Designed for Kockums Type 471. Collins Class.

Hedemora Diesel AB has over 30 years experience in the development and delivery of turbo-charged diesel engines for submarines.

We are very proud to have been selected to supply engines for the world's most advanced diesel-driven submarines for the Royal Australian Navy.

Amongst the important aspects of the Australian submarine concept, specifically highlighted by the assessment team, are the low noise and vibration levels – aspects directly attributed to Hedemora engines. Other major features are the engines' well-proven reliability and good serviceability.

Hedemora diesel engines are manufactured in two proven basic versions, with 6 - 18 cylinders and an output range of 500 - 2800 kW. Their compact and sturdy design makes the engines ideal for submarine applications.

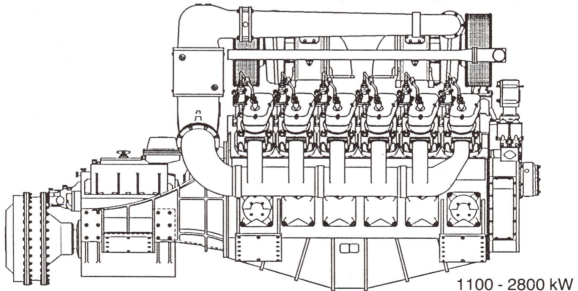

1100 - 2800 kW

ADLINK

HEDEMORA DIESEL AB

S-776 00 Hedemora, Sweden. Phone +46 225 155 40. Fax +46 225 154 34.

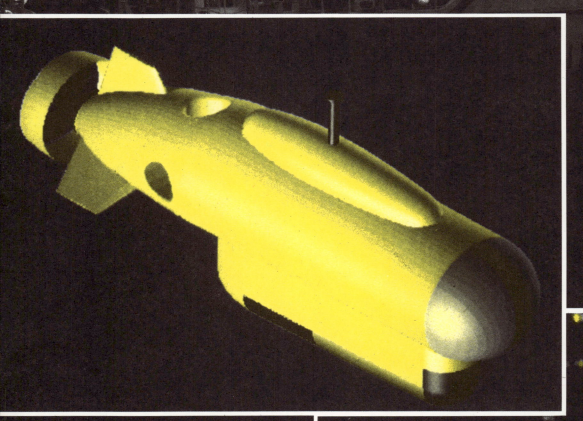
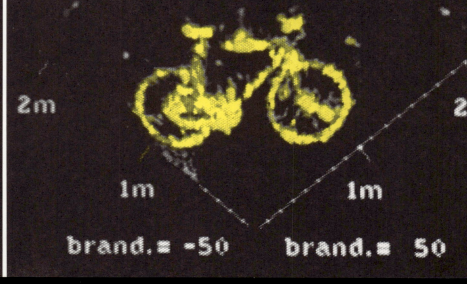

JANE'S UNDERWATER WARFARE SYSTEMS

FOURTH EDITION

EDITED BY
ANTHONY J. WATTS

1992-93

ISBN 0 7106 0984 1

JANE'S DATA DIVISION

"Jane's" is a registered trade mark

In the USA and its dependencies
Jane's Information Group Inc, 1340 Braddock Place, Suite 300, Alexandria, VA 22314-1651, USA

British Library Cataloguing-in-Publication Data.
A catalogue record for this book is available from the British Library.

Printed and bound in Great Britain by Butler and Tanner Limited, Frome and London.

LAUNCHING UNIQUE UNDERWATER SYSTEM SOLUTIONS.

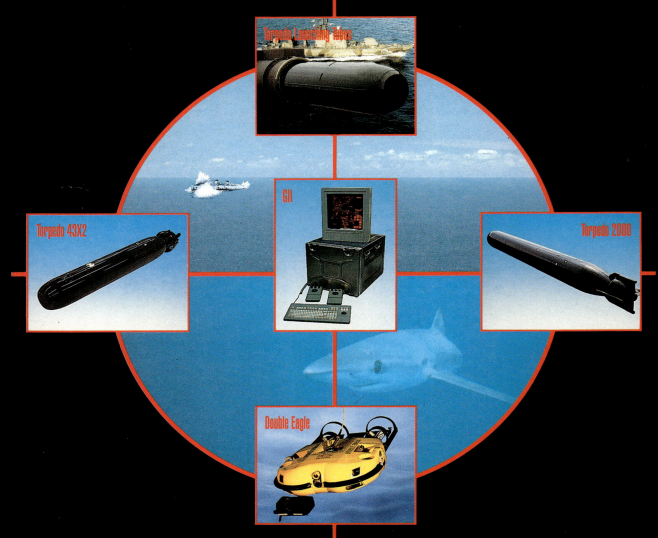

Torpedo 2000, the new heavyweight dual purpose torpedo, offers Longer Range-, Higher Speed-. Silent and deep running advantages.

Torpedo 43X2, the only wireguided lightweight ASW torpedo, capable of detecting and tracking submarines not only in blue waters, but what is more, also in shallow and narrow waters.

Double Eagle, with 360 degrees manœuvrability in roll, pitch and heading is the ultimate MCM ROV.

GII, the Graphic Interactive Indicator, a flexible C³I system with a great variety of applications.

Torpedo Launching Tubes, used also as transport containers.

A new range of products, from which unique underwater systems can be tailored to meet the Navy-requirements well into the next century.

Contents

Alphabetical list of advertisers

Naval Technology...

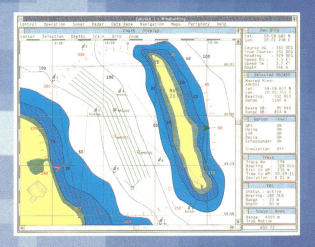
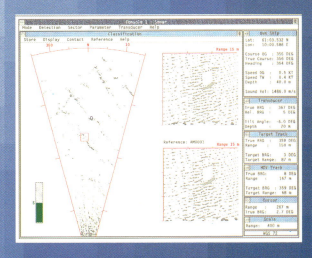

Classified list of advertisers

The companies listed advertising in this publication have informed us that they are involved in the fields of manufacture indicated below:

Acoustic countermeasure systems: submarine/surface ship
Thomson Sintra/ASM

Acoustic monitoring equipment
Thomson Sintra/ASM
WELSE

Acoustic ranges & targets: static/portable
Thomson Sintra/ASM
WELSE

Acoustic trials equipment
Thomson Sintra/ASM
WELSE

Airborne ASW dipping sonars
Thomson Sintra/ASM

Airborne ASW receiving & processing
Thomson Sintra/ASM

ASW fire & control systems
Swedish Ordnance-Underwater
Thomson Sintra/ASM
WELSE

Degaussing systems: static portable
Thomson Sintra/ASM

Divers' hand-held equipment
Thomson Sintra/ASM

MCMVs
Consorzio SMIN

Mines & depth charges: air-launched
SA Marine

Mines & depth charges: submarine-launched
SA Marine
Thomson Sintra/ASM

Mines & depth charges: surface-launched
SA Marine
Thomson Sintra/ASM

Minehunter propulsion systems
Hedemora Diesel

Minesweeping equipment
SA Marine
Thomson Sintra/ASM
WELSE

Naval oceanographic systems
Thomson Sintra/ASM

ROVs
Consorzio SMIN
Swedish Ordnance-Underwater

ROV sonar systems
Consorzio SMIN
Thomson Sintra/ASM

Ship sonar systems: hull-mounted
Thomson Sintra/ASM

Ship sonar systems: variable depth
Thomson Sintra/ASM

Sonobuoys: active/passive
Thomson Sintra/ASM

Sonobuoys: telemetry systems
Thomson Sintra/ASM

Static detection systems
Thomson Sintra/ASM

Submarine early warning systems
Thomson Sintra/ASM
WELSE

Submarine propulsion systems
Hedemora Diesel

Submarine sonar systems: variable depth
Thomson Sintra/ASM
WELSE

Submarine weapon discharge equipment
WELSE

Submarine weapon handling equipment
WELSE

Torpedoes: air-launched
Swedish Ordnance-Underwater
Thomson Sintra/ASM

Torpedoes: submarine-launched
Swedish Ordnance-Underwater
Thomson Sintra/ASM
WELSE

Torpedoes: surface-launched
Swedish Ordnance-Underwater
Thomson Sintra/ASM

Torpedo tubes
Swedish Ordnance-Underwater

Towed arrays & equipment
Thomson Sintra/ASM
WELSE

Towed array handling equipment
Thomson Sintra/ASM
WELSE

Underwater communication: ship/submarine
Thomson Sintra/ASM
WELSE

Underwater communication: divers' systems
Thomson Sintra/ASM

Weapon guidance systems
NFT
Swedish Ordnance-Underwater
Thomson Sintra/ASM
WELSE

At Thomson Sintra ASM, Our Sphere Is The Sea.

Our mission at Thomson Sintra ASM is to supply the advanced know-how and cutting-edge technology necessary to keep the seas safe. With over forty years' experience to our credit, a technological and industrial base second to none and cooperative programs under way with partners and new subsidiaries on five continents, we're one of the world's leading companies in the field.

By making it our business to develop the most effective technology to counter the threat at sea — on it, beneath it or above it — we have become the leading supplier of sonar systems to the world's navies, No. 1 worldwide in minehunting, a leader in mine warfare, and the world-class specialist in the processing of sonar data. A global player producing combat and detection systems for every aspect of antisubmarine warfare, for every kind of platform.

THOMSON-CSF
World-Class Electronics

CORP ASM-75

THOMSON SINTRA ACTIVITES SOUS-MARINES - 525 Route des Dolines - B.P. 138 - Parc Sophia Antipolis - 06561 Valbonne Cedex - France - Tel.: (33) 92 96 30 00 - Telex: THOM 616780F.

A MIN Mk 1 remotely operated underwater vehicle aboard an Italian Navy 'Lerici' class MCMV. Four of the Mk 1 vehicles are in operation. The modified and upgraded Mk 2 vehicle is being procured for the Italian Navy's 'Gaeta' class MCMVs.

JANE'S UNDERWATER WARFARE SYSTEMS 1992-93

Please note that because of publishing time scales and the uncertainty over the future of the structure of the former Soviet Union the term 'Union of Sovereign States' (USS) has been used throughout this edition of Jane's Underwater Warfare Systems. This term was replaced by 'Commonwealth of Independent States' (CIS) at the beginning of 1992. The CIS is led by Boris Yeltsin, President of Russia, and includes all the countries previously in the former USSR except for Georgia, Lativa, Estonia and Lithuania.
10 March 1992

Jane's Information Group Limited, Sentinel House, 163 Brighton Road, Coulsdon, Surrey CR5 2NH, UK
Jane's Information Group Inc, 1340 Braddock Place, Suite 300, Alexandria, VA 22314-1651, USA

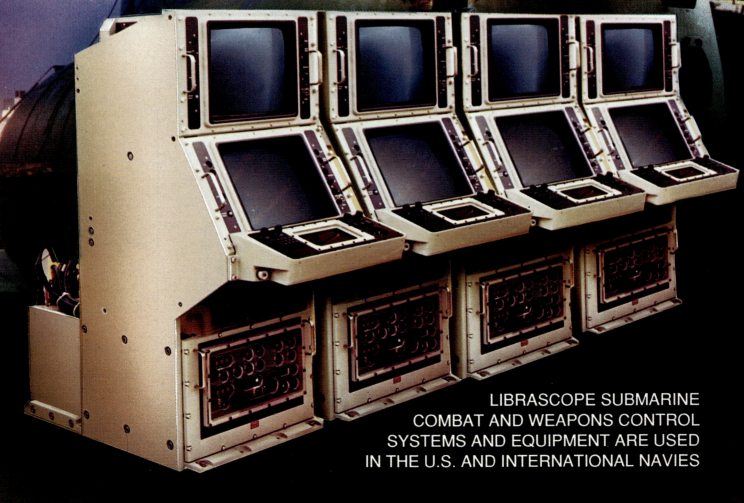

Foreword

The World Scene

In the foreword to last year's edition I commented that "... the world at the present time seems to be in greater turmoil than it has been for many years". That turmoil has in no way abated during 1991, in fact just the opposite. We are now living in an age in which uncertainty and unease stalk the corridors of power like Old Father Time.

Although the Gulf War ended with a crushing victory for the Coalition Forces, it was not quite as clear cut as many might have hoped. President Saddam Hussein remains in power, and although required by the United Nations to destroy Iraq's arsenal of nuclear and chemical weapons under UN supervision, he has tried to make things as difficult as possible for those inspecting Iraqi installations. Furthermore, there is growing evidence and concern that Iraq is sidestepping the investigators, and hiding away much of what survived the war, ready, some consider, to fight another day.

But it was to Eastern Europe that most eyes were turned throughout the latter part of 1991, and will continue to be turned during 1992. Events in the Soviet Union and the former Warsaw Pact developed with alarming rapidity in 1991. Following the throwing-off of the yoke of Communism by the former Eastern Bloc countries, as noted in last year's Foreword, even the Soviet Union herself cast off the Red mantle. And it is this which has led to feelings of considerable unease around the world, and possible instability as well.

Never before has the world witnessed the demise of a Super Power which might have held it to ransom, and which has left in its wake a power vacuum of enormous proportions. There is no parallel in recorded history on which politicians, economists and the military can draw for reaction to such an event. World leaders are having to feel their way very carefully through not only a political minefield, but a military and a growing economic one as well. In place of the Soviet Union came the Union of Sovereign States (The USS. Because of publishing time scales and the uncertainty over future developments concerning the former Soviet Union, this is the term which has been used throughout this edition for the sake of continuity), a grouping formed under President Gorbachev, which itself collapsed at the beginning of 1992. This in its turn was replaced by the Commonwealth of Independent States (CIS), each with its own president and held together by what are now seen to be extremely tenuous links.

The leader of the CIS is Boris Yeltsin, President of Russia. But even his power base is uncertain. Following the mass demonstration in Moscow during the weekend of 8-9 February 1992, just before this Foreword was written, there is now uncertainty as to how long he may be able to hold on to power. Many of those who only a few months ago hailed Yeltsin as their saviour, are now seeking to cast him out along with Gorbachev. Yeltsin's economic reforms aimed at turning Russia into a market economy have wrought havoc with the purchasing power of the Russian man in the street, who is not at all happy about the turn in events. Many are living literally on the bread line. Even those who under the old regime were comfortably well off with reasonable salaries – the scientist, schoolteacher, doctor, dentist and so on – now rank with the poorest in Russian society.

And in the background lies the military, its future also in grave doubt as swingeing cuts are proposed for the armed forces. But are the armed forces in fact being cut to the extent that Western observers are being led to believe? Again, there are those who doubt that quite so much equipment has been scrapped or destroyed as the CIS would have the West think. A good indicator of the state of the armed forces is the Navy, and recently there have been many pictures taken around former Soviet bases, principally in the Black Sea, which show all too clearly the potential might of the naval forces. True, many ships have been scrapped, but these have been mainly obsolete units of little or no fighting value. There have also been reports that some units currently under construction may be put up for sale, and again it has been reported that India and others may be interested in purchasing the aircraft carrier under construction. But so far there is no real evidence that units of any size or in any numbers are being sold off or scrapped. Of course the disruption caused by the break-up of the Soviet Union, the acute lack of finance and so on, together with lack of co-ordinated control over shipyards which are situated among the various Republics, may well cause interruption to the completion of any new units under construction. This will be felt more acutely in the complex area of submarine construction. It may, therefore, be some time before it becomes clear as to what is happening to shipbuilding programmes, and in particular the rate of submarine construction.

Likewise, there is little real evidence that the organisation of the Navy itself has undergone any major restructuring in the post-Soviet Union era, although it is early days to forecast what might happen. One thing is certain, however, and that is that further cracks in the already shaky Commonwealth are already showing over how existing military assets will be shared out among the republics of the CIS. The Ukraine is claiming sovereignty over the Black Sea Fleet, while Yeltsin claims it should remain under Russian control. And the Navy apparently appears to be supporting President Yeltsin in this stand. It is perhaps worth remembering that in all her turmoils throughout history it has been the Navy which has always been at the centre of major changes to the infrastructure of Russia/Soviet Union/USS/CIS.

With so many questions left unanswered, the nations of the West are having considerable difficulty in determining their response to events in the former Soviet Union. While it has been relatively straightforward recognising the various republics and their presidents, it is not so easy trying to determine who will ultimately be responsible for the military, and in particular the naval forces, and hence with whom one should communicate and, ergo, what the West's response to any discussions should be.

The clamour in the West for the so-called 'Peace Dividend' has been enormous, and many have responded to it with alacrity. But the time has come for a pause, and no-one is quite sure what sort of a response is now required in the face of the uncertain future regarding former Soviet military forces, and changes to the world order. One thing is certain, there is no longer any need for such a vast arsenal of nuclear weapons, and already the United States has indicated that massive cuts in its nuclear capability are being planned and implemented.

However, there is an uneasy feeling in some quarters that it would not be wise, certainly for the present, to completely negate the West's response to what was the Soviet Union's nuclear capability. In the wake of the collapse there has been much speculation over what will happen to the scientists, the military and all the civilians involved in the former Soviet Union's nuclear programme. Already there are signs that some may well be lured overseas by the prospects of alternative, higher salaries, and they may not be too bothered about the politics of those countries, or even the possible threats which those countries may wish to pose with nuclear weapons.

If such is the case, how should the nuclear capabilities of NATO and the West be realigned? Will it be possible to prevent some of the former Soviet Union's nuclear arsenal

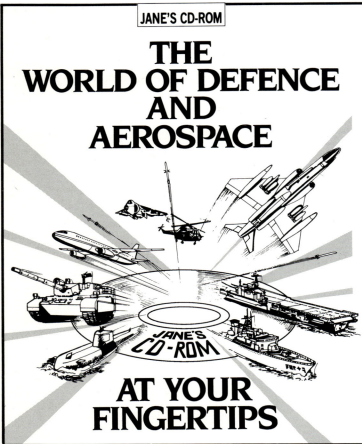

from being spirited abroad to a foreign country willing to pay a high price for such hardware? What will happen to the former Soviet Union's naval nuclear capability? Will this too be up for grabs so to speak, or if not how can its destruction be supervised? These are the thorny questions which now have to be addressed.

On the conventional war front huge cuts in the armed forces of many western nations are already being implemented. To date, most of these cuts have fallen more heavily on land- and air-based forces. Naval forces, while suffering cuts, mainly among personnel, have had their infrastructures left more or less intact. Many of the planned reductions in naval forces will affect future programmes. In other words there is likely to be a slowing down rather than outright cuts in naval forces. Warship construction has been slowing down for some years, now it is time for equipment to be reduced, and already some programmes are being hard hit by financial reductions, or abandoned altogether. Most of the cuts will relate to new systems either in the process of acquisition, or under research and development. The outcome of this is likely to be that existing systems will be required to remain in operation for much longer than hitherto, with greater reliance on upgrading and modernising, rather than outright acquisition of new equipment.

The Naval Scenario
Not for nothing are naval forces often referred to as 'The Silent Service'. At present many have retreated behind closed doors to debate their future stance under the new world order. What is likely to be their response? Some indications are becoming clear. Within NATO there is much talk of 'Rapid Reaction Forces' which, by the very nature of its terminology, demands a maritime response. Within this context too, it seems clear that such forces will not necessarily be confined within the boundaries of NATO, as they formerly were in the face of the threat from the Soviet Union. If so-called 'out-of-area' responses are being potentially considered, then obviously amphibious warfare must play a major role in any such plans.

In the face of slowly but surely expanding submarine forces worldwide, it will be essential for rapid reaction forces to be adequately protected against submarine attack. Furthermore, operations in the Gulf in 1991 showed yet again how deadly the mine threat is, even from unsophisticated horned mines! Again naval forces must ensure that they have adequate mine countermeasures assets available to meet any contingency.

But mines and submarines can also be used as defensive weapons, a factor which has been exploited to the full by Scandinavian countries. In fact these countries probably lead the world in the use of the mine and submarine as a defensive weapon. In the light of the potential development of rapid reaction forces, many countries may now seek to build up their submarine and mine warfare arms for defence against the possible threat of attack from a rapid reaction force.

Thus, while as noted above there may be a slowing down in procurement programmes as a result of the demise of the Soviet Union, there is no such indication that nations have completely abandoned the view that ASW and mine warfare are potentially the two most important aspects of naval warfare. As such, therefore, ASW and MCM will probably remain at the top of the agenda for most naval forces, but perhaps at a less intense level than was prevalent during the height of the Cold War. In the case of mine countermeasures this will not be altogether welcome, for mine countermeasures has always tended to be disregarded, until it is too late!

Anti-Submarine Warfare
Any navy which has to ensure the safety of merchant shipping and secure sea lines of communication must possess an adequate ASW capability. Throughout its history the submarine has proved its effectiveness as an offensive weapon when used against commerce on the high seas. Even against naval units it has proved to be a lethal weapon and one which poses a major threat to the movement of naval forces. This threat has now been further increased by the submarine's capability to launch both tactical as well as strategic ballistic missiles. During the Gulf War an American submarine stationed in the Red Sea fired TOMAHAWK cruise missiles against Iraq.

In the future the capability of the non-nuclear submarine will be further enhanced by the adoption of air-independent propulsion, offering smaller navies and non-nuclear powers the ability to conduct underwater operations of virtually unlimited endurance.

The submarine is an extremely potent weapon with a capability for destruction and an ability to dictate strategy and tactics far outreaching its size and infrastructure. The effectiveness of the submarine lies in its stealth features, which today are being further increased by the use of various noise reducing characteristics, both through features built into its design, the choice of materials used in its construction, and by the measure of treatment to its hull and onboard systems. Added to this is an increasing knowledge and build-up of data on the ocean environment in which the submarine operates and which it can use to its advantage. Hence it is exceedingly difficult for either surface or sub-surface forces to be absolutely certain that they have located all the underwater targets which may affect friendly maritime operations and national strategy.

The initial detection of a submarine is a very exacting and time consuming operation which ties up vast numbers of ASW assets, including submarines, surface forces and fixed-wing aircraft and helicopters. Much of the effectiveness of the initial detection relies firstly on intelligence concerning the enemy, his way of thinking, modus operandi and so on, and then on the use of very long range passive sonar to provide the first indication of potential targets. Even assuming one has detected and can then effectively track and localise a target, it still has to be attacked and destroyed, and this is no easy matter either. Future technology will confer on the submarine the ability to deploy effective countermeasures against anti-submarine torpedoes.

Apart from developments in the area of combat information, ASW sensors are also undergoing continual development. Among areas of increasing importance are those related to the development of active towed array sonars and the introduction of thin line towed arrays, some of which will make use of fibre optic links for the transfer of data.

With the stealth features now being incorporated in submarine designs, the need to provide highly accurate data on potential targets much earlier will require the deployment of long-range active towed array sonars. In the face of improved stealth techniques long-range passive towed array systems will find it increasingly difficult to provide the necessary volume of accurate data from which to compile a detailed tactical picture.

The use of lightweight thin line arrays will enable much smaller ASW vessels to deploy towed array sonars, and the array itself will be able to be deployed at much greater distances behind the towing platform, removing it even further from areas of unwanted noise produced by the towing platform, thus increasing its effectiveness. In addition, the use of fibre optics will enable a much greater volume of data to be passed back to the towing ship for processing. The use of newly developed microchip technology will enable some pre-processing to be carried out in the arrays themselves, further enhancing the capability of the systems.

In the face of such developing technologies and new

systems, albeit entering service at a possibly slower rate than hitherto, it ought to be emphasised that now is not the time for navies to allow their ASW capability and assets to stagnate, much less for them to be reduced.

Underwater Weapons

Stealth will also be the main area of development in underwater weapons. This will enable the platform deploying the weapon to retain its anonymity, and also for the weapon itself to remain undetected.

For torpedoes the main objective will be the need to develop quieter propulsion systems with increased endurance, enabling weapons to follow more closely targets manoeuvring at high underwater speed and to carry out a number of re-attack procedures in the event of lost contact. Secondly it will be necessary to continue development of warheads and in particular charges and fuzes, in order to ensure that weapons remain effective against modern submarines. Finally, as anti-torpedo defences are developed, so the homing capabilities of the torpedo will need to be enhanced in order to overcome ECM and ECCM capabilities.

Mines too will need to exhibit increased stealth capabilities. Modern minehunting sonars and remotely operated vehicles fitted with high resolution sonars, TV cameras and so on are conferring on mine countermeasures forces much greater effectiveness against the latest mines. Mine design will need to concentrate more on the ability of the weapon to remain invisible to modern sonars, a capability for ground mines to merge more with the background, and even, where conditions are appropriate, the ability to bury themselves in the sea bottom out of range of bottom-penetrating sonars.

To counter ASW aircraft, developments are now in hand which will enable the submarine to detect aircraft while still submerged, and to deploy anti-air weapons, primarily against the helicopter, from a submerged position. As such developments will reduce much of the effectiveness of the helicopter as an ASW platform, there will be a need for some form of shipborne long-range weapon capable of delivering an ASW torpedo within suitable range of a hostile submarine.

Acknowledgements

The easiest and most pleasant part of any editor's foreword is to acknowledge the help that has been given from other editors and contributors. Without the assistance of these friends and colleagues the task of producing a yearbook would be very much more difficult.

I am most grateful to Bernard Blake, Keith Faulkner and Ruth Simmance whose guidance and help have enabled me to rearrange the contents this year to provide what I hope will be more comprehensive coverage of the various aspects of underwater warfare. This has not been an easy task, and I hope that the new arrangement will meet with readers' approval.

Secondly I must acknowledge the value of the very extensive library of data and photographs compiled over many years by Maritime World for use in its publication *Navy International*. Again, without the backup of these records the work of producing this yearbook would have been very much harder.

The support of the excellent backup team at the Jane's Information Group headquarters has also greatly eased the work of editing this yearbook. Under the leadership of Keith Faulkner, the in-house editorial team has performed an immense amount of work. In particular my thanks must go to Ruth Simmance and her computer team Sarah Erskine and Christine Varndell and to Lynn Morse and Kevin Borras who have had the thankless task of checking the spelling, phraseology and general layout of the book. Together they have helped me enormously in standardising the many entries, and wherever possible to simplify and explain the intricacies of what is an extremely complex subject.

Many other people in Jane's have also been of help, in particular Captain Richard Sharpe, the editor of *Jane's Fighting Ships*. The Jane's publications *International Defense Review* and *Jane's Defence Weekly* have been invaluable as reference sources.

Thanks are also due to the staff at Butler and Tanner Limited who are responsible for the finished product.

Finally, I must acknowledge the support and assistance of my wife Mary.

Because of the rate of change within the defence industry, which has accelerated enormously since the end of the Cold War, updating is a continuous process, and necessarily a yearbook is only as accurate as the day it was passed for press. A copy of the current entry is sent to the appropriate contractor each year with a request for amendments. It is essential that these are returned promptly to ensure that the yearbook is kept correctly up to date. It is always difficult and sometimes impossible to incorporate material which is received late. New material for updating information or describing new equipment and new photographs and so on are always gratefully received at any time of the year. I would like to acknowledge all those who have returned material together with amendments, and new information.

Anthony J Watts
Newdigate
February 1992

Glossary of acronyms and abbreviations

AA   active adjunct (sonar)
ACINT   acoustic intelligence
ADCAP   advanced capabilities
ADP   automatic data processing
AF   audio frequency
AFC   automatic frequency control
AGC   automatic gain control
AI   artificial intelligence
ALCS   airborne launch control system
ALFS   airborne low frequency sonar
ALWT   advanced lightweight torpedo
AM   amplitude modulation
AMCM   advanced mine countermeasures
ARE   Admiralty Research Establishment (UK)
ASROC   anti-submarine rocket
ASW   anti-submarine warfare
ASW/SOW   anti-submarine warfare/stand-off weapon
ATE   automatic test equipment

BITE   built-in test equipment
BT   bathythermograph

CASS   command active sonobuoy system
CCIS   command and control information system
CCTV   closed-circuit television system
CEP   circular error, probable
CFAR   constant false alarm rate
CIU   control interface unit
CPA   closest point of approach
CRT   cathode ray tube
CSDT   control for submarine discharge torpedo
CSLT   control for surface-launched torpedoes
CSU   control selection unit
CW   continuous wave
CZ   convergence zone (sonar)

DARPA   Defence Advanced Research Projects Agency (USA)
DCN   Direction des Constructions Navales (France)
DEMON   demodulated noise (sonar processing)
DF   direction finding
DHS   data handling system
DICASS   directional command active sonobuoy system
DIFAR   direction frequency and ranging
DMT   deep mobile target
DoD   Department of Defense (USA)
DPM   digital plotter map
DRA   Defence Research Agency (UK)
DRTS   detecting, ranging and tracking system

ECAN   Etablissement des Constructions et Armes Navales (France)
ECCM   electronic counter-countermeasures
ECM   electronic countermeasures
ECR   embedded computer resources
EDM   engineering development model

EHF   extremely high frequency
ELF   extremely low frequency
EPROM   erasable programmable read-only memory
ERAPS   expendable reliable acoustic path sonobuoy
ESM   electronic support measures
EW   electronic warfare

FCS   fire control system
FFT   fast Fourier transform
FLIR   forward looking infra-red
flops   floating point operations per second
FRAS   free-rocket anti-submarine
FSD   full-scale development
FSK   frequency shift keying

GPS   Global Positioning System (formerly NAVSTAR)

HE   high explosive
HF   high frequency

IF   intermediate frequency
IFCS   integrated fire control system
IFF   identification, friend or foe
IFM   instantaneous frequency measurement
II   image intensifier
INS   inertial navigation system
IR   infra-red
IUSS   integrated undersea surveillance system

LAMPS   light airborne multi-purpose system
LANTIRN   low-altitude navigation and targeting infra-red night system
LAPADS   lightweight acoustic processing and display system
LED   light emitting diode
LLTV   low-light television
LOFARGRAM   low frequency analysis and recording gram
LRAAS   long-range airborne ASW system

MAD   magnetic anomaly detector
MCM   mine countermeasures
MCMV   mine countermeasures vessel
Megaflops   million floating point operations per second
MF   medium frequency
MNS   mine neutralisation system
MoD   Ministry of Defence (UK)
MoU   memorandum of understanding
MPA   maritime patrol aircraft
MSK   minimum shift keying
MSO   ocean-going minesweeper
MTBF   mean time between failures
MTI   moving target indication
NAVSTAR   navigation satellite timing and ranging
NSR   naval staff requirement (UK)

OTPI   on top position indicator

PCU   power control unit

PPI   plan position indicator
PRF   pulse repetition frequency
PSP   programmable signal processor

RDSS   rapid deployable surveillance system
RDT&E   research, development, test and evaluation
RF   radio frequency
RFS   radio frequency surveillance
ROV   remotely operated vehicle
RWR   radar warning receiver

SACU   stand-alone digital communications units
SEPADS   sonar environmental prediction and display system
SEWACO   Sensor Weapon Control and Command
SHF   super high frequency
SINS   ship's inertial navigation system
SLBM   submarine-launched ballistic missile
SLCM   submarine-launched cruise missile
SLMM   sea-launched mobile mine
SOSUS   sound surveillance system
SPAT   self-propelled acoustic target
SSB   conventional-powered ballistic missile submarine or single sideband
SSBN   nuclear-powered ballistic missile submarine
SSE   submerged signal ejector
SSK   ASW submarine (non-nuclear)
SSKP   single shot kill probability
SSN   nuclear-powered attack submarine
SUBROC   submarine-launched rocket
SURTASS   surveillance towed array sensor system
SUT   surface and underwater target

TACTAS   tactical towed array sonar
TMA   target motion analysis
TWS   track-while-scan (radar)
TWT   travelling wave tube

UAV   unmanned air vehicle
UCS   underwater combat system
UHF   ultra high frequency
USB   upper sideband
UV   ultraviolet

VDS   variable depth sonar
VHF   very high frequency
VHSIC   very high speed integrated circuit
VLA   vertical launch ASROC
VLF   very low frequency
VLSI   very large scale integration (electronic circuits)

XBT   expendable bathythermograph
XSV   expendable sound velocity system

Introduction

The objective of *Jane's Underwater Warfare Systems* is to cover the complete scenario of underwater warfare. This year it has been decided to completely revise the order of contents to cover in a more meaningful way systems associated with specific forms of underwater warfare. Hence the book is arranged in four broad areas covering: Anti-submarine Warfare, Underwater Weapons, Mine Warfare and Associated Underwater Warfare Systems. These main areas have been subdivided into primary categories and these have been further subdivided into secondary categories. In this way particular types of system related to one specific form of underwater warfare are all grouped together, making reference and comparison much easier.

ANTI-SUBMARINE WARFARE

Command and Control and Weapon Control Systems – includes submarine combat information systems, weapon control systems and airborne acoustic processing systems.

Sonar Systems – includes both surface ship and submarine sonars covering hull-mounted systems, variable depth sonars and towed arrays. It covers sensors, arrays, processing and display systems and handling gear in the case of towed arrays. Airborne systems include dipping sonars and various types of active and passive sonobuoys and their receivers. Static detection systems covers all types from the large systems such as the United States SOSUS to the smaller equipments used for harbour protection, offshore platform protection and so on.

Countermeasures – covers submarine and surface ship acoustic countermeasures and noise projectors and submarine electronic warfare equipment.

Underwater Communications – in this section systems used for submarine-to-submarine, ship-to-submarine, aircraft-to-submarine and satellite-to-submarine communications and including underwater telephones and communications buoys are covered.

Electro-optical Sensors – covers periscopes and optronic masts and miscellaneous optical and infra-red sighting devices.

Magnetic Anomaly Detection Systems – covers those systems installed in aircraft and helicopters to assist in tracking submarines by detecting variations in the magnetic field.

UNDERWATER WEAPONS

Torpedoes, Guided Weapons, Rockets, Mines and Depth Charges – cover the complete range of underwater weapons, including submarine, surface ship and air-launched torpedoes, guided ASW weapons, rockets and weapon discharge systems. Also included is the full range of mine types, including moored, ground influence, tethered and controlled, and depth charges.

MINE WARFARE

Command and Control and Weapon Control Systems – covers command and control systems and precise navigation and positioning systems for mine countermeasures vessels.

Sonar Systems – this section includes surface vessel sonars used for mine countermeasures, including hull-mounted, variable depth and towed side-scan sonars. It also includes ROV sonar equipment.

Mine Disposal Vehicles – all types of ROVs used for underwater warfare and defence and associated ancillary equipment.

Minesweeping Systems – minehunting and minesweeping systems and sea mine clearance and ordnance disposal systems.

Divers' Systems – covers all types of equipment associated with divers and their operations, including lights, cameras, communications equipment, hand-held sonars, pingers and small underwater vehicles used to transport divers.

ASSOCIATED UNDERWATER WARFARE SYSTEMS

Acoustic Management Systems – covers probes and associated buoys such as telemetry systems and those used to measure various ocean parameters.

Hydrographic Survey Systems – various systems employed in underwater survey include echo sounders, sonars and ocean survey systems.

Signature Management – in this section acoustic and magnetic ranges, degaussing systems and acoustic control and measurement systems are covered.

Training and Simulation Systems – training and simulation systems for all aspects of underwater warfare, including command team trainers, equipment trainers and acoustic targets are covered.

Other systems covered in this section are **Navigation and Localisation Systems, Submarine Radar, Consoles and Displays, Transducers, Miscellaneous Equipment** and **Addenda**.

Analysis – quick reference tabular information on the systems covered in the above sections and also those systems for which insufficient, non-restricted information is available to justify an entry. For reference purposes a number of obsolete systems are also included.

Anti-submarine warfare

Any country with a littoral, and even more so islands such as the United Kingdom, is almost entirely dependent on its sea lines of communication for its continued existence. The protection of those sea lanes against underwater attack in particular is vital to ensure the safe carriage of imports and exports (food and materials in peacetime and during hostilities, reinforcements as well to support the war effort), as was evidenced during the First and Second World Wars. It is also vital to a continental power such as the United States since, if it were to become engaged in hostilities, it would need to transport vast quantities of troops and their supplies by surface vessel to wherever its forces were engaged. That such would be the case was highlighted by the vast movements of troops and supplies prior to and during Operation Desert Storm, the war with Iraq in 1991. Although the demise of the former Soviet Union and the current indecision over the control and possible dismemberment of the former Soviet Navy lies unresolved, a submarine force of some considerable size still exists and at present it is not at all certain as to what the fate of that force will be, nor, if it is retained, what its main role is likely to be. In fact the whole situation is so uncertain that some of the former Soviet Navy's submarines may well be sold off and the question then arises – 'To whom will they be sold?'

Furthermore, all evidence goes to show that submarine forces worldwide are expanding rather than contracting. New forces are being organised and those with obsolete boats are seeking to acquire replacements.

By far the greatest problem faced by ASW forces is finding submarines in the first place and then tracking them, not sinking them. The range of sonar devices, although limited, has improved in recent years, particularly with the advent of the long-range passive sonar. But over the last few years the underwater speed of submarines has increased enormously. Modern submarines are also extremely quiet, particularly when running at slow speeds, and the latest types use acoustic cladding to further reduce their 'signature'. Many types of sonar have been developed for submarine detection, including large static moored systems, hull-mounted and variable depth sonars, towed arrays, sonobuoys and airborne dipping sonars, plus other devices such as magnetic anomaly detectors and wake detection systems. Despite the great variety of available systems, employing the latest technology in sound detection and data processing, the detection of a hostile submarine is still an extremely difficult and slow process and ties up large numbers and forms of ASW assets.

The large static systems moored on the ocean bed are operated mainly by the United States and the former Soviet Union. They tend to be used in the most strategically important and sensitive areas and in particular at choke points, that is those areas which are limited in width by land masses, but which serve as entry and exit lanes for submarines and in particular the strategic missile submarines. These areas are also patrolled by attack submarines whose role is to detect hostile submarines and in certain tactical scenarios provide data for nuclear-powered submarines to attack the hostile boats, or carry out an attack themselves. Should a hostile submarine escape this first line of defence, it must then be detected and attacked by a second line of defence such as surface ship and aircraft using towed array, hull-mounted or variable depth sonar, dipping sonar, sonobuoys, magnetic anomaly detectors and wake detection systems and torpedoes and depth charges.

Although great strides have been made in the design of transducers and other sensing devices, the greatest advances in conventional sonars today are those related to the processing and display systems and their integration into a combat system. As with radar, the advent of digital computers, microprocessors and even more sophisticated electronic components, has meant that large amounts of information can be extracted from relatively weak signals.

Both hull-mounted and variable depth sonars are used for passive and active detection and towed arrays are employed for longer range passive surveillance. The towed array has many advantages over the hull-mounted system in that it is deployed a considerable distance behind the ship and is thus not affected by any noise emanating from the towing vessel. Towed arrays also achieve very long detection ranges by operating at very low frequencies where propagation losses are lower, enabling the low frequency sound emanating from propeller cavitation and machinery of a hostile boat to be detected. The towed array does, however, suffer from a number of disadvantages, for example being unable to determine the range of a contact, ambiguity in bearing, directional uncertainty because of sideways movement of the array and the towing cable, flexing of the hydrophone array and a number of other physical factors. These, however, are now being overcome. In addition a large winch and handling gear is required to deploy the array. This is not a major problem on surface ships (except from a weight point of view) but is impractical in all but the largest submarines. In some cases the array has to be clamped on after the submarine leaves harbour and removed just before re-entering, making the boat extremely vulnerable when this operation is being carried out, although some submarines are able to stow the array on the side of the hull. However, despite its disadvantages the towed array is of immense value in long-range detection and a considerable amount of development effort is currently being devoted to the design of thin line towed arrays to provide a much lighter and more manoeuvrable system. Development of active towed arrays, usually by combining the receiving array with a hull-mounted transmitter and resolution of the ambiguity factor, is also proceeding apace.

The primary airborne detection system is the sonobuoy which is produced in its tens of thousands for deployment from fixed-wing aircraft and helicopters. Since they are expendable, sonobuoys have to be relatively cheap to manufacture but reliable in operation. Large numbers of these devices are in current service and include both passive and active buoys, directional and non-directional, large size and small size, all of which transmit information back to the aircraft for processing and display. Another underwater threat which is of increasing importance is the free swimmer and small submersible. Harbours and offshore platforms are particularly vulnerable to this form of attack and a number of static sonar systems have been developed to detect such intruders. The sensors of these systems are usually bottom or cable moored and connected to data processors and displays in a shore-based centre. In many cases an integrated system using a variety of radar, acoustics, electro-optical and/or TV-based sensors is employed.

Torpedoes

The torpedo is the primary ASW weapon being used by surface ships, aircraft and submarines.

Its importance is such that it is worth looking briefly at its history, which dates back to 1866 when Robert Whitehead developed the first real torpedo in his factory at Fiume. Driven by cold compressed air at a speed of about 7 kts it was successfully demonstrated to a number of interested countries and by the early 1880s had been developed to run at speeds of up to 30 kts for distances of about 1 km. This method of propulsion continued to be developed until the early part of this century when experiments with heated air in the UK produced a dramatic increase in performance. In this system paraffin, water and air were mixed, the result being a steam-air mix driving a radial engine. In the United States a turbine-powered weapon was developed using alcohol as the fuel. By the middle of the First World War, the torpedo had become a major reliable and effective weapon, despite problems with the fuzing. Indeed the British Mk 4 torpedo could run for distances of around 5 km at speeds up to 40 kts, figures that, with the exception of a Japanese torpedo in 1942, were not exceeded until a few years ago.

Between the First and Second World Wars a great deal of development continued in the field of warhead, fuzing and propulsion systems. One of the most remarkable and certainly the most long-lived products of this period was the British Mk 8 which was developed in the early 1930s and remained in service with the Royal Navy until 1986, more than 50 years of service. Indeed it is believed that a few examples of this straight running, non-homing torpedo are still in service in a number of navies. It was, however, Germany who revolutionised the torpedo by introducing an electrically propelled weapon in 1939, just in time for the Second World War. Powered by lead/acid batteries it could travel at 27 kts over ranges up to 8 km. Although slower than the thermally powered torpedo, this was of little consequence to the German Navy which concentrated on the destruction of relatively slow speed merchant shipping. The German Navy also produced the first acoustic homing torpedo in 1943, another major milestone in torpedo development, as well as the wire-guidance system which is used on all modern heavyweight torpedoes, and also wake-following techniques.

By this time it had also been realised that not only were torpedoes essential for use against surface shipping, but they were even more necessary to combat hostile submarines. Indeed this latter capability has now become the main requirement for the modern torpedo.

After the end of the Second World War development continued apace, mainly in the field of electrically propelled weapons with acoustic homing systems, until the increasing speed and depth capabilities of submarines necessitated the return to a thermally powered torpedo. The latest torpedoes are the thermally powered US Mk 50 which has now reached the end of nearly 20 years of development and the Swedish Torpedo 2000. Torpedo development takes a very long time.

Most countries with a history of torpedo development have ongoing development programmes. However, in the present economic climate it is proving increasingly difficult to justify funds for development of a weapon which may take years to perfect. In the light of current situations it is perhaps not surprising to find increasing interest in international collaboration. The latest such venture is that between France and Italy for the development of a new lightweight weapon based on experience with national programmes involving the Murène and A290 weapons.

Whatever the type of torpedo, it has to defeat the quiet running, high speed modern submarine. The torpedo must therefore be faster, must operate passively so as not to betray its presence until within close range when in can switch to active homing, must have sufficient capability to re-attack if it misses first time, must have a highly efficient fuzing system and have an adequate warhead to penetrate the double-hull structure of most submarines. In many cases it must also have a dual capability to operate equally efficiently against both surface ships and submarines – space on board submarines is too precious to carry two types of torpedo. It has to pick up the noise of a quiet running submarine against the background of continuous natural noise, as well as sophisticated acoustic countermeasures. As anyone who has listened to a sonar system will verify, the ocean is a very noisy place because of the sea itself, underwater creatures and man-made traffic. A homing torpedo has to overcome all this, as well as its own flow noise, which at high speed is quite considerable. The result of all these requirements for a modern torpedo involves many years of design and development, not only for the design of the weapon's physical characteristics and propulsion system but, even more importantly, the development of the electronics software and hardware to provide a highly sophisticated onboard sensor, computer and processor. The torpedo of the 1990s is really an underwater guided missile.

There is always a great deal of interest in the composition of the warhead of a modern torpedo and many people suspect that a nuclear charge is often incorporated. Certainly the United States experimented with nuclear warheads on the Mk 45 torpedo, which was subsequently withdrawn from service, and the former Soviet Union equipped some of its weapons with nuclear warheads. Generally, however, the warhead comprises conventional explosives configured in the form of a shaped charge which fires a 'red-hot slug' on impact to penetrate the hull, leaving a relatively large hole. Water pressure does the rest.

Underwater Ordnance

The sea mine is one of the most cost-effective of weapons, being small and cheap in comparison with other weapon systems. It is easy to deploy, easy to hide and creates a physical and psychological effect out of all proportion to its cost and size. Events in the Gulf in 1991 again highlighted the threat that the sea mine poses, as well as the immense efforts that are needed to combat the threat. Modern mines are far removed from the Second World War horned variety, although there are still large numbers of this type around which are just as deadly, again witness the Gulf! However, the most common types are the ground influence mines which are designed to react to either magnetic, acoustic or pressure signatures or a combination of these.

Mines are covered in greater depth in the *Underwater Weapons* section of this book.

Mine countermeasures are carried out mainly by specialist mine countermeasure vessels equipped with mine detection sonars, minesweeping equipment and/or remote operating vehicles which can detect mines, lay charges against them and detonate them from a distance. Divers are also employed in mine disposal operations. One recent innovation in mine detection is the Craft-of-Opportunity (COOP) technique which uses lightweight, towed side-scan sonars that can be fitted to practically any small boat. Again the subject of mine countermeasures is covered in depth in the introduction to the *Mine Warfare* section in this book.

Other areas of prime importance in underwater warfare are: communications, oceanography and hydrography, and signature management.

Communications between a submarine and the surface have relied for many years on the submarine coming to the surface, or deploying an antenna at the surface to receive conventional radio transmissions; sitting just below the surface to receive VLF/ELF communications; or using underwater telephony (that is acoustic communications) for surface ship/submarine transmissions. More recently much development effort has gone into the design of aircraft-to-submarine, or satellite-to-submarine laser communications. These systems have met with a fair measure of success and trials are still proceeding. Communications between land-based/shipborne/air and satellite-based systems and a submerged submarine tend to be rather one-way; for obvious reasons the submarine is most reluctant to transmit.

Acoustic and magnetic ranges and degaussing systems are an essential element in the development of an effective underwater warfare capability. There are many acoustic ranges throughout the world which are used for trials purposes to test torpedoes, countermeasures, sonar systems and practically every type of underwater weapon or device. These vary from large static underwater systems with arrays of hydrophones to small portable systems which can be carried on board ship and deployed when required. Because of the highly sensitive nature of their work, very few details emerge regarding their operation or the results achieved. Degaussing is really a subject in its own right and a number of both static and portable types of system exist around the world. The object of these systems is to assess the magnetic signature of a vessel and neutralise this so that magnetic mines are unable to detect the signature.

Oceanography and hydrography is a vast subject and covers both naval and commercial applications. We have tried to give an overall picture of the subject when viewed from the naval standpoint, but it is such a grey area that it is very difficult to separate the commercial from the naval. (The same applies to underwater remotely operated vehicles.) Many systems which are used in commercial applications relate to currents, tides, sea temperatures, sea bottom geography and formations, movements of marine creatures and many other aspects that are vital to the knowledge of sea conditions and which affect very considerably underwater acoustic and magnetic conditions. Most of these are just as important to the designer of a sonar system or an underwater weapon as they are to an oceanographer.

Conclusion

The field of underwater warfare is a vast subject that has come to the fore over, perhaps, the last 50 years, and certainly the past two or three decades have seen vast strides in design and development of various systems. Efforts have been concentrated in two main areas, on the one hand the submarine, its weapons, detection devices and its self-protection measures; and on the other hand the countermeasures necessary to detect and neutralise them by surface ships, aircraft, static systems and other submarines. This is an ongoing battle between one side and the other. As fast as development on one side allows the building of better, faster, deeper diving submarines with improved weapons, so the other develops better detection devices and more efficient weapons to hunt and destroy the submarine.

Anti-Submarine Warfare

Command and Control and Weapon Control Systems
Submarine Combat Information Systems
Weapon Control Systems
Airborne Acoustic Processing Systems

Sonar Systems
Surface Ship ASW Systems
Submarine Sonar Systems
Airborne Dipping Sonars
Sonobuoys
Static Detection Systems

Countermeasures
Acoustic Decoys
Electronic Warfare

Underwater Communications
Submarine Communications Systems
Underwater Telephones
Communications Buoys

Electro-Optical Sensors
Periscopes
Miscellaneous Electro-Optical Sensors and Systems

Magnetic Anomaly Detection Systems

COMMAND AND CONTROL AND WEAPON CONTROL SYSTEMS

Considerable advances have been made in submarine fire control and combat information systems. These have resulted principally from major developments in the field of electronics and the miniaturisation of components. The overall general effect has been to considerably improve weapons capability, which in turn has demanded greatly improved assimilation and correlation of data from the various sensors, which themselves have been greatly improved in capability due both to the electronic revolution, and to newly developed materials.

The sum total of these developments has had a considerable effect on the design and capability of both weapon control systems and combat information systems. The constant reduction in size and weight of individual systems has a two-fold effect on the submarine. Firstly it has enabled the smaller displacement coastal type submarine to be equipped with fairly sophisticated fire control systems, not always possible previously because of weight and space restrictions. Secondly it has enabled the larger ocean-going submarine to be fitted with an extremely comprehensive fire control and combat information system, considerably enhancing its capability, although not necessarily with any saving in weight or space, for what has been saved by miniaturisation has been put to use in cramming in more data processing capability to improve the overall system effectiveness.

The advent of passive ranging sonar (broadband, wideband and narrowband), intercept sonar, passive cylindrical bow hydrophone arrays, flank array sonar, active sonar, low frequency sonar and towed array sonar all help to provide the submarine with an incredibly large database from which to draw for attack and defence. Other sensors have also developed and been improved in quality such as ESM, radar and optical systems, many of which are now equipped with IR imaging, laser rangefinding, low light TV and line-of-sight stabilisation.

Another area of considerable importance which affects the overall mission capability of the submarine is its navigation equipment. Many submarines now fit inertial, satellite and Omega navigation systems as well as the usual gyro-compass and log. All these help to provide a highly accurate position-finding capability – essential if modern weapons are to be directed accurately onto their target.

This huge bank of data must be assimilated and processed at high speed to generate an accurate picture of the complete underwater and surface environment. It requires programmable high capacity computers, which have the ability to handle such vast amounts of data for a number of targets simultaneously, retrieving the information from different sources, and despatching it to different sources.

The best and most effective way of handling these data so that they are available to as many systems as require them, is undoubtedly via the databus. The ability to have all these data passing round a system has enabled the design of the multi-purpose console which provides enormous flexibility and redundancy, each console being able to take over a variety of different tasks at the command of the operator, depending on the priority and nature of the breakdown.

With such enormous potential and complexity, designers have now moved towards the concept of a totally integrated combat system for the submarine, combining in one total package sensors, combat information and weapon control. To treat these three prime areas as a single system is of considerable value as it overcomes many of the problems of interfacing various systems which have occurred in the past.

With the advent of the fully integrated combat system, the increasing use of automation, and the potentially enormous advantages of a submarine's CO being fully aware and in complete control of everything around him, submarine designers are now integrating all command and control functions within the confines of the control room.

This arrangement is now typical of control rooms which are under consideraton for future generations of submarine. Such a concept is, however, only possible with a high degree of automation and with systems which are completely reliable.

Coincident with these developments is a revolution in the development of masts and sensors. Considerable effort is now being devoted to the development of non-penetrating modular masts for all types of submarine. These will primarily be used to carry an optronics pod to replace the conventional periscope and to carry a TV/visual window, thermal imager, laser system, EW and possibly communications antenna as well. This will enable the number of masts to be reduced, leading to a smaller sail and improved hydrodynamic performance of the boat. The operations room will be free of all mast encumbrances, resulting in an entirely new concept of control room layout and command team relationship. The use of fibre optic links between mast and control will also enable the sail to be re-sited in a much better position on the hull for hydrodynamic purposes.

SUBMARINE COMBAT INFORMATION SYSTEMS

DENMARK

SUB-TDS

Type
Tactical data system for submarines.

Description
The Terma Tactical Data System for submarines (Sub-TDS) is designed to meet modern requirements of combat information systems for coastal submarines.

The Sub-TDS modules are housed in water-cooled cabinets shaped to fit into the narrow space of the hull.

The system provides all functions needed for picture compilation, situation assessment, tactical manoeuvre calculations and weapons employment. It provides a standard tactical data link to exchange track and ESM information as well as plain text messages, utilising existing onboard radio equipment, but may also be supplied with NATO Link 11 or other customer-specified data link.

The Sub-TDS supports the Terma Torpedo Fire Control System (TFCS), which has its own processor cabinet and torpedo control panels. The system provides control of up to four simultaneously fired torpedoes with wire guidance and homing facility. It handles readiness status reports from the tubes and torpedoes, and controls run-up and firing as well as course and depth guidance when under way. The TFCS also supports fire control solutions for straight running torpedoes.

The Sub-TDS consists of the following basic modules which may be added in any number to suit the specific submarine type:
(1) the Graphic Display (GD), which is a 20 in high resolution raster scan colour monitor, portrait oriented, the lower part being utilised for information readout. The GD can display three different types of picture:
  (a) GOP (General Operational Plot), which compiles and displays a complete situation picture with maps, tracks, ESM data, and an extended selection of tactical patterns and tools. The entire system relates to own ship's navigational position and readout positions in Latitude/Longitude, Colour Grid, or true or relative bearing/range as desired. The GOP has a wide scale of editing tools, and may off-centre and zoom to view any local geographical position
  (b) CEP (Contact Evaluation Plot), which plots contacts as a function of time and allows the submarine to produce a situation picture and an attack evaluation based on inputs from passive sensors
  (c) ATP (Attack Plot), which extracts information from the two above mentioned plots, and produces the local attack situation; it also includes all the necessary vector diagrams, forecasts and recommendations for torpedo attack on selected targets.
(2) the System Terminal forms the man/machine interface with a keyboard containing a 'QWERTY' keypad and 60 dedicated function keys, and a high precision rollerball.
(3) DPU (Data Processing Unit), which serves as the main TDS data co-ordination unit and holds all interfaces to the internal TDS modules as well as external interfaces like navigational fix systems, ESM, communications, sonars, hydrophones, periscopes, radar and depth gauge.
(4) TGU (Torpedo Guidance Unit), which is a processor that handles all calculations for launch, guidance and torpedo position determination. The unit interfaces torpedoes and tubes to the entire combat system, and also handles the interface of the Torpedo Fire Control Panel, Tube Order Panel, and the Torpedo Distribution Unit, which are peripheral equipments to the TGU.
(5) Peripheral equipment encompasses recording to tape cassettes or disks, printer and colour plotter. The on-line functions are the recording of all events within the system and the loading of maps for display. Off-line the system may produce playback to the screens for lessons learned, and printed records of various selections and formats, such as track records or narratives.

Operational Status
In production and sold to the Royal Danish Navy for an update programme of the 'Kobben' class and the 'Narhvalen' class submarines.

Contractor
Terma Elektronik AS, Lystrup.

FRANCE

SUBICS

Type
Submarine integrated combat system.

Development
SUBICS is a modular family of integrated combat systems developed to comply with the performance requirements of most navies and to provide growth potential for any new requirements.

Description
SUBICS is a fully integrated system combining all available information from sonars, optronics, ESM, radar, data links and so on using common standard processors and Colibri multi-function consoles. The consoles feature large (up to 19 in) TV screens, high resolution, ergonomic design and reduced space requirements for use in submarines. SUBICS features a very high level of automation with computer assistance. It incorporates a broad-base design of acoustic arrays covering a variety of performances and sizes offering a wide range of functions.

The multi-array sonar and multi-processing techniques enable automatic sonar detection and tracking to provide expanded range, bearing and frequency spectrum coverage. Fully automatic processing of the high number of contacts resulting from the long detection ranges reduces the operator's workload and enables him to focus his attention on non-recurrent tasks. The system also features anti-jamming techniques. Target localisation is achieved using a wide range of highly efficient operating modes including discrimination improvement with direct passive ranging. Automatic Target Motion Analysis (TMA) is applied to all sonar tracking channels and interactive TMA to operator-designated tracks. Identification is achieved using radiated target noise and sonar pulse analysis by hi-fi audio and spectrum analysis with computer database comparison of pre-recorded signatures. ESM and optronics classification data are also used when available.

Other features include automatic sonar multi-array multi-processing track association and data fusion using computer-aided procedures to process acoustic and non-acoustic data for operational decision making. Command and control functions feature a comprehensive overview of the tactical situation on a console and a large screen command plot. Threat assessment is provided including torpedo alert. A wide range of computer-aided procedures assists the command in decision making, improving submarine covertness and performance and enabling effective attack or escape planning to be implemented, together with weapon engagement. Engagement facilities include the adaptability to different types of torpedo and missile, tube/target assignment, weapon loading and computing preset parameters, launch sequence and torpedo wire guidance with availability of concurrent missile/torpedo salvos.

SUBICS ensures maximum performance levels for the submarine platform characteristics (size, speed and noise), a reduced operating crew, high flexibility and resistance to failure, and a design tailored to the submarine's size and requirements.

Emphasis has been put on this facility to operate the combat system through menus and user-friendly dialogues.

Currently under development is a version of SUBICS that is adapted to meet the specific requirements of 'mini' submarines.

Operational Status
Completing development; subassemblies are undergoing trials at sea.

Contractor
Thomson Sintra Activités Sous-Marines, Valbonne Cedex.

Thomson Sintra integrated combat system in a mock-up submarine control room

GERMANY

OSID

Type
Integrated submarine combat system.

Description
OSID has been designed to provide small- and medium-sized submarines with an integrated combat suite for surveillance, target tracking, target analysis and classification and weapon control. The system integrates the following sensors: cylindrical or conformal arrays; intercept arrays; own noise hydrophones and accelerometers. These sensors and computer algorithms for data evaluation and weapon control provide target and situation data through performance of the following functions: broadband detection over 360° with high bearing accuracy; sonar pulse detection over 360° in a very broad frequency band; detection of self-generated noise; automatic and interactive tracking of targets; audio analysis of broadband noise and sonar pulses, transposed into the optimum audio band; DEMON analysis of broadband noise; target classification aided by a computer database; sonar signal analysis with respect to carrier frequency, modulation, pulse length and pulse repetition time; launching suggestion for weapons; control of wire-guided torpedoes and display of target/weapon situation.

Contractor
Krupp Atlas Elektronik, Bremen.

ISUS

Type
Integrated combat system.

Description
ISUS is an integrated submarine combat system designed to acquire, process, analyse and display data from all sensors fitted in the boat. This interactive process leads from the display of raw sensor information to an easy-to-read display of the combat situation, providing the command team with optimum assistance in the decision making process.

The system integrates a wide variety of acoustic arrays including: cylindrical or conformal; flank; towed; passive ranging; cylindrical transducer; and intercept, as well as own noise hydrophones and accelerometers; Doppler and EM log; underwater telephone; radar; ESM; optic and optronic sensors and navigation sensors.

The system performs the following functions on data either sequentially by one operator or in a task sharing concept simultaneously by several operators: broadband detection over 360° with high bearing accuracy; sonar pulse detection over 360° in a very broad frequency band; detection of low frequency spectral lines over a long range with high accuracy; transient detection and storage; accurate range measurement in two 120° sectors; moored mine location; detection of own hydrodynamic and machinery noise; location of targets by single pulse active sonar; navigation; integration of periscope and radar displays into the multi-function consoles; automatic and interactive tracking of targets (the target track is controlled by one main sensor and supported by auxiliary sensors); audio broadband and narrowband analysis (the analysis leads to computer-aided classification suggestion); threat analysis; launching suggestion for weapons; integration of the torpedo sonar information from wire-guided torpedoes; control of wire-guided torpedoes and display of the target/weapon situation.

Contractor
Krupp Atlas Elektronik, Bremen.

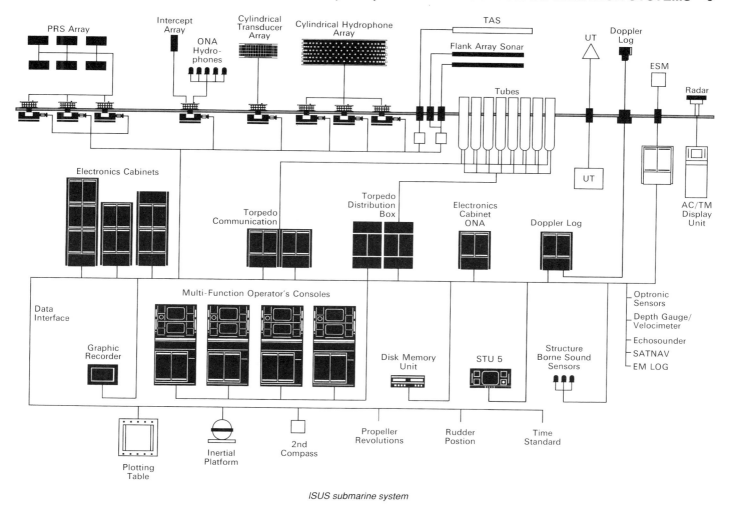

ISUS submarine system

ITALY

MM/BSN-716

Type
Submarine action information system.

Development
SACTIS is the abbreviated title given to the submarine action information system developed for Italian Navy submarines of the 'Sauro' class by SMA in collaboration with Datamat. The former concern was selected as prime contractor for this project in 1981 when it was decided to equip this class of submarine with a new combat information system. Two versions are produced, known as SACTIS 1 and 2, or by the Italian Navy designations MM/BSN-716(V)1 and MM/BSN-716(V)2, respectively. The two versions are intended for fitting all the 'Sauro' class boats.

Description
In both versions the principal functions include:
(a) automatic data acquisition from all sensors (for example, hydrophone, sonar, passive ranging system, search and attack periscopes, radar, ESM, navigation system, depth sounder)
(b) manual data input
(c) real-time computation of ship's position and target position (from bearing data only)
(d) data display (raw and processed) with four different presentations: unfiltered situation; tactical situation; time/bearing display; tactical operation tabular evaluation
(e) calculation and display of data for typical manoeuvres such as screen or barrage penetration, evasion, target interception or collision course, approach and divergence routes, closest point of approach (CPA)
(f) data and event recording
(g) playback for training and analysis.

The hardware comprises: two or three display groups which have vertical CRT displays and keyboards for data input and system management, with separate alphanumeric 'tote' displays above the situation display CRTs; a central processor unit based on a Rolm MSE 14 digital computer; disk memories; a printer; a specially designed interface unit for input/output between SACTIS and the various sensors. The displays are arranged side-by-side to form an operating console and up to 30 separate targets can be displayed, from which individual track histories can be selected for examination and 10 of them presented as filtered targets. Main sensors connected to the system include the BPS-704 radar, IPD-70S sonar system and the Thetis electronic warfare system.

The later SACTIS 2 version includes provision for connection with the submarine's A184 torpedo fire control system, inclusion of the ELT/810 sonar prediction system and connection to a Link 11 receiver. There is also extended computer capacity and a third operator position at the SACTIS console.

Operational Status
Italian 'Sauro' Class submarines are fitted with the MM/BSN-716 system.

Contractors
SMA SpA, Florence (main)
Datamat SpA, Rome (software).

Three-console SACTIS system

SICS

Type
Submarine integrated combat system.

Description
The Italian Submarine Integrated Combat System (SICS) has the following functions:
(a) sensor and tactical data handling
(b) fire control
(c) command
(d) common services.

The system hardware is designed in a compact, modular fashion to permit adaptation or expansion as required. Data and video distribution network, advanced computer technology and sensor data processing for optimum weapons utilisation are included also. Growth potential was considered essential to permit changes in sensor and weapons outfit as modifications to operational procedures are unavoidable during the life of a modern submarine.

The system uses standard consoles which are completely interchangeable for operational functions. This allows consoles to be easily reconfigured to accommodate changing role requirements, and also permits consoles to change their roles while the boat is 'on patrol', according to varying operational

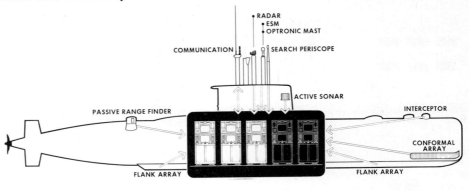

Outline of typical SICS arrangement

circumstances, to permit a reduction in manning requirements whenever possible.

The system's sensor and tactical data handling controls the processing of inputs from sonar, periscope, ESM, radar, log and compass, and other navigational equipment.

The FCS function provides for TMA solutions, weapon status monitoring, computation of optimum target engagement and the computation of weapon presettings before launch.

The co-ordination, evaluation, designation and other command functions are carried out by the system, together with comprehensive monitoring and display facilities, using data directly from sensors to produce a filtered situation picture.

Operational Status
Advanced development.

Contractors
Sistemi Subacquei WELSE SpAC, Genoa.

FCS Mk 3

Type
Submarine command and control systems.

Description
The *Società di Elettronica per l'Automazione* (SEPA) member of the FIAT Group, in addition to participating in the WELSE underwater warfare consortium and contributing to the SICS programme, independently offers naval fire control systems (FCSs) for various other applications. Such digital systems are mostly based on the use of SEPA militarised mini- and microprocessors of the company's ULP Series. For example, a developed version of the torpedo and weapons control subsystem for SICS is available separately as the FCS Mk 3 for submarine command and control, which will be used in the Improved 'Sauro' class submarines.

The FCS Mk 3 is a fully integrated submarine command and control system having as its main functions:
(a) acoustic sensor performance prediction
(b) prediction of counter-detection range
(c) target motion analysis
(d) threat evaluation
(e) target designation for attack
(f) weapons control (torpedoes and missiles)
(g) countermeasures control.

Sensor inputs to the system include radar, navigation and attitude information, periscope and acoustic data from the integrated active/passive sonar system.

Target and ship's own data from these sources are used to calculate and display target positions and target vectors, impact point predictions, tactical situation, a launch and guidance display, and wire guidance signals. The local control and switching box connects the signals from the FCS to the selected torpedoes, provides launching tube control, and feeds electric power to torpedoes in the tubes. Within the torpedo, the electronic control unit carries out wire signal interfacing, torpedo steering in accordance with

FCS commands, computation of target data from the torpedo acoustic head, and torpedo homing.

Inputs to the system include: active and passive sonar data, navigational information, radar and ESM/EW sensors, and periscope. Weapons include A184 torpedo tubes, which are connected via a local control and switching box.

Operational Status
In production.

Contractor
SEPA, Turin.

Consoles for the SEPA FCS Mk 3

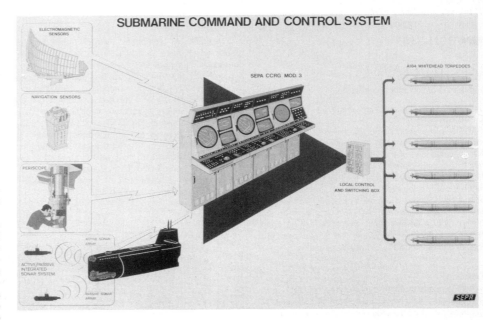

Schematic diagram of FCS Mk 3 for submarines

NETHERLANDS

GIPSY

Type
Submarine data handling and weapon control system.

Description
Gipsy is an automated data handling and weapon control system designed for the 'Walrus' class submarines of the Royal Netherlands Navy. The system forms the link between the submarine's sensors and its weapons and consists of seven identical display and computer consoles (DaCCs). The built-in computer is of the SMR-MU type; the 16 in (406 mm) plan view display (PVD) provides a high-load, synthetic picture together with compressed radar video or sonar video, and the control panel is a multi-purpose unit. The use of identical DaCCs offers maximum flexibility.

A central control unit (CCU) is used to regulate the mutual data transfer between the DaCCs and the sensors and weapons. For this function two SMR-MU computers are provided, one active and the other a hot standby machine. The large amount of data from the sonars is handled by two extra SMR-MUs, also housed in the CCU. Three types of sonar are fitted (towed array, flank array and circular array), and other sensors and data sources include a noise analyser, ESM facilities, radar, periscopes and position finding equipment.

The weapon control system can control a mixed load of weapons for sub-surface and surface engagements, and the ship's launching system consists of two mutually independent sections, each of which is controlled from a launching system control panel (LSCP). The interface with the weapons (modern sub-surface missile) is formed by two identical distribution cabinets. All hardware necessary for the integration of these weapons is included in the distribution cabinet, avoiding the necessity for additional equipment.

The main functions of Gipsy are:
(a) sonar display and control. This makes it possible to omit the original sonar displays and controls
(b) contact evaluation. This entails displaying information obtained from all sensors in a time/bearing format
(c) classification, where ESM, ASM and noise information is compared with cassette tape stored libraries to provide a rapid classification of the target
(d) contact motion analysis. Modern tracking filters are used for the automatic determination of target movements
(e) tactical plot and general plot functions, giving an up-to-date survey of the tactical and/or navigational situation or an historical situation survey
(f) weapon control, whereby the weapon systems are provided with the requisite aiming data
(g) simulation and test to verify the system's operability and to aid in operator training for the functions (a) to (f).
Additional functions can be inserted easily due to the fact that the system is highly software oriented.

Operational Status
Four systems have been delivered for fitting in the boats of the 'Walrus' class for the Royal Netherlands Navy, together with a reduced system for shore-based training.

Contractor
Hollandse Signaalapparaten BV, Hengelo.

SINBADS

Type
Submarine integrated battle and data system.

Description
SINBADS is a compact data handling and weapon control system suitable for use in submarines of the small coastal type and up to the larger, ocean-going type. It succeeds the Signaal M8 Series of torpedo FCSs for submarines. The computer used is the Signaal fourth-generation general-purpose machine SMR-MU, and the complete system combines the weapon control and data handling functions.

The data handling function covers sensor display and sensor selection for track initiation. The raw sensor data of all sensors can be displayed. Each individual sensor is indicated with a unique label. The development of SINBADS is based on the application of modern passive sonars, capable of automatic tracking of targets. The tracking algorithm is based on modern filtering techniques and the tracking system basically includes as its main mode of operation a 'bearing-only' analysis which, however, also accepts other inputs of target information. The complete tracking system functions as an interactive system in which the command can intervene in the tracking process. For this purpose there are four display formats available.

SINBADS can handle five targets and three torpedoes simultaneously. These can be guided, unguided or a mixture thereof. The system also includes data recording for on- and off-line use, weapon simulation for training, on-line failure monitoring, and an emergency mode for which hand controls are provided to enable torpedoes to be set, fired and controlled in case of a computer or display failure.

Operational Status
Produced for the following navies: Argentina, six for Santa Cruz (TR-1700) type; Greece, four for 'Glavkos' (209) class; Indonesia, two; Peru, four for 'Casma' (209) class; Turkey, four for 'Atilay' (209) class.

Contractor
Hollandse Signaalapparaten BV, Hengelo.

SPECTRUM I AND II

Type
Submarine combat information system.

Description
SPECTRUM I is a fully integrated combat system designed to carry out all tactical functions for small displacement and larger patrol submarines. A particular feature is the full Fourier frequency analysis on all pre-formed beams and the use of this information for spectral and spatial normalisation of sonar information before narrow and broadband detection. This provides a digital system that automatically optimises detection, analysis and contact tracking under all conditions.

In its full configuration, SPECTRUM I includes a multiple array broad and narrowband sonar suite (SIASS), a command system including interchangeable multi-function display control consoles, a ship's integrated communication system (SINCOS) and a weapon control system compatible with modern wire-guided torpedoes and air-flight weapons.

A navigation radar (ZW07) also forms part of the complete system configuration. SPECTRUM I subsystems are fully integrated and incorporate a software-based multi-processor modular design.

Data collected by SPECTRUM's sonar suite and other sensors (such as ESM and radar) are processed to localise and identify contacts. The search results from the multiple sensor fit are correlated and displayed on the interchangeable consoles to generate a concise tactical picture. From this information decisions are made about target designation and priorities, threat assessment, weapon assignment and tactical manoeuvring of own ship. Weapon employment from warm-up, presets through launch and post-launch guidance are accomplished by the weapon control function. SPECTRUM I also incorporates a tactical data handling function and functions for navigation, data logging and crew and operator training and is capable of integration with electro-optical sensors (periscopes) and electronic navaids.

Sonar suite (SIASS)
The sonar suite consists of:
(a) a Cylindrical Array Sonar acting as the primary attack sonar. The combination of narrow and broadband processing is to extend detection ranges, to enhance tracking and to provide quick and accurate classification of contacts
(b) a Flank Array Sonar, using 24 acceleration cancelling type hydrophones on either side of the boat, serves as long-range detection and tracking sonar especially for the lower frequencies
(c) a Passive Ranging Sonar with three arrays on either side that determining target ranges. The passive ranger's signal processing uses a method of frequency-domain correlation to measure wavefront curvature and thus provide range information. The use of variable tracking bandwidths reduces the influence of unwanted noises and maximises the signal-to-noise ratio
(d) an Intercept Sonar providing coverage up to 90 kHz and employing full digital processing utilising fast Fourier transform techniques to detect, correlate and classify intercepts of sonars and torpedoes. Automatic alarms are given for a number of preset threats.

Contacts in automatic track are automatically subjected to passive bearings-only Target Motion Analysis (TMA). A new generation TMA algorithm is employed which has demonstrated speed and accuracy at sea trials. An electronic classification library is included with a capacity of up to 300 specific platforms.

Weapon control
Two identical weapon interface cabinets are incorporated in the combat system, each containing interfacing and processing equipment required for the control of the weapons in one half of the launching system. Normal and emergency modes are provided to ensure high operational availability.

The weapon control function is compatible with various submarine launching systems and submarine-launched weapons with three firing modes provided. The system is capable of simultaneous launch and control of four modern acoustic wire-guided torpedoes and four submarine-launched missiles.

Tactical data handling
The tactical data handling complex of SPECTRUM I comprises two redundant computers, interfaces to own ship's sensors and a data logging capability. The complex gathers data from all subsystems to compile a comprehensive picture of the tactical situation, correlates data from the various subsystems, supports threat assessment, threat priorities, weapon assignment and tactical manoeuvring, provides aids to navigation such as CPA calculations, stationing and position fixing, and carries out environmental predict and system control and monitoring.

SPECTRUM II is an advanced submarine combat system based on SPECTRUM I. The system includes all of SPECTRUM I system functions using standardised processing and display components recently developed for the 'Karel Doorman' class frigate programme of the Netherlands. This has resulted in the following major upgrades:
(a) the multi-function operator consoles are equipped with high resolution (1376 × 1024 pixels) colour raster scan displays capable of showing brightness modulated sonar pictures and radar video
(b) system data processing is distributed between

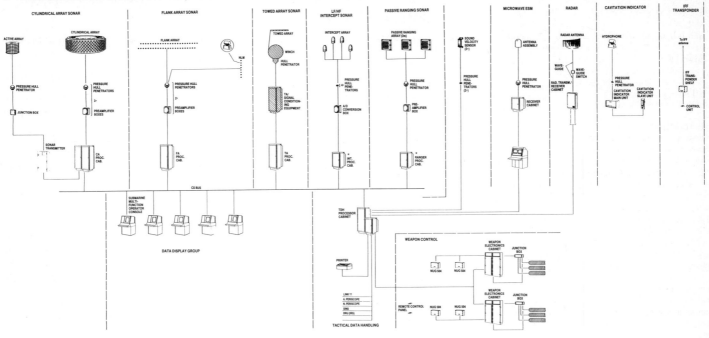

SPECTRUM II main elements and sensor systems

several 68020 family-based general-purpose processors. Distribution of processing tasks and resources is shared co-operatively

(c) an Ethernet-based databus is used to interconnect the various processors within the system. It transmits information over two mutual independent triaxial cables at a rate of 10 Mbits/s. Distributed control of the network is fundamental to the Signaal databus architecture

(d) for storage and retrieval of program, library and

mission data use is made of two digital optical recorders capable of storing up to 250 Mbits of data on compact disc.

All major processing electronics are housed in a newly developed standard 19 in cabinet. Features of the new design include Eurocard construction format, standardised power supplies and a flexible cooling system. The cabinets can be delivered in three different heights for optional accommodation on board submarines.

Operational Status
SPECTRUM I is operated in the 'Seadragon' class submarine. SPECTRUM II is on offer to various navies.

Contractor
Hollandse Signaalapparaten BV, Hengelo.

NORWAY

MSI-90U

Type
Submarine basic command and weapons control system.

Development
MSI-90U has been designed to meet the requirements of the Royal Norwegian Navy and the German Navy for their next generation of submarines, the Norwegian 'Ula' class and German 'U212' class.

Emphasis has been placed on achieving a highly flexible and capable system with extensive redundancy and independent operation of the different subsystems. Careful attention to standardisation in the highly modular system should considerably assist in limiting through-life support costs.

An MSI-90U evaluation system has been undergoing rigorous testing in a Factory-based Test Stand (FTS) since 1986. This FTS consists of a complete MSI-90U system and utilises VAX computers for simulation of all onboard sensors, torpedoes and tactical scenarios. The FTS is also used for education and training of submarine crews in system operation and maintenance.

Description
MSI-90U is a software-based command and weapons control system which uses distributed processing, a high capacity serial data transmission system (LAN), and multi-function operator consoles to achieve a high degree of capability, flexibility and availability. The main features of the system are:

(a) distributed data processing using the 32-bit KS-900 general-purpose computer designed for real-time processing, based on the commonly used Motorola 68000/68020 range of microprocessors. The modular KS-900 is designed for programming in Ada, Pascal and C

(b) a high capacity Local Area Network (LAN) used for data communication between subsystems of

MSI-90U in operation aboard HNoM Ula

MSI-90U, between MSI-90U and sensors, and sensor to sensor communication

(c) multi-function operator consoles allowing every operator to have access to all information in the system as well as permitting the number of operators to be adjusted to suit the current tactical situation. This means that one operator can control the complete system with one console in, for example, a patrol situation. In the standard

MSI-90U configuration four identical multi-function operator consoles are used, which can be configured for different tasks. The number of consoles can be adapted to customer requirements

(d) built-in redundancy allowing subsystems to operate separately in graceful degradation fallback mode.

The LAN, named BUDOS, uses standardised interfaces which comply with NATO STANAG 4156.

Standard interfaces such as RS 422 are also available. Via BUDOS any subsystem can communicate with any other connected subsystem. Connection of new subsystems can easily be made by means of spare interfaces on the existing BUDOS multiplexers, or by adding more multiplexers.

Adequate redundancy is assured by multi-function operator consoles (each with its own KS-900 computer), three main KS-900 computers and two or three KS-900 weapon computers (depending on weapon type and customer requirements).

Information is presented on two high resolution colour raster scan displays. The man/machine interface is via a programmable entry panel (a plasma display with touch sensitive overlay matrix), tracker-ball with associated control buttons and standard alphanumeric keyboard. The concept of the MMI, including layout of display pictures, has been subject to extensive MMI studies and trials in co-operation with the Norwegian Defence Research Establishment and the Royal Norwegian Navy.

The system is designed to carry out sensor integration covering target motion analysis, classification and identification, and weapons assignment and control. A number of supplementary facilities are also provided, such as tactical evaluation and navigation, threat evaluation, engagement analysis, pre-programmed movements, sound trajectory calculations and presentations, predicted sonar ranges for own ship and hostile ships, presentation of geographical fixed points and areas, data recording, and simulation for training purposes and so on.

Operational Status

Delivery of MSI-90U for the Norwegian 'Ula' class submarines has been completed, six systems having been delivered between 1988 and 1991. After installation each system has successfully undergone Harbour Acceptance Trials (HATs) and Sea Acceptance Trials (SATs). The SATs also included successful firings of DM2A3 torpedoes. The lead ship of the 'Ula' class entered service with the Royal Norwegian Navy in April 1989 and the remaining five boats will be in service by the end of 1992.

Contractor

Norsk Forsvarsteknologi A/S, Kongsberg.

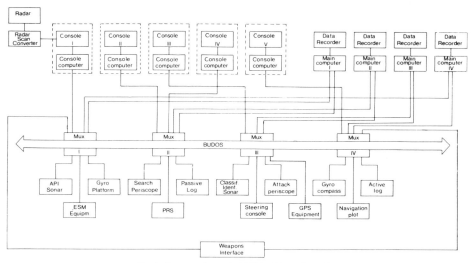

Block diagram of a typical MSI-90U configuration

SWEDEN

NEDPS (NIBS)

Type

Submarine action information and fire control system.

Description

The NEDPS (Näcken Electronic Data Processing System (in Swedish, NIBS)), was developed for use in the 'Näcken' class of submarines of the Swedish Navy.

This is a fully integrated system designed to acquire, process and display information for tactical evaluation and to form a basis for decisions regarding selected targets and torpedo guidance. The system includes complete hardware and software for controlling wire-guided homing torpedoes. Also included is computer control and monitoring of ship's functions, such as propulsion, steering, depth keeping, trim and storage battery condition.

The system has a dual computer configuration, the information being gathered through an extensive data collecting system from the different surveillance and weapon systems as well as from the navigation system. The information is presented to the operators on two separate displays, each of which is equipped

The Mk 2 action information/fire control system consoles as fitted in the 'Västergötland' class submarines

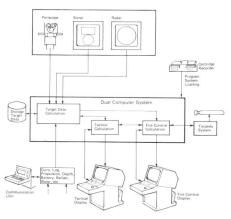

with input facilities in the form of special keyboards and tracker-balls. Each operator console incorporates three weapon panels in the upper part of the console together with an alphanumeric display, which presents supplementary information such as data link messages and ESM data. After target acquisition the tracking is carried out automatically and relevant fire control data are calculated continuously. Upon firing, the torpedoes are normally controlled fully automatically with graphic and alphanumeric presentation of all relevant information. The operators at the tactical and fire control displays can at any time take over or adjust the procedure. The operational plot provides an extensive range of supplementary information, including: own submarine operational area; own submarine movements together with navigation reference points and times; operational areas for other submarines; minefields; coastal radar stations with radar parameters. The plot also displays up to 50 internal labelled targets, 10 target channels for target-motion-calculated targets; 240 external (via data link) targets with target numbers and identities; and 32 ESM lines.

The tactical plot is used for the passive sonar TMA calculations on up to 10 targets simultaneously. On the 'Näcken' class this plot is manned by the executive officer who supervises the fire control operator as well as the sonar, radar and ESM operators. He also handles the periscope observations made by the commander.

Operational Status

The Mk 1, or NEDPS, is in service with the 'Näcken' class submarines while the Mk 2, which uses software developed from the Mk 1 with a new single multi-function console, is in service with four 'Västergötland' class submarines and three 'Sjöormen' class submarines and is being delivered for retrofitting the 'Näcken' class submarines. Retrofits have been studied for HDW Type 205 and Type 209 submarines. One system has also been installed in a training simulator at Berga Naval College.

NobelTech also has in production a new generation of command system for the next generation of Swedish submarines, the 'A-19' class. This will use state-of-the-art hardware from the base system 2000 (also being used in the ANZAC frigates and so on) and further developed software written in Ada.

Contractor

NobelTech Systems AB, Järfälla.

The Mk 2 action information/fire control system fitted as part of the mid-life update on the 'Sjöormen' class submarines

UNITED KINGDOM

DCB

Type
Submarine tactical data handling and integrated fire control system.

Development
To reduce the manpower requirements of attack teams the fitting of digital computer systems in the Royal Navy's nuclear submarine fleet began in the early 1970s with the combat information System DCA. System DCB was developed from System DCA in the late 1970s when a second digital computer, for fire control, was integrated into the system. Development of DCB has continued with regular software updates to incorporate new sensors, weapons and tactics. System DCB is in service in all existing nuclear-powered Royal Navy submarines.

Description
System DCB is a well proven tactical data handling system with integrated fire control facilities. The system comprises a two-position combat information console and a two-position fire control console installed in the control room; with weapon interface and battery monitoring equipments installed in the weapon compartment; and the combat information and fire control computers (two FM 1600B computers) and associated electronics installed in the computer room. All console positions are provided with two CRT displays, one pictorial and the other for tabular tote information. The main operator interaction with the system is through light pen selections on the displays, supplemented by dedicated switches.

The submarine's sonars, periscopes, ESM and navigation equipments are linked to the combat information computer by S^3 (Serial Signalling System) interfaces, and synchro and resolver inputs which are converted in computer peripheral units and input as digital data. The computer processes the received sensor data into tracks. Control of track association can be exercised by the operator. Additional information such as classification and intercept data can also be input by the operator in coded digital form from the keyboards. Target motion analysis, using bearing-only techniques, is carried out for all contacts. The tracking solutions and the sensor data received are compiled into a tactical picture that is shown in plan, time/bearing or time/frequency form on the combat information console pictorial displays. To assist in the evaluation of the tactical picture a variety of tactical calculations can be carried out at operator request. The results of these calculations, together with amplifying data on contracts, can be presented on the tote displays. Contact data are also output to the Automatic Contact Evaluation Plotter (ACEP). Digital data on selected targets are passed to the fire control computer where fire control solutions for either wire-guided torpedoes or anti-ship air flight missiles are calculated. The solutions are shown in plan form on the pictorial displays at the fire control console while the weapon settings are shown in tabular form on the tote displays. Weapon settings, tube orders and weapon orders are passed to the weapon interface

System DCB console

equipment as digital words on duplicated serial highways. The weapon interface equipment decodes the digital data and applies power, switch signals, discharge gear drive pulses and guidance commands to the weapons and tubes as necessary.

After firing, data feedback received from the wire-guided torpedoes is passed to the fire control computer on duplicated serial highways. The fire computer decodes the feedback data and generates steer commands to the weapon if required as well as updating the fire control information on the displays.

The duplication of the serial highways for weapon data and the cross-linking between the weapon compartment units provides a choice of fallback modes to maintain a weapon firing capability.

Operational Status
In service with all current Royal Navy nuclear submarines. Hardware update with F2420 computers is well in hand and all FM1600B computers will be replaced by the end of 1992.

Contractors
Ferranti Computer Systems Limited, Naval Command and Control Division, Bracknell.

Dowty Maritime Command and Control Systems Limited, Feltham.

DCC

Type
Submarine tactical data handling and integrated fire control system.

Development
System DCC is the tactical data handling and fire control system developed for the 'Upholder' class SSKs for the Royal Navy. The system is derived from System DCB which is in service with Royal Navy nuclear submarines.

Description
System DCC comprises a three-position console (with electronics in the lower compartments) in the control room, supported by computer equipment in the

computer room and weapon interface equipment in the forward part of the submarine. The main system computer (FM 1600E) is redundant; a second computer being provided to run the system should a fault occur in the primary computer, thus ensuring that the facilities of DCC are maintained.

Each operator position provides two circular 12 in cursive display units, one for pictorial displays and the other for tabular tote data, a light pen and a fire control order and status unit keyboard. The main human computer interface with the system is provided by the light pens, but certain selection and switching facilities for fire control are provided on the console keyboards. Each operator position provides full combat information and fire control facilities, allowing increased flexibility in the manning of the system, such that a single operator can control the complete system. There is close integration between the

sensors and System DCC with sensor data being passed to DCC via a digital databus. The sensors passing tactical and environmental data to DCC include: long-range passive sonar, medium-range passive/active sonar with cylindrical array located in the bow, sonar intercept, passive ranging sonar using three aligned arrays on each side of the hull, underwater communications, bathythermograph, and sound speed-measuring system. The periscope masts are provided with facilities for fitting a variety of electronic and electro-optical devices. A Type 1006/7 surface surveillance and navigation radar and omnidirectional and directional warning antennas for the ESM are carried on a separate mast or incorporated in the search periscope.

The sensor data are collated, processed and presented in plan, time/bearing or time/frequency form on the pictorial displays of the console, with supplementary or amplifying data being shown on the tote displays. Contact data are also output to the Automatic Contact Evaluation Plotter (ACEP). The processing, which includes target motion analysis and tactical calculations, is provided to support the command in the assessment of the tactical situation and the decision making processes.

The weapon fit for the 'Upholder' class includes dual purpose wire-guided torpedoes (anti-submarine and anti-ship), anti-ship air flight missiles, and mines. The fire control facilities for these weapons are inte-grated into the system and include target designation, weapon and tube preparation, weapon launch and, where appropriate, post-launch guidance. Weapon settings, tube orders and weapon orders are passed to the weapon interface equipments which apply power and switch signals to the tubes and weapons during the preparation and launch phases. After firing, the data feedback from the wire-guided torpedoes is passed to the central computer which generates steer commands as required, and also updates the fire control information on the displays.

Operational Status
Four systems have been delivered for fitting in the four 'Upholder' class SSKs. The systems fitted in the first three submarines are undergoing acceptance trials.

Contractors
Ferranti Computer Systems Limited, Naval Command and Control Division, Bracknell.

Dowty Maritime Command and Control Systems Ltd, Feltham.

KAFS

Type
Submarine combat information and fire control system.

Development
KAFS is the submarine system variant of the Ferranti Modular Combat System range developed for the export market in the mid-1980s. The development of KAFS reflects Ferranti's experience in submarine combat information and fire control systems, gained from the development of Systems DCA, DCB and DCC for the Royal Navy.

Description
KAFS is suitable for conventional submarines of all sizes. The system uses the latest processing tech-niques and software algorithms to provide a wide range of operational facilities including sensor data handling, target motion analysis, picture compilation, fire control, tactical calculations and data recording. A comprehensive onboard training facility enabling training in both combat information and fire control procedures to be carried out in harbour or at sea is also included. The system hardware and software has been specifically designed in a compact, modular manner to provide a range of facilities which can be adapted or expanded as necessary to meet individual customer requirements.

The equipment configuration is a console in the control room linked by a MIL-STD-1553B databus to the weapon control equipment in the torpedo room. The torpedo room equipment comprises a local control panel which provides common services together with two weapon interface units and two tube switching units. The weapon interface and tube switching units are provided on a sided basis, port and starboard, so that in the event of a failure in the equipment of one side the other remains operational. Data from the sonar, search and attack periscopes, ESM, radar and navigation sensors are also passed to the system via the databus.

The control room console provides two identical operator positions mounted above two electrical sec-tions which house the central computer and all the associated electronics. Full combat information and fire control facilities are provided at each console operator position and either operator can control the whole system thereby allowing one position to be shut down in patrol state. Each console position has two CRT displays, two command data panels, and a weapon control and system status panel. The two display areas provide the operator with a label plan, time/bearing or vertical section display and a tote display providing pages of data and selections to enable him to interact with the system using a light pen. Track data, intercept alarms and tube and weapon status are displayed on the combat informa-tion and fire control command data panels. Tube and weapon preparation orders are passed to the torpedo room using the weapon control and system status panel, which also has facilities for manual guidance

KAFS console

post-launch and system control as well as status indications.

The weapon control equipment in the torpedo room, which is microprocessor-based, provides the interfaces to the tubes and weapons. Tube and weapon status is displayed on the local control panel which also provides weapon discharge and control facilities in fallback mode, independent of the central processor and facilities for weapon battery heating together with monitoring to detect hazardous condi-tions. The weapon interface and tube switching units apply switch settings and power supplies to the tube and weapons during the preparation and launch phases. After-launch guidance commands, if appro-priate, are also routed to the weapons via these units.

Operational Status
The Brazilian Navy placed an order for two systems and a shore training system in 1984 followed by an order for two further systems in 1987; the systems are being fitted in HDW Type 209 submarines. The first boat system is now fully operational – including suc-cessful torpedo firings — and the shore system has been installed and accepted in Brazil. The three remaining systems which were built under subcon-tract by SFB Informatica in Rio de Janiero have been delivered to the Mariuha do Brasil.

Contractor
Ferranti Computer Systems Limited, Naval Command and Control Division, Bracknell.

DCG

Type
Submarine tactical data handling system.

Development
DCG is a tactical data processing system which has been designed, developed and built to support the command and control system in UK submarines. The system exploits commercially available computing units to achieve high performance and high reliability within a low lifetime cost and short procurement lead times.

Description
System DCG can operate independently although it is linked to the main submarine command (action infor-mation) computer from which it derives data describ-ing the submarine's own movement and the perceived environment. System DCG subjects these data to rigorous and intensive analysis to produce a more accurate picture of the tactical situation.

While the system was developed originally for UK submarines, connections to any automated command system may readily be achieved using standard interfaces. Alternatively, the manual data entry capability enables System DCG to be used independently.

Outfit DCG consists of:

(a) an array processor which is able to perform high speed matrix mathematics. It executes up to eight million floating point operations per second: about one hundred times the speed of a conventional minicomputer

(b) a mass memory device to hold long timespans of data necessary to detect trends in slow-moving events

(c) sophisticated graphics terminals to present the data in easily assimilated picture form

(d) lightweight keypads with an ergonomic layout specifically designed for DCG interaction

(e) a general-purpose minicomputer to organise the data and control the devices

(f) a high resolution colour graphics printer.

The skill of the system integrator consists in selecting matched units which, when combined, deliver the system capability and the system reliability demanded.

System DCG supports an extensive catalogue of tactical software, and the operator terminals are able to perform different functions simultaneously.

These command aids, tracking and combat functions are accessed through a menu driven operator interface, with emphasis on a common style of interaction, data compatibility between functions and ease of extension.

Operational Status
A total of 37 systems in service with the Royal Navy.

Contractor
Dowty-Sema Ltd, Esher.

Console configuration of System DCG

COMKAFS (OUTFIT DCH)

Type
Submarine combat information and fire control system.

Development
COMKAFS is a range of compact submarine combat information and fire control systems suitable for all classes of conventional submarine, but the flexibility in equipment configuration makes it especially suitable for installation in smaller submarines. COMKAFS is a variant of the KAFS system fitted in a number of submarines, and reflects the design experience gained from the supply of Systems DCB and DCC to the Royal Navy. The Royal Navy version of COMKAFS is known as Outfit DCH, which is being fitted as part of the 'Oberon' class update programme.

Description
The use of advanced computer technology, the latest distributed processing techniques and modular software results in a compact and powerful system. The operational facilities provided include: sensor data handling; target motion analysis; picture compilation; tactical evaluation; target indication; fire control; navigation and sonar ray path analysis. COMKAFS can engage surface and subsurface targets and assist the command during the participation in ASW exercises and in covert operations such as minelaying, periscope reconnaissance and intelligence gathering. Onboard training facilities, built-in status monitoring and fallback modes of operation are included.

The equipment fit can be varied to suit the particular class of submarine, as an example the DCH system architecture is shown. The control room equipment is linked to the weapon interface equipment by a MIL-STD-1553B databus, which can also be used to interface the submarine's sensors to the system. COMKAFS can be interfaced to the full range of modern submarine sensors. The processing units employ microprocessors in single shelf, self-contained enclosures which can greatly ease the difficulties of ship fitting in smaller submarines.

The display positions use a monochrome raster display although colour raster is available as a customer option. The screen formats include: labelled plan display; time/bearing display; time/frequency

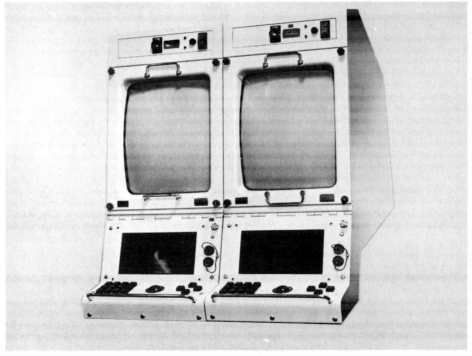

Double console arrangement of Outfit DCH

display; vertical section display and fire control display combined with a display of sitrep data or tube and weapon status. The display positions contain a plasma panel, with touch-sensitive overlay, that provides the main operator interaction medium using a series of prompt driven menus. A high level of automation is employed in COMKAFS which minimises the need for operator intervention and enables a single operator to control the complete system.

COMKAFS is designed to interface to dual purpose wire-guided torpedoes (anti-submarine and anti-ship) and anti-ship air flight missiles. The fire control facilities for these weapons are integrated into the system and include target designation, weapon and tube preparation, weapon launch and where appropriate post-launch guidance.

COMKAFS is also available fully integrated with the Triton submarine system which consists of a coupled federation of sonar sensors designed and developed for the 'Oberon' update programme by the three major UK sonar manufacturers: Ferranti, Marconi and Plessey. Further options include an interface to a plotting table and an automatic contact evaluation plot.

Operational Status
In production for the Royal Navy 'Oberon' class update programme. System DCH has been successfully integrated with the new 2051 sonar suite and ESM, and sea trials are under way following the completion of onshore tactical weapon system trials. Installation of DCH hardware began with HMS *Opossum* during a normal refit. She carried out successful test firings and system acceptance in 1990. 'Oberon' class submarines remaining in service with the Royal Navy were subsequently fitted with System

DCH and the system saw service in the Gulf War in 1991.

Contractor
Ferranti Computer Systems Limited, Naval Command and Control Division, Bracknell.

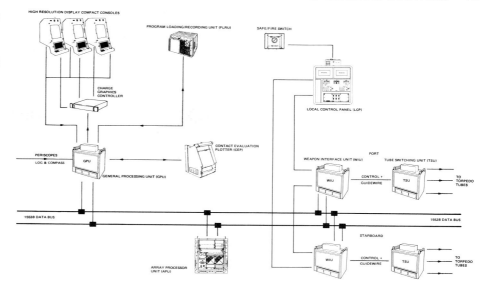

Overall system architecture of COMKAFS

FISCS

Type
Submarine integrated combat system.

Development
The Ferranti Integrated Submarine Combat System (FISCS) is the latest system of this type to be produced by the company. It incorporates the necessary system architecture, processors and display technology to provide a combat suite combining 'state-of-the-art' sonars with advanced data handling and weapon control facilities.

Description
FISCS accepts multi-sensor inputs and presents a comprehensive tactical picture by the integration of target data from all sensors. Tactical evaluation facilities include relative velocity calculations, background data manipulation, and command advice. Weapon control capabilities extend to salvo fire of wire-guided torpedoes and submarine-launched missiles. Other features include comprehensive onboard data link facilities, a minelaying package, and automatic deployment of decoys and countermeasures. Periscope and radar video may be fully integrated within the combat suite.

FISCS comprises a six-position operator console with 17 shelves of electronic equipment in the control room, a control panel and seven shelves of electronic equipment in the torpedo room, and the submarine's complete fit of sonar arrays and equipment.

The system draws its information from a variety of sensors, but the heart of FISCS is the advanced sonar capability which provides the acoustic information for subsequent operational activities by the submarine, both inshore and in deep ocean environments. It is optimised against surface vessels, submarines, ASW aircraft and their associated weapon systems. Acoustic detection, signal processing and analysis are employed to maximise detection ranges, provide an advanced warning capability, and capitalise on the latest methods of target analysis and classification. A comprehensive transducer suite facilitates all-round coverage over selectable or fixed frequency bandwidths, and includes towed, flank, bow, intercept and specialist array systems such as passive ranging array groups.

The received information is processed to provide wideband, broadband, narrowband and specialist data for subsequent analysis and use. Nine-core elements operate either individually or in conjunction with other elements, and are integrated together for

Ferranti Integrated Submarine Combat System

presentation on any or all of the display screens. Information is readily interchangeable between sonar elements or with the tactical data handling and weapon control subsystems, and the latest ergonomic and man/machine interface designs are incorporated.

Contractor
Ferranti-Thomson Sonar Systems UK Ltd, Cheadle Heath, Stockport.

SMCS

Type
Submarine command system.

Development
To manage the increased data processing demanded by improved sensors, SMCS (Submarine Command System), based on the Successor architecture, has been selected as the new command system for fitting or retrofitting to the Royal Navy's 'Vanguard', 'Trafalgar', 'Swiftsure' and 'Upholder' class submarines. It will replace DCB or DCC in existing submarines.

Description
SMCS interfaces to the submarine's sensors and tactical weapons providing the control processing and functionality of the combat suite. Functions include: own ship data handling; TMA; tactical picture compilation; oceanographic data analysis; weapon management; weapon command tactical aids; simultaneous onboard training and exercise analysis.

SMCS is a high capacity open architecture distributed processing command system based on a dual redundant fibre optic Local Area Network (LAN). The processors feature INTEL 80386 32-bit chips and INMOS transputers on a Multibus II backplane. Operator interface is by high resolution multifunction colour graphics displays with interactive plasma panels and a 'puck' (similar to a mouse). By capturing the latest technology this system has many times the processing power of previous systems; it has the capacity for significant expansion and update. Except for certain highly specialised areas, software is written in Ada.

Operational Status

First submarine fits being undertaken. Equipment harbour trials successfully completed. Software under development and scheduled for phased deliveries.

Contractor

Dowty-Sema Ltd, Esher.

SMCS console

COMBAT SYSTEM INTEGRATION

Type

Integrated combat system facilities.

Description

VSEL has pioneered the concept, design, build and operation of shore-based combat system integration facilities in the UK. The idea has resulted in major contributions to the effectiveness of total submarine development.

The Shore Integration Facilities (SIFs) operated by the Combat Systems Division offer: parallel development of submarine and combat systems; reduced overall development times; reduced build time; reduced risks to programme and cost; simpler onboard installation and system proving; earlier operator training and post-trials analysis. There are currently two SIFs in the UK, designed for the UK 'Upholder' and 'Vanguard' submarine classes. Both are located at VSEL's Barrow works, alongside construction of the first of class vessels.

Of particular advantage to operators of the SIF system is the ability to set the combat system to work progressively. Starting with the combat system databus, the facility continues with phased integration of other equipment, leading to a completed functional system.

This progressive working uses a scenario generator which exercises and demonstrates the system in an operational environment, using a variety of stimulation and simulation facilities as necessary. The design and production of software for the scenario generator is handled by VSEL. Complementing the work of the SIFs are stimulation facilities carried on board submarines for combat system integration trials. Validation is achieved by means of advanced software-controlled trials data recording and analysis facilities.

Contractor

Vickers Shipbuilding and Engineering Ltd, Barrow.

UNITED STATES OF AMERICA

Mk 113

Type

Submarine combat system.

Description

The FCS Mk 113 Mods 6, 8 and 10 are used in SSN attack class submarines of the US Fleet. Target bearings from the BQQ-5 sonar suite are pre-filtered, and together with own ship motion data, are processed in the Mk 130 digital computer, the first electronic digital computer to be used aboard attack class submarines. With input of own ship state vector information, the target range, course and speed can be computed from passive bearings only. The operator interacts with the target motion analysis (TMA) function at the analyser console Mk 51.

The FCS Mk 113 consists of the following equipments:

(a) the attack director Mk 75 receives own ship and target information to compute torpedo ballistics and wire-guide controls, and performs position-keeping on both target and torpedo. The Mk 75 also supplies necessary data to levelling computer Mk 129 for SUBROC ballistic computations

(b) the attack control console Mk 50 is the system

Attack control console of Mk 113 fire control system

firing panel, tactical display, and torpedo room status panel

(c) the levelling computer Mk 129 performs SUBROC (submarine-launched rocket) ballistic computations, and in conjunction with the reference sensing element Mk 1, provides signals for levelling the SUBROC (submarine-launched rocket) inertial platform.

The addition of two torpedo control panels Mk 66, two tone signal generators Mk 47, together with major modifications to existing Mod 2 equipments, resulted in FCS Mk 113 Mod 6 and Mod 8 which accommodate the Mk 48 torpedo. Torpedo control panel Mk 66 (which represented the first application of MSI/LSI circuit technology on submarines) acts as a preset panel while TSG Mk 47 generates the signals for transmission of data to the torpedo. The Mod 6 is a field-modified Mod 2 while the Mod 8 comes directly from the factory with the modifications built in.

Operational Status

The FCS Mk 113 has been operational since 1962. It is the weapon control system deployed aboard the US Navy nuclear attack class and, with the addition of an analyser console Mk 78 (FCS Mk 113 Mod 9), SSBN fleet ballistic missile submarines. In the late 1960s additional equipment required for the Mk 48 torpedo was developed. During the past 10 years, the attack class submarines have been upgraded with the FCS Mk 117 and Combat Control System (CCS) Mk 1 equipments. Only four FCS Mk 113 systems are currently operational on SSN attack class submarines. However, approximately 30 FCS Mk 113 Mod 9 systems remain operational on SSBNs.

The Fire Control System Mk 113 Mod 9 is the heart of the defensive weapon system of the older

'Lafayette' and 'Benjamin Franklin' classes of Fleet Ballistic Missile (FBM) submarines. The system controls preparation, status, launch and guidance of the Mk 48 heavyweight torpedo, the primary defensive ASW weapon aboard the FBM submarine.

SCCS Mk 2

Type
Submarine combat control system.

Description
This is an improved version of the submarine fire control system (SFCS) Mk 1 that was initially produced for the Royal Australian Navy's modernised 'Oberon' class submarines and subsequently fitted in boats of the Canadian and Indian navies. The new system is mainly for export applications, although its use in certain classes of US ship, or surface vessel fittings, should not be ruled out. The original Mk 1 SFCS Mod 0 was described in the Australian section of this part of *Jane's Weapon Systems* in previous editions.

The Mk 2 combat control system currently in production is a completely integrated action-information system for submarines, corvettes and frigate-type ships. It provides comprehensive processing and evaluation of all combat data and co-ordinated operational control of all sensors, weapons and combat manoeuvres. The system's unique and innovative architecture has resulted in lower costs, higher reliability, and easier operation and maintenance than its predecessors.

The system provides for surveillance of a large number of contacts and the simultaneous engagement of multiple threats. The surveillance process is highly automated due to the large amount of navigation, contact tracking, communication link and weapon telemetry surveillance data that must be processed and evaluated. The results of the automatic processing are presented to system operators in user-friendly pictorial and natural language forms. The highly automated approach frees the operators from perfunctory tasks and allows them to manage by exception. This results in a reduction in the number of operators required for surveillance and allows the assignment of additional operators to the prosecution of contacts which are a threat to the ship's mission. The contact prosecution process, while operator intensive, includes the display of tactical options and recommendations to aid operators in attacking threats using torpedoes and missiles. In addition, the system provides support for performance monitoring and fault localisation, combat data recording and reconstruction, and operation and maintenance training.

The system hardware is completely redundant and can therefore sustain single failures of any type without any loss of tactical capability. The equipment is very compact and will fit down a 650 mm hatch. Central to the system are a set of distributed computers which are embedded in each of the hardware elements. The computer design is modular and therefore expandable by inserting additional processing and memory cards. The system is comprised of the following hardware elements: a sensor data converter/

processor with two computers; multiple combat control consoles with one computer each; and a weapon data converter/processor with two computers, all interconnected by a dual combat databus.

The sensor data converter/processor provides:
(a) interfaces with the navigation sensors, the tracking sensors (including passive/intercept/active sonar, radar, electronic support measures, TV, and periscopes), and the communication link
(b) automatic surveillance and support processing
(c) combat data recording.

Combat control consoles provide a truly multi-function display of sensor video and processor generated graphics and alphanumerics. The main display surface is a 45 Hz non-interlaced raster scan colour CRT with 1280 × 1024 picture element resolution. Each console independently provides for the simultaneous display of video from any two sensor sources plus the overlay of graphic and alphanumeric annotation. Radar, navigation charts and annotation may be combined for use in combat navigation. The console contains a 17 × 8.5 in (430 × 22 mm) touch interactive plasma display which is used as a tactical summary display and for interaction with the weapon launch compartment personnel during preparation for a launch.

The weapon data converter/processor provides:
(a) automatic contact prosecution and support processing
(b) an interface with the launch tubes and weapons
(c) emergency weapon setting, launching and control.

A plasma display like the one in the combat control console is included and is used for interaction with the console operators during preparation for a normal launch and for local control during emergency operations.

The combat databus provides communication between the other elements of the system at a one million bit/s serial rate over each half of a dual bus in the MIL-STD-1553B format. Data are broadcast on the bus from each computer to all other computers simultaneously so that they all receive the data simultaneously. Each half of the bus has the capacity to provide all of the required system communication. The bus is available in both copper and fibre optic forms.

The sensor and weapon data converter/processor performance monitoring tasks combine to provide a double check on system failures and in conjunction initiate automatic switching to redundant hardware/software in the event of a single point of failure. Switching processors is simplified as program loading and database updates are not required. Switching halves of the databus is also simplified by switching all communication from one half to the other in the event of any failure on the half that is providing system communication.

The Mk 2 combat control system is a completely integrated action-information system which can

useful life of the Mk 113 Mod 9 system through the 1990s.

Contractor
Librascope Corporation, Glendale, California.

The first of the militarised production control consoles for the Librascope SCCS Mk 2 system

survive any single point of failure without any loss of tactical functions. It centres around multi-function colour display consoles, and the cost is low due to the simple and innovative design.

Operational Status
The preceding Mk 1 version was initially supplied to the Royal Australian Navy for fitting in modernised 'Oberon' class submarines, and was later adopted for Canadian Navy boats of the same class. It was procured for German-built 1500-ton submarines for the Indian Navy. The Mk 2 SCCS is currently in production. Librascope's next generation submarine combat control system Mk 3 is currently in the development phase and is scheduled for installation on the Royal Australian Navy's new construction Swedish Type 471 submarines.

Contractor
Librascope Corporation, Glendale, California.

Librascope's current system refurbishment contract makes provisions for an embedded AN/UYK-44 computer and a new state-of-the-art CRT electronics section for the Analyzer Console Mk 78, which computes TMA data. These modifications will extend the

---

COMBAT CONTROL SYSTEM MARK 2 (CCS Mk 2)

Type
Submarine integrated weapon control system.

Development
The CCS Mk 2 is a major US Navy upgrade that will reduce the many submarine combat control system variants currently in the US fleet to four.

In late 1988, Raytheon won the competition to develop and manage the CCS Mk 2 upgrade. The first system is scheduled to become operational in 1992. As prime contractor for much of the currently deployed CCS Mk 1 system, Raytheon will use its experience to retain approximately 80 per cent of the CCS Mk 1 functionality in the CCS Mk 2 system.

CCS Mk 2 will provide commonality across combat systems on the 'Los Angeles' class attack submarines and 'Ohio' class ballistic missile submarines. The system will integrate data from advanced submarine sensors and will control weapon targeting and launch. The system will also have significant potential for growth to handle additional submarine mission requirements.

The CCS Mk 2 programme will also develop and implement a common in-service support programme for both the 'Los Angeles' and 'Ohio' class platforms. Through the adoption of common systems, CCS Mk 2 will significantly reduce support costs over the next five years.

Description
The upgrade programme is a major step toward the US Navy goal of evolving fire control equipment systems into comprehensive combat control systems.

The effort involves consolidating the functions of sonar, target motion analysis, tactical displays, weapon order control and submarine navigation and communications.

The improvement will provide common equipment and software in the 'Ohio' class and 'Los Angeles' class submarines, including those with vertical-launch capability. The upgrade will also be installed aboard new 'Los Angeles' class submarines equipped with the advanced AN/BSY-1 sonar system.

With the installation of CCS Mk 2, the many different combat systems currently in the fleet will be reduced to four variants, which will achieve 90 per cent hardware and software commonality. In replacing the AN/UYK-7 standard computers used in older combat control systems with the AN/UYK-43 model, Raytheon will provide increased processing performance with reserve capacity for future enhancements. In addition, the CCS Mk 1 OTH processor will be replaced by the AN/UYK-44 and a parallel processor. The parallel processor will provide a significant increase in responsiveness for the track/correlation function.

CCS Mk 2 will also include an upgrade of all cursive displays to modern raster displays. Based on commercial technology from Silicon Graphics Inc, the display system includes a sophisticated graphics engine and a high speed display processor. With a local processing capability of eight million instructions per second (MIPS), the display processor decreases dependency on the central processing unit by more than 50 per cent, providing significant processing reserve for future fire control system enhancements.

The design of the tactical system will be paralleled by the development of a common logistics and in-service support programme. This support programme will realise efficiencies and reduce the overall cost of the support programme. Due to the commonality of the combat systems and the common support programme, the training programmes for the submarine crew will be combined and streamlined.

Operational Status

Raytheon Company won a $405 million competitive contract for the system upgrade in September 1988 from the US Navy's Naval Sea Systems Command. As prime contractor, Raytheon's Submarine Signal Division leads the programme's management and technical effort, which includes a major software upgrade. The company's Equipment Division, based in Marlborough, Massachusetts, is providing the raster displays. A Raytheon Service Company support facility will integrate and inspect the system prior to ship installation and will be responsible for final assembly of console upgrade kits and new consoles.

Contractor

Raytheon Company, Submarine Signal Division, Portsmouth, Rhode Island.

AN/BSY-1(V)

Type

Submarine combat system.

Development

The BSY-1(V) integrated combat system is being developed as part of the two part programme involving the BSY-2 system, initiated in May 1985 as a result of restructuring the SUBACS programme. As part of the redesigned SUBACS the fibre optic databus distribution system was replaced by the traditional copper cable together with already developed hardware in order to accomplish data distribution. The system is designed to equip the 'Los Angeles' class attack submarines. Initially the system will equip later units of the class, retrofitting earlier units as they become due for mid-life update.

Description

The system is the first submarine combat system in the US Navy to combine the functions of sonar and weapon/fire control systems. BSY-1(V) integrates the medium to low frequency bow-mounted Submarine Active Detection System (SADS) sonar used for the detection and fire control solutions on hostile subsurface to surface targets. This operates over 360° in long-range search, and provides a passive listening mode. Also integrated in the system is the Long Thin Line Towed array (TB-23), which is an improved towed array for long-range submarine target detection. The system also integrates the high frequency, active Mine and Ice Detection Avoidance System (MIDAS) mounted in the sail and which is used for close-range detection of mines and polar navigation.

The principal subsystem is the stand-alone Wide Aperture Array (WAA) installed in 'Los Angeles' units ordered in FY89 and subsequent programmes. This passive flank array is mounted on both sides of the submarine's hull. Finally the system integrates two towed passive arrays: the TB-16 equipped with the AN/BQQ-5 sonar and the lightweight, long thin line TB-23 towed array with the AN/BQQ-5D sonar, the latter providing long-range submarine target detection. The TB-23 can be reeled into the submarine's ballast tanks.

The integrated combat system will carry out target motion analysis and calculate a fire control solution for Mk 48 Adcap torpedoes, Harpoon and Tomahawk cruise missiles.

The system is designed to improve data processing and management capabilities using new and more capable computers, new data displays and additional software and increased automation in areas such as surveillance, detection and tracking of targets. This will enable operators to perform multiple tasks and handle multiple targets simultaneously. The aim is to reduce the response time between initial detection and launching of the weapon.

Operational Status

The first unit designated to receive the BSY-1(V) system was the 'Los Angeles' class submarine *San Juan*. It was found, however, that when completed the system was too large for installation in the submarine. Consequently some of the system hardware had to be modified and some internal spaces in the submarine re-designed to enable the system to be fitted. The system was then installed during the Summer of 1987. Coincident with the first system going to sea, the US Navy commissioned a shore trainer. Since then four submarines (SSN 751 to SSN 754) have received preliminary basic systems providing them with limited self-defence capabilities (acoustic, safety and weapon firing functions) necessary to operate until the systems are upgraded later with the Navy's new standard AN/UYK-43 computer (which will replace the old AN/UYK-7), improved software, and extra hardware for signal processing and polar ice cap operations to provide full offensive capabilities. The second version is the full performance capability system which will equip the remaining 20 'Los Angeles' class submarines, commencing with SSN 755. Due to continuing problems arising from design changes to the system, delivery to the navy of the first nine submarines fitted with AN/BSY-1(V) will be about 17 months late. In December 1987 two contracts were awarded for BSY-1(V) systems, one for four systems to fit SSN 760 to SSN 763, with delivery due in July 1991, and a second contract to deliver four systems for SSN 764 to SSN 767 for delivery in July 1992. A total of 23 AN/BSY-1(V) combat systems will be purchased (a 24th system was cancelled following the FY90 appropriation decisions).

Contractor

IBM, Federal Systems Division, Bethesda, Maryland (production).

Martin Marietta, Glen Burnie, Maryland (towed array).

Raytheon Company, Submarine Signal Division, Portsmouth, Rhode Island (hull array).

General Electric, Moorestown, New Jersey (production).

EG&G, Rockville, Maryland (system engineering).

AN/BSY-2

Type

Advanced submarine combat system.

Development

The BSY-2 combat system developed out of the SUBACS programme which was cancelled in FY86. It is being developed for the new SSN-21 *Seawolf* nuclear powered attack submarine of the US Navy. The development of BSY-2 is considered to be one of the largest computer software efforts ever undertaken for a submarine. Very considerable effort will be devoted towards developing and integrating a massive amount of software involving some 3.6 million lines of code, the majority of which will be written in Ada (approximately 2.2 million lines of code). Responsibility for developing the software is to be shared among seven development organisations under the direction of the prime contractor. Four of these form part of the prime contractor and three are subcontractors.

Description

BSY-2 is an integrated sensor and fire control system combining active sonar, passive flank array sonar, passive towed array sonar and combat control system with an advanced computer system which is designed to detect, classify, track and launch weapons at enemy subsurface, surface and land targets. The boat's sensors will include a wide aperture array, large spherical array, tactical situation plotter, combat system display consoles, transmit group, a large vertical display screen and a large horizontal display screen, multi-function combat system display consoles, weapon launch system, multi-array conditioner, long thin line towed array (TB-23), a hemispherical transmit array, a standard towed array (TB-16), and a Mine and Ice Detection Avoidance Sonar (MIDAS).

The system is being developed to counter the Soviet submarine threat of the 21st century, and as such is being designed to enable the submarine to: detect targets in a much shorter time than is currently possible; allow operators to perform multiple tasks and handle multiple targets simultaneously; and to greatly reduce the time between detecting a threat and the launch of weapons.

The new UYS-2 acoustic processor, which has several times the capacity of the UYS-1, will be used as standard, and the BSY-2's distributed architecture is specifically designed to cope with the increased processing requirements of the acoustic array suite. The system will comprise about 200 Motorola 68030 32-bit processors which will carry out the various functions within the combat suite. The design specification requires the processors to be grouped in four task areas: acoustics, command and control, weapons and display. Redundancy will be catered for in various ways. In the weapon cluster a designated backup spare will be available for each of the four processors. Thus, if any processor fails it will be automatically replaced by its spare, providing 100 per cent redundancy. The remaining three clusters will not, however, achieve the same level of redundancy. For these, various backup support options will be available should one or more processors in a cluster fail. At the first level of redundancy two or three spare processors will be available for electronic substitution in the event of failure. As a second line of defence, other processors in the affected cluster could be available for use as spares, if they are not being used for their own function at the time. Finally, if all available processors are in use and failure occurs, then the system will suffer gradual degradation with some loss in operational capability. If operational capability is lost a prioritised list of functions will be implemented, assuring that minimum defensive functions (self-protect functions) are maintained at all times. These will afford the submarine protection against hostile threats, but will not provide it with the full capability required to carry out an attack mission. Finally, should a cluster begin to lose operational capability, then spare processors carried onboard can be manually inserted into the system, but this takes time which in certain threat situations may not be available.

Operational Status

Contract for full-scale development and production of first three systems awarded March 1988. Total development and procurement costs for 28 planned systems estimated to be $9.1 billion in March 1989. Integration and testing of the system is taking place between 1989 and 1994, the first system being due to be delivered in November 1993. However, this first system will not possess full capabilities as additional software needed to increase performance to full operational capability is not due to undergo testing and integration until April 1993, for delivery in June 1994.

Contractor

General Electric, Syracuse, New York.

SOAS (SUBMARINE OPERATIONAL AUTOMATION SYSTEM)

Type
Submarine command support software programme.

Description
The demand to process and present the growing volume of constantly changing data provided by current and next generation advanced sensors, covering both the tactical and ocean environment as well as the mass of data on own ship systems, and the increasingly lower acoustic signatures generated by modern submarines, is constantly increasing. To meet this demand the US Defense Advanced Research Projects Agency (DARPA) is studying software requirements for the next generation of submarine command systems.

Referred to as SOAS (Submarine Operational Automation System), the initial phase of the programme involves the development of a series of prototype software programs. These will form the key element in the next generation of software architecture for use in submarine command systems designed to provide highly integrated information and data management to assist the commander in tactical and ship control decision making. Using advanced computer hardware designs based on a series of parallel computers capable of carrying out millions of calculations simultaneously, and incorporating neural networks and artificial intelligence-based programs, SOAS will integrate signature and vulnerability management, ship monitoring and control, situation assessment, tactical planning, and the ship-crew interface.

Unlike the development of the BSY-1(V) system which was designed to meet a specific set of operational requirements, SOAS adopts a 'system extension' approach in which operational requirements are added as development of the system proceeds.

To meet this requirement General Electric has been awarded the contract to develop six software prototypes (P1 to P6) in which programs are developed to satisfy an increasingly demanding tactical scenario. The first prototype, P1, completed in December 1990, covered the requirement for a submarine to carry out surveillance of hostile forces and reconnaissance in an area of low hostile activity. A land-based test system has been established at General Electric's facility at Moorestown, where elements of the software undergo tests to prove the concept and where new concepts are evaluated and successful prototype designs are then integrated with the established architecture.

Development of P2 now under way envisages a scenario where the number of contacts is significantly increased and will incorporate stealth signature management programs being developed at the Applied Physics Laboratory of John Hopkins University.

P3 will address requirements for the submarine to penetrate a hostile screen and enter an area of intensive hostile ASW activity. This program will include software models of ASW tactics likely to be employed by enemy attack submarines, surface ships and aircraft.

P4 will incorporate facilities enabling the submarine to approach the target area avoiding enemy defences such as minefields, identify the target, and position itself for weapons launch. P5 software will cover target strike/attack requirements while P6 will enable damage assessment on the target to be carried out.

Operational Status
Phase 1 contracts for development plans and demonstration of capabilities were awarded to General Electric, Lockheed-Georgia, McDonnell Douglas and Martin Marietta in January 1989. In February 1990 General Electric won the Phase 2 development contract, the design phase for which was completed in May 1991. Software coding was completed during the Summer of 1991 and full software integration was scheduled for October/November. Demonstration of the integrated P1 and P2 software architectures was due in December 1991. General Electric plans to develop and integrate one prototype design per year, the programme being scheduled for completion in 1995.

Contractor
General Electric, Moorestown, New Jersey.

WEAPON CONTROL SYSTEMS

FRANCE

DLT-D3

Type
Submarine torpedo fire control system.

Description
The DLT-D3 torpedo fire control system is used in French Navy submarines. All types of torpedo employed by the French Navy can be launched, including wire-guided. The system may also be expanded for anti-surface missile applications.

Target data are fed to the system from onboard sensors, which comprise fore and aft sonars, acoustic rangefinder, and attack and surveillance periscopes.

The system employs a general-purpose digital computer, associated with a CRT display terminal. The following functions are performed by the system:
(a) updating of the tactical situation from the data delivered by the various sensor systems fitted and the navigation equipment
(b) assistance in calculating target components by the use of special recorded programs
(c) weapon control; computation of the firing path, remote setting of torpedoes, and firing sequence control. The system is designed for launching any type of torpedo
(d) maintaining a chronological record
(e) maintenance assistance by means of test programmes.

The DLT-D3 operating programs have been compiled to permit tracking of eight targets, simultaneous guidance of two wire-guided torpedoes and preparation of a third for launching. Each of the three displays of the terminal is dedicated to the presentation of the following data, in accordance with the program implemented: tactical situation, firing path, and alphanumeric display of parameters (tote) and decoding of the designations and functions of the two common keyboards. Conversation between the operator(s) and the computer are by means of these keyboards.

DLT-D3 submarine torpedo fire control system

Emergency launch of torpedoes is possible from either bow or stern station.

The system is designed to be served by one or two operators at the operations centre, and one operator at the bow station with possibly another at the stern station. The equipment arrangement, typically, is one CIMSA 15M125 digital computer (QTD), a monitoring and control console (VIC), an azimuth relay (RZ), a true-bearing diagram (GZ) in the operations centre, a tube selection panel (PAT), and tube servicing station (PST) in the bow and/or aft torpedo tube compartments.

Operational Status
This system is in service in 'Daphné', 'Agosta' and 'Rubis' class SSNs of the French Navy, and is in production for Spanish and Pakistan 'Agosta' class submarines.

Contractor
Thomson Sintra Activités Sous-Marines, Valbonne Cedex.

INTERNATIONAL

KANARIS

Type
Submarine fire control system.

Development
This system is being developed by Unisys in collaboration with the Hellenic Navy's Research and Development Centre (GETEN) for half the Greek Navy's 'Glavkos' class (Type 209) submarines.

Description
Each system will consist of a two-cabinet firing distribution unit and the main control unit with two consoles. The latter are the Unisys Tactical Modular Displays (TMDs) which feature embedded AN/UYK-44 computers permitting the system to control six weapons simultaneously.

Operational Status
Five systems have been ordered: four for shipborne use and the fifth as a land-based training unit.

Contractors
Unisys Defense Systems, New York.
GETEN (Hellenic Navy Research and Development Centre).

ITALY

WEAPON CONTROL – SHIPBORNE

Type
Shipborne torpedo control systems.

Description
These systems allow the remote selection, presetting and launch of lightweight torpedoes such as A 244/S, A 244/S Mod 1, Mk 44, Mk 46, A290 and others. The systems can be supplied in three different configurations:

Configuration 1 – the system is manually controlled by the operator by means of a remote-control panel installed in the ship's operations room;
Configuration 2 – the remote-control panel is interfaced with the combat information system. In this case the optimum presetting values which vary with the tactical situation are determined and updated by the combat information computer which automatically sends them to the selected torpedo through the control panel;
Configuration 3 – is intended for light vessels not fitted with a combat information system. The remote-control panel is directly interfaced with the sonar: it

determines and updates the target's course and speed and, on the basis of these data, computes the optimum presetting parameters which are sent to the selected torpedo. A display is incorporated in the control panel for re-presetting the tactical situation and the recommended attacking manoeuvre.

Operational Status
In production. In service with the Italian Navy and several other navies.

Contractor
Whitehead, Livorno.

WEAPON CONTROL – AIRBORNE

Type
Airborne torpedo control systems.

Description
These miniaturised systems are designed to allow remote selection, presetting and launch of the A 244/S Mod 1 torpedo and other existing types of lightweight torpedo, from any type of helicopter or fixed-wing aircraft. The systems are normally supplied in one of two configurations:

Configuration 1 – for helicopters fitted with dunking sonar. The torpedo control can be carried out either in hovering or in forward flight. In the former case, data from the dunking sonar are processed by the computer unit of the system to determine the target course and speed, and the computer then determines the optimum torpedo presetting values which are sent automatically to the selected torpedo. The hit probability is also displayed on the control panel to assist in decision making.

Configuration 2 – for fixed-wing aircraft and helicopters not fitted with dunking sonar. This con-figuration is intended for torpedo launch in forward flight rather than both the hovering and forward flight conditions.

Operational Status
In production. In service aboard almost any type of ASW aircraft and helicopters.

Contractor
Whitehead, Livorno.

APS SERIES CONTROL SYSTEMS

Type
Torpedo control systems.

Description
These miniaturised systems are designed to allow remote selection, presetting and launch of the A 244/S torpedo, and other existing types of light-weight torpedo, from any type of helicopter or fixed-wing aircraft. The systems are normally supplied in one of two configurations:

Configuration 1 – for helicopters fitted with dunking sonar. The torpedo control can be carried out either in the hover or in forward flight. In the former case, data from the dunking sonar are processed by the control panel computer to determine the target course and speed, and the computer then determines the optimum torpedo presetting values which are sent automatically to the selected torpedo. The hit probability is also calculated and displayed on the control panel to assist in decision making.

Configuration 2 – for fixed-wing aircraft and helicopters not fitted with dunking sonar. This configuration is intended for torpedo launch in forward flight rather than both the hover and forward flight conditions.

Operational Status
In production.

Contractor
Whitehead, Livorno.

Panel of APS 102 airborne torpedo control system for A 244/S torpedoes

NETHERLANDS

M8

Type
Submarine torpedo fire control system.

Development
The M8 Series originated in a Royal Netherlands Navy contract awarded in 1955 for the development of a torpedo FCS for the 'Walrus' class submarines. West German interest in the system gave additional impetus to subsequent development and led to the successful development of systems for use in both submarines and surface ships. Two types of torpedo were involved: the AEG Seal and Seeschlange, for use against surface vessels and submarines respectively. This successful collaboration with the German industry continued and M8 Series systems are standard fitting on all submarines produced by Howaldswerke Deutsche Werft in Kiel. The ultimate development is the SINBADS system, but other related developments are the M9 systems in the German 'Köln' and 'Thetis' class vessels, and the M11 systems for two new Argentine fast patrol boats.

Description
The M8 is a digital computer-based fire control system for use in submarines for the direction of torpedoes against either surface shipping or submerged targets. It was produced in several versions with designations ranging from the M8/0 of the mid-1950s prototype to the latest. M8 is now out of production and succeeded by SINBADS, Gipsy and Submarine SEWACO. The basic system comprises a torpedo display control and computer console, a sound path display unit, amplifier and supply unit, distribution box, local control panel(s), and gyro angle setting units. Complete system weight is about 900 kg. The system may be operated by one man, or two men if the submarine is operating with consorts.

The system will accept target data inputs from a range of sensors which includes radar, sonar, passive sound detection systems, periscope observation and consort reports. Ship's own navigational data are also fed into the M8 computer. The display, which has range scale settings for 20, 10 and 5 km, presents the positions of all contacts from all sensors simultaneously. One or more sensors may be connected to the computer for torpedo engagement, and up to three targets may be attacked simultaneously. The computer is programmed to provide firing data for wire-guided, programmed, conventional, and other types of torpedo, and performs automatic calculation of target position, course and speed. The CRT display can give true motion, relative motion or off-centred presentation of the tactical situation.

Operational Status
The following list is believed to record accurately the known installations.

Argentina	Type 209 (2)
Colombia	Type 209 (2)
Ecquador	Type 209 (2)
Germany	Type 205 (5)
	Type 206 (18)
Greece	Type 209 (4)
Netherlands	'Potvis' class (1)
	'Zwaardvis' class (2)
Norway	Type 207 (14)
Peru	Type 209 (2)
Turkey	Type 209 (2)
Venezuela	Type 209 (2)

Contractor
Hollandse Signaalapparaten BV, Hengelo.

UNITED KINGDOM

TORPEDO LAUNCH CONTROLLER

Type
Electronic launch controller for submarine discharge systems.

Description
The launch controller is a microprocessor-based system which monitors and controls the torpedo firing sequences from loading to discharge. It also integrates the discharge instrumentation system with comprehensive self-diagnostic and fault tolerant interfaces to the torpedo tube. This system allows remote-control and monitoring of the torpedo tubes from the command centre without operator assistance, as well as fault annunciation.

Operational Status
In service or under development for 'Upholder' and 'Vanguard' class submarines (UK).

Contractor
Dowty Maritime Command and Control Systems, Feltham.

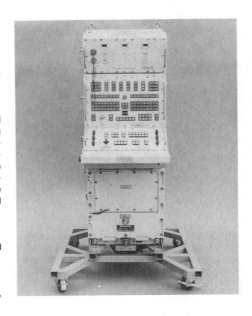

Torpedo launch controller

UNITED STATES OF AMERICA

TRIDENT FCS

Type
Submarine-launched strategic missile fire control system.

Development
General Electric Ordnance Systems Division has received a contract award for the development of US Navy fire control systems and guidance support equipment. The contract covers work extending over a four-year period for design and initial manufacture of fire control systems for the *Trident* submarine and for the modifications required for Poseidon fire control systems currently deployed in order to accommodate a larger C-4 missile.

Description
As a key subsystem of the Trident strategic weapons system, the GE fire control technology will serve the primary purpose of preparing the missile guidance system for flight and controlling the missile launch sequence.

Support equipment will consist of: guidance system test equipment for shore-based activities and nuclear submarine strategic weapons facilities; guidance system containers and handling equipment; and equipment to support fire control and guidance operations at all maintenance levels. Deliveries of fire control systems and support equipment are continuing. No further details were available at the time of publication.

Contractor
General Electric Company, Ordnance Systems Division, Pittsfield, Massachusetts.

AIRBORNE ACOUSTIC PROCESSING SYSTEMS

CANADA

AN/UYS-503 SONOBUOY PROCESSOR

Type
Multi-sonobuoy processing equipment.

Description
The AN/UYS-503 sonobuoy processor is a small, lightweight, low-cost multi-sonobuoy equipment designed for a variety of ASW platforms. It employs a new concept in processor architecture and a digital demultiplexer that provides processing for all sonobuoys, and growth capability for current and future sonobuoys.

The AN/UYS-503 is made from identical, independent parallel processing modules. The processing capability of each module when used with passive sonobuoys is seven bands from DIFAR (one surveillance and six vernier), and 14 bands from LOFAR (two surveillance and 12 vernier). All bands are processed simultaneously and stored continuously for real-time viewing in the active mode; each module processes one DICASS (CW and FM) or one RO (CW) buoy.

A single AN/UYS-503 unit system can simultaneously process eight LOFAR, or four DIFAR, or four DICASS, or four BATHY, or four VLAD, or any combination of these. Multiple AN/UYS-503 units can be combined to provide 16 or 32 sonobuoy systems with one or two operators as required. The system has excellent reliability (MTBF greater than 1600 h), and graceful degradation is inherent in the design since each processing module functions independent of the others.

Display formatting is designed to enable the operator to view several of the processing data streams simultaneously. Display formats include: GRAM, ALI, BFI for passive; and range-amplitude, range-Doppler and range-bearing for active. DIFAR processing includes both cardioid and omni-modes.

The system is RS 232 and 1553B compatible and uses RS 343 composite video output to any standard 875-line TV screen.

AN/UYS-503 sonobuoy processor

Operational Status
Production deliveries to the US Navy, Royal Swedish Navy and Royal Australian Navy are complete. Further deliveries to the US Navy are ongoing.

Specifications
Frequency range: 0-2560 Hz
Input channels: Multiple of 8 standard sonobuoy receivers
Size: 1 ATR short–22.9 × 19 × 30.5 cm (per eight-buoy unit)
Weight: 20 kg (per eight-buoy unit)

Contractor
Computing Devices Company (a division of Control Data Canada Ltd), Ottawa, Ontario.

AQA-801 BARRA SIDE PROCESSOR

Type
Barra sonobuoy processing equipment.

Description
The AQA-801 Barra Side Processor (BSP) is the latest Computing Devices product for processing the AN/SS8-801 Barra sonobuoy. The BSP provides ASW operators with the capability to process, control and display Barra sonobuoy data without modifying existing onboard sonobuoy processors.

Barra is a digitally telemetered, broadband passive array sonobuoy that provides unmatched passive bearing accuracies enabling ASW platforms to conduct submarine attacks while remaining totally passive. The BSP recovers the acoustic data from the Barra array, processes and extracts directional broadband and spectral intelligence to support broadband detection, tracking, spectral analysis and passive

AQA-801 Barra Side Processor

attack ASW operations. A single BSP unit processes up to four Barra buoys simultaneously and outputs data for presentation via the onboard sonobuoy processor in a GRAM or ALI Bearing Time History plot, or a P-Theta format.

Like the AN/UYS-503 sonobuoy processor, the BSP uses modular system architecture and digital multiplexers to ensure multi-buoy real-time expanded system capability and growth.

Operational Status
In operational service with the US Navy. Production deliveries to Sikorsky for Royal Australian Navy Seahawks.

Specifications
Size: 1 ATR short–22.9 × 19 × 30.5 cm (per four-buoy unit)
Weight: 20 kg (per four-buoy unit)

Contractor
Computing Devices Company (a division of Control Data Canada Ltd), Ottawa, Ontario.

FRANCE

SADANG ACOUSTIC SYSTEM

Type
Sonobuoy processing and display system.

Description
Installed aboard fixed- or rotary-wing aircraft, this equipment processes and displays signals received from various types of sonobuoy. Modular design permits optimum system configuration to suit all sonobuoy types and the specific number to be processed simultaneously. All signals received are permanently recorded before being processed to allow more extensive utilisation of data at shore-based processing stations. Categories of sonobuoy whose data can be processed include: passive buoys (directional and omnidirectional); active buoys (directional and omnidirectional); and environmental buoys such as bathythermographic, sea noise measuring and so on. Numerous buoys of the same or several different types can be handled simultaneously.

Three main elements comprise the SADANG equipment:

(a) communications unit, in which will be found a 99-channel VHF receiver, a UHF transmitter/receiver for active sonobuoy remote-control, and ASSG overall test systems. Each set is associated with a control box for the system operators

(b) signal processing and display unit. This carries out analogue pre-processing and full processing functions in the active and passive modes using digital processors and appropriate software. Information is presented by a high density graphic recorder and a CRT display. The latter is associated with a high-capacity bulk storage unit and numerous image formats are available by means of elaborate software appropriate to the specific types of buoy and signal processing. The dual processor configuration is normally operated by two operators using an alphanumeric keyboard

SADANG airborne acoustic system installed on French Navy Atlantique maritime patrol aircraft

and trackerball. A digital databus permits dialogue between the operators and the tactical co-ordinator, via the tactical computer. Several signal processing and display units can be associated aboard the same aircraft so that system performance can be enhanced by increasing the number and types of buoys which can be processed simultaneously. An on-top position indicator is used in conjunction with a radio compass for homing towards the sonobuoy

(c) magnetic recording unit. This multi-track recorder records all the sonobuoy data at the analogue processing level for subsequent analysis ashore.

Other optional equipment includes an acoustic signal generator simulator for test purposes, and remote-control of active sonobuoys.

Operational Status
In production.

Contractor
Thomson Sintra Activités Sous-Marines, Valbonne Cedex.

LAMPARO PROCESSING EQUIPMENTS

Type
Sonar signal processing and display systems.

Description
Derived from the SADANG equipment (see separate entry), these airborne sonar signal processing and display systems are designed as modern digital equipments for ASW fixed- or rotary-wing aircraft.

The processing capabilities of these systems are based on a single processing unit in a light and compact packaging manned by a single ASW operator. The same wide panel of facilities and processing modes is available.

The system is orientated towards the processing of omnidirectional passive and active buoys.

The signals are displayed on a CRT in TV format and a hard copier can be connected for permanent recording. An on-top position indicator is used in conjunction with a radio compass for homing on the sonobuoy. Weight of the complete system is 70 kg.

Operational Status
In operational service.

Contractor
Thomson Sintra Activités Sous-Marines, Valbonne Cedex.

Lamparo sonobuoy processing system

HS 312 ASW SYSTEM

Type
Passive and active ASW system.

Description
The HS 312 is an acoustic system for helicopters incorporating the facilities of the HS 12 system and the Lamparo sonobuoy processing equipment. The equipment functions in both passive and active modes (CW and MF) and has a longer range than the HS 12. The acoustic subsystem will process four or eight sonobuoys simultaneously with the dipping sonar.

Only a single operator is required and the light weight and compactness of the system mean that the HS 312 can be fitted to any type of light or medium size helicopter. The performances of the basic components have been improved by integration of the processing units, integration of a standardised keyboard and the use of only one display screen.

Operational Status
In series production. Three systems were delivered to China in 1987.

Contractor
Thomson Sintra Activitiés Sous-Marines, Valbonne Cedex.

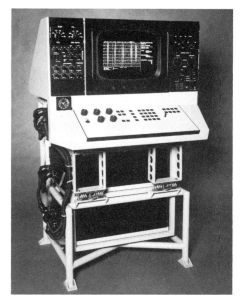

HS 312 helicopter acoustic system control/display

UNITED KINGDOM

AQS-901 PROCESSING SYSTEM

Type
Sonobuoy processing and display system.

Description
The AQS-901, which first entered service in 1979, processes data from all NATO inventory sonobuoys and exploits the considerable advantages gained from advanced buoys such as Barra, CAMBS and VLAD. It is claimed to be the most powerful and comprehensive system in service.

The aircraft contains two systems, each with two CRT displays and two hard copy displays. A fifth CRT, which is able to show data from either system, is provided for system management. Information is displayed via both hard copy and CRT display media, and is controlled and interrogated by the operator using a keyboard, trackerball and cursor.

Operational Status
In production and service. The AQS-901 is fitted to over 30 RAF Nimrod MR Mk 2s and 20 RAAF P-3C Orions. Deliveries were completed in Autumn 1986. The current software and hardware product improvement programme will continue for several years.

Contractor
GEC Avionics Limited, Maritime Aircraft Systems Division, Rochester.

AQS-901 sonobuoy processing and display system

AQS-902/AQS-920 SERIES PROCESSING SYSTEMS

Type
Sonar processing and display systems.

Description
The AQS-902 and its export variants, the AQS-920 series systems, have been designed and developed for installation in ASW helicopters, maritime patrol aircraft and small ships. This range of lightweight and flexible systems will handle data from all current and projected NATO inventory sonobuoys and will interface with a wide range of 31 or 99 RF channel sonobuoy receivers. A maximum of eight sonobuoys may be processed, while for ASW helicopters these systems offer the additional advantage of simultaneously displaying and processing dipping sonar data.

The modular structure of the AQS-902/920 series acoustic processors provides the capacity for every configuration to be tailored to the aircraft, operational role and specific requirements of each customer. As a result the equipment is suitable for both mid-life

updates of existing ASW aircraft and equipment fits in aircraft having no current sonobuoy capability. Installation of AQS-920 on a small ship provides a flexible, cost-effective, variable depth ASW capability without requiring the modifications dictated by hull-mounted sonar systems.

The installation comprises four basic units, each of which contains the dedicated modules which allow construction of the selected system configuration at minimum weight and cost. These allow subsequent expansion and modification of the system to suit changing requirements.

The processor unit converts the received sonobuoy signals into digital form, carries out the necessary filtering and analysis, and processes the data into a form suitable for display. It is packaged into a short 1 ATR case.

The display unit may be in the form of a CRT display or a chart recorder, with configurations using either or both of these as required. All CRT and hard copy displays are fully annotated. The operator uses a simple control panel to specify processing modes and parameters, and a trackerball or stiff stick control and cursor to interrogate data on the CRT display.

AQS-902 provides numerous standard processing and display facilities, including: simultaneous passive and active processing; wide and narrowband spectral analyses; broadband correlation; CW and FM active modes; passive and active auto alerts; and auto calculation and tracking of line frequency, target bearing and range, signal to noise ratio, Doppler fixing, CPA analysis, bathythermal processing and ambient noise measurement.

A unique feature of the systems is the acoustic localisation plot which presents the operator with a geographical plot of sonobuoy and target positions.

In May 1987 GEC Avionics was awarded a contract worth in excess of £40 million to supply more than 150 AQS-902G/DS systems for the Royal Navy's Sea King ASW helicopters. These systems, which are now entering service in the Sea King Mk VI, feature new CRT display facilities and also control, process and display data from the 2069 dipping sonar.

Operational Status
Configurations of this range of equipment are in service with the Royal Navy, the Royal Swedish Navy and the Indian and Italian Navies. Further variants are

in production for Grumman S-2 Turbo Tracker aircraft.

Contractor
GEC Avionics Limited, Maritime Aircraft Systems Division, Rochester.

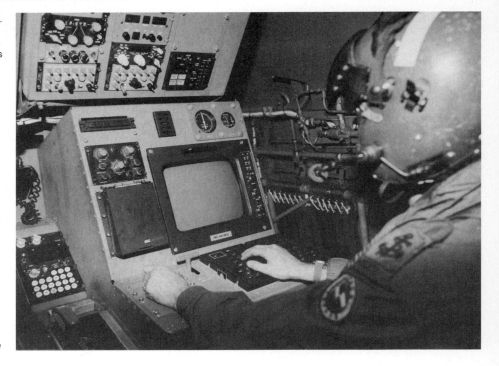

AQS-902G/DS installation in the RN Sea King Mk VI

AQS-903/AQS-930 SERIES PROCESSING SYSTEMS

Type
Sonar processing and display systems.

Description
The AQS-903 and its export variants, the AQS-930 series, are powerful lightweight acoustic processors currently under development for future helicopter and fixed-wing applications. These systems will process data from a basic eight to 32 sonobuoys in any combination. This flexibility is derived from the design based around a number of processing 'pipelines', each able to handle a certain quantity of passive and/or active sonobuoys. Four of these pipelines will provide a processing power eight times that of the AQS-901 at a fraction of the size and weight.

AQS-903/930 series systems make extensive use of distributed processing with the systems incorporating the latest component technology to provide the operator with a wealth of information. Data from up to eight sonobuoys can be processed simultaneously, providing full 360° surveillance. Sector coverage offers additional classification and/or tracking analysis to allow smooth transition from one mission phase to another, irrespective of the type of buoy in use. Analysis and display cues are in plain language, requiring a minimum of operator training. Control of the twin CRT displays is achieved by the operator using a simple control panel and multi-function keys.

AQS-930 systems will interface to other systems responsible for sonobuoy launching, navigation, tactical tracking and weapon release.

Operational Status
A configuration from this range is currently under development for the Royal Navy Merlin ASW helicopter, due to enter service in the 1990s. The basic configuration will process data from the complete range of NATO inventory sonobuoys. Systems have been delivered to Westlands and flight trials in the Royal Navy EH 101 Merlin have been completed.

Contractor
GEC Avionics Limited, Maritime Aircraft Systems Division, Rochester.

AQS-903 acoustic processing rack

RN Merlin ASW helicopter

ASN-902, ASN-924 AND ASN-990 TACTICAL PROCESSING SYSTEMS

Type
Tactical processing and display systems.

Description
The ASN-902 Tactical Processing System (TPS) provides a means of correlating and processing data for display from the wide range of sensors and navigation subsystems in the modern maritime patrol aircraft and helicopter.

The system enables the tactical co-ordinator to display data in an easily assimilated form and assist in the solution of complex navigation, intercept and attack problems. ASN-902 and -924 systems, which are based on proven AQS-902/920 hardware, have a flexible design which makes them readily adaptable

for use as the central element in an integrated mission management system. It can replace the variety of individual sensor system control and display units with common integrated units, providing flexiblity of operation. Standard ARINC 419, 429 and MIL-STD-1553B data interfaces allow installation as original equipment or as a retrofit.

The ASN-902 system, in service in Sea King Mk 42B helicopters, has a monochrome display, whereas the ASN-924, which is in production, has a colour display to enhance the information presented to the tactical co-ordinator. A totally new series, designated ASN-990, is now under development. These systems will use Motorola 68000 series microprocessors and be programmed in Ada.

Systems integration considerably improves the efficiency and flexibility of the mission avionics suite, minimises the weight of combinations of multiple sensors and simplifies logistic and training problems with common control units and multi-purpose displays. The company's Maritime Aircraft Systems Division has carried out systems integration for the Advanced Sea King Mk 42B. This programme incorporates an AQS-920 series acoustic processor and as its core element, an ASN-902 tactical processing system.

Operational Status

The ASN-902 Tactical Processing System is in service in the Advanced Sea King Mk 42B. The ASN-924 is in production and the ASN-990 is under development.

Contractor

GEC Avionics Limited, Maritime Aircraft Systems Division, Rochester.

Integrated mission avionics suite in an Indian Navy Sea King incorporating AQS-920 series system (foreground) and TPS (left)

SP 2104 PROCESSOR

Type

High speed acoustic processor.

Description

The SP 2104 has been designed to reprocess at high speed the acoustic tape recordings made during airborne anti-submarine operations. The equipment is able to process recorded signals at up to eight times normal time. It accepts data from four Jezebel-type sonobuoys. These data are processed in up to eight processing channels, each of which is independently configurable to a comprehensive set of processing

Fast time acoustic processor

parameters. Control of the system is achieved via an interactive display incorporating a logical menu system to simplify initialisation procedures and subsequent parameter changes.

The system is used to find targets missed during operational analysis, to confirm the presence and extract more information from specific targets for classification purposes, to provide operator assessment and highlight training requirements and to extract data for library purposes.

The SP 2104 system is modular to allow easier configuration of equipments to meet individual user requirements. Standard interfaces have been used throughout.

Contractor

Marconi Underwater Systems Limited, Croxley.

UNITED STATES OF AMERICA

LAMPS Mk III

Type

Integrated helicopter ASW system.

Description

The USN LAMPS (Light Airborne Multi-Purpose System) has been designed to extend and enhance the capabilities of surface ships, particularly in anti-submarine warfare (ASW) and increase effective operational range for the weapon systems fitted in surface vessels of the US fleet. This system extends the electronic and acoustic sensors and provides a reactive weapon delivery capability for destroyers and frigate class ships by operating manned helicopters from them. Sensors, processors and display capabilities aboard the helicopter enable the three-man crew to extend the ship's tactical, decision-making and weapons delivery capabilities. Classic line-of-sight limitations for surface ships and limitations to underwater acoustic detection, classification and localisation are thus mitigated through the use of the manned LAMPS aircraft.

LAMPS is being acquired in two phases. The first, Mk I, involved installation of shipboard equipment and conversion of H-2 helicopters already in the inventory to an SH-2 configuration. Mk I became operational in 1972. The Mk III includes the integration of improved avionics and shipboard systems to be used with the SH-60B Seahawk, a modified version of the Army UH-60A Blackhawk helicopter.

In 1974 the USN selected IBM as the system prime contractor to work with the Naval Air Systems Command in developing LAMPS Mk III. By late 1976, after extensive developmental testing, system capabilities were demonstrated in deep water using two modified helicopters with representative avionic and shipboard electronics. Two open-ocean tests concluded the validation of the LAMPS Mk III mission performance. The first was conducted jointly by the Naval Air Systems Command and IBM. After the data reduction was completed and the technical performance was assessed, the test was judged highly successful. The second test was conducted under the direction of the commander of the operational test and evaluation force to verify the operational

LAMPS helicopter flying over USS McInerney

suitability and performance of the LAMPS Mk III concept. The Navy was given permission to continue with full-scale development by the Department of Defense in 1978, with IBM as system prime contractor, Sikorsky Aircraft as the air vehicle contractor, and General Electric as the engine manufacturer for the new Mk III air vehicle.

The LAMPS Mk III weapon system embodies the integration of the parent ship (frigate, destroyer, cruiser) and the manned aircraft (SH-60B Seahawk) operating from that ship for both ASW and anti-ship surveillance and targeting (ASST) missions. As an adjunct of the sensor and attack systems of the parent or similarly equipped ships, the aircraft extends the detection, classification, localisation, and attack capabilities where needed to increase line-of-sight and range capabilities of the parent ship. Radar and ESM sensors extend the ship's line-of-sight against surface threats, whereas sensors that operate relative to the ocean medium (such as sonobuoys and magnetic anomaly detectors) enable the ship to engage the underwater target.

The primary control of aircraft tactics and selection of sensor and weapon modes remain with the parent ship for both the ASW and ASST missions, with an equally important, independent mode of LAMPS aircraft control provided as a backup.

In an ASW mission, the LAMPS aircraft is deployed

from the parent ship when a suspected threat has been detected by the ship's towed array sonar, hull-mounted sonar, or by other forces. It proceeds to the estimated target area, where sonobuoys are dropped into the water in a pattern designed to entrap the target. The sounds (acoustic signatures) detected by the buoys are transmitted over a radio frequency link to the aircraft where they are analysed, codified and retransmitted to the ship for interpretation and analysis. When the location of the threat has been determined with adequate precision, the aircraft descends near the ocean's surface for final confirmation. This may be accomplished using active sonobuoys, passive directional sonobuoys or by trailing a magnetic anomaly detector behind the aircraft. On final confirmation by any of these methods, a torpedo attack can be initiated. The extension of the ship's sensor, tactical control and attack capabilities is achieved by using a duplex data link that transfers acoustic, electromagnetic and command data and voice from the Seahawk, and command data and voice to the aircraft. Data are processed and evaluated aboard the parent ship where tactical and command decisions are made, and tactical instructions, weapon delivery information and transmission of processed data are linked back to the SH-60B Seahawk.

Data transmitted to the parent ship (or other

LAMPS-equipped ship) are processed by digital computers and specialised processors, distributed, and evaluated by the parent ship's operators: the LAMPS air tactical control officer; acoustic sensor operator; remote radar operator and ESM operator. The combat information centre evaluator uses the data in context of the overall tactical situation to determine what actions to take in attacking. Operation of the LAMPS Mk III system was described in detail in previous editions of *Jane's Weapon Systems*.

Operational Status
Introduction to the US Fleet of the SH-60B LAMPS Mk III began in 1984, and such helicopters are being deployed aboard about 100 US Navy surface vessels, including DD-963 and DDG-993 destroyers, CG-47 cruisers, and FFG-7 frigates. The earlier LAMPS Mk I Seasprite will continue to serve in some older FFG-7s. A total of 204 SH-60Bs is the US Navy planned requirement, and a substantial number are already in service.

Contractors
International Business Machines (IBM) Corporation, Federal Sector Division, Bethesda, Maryland (system prime contractor).

Sikorsky Aircraft (SH-60B Seahawk helicopter).
General Electric Company (engine contractor).

AN/AYA-8B PROCESSING SYSTEM

Type
ASW data processing system for the P-3C.

Description
Since 1968, GE has been producing the P-3C data processing system (DPS). The DPS equipment constitutes a major portion of the P-3C anti-submarine aircraft avionics.

The purpose of the AN/AYA-8B DPS is to provide an interface between the central computer and other aircraft systems. The main aircraft systems consist of the tactical co-ordinator (TACCO) station, non-acoustical sensor station, navigation communications (NAV/COM) station, acoustical sensor stations, pilot station, radar interface unit, armament/ordnance system, navigation systems, ARR-72 receivers, submarine anomaly detector, and OMEGA. The DPS consists of four major logic units, as well as a selection of manual keysets and panels.

Logic Unit 1 provides an interface to four types of peripheral information systems:
(a) manual entry
(b) system status
(c) sonobuoy receiver
(d) auxiliary readout display.

The manual entry subsystem provides the communication between the operator stations and the central computer of the man/machine interface. Each operator has a complex of illuminated switches and indicators by which he communicates with the central computer.

System status is received and stored by the status logic subunit of Logic Unit 1 and transmitted to the central computer. The status words are transmitted whenever any status bit changes or upon interrogation by the computer.

The sonobuoy receiver logic provides for the digital tuning of the sonobuoy receiver subsystem by the central computer. Each of the 20 receiver processor channels can be tuned to one of 31 RF cells by the central computer or manually by the operator.

The auxiliary readout display logic provides the digital interface between the computer and the combat information displays located at the TACCO station and the NAV/COM station.

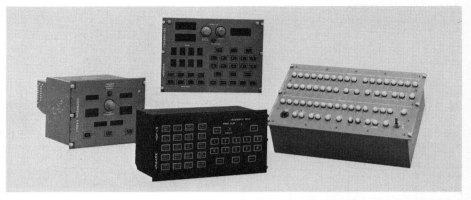
AN/AYA-8B pilot keyset (front) with (left to right) ordnance panel, universal keyset and armament/ordnance panel

Logic Unit 2 provides digital communications to three major aircraft subsystems:
(a){ind1}navigation
(b) armament/ordnance
(c) magnetic tape transports.

Logic Unit 3 provides the interface between the computer and three display subsystems:
(a) TACCO multi-purpose display
(b) sensor multi-purpose display
(c) pilot display.

Logic Unit 4 provides for the expansion of the computer input/output capability by means of the data multiplexer subunit (DMS), furnishes an increased memory capacity by means of the drum auxiliary memory subunit (DAMS) and provides an interface between the computer and two aircraft subsystems:
(a) OMEGA
(b) auxiliary display.

Keysets and Panels
Three universal keysets provide the navigation/communication operator and the two acoustic sensor operators with the capability of entering information into and receiving information from the computer program. The pilot keyset is used by the aircraft pilot to control information presented on his CRT display, to enter navigation stabilisation information into the computer, to drop or cause the dropping of smoke floats and weapons, and to enter information on visual contacts.

The ordnance panel is used to display commands from the computer to the ordnance operator concerning search stores, that is, bin and chute number, and status information.

The armament/ordnance test panel provides the capability of monitoring each output to the aircraft armament and ordnance systems from the armament output logic unit and ordnance output logic subunits of Logic Unit 2.

Operational Status
In production.

Contractor
GE Aerospace, Utica, New York.

AN/AYA-8B modernised logic units for P-3C aircraft

BIRADS

Type
ASW processing and display system.

Description
BIRADS (BIstatic Receiver And Display System) is a portable equipment which has been developed to upgrade existing ASW electronics suites. By using off-the-shelf computer and display components, the BIRADS system integrates matched filtering, pulsed energy detection and constant wave energy detection. It is intended to work alongside existing hardware to improve real-time, active sonar data analysis and interpretation. It can be used as an aircraft or a towed array active ASW receiver.

BIRADS uses a Concurrent (MASSCOMP) 6650 computer to digitise and process 32 channels of acoustic sensor data from different sensor types. For coded active pulses, the system can store the signals in raw form, perform optional beam forming, and conduct matched filtering. For CW signals the system executes a high resolution frequency analysis through a user-defined configuration file.

Processed signals are presented on a Sun 4 SPARCSTATION display as either A-scan traces or colour/intensity modulated sonograms. Operators can pick out signal peaks manually, using a mouse or cursor control, or set a threshold for automatic detection. On the display a split-screen approach allows the operator to see not only the signals themselves, but also observe the detected peaks mapped out on a geographic display. These displays provide information on bathymetric contours, convergence zones,

sensor fields, planned source and target tracks, and transmission loss coverage. The data for all peak detections are stored for subsequent track processing.

For post-mission analysis BIRADS can play back multiple-platform sensor data, re-analyse processed data and reprocess raw data as required. Special features include analysis of echo features and manipulation of the measurement database. BIRADS can also be used in a simulation mode for training purposes.

Operational Status
In production.

Contractor
BBN Systems and Technology Corporation, Cambridge, Massachusetts.

SPARTON SONAR PROCESSING EQUIPMENT

Type
Range of airborne sonar processors and displays.

Description
Sparton's modular airborne sonar processing systems and components are a flexible, lightweight, low-cost means to process and display data from sonobuoys and similar submarine tracking devices. The modular approach allows individual units to be used separately or combined to form a compact system capable of processing and displaying signals from most of the sonobuoy types currently in use or being developed. Typical hardware is described briefly in the following paragraphs.

Processing System
The processing system consists of five modules: command signal transmitter; command signal generator; active processor; sonar display and power supply. The active processor system is capable of simultaneously processing and displaying two AN/SSQ-47 or AN/SSQ-522 active sonobuoys. The modifications provide the matched filter processing required for the additional CW pulse lengths and an FM processor utilising replica correlation. The system can be further expanded to process and display the AN/SSQ-62 or proposed DICASS directional command-active sonobuoys by the addition of a sixth module. This module contains a digital arc tangent computer that calculates and displays bearing with a direct readout in degrees.

Command Signal Generator
The command generator is the active sonobuoy's information source. It is a dual channel control unit that generates the sonar pulse information for two sonobuoys. Selected channels from the sonobuoy receiver are routed to the active processor. The command signal generator also provides the active processor with trigger pulses and other digital control information.

Active Processor
The active processor is a lightweight, dual channel unit capable of simultaneously processing two command-active sonobuoys. It automatically analyses information for the detection and localisation of submarines. It is completely solid-state, with integrated circuits used extensively in the analyser portion of the unit. The processed signals from two sonobuoys can be displayed in a split-screen mode on the sonar display.

Sonar Display
The sonar display can be used with the active processor to provide target range and Doppler information in the CW mode and target detection and range information in the FM mode. The Sparton sonar display utilises a variable-persistence cathode-ray tube that

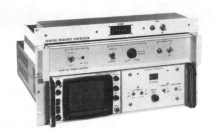

TD-1135/A demultiplexer processor/display

provides bright, high quality readouts. In use, the screen is divided vertically into two displays where a combination of A scans or B scans can be displayed for either one of two sonobuoys simultaneously. Persistence time is adjustable from zero to full storage. Stored spot resolution is equivalent to 20 lines/cm.

Specifications
Display: Two channels, A and/or B scan
CW: Target range and Doppler
FM: Target detection and range
Display integration: Variable-persistence CRT for time integration
Controls
Threshold: Provides capability to optimise target detection

TD-1135/A Demultiplexer Processor/Display
The TD-1135/A is a high quality instrument for testing directional frequency analysis and ranging (DIFAR) and directional command-active (DICASS) sonobuoys. The TD-1135/A accepts a composite DIFAR/DICASS signal from any standard sonobuoy receiver, demultiplexes it, and displays the output as a radial vector on a variable persistence CRT calibrated in degrees from magnetic north. North-south, east-west and omnidirectional outputs are provided for computer analysis if greater accuracy is required.

Designed for use with standard sonobuoy monitoring equipment, the TD-1135/A is compatible with receiver types AN/ARR-52, AN/ARR-72, and AN/ARR-75, and with analyser display group types AN/AQA-3, AN/AQA-4 and AN/AQA-5. Included with the TD-1135/A is a Sparton DFS-101A digital frequency synthesiser, with crystal oscillator reference, which is tunable in discrete steps over the entire DIFAR/DICASS frequency band. If an AN/AQA -3, -4 or -5 is not available, the TD-1135/A and frequency synthesiser can be modified by adding a spectrum analysis capability.

The TD-1135/A demultiplexer is available as a separate unit for operation with DIFAR or DICASS sonobuoys.

Specifications
Frequency range: Broadband DIFAR audio
Input impedance: 10 kΩ (min)
Output impedance: 10 kΩ (max)
Variable gain: 48 dB (manual or remote)

Time constants: 0.1, 1 and 10 s
Phase pilot bandwidth: 1 Hz (max)
Processor analysis bandwidth: 1 Hz (max)
MTBF: 1500 h (min)
Outputs available
Demultiplexer: N-S, E-W and omni
Processor: N-S, E-W and omni

Spectrum Analyser
The Sparton spectrum analyser is a low-cost, special-purpose digital processor that performs real-time spectral analysis. The signature of the target being analysed is displayed simultaneously using both channels of the Sparton sonar display. Discrete Fourier transform technology is used with circuits operating at optimally low bit rates. Features include linear integration of spectral estimates, manual or automatic modes, internal cursor for accurate frequency identification, and total spectrum coverage.

Specifications
Frequency range: 14-912 Hz
Frequency bands: 14-114 Hz (0.25 Hz resolution); 112-912 Hz (2 Hz resolution)
Number of spectral estimates per band: 400
Linear integration of spectral estimates
Manual: 1-100 samples
Automatic: 100 samples
Input amplitude dynamic range: 80 dB

DICASS Command Signal Monitor
The Sparton DICASS (directional command-active sonobuoy system) command signal monitor is used to monitor valid command signals during testing of the AN/SSQ-62. The unit is self-contained and requires a UHF antenna and 115 V AC for operation. It provides a permanent printed record of the VHF channel command, identification of command function, RF signal level during command, time of signal transmission, length of command, and RF signal level prior to command.

Sea Noise Directionality and Level Indicator (ID-1872/A MOD)
The ID-1872/A MOD adds the ambient sea noise directional estimator (ANODE) function to a standard ambient sea noise indicator. It maintains the capability of the ASNI ID-1872/A, which measures and displays the omnidirectional sea noise level sensed and transmitted by an AN/SSQ-57 sonobuoy and adds the ANODE function.

Specifications
Analysis centre frequency: 50, 100, 200, 440, 1000, 1700 Hz
Analysis bandwidth: 25 Hz
Weight: 3.18 kg

Contractor
Sparton Corporation, Electronics Division, Jackson, Michigan.

SONAR SYSTEMS

This section covers all types of sonar used for ASW and deployed by surface ship, submarine and helicopter, as well as sonobuoys and static sonar systems. Shipborne sonars used for naval hydrographic purposes and training systems and simulators are described later in this book.

Sonars are all affected by the same environmental factors; for example, temperature, salinity and pressure, and seasonal variations in these factors as well as the geographical area in which operations are to be conducted.

In carrying out ASW, sonars are assigned to the following tasks: detection – medium-range passive; classification – medium-range active/passive using high speed, highly accurate signal processing, Doppler correlation and digital analysis; localisation – shorter range active/passive for tracking and final computations for directing strike action.

Detection Problems

Passive sonars rely for detection on flow noise and cavitation caused by movement in the water of hull and propellers which creates noise over a wide frequency band; on regular low frequency sound created by machinery; and on hostile active sonars.

Active sonar is bound by the same physical laws which affect passive sonars, except that the effect is doubled, for the sound has to travel in two directions – out and back – before signal processing can be undertaken. Active pulses also lose energy because of the increased distance they have to travel and by absorption in the hull of the target. Active sonars therefore suffer high propagation losses and their effective range is not as great as a passive sonar. The range problem can be overcome by transmitting a low frequency pulse (typically 1 kHz with a wavelength of approximately 1.6 m), but it is difficult to focus the beam accurately down a single point bearing. The spread of the pulse at low frequency also reduces accuracy by which range can be determined. Range can be improved by increasing transmitting power, but the relation is not linear. Advantages are that low frequency pulses can penetrate anechoic coatings. It is also more difficult to detect low frequency pulses at great distances (typically in the order of hundreds of miles).

Data Handling

On the data processing side the future will see significant increase in automated detection, classification, tracking and so on and the presentation of data in synthetic form. Processing power is available in abundance to cope with the demands that will be made on it in the future. The problem will be knowing what to do with the vast amount of information gathered in order to process it in the right way, to present the command with the right type of picture best suited to

particular functions, and from which accurate assessments can be made as to future tactics.

In the future, command systems must be fully integrated, which includes the sonar plot. This integration can be achieved in one of two ways:
(a) by passing fully processed data on all contacts to the command system for evaluation, a task which will require the application of artificial intelligence
(b) by relegating some functions of the command system back to the sonar processing system so that the command system receives heavily processed track data via a distributed system.

Factors Affecting Detection

Sonars suffer from a number of disadvantages, and their effectiveness depends on a number of factors. Efficiency depends on whether one is operating in passive or active mode. All ships create self-generated noise from their movement through the water (cavitation) and from their machinery as noted above. Machinery-generated noise is more of a problem when the sonar is operating in passive mode, while cavitation tends to dissipate the energy of the transmitted pulse. In active mode effectiveness can be further curtailed by temperature variations which may restrict range capability.

These problems can be overcome in two ways, either singly or combined, depending on the requirements. One is to increase the transmitting power. This requires increased volume and weight capability in the ship together with the necessary generating power, and a larger transducer array. The second solution is to mount the transducer as far away from noise sources as possible; for example, it would have to be a towed, variable depth or bow-mounted array. Bow arrays also suffer disadvantages, for it is not unknown for ships to be manoeuvred forgetting the extra projection at the bow, resulting in damage to the protective cover and even the array itself. To further improve the efficiency of the sonar it is necessary to ensure that the ship is as quiet as possible. This can be achieved by providing a self-measuring noise diagnostic system which will enable the signal processing to cancel out self-generated noise frequencies.

Solutions to temperature variations and self-generated noise in passive mode are overcome by using variable depth sonars and the latter by towed arrays. All sounds possess their own very distinctive characteristics or 'footprints', for example frequency, amplitude and so on. Classification of targets can therefore be greatly aided if received echoes can be matched against a known library of sounds.

Taking advantage of the various properties of the medium through which the sound passes in order to extend and improve the capabilities of the sonar, requires various ancillary systems which can measure

environmental factors and provide propagation predictions. Using this equipment an operator can decide how best to use his sonar to obtain more accurate and positive results. For example, under the right conditions, range can be extended by use of bottom bounce, but it suffers in that the beam cannot be focused after reflection, thus scattering and divergency occurs.

Alternatively the convergence zone mode of operation can be used which does not require so much power, but which may not always give such an extended range. The mode of operation is selected according to conditions and the tactical situation, but smaller sonar sets may not have the capability to operate on the bottom bounce mode.

Sonars usually operate on a variety of frequencies. For example, some older types of sonar work on a short range of about 2750 m using two narrow beams with phase shift covering the 17-30 kHz band. Medium-range sets use a wider beam and operate at a lower frequency to overcome the problem of attenuation and so increase range.

Modern sonars use electronic scanning to provide a panoramic picture (except for the small area round the stern which would be blanked by propeller noise). These are usually low frequency sets with very high power to achieve long-range detection. Torpedo detection is now an essential and vitally important function of any sonar, for without it a ship is virtually defenceless against torpedo attack.

The section on sonobuoys covers both passive and active, directional and omnidirectional buoys designed for the detection, tracking and localisation of underwater targets.

Probes, profiling systems and similar devices designed to provide acoustic propagation data are covered later in this book.

The section on static systems covers both large static detection systems such as SOSUS, and the smaller systems used for harbour protection, oil/gas rig protection and so on. The very large fixed systems, for both long-range surveillance and as 'trip-wires' at specific locations, are mainly deployed by the USA and the USS. They are relatively widespread but their security classification remains very high. Consequently, very little specific information is freely available.

The United States is carrying out a comprehensive development programme on underwater target surveillance systems and on the engineering of acoustic search sensors. There seems little doubt that the USS has been carrying out a similar programme. In both cases security restrictions preclude the access of information.

SURFACE SHIP ASW SYSTEMS

AUSTRALIA

MULLOKA SONAR

Type
Hull-mounted scanning sonar system.

Description
Mulloka is a lightweight, relatively high frequency active hull-mounted scanning sonar for installation in Royal Australian Navy anti-submarine escorts and designed for operation in coastal waters around Australia. It is a solid-state forced air-cooled equipment of high reliability which has been designed using a standardised packaging approach to aid maintainability and keep production simple.

The equipment includes a dedicated on-line computer which is used to control transmission parameters, perform automatic signal processing functions

and conduct checkout procedures. All signal interface equipment with other ships' systems has been grouped at a single location so that modifications necessary to interface Mulloka with other ship equipment configurations are confined to one unit.

The sonar beam forming process involves the development of a large cylindrical transducer array consisting of 96 staves with 25 transducers per stave, driven in differing combinations and phases under computer control to form the required sonar beams electronically. The hydrophone elements of the array are driven by 480 power amplifiers during the transmit period, the length of which is a function of the pulse length of the transmitted signal. The transmit period is selectable over the range of approximately one to four seconds. The transmitter beams are formed sequentially during this period and cover the ocean for 360° in azimuth. The formation of the beam transmit

pattern is random in nature from one transmission to the next in order to prevent the submarine commander having any indication of detection. Immediately following the transmission, the sonar reverts to the listening mode for a period determined by the range scale selected by the operator. Range is about 10 000 yd.

Operational Status
In 1979 a prototype Mulloka sonar was formally handed over to the RAN after undergoing trials in HMAS *Yarra* since 1975. A production programme to build systems to equip six RAN 'River' class destroyer escorts has been completed.

In April 1980 it was announced by the Australian Ministry of Defence that an initial contract had been awarded to Honeywell Inc in America to provide a transducer array for the Mulloka sonar. Two

Australian companies, GEC Marconi Systems Pty Ltd and Dunlop Industrial and Aviation, were also participating. A contract for the electronics of the system was let to THORN EMI (Australia) Ltd in 1979.

In early 1982, two production contracts were awarded for the 'River' class ships to be fitted with Mulloka sonars. One contract for A\$16.1 million was awarded to THORN EMI (Australia) Ltd and the other

to Honeywell, USA, for A\$9.4 million. The THORN EMI contract was for six systems delivered in the period 1983-85 and the Honeywell contract for five transducer arrays delivered in 1983/84.

In 1985, contracts were awarded for two systems to be fitted to the 'FFG 7' class frigates being built in Australia. As a result of the transferred technology, Plessey Australia was contracted to establish a

production facility and to manufacture two transducer arrays and spares which were delivered over the period 1987 to 1989. In mid-1987, THORN EMI (Australia) Ltd was awarded a A\$6 million contract to upgrade the system electronics, including a new display system. The Mulloka sonar will also equip Australia's ANZAC frigates.

KARIWARA TOWED ARRAY SONAR

Type
Thin line towed array sonar project.

Description
The Australian Defence Research Centre has refined a towed array sonar for surface ship or submarine use to what is claimed to be the most compact system available. Code named Kariwara, the array carrier is only about 19 mm thick, with cigarette-sized

sensors spaced along it. It is stated to be suitable for ships down to small patrol boat size. The system has a thick rubberised coating to make it less vulnerable to predators. No other details are available, but the sonar is understood to be in the early development phase.

Operational Status
As at the beginning of 1990, the Kariwara project was still in the development phase. It was hoped that it would be sufficiently developed for installation in the two 'FFG 7' class frigates being built in Australia, but this did not transpire. Kariwara has been trialled on

one of Australia's 'Oberon' class submarines and is to equip the new 'Collins' class.

The Australian Department of Defence was scheduled to begin a competition in mid-1989 for an Advanced Surface Ship Towed Array Surveillance System (ASSTASS) as the surface surveillance application of Kariwara. The surface ship designation is Project 1300; the submarine designation is Project 1100. Problems have been encountered with the submarine towed array and it is possible that an alternative will be fitted.

CANADA

HS-1000 SERIES SONAR

Type
Lightweight hull-mounted and VDS sonars.

Description
The HS-1000 series of sonars is a family of lightweight omnidirectional ASW sonars which have been developed to give small naval vessels a modern sonar capability in search, detection and attack. The numbering of each sonar designates transducer mounting and/or sonar power:

(a) HS-1001: lightweight variable depth sonar (VDS)
(b) HS-1002: lightweight hull-mounted (HM) sonar
(c) HS-1001/2: lightweight combined VDS/HM sonar with single electronics and a transfer switch
(d) HS-1007: medium-sized sonar available in either the HM or VDS configuration featuring higher power
(e) HS-1007/7: medium-sized combined VDS/HM sonar with single electronics and a transfer switch.

The HS-1001 and HS-1002 sonars are intended for use on patrol craft and corvettes. The HS-1007s are intended for use on corvettes, frigates and destroyers.

Operational Status
The HS-1000 series sonar is available on a current production status and has been delivered in the VDS configuration and in the combined VDS/HM configuration.

Contractor
Westinghouse Canada Inc, Burlington, Ontario.

AN/SQQ-504 TOWING CONDITION MONITOR

Type
Monitoring system for VDS sonar systems.

Description
The AN/SQQ-504 towing condition monitor is an aid to ship's personnel for successful launch, tow and recovery of the variable depth sonar (VDS) body. With the introduction of this type of equipment, the command and sonar operations are provided with a badly needed means to optimise operations and minimise risk to VDS equipment due to the variability of sea conditions, ship's speed and motion and so on.

The monitor provides VDS tow cable strain information (actual strain when the body is under tow, or

AN/SQQ-504 towing condition monitor, showing primary indicator (left) and remote indicator (right)

predicted strain when the body has not been launched or is to be towed at a different cable length) to the operator and ship's command. The system also

provides information by which the VDS operator can judge the performance of the VDS hoist system and provides information to enable personnel to monitor the degrading effect of ship's motion on the VDS sonar detection.

The system consists of a group of data sensors, a processor unit, two recording units, three identical remote display units, and a VDS operator's control and display unit. These items perform the functions of data acquisition, signal conditioning, data processing, display and recording. Sensors include boom and deck accelerometers, a cable strain gauge, a boom switch, a cable length synchro and ship's speed data.

Operational Status
The AN/SQQ-504 is fitted to Canadian Navy destroyers.

Contractor
Westinghouse Canada Inc, Burlington, Ontario.

AN/SQS-505 SONAR

Type
Hull-mounted and VDS search and attack sonar.

Description
The AN/SQS-505 is an evolving family of 7 kHz medium-size search and attack sonars intended for use on vessels of 1000 tons and above. It may be used in a hull-mounted and/or variable depth application.

In the Canadian ships, each vessel has two separate but identical SQS-505s on board. One system is variable depth, the other is hull-mounted retractable. Other navies have used the SQS-505 in hull-mounted only applications, with and without hull outfit retracting mechanisms. It is possible to use the 505 in direct communication with a digital computer, either for the transfer of range and bearing information or for assistance to the operator in detection and tracking.

Mounted on major platforms, such as frigates, corvettes and destroyers, the AN/SQS-505 is a deep water sonar surveillance system, proven in operational service and tested to military specifications. It is an omnidirectional sonar, also able to operate in a directional mode that can maintain search ability while tracking defined targets. Three selectable frequencies are available so that several ships can survey an area in close proximity without cross-talk. In high sea states which result in excess roll and pitch of the ship, the AN/SQS-505 utilises transmit beam stabilisation. This circuitry delays transmission until near optimum roll and submergence conditions prevail, thereby minimising chances of the target being missed altogether.

Hull-mounted Hull Outfits
The SQS-505 hull-mounted transducer can be used in any one of the following applications:
(a) fixed transducer and dome
(b) fixed dome/retractable transducer (E-5 hull outfit)
(c) bow dome.

The requirement for these various alternatives would depend upon the configuration of the ship and the location available for the mounting of the transducer.

Operational Status
The SQS-505 is currently in production. Substantial advancements in the sonar's capability have been incorporated since the system was first installed in ships in the early 1970s.

In addition to the seven Canadian ships outfitted with dual SQS-505 sonars (VDS/HM), four hull-mounted systems have been installed on the latest Belgian 'Wielingen' class frigates, and the Royal Netherlands Navy has fitted bow dome systems to six of its 'Kortenaer' class frigates. The Hellenic Navy has two 'Kortenaer' class frigates fitted with AN/SQS-505 systems.

The latest AN/SQS-505(V)6 sonar system configuration is to be fitted to the 12 new Canadian patrol frigates of the 'Halifax' class, with fixed, faired dome and retractable transducer.

Specifications

Coverage: 360° (azimuth)
Range: 200 to 3200 yd
Doppler range: ±40 kts (7.2 kHz)
Displays: Range (A Scan); Bearing (B Scan); Doppler
Frequencies: 6.4, 7.2 and 8.0 kHz
Source levels: From 224.233 dB/μPa/m (transmission mode dependent)

Initial detection: 12 ping history of returns
Bearing accuracy: 1 RMS
Range accuracy: ±1% range reading
Transmission modes: Omni, TRDT, DT, ASPECT
Power level: High and low (10 db reduction)

Contractor

Westinghouse Canada Inc, Burlington, Ontario.

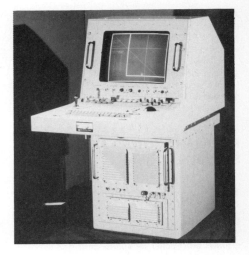

AN/SQS-505(V)6 (AN/UYQ-501(V) SHINPADS) control and indicator console

---

AN/SQS-509 SONAR

Type

Hull-mounted and VDS search and attack sonar.

Description

The AN/SQS-509 is a lower frequency version of the SQS-505 intended for use on vessels of 2500 tons and above. It has been derived from the SQS-505 by changing the frequency-dependent items, such as the transducer, and consequently there is considerable logistic commonality with the SQS-505. Apart from a lower operating frequency (5.4 kHz), the SQS-509 has the same operational features as the SQS-505 described in the previous entry.

The transducer is approximately 1.6 m in diameter and 1.6 m high. It consists of 360 individual elements, each of which can be removed while the array is under water. The transmitter group consists of two cabinets: the power supply cabinet and the transmitter cabinet. The power supply converts the ship's prime AC power to DC, required for the operation of the transmitter. The transmitter contains 36 power output modules, each driving one stave of the transducer. It is completely solid-state and contains its own integrated test equipment. The principal transmission modes are Omni and TRDT.

The receiver is contained in three cabinets of the standard SQS-505 size. It is completely solid-state with its own automatic built-in test equipment which enable the operator to determine the location of a fault down to the replaceable module level.

The control and indicator console consists of a display and controls which enables one operator to have complete control over the operation of the sonar and the ability to detect, classify and track targets. In addition, via separate displays in the console, the operator is able to maintain continuous independent surveillance of the passive mode. Because of the larger transducer size, the preferred mounting for the transducer is in a bow dome.

Operational Status

The SQS-509 is fitted in four 'Kortenaer' class and two 'Jacob van Heemskerck' class frigates of the Royal Netherlands Navy.

Contractor

Westinghouse Canada Inc, Burlington, Ontario.

TYPE 5051

Type

Surface ship search/attack sonar.

Description

The compact, modular Type 5051 sonar is designed for use on vessels with a displacement of 100 tonnes and above. It uses a wide operating frequency range of 1 to 15 kHz, and with its modular design can be readily adapted to numerous array configurations and operational roles. The system can be acquired as a turnkey system or as replacement electronics for existing sonars.

The transmitter, receiver, T/R switch and power supply is housed in two electronics cabinets, and the system also features a fault tolerant signal processor and BITE.

Three different frequencies and array sizes (VDS in two), are currently available for the 5051: a 36 stave 5 kHz; a 36 stave 7 kHz and a 24 stave 10 kHz. The system can also be adapted to other existing array configurations, including towed arrays.

The transmitter uses high efficiency linear amplifiers to faithfully reproduce digitally synthesised waveforms for precise beam forming. The receiver is built around a Westinghouse Canada proprietory digital signal processor design. Its modular design allows for future system upgrade or modification through the use of additional processor cards. Data are presented on a variety of display formats which include:

(a) A-scan for initial active detection
(b) B-scan for active localisation
(c) PPI for active localisation
(d) Zoom B-scan for active classification
(e) Doppler for active classification
(f) ASPECT for active classification
(g) Bearing/Time for passive detection/DOA estimation
(h) SPECTRUM for passive detection/classification.

Operational Status

In development.

Specifications

Transmitter
Beam: Omni/TRDT/DT
Modulation: CW, LFM, Hyperbolic FM, CW-FM Hybrid
Pulse length: Typically 2.5, 40, 100 ms but can be varied
Power output: Up to 1.5 kW per stave ± 36 stave drive capability
Receiver
Beam forming: 24 or 36 pre-formed beams with 1 steered beam
Active processing: FFT for CW, correlation for FM
Passive processing: FFT for spectral analysis and short and long term integration for bearing/time display

Contractor

Westinghouse Canada Inc, Burlington, Ontario.

SCAN 500 SERIES SONARS

Type

Shallow water sonar for patrol boats.

Description

Designed with shallow water operation in mind, the SCAN series of sonars meets the mine detection and anti-submarine warfare mandate of today's patrol boat fleets. Available in 22 kHz, 45 kHz and 120 kHz models, SCAN can operate as a stand-alone search and attack sonar, or in support of other search methods such as a stern-rigged dipping sonar or a variable depth sonar. The new Westinghouse SCAN VDS system features computer control, variable tow depths, terrain avoidance and ray plot capability with a total system 'dry weight' of 2500 kg. All configurations of SCAN are easily adaptable to vessels down to 35 m in length and 100 tons displacement.

Search and attack modes are available with outputs of range, bearing, target depth, true audio, VDS location and Doppler presented to the operator. SCAN features include FM and CW processing, weapons system interface, three-ping target correlation, PPI and data displays, true motion, true north or ship's orientation, reduced power transmission, transmit delay, RS232 outputs, BITE and vertical beam steering.

Operational Status

In production.

Contractor

Westinghouse Canada Inc, Burlington, Ontario.

AN/SQR-501 SONAR SYSTEM

AN/SQR-501 CANTASS

Type
Surface ship towed array sonar system.

Description
The AN/SQR-501 Canadian Towed Array Sonar System (CANTASS) equipment is a critical angle surface ship towed array which uses two Computing Devices standard Canadian Navy products: the AN/UYS-501 array processor and the AN/UYQ-501 SHINPADS display (see separate entry). CANTASS employs the Martin Marietta AN/SQR-19 towing winch and hydrophone subsystem, and modular interfacing and data management electronics. This provides a capable, cost-effective, long-range passive detection and surveillance system for surface ships.

The system provides frequency and bearing analysis of acoustic emissions from long ranges, in particular for nuclear submarines. Data are presented to operators on programmable CRT displays in a number of alternative formats. A variety of processing algorithms have been developed by the Canadian Navy Defence Research Establishment and Computing Devices Company.

The implementation of the system processing and data manipulation algorithms is entirely in software, providing the system with a growth capability and the adaptability to meet changing needs. The modular physical configuration and the use of standard components allows the system to be adapted to other platforms, such as submarines.

Operational Status
A contract worth $89 160 712 for 15 systems was placed in August 1990.

Contractor
Computing Devices Company (a division of Control Data Canada Ltd), Ottawa, Ontario.

AN/SQS-510 SONAR SYSTEM

Type
Hull-mounted active sonar system.

Description
The AN/SQS-510 is a new digital sonar system that integrates control processors with the DND AN/UYS-501 digital signal processor and the AN/UYQ-501 dual screen colour display system, both produced by Computing Devices Company, to replace the receiving and display elements of the AN/SQS-505 sonar (see separate entry). It is intended to provide surface units with a hull-mounted or VDS active detection and localisation system, either as prime sensor or as a complementary sensor to the AN/SQR-501 CANTASS equipment.

Operational Status
In production for ships of the Canadian and Portuguese navies.

Contractor
Computing Devices Company (a division of Control Data Canada Ltd), Ottawa, Ontario.

AN/SQS-510 sonar advanced development model

AN/UYS-501 PROCESSOR

Type
Digital signal processor.

Description
The AN/UYS-501 is a militarised digital signal processor that forms part of the DND AN/SQR-501 Canadian Towed Array Sonar System (CANTASS) and AN/SQS-510 digital sonar system. The signal processor meets the latest fast computation-intensive requirements through its high bandwidth, internal busses and innovative memory addressing techniques.

The AN/UYS-501 uses an M68020-based CPU set and will support up to six I/O devices which can be software configured to be a NATO STANAG 453, Computing Devices Company 16-bit parallel (bi-directional, maximum data rate of 2 million 15-bit words/s) or Computing Devices high speed serial (unidirectional, maximum data rate of 10 million bits/s) interfaces.

The AN/UYS-501 will execute IEEE 754-1985 floating-point complex word (32-bit real, 32-bit imaginary) or 32-bit real word arithmetic. It contains 4 194 304 complex words of working or main memory, with an expansion capability to 67 108 864 complex words. It also contains a total of 32 768 complex words of cache memory with an expansion capability to 65 536 complex words. Dedicated hardware optimises the processor for FFT heterodyning, beam forming and data recording. With its eight arithmetic units configured in parallel, the throughput is 32 Mflops, and it will compute a 1024 length complex FFT in 160 μs.

The processor is provided with real-time multitasking, development and debug software. Power up/on-line/off-line diagnostics simplify fault finding and isolation to a single card. It is also designed to support the signal processing high level language (Fourier).

The AN/UYS-501 is contained in a single cabinet with dimensions 65 in high × 26.8 in wide × 26 in deep (165 × 68 × 66 cm). It is air-cooled using ambient air and operates from 110/120 V, single phase, 60 Hz, 2 kVA supply.

Operational Status
A contract to provide an engineering development model (EDM) was awarded in December 1985. Scheduled completion of the EDM was due in March 1989. A production contract, as part of the production AN/SQS-510 system, was awarded in April 1988. A further Canadian Department of National Defense production contract, as part of the EDM and production AN/SQR-501, was expected in the early part of 1989.

Contractor
Computing Devices Company (a division of Control Data Canada Ltd), Ottawa, Ontario.

CTS-36 LHMS OMNI SONAR

Type
Small ship hull-mounted sonar.

Description
The CTS-36 LHMS OMNI sonar is a hull-mounted small ship system for use in surveillance, anti-submarine warfare and mine countermeasures support roles in coastal waters. It has been optimised for shallow water operation. The CTS-36 is a scanning and multi-beam sonar, using narrow receiving beams, high speed scanning and digital processing techniques, to provide instantaneous detection, classification and target data in a full 360° field-of-view. Simultaneous video and audio information is available in both active and passive modes. A minimum of three tracking channels is provided with target and ship's track, target range, depth, bearing, speed and heading instantaneously displayed.

The CTS-36 also provides:

(a) detection of torpedoes and divers at short range
(b) detection of targets such as mines, wrecks and rocks on the seabed
(c) detection, location and classification of large targets at long ranges
(d) electronic tilt and stabilisation of transmit and receiving array
(e) interface to ship's own command/communications systems
(f) a high resolution control/display unit.

Operational Status
The Royal Danish Navy has signed a contract for the supply of CTS-36 RDN OMNI sonar systems for its IS-86 inspection ships.

Specifications
Operating frequency: CW mode; nominal 36 kHz (33 kHz optional) FM chirp; 1 kHz centred at 36 kHz
Beam tilt/stabilisation: Electronic tilt continuously adjustable from 8° above the horizontal to 24° down. Stabilised for roll and pitch
Receiving beamwidth: Horizontal and vertical: 6° transmitting beam omni or directional sector
Transmitter: 270 channels; power output/channel 50 W max
Source level: 220 dB/μPa/m

Modes of operation
Omni active: Simultaneous audio and video selectable, omni or directional CW transmission. Automatic detection mode in which area on display is automatically outlined where signal reception exceeds operator-selected threshold. Audio continuously steerable through 360°. Transmit sector width and position selectable
Omni passive: Reception system as above
Tracking: FM chirp of CW. Each of three independent receive beams may be manually locked on targets through 360°. Omni video and/or manually positioned audio
Surveillance: Operator selectable transmit sector width and position, 12° receive beam stepped through operator selected sector. Video and audio
Search MCC: Omni transmission. Broad 30° depressed receive beam facilitates maintenance of contact on close-in targets

Contractor
C-Tech Ltd, Cornwall, Ontario.

CMAS-36

Type
Mine avoidance sonar.

Description
Derived from the CTS-36 mine avoidance sonar, the CMAS-36 has been developed with the emphasis towards locating tethered mines in shallow waters, and torpedoes and divers at short range. Like the CTS-36, this sonar is a lightweight, high speed scanning, multi-beam, hull-mounted sonar using narrow receiving beams and digital processing techniques to provide instantaneous detection, classification and target data. Video presentation on a 16 in colour display in either PPI or B-scope is available in both active and passive modes, a trackerball controlled cursor providing target range, depth, bearing, speed and heading for instantaneous display.

Selectable omni or directive CW transmissions are available in the active mode with selectable transmit sector width and position. The automatic detection mode outlines one or more of approximately 100 detection zones in which signal reception exceeds the operator-selected threshold. The surveillance zone is fixed at 72° centred on the bow.

Specifications
Frequency: 36 kHz (30 kHz optional) in CW mode
Beamwidth: Receive – (horizontal and vertical) 6°; transmit – omni or selectable sector
Transmitter: 270 channels; power output/channel 50 W

Contractor
C-Tech Ltd, Cornwall, Ontario.

INDAL TECHNOLOGIES SONAR HANDLING EQUIPMENT

Type
Sonar towed bodies, handling equipment and sonar domes.

Description
Indal Technologies specialises in the provision of sonar towed bodies, handling equipment and sonar domes for underwater defence and commercial applications.

For cable handling and towing systems applicable to variable depth sonar (VDS) systems, Indal has a range of equipment which is normally mounted at the aft end of the ship and is self-contained, apart from electrical power requirements. The range includes the 6-200, 9-330, 9-600, 15-750 and 10.5-600 systems.

Indal Technologies produces fully integrated systems and subsystems for ASW and countermeasures requirements, including tactical towed line array handling systems as well as helicopter lightweight dipping sonar winches and torpedo decoy handling subsystems.

The Towfish and Rayfish are general-purpose towed bodies designed to suit a wide range of applications. They are used to house water sampling, seismic side-scan, sub-bottom profiling, video, magnetometer and fish-finding sonar systems. Both bodies have low flow noise characteristics and access to instrumentation through easily removable shells.

Indal also designs, develops and manufactures a range of sonar domes. These are configured by optimising acoustic, hydrodynamic and structural parameters for specific operational frequency and bandwidth, and incorporate a stainless acoustic window of double skin design.

Operational Status
In production.

Contractor
Indal Technologies Inc, Mississauga, Ontario.

FRANCE

DSBV 61A SONAR

Type
Surface ship towed array sonar.

Description
The DSBV 61A is a very low frequency passive towed array sonar system consisting of a linear array, a towing cable which is also used for data transmission, a winch for handling and storage of array and cable, and the necessary signal processing and data extraction electronics. The latter is made up of five cabinets, plus two display consoles associated with two TOTES. The DSBV design allows two selectable towing modes: critical angle towing or towing behind the DUBV 43C VDS body.

The display consoles feature three-colour display of target information and can display on request, the various sonar memory pages. A built-in fault display allows a quick reconfiguration and/or rapid maintenance of the system.

DSBV 61A linear towed array sonar (serial equipment) on board the Primauguet *C-70 type ASW destroyer*

Operational Status

In series production. Operational on the C70 type ASW corvettes. A derivative known as Anaconda is described briefly under a separate entry.

DUBA 25 SONAR

Type

Hull-mounted active panoramic sonar system.

Description

The DUBA 25 (the service designation for the Tarpon TSM 2400) is a powerful attack sonar fitted in the French 'Aviso' class ships and similar anti-submarine vessels. It is a medium frequency active sonar with a panoramic transducer array and assembly with sonar dome supplied by the French Navy. Although the sonar is suitable for use as towed VDS equipment, in the DUBA 25 configuration it is hull-mounted, and provides 'Aviso' class ships with all-round surveillance, acquisition and attack facilities for ASW operations. The power employed is sufficient for operating ranges of several kilometres. The transducer array consists of 36 staves arranged to form a cylinder 110 cm in diameter and provided with roll stabilisation.

Contractor

Thomson Sintra Activités Sous-Marines, Valbonne Cedex.

Operational Status

Operational since 1975. No longer in production. More detailed information is given in earlier editions of *Jane's Weapon Systems*.

Contractor

Thomson Sintra Activités Sous-Marines, Valbonne Cedex.

DUBV 23D SONAR

Type

Surface ship low frequency search sonar.

Description

The DUBV 23D is a bow-mounted, low frequency, panoramic sonar for anti-submarine operations. The 48-column transducer array is housed in a bulb at the fore part of the ship, the bulb being of streamlined design to reduce parasitic noises to permit listening at high speeds. The panoramic sonar is intended for both search and attack roles.

In addition to the transducer array, the equipment includes the transmitter/receiver unit, a computer section for the processing of data being fed to weapons, and control and display consoles at the anti-submarine attack station.

The DUBV 23D is of identical design to the towed sonar DUBV 43B/C and in French vessels the two sonars are used together for anti-submarine warfare.

Operating modes provide for: panoramic surveillance, sector surveillance, step surveillance, passive surveillance at sonar frequency, panoramic attack transmission, or 'searchlight' attack transmission. In addition to the system's own display devices, the DUBV 23D provides for target data outputs to other ships' systems and repeater PPIs.

Operational Status

The DUBV 23D is in service in converted anti-submarine escorts Types T53 and T56, frigates of the 'Suffren' class and Type C67, and Type C70 corvettes of the French Navy. The system is no

DUBV 23D sonar control console

longer in production, but systems in service will be upgraded through the SLASM programme (see separate entry).

Specifications
Transmitter
Frequencies: four operating frequencies in the neighbourhood of 5 kHz, of which two are operational
Power: 96 kW (2 × 48 kW)
Type: FF (fixed frequency), FM (linear frequency modulation with non-coherent data processing at reception)
Duration: 4, 30, 150 or 700 ms

Scatter echo: With or without rejection
Doppler effect correction: On all 48 channels
Cadence: Adjustable step by step from 1500 to 48 000 yd
Receiver: Panoramic, directional, passive listening in sonar band

Contractor
Thomson Sintra Activités Sous-Marines, Valbonne Cedex.

DUBV 43B/C SONAR

Type

Variable depth sonar system.

Description

The DUBV 43B/C low frequency variable depth sonar (VDS) comprises a streamlined towed body containing the sonar transducer array. This is towed from the stern of the parent vessel and can be set to run at depths between 10 and 200 m in the 'B' version and up to 600 m in the 'C' version. It is equipped with stabilisers providing for control in roll, pitch and depth. Dimensions of the 'fish' are length 550 cm, width 170 cm; and submerged weight is 7.75 tonnes. The towing cable also incorporates 48 pairs of signal conductors. The range of towing speeds is 4 to 30 kts, and detection ranges of up to 25 km are quoted.

Operational Status

The DUBV 43B is in service in converted anti-submarine escorts Types T53 and T56, frigates of the 'Suffren' class and Type F67. The DUBV 43C is in service with Type C70 corvettes of the French Navy.

Two updates of the system are in progress:
(a) UTCS 1B which includes improvement of the signal processor
(b) a torpedo alarm receiver is soon to be added to sonars in service.

DUBV 43C VDS towed transducer and towing gear aboard the French corvette Georges Leygues

Contractors
Direction des Constructions Navales, Paris.
Thomson Sintra Activités Sous-Marines, Valbonne Cedex.

ANACONDA TACTAS SYSTEM

Type
Hull-mounted and towed array tactical sonar.

Description
Thomson Sintra Activités Sous-Marines has recently been contracted to supply a system of hull-mounted and tactical towed arrays, known as Anaconda. This system configures a digital wet end, currently in service with the French Navy (DSBV 61A), with an advanced technology sonar receiver built around the powerful Mangouste Mk 2 processor. This latter equipment was developed by Thomson Sintra Activités Sous-Marines to increase the performance of all sonars, particularly in automatic detection, narrowband analysis, automatic tracking, adaptive beam forming and self-noise cancellation.

Operational Status
Ordered for the Royal Netherlands Navy 'M' ('Karel Doorman') class frigates.

Contractor
Thomson Sintra Activités Sous-Marines, Valbonne Cedex.

DIODON SONAR/TSM 2633 MF

Type
Hull-mounted and variable depth sonar systems.

Description
Diodon is a panoramic medium frequency sonar for small and medium sized surface ships. It offers high level performance in signal and data processing, target classification, automatic detection and tracking, and man/machine interface.

The sonar is well balanced, being as efficient in shallow waters as in deep, and is very easy to use. Control is exercised through a one-man console using a dual screen high resolution colour display. The array can be fitted within a fixed or retractable dome, or supplied as a variable depth towed sonar (VDS); in case of a combined system, the same electronics suite can operate both the hull-mounted and VDS arrays. Embedded facilities include passive listening, sound ray tracer, integrated onboard simulator, an acoustic propagation prediction for the day and, optionally, a video tape recording system.

A VDS version is available, and Diodon can be combined with other active or passive TSM sonars.

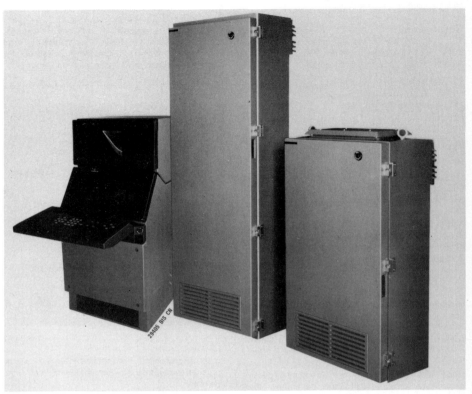

Diodon sonar system, showing (left to right) display console, receiver and power supply

Diodon sonar transducer array

Operational Status
In production and operational in more than seven European or overseas navies.

Specifications
Acoustic array: 24 identical staves; weight 450 kg
Frequency: 12 kHz (approx)

Transmitter: Omnidirectional transmission in CW and FM modes
Receiver: 36 pre-formed beams; advanced processing with replica correlator in FM and spectrum analysis (FFT) in CW

Video processing: Data processing in FM and CW modes: automatic detection and tracking; simultaneous multi-track extraction (up to 64); target analysis (apparent length, aspect angle, Doppler measurement). The sonar is easy to link to the own ship combat system
Display: High resolution flicker-free colour TV monitors
Total weight: 1500 kg (additional 8 tonnes for the winch and handling system in the VDS version)

Contractor
Thomson Sintra Activités Sous-Marines, Valbonne Cedex.

SPHERION SONAR/TSM 2633 LF

Type
Hull- or bow-mounted sonar systems.

Description
Spherion is the primary system in the TSM series of active sonars, which also encompasses the Diodon and Salmon systems.

Spherion is a long-range active sonar for ASW surface combatants. It comprises a spherical hull-mounted array with an electronic processing system that allows for real-time computation in transmission and reception modes; three-dimensional stabilisation to compensate for pitch, roll and yaw in the parent vessel; and a beam-tilting capability to counter adverse sound velocity profiles and strong reverberation.

The sonar can be combined with the medium frequency Diodon, sharing the same electronic cabinets and console.

Operational Status
In production and operational at sea. Spherion has been chosen by the Royal Norwegian Navy to equip the 'Oslo' class frigates and the 'Sleipner' class corvettes, and also by the Australian and New Zealand navies to equip the new ANZAC class frigates.

Specifications
Acoustic array: Spherical with 160 individual transducers
Frequency: 7.5 kHz (approx)
Transmitter: Omnidirectional transmission in CW and FM modes
Receiver: 36 pre-formed beams stabilised in azimuth and elevation; advanced processing with replica correlator in FM and spectrum analysis (FFT) in CW
Video processing: Data processing in FM and CW modes: automatic detection and tracking; simultaneous multi-track extraction (up to 64); target analysis (apparent length, aspect angle, Doppler measurement). The sonar is easy to link to the own ship combat system

Display: High resolution, flicker-free colour TV monitors
Total weight: 4000 kg

Contractor
Thomson Sintra Activités Sous-Marines, Valbonne Cedex.

Hull-mounted array of Spherion/Diodon sonar

LAMPROIE SONAR

Type
Surface ship or submarine towed array sonar system.

Description
The TSM 2930 Lamproie linear towed array is the latest development in the range of linear towed arrays designed and produced by Thomson Sintra. The array is used for passive detection, classification and position fixing of underwater targets. Panoramic broadband and narrowband analysis of the detected signals is performed continuously and displayed on two high resolution colour TV screens.

The array is designed to be towed by either a submarine or surface ship. When used by a surface ship the array can be towed either behind a long, heavy cable (critical angle tow), or behind a composite system. In the latter case the array is towed by a long, neutrally buoyant cable connected to a depressor, which itself is then connected to the towing vessel through a short, heavy cable. For submarine use the array is 'clipped-on'.

Lamproie is able to track several targets concurrently, either in broad or narrowband, with a very fine bearing accuracy. Tracking can be automatic or manual. An automatic rejection system for narrowband countermeasures protects trackers from sonar jamming. Narrowband analysis can be performed concurrently on several beams, steered either under automatic tracking or manually. LOFAR analysis includes refined vernier capabilities. DEMON analysis demodulates signals for cavitation detection. A high fidelity audio channel improves target identification.

Operational Status
In production.

Contractor
Thomson Sintra Activités Sous-Marines, Valbonne Cedex.

SALMON SONAR/TSM 2643

Type
Hull-mounted and variable depth sonar systems.

Description
The Salmon sonar is a compact, medium-range, active sonar designed for light patrol vessels conducting ASW operations mainly in shallow waters. It offers high detection probability in severe noise environment and includes sophisticated signal and data processing, automatic detection and tracking functions, and operator aids.

The basic version consists of an acoustic array of 24 staves housed in a streamlined fish, a towing winch, a transmitter and a receiver unit, and an operator console equipped with a multi-colour display. Embedded facilities include passive listening, sound ray tracer, integrated onboard simulator, and, optionally, an acoustic propagation prediction for the day and video tape recording system.

A hull-mounted version is also available, and Salmon can be combined with the other active or passive TSM sonars.

Operational Status
In production. Two systems have been installed on the 'Stockholm' class, and four of the improved variant on the new 'Goteborg' class corvettes of the Royal Swedish Navy. Salmon systems are also being supplied to the Royal Danish Navy for fitting on board the Stanflex 2000 'Thetis' class fishery patrol vessels, and the Stanflex 300 'Flyvefisken' class patrol vessels.

Specifications
Acoustic array: 24 identical staves; weight 260 kg
Frequency: 19 kHz (approx)
Transmitter: Omnidirectional transmission in CW and FM modes
Receiver: 36 pre-formed beams; advanced

A Salmon VDS sonar fitted on the Danish frigate Thetis. *The illustration shows the vessel before having her stern modified to a transom configuration*

processing with replica correlator in FM and spectrum analysis (FFT) in CW
Video processing: Data processing in FM and CW modes: automatic detection and tracking; simultaneous multi-track extraction (up to 64); target analysis (apparent length, aspect angle, Doppler measurement). The sonar is easy to link to the own ship combat system
Display: High resolution flicker-free colour TV monitors
Total weight: 850 kg (additional 7 tonnes for the winch and handling system)

Contractor
Thomson Sintra Activités Sous-Marines, Valbonne Cedex.

SLASM

Type
High performance integrated sonar system.

Description
SLASM is a high performance integrated sonar system for large ASW surface ships (frigates or destroyers), comprising the following main sub-systems or functions:
(a) a very low frequency, deep diving, activated towed array sonar
(b) an active low frequency hull-mounted sonar
(c) an active low frequency variable depth sonar
(d) a passive very low frequency towed array sonar
(e) a torpedo alert system, using some of the previously mentioned arrays.

The system consists of:
(a) a large, deep diving, towed fish housing the very low frequency transmitter array, and the low frequency transmitter/receiver arrays
(b) a long linear array, to be towed clipped on to the towfish, or towed independently
(c) a hull-mounted low frequency transmitter/receiver array
(d) a large towing and handling system, a by-product of the existing DUBV 43 system.

Signal and data processing are performed through three identical cabinets, each being able to process any function of the system; two more cabinets house the recording and simulator system, the post-processing and the system controller.

The man/machine interface consists of four multi-function consoles and two dedicated monitors.

The SLASM has been ordered by the French Navy for upgrading the F67 ASW frigates; development is in progress.

A second version is foreseen, probably through a collaborative venture with some other European navies; this second batch is to be installed in the French Navy aboard the C70 ASW frigates, by the end of the century.

Contractor
Thomson Sintra Activités Sous-Marines, Valbonne Cedex (industrial prime contractor).

ATAS SONAR

Type
Active towed array sonar system.

Description
ATAS is an active towed array sonar being developed by Thomson Sintra ASM (France) and BAeSema (UK) for advanced ASW operations.

For more details, refer to United Kingdom later in this section.

Contractor
Thomson Sintra Activités Sous-Marines, Valbonne Cedex, France.

BAeSema Marine Division, Bristol, UK.

ALBATROS

Type
Torpedo alert system.

Description
The Albatros torpedo alert system is an advanced modular sonar system specially designed for torpedo detection and classification.

The basic system comprises:

(a) a short special towed array
(b) a small size Angler-type towing and handling system
(c) a processing cabinet
(d) a display.

The main characteristics of the system are:
(a) large azimuth coverage
(b) permanent surveillance
(c) automatic pre-alert

(d) bearing warning without port/starboard ambiguity
(e) final classification and decision under operator control.

Operational Status
Under development: parts have yet to be delivered for trials.

Contractor
Thomson Sintra Activités Sous-Marines, Valbonne Cedex.

U/RDT-1A PASSIVE SONAR

Type
Wideband passive torpedo detecting sonar.

Description
The U/RDT-1A is a pre-formed beam passive sonar designed specifically for the detection of the noise level radiated by torpedoes. The level and frequency spectrum is determined by an operator with a colour video screen for panoramic surveillance, a monochrome classification video screen and an audio system for received-noise listening. An automatic alarm alerts the operator when a new source is detected. The system normally uses the array of the active sonar (VDS on hull-mounted sonar), but can use an independent passive array. The system detects torpedo-generated noise at ranges up to 10 km. It can be adapted to most existing types of array:

(a) 24, 36, 48 (and so on) staves
(b) 1 to 10 kHz bandwidth

Operational Status
In service with the French Navy.

Contractor
Safare Crouzet, Nice.

SWAN

Type
Towing and handling systems.

Description
Swan towing and handling systems constitute a family of equipments adapted to various sizes of VDS. Swan systems include:
(a) towed bodies, adapted to sonar array sizes
(b) towed faired cables
(c) winch assemblies (winches, poles, hydraulic power units)
(d) winch operator consoles.

Main characteristics of the Swan family are:
(a) active stabilisation of the towing system, which compensates for the ship's stern movements. In conjunction with a highly stable towed body, it enables the VDS to operate in rough seas
(b) high survival towing speed
(c) safe and easy operations, with secured launching and recovery sequences (only one man is required for operations)
(d) compact size.

To date, the Swan family consists of three models: Swan 500, Swan 1000 and Swan 2000, respectively adapted to Gardon, Salmon and Diodon sonars. Derivatives are possible for other sonar types.

Three versions have been ordered or delivered up to now, in order to meet own ship characteristics:
(a) a turntable version, to be installed upon the deck
(b) a sliding or railed version, to be installed below the helicopter deck
(c) a containerised version.

Operational Status
Operational at sea.

Contractor
Thomson Sintra Activités Sous-Marines, Valbonne Cedex.

ANGLER

Type
Towing and handling systems.

Description
Angler towing systems constitute a family of equipments designed to be adapted to various types of linear towed arrays.

Main characteristics are:
(a) compact and lightweight design
(b) system operated by a single operator most of the time
(c) easy and fast onboard installation
(d) a slipring allows permanent sonar operation during handling.

The Angler family comprises different models adapted to the various towed sonars (cables, arrays),

one of them being the TSM Lamproie passive towed array sonar.

Operational Status
Operational at sea.

Contractor
Thomson Sintra Activités Sous-Marines, Valbonne Cedex.

GERMANY

DSQS-21 (AS083–86) SERIES SONARS

Type
Hull-mounted and variable depth sonar systems.

Description
The DSQS-21 series sonars (also known as the AS083 to 86 series) are designed for operation in surface ships as part of an anti-submarine weapon system. For operations against submarines below the thermal layer, the sonar can be supplemented with a towing system providing a variable depth sonar (VDS) capability.

Computer-aided detection techniques are used for classification and tracking and the information is presented on colour CRT displays to permit Doppler coding and the discrimination of data on the integrated displays. A 'Z' in the nomenclature indicates variants equipped with electronic stabilisation to minimise the effects of ship's motion.

The VDS is designed to alternate with the hull-mounted sonar, and the displays of the two sonars

Specifications

Sonar		Diameter of Transducer Arrays		No of Pre-formed Beams		Class of Ship
		VDS	hull-mounted	VDS	hull-mounted	
DSQS-2	B/BZ	1.8 m		64		Corvettes, frigates
	B/VDS	1.8 m	1 m	64	32	Corvettes, frigates
	BZ/VDS	1.8 m	1 m	64	32	Corvettes, frigates
DSQS-21	C	1 m		32		Corvettes, frigates
	CZ	1 m		32		Corvettes, frigates
	C/VDS	1 m	1 m	32	32	Corvettes, frigates
	CZ/VDS	1 m	1 m	32	32	Corvettes, frigates
DSQS-21	D	1 m		32		Up to corvettes
	DZ	1 m		32		Up to corvettes
	D/VDS	1 m	1 m	32	32	Up to corvettes
	DZ/VDS	1 m	1 m	32	32	Up to corvettes

are integrated. Fitted with automatic depth control and heave compensation for stability of the 'fish', the VDS can be handled automatically or manually.

Operational Status
In service. Fitted to destroyers and frigates of the German Navy. More than 50 systems have been delivered to 15 countries. A number of systems were sold to Taiwan. The DSQS-21 is now being built under licence in Taiwan and is being fitted to most surface vessels of the Taiwan Navy.

Contractor
Atlas Elektronik GmbH, Bremen.

ASO 90 SERIES

Type
Hull-mounted, variable depth and towed array sonar systems.

Development
The ASO 90 series (comprising versions ASO 91–ASO 96) has been developed from experience with the DSQS-21 systems which were built in various versions, including VDS. The ASO 90 has been designed to equip the new Type F123 frigates of the German Navy which are due to enter service in 1994.

Description
The ASO 90 systems feature electronically stabilised (optional) cylindrical or conforming transmit and receive arrays with a large number of transducers. Arrays are either hull-mounted or variable depth types and are variously designed for low, medium and high frequency applications, according to ship type. Submerged arrays, either in a combination with hull-mounted or as stand-alone, may be used in the event of adverse sound propagation conditions. The use of switched-mode power amplification provides for optimum efficiency.

Advanced detection techniques for parallel, active and passive mode operation over 360° azimuth combined with spatial and temporal signal processing using the latest technology enable the detection of submarine echoes even in difficult waters. The usual techniques of Doppler filter bank (continuous wave) and pulse compression (frequency modulation) are supplemented by special pre- and post-processing of the signal data for full colour presentation and raster display on consoles specifically designed to reduce operator workloads. Incorporating automatic fault detection and location functions, complete stand-alone or integrated assemblies can be configured for interface to other ships' systems including onboard or ship-based simulators as well as recording/replay units.

The ASO 90 is designed for computer-aided detection, localisation and classification over an entire 360° azimuth. The modular design of the systems enables them to provide, as required: underwater telephone; automatic target tracking; classification; ray trace; intercept; torpedo warning; vertical stabilisation; simulation; interface to own-ship's fire control system; interface to towed array sonar; underwater survey display and torpedo command functions.

The VDS is fitted with automatic depth control and heave compensation for stability of the towed body. The VDS can be handled automatically or manually. The design of the winch allows the towed array sonar to operate with different cable and streamer lengths.

Specifications

Sonar system	HMS transducer array	VDS transducer array	Frequency	Ship
ASO 96	2.5 m	1.0 m	Low	Destroyers, frigates
ASO 95	1.8 m	1.0/0.58 m	Medium	Destroyers, frigates
ASO 94	1.0 m	1.0/0.58 m	Medium	Corvettes, frigates
ASO 93	0.58 m	0.58 m	High	Patrol boats, corvettes
ASO 92	Conformal array 0.7 × 0.3 m	Conformal array 0.7 × 0.3 m	High	Patrol boats, corvettes

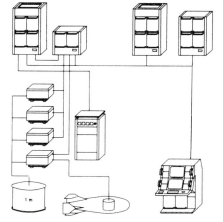

ASO 94 with VDS, stabilised version

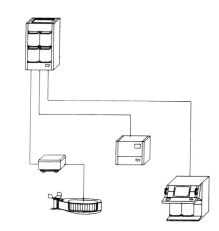

ASO 92/VDS

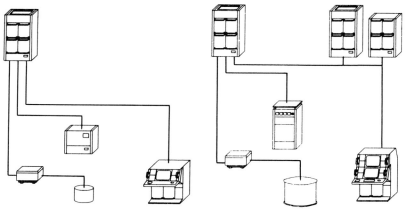

ASO 93   *ASO 94*

Operational Status

The first ASO 90 series sonars will be installed on the German F123 Type frigates by 1992.

Contractor

Atlas Elektronik GmbH, Bremen.

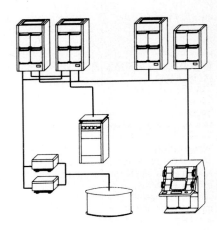

ASO 95

TAS 90 SERIES

Type

Passive towed array system.

Description

The TAS 90 series is designed for installation on surface ships and operates in the very low frequency band.

The sonar uses the latest spatial and temporal signal processing techniques and can detect extremely weak signals from submarines, torpedoes and surface ships even in difficult acoustic conditions. Data extraction and presentation is on full colour, raster display, and provides a permanent overview of the tactical situation. Detailed analysis of detected targets is carried out using LOFAR, DEMON and SPECTRUM techniques. Detection, tracking and analysis operations by the operator are facilitated by special features such as surveillance channels, independent broadband/narrowband facilities, frequency/time records with long time history, and magnifier and transient detection. The TAS 90 series is fully compatible with the ASO 90 active sonars either as a stand-alone system or an integrated version.

Various combinations in lengths of cable and streamer, including a long acoustic aperture with a large number of pre-formed beams, stored on an easy to operate winch, are provided.

Contractor

Atlas Elektronik GmbH, Bremen.

INTERNATIONAL

GETAS SONARS

Type

Towed array passive sonar system.

Description

The GETAS series of passive sonars has been developed and proven against quiet submarine and surface targets in both a towed array and bottom-laid version. The systems provide a fully integrated comprehensive solution to passive sonar detection, tracking and classification. They are designed for single operator manning.

Marconi Underwater Systems provides the 'dry end' processing and display equipment and Geophysical Company of Norway (GECO A/S) provides the 'wet end' components.

The onboard or shore-based equipment comprises a signal conditioning unit, broad and narrowband data processors, touch-sensitive control panels and waterfall displays housed in a single cabinet. A number of features have been incorporated including BITE, automatic track followers, remote hard copy and tape recording outputs. In addition, real-time clock and date, array heading, depth and tension and seawater temperature and salinity are available on the display.

There are two variants of the shipborne 'wet end', each consisting of a winch and cable with acoustic vibrator, isolator and instrumentation array sections. The all-up weight of the standard system is 8550 kg and of the small ship version it is 2155 kg.

A submarine version is also available. The bottom-laid version is similar to the standard but modified for abrasion resistance and negative buoyancy. All systems are easily deployed and recovered.

Specifications

Acoustic section
Overall length: 100 m
Operational speed: 8-12 kts
Operating depth: 200 m
Survival depth: 1200 m
Tow cable
Length: 1000 m
Processors
Broadband: Display mode waterfall; X-axis bearing, Y-axis time
Narrowband: Waterfall display

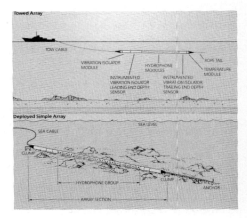

Contractors

Marconi Underwater Systems Ltd, Waterlooville, UK.
GECO A/S, Kjorbokollen, Norway.

ITALY

SELENIA ELSAG SONAR SYSTEM

Type

DE 1167 sonar manufactured under licence.

Description

The DE 1167 sonar is manufactured by Selenia Elsag Sistemi Navali under a licence agreement with Raytheon (see separate entry in the USA section). The system configuration includes 7.5 kHz hull-mounted 36-stave transducer arrays, a transducer array dome, four cabinets and a single operator console.

The hull-mounted DE 1167 is an active/passive pre-formed beam, omni and directional transmission sonar which uses three non-interfering 600 Hz wide FM bands centred at 7.5 kHz and a spatial Polarity Coincidence Correlation (PCC) receiver. The passive mode, which is selected automatically when transmissions are halted, is primarily useful for torpedo detection. A half-frequency simultaneous PCC receiver (bandwidth 3 to 8 kHz) is also part of the system. Optional items include a performance prediction subsystem, an auxiliary display, a remote display, and a training/test target remote-control unit. Signal reception and beamforming are accomplished by broadband analogue circuitry followed by clipper amplifiers for perfect data normalisation. Detection processing, display processing system control/timing and waveform generation are carried out digitally.

Two microprocessors perform display ping history, cursor ground stabilisation and target motion estimation functions for torpedo direction. The modular air-cooled 12 kW transmitter uses highly efficient class A/D power transistor techniques.

Several operational features are unique to a sonar of this size and range:
(a) the display processing incorporates a ping history mode through which the sonar data obtained in three of the previous ping cycles may be retained on the viewing surface, allowing the operator to differentiate readily between randomly spaced noise events and geographically consistent valid acoustic reflectors
(b) the clipped PCC processing permits accurate thresholding of all signals such that the false alarm rate, or number of random noise indications on the screen remains relatively low and constant over all variations in background noise and reverberation levels, further facilitating contact detection
(c) ground stabilised cursors and target motion analysis permit rapid determination of contact motion over the bottom, an excellent clue as to the nature of the contact.

These three operational features, combined with extremely accurate tracking displays, a built-in fault detection/localisation subsystem, performance verification software and test/training provides ease of maintenance, operation and support.

Operational Status

Eight systems are in operational service with the Italian Navy, with one system at the training centre. Four systems are under contract for the Italian Navy.

Specifications

Centre frequency: 7.5 kHz
Receiver type: Spatial polarity correlator between 36 pairs of half beams
Beam characteristics: 36 sets of right and left half beams for active and passive detection. Selectable 15° horizontal × 12.5° vertical sum beam for audio listening. 1.25° bearing interpolation for display. Selectable 24° horizontal × 21° vertical passive sum beam for audio listening. 2.5° bearing interpolation for display
Active display: 300 range cells, 288 bearing cells; single and multiple echo history; 4 intensity levels
Passive display: Electronic bearing time recorder with medium time averaging. LTA/STA with passive receiver
Track displays: Sector scan indicator (1000 yd × 10°) and target Doppler indicator (1000 yd × ± 60 kts)

Target data: Range 8 yd (6.1 m) resolution. Bearing 1.25° resolution plus active search display 0.1° and 3.3 yd on SSI display.± 60 kts of Doppler at 1 kt steps on the TDI
Data format: Standard: NTDS slow (digital)
Power requirements: Active 20 kVA (pulse) at 15% (max) duty cycle 400 V 60 Hz 3-phase
Weight: Hull-mounted 7.5 kHz transducer 1600 kg; GRP dome 750 kg

Contractor

Selenia Elsag Sistemi Navali, Rome.

JAPAN

OQS-4 SURFACE SHIP SONAR

Type

Hull-mounted sonar system.

Description

Nearly all Japanese ASW ships have been fitted with various US types of sonar systems. However, the 'Hatsuyuki' class destroyers have been fitted with a new Japanese-designed low frequency sonar system; the OQS-4. No published data are available, but the OQS-4 is a solid-state, low frequency design which has been developed from the OQS-3 system. This latter sonar is the Japanese designation for the United States AN/SQS-23, built under licence since the late 1960s.

Operational Status

In operational service with the 'Hatsuyuki', 'Asagiri' and 'Hatakaze' class destroyers of the Japanese Maritime Self-Defence Force. The sonar also equips the destroyer *Murakumo* and frigates of the 'Abukuma' and 'Yubari' classes. Will eventually be replaced by the AN/SQR-19 TACTAS towed array on these ships.

NETHERLANDS

PHS-32 SONAR

Type

Hull-mounted search and attack sonar system.

Description

The PHS-32 is a medium-range high performance search and attack sonar in which the newest technological developments for signal processing and operation are employed and aided by the use of a general-purpose computer. The computer yields a compact, lightweight sonar for corvettes from 200 tons up to frigate-size ships.

The signal processing facilities provided include fast Fourier transformation processing of all pre-formed beam receiving channels. All data are presented on a single TV-type display, while operation has been much simplified by the use of light pen control. Automatic tracking of up to four targets is provided. These features, combined with a high accuracy, make this sonar very useful in an attack.

A circular transducer permits all-round coverage in various modes of transmission such as: Omni, TRDT, MCC (wide vertical beam), LISTEN (passive with time/bearing recorder presentation). An audio beam is also available. The system can be delivered in a fixed, retractable or VDS-dome outfit.

Operational Status

In series production and operational service on board ships from fast patrol boats to frigates.

Specifications

Features: single operator control; display data processing for continuous presentation and memory mode (ping history); four-target automatic tracking; built-in energy storage for lower peak power demands on ships' mains; integrated on-line and off-line test systems.
Frequencies: 3
Pulse lengths: 12.5, 25, 50, 100 ms and 400 ms long pulse (CW or FM)
Detection range: >10 000 yd
Vertical beamwidth: 12 or 20°
Bearing accuracy: 1° RMS
Range accuracy: ±0.5-2%
Notch filtering: Selectable rejection bandwidth
Own Doppler correction: On all 60 channels
Roll and pitch performance: Automatic co-ordination of transmitting pulse and vertical beamwidth in rough weather
Weights
Electronic cabinets: (Sonar console, duplexer and amplifier cabinet, transmitter cabinet) 1041 kg
Retractable hull-outfit including transducers: 7900 kg
Fixed hull-outfit including transducer: 2500 kg
Variable depth system: 5000-8000 kg

Contractor

Hollandse Signaalapparaten BV, GD Hengelo.

Lightweight modular transducer of PHS-32 sonar

PHS-36 SONARS

Type

Hull-mounted search and attack sonar systems.

Description

The PHS-36 series of sonar systems is modular in design and intended for worldwide operation on any type of surface ship. The PHS-36 which will be fitted in the M-frigates of the Royal Netherlands Navy is stated to be the first system in the world which is fully integrated into the SEWACO (Sensor Weapon Control and Command) system.

The PHS-36 provides information for the data handling functions of active sonar, passive sonar, acoustic support and track management and other functions such as equipment monitoring. The sound velocity processing and recording system also provides data for acoustic support.

A colour-coded presentation on a high resolution display gives a graphic picture of sonar information.

The dual processing system performs FM and CW processing simultaneously to cope with both noise-limited and reverberation-limited conditions, as well as with a variety of submarine running conditions as provided for ASW scenarios. The main functions performed by the sonar processor are auto-detection, auto-track initiation and auto-track processing.

The flexibility of the modular PHS-36 sonar processor allows easy add-on of a towed line array with or without active adjunct or other types of ASW sensor.

Operational Status

The sonar, which was a result of a joint development programme with the Royal Netherlands Navy, is in production. Nine systems were ordered for the Royal Netherlands Navy in December 1987 for fitment to the new 'Karel Doorman' class frigates.

Specifications

Transmission
Frequencies: Two frequencies around 7 kHz
Pulse type: CW, FM, FM + CW (a sequential transmission per interval)
Pulse length: 75-1200 ms depending on range-scale/pulse type selections
Reception
Pre-formed beams: 32
Bandwidth passive mode: Band I 1.5-3.5 kHz; Band II 5.8-10.5 kHz
Accuracy
Range: CW (300 ms) <70 yd; FM <20 yd
Bearing: <2°
Doppler speed: CW (300 ms) <0.2 kts
Vertical beamwidth: 13° (narrow beam); 24° (wide beam)

Contractor

Hollandse Signaalapparaten BV, GD Hengelo.

NORWAY

SS105 SCANNING SONAR

Type
Hull-mounted active sonar system.

Description
This sonar is intended to fulfil the sonar requirements of modern ocean-going coastguard vessels. It is a 360° scanning sonar with 48 pre-formed receiving beams with 11° beamwidth and 'split beam' processing in each beam. The working frequency is 14 kHz. The sonar system consists of an operator's console, transmitter unit, receiver unit, hydraulic power unit and hull unit.

The transmitter consists of 48 switching type amplifiers, 600 W each. Total output power is approximately 15 kW. Both Omni, Single RDT and Triple RDT are available transmission modes. The main display has a CRT PPI, 280 mm in diameter, with scale ranges from 2 to 16 km. Markers provided are target cursor, stern cursor, transmitting and receiving sectors. An LED display shows target range, relative/true target bearing and ship's speed. Target data are transmitted digitally to other systems on board as required.

The 48 stave transducer is installed in a streamlined retractable sonar dome which will take ship's speed up to 25 kts. A fixed dome arrangement is also possible.

Operational Status
The SS105 is fitted to coastguard vessels of the Royal Norwegian Navy and the Finnish Coastguard service. This sonar is no longer in production.

Specifications
Transmitter
Type: Class S amplifier
Number of channels: 48
Max power output per channel: 600 W
Pulse lengths: 10, 30 and 60 ms
Frequency: 14 kHz
Transducer
Number of staves: 48 (circular-mounted)
Active face per stave: 225 cm²
Beamwidth vertical plane: 12 ± 1°
Resonance frequency: 14 kHz
Tilt: 6° (mechanical)
Transmitting modes and performance
Omni
Beamwidth vertical plane: 12 ± 1°
Beamwidth horizontal plane: 360°
Max output level: 219 dB/μPa/m
Directivity index: 9.5 dB
SRDT
Beamwidth vertical plane: 12 ± 1°, one beam
Beamwidth horizontal plane: 8.5 ± 1°, scanning a sector variable from 10 to 115°
Directivity index: 25 dB
Max output level: 230 dB/μPa/m
TRDT
Beamwidth vertical plane: 12 ± 1°
Beamwidth horizontal plane: 8.5 × 1°, 3 beams each scanning a sector of 120°
Max output level: 230 dB/μPa/m
Directivity index: 25 dB
Receiving performance
Number of simultaneous beams: 48
Bandwidths: 400 and 800 Hz
Beamwidth vertical plane: 12 ± 1°

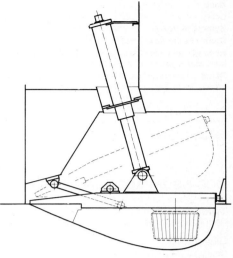

SS105 hull unit

Beamwidth horizontal plane: 11 ± 1°
Directivity index: 26 dB

Contractor
Simrad Subsea A/S, Horten.

SS245 SONAR SERIES

Type
Series of multi-beam sonar systems.

Description
The SS245 series is a family of active high resolution multi-beam sonars designed for naval vessels. The relatively low operating frequency of 24 kHz and a high source level ensures long-range detection. The transmitting modes are omnidirectional, or directive with a single beam. In all modes both transmission and reception beams are tiltable from +10 to −20° in 16 steps. In addition the system contains an MCC-mode (Maintenance of Close Contact) with a broad vertical beam in both transmission and reception. Signal processing and beam forming is carried out by a high speed digital processor using the full dynamic range of the signals. The echoes are presented in 64 beams, interpolated to 128 on the display unit.

The sonar transducer is a cylindrical array arranged in seven circles with 27 elements in each. The geometry is arranged in a lattice pattern in order to keep the spatial sampling of the elements sufficiently dense. The active transducer array is 432 mm diameter × 280 mm high, and the 24 kHz operating frequency gives good angular resolution. The weight of the array is approximately 300 kg.

The various sonars of the SS245 series have different hull units. The options are:
SS245 – hoistable transducer without dome
SS246 – hoistable transducer with GRP dome
SS247 – fixed transducer with GRP dome

SS248 – hoistable transducer and dome
SS249 – hoistable transducer with inflatable rubber dome.

The SS245 and SS246 hull units are equipped with a motor-driven screw to hoist and lower the transducer in and out of its operative position. The SS248 and SS249 use a hydraulic system for raising and lowering of the transducer and dome.

All command and control functions are performed on the control and display unit which incorporates a 20 in raster scan high resolution colour display, an operator panel and a rack assembly with a set of circuit boards and a power supply.

The man/machine interface uses dedicated control buttons on the CDU keyboard which offer direct access to primary sonar functions such as range, gain, tracking, etc. Secondary functions are menu controlled. The menu is displayed alongside the sonar picture and commands are selected via a joystick-controlled cursor on the display.

Two main display modes are available: relative or true motion.

Zoom and aspect scan facilities are available in each mode and Doppler analysis is also available in each mode. Other options available include passive DEMON analysis, ray trace based on a temperature probe input and video recording.

Operational Status
In production.

Contractor
Simrad Subsea A/S, Horten.

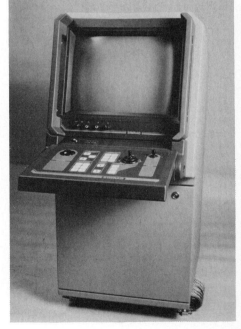

Control/display unit of the SS245 sonar system

SS575 SONAR SERIES

Type
Medium-range active sonar systems.

Description
The SS575 series of active high resolution sonars is designed for medium-range detection and tracking of submarine targets. Detection is achieved using a medium frequency (57 kHz) combined with a high source level and FM signal processing. The transducer is mounted in a spherical array offering unlimited possibilities for stabilised beam configurations under and around the vessel.

Transmission and reception modes vary from one to 64 beams covering 360° while the transmitting beam can be tilted between +10° and −90°. Displays are presented either in relative motion on the full screen or in a combination of modes on a split screen. Within the different modes the operator can select zoom, aspect scan or Doppler analysis submodes. Signal processing and beam forming are performed by a high speed digital processor using the full dynamic range of signals. Echoes are presented on a 20 in high resolution colour CRT.

Two types of hull transducer mounting are available, each with its own designation; SS575 with a standard length hoistable transducer, and SS576 with an extended hoistable transducer.

The sonars comprise the following elements: control and display unit (CDU); transceiver unit (TRU); hull unit with transducer array; hoist control unit (HCU); transducer matching unit (TMU) and vertical reference unit (VRU).

All command and control functions are performed on the CDU with functions such as range, gain, tracking and so on having dedicated control buttons on the CDU keyboard. Supplementary functions are menu-controlled, the menu being divided into primary and secondary parts. The menu is displayed alongside the sonar picture, with commands being selected by a joystick-controlled cursor on the display.

Apart from selecting display modes, the operator can also select functions such as target tracking, position tracking and transference of data to weapon control systems, the latter being an optional feature.

Specifications
Transceiver unit
No of individual transmitters: 256
Power output: 25 W per transmitter
Transmitter modes: CW short, CW normal, CW long, FM-0 (1 pulse), FM-1 (2 pulses), FM-2 (4 pulses), FM-3 (8 pulses), FM-Auto
Pulse length CW: 0.6 to 160 ms depending on range (60 ms in omni mode)
Pulse length FM: 0.6 ms, 2.5 ms, 10 ms, 40 ms
Bandwidth FM: 1.78 kHz
Beam forming: 64 digitally formed beams

Ranges: 250 m, 500 m, 1000 m, 1500 m, 2000 m, 4000 m, 6000 m
Transducer array
No of elements: 256
Operating frequency: 57 kHz
Beamwidth (single transmission): Horizontal 11°; vertical 11°
Beamwidth (omni): Horizontal 360°; vertical 11°

Contractor
Simrad Subsea A/S, Horten.

ST240 TOADFISH

Type
Dipping and variable depth sonar.

Description
The ST240 Toadfish is an active, multi-beam, single frequency, high resolution sonar system designed for ASW and MCM and for the detection of other submerged targets. The transducer array is located in a submerged array which may either be towed or dipped.

The system comprises a display and control unit, power and beam forming unit, towed body (Toadfish) and a crane and hydraulic winch system. The ST240 is an omni and sector transmitting system with multi-beam reception covering a sector of 360°. Operating modes include omni, sector and passive. Tracking is implemented in the active modes. Toadfish is fitted with a CTD sonde to present a real-time ray trace on the screen. Transmission modes are CW and FM, with the possibility of Doppler analysis in CW. The power supply and beam forming unit receives pre-processed digital data from the towed body and performs signal processing (detection, normalisation and ping-to-ping correlation) with a fast digital signal processing system comprising a total of four single board processors using the full dynamic range of the signals. The unit also interfaces with the single operator control and display unit in which presentation is performed on a raster scan, high resolution 20 in colour monitor. Interfaces to Doppler log and gyro compass are provided, and the towed body itself contains several sensors.

Winch assembly and towed body of ST240

Specifications
Frequency: 24 kHz
Beam forming: 64 digitally formed beams
Beamwidth transmission:
  Single beam: 8° (horizontal); 12° (vertical)
  Sector: 30°, 60°, 120° (horizontal omni); 12° (vertical)
Bandwidth FM: 1900 Hz
Pulse length CW: 1.3 to 200 ms depending on range
Pulse length FM: 1.3 ms, 2.6 ms, 5.2 ms, 10.4 ms
Ranges: 500 m, 1000 m, 2000 m, 4000 m, 6000 m, 12 000 m

Operational Status
In production.

Contractor
Simrad Subsea A/S, Horten.

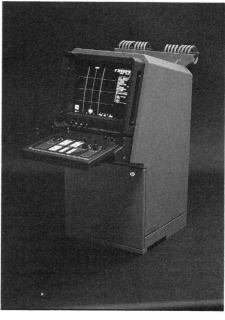
Single operator display console of ST240

UNION OF SOVEREIGN STATES

Because of security constraints, and the generally inaccessible location of sonar equipment, it is not possible to treat USS underwater equipment in the same manner as has been adopted for that of other nations. Instead, the following notes summarise what little has been gathered, and which is free of Western security restrictions. While this inevitably falls well short of the ideal treatment, it is hoped that readers will find the ensuing paragraphs of some help. The table in this entry gives a list of the types of sonar carried by various USS surface ship classes.

Underwater acoustic experiments by what was formerly the Soviet Union, initially in relation to communications applications, are widely agreed to date back to the years immediately prior to the First World War, but the earliest references to anything resembling what is now known as sonar occur in the 1930s when research into hydrophones for submarines is mentioned. At this general time, the possibilities of thermal detection devices for both aircraft and surface shipping were pursued. In the same period, the USSR as it then was is credited with the production of passive seabed acoustic detector equipment. Early Soviet records claim that at the start of the Second World War, the naval forces had a variety of sonar equipments available for shore and ship installations.

Most, if not all of those reported were apparently passive devices, those for submarine fitting consisting of an elliptical array made of 8, 12 or 16 hydrophone elements. There were also passive sonar sets for surface ships, one of these being named Tamir which began sea trials in 1940. This equipment is stated by the former Soviet Union to have become the standard

sonar employed in the anti-submarine campaign of that time. Neither the then Soviet Union's former allies, nor her enemies, appeared to have been unduly impressed by the results achieved, according to historians, official and unofficial. By the post Second World War period, the nation had gained access to sonar technology originating in America, Britain and Germany, either by gift or as booty.

Since then, there has been steady and impressive growth of Union interest in all aspects of submarine warfare, and it is reasonable to suppose that sonar equipment is accorded a priority within these activities at least as great as that given it by the Western navies. The advent of ASW helicopter and aircraft carriers, such as the *Moskva*, *Leningrad* and *Kiev*, in advance of comparable ships being commissioned in Western navies, might imply a higher priority.

The appearance of these ships was preceded by smaller vessels designed for ASW operations and special-purpose aircraft and helicopters for naval duties, the latter being deployed in both land-based and embarked formations. These developments occurred in the mid-1950s to mid-1960s period.

Among the ASW ships, 'Udaloy' class destroyers have bow or hull-mounted sonar suites and variable depth sonars. The 'Petya' and 'Mirka' class vessels have hull-mounted sonars and some of the earlier 'Petya' class have been retrofitted with a variable depth sonar which replaces the after torpedo tubes. The Kresta I ASW cruisers have hull-mounted sonar and can carry 'Hormone' helicopters which have been equipped with dipping sonar and a magnetic anomaly detector (MAD), and can deploy sonobuoys. A bow-

mounted sonar is fitted in the later 'Kresta II' class, and the still more recent 'Kara' class ships have a variable depth towed sonar. The 'Krivak' class ships, which appeared at sea in 1971, have both bow sonar and a variable depth sonar.

Information is also available on a sonar system fitted to the 'Kashin' class destroyers. This is listed as a keel-mounted HF sonar, which includes Tamir, Pegas and Hercules types. The first two of these systems are very old and it is likely that the system fitted to the 'Kashin' class is the Bull Horn, probably operating in the 20 to 30 kHz range. The sonar is a unit of four pull-out chassis, with a small (77 mm) CRT behind an orange filter screen. The lower three racks feature simple on/off toggle switches and push-buttons, only two multiple selector switches being visible. The only outputs seem to be to the small CRT and a meter which implies a simplistic sonar technology from the 1960s.

A medium frequency sonar, NATO designation Bull Nose, is fitted to the 'Kara', 'Kresta II' and 'Krivak' class ships. Often associated with Bull Nose is the Elk Tail VDS sonar. Other systems include the Horse Jaw LF hull-mounted sonar fitted to the 'Kirov' battle cruisers and the 'Udaloy' destroyers, and the Mare Tail, Rat Tail, Lamb Tail, Horse Tail and Foal Tail VDS sonars fitted to various ships. Moose Jaw is a very large low frequency hull-mounted sonar fitted to the 'Moskva' cruisers and the 'Kiev' carriers.

In addition to the 'Hormone' helicopters, the earlier 'Hound' is believed to be capable of operating a dipping sonar or possibly a towed magnetic anomaly detector (MAD). The Mi-14 'Haze' shore-based

This is probably the most detailed unclassified view of a modern four-unit group of a low frequency hull search and probable medium frequency VDS sonar. The ASW fire control computer and switchboard are at the far end of the control space. This arrangement could be typical of the sonar layout found on 'Kara' class cruisers or 'Krivak' class destroyers (Jane's Intelligence Review)

helicopter also carries MAD. Other fixed-wing maritime aircraft, such as the Be-12 'Mail' and the Il-38 'May', are equipped with MAD, and the latter type can also deploy sonobuoys and process data obtained by this means. 'Mail' has been followed into service by an ASW version of the Tu-20 'Bear-F', which is understood to employ sonobuoys for submarine target detection.

Many submarines of the fleet prominently feature large sonar arrays and it has been estimated that between 60 and 65 nuclear attack submarines have been commissioned, supported by perhaps as many as 95 diesel attack submarines. These are in addition to the large classes of nuclear and diesel ballistic and cruise missile submarines, all of which have sonar installations.

Much more recently, a certain amount of information has become available on acoustic and non-acoustic ASW systems. The USS appears to be making a considerable effort to develop non-acoustic sensors, including surface signature detection systems which sense the disturbance on the ocean surface caused by a passing submarine.

Wake detection systems that sense the turbulent wakes, internal wave wake, or contaminant (radioactive or chemical) wake of submerged submarines are also believed to be under development, as are magnetic or electric field detection systems that pick up the submarine or the subsurface disturbances caused by its passage. These wake and disturbance detection systems must operate at high altitude to offer effective surveillance, and must therefore be either airborne or spaceborne.

In the traditional acoustic sensing field the USS is understood to be continuing to deploy 'Cluster Lance' planar acoustic arrays in the Pacific waters near the mainland where broad area surveillance is required. It is believed that they may also be deploying static barrier arrays at entry and exit points in the Barents, Greenland and Kara seas. The USS is also believed to be evaluating a surveillance towed array, similar to the US SURTASS system.

The table of surface ship sonar types represents the latest information available. Information on submarine sonars is given in the *Submarine Sonar Systems* section of this book.

The Analysis section contains a list of current sonar systems under their NATO designations.

Surface ship class	Sonar system
'Kiev' aircraft carriers	Hull-mounted, active search and attack, low/medium frequency. VDS active search, medium frequency.
'Kirov' battle cruisers	Hull-mounted, active search and attack, low/medium frequency. VDS active search, medium frequency. The hull-mounted sonar is of a fairly new and powerful design, while the VDS may have a depth capability down to 150 to 200 m, depending on speed.
'Moskva' helicopter cruisers	Hull-mounted active search and attack, low/medium frequency. VDS active search, medium frequency.
'Slava' cruisers	Hull-mounted active search and attack, low/medium frequency. VDS active search, medium frequency.
'Kara' cruisers	Hull-mounted active search and attack, low/medium frequency. VDS active search, medium frequency.
'Kresta I' and 'Kresta II' cruisers	Hull-mounted active search and attack, medium frequency.
'Kynda' cruisers	Hull-mounted active search and attack, medium frequency.
'Sovremenny' destroyers	Two hull-mounted, active search and attack, medium frequency.
'Udaloy' destroyers	Hull-mounted, active search and attack, low/medium frequency. VDS active search, medium frequency.
'Kashin' destroyers	Hull-mounted, active search and attack, medium frequency. VDS search, medium frequency (modified 'Kashin' class ships only).
'Kildin' destroyers	Hull-mounted, active search and attack, high/medium frequency.
'Sam Kotlin' destroyers	Hull-mounted, active search and attack, high frequency.
'Kanin' destroyers	Hull-mounted, active search and attack, medium/high frequency.
'Kotlin' destroyers	Hull-mounted, active search and attack, medium frequency.
'Skory' destroyers	Hull-mounted, active search and attack, high frequency.
'Krivak' frigates	Hull-mounted, active search and attack, medium frequency. VDS active search, medium frequency (the *Zharky* has a new type of VDS)
'Koni' frigates	Hull-mounted, active search and attack, medium frequency.
'Grisha' frigates	Hull-mounted, active search and attack, high/medium frequency. VDS active, high frequency, similar to the 'Hormone' helicopter dipping sonar.
'Mirka' frigates	Hull-mounted, active search and attack, high/medium frequency. Dipping sonar is also fitted in some ships in the transom, and in others abreast the bridge.
'Petya' frigates	Hull-mounted, active search and attack, high/medium frequency. VDS active search, high frequency (in some ships)
'Parchim' frigates	Hull-mounted active search and attack, medium frequency. Helicopter type VDS, high frequency.
'Turya' fast attack craft	VDS active search and attack, high frequency (similar to 'Hormone' helicopter type).
'Pauk' fast attack craft	Appears to have a 2 m extension at the stern for dipping sonar.
'SO1' large patrol craft	Tamir 2, hull-mounted, active attack, high frequency.
'Stenka' fast attack craft	VDS, high frequency. 'Hormone' type helicopter dipping sonar.
'Slepen' fast attack craft	Dipping sonar (probably the 'Hormone' helicopter type).
'Natya' minesweepers	Hull-mounted active minehunting, high frequency.
'Yurka' minesweepers	Hull-mounted active minehunting, high frequency.
'T43' minesweepers	Hull-mounted active minehunting, high frequency
'Yevgenya' minesweepers	A small sonar is fitted and lifted over the stern on a crane.

UNITED KINGDOM

TYPE 2016 SONAR

Type
Hull-mounted fleet escort search and attack sonar with bow or keel-mounted variants.

Development
Design was initiated in the late 1960s and an experimental equipment underwent tests in HMS *Matapan*. A development contract was awarded in 1973 and the first prototype was completed in 1978 with fleet weapons acceptance in 1983. A bow-mounted variant, incorporating electronic beam stabilisation, was developed in 1983.

Description
The Royal Navy's Type 2016 fleet escort sonar is a replacement for the existing RN sonars Types 177 and 184. It employs computer-aided techniques which are operator interactive, have automatic detection and tracking, enhanced signal processing and displays complemented by an information storage facility.

The Type 2016 is a hull-mounted panoramic surveillance and attack equipment with facilities for classification and multiple target tracking. Interference between nearby vessels using the Type 2016 sonar is largely eliminated by the use of a new type of broadband transducer. The sonar display console is designed for manning by a single operator under normal cruise conditions, this crew member being able to initiate the preparatory actions necessary for urgent action.

Digital data processing facilities are based on use of Ferranti computer equipment, with other subcontractors including Marconi Radar.

The Type 2016 system comprises the following main elements: four active receiver cabinets, one passive and control cabinet and one T/R switch and beam former cabinet, all housed in Marconi MC70 type cabinets. The Ferranti computer suite consists of three D.811A cabinets and a separate cabinet for computer spares. The two transmitters are housed in two standard RN/AUWE cabinets. The solid-state electronics are based on a modular system approach using medium scale integrated devices, printed circuit back planes with wire wrap connections and standard line-drive receivers for all inter-cabinet signal connections. Extensive use is made of hybrid circuit techniques to reduce volume.

The Type 2016 array system, which is roll stabilised, is fitted in a ribless monocoque glass-reinforced plastic sonar dome within a fixed hull outfit. Depending on the ship fit, the array is mounted either in a keel or bow dome; array stabilisation is either hydraulic or by electronic beam steering.

The display console contains three main displays and a top-mounted versatile console system (VCS). The right-hand display is dedicated to passive and auxiliary data, the passive data are displayed with two different integration times enabling both torpedoes and more distant targets to be tracked. The auxiliary data are used for data logging and system information including the display of such items as course, speed, weather, sea bottom conditions and bathymetric information. System monitoring information is also provided here and on a maintainer's teleprinter or intelligent terminal, which enables faults to be pinpointed down to board level. The other two displays

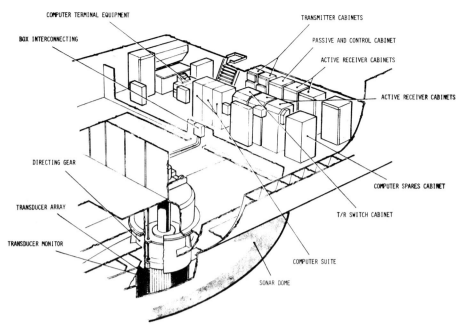

Arrangement of Type 2016 fleet escort sonar equipment and hull outfit compartment

Console of Type 2016 sonar aboard an RN frigate. Three displays (left to right) are for classification, surveillance, and passive presentation

can be used to view surveillance, classification or target history information and are fully interchangeable. Special keyboards are provided at both positions for interaction with the automatic tracking process, display selection and data transmission to the command/combat information system.

Operational Status
In service with the Royal Navy in Type 22 frigates, Batch 3 'Leander' frigates, 'Invincible' class aircraft carriers and Type 42 destroyers. Production was completed in 1987.

Contractors
Marconi Underwater Systems Ltd, Templecombe (prime contractor), with Ferranti Computer Systems, Marconi Radar Systems and other subcontractors.

TYPE 2031I SONAR

Type
Passive towed array sonar system.

Development
The Type 2031I sonar was developed and manufactured, in conjunction with the Admiralty Research Establishment, by GEC Avionics (now part of Marconi Underwater Systems Ltd) who provided the signal processor and display system. Plessey Naval Systems provided the array and terminal unit, STC

the tow cable, and NEI Clarke Chapman the winch and handling gear.

Description
The Type 2031I is a passive towed array sonar system fitted to Royal Navy frigates. It consists of a multi-octave towed array, tow cable with handling gear, and an inboard processor and display system. The inboard electronics are housed within a self-contained cabin located below-decks.

The sonar provides all-round cover in surveillance and classification mode with narrowband, broadband

and demodulation analysis facilities. It has extensive data recording and audio facilities together with display of array parameters including depth, heading and cable tension.

The Type 2031I may be operated by up to three operators, with three stations each including an acoustic display, touch-screen, cursor control and audio output. Data are displayed on three acoustic displays, with comprehensive cursor and annotation facilities. Additional facilities for driving remote displays (such as those in the Operations Room) or external video recorders are provided. There is a comprehensive built-in test equipment, and the processing is reconfigurable in the event of a fault.

An export variant, the SP 2110, is also available and is described under a separate entry.

Operational Status
In operational service with the Royal Navy.

Contractors
Marconi Underwater Systems Ltd (was GEC Avionics), Croxley (prime contractor).

STC Cable Systems Division, Newport, Gwent.

NEI Clarke Chapman Ltd, Gateshead.

Type 2031I passive towed array sonar

TYPE 2031Z SONAR

Type
Towed array sonar system.

Description
The Type 2031Z is a towed array sonar system currently being fitted to all new ASW frigates of the Royal Navy. Using the innovative 'Curtis' architecture developed at the UK Admiralty Research Establishment in Portland, Dorset, the 2031Z electronic signal processing suite has, because of its compactness, a considerably lower installed cost than other systems, and yet provides a multi-octave broad and narrowband analysis facility capable of tracking the most elusive threats.

The array itself is towed behind the ship and has a very low self-noise characteristic and is considerably longer than that used by any other self-contained sonar. It provides good bearing accuracy, even down to the very low frequency end of the acoustic spectrum. The combination of an advanced signal processing architecture and a high performance towed array ensures that the system is a dual capability sensor, able to fulfil both the tactical and surveillance role. A 2031Z equipped frigate can, therefore, act as an ASW picket or close escort.

The modular nature of the 2031Z, both as regards the digital signal processing electronics and the towed array itself, makes it possible to configure other types of sonar systems from the same range of modules. As an example, a variant of the 2031Z signal processing, suitable for submarine applications, is also available and has been delivered to the Royal Navy.

Operational Status
The Type 2031Z is now in series production for all Royal Navy Type 22 (Batches 2 and 3) and all Type

Sonar 2031Z units, including dual display consoles, non-acoustic data displays and hard copy recorders

23 frigates. A total of 20 systems had been ordered by the end of 1989.

Contractors
Dowty Maritime, Weymouth (manufacturer of the inboard electronics suite).

Marconi Underwater Systems Ltd, Templecombe (towed array).

NEI Clarke Chapman Limited, Gateshead (winching equipment).

TYPE 2050 SONAR

Type
Hull-mounted active search sonar system.

Description
The Type 2050 is the new hull-mounted active sonar for the Royal Navy and is being fitted to the Type 23 and retrofitted to Type 42 and Type 22 frigates. It is a successor to the Type 2016 and is compatible with both bow and keel variants of the Type 2016 array. The equipment has been developed from the Ferranti FMS series (see separate entry), and includes digital signal processing and distributed data processing using Ferranti Digital Signal Processing (DSP) cards and multiple Ferranti Argus M700 processors.

The provision of extensive data processing facilities and improved man/machine interfaces allows one operator to control the complete system. The equipment will interface to the combat system data highway feeding data to the Ferranti CACS action-information system. The suite consists of five processing cabinets and the system uses the same display consoles as will be used in the Type 2054 equipped Trident submarines. These are air-cooled and in monochrome, but colour is available in the system for possible export applications.

Operational Status
The UK MoD has ordered 27 Type 2050 sonar systems for the Royal Navy, and these are being fitted in Type 23 frigates, and retrofitted into Type 22 frigates and some Type 42 destroyers. An export version, the FMS 21, is described in the FMS series entry.

Contractors
Ferranti-Thomson Sonar Systems UK Ltd, Stockport (prime system contractor).
   Marconi Command and Control Systems Ltd (sub-contractor for the transmit/receive function).
   Marconi Underwater Systems Ltd, Templecombe (dome and array modules).

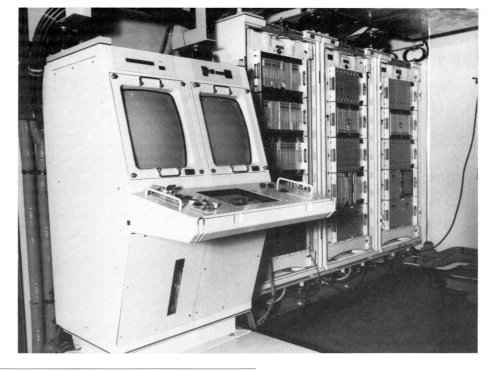

Type 2050 hull-mounted sonar system

TYPE 2057 SONAR

Type
New passive towed array sonar project.

Description
The Type 2057 is the new passive towed array sonar for the Royal Navy. It is intended for use in a variety of both surface ships and submarines. Engineering features are likely to include arrays, vibration isolation modules and non-acoustic sensors, winch cables, inboard signal and data processing and displays, and recording and systems control.

Operational Status
In September 1989, the UK MoD selected Ferranti Computer Systems and Plessey Naval Systems (now Marconi Underwater Systems) as lead companies of two industrial teams to work on a project definition phase. Each study is funded to around £4 million and covers system definition and risk reduction in all areas, resulting in bids for full development and initial production. A number of technology demonstrators are envisaged during this project definition phase, including the production of prototype hardware for preliminary sea trials. After this phase a single consortium will be chosen for full development and initial production.

Contractors
Ferranti-Thomson Sonar Systems UK Ltd, Stockport.
   Marconi Underwater Systems Ltd, Templecombe.

PMS 26 and PMS 27 SONARS

Type
Hull-mounted passive sonar systems.

Description
Hull-mounted ship sonars designed for single operator control in both the surveillance and attack roles, the PMS 26 and PMS 27 are also used as 'dunking' sonars in certain helicopters under the designation Type 195.

The PMS 26 is a self-contained system for ships and patrol craft down to 150 tons. It provides full 360° coverage in four steps of 90° and may be manually controlled to cover a particular sector, or set to carry out automatic search procedure. It incorporates a 'maintenance of close contact' facility for tracking close or deep targets. The single operator controls the sonar through a special console. He is provided with three sources of sonar information: audio, visual Doppler, and visual sector.

The Doppler facility provides increased initial detection range and improved classification capabilities compared with conventional small ship sonars. The PMS 26 transducer array is mounted within a hull outfit with a glass-reinforced plastic dome.

The PMS 27 differs from the PMS 26 only in its associated hull outfit. The PMS 27 transducer array is mounted in a Royal Navy hull outfit 19 or similar, which makes the equipment suitable for installation in small escorts down to about 650 tons displacement. The PMS 27 can be used as a surveillance sonar in association with a separate fire control sonar within the same hull outfit.

This enables the ship to continue surveillance for new threats whilst engaging a target already detected.

Operational Status
In operational service but no longer in production.

Contractor
Marconi Underwater Systems Ltd, Templecombe.

HYDRA SONARS – PMS 25/26/27 MOD 1

Type
Modernised surface ship sonars.

Description
The PMS 26 and 27 surface ship sonars have been modernised to form part of the Hydra family of sonar systems. They include a number of features of the Hydra PMS 56 sonar providing fully automated hull-mounted sonar systems designed for single operator control in both surveillance and attack roles. The modernisation retains the existing PMS 26/27 hull outfit and introduces a new array and inboard console. The operating frequencies remain the same, allowing for underwater telephone function if required. The sonars offer a high degree of flexibility, allowing features to be tailored for individual operational needs without special development programmes.

The modernised sonar systems have four modes:
(a) active replica processing with linear pulse modulation which provides good performance at all target speeds with an improved noise-limited range
(b) active Doppler processing with CW pulses which is used against high speed vessels to provide target motion information
(c) active energy detection using short CW pulses to provide close-in cover. The minimum range in this mode is less than 50 m
(d) a passive mode which offers an alternative to the active mode. It gives all-round passive cover with variable integration lines on the display to increase detection capability.

The single operator, single display console has a high resolution, semi-ruggedised colour display with sealed membrane control interfaces.

The set may be pre-programmed into the four processing modes. Switching between modes can be carried out rapidly by means of the control cursor keys and control inject button.

Display formats for each mode of operation give suitable displays for both detection and tracking. The displays are stabilised to a given bearing for improved detection of moving targets against background echoes.

A number of optional enhancements are available, including:
(a) wide LPS pulse bandwidth, for improved performance in high reverberation areas
(b) automatic passive tracking
(c) 20 kHz operation
(d) underwater telephone mode
(e) built-in training
(f) command system interface
(g) selection of arrays including longer hull-mounted, variable depth and towed body.

Operational Status
In production.

Contractor
Marconi Underwater Systems Ltd, Templecombe.

PMS 56 – HYDRA ASW SONAR

Type
Small ship hull-mounted or variable depth sonar.

Description
PMS 56 is a lightweight, compact shipborne sonar, one of the Marconi Hydra range of modular sonar systems. PMS 56 provides all the facilities required for effective ASW operations by small ships and patrol craft. It is available with fixed, retractable or variable depth arrays for installations on vessels of up to 300 t displacement and with maximum speeds of up to 30 kts.

The system includes a high resolution colour display which provides the single operator with facilities for search, detection and classification. The man/machine interface, provided by pull down menus, is designed for ease of use and reduced training requirements.

The lightweight panoramic array uses transmit and receive hydrophones for maximum efficiency. Incorporating the latest technology, all the transmit and receive electronics are contained in two lightweight air-cooled shelves.

Operational Status
Currently being evaluated by several navies.

Specifications
Modes: CW, FM, passive, aural
Range scales: 625-20 000 yd
Pulse lengths: 5-1000 ms
Weights: 220 kg (console); 50 kg (transmitter rack); 385 kg (hull outfit and array)

Contractor
Marconi Underwater Systems Ltd, Templecombe.

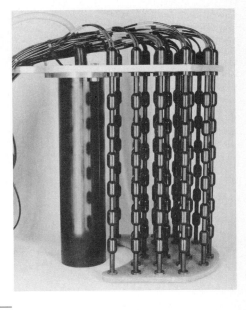

PMS 56 lightweight panoramic array

COMTASS – HYDRA TOWED ARRAY SYSTEM

Type
Compact towed array sonar system.

Description
COMTASS is a cost-effective compact towed array system belonging to the Marconi Hydra family of modular sonar systems. It can be installed in ships or submarines with the minimum of structural changes. The high performance array has been designed to withstand high speed towing, and would typically be installed on a surface ship in less than half a day. The processing and electronics are contained in two 19 in units, each with power supplies and cooling fans. The total modularity of the display and processor gives scope for additional processing if required.

The standard console, which requires only one operator, allows for the simultaneous display of broadband and narrowband passive data for the long-range surveillance of surface and submarine targets. It is also possible to integrate active sonars into the COMTASS console, with flexible control of the displayed data.

The console consists of two high resolution, colour, refreshed raster-type displays in a variety of formats, according to the customer's specification.

A turnkey system design leads to simple on-deck and below-deck installation, and the critical angle tow lends itself to ease of handling, deployment and retrieval. The system provides passive detection and tracking of targets on a multi-function operator's console with detection ranges given as 30+ miles, with an extensive over-the-horizon ships surveillance capability.

The handling system, consisting of winch and control panel, is provided by Strachan & Henshaw. The system is designed for minimum maintenance and semi-automatic deployment. To provide further flexibility it can be palletised or containerised for easy removal or transfer between vessels, and for quick fitment to reserve or auxiliary vessels.

Operational Status
COMTASS has been evaluated by a number of navies and has been fully validated by the UK defence authorities.

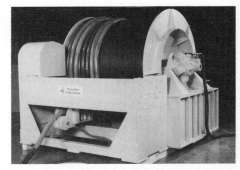

COMTASS lightweight array and handling system

Contractors
Marconi Underwater Systems Ltd, Templecombe.
Strachan & Henshaw, Bristol (handling system).

MARCONI TOWED ARRAYS

Type
Towed array sonars.

Description
Marconi has a unique background in towed array sonars. Over the past two decades the former Plessey (now Marconi Underwater Systems), in close collaboration with the UK MoD, played a major role in developing every indigenous 'wet end' for the Royal Navy towed array sonars (Types 2024, 2026, 2031I, 2031Z, 2046, 2054 and 2062) covering all the current service equipments. The Type 2023, the earliest towed array sonar for SSBMs, was procured from the USA, replaced on an interim basis by the Type 2062 and ultimately by the Type 2046. Similarly, Marconi developed tow cable terminations and the 'clip-on' systems for submarines. Winch equipments were developed by NEI Clarke Chapman for the Type 2031 surface ship arrays and STC, under subcontract from Marconi as prime contractor, for the Type 2054 submarine system. With the exception of the Types 2031Z and 2046 processors, which were developed under contract from the MoD, Marconi has developed all the terminal units, beamformers and processing for Types 2024, 2026, 2031 and 2054, in addition, it has been either prime contractor or co-ordinating design authority for these systems.

This total system involvement with Royal Navy sonars is being continued in the Type 2057 programme where Marconi Underwater Systems Ltd is the prime contractor and design authority for the 'wet end' and the processor, and Strachan & Henshaw the design authority for the handling systems. The

Hydra towed array system fitted on a trials ship

Type 2057 will in time provide surface ship and submarine towed array sonars with common modules for use throughout the fleet.

Hydra Modular Systems
In terms of private venture, the experience gained in MoD and other programmes has been applied to

generate a new family of modular sonar systems known as Hydra (see separate entries). These systems consist of a large number of mutually compatible modules covering beamformers, signal and data processing, man/machine interfaces, data fusion and arrays (helicopter dunking to VDS, flank, passive, active/passive and conformal as well as towed arrays).

Combinations of these elements of Hydra sonar systems provide cost-effective optimum solutions for all surface ship and submarine applications. COMTASS and GETAS are examples of Hydra towed array systems. Other combinations cover requirements from the equivalent of a full Royal Navy fit to a modest single octave trainer. They can all be tailored to suit individual technical specifications and budgets. Research into fibre optic technology has been progressing for some time and technology demonstrator towed array systems are now on trials at sea.

Operational Status
Towed array sonars in the Royal Navy are deployed as follows:

Type 2024	'Oberon' class submarines
Type 2026	Some 'Trafalgar' class submarines
Type 2031I	'Leander' class frigates
Type 2031Z	Type 22 Batch 2 & 3, Type 23 frigates
Type 2046	'Churchill' class submarines
	'Resolution' class submarines
	'Swiftsure' class submarines
	Some 'Trafalgar' class submarines
	'Upholder' class submarines
	'Valiant' class submarines
Type 2054	'Vanguard' class submarines
Type 2062	This was superseded by the Type 2046.

Contractor
Marconi Underwater Systems Ltd, Templecombe.

FMS SERIES SONAR SYSTEMS

Type
Family of passive and active sonar systems.

Description
The Ferranti Modular Sonar (FMS) series is an integrated range of low-cost, compact, high performance systems, developed by the company as a private venture to meet the needs of a wide variety of platforms and operational roles, and is based on the new generation of sonar designs selected by the Royal Navy. The series has been designed to optimise performance, simplify installation and in-service support and reduce through-life costs.

The systems within the series consist of:
FMS 12 - passive narrowband (2 octave)
FMS 13 - passive narrowband (3 octave)
FMS 15 - passive narrowband (5 octave)
FMS 20 - active (4 ft hull array with VDS option)
FMS 21 - active (6 ft hull array with VDS option)
FMS 30 - passive broadband (single shelf - 3 ft array)
FMS 31 - passive broadband
FMS 52 - navigation and obstacle avoidance sonar.

All FMS systems employ the latest advanced distributed digital signal and data processing techniques to provide significant cost, space and performance improvements over the majority of current equipments. These techniques are subject to continuing development to further improve these factors.

Signal processing, based on advanced 'Curtis' techniques, is achieved using programmable Ferranti digital signal processing modules which can be easily configured to meet the optimum performance requirements of specific array outfits and for easy through-life enhancement.

The data processing equipment is based on multiple M700 processors, arranged in distributed arrays. The use of large numbers of these interconnected in the data processing system gives very high levels of computer assistance and automatic operation. Features such as automatic detection, tracking and classification of multiple targets are provided. Standard programmable modules for both signal processing and data processing functions enable significant reduction in spares holding, simplify maintenance and reduce overall cost.

The ergonomically designed operator consoles allow single operator control of the complete system. Each contains two 1000-line high resolution raster displays and an operator's desk. The man/machine interface includes a trackerball, nudge keys and a touch-sensitive plasma panel. From the plasma panel the operator can select various displays, focus on selected beams, scroll around beams, set verniers, annotate features on the display and select a variety of cursors. The trackerball and nudge keys are used for fast and accurate positioning of the cursor to provide specific displays of frequency, bearing, time and harmonic ratio information.

Each sonar in the FMS series is configured by selecting from a range of standard modules. This allows active or passive sonars for ships and submarines, large or small, to be proposed to suit customers' specific requirements. The modules are the same as, or closely related to, equipment in or ordered for RN service. This capability enables Ferranti to offer low-cost and risk solutions to most sonar requirements.

The processing equipment can be housed in air- or water-cooled equipment practice cabinets to meet customers' individual requirements.

FMS 12/FMS 13/FMS 15
The FMS 12, 13 and 15 are high resolution narrowband surveillance sonars designed to derive signal inputs from either towed or hull-mounted line arrays. Each system consists of a single electronics cabinet which weighs 300 kg and a single operator's console weighing 350 kg.

FMS 12 provides frequency cover over two octaves, FMS 13 over three octaves and FMS 15 over five. Features common to both systems include: 360° bearing cover, 32-beam resolution and multiple beam surveillance display. They also provide high resolution vernier analysis classification, with sonar aural and audio communications interface. Both systems have built-in test equipment, an air-cooled operator console and electronics cabinet.

Operational Status
An enhanced version of FMS 12 is included in Project Triton — the Royal Navy's Type 2051 sonar update programme for the 'Oberon' class submarines. FMS 12 also provides the basis for the RN Type 2046 passive sonar system, 26 of which have been ordered for updating the current and next generation of nuclear powered attack submarines, and for inclusion in the new 'Upholder' class of conventional submarines.

The Royal New Zealand Navy ordered an FMS 15/2 system in March 1989, and which Ferranti supplied in January 1990.

FMS 20
FMS 20 is an active sonar system which combines the range resolution of an FM sweep with the Doppler discrimination of a CW pulse. It also has a passive capability for a built-in torpedo alarm facility. In its most comprehensive form, FMS 20 consists of four electronic cabinets, each with a weight of 765 kg and a two-display operator console weighing 340 kg. The associated transducer arrays may be hull-mounted (4 ft diameter) with or without an associated variable depth array.

The equipment provides a 32-beam resolution, electronic array motion compensation, 360° bearing cover, variable range scales and a variable centre frequency of operation. The display formats available to the operator as standard features include active FM and CW information, classification of targets, track and history totes, environmental and ray tracing, and monitoring of the equipment's status.

With Ferranti's concept of using the most modern shelf-based modular technology, the number of electronic cabinets required to drive the FMS 20 system can be reduced to two, albeit with some reduction in overall capability.

Operational Status
FMS 20 is currently being considered by several NATO and other navies for inclusion in both new ship and update programmes.

FMS 21
This system is derived from, and the export variant of, the Royal Navy's new Type 2050 sonar. It is an active sonar system which, like FMS 20, combines the range resolution of an FM sweep with the Doppler discrimination of a CW pulse. Associated transducer arrays may be hull-mounted (6 ft diameter), with or without an associated variable depth array, and provide a 64-beam resolution.

The four cabinets which contain the electronics for the FMS 21 system and its two-display operator console (providing single operator control) weigh 765 kg and 340 kg respectively. Equipped with a variety of display formats as standard features, the operator has active FM and CW information, classification of targets, track and history totes at his fingertips, as well as environmental, ray tracing and system status information. FMS 21 also has a passive capability with an automatic torpedo warning alarm.

Operational Status
FMS 21 is currently being considered by the Pakistan and Hellenic Navies for inclusion in their new frigate programmes, whilst 17 systems of Type 2050 have been ordered by the RN.

FMS 52
FMS 52 is a high frequency, active navigation and obstacle avoidance sonar which minimises the chances of collision with all types of sub-surface obstacles, including moored mines. The 64-element transducer array can be either hull-mounted (for surface vessels), fin or hull-mounted (for submarines), or have fixed installation for harbour surveillance. The

A purpose-built combination of the FMS 15 and FMS 21 sonar systems

use of electronic beam steering in preference to mechanical methods increases the accuracy and data update of the system.

Other features of the FMS 52 include:
(a) steerable transmitter beams, designed to ensure that transmission patterns can be optimised for searches at different depths and in varying water conditions
(b) bearing resolution of 3°
(c) expansibility—up to four arrays giving full 360° coverage
(d) capability of unmanned operation.

Operational Status
The FMS 52 has been selected for the German Navy's new generation Type 212 submarines.

Contractor
Ferranti-Thomson Sonar Systems UK Ltd, Stockport.

SP 2110 SONAR

Type
Passive towed array sonar system.

Description
The SP 2110 is an extension of the GEC Avionics Type 2031I and 2026 sonars which are in operational service with the Royal Navy. The SP 2110 is smaller with increased processing power and is stated to be considerably cheaper. It is modular and consists of the array and handling system, the array terminal unit, and the processing and display console. Additional equipment can be supplied and interfaced to provide a complete towed array system providing narrowband surveillance and tracking, broadband, vernier, DEMON and aural processing. The system can be configured to meet customer requirements.

Contractor
Marconi Underwater Systems Ltd (was GEC Avionics), Croxley.

TYPE 184M/P SONAR

Type
Active and passive hull-mounted search sonar.

Description
The Type 184M is one of the Royal Navy's primary surface ship anti-submarine search and attack sonars.

It is a 360° scanning sonar incorporating both active and passive modes of operation; it provides range, bearing and target Doppler data to the fire control computer. The Type 184M was designed in association with the Admiralty Underwater Warfare Establishment (AUWE), and is fitted in many RN surface ships. The equipment has also been supplied to a number of other navies.

Few technical details have been cleared for publication, but the equipment provides a dual frequency transmission and three receiver systems. The latter comprise:
(a) an all-round search and tracking PPI system with eight receiving channels
(b) a 4-beam sector search Doppler system with B-scan display
(c) a continuous torpedo warning system with its own display.
A circular 32-stave transducer array is employed.

Solid-state modernisation kits (184P) to improve the performance and reliability of this sonar are now in production for the Royal Navy. This programme is being implemented in three phases. The first phase introduced a 16-channel PPI system and replaced all electromechanical selectors and timers with solid-state electronics. The second phase offers improved

Type 184 sonar control console

detection capability against moving targets. The 31 digital filters in each of the four Doppler beams measure up to ± 40 kts of target speed with a resolution of better than 3 kts. The third phase provides separate Doppler, PPI and hydrophone effect geographically stabilised memory colour displays, with special attention paid to man/machine interfaces.

The PPI display offers improved detection through noise discrimination; 4-beam sector search Doppler display provides moving target detection capability; and the hydrophone effect display in the passive mode presents the output of the hydrophone effect processor on the periphery of a circular plot with maximum noise signal directed towards the centre.

Operational Status
In production. Operational with RN and other navies. RN fittings include Type 21 frigates, Type 42 destroyers and 'Leander' class ships. The G 750 medium-range sonar is a solid-state, improved version.

Contractor
Graseby Marine, (Graseby Dynamics Ltd), Watford.

G 750 SONAR

Type
Active and passive hull-mounted search sonar.

Description
The G 750 is a search and attack all-round sonar for ships of frigate size and above. It is a modernised and improved version of the Type 184M, providing accurate fire control data from two independent automatic tracking systems whilst simultaneously maintaining all-round surveillance, independent Doppler search, and classification and continuous torpedo warning.

Three separate processing and display systems are incorporated for:
(a) all-round search and tracking (PPI)
(b) Doppler
(c) passive search.

Digital readout displays are used extensively and visual displays are supplemented by audio in all three systems. Nine cabinets are required for the complete electronics. Three are display consoles: three contain transmitter and control equipment, two house receivers, and one is used for monitoring purposes. For installation in smaller ships or submarines, single passive or PPI display systems with smaller transducers are produced. In the standard G 750 system, an improved version of the Type 184M cylindrical 32-stave transducer is used to form a large number of beams. Range and bearing measurement is by in-beam scanning, obviating the need for separate expanded displays.

Target range and bearing are measured by either of two methods:
(a) by means of a manually positioned marker technique, in which the circular markers are placed over the target echo it being the marker position that is measured; or
(b) by echo gating, where the echo range and bearing are measured directly. This form of tracking is normally on acquisition of a possible target and is available on either computer-assisted or non computer-assisted basis. Marker boxes representing the gated area replace the circles. In both methods the targets are individually identified by dotted or full markers.

Provisions are made for interfacing the G 750 sonar with computer-based ASW and tactical information systems to provide a source of range, bearing, and target Doppler data. Operational capabilities of the equipment include active detection of submarines throughout 360° and the ability to track two targets simultaneously. Similar detection coverage and tracking capabilities are available in respect of torpedo targets, with suitable outputs to torpedo warning and avoidance systems. There is also an active search capability over a 45° arc with facilities for determining target Doppler. Four 11° beams forming a 45° 'searchlight' are providing a 360° search facility capable of measuring target Doppler up to ± 40 kts. The system is designed for operation in adverse reverberation conditions, with ripple or omnidirection dual frequency modulated and Doppler CW transmission modes. The system can be applied as an HF or

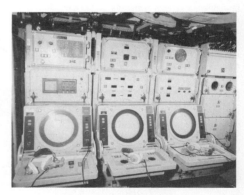

Type G 750 sonar control console

LF variant. Targets can be tracked to a maximum range of 20 km.

Operational Status
In production. Three systems delivered to Royal New Zealand Navy and six systems to the Indian Navy. The system is likely to be updated with new displays and improved signal processing.

Contractor
Graseby Marine (Graseby Dynamics Ltd), Watford.

SEA HUNTER SONAR SYSTEM

Type
Medium-range small ship sonar.

Description
Sea Hunter is a compact sonar system for small craft of over 150 t. It is a medium-range, high performance equipment which employs a conventional PPI display for all-round active and passive surveillance, with both manual and auto-tracking facilities. Sea Hunter consists of a single display cabinet, two electronics cabinets and a lightweight 24-stave, hull-mounted transducer array. A computer-aided colour display and the latest signal processing techniques are employed.

Major features of Sea Hunter include:
(a) all-round active/passive surveillance with manual and auto-tracking facilities
(b) single computer-aided display for improved detection performance
(c) variable sonar parameters to suit prevailing environmental and operational requirements
(d) lightweight hull-mounted circular array
(e) automatic monitoring of systems faults
(f) optional onboard simulator/trainer.
The sonar provides automatic tracking on up to four targets with 5-ping history.

Operational Status
In advanced stage of development.

Specifications
Transmission: Random rippled directional; omnidirectional
Frequency: 10.5 kHz nominal
Pulse length: 20, 40 and 100 ms
Modulation: FM
Source level: 220 dB/μPa/m maximum
Range scales: 2000, 4000, 8000 and 16 000 yd/m
Data processing: Auto-tracking of up to four targets; manual tracking; 5-ping history
Range accuracy: 1% of range selected
Bearing accuracy: ± 1° RMS

Contractor
Graseby Dynamics Ltd, Watford.

SEA SEARCHER SONAR SYSTEM

Type
Active and passive ASW sonar system.

Description
Sea Searcher is a new high performance sonar system for active and passive submarine detection, and is suitable for installation on all ships over 1500 t. It is designed to measure target range, bearing and Doppler to the high degree of accuracy required by modern weapon systems. The transducer array consists of 32 shock-hardened staves assembled on a circular frame. The electronics and display are housed in seven fully ruggedised cabinets.
Major features of the Sea Searcher include:

(a) all-round 360° surveillance with simultaneous auto-tracking of up to four targets
(b) choice of two or three separate geographically stabilised colour displays
(c) multi-ping back history of all potential targets
(d) numerical display of speed and course of selected targets
(e) variable threshold setting to improve maximum range detection in changing environments
(f) variable pulse length to overcome high reverberation environments
(g) rippled directional transmissions to provide maximum source level.

Operational Status
In advanced stage of development.

Specifications
Transmission
Type: Random rippled directional; omnidirectional
Frequency: Between 6 and 9 kHz
Pulse length: 20, 40 and 100 ms
Modulation: FM; shaped CW
Range scale: 5000, 10 000, 20 000 and 40 000 yd/m
Reception
No of channels: 16
Auto-tracking: Up to four targets
Range accuracy: 1% of range selected
Bearing accuracy: ±1° RMS
Doppler: Range ±40 kts; resolution 2 kts

Contractor
Graseby Dynamics Ltd, Watford.

MODEL 3025 SEARCH SONAR

Type
Search sonar for collision avoidance.

Description
The Model 3025 search sonar is a new product designed to prevent forward collision of either surface craft or remotely operated vehicles. It has an operating range of 500 m and a vertical beamwidth of 12°.

The sonar will easily detect any buoyant or mid-water targets.

The 3025 removes distortion found in other long-range single beam sonars by electronically sweeping eight 3° receive beams through a 24° horizontal sector. The transducer array is moved within a scan sector by a powerful brushless DC motor. Scan rates vary from 30 to 170°/s, depending on the selected range.

The surface display has a true zoom expansion box which uses the raw sonar data, applies the selected magnification factor and displays the result on the screen. This facility is capable of a range resolution of 120 mm.

Operational Status
One system has been supplied to the UK MoD for research purposes.

Contractor
Ulvertech (a subsidiary of MSI Defence Systems Ltd plc), Ulverston.

ATAS SONAR

Type
Active towed array sonar.

Description
ATAS is an active towed array sonar which is being developed as a collaborative programme between BAeSema and Thomson Sintra ASM for ASW operations. It incorporates the best features of active variable depth sonars and passive towed arrays. The system provides both active detection of targets and simultaneous torpedo warning capability. Provision is also made to permit the addition of a conventional, low frequency passive towed array and integration with Thomson Sintra Spherion hull-mounted sonars. The ATAS system comprises a high power, low frequency transmitter, operating in the 2.4 kHz band, which can be deployed at up to 900 m behind the towing platform, together with an in-line receiver array towed a further 300 m behind the transmitter all of which may be towed at depths suited to prevailing oceanographic conditions. The port/starboard ambiguity normally associated with a towed array is automatically and instantaneously resolved within the system without the need for ship manoeuvre. ATAS can be operated at depths of 235 m.

The transmitter has a high acoustic sound source consisting of a vertical stave of low frequency flextensional transducers. Transmissions are omnidirectional in the horizontal plane with a vertical beamwidth of 25°. The receiver array is a 30 m long flexible tube housing a line of hydrophones and associated electronics to condition and digitise hydrophone signals. The receiver gives full azimuth cover and in the broadside direction a bearing resolution of up to 0.5°.

The processing equipment uses current technology

ATAS sonar showing the deployment mechanism and the transmitter

to provide high processing capability within a single compact unit.

The compact nature of ATAS permits it to be deployed from non-specialised ships of 200 tonnes and above, including ships of opportunity or ships taken up from trade, with a minimum of modification to the platform. Hull penetration is eliminated and a space of only 1.8 m by 5.6 m is required on the stern of the ship for the self-contained winch and handling equipment which weighs 4.7 tonnes (excluding in-water equipment). Other shipborne equipment comprises two electronic cabinets and a display unit incorporating dual high resolution colour monitors. The display console can be installed remotely from all other equipment, it displays active and passive search, localisation and classification information in addition to performance of the day and performance monitoring data. All system equipment can be supplied in containerised form.

Operational Status
In an advanced stage of development. Successful sea trials of the ATAS system have been carried out. ATAS was evaluated for the US Coast Guard under a US Navy contract during the Summer of 1990.

Contractors
BAeSema Marine Division, Bristol.
Thomson Sintra Activités Sous-Marines, Valbonne Cedex.

UNITED STATES OF AMERICA

AN/SQR-15 SONAR

Type
Towed array sonar system.

Description
The AN/SQR-15 is a surface ship towed array sonar system which has been in operational service for several years. It provides a mobile passive surveillance capability for escort vessels against enemy submarines under most environmental conditions, and is optimised for long-range surveillance. The system is an improved and redesigned version of the earlier AN/SQR-14 equipment.

Operational Status
In operational service.

Contractor
Martin Marietta, Glen Burnie, Maryland.

AN/SQR-17A SIGNAL PROCESSOR

Type
Integrated shipboard submarine detection, classification, signal processing and display system.

Description
The AN/SQR-17A is a totally integrated submarine detection and classification system for shipboard applications. The system configuration includes the AN/SQR-17A sonar signal processor and display unit, the RD-420B tape recording system, the AN/ARR-75 sonobuoy receiver interface unit and the antenna unit.

The AN/SQR-17A can process signals detected by DIFAR, DICASS, VLAD, LOFAR, BT, ANM and RO sonobuoys, and displays raw acoustic in either A- or B-scan format. Data are then enhanced, incorporating alphanumeric designations and graphics with human-factor considerations, to aid personnel in target analysis and tactical co-ordination.

The AN/SQR-17A has a 19 in acoustic data display and two 9 in auxiliary video-monitor displays for menu prompting, for selection of the appropriate system operating mode and for monitoring the status of multiple acoustic sensors.

The high density solid-state mass-memory display provides full multi-target detection, tracking and classification for ASW encounters. The display uses real-time, high resolution, TV raster formats and is designed modularly with electrographic hard copy capabilities in standard-gram format.

The AN-SQR-17A has gone through a series of upgrades and improvements through the (V)1, (V)2 and (V)3 versions.

Operational Status
Currently under contract to the US Naval Sea

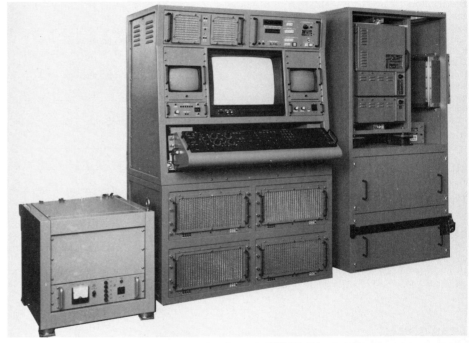

(Left to right) AN/ARR-75 sonobuoy receiver/interface unit, AN/SQR-17A sonar signal processing system, RD-420B recorder/reproducer

Systems Command. To date, 97 shipboard units have been procured. In addition, 16 modified units are being procured for mobile inshore undersea warfare applications.

Contractor
Diagnostic/Retrieval Systems Inc, Oakland, New Jersey.

AN/SQR-18A(V) TACTAS SONAR

Type
Towed array passive sonar system.

Description
The AN/SQR-18A(V) tactical towed array sonar (TACTAS) has been designed to enhance the air/sea warfare capabilities of the US Navy surface ship fleet. It provides long-range passive detection and classification of submarine threats. The system is a high technology, single operator equipment providing full azimuth coverage in both narrowband and broadband search modes.

The 800 ft long array consists of an acoustic section, a vibration isolation module and a rope drogue. In the short tow version it is towed behind the variable depth sonar (VDS) body of the AN/SQS-35V (see separate entry). Pre-amplifiers are in the array, while signal processing and other electronics are on board the ship.

The original AN/SQR-18A has now been replaced by the AN/SQR-18A(V)1 version, which with a new low-noise array provides increased detection range and allows effective operation at high own ship's

speed. An adaptive processor provides interference cancellation, making the system virtually immune to tow ship noise, and allows operation near other noisy ships. An improved tracker provides accurate bearing data for effective target motion analysis solutions, even on weak targets. The most significant improvement made to the basic system has been the installation of an operator auto alert capability which provides target detection for operator selected frequencies. The system is capable of providing acoustic data to an external recorder and playing back recorded data.

The AN/SQR-18(V)2 critical angle towing version has its own towing and handling capability permitting it to be used on ships without a VDS. This variant has been recently completely modernised and utilises an Advanced Modular Signal Processor (AMSP), a programmable system using interactive software design. The new processor provides additional processing capability used in conjunction with an improved operator automatic alert system which can be optimised for each individual threat and alert level.

In the US Navy the AN/SQR-18A(V)1 and (V)2 systems are integrated with the AN/SQR-17 sonobuoy processor and AN/SQS-26 hull-mounted sonar. This combination provides an effective combat

suite for the LAMPS Mk-1 ASW suite. The AN/SQR-18A(V)1 or (V)2 are capable of effective operation as a stand-alone system or can be integrated with other combat information systems.

The AN/SQR-18A(V)1 system electronics consist of three major and five small units while the (V)2 consists of two major and six small units.

A signal conditioner provides gain, equalisation, interfacing, auto-ranging gain control and readies the hydrophone signal for digitisation. This is the only analogue part of the AN/SQR-18A(V).

An embedded trainer is included which provides realistic targets in dynamic interactive scenarios with high fidelity classification clues that allow onboard training of both the operator and the target localisation team.

Digital signal processing techniques are used in both variants to present narrowband and broadband low frequency information. The system may be operated in both narrow and broadband simultaneously. Tracking is provided by an automatically stabilised tracker beam. The automatic target following circuits keep the beam nulled on a target through own ship and target manoeuvres. All processed data are stored in a modular electronic solid-state memory.

Narrowband outputs are presented on the CRT screens in LOFAR format. The AN/SQR-18A(V)1 provides two independent CRT screens while the (V)2 provides an additional CRT to present situation summary information. Controls are provided to measure the frequency and amplitude of tonals, and harmonic ratio of target frequencies. Results of measurements, bearings of beams displayed and tag numbers are displayed in alphanumeric format.

Operational Status

In use with the US Navy and with the Royal Netherlands and Japanese navies. Forty AN/SQR-18A systems have been delivered to the US Navy

and have been updated to an AN/SQR-18A(V)1 standard. Seven AN/SQR-18A(V)2 systems have been delivered with these being designated for the US Navy FF 1052 frigates not equipped with VDS, as well as some FFG-7 class frigates.

EDO is currently under contract to develop additional improvements to these equipments.

Contractors

EDO Corporation, Government Systems Division, College Point, New York.

Martin Marietta, Glen Burnie, Maryland (towed array and handling and storage subsystem for the (V)2).

AN/SQR-18A console/display

AN/SQR-19 TACTAS SONAR

Type

Passive towed array sonar system.

Description

The AN/SQR-19 is a passive towed sensor array designed to give surface vessels long-range detection and tracking of both submarines and surface ships. It includes: a towed sensor array with improved ranging ability and reduced self-noise; a handling system which improves system performance by maintaining array position and depth; and a sophisticated signal processing system providing more sonar information with less operator workload. The 'wet end' consists of a linear array and tow cable, plus the associated handling and stowage equipment. Coaxial conductors within the tow cable carry the flow of telemetred acoustic signals, plus heading depth and temperature data from the array to the ship, and also handle the

commands to the array. Located below decks are the signal and data processing equipments, plus the display and control consoles. The displays are shared with the AN/SQS-53 hull-mounted sonar and the LAMPS III ASW helicopter, so that the ship is provided with a single integrated ASW system (the AN/SQQ-89).

The AN/SQR-19 provides omnidirectional long-range passive detection and classification of submarine threats at 'tactically significant' own ship speeds in seas up to state 4, using an 82 mm diameter array towed on a 1700 m cable to provide tow depths down to 365 m.

Operational Status

The first operational system was installed on board the USS *Moosbrugger* early in 1982 and has completed both technical and operational evaluation. Installation aboard six US Navy 'Spruance' class destroyers is underway. It is intended to install the AN/SQR-19 aboard all US Navy DDG-993 and DDG-

51 destroyers, 'Ticonderoga' class cruisers, and the 'Oliver Hazard Perry' class frigates. Martin Marietta has received a number of contracts for the system, including one from the Canadian Navy for the towed array portion, and another in December 1987 valued at $100 million for the US Navy and Spanish Navy. As far as the US Navy is concerned, procurement of the AN/SQR-19 is incorporated with the AN/SQQ-89 integrated ASW system, of which it forms a part. Allied-Signal, as a member of the Westinghouse AN/SQQ-89 team, received an order for seven systems in 1990.

Contractors

GE Electronic Systems Division, Syracuse, New York.
Allied-Signal Aerospace Co, Sylmar, California.
Westinghouse Electric Corporation, Sykesville, Maryland.
Martin Marietta, Glen Burnie, Maryland.

AN/SQS-35, AN/SQS-36, AN/SQS-38 SONARS

Type

Hull-mounted and variable depth sonar systems.

Description

These systems have been designed to detect submarines at medium ranges in both deep and shallow waters. The AN/SQS-35 and the AN/SQS-38 sonars have been in service with the US Navy since the mid-1960s.

The US systems are improved miniaturised solid-state versions of previously developed vacuum-tube equipments manufactured for the US, Italian, Norwegian and Japanese Navies.

Some versions combine both variable depth capability and hull sonar capability, selectable by the sonar operator at the control console.

Weapons associated with the system are the Mk 44 torpedo and ASROC (USA), Terne (Norway) and Lanciabas (Italy). Over 75 systems have been delivered.

Operational Status

In service with the US Navy.

Contractor

EDO Corporation, Government Systems Division, College Point, New York.

AN/SQS-35(V) variable depth sonar seen aboard escort ship USS Francis Hammond

---

AN/SQS-53/26 SONARS

Type

Hull-mounted passive and active sonar system.

Description

Claimed to be the most advanced surface ship ASW sonar in the US Navy inventory, the AN/SQS-53 is a

high power, long-range system evolved from the AN/SQS-26CX. Functions of the system are the detection, tracking, and classification of underwater targets, and underwater communications, counter-measures against acoustic underwater weapons and certain oceanographic recording uses. Target data obtained by the sonar are transmitted to the ship's Mk 116 digital underwater fire control system. The latter translates target range, bearing and depth data sent

from the sonar to the ship's central computers into signals controlling the launch of ASW weapons (ASROC or torpedoes).

The AN/SQS-53 can detect, identify and track multiple targets and is the first USN surface ship sonar designed specifically to interface directly with a vessel's digital computers. The system has a cylindrical array of 576 transducer elements housed in a large bulb dome below the water line of the ship's

Bow housing of AN/SQS-26 sonar

bow. There are 37 cabinets of signal processing, transmitting and display equipment. Passive and active operating modes are possible.

There are three active modes:
(a) surface duct
(b) bottom bounce
(c) convergence zone.

The surface duct mode depends upon sound energy being transmitted essentially in the horizontal plane. Due to the high level of noise introduced into the return signals near the surface, this mode is useful only for relatively short distances. Nevertheless, this method is conventional for many surface ship sonars and the high transmission power of the AN/SQS-53 is stated to provide longer range capability in the surface duct mode than previous sonars.

In the bottom bounce mode the sound energy is directed obliquely toward the seabed. The energy is reflected upward from the ocean floor toward the surface at considerable distances from the ship. Submarine echoes are received via a similar return path. This method is useful in waters of more than certain minimum depths and where the seabed has the requisite favourable characteristics.

Convergence zone mode operation takes advantage of the characteristics of very deep water. The sound energy is refracted downward due to the temperature and pressure conditions near the surface, but, as depth increases, these physical effects change and the sound path alters direction to cause the energy to return to the surface in a coarsely focused convergence zone. This zone can form at great distances from the ship, and provides the longest range of coverage for the sonar when the water conditions are favourable for this mode of operation.

Passive detection gives the bearings of targets

Bow dome cutaway showing partial view of AN/SQS-53 cylindrical transducer

based on their own noise generation, rather than by echo location. It has proved a valuable method on the currently deployed AN/SQS-26 equipped ships, especially at low speed. The *Spruance* and her successors are expected to have improved passive detection capabilities at higher speeds due to improved noise suppression measures. The passive mode can be operated simultaneously with the active modes.

Operational Status

The AN/SQS-26 is fitted in 10 'Garcia' class frigates; 6 'Brooke' class missile frigates; and 46 'Knox' class frigates. The AN/SQS-53 is being fitted in 'Spruance' class destroyers as they are built and a total of 30 ships is planned. The five 'Virginia' class nuclear cruisers are fitted with AN/SQS-26CX or AN/SQS-53A systems. An updated version, known as the AN/SQS-53B, is in service and $104 million contracts have been awarded to Hughes Aircraft Company by the US Navy for 30 AN/SQS-53B digital display subsystems for 'Spruance' and 'Ticonderoga' class ships. In September 1986 a contract valued at $39 million was awarded to General Electric, followed by a $34 million contract in May 1987.

A new version of the system, the AN/SQS-53C Battle Group Sonar, has been developed by General

Electric. The AN/SQS-53C is a much more advanced system which uses 22 fewer cabinets with a consequent reduction in cabling and so on. The new version uses seven Sperry AN/UYK-44(V) computers linked together in a multiple embedded configuration. The AN/SQS-53C is scheduled to be retrofitted into a number of ships currently equipped with earlier versions.

The SQS-53C incorporates a number of improvements, including a better performance in the active mode, simultaneous operation in active and passive mode, multiple target automatic tracking and improved displays. The first fitment was in USS *Stump* in August 1986 for evaluation purposes. This was completed in mid-1989.

In 1990 Westinghouse was awarded a contract for eight SQS-53C systems as part of their SQQ-89 integrated ASW combat system award.

Contractors

General Electric Company, Electronic Systems Division, Syracuse, New York (Types 53A and 53C).

Hughes Aircraft Company, Systems Group, Fullerton, California (Type 53B).

Westinghouse Electric Corporation, Sykesville, Maryland (Type 53C).

AN/SQS-56/DE1160 SONAR

Type

Hull-mounted active and passive sonar system.

Description

The AN/SQS-56 is a modern hull-mounted sonar developed as a company funded product by Raytheon's Submarine Signal Division for the US Navy's 'Oliver Hazard Perry' class frigates. The US Navy has provided the AN/SQS-56, via the Foreign Military Sales (FMS) programme, to Saudi Arabia for its PCG ships, to Australia for its 'FFG-7' class frigates and to Turkey for its MEKO 200 programme. Versions commercially exported with the designation DE1160 are operational on the Italian Navy's 'Lupo' class frigates and the Spanish, Moroccan and Egyptian navies' 'Descubierta' class frigates and are being considered for installation in new construction ASW ships of several other navies. The DE1160, when configured with 36 kW transmitters, is identical to the AN/SQS-56. Outfitted with a VDS array and handling subsystem, it becomes Raytheon's DE1164 sonar (see separate entry) and is installed in the Italian Navy's 'Maestrale' and 'Alpino' class ships.

The DE1160, when equipped with a larger, low frequency transducer array and three additional

transmitter cabinets, is designated DE1160LF and is capable of convergence zone performance. The DE1160LF was delivered to the Italian Navy for the helicopter carrier *Giuseppe Garibaldi* early in 1984. This system has also been ordered by the Spanish Navy and will upgrade its 'Baleares' class frigates.

A VDS version of the DE1160LF will combine the convergence zone performance of the *Garibaldi* sonar with the environmental adaptability of the DE1164 under the denomination DE1164LF, a sonar system for major ASW combatants. This system is installed on the Italian Navy's 'Animoso' class ships.

The AN/SQS-56 sonar features digital implementation, system control by a built-in mini-computer and an advanced display system. Digital implementation allows packaging of the complete multi-function active and passive sonar in five medium-size electronic cabinets and one operator's console. Computer-controlled functions provide a system which is extremely flexible and easy to operate. The computer is also used to provide automated fault detection and localisation and a built-in training capability. The human-engineered display ensures proper interpretation by operators, even by those with relatively low levels of training.

The sonar is an active/passive, pre-formed beam, digital sonar providing panoramic echo ranging and panoramic (DIMUS) passive surveillance. All signal

processing, except transducer received signal amplification and linear transducer transmit drive, is accomplished in digital hardware, most of which is implemented using US Navy SEMP (standard electronic module programme) components in compact water-cooled cabinets small enough to allow installation through standard size hatches. All visual data are presented on flicker-free, digitally refreshed television type raster scan CRT displays. Complete symbol and alphanumeric facilities are included. System timing, control and interface communication are accomplished by a general-purpose mini-computer which is a component of the basic system. Both 400 Hz synchro and MIL-STD-1397 Type A or C, category II digital interfaces are available in the basic sonar system. Except for the 400 Hz synchro reference power, the entire sonar operates from 440 to 480 V, three-phase 60 Hz ship's power.

The basic system includes the transducer array, transducer junction box, five electronic cabinets (array interface, transmitter(s), receiver and controller), operator console, sonar dome and control unit. Options in production include a loudspeaker/intercom, a water-cooling unit, a remote display and a performance prediction subsystem.

A single operator can search, track, classify and designate multiple targets from the active system while simultaneously maintaining anti-torpedo surveillance on the passive display. Computer-assisted system control permits the operator to concentrate on the sonar data being displayed rather than on the system control.

Operational Status

Approval for service use was issued early in 1980 by the US Navy following successful completion of final operational test and evaluation. By November 1989 systems ordered for the AN/SQS-56/DE1160 totalled more than 100 and included 53 systems and 10 trainers for the US Navy; four systems for Australia (AN/SQS-56); seven systems for Saudi Arabia (AN/SQS-56); six systems for Turkey (AN/SQS-56); six systems for Italy (four DE1160, one DE1160LF and one trainer); 18 systems for Spain (six DE1160, five DE1160LF, six AN/SQS-56SP and one trainer for 'Oliver Hazard Perry' class ships); one DE1160 for Morocco; and four DE1160 HM/VDS for Greece (for the MEKO frigates).

In January 1990 Raytheon announced a contract to supply AN/SQS-56 hull-mounted sonar systems to Taiwan for installation on new patrol ships. Six systems have been ordered with first deliveries due in early 1992. Elsag of Italy has completed delivery of a MAST trainer to the Italian Navy (see *Training and Simulation Systems* section).

Contractor

Raytheon Company, Submarine Signal Division, Portsmouth, Rhode Island.

A US Navy sonar operator using the AN/SQS-56 dual sonar display

DE1164 SONAR

Description

The DE1164 sonar consists of the Raytheon DE1160 hull-mounted sonar augmented by a fully integrated variable depth sonar subsystem. All sonar functions of the DE1164 are identical to those of the DE1160. However, the DE1164 provides transmission and reception via various combinations of the hull-mounted and/or the towed variable depth sonar (VDS) transducer arrays. Addition of the VDS subsystem improves overall sonar operational flexibility, and allows the VDS transducer array to operate at acoustically favourable depths and in a much quieter environment.

In addition to the components of the full DE1160 hull-mounted system, the DE1164 includes one extra cabinet of electronics, a VDS towed body with associated cable, and the electrohydraulic mechanism associated with launching, towing, and retrieving the VDS body. The VDS handling equipment provides for one-man operation for launching and retrieving and unattended no-power towing. For reliability two independent hydraulic power supplies are provided, either of which may support the entire operation; an emergency retrieval system is available as an option. VDS body weight, cable size and length and careful attention to drag provide a VDS depth capability greater than 200 m at 20 kts of ship's speed.

Both the hull-mounted and VDS arrays use the common set of DE1160 transmitting, receiving, and display electronics. Selection of the particular combination of transmit/receive array functions is ordered by the operator via the sonar console input keyboard; the system computer then sets up the required sequence. Alphanumeric symbols on the display inform the operator about which particular array is in use during any specific ping-cycle.

During the normal operation the power requirements of the DE1164 are identical to those of the

VDS configuration of DE1164 sonar on the Libeccio *of the Italian Navy*

DE1160C. A maximum of 74 kW additional power is required from the 440 V, three-phase, 60 Hz power mains during VDS retrieval or launching. The hull-mounted sonar may be operated as a DE1160 when the VDS is stowed or being launched/recovered.

Operational Status

A total of 10 DE1164 systems have been delivered to the Italian Navy and are in operational service on 'Maestrale' and 'Alpino' class ships. On order is one VDS trainer and two DE1164LF systems for use with the Italian Navy new construction 'Animoso' class

guided missile destroyer. The primary electronics system for this latest DE1164LF award will be built in Italy under co-production and licence agreements. The mechanical hoist will be built by Fincantieri Naval Shipbuilding Division in Genoa, and part of the transmitter and transducers will be built by Elsag and their subsidiaries.

Contractor

Raytheon Company, Submarine Signal Division, Portsmouth, Rhode Island.

DE1167 SONAR

Type
Hull-mounted and variable depth sonar systems.

Description
Raytheon's DE1167 family of sonars implements the proven features of the AN/SQS-56/DE1160 systems using advanced microprocessor architecture and state-of-the-art display and transmitter technology. The DE1167 series is based on large size modules and air-cooled cabinets, which permit the production of smaller, simpler, cheaper sonar systems, and facilitate in-country manufacturing/repair participation where required. The DE1167 family is designed to satisfy the requirements of most ASW platforms and missions. Configurations include hull-mounted, VDS and integrated HM/VDS systems featuring a 12 kHz VDS and either 12 kHz or 7.5 kHz hull-mounted 36-stave transducer arrays. Configurations using 48-stave arrays or a frequency lower than 7.5 kHz are possible.

Like the Raytheon DE1160B/C and DE1164 sonars, the DE1167 features primarily digital electronics and an advanced control and display system. The standard inboard electronics consist of three cabinets and a single operator console. Outboard units consist of a transducer array and 2.74 m long dome for the hull-mounted 12 kHz installation (4.17 m long dome for 7.5 kHz) and/or the VDS winch, overboarding assembly, control station, hydraulic power supply, faired cable and towed body for the 12 kHz variable depth subsystem.

The basic DE1167 HM is an active/passive, pre-formed beam, omni and directional transmission sonar which uses three non-interfering 600 Hz wide FM transmission bands centred at 12 or 7.5 kHz and has a spatial polarity coincidence correlation (PCC) receiver. The passive mode, which is selected automatically when transmissions are stopped, is primarily useful for torpedo detection. Optional items in production include a performance prediction subsystem, an auxiliary half-frequency passive receiver, an auxiliary display, a remote display and a training/test target remote-control unit. Signal reception and beam forming are accomplished by broadband analogue circuitry followed by clipper amplifiers for perfect data normalisation. Detection processing, display processing, system control/timing and waveform generation are done digitally. Two microprocessors perform display ping history, cursor ground stabilisation, and target motion estimation functions for torpedo direction. The modular air-cooled 12 kW transmitter uses highly efficient class A/D power transistor techniques.

Several operational features are unique to a sonar of this size/range. Firstly, the display processing incorporates a ping history mode through which the sonar data obtained in three of the previous ping-cycles may be retained on the viewing surface, allowing the operator to readily differentiate between randomly spaced noise events and geographically consistent, valid acoustic reflectors. Secondly, the clipped PCC processing permits accurate thresholding of all signals such that the false alarm rate, or number of random noise indications on the screen,

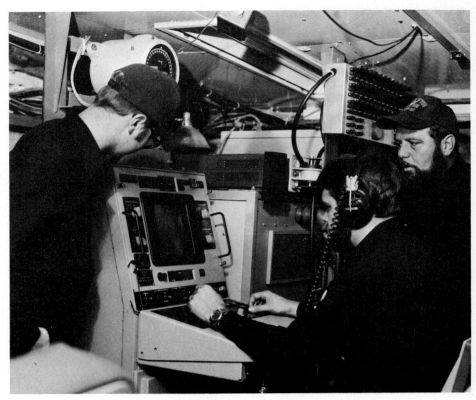

Operator's display/control console of Raytheon DE1167 sonar

remains relatively low and constant over all variations in background noise and reverberation levels, further facilitating contact detection. Thirdly, ground-stabilised cursors and target motion analysis permit rapid determination of contact motion over the bottom, an excellent clue as to the nature of the contact. These three operational features, combined with extremely accurate tracking displays, a built-in fault detection/localisation subsystem, performance verification software and test/training, result in a high performance system which is easy to operate, maintain and support.

Operational Status
As of November 1988, 41 systems were under contract, including 26 DE1167 for the Korean Navy, two DE1167LF/VDS systems for Spanish-built corvettes for the Egyptian Navy, nine DE1167LF systems for the Italian Navy and four DE1167LF systems for Japan. ELSAG has implemented a licence agreement with Raytheon for manufacturing DE1167 sonars for the Italian Navy corvette programme.

Specifications
Centre frequency: 12 kHz (HM and VDS), 7.5 kHz (LF and HM)
Source level: TRDT 227 dB (HM), omni 217 dB
Pulse type: 600, 2000 Hz FM sweep; 100, 200, 50, 6 ms pulse lengths
Receiver type: Spatial polarity coincidence correlator (PCC) between 36 pairs of half-beams

Beam characteristics: 36 sets of right and left half-beams for active and passive detection. Selectable 10°H × 13°V sum beam for audio listening. 1.25° bearing interpolation for fine search display
Active display: 300 range cells, 288 bearing cells; single and multiple echo history; 4 intensity levels; flicker-free, bit-image memory technique
Passive display: Electronic bearing time recorder (EBTR) with medium time averaging. DIMUS-type LTA/STA with optional passive receiver
Track displays: Sector scan indicator (1000 yd × 10°) and target Doppler indicator (1000 yd × ± 60 kts)
Target data: Range: 8 yd (6.1 m) resolution. Bearing 1.25° resolution plus active search display — 0.1° and 3.3 yd on SSI display. ± 60 kts of Doppler at 1 kt steps on the TDI
Data format: Standard: NTDS ANEW (digital). Optional: NTDS slow, Fast Serial D/S synchro converters
Power requirements: Passive 800 W. Active 20 kVA (pulsed) at 10% (max) duty cycle 440 V, 60 Hz, 3-phase, VDS launch/retrieve 75 kVA (max) 440 V, 60 Hz, 3-phase
Weights: Hull-mounted 12 kHz; 1500 kg (nominal); VDS 10 000 kg (nominal) with 200 m cable length

Contractor
Raytheon Company, Submarine Signal Division, Portsmouth, Rhode Island.

DE1191 SONAR MODERNISATION

Description
The DE1191 sonar upgrades and improves the performance of older AN/SSQ-23 and AN/SQS-23 sonars which are being used by navies worldwide aboard numerous FRAM II destroyer class ships.

The DE1191 features advanced inboard electronics which, when coupled with the existing AN/SQS-23 dome and array, comprise a newly configured, lightweight system with the capability to outperform most current modern surface ship sonars. The AN/SQS-23 is a sonar of 1950s design. The system has the potential for excellent performance, with a large low frequency transducer array situated in a quiet location and on a relatively quiet platform.

Replacement of the AN/SQS-23 dated, inboard electronics with DE1191 hardware, a solid-state transmitter and a modern receiver/display subsystem, eliminates weight totalling 12 tons and reduces onboard space requirements by 450 cu ft.

This massive equipment reduction greatly

enhances long-term reliability, maintainability and logistic support for both transmitter and receiver/display functions. Furthermore the DE1191's computer-assisted receiver/display, a slight modification of Raytheon's DE1167 receiver/display, has notable superior detection range and operability features, resulting in an impressive 14 dB improvement in sonar performance.

DE1191 inboard electronics are solid-state, ensuring a longer and more economical life cycle as compared to the vacuum tubes currently used by AN/SQS-23 systems. Improved equipment and the reduction in hardware minimise the need for large and costly spare part inventories, both on board and at depot. In direct contrast to the AN/SQS-23, where spare inventories are soon to be completely phased out by the US Navy, the DE1191 equipment is common to several other Raytheon sonars and availability is guaranteed for 20 to 30 years.

Modernisation can be achieved incrementally in two major stages: the solid-state transmitter, or SST, which is always installed first, and later the receiver/display – or all at once, as time and funding

permit. The DE1191's streamlined configuration will reduce both time and expense necessary to conduct maintenance and operator training.

The modern integrated circuits of the DE1191, proven system techniques and high volume production have all contributed to low-cost equipment that is economical to support. It is available to the international market directly through Raytheon, or via foreign military sales arrangements with the United States government.

Operational Status
In production. In February 1990 Raytheon reached agreement with the Greek Navy for production of three DE1191 systems for use aboard former US Navy destroyers. These will use the existing AN/SQS-23 dome and array, but will employ new lightweight electronics.

Contractor
Raytheon Company, Submarine Signal Division, Portsmouth, Rhode Island.

RAYTHEON SOLID-STATE TRANSMITTER (SST)

Type
Sonar transmitter subsystem.

Description
The Raytheon SST provides sonar systems with the benefits of modern, solid-state technology, offering greatly improved reliability and maintainability, plus very significant space savings relative to vacuum tube transmitters. This transmitter subsystem is made available for system modernisation or for new system applications.

Configured to meet a specific system's requirements, the SST consists of multiple 1 kW modules with associated power supplies. As configured for the AN/SQQ-23 and AN/SQS-23 sonars upgrade, the subsystem consists of two cabinets of 24 modules each and a performance monitor/system interface cabinet. It replaces more than 30 units of the previous transmitter, including all of the energy storage motor generators.

Operational Status
Current production includes transmitters for seven navies. Similar transmitters are used on the AN/BQS-13, AN/SQS-56, DE1160 series and the AN/SQQ/SQS-23 sonar systems. An air-cooled version of the SST has been delivered in a programme to update the AN/BQS-4 submarine sonar. Some 72 transmitter subsystems had been sold by November 1988.

Specifications
Standard 1 kW module (1 per channel)
Power output: 1 kW
Operating frequency: ± 1.5 kHz (nominal)
Distortion: 6% max
Load: Nominal ± 100%
Duty cycle: 15%
Linearity: ± 1 dB from 0 to −12 dB
Gain adjustment: 3 dB
Protection: Short circuit and over-temperature
Mean time to repair: 3 mins
Weight and size: 1651 kg in 1.2 × 1.8 m of deck space
SST as manufactured for AN/SQQ/SQS-23
Power requirements: 115V AC, 60 Hz 1-ph, 0.37 kVA, 0.92 PF; 440V AC, 60 Hz, 3-ph, 128 kW (Max 0.9 PF)
Modes of operation
(a) Transmit: same as any associated system such as omni, sector, FM or CW
(b) Performance monitoring: automatic and manual
(c) Self-test: receiving system not required.

Contractor
Raytheon Company, Submarine Signal Division, Portsmouth, Rhode Island.

Raytheon standard sonar transmitter

MODEL 610 SONAR

Type
Hull-mounted active sonar system.

Development
Model 610 was developed by the EDO Corporation as a private venture starting in 1965. The first prototype was completed in 1966 and the first production model completed its sea trials in 1969. The Model 610 has been continuously improved, and the 610E model is of all solid-state construction.

Description
Designed for the long-range detection of submarines in deep and shallow water, the EDO Model 610 scanning sonar has two active consoles, enabling it to perform a search-while-track function. Facilities offered include a search capability in three 120° sectors, passive correlation, and reverberation processing. The transmitter and receiver beams are pre-formed. Output is available for a fire control system. All mode changes and range scale changes are controlled by console push-buttons, and displays include a Doppler display on each of the active consoles and a passive sonar bearing time recorder display.

Operational Status
No longer in production. Model 610 systems are in operational service with the Royal Netherlands Navy (on the 'Tromp' class as the CWE 610), the ex-Netherlands 'Van Speijk' class frigates of the Indonesian Navy (as the CWE 610), the Peruvian 'Lupo' class frigates, the Venezuelan 'Lupo' class frigates (in a modified version as the SQS-29), and on the Brazilian 'Niteroi' class frigates.

Contractor
EDO Corporation, Government Systems Division, College Point, New York.

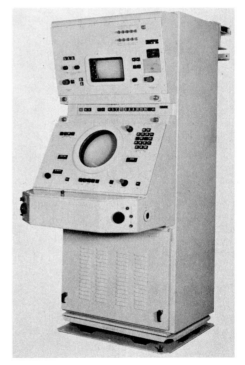

Model 610E sonar control console No 1

MODEL 700 SERIES SONARS

Type
Hull-mounted and variable depth sonar system.

Development
Like the Model 610, the Model 700 series was developed by the EDO Corporation as a private venture for sale on the international market.

Description
Latest equipment of this range of single operator sonars is the model 700E medium-range hull-mounted sonar. The Model 700/702 hull-mounted VDS uses common 700E electronics with a hull-mounted transducer and a lightweight VDS hoist. Selection of hull-mounted or VDS operation is by push-button on the operator's console. The Model 700/701 is a VDS which provides a capability to detect deep targets when bathythermal conditions are unfavourable for hull-mounted sonars.

The basic equipment has a 254 mm panoramic CRT display and a Doppler display. All mode and range scale changes are made by push-button controls on the operator's console.

Operational Status
No longer in production. Model 700 sonars are in operational service with the Brazilian Navy (on the ASW frigates of the 'Niteroi' class), and the ex-Netherlands 'Van Speijk' class frigates now in service in the Indonesian Navy.

Contractor
EDO Corporation, Government Systems Division, College Point, New York.

EDO 700E VDS in stowed position aboard Brazilian Mk 10 frigate Niteroi

MODEL 780 SERIES SONAR

Type
Hull-mounted and variable depth sonar systems.

Development
The Model 780 Series has been developed by the EDO Corporation for sale on the international market.

Description
The EDO Model 780 Series sonar is a family of high performance computer-based sonars designed to maximise the ASW capability of ships ranging from high-speed patrol craft to ASW frigates. Configured as a modular system, the Model 780 Series can be assembled to match ship-size constraints and required ASW capability for hull-mounted or variable depth sonar operation. The following models are available:

780:	13 kHz variable depth sonar (VDS)
786:	13 kHz hull-mounted
795:	5 kHz hull-mounted
796:	7 kHz hull-mounted
7860:	Combined 13 kHz hull-mounted with 13 kHz VDS
7867:	Combined 13 kHz hull-mounted with 7 kHz VDS
7950:	Combined 5 kHz hull-mounted with 13 kHz VDS
7960:	Combined 7 kHz hull-mounted with 13 kHz VDS
7967:	Combined 7 kHz hull-mounted with 7 kHz VDS.

The system features simultaneous active and passive search, tracking and classification. The operator can conduct 360° active/passive search operations while automatically tracking and analysing targets of interest. Display formats include panoramic B-scan detection, panoramic B-scan tracking and identification, expanded range versus speed target Doppler indicator, expanded range versus bearing sector scan indicator, 360° passive bearing versus time search, expanded passive bearing versus time bearing deviation indicator, passive analysis target noise indicator and acoustic performance raypath plots. Alerts of hostile sonar emissions detected by the integral acoustic intercept receiver are prominently displayed on the CRTs. A variety of transmission modes allows the operator to select the appropriate transmission mode for the tactical situation.

Recent at-sea tests of the new linear FM coherent signal processor have resulted in exceptional long-range detections under difficult high reverberation, shallow water conditions.

The Model 780 Series' electronics are flexibly designed. The latest Model 796 Mod 1 sonar has been adapted to use a customer-furnished standard console in place of the original control console. This system incorporates the latest improvements, among which are automatic multiple target tracking, multi-colour display, increased target capacity and flexible software-based display formats.

The VDS versions have exceptionally lightweight handling systems which permit installation on ships as small as 250 tons. The Model 787 features a new design compact 7 kHz transducer which has the

780 series VDS

Model 780 sonar system equipment comprising sonar control console, data storage and control, transmitter/power supply and sonar receiver

same weight and dimensions as the 13 kHz Model 780. This provides for greater detection ranges without the weight and space penalties associated with conventional 7 kHz VDS systems. The Model 780/787 VDS tow body is very similar to the AN/SQS-35 tow body and is therefore capable of towing the AN/SQR-18A(V)1 tactical towed array system.

The combined hull-mounted/VDS systems are also single operator systems. The combined electronics allow the one operator to simultaneously operate both the hull-mounted and VDS sonars including VDS depth control.

Operational Status
Hull-mounted and VDS versions are currently deployed operationally. The Model 796 Mod 1 is currently in production for an international customer.

Contractor
EDO Corporation, Government Systems Division, College Point, New York.

MODEL 4200 SYSTEM

Type
Obstacle detection sonar system.

Description
The Model 4200 obstacle detection sonar system with an improved soundhead, is a small DC-powered, high resolution sonar capable of detecting underwater targets such as buoys, pilings, piers and other underwater objects. The sonar set consists of two units, the soundhead and the control electronics. These are mounted in separate pressure housing and are connected by a single cable.

Electronic improvements have resulted in lower receiver self-noise, narrower receiver bandwidth and improved beam forming. The implementation of both has resulted in better sensitivity and better target definition for all horizontal and vertical beams.

Operational Status
In production.

Contractor
EDO Corporation, Electro-Acoustics Division, Salt Lake City, Utah.

Model 4200 obstacle detection sonar system

MODEL 984 SONAR SYSTEM

Type
Precision altitude sonar system.

Description
The Model 984 precision altitude sonar system is designed to track the altitude of a submersible or towed vehicle and to provide precise information for input to devices such as automatic altitude control systems or collision avoidance systems. The Model 984 is capable of providing depth-off-the-bottom information with an accuracy of 0.2 per cent, and a resolution of 0.1 ft over a range of 5 to 450 ft. Thirteen bits are output serially at a rate of one per second. The equipment operates with either the internal clock keying the transducer and updating the altitude data once per second, or on an internal key input at a maximum of eight per second. The Model 984 receiver contains the automatic time varying gain circuitry to compensate for the spreading loss of sound in water.

Model 984 consists of the transducer assembly and the control electronics assembly. The transducer assembly contains the tuning elements and is connected to the electronics assembly by a 10 to 100 ft cable.

Operational Status
In production.

Contractor
EDO Corporation, Electro-Acoustics Division, Salt Lake City, Utah.

Model 984 sonar system transducer

5951 SUBMARINE CLASSIFICATION SONAR SYSTEM

Type
Side-scan sonar system for detection of bottomed or hovering submarines.

Description
Klein side-scan sonar systems provide the ASW operator with the capability to classify bottomed or hovering submarines. The sonar provides high resolution images of the target area permitting high probability classification of targets that standard low frequency ASW sonars do not have the resolution to resolve.

The 5951 side-scan sonar system provides the operator with sufficiently high resolution images of the target area to resolve bottomed or hovering submarine targets from geological or biological targets without difficulty.

The system consists of a lightweight variable depth sonar (VDS) transducer which is towed in the vicinity of the target to be classified. Sound energy is projected outward from the transducer in a very narrow horizontal beam pattern, 1° or less depending on frequency of operation, to provide extremely high along-track resolution with a resulting high definition image of the target. The vertical beam pattern is large, typically 40° to provide the maximum insonification of a large volume of the sea.

The 5951 sonar system is lightweight (weighing less than 100 kg (225 lb) complete, including the VDS transducer), operates on 110/220 V AC or 24 V DC and consumes less than 200 watts of power, permitting operation from various platforms as large as destroyer type vessels down to as small as 10 m patrol craft. Due to the low weight and power requirements, installation of the sonar system will have minimal impact on the stability of the vessel or the power generation system.

Various operating frequencies are available, including 50 kHz for long-range, medium resolution

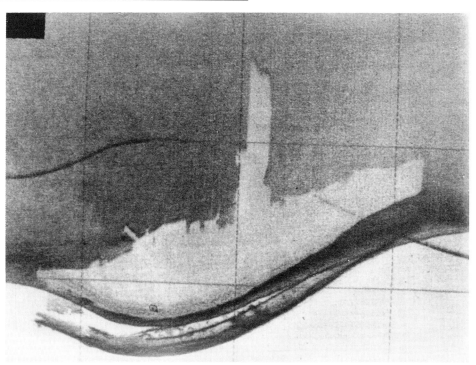

Record of a US Guppy I submarine showing stern quarter aspect (stern to the left of photograph) from a Klein 500 kHz side-scan sonar. Note the conning station on the forward (right) edge of the conning tower, and possible antenna bracket on rear of conning tower. The picture was gathered at a range of 50 m (164 ft) and depth of 35 m (115 ft)

operations, 100 kHz for high resolution operations, 500 kHz for very high resolution and simultaneous 100/500 kHz operation for multi-spectral imaging for resolution of the more difficult targets.

In addition to the ASW applications, the system can provide a secondary capability when used for the detection of mine and mine-like targets.

Operational Status
All versions of the sonar are currently in production and are in service with navies worldwide for anti-submarine warfare, mine countermeasures and oceanographic-related applications.

Contractor
Klein Associates Inc, Salem, New Hampshire.

SURTASS

Type
Mobile towed array surveillance system.

Development
SURTASS commenced full-scale development in 1976. Testing was carried out at sea in 1977. Technical evaluation of SURTASS was completed in March 1978 and operational evaluation took place in May and June 1978. The first full tests of the system, including the satellite link and data processing, took place in July 1979, and the system was declared ready for operational evaluation in March 1980. Introduction to US fleet service began in late 1984.

Description
SURTASS is the successor system to the AN/BQR-15 towed array surveillance system (TASS) already in operation with the US fleet. This mobile surveillance system complements fixed networks by providing the essential flexibility to respond to changes in Soviet submarine deployment patterns and by extending coverage to remote ocean areas not monitored by fixed systems. It can also serve as an emergency reserve facility if fixed networks are disabled. There are two shore-based data processing centres (one on the east coast of the USA and one on the west coast) to which acoustic target information gathered by the towed arrays is relayed by satellite. After processing, target information is transmitted back to operational ships at sea. Several ships are planned for each of the two processing centres, operating in waters to the east and west of the North American continent respectively.

Whereas the AN/BQR-15 is towed by submarines, slow surface ships will tow long SURTASS arrays back and forth over designated patrol lines. Funds have been set aside for a special type of platform to operate SURTASS, designated T-AGOS. More detailed information on these ships is given in *Jane's Fighting Ships*.

The basic portion of SURTASS is a hydrophone array at the end of a cable more than 1500 m long. Five AN/UYK-20 data processors and an AN/UYS-1 signal processor handle data processing on board the T-AGOS ships before the data are relayed via the satellite. The sonar element also includes the winch, handling equipment and associated electronics. The other three major elements include: the communication/navigation system for acoustic data transmission, command and control, and vessel navigation; the T-AGOS ships themselves and the shore processing stations. It is intended that the T-AGOS ships will operate worldwide on 90-day missions.

During the past five years a number of product improvements has been continued and introduced into the product line, and during 1984 a product improvement programme was implemented. This applied to the AN/UYS-1 processor and other classified units, and during 1985/86 specifications and designs for classification improvements for the AN/UYS-2 enhanced modular signal processor were developed.

Details concerning the Low Frequency Active Transmit Subsystem (LTS) can be found in the section on *Transducers*.

Operational Status

A total of 18 T-AGOS ships are now operational. In mid-1990 Hughes was awarded a $24 million contract for block upgrades to the basic SURTASS equipment. The US Navy called for a design revision in the T-AGOS programme in 1986 and the new 'Victorious' class ships are now completing to a Small Waterplane Area Twin Hull (SWATH) design. In mid-1990 Hughes was awarded a $45 million contract to provide additional capabilities to additional SURTASS systems on the new SWATH ships. In addition, five more ships, known as T-AGOS-23 types, are projected. Two additional T-AGOS ships are to be procured by the Japanese Maritime Self-Defence Force and Hughes has been awarded a contract to supply electronics equipment for these ships. They will be fitted with a complete SURTASS package and manned by a joint Japanese/US crew. They are scheduled to be operational in 1992.

Contractors

Hughes Aircraft Company, Fullerton, California.
TRW Systems, McLean, Virginia.

ASPRO COMPUTER

Type

Parallel/associative computer.

Description

ASPRO is a fully militarised parallel processing computer with a unique architecture which allows high speed computing capability. While simultaneously performing sophisticated tracking, correlation and data fusion algorithms, ASPRO can also process doctrine management statements that alert the operator of situations of interest. Doctrine management, a high speed function when implemented on ASPRO relieves operator workloads in complex situations.

The associative search capability of ASPRO allows complex searches, based on content, of its entire database (up to 32 Mbytes) in only a few instruction cycles. In addition, the parallel architecture allows thousands of operations to be performed simultaneously, dramatically reducing processing times.

ASPRO also contains a high performance MIPS R-3000 RISC sequential processor to perform independently very complex non-parallel tasks at a rate of 15 million instructions per second. This processor contains high speed instructions and data caches, 4 Mbytes of main memory plus Ethernet and dual RS-232 interfaces.

Operational Status

In operational service on 'Los Angeles' class submarines as part of the command and control system, and on the US Navy's E-2C maritime surveillance aircraft.

Contractor

Loral Defense Systems, Akron, Ohio.

ASPRO associative parallel processor

ASPRO-VME

Type

Parallel/associative computer.

Description

The ASPRO-VME open architecture parallel processor follow-on to the ASPRO associative processor has been developed in response to the dramatically increased need for higher throughput in command and control applications.

The ASPRO-VME fourth-generation parallel processing computer is a modular, open architecture, VME-compatible card set. It is capable of performing between 150 Mflops and 2.4 Gflops. The basic three-module 512-processor configuration can be expanded from 512 processor increments to 8192 processors. Each of the parallel processors, called processing elements (PEs), contains a full 32-bit IEEE floating point processor and a bit-serial processor.

A module occupies a single VME card slot and consists of two printed circuit boards attached to a cold plate. The entire ASPRO-VME in its basic configuration requires three VME slots.

Programmable in Ada and C, ASPRO-VME is supported by a powerful software development tool set which allows application programs to be developed easily in a high order language.

Because of ASPRO's single-instruction multiple-date stream architecture, modular expansion (512 PEs to 8192 PEs) can be accomplished without rewriting software.

The architecture delivers high speed, flexible computational capability. While performing sophisticated tracking, correlation and data fusion algorithms, ASPRO can also process doctrine management statements that alert the operator to situations of interest.

ASPRO-VME is available in a rugged and full MIL-SPEC configuration. It can operate in a stand-alone mode to be directly embedded in commercial and rugged workstations with 6 VME slots.

Applications for ASPRO-VME include: command and control; correlation and tracking; data fusion; database management; signal processing; artificial intelligence; expert systems; neural networks and image processing.

Operational Status

Pre-production units available Autumn 1991. Production begins Spring 1992 with first units available by mid-Summer.

Contractor:

Loral Defense Systems, Akron, Ohio.

CONCURRENT MODEL 6450 COMPUTER

Type

Onboard multi-processing computer system.

Description

The dual processor Concurrent Model 6450 computer acquires data from hydrophones in real time, processes them in a shipboard installation, and provides immediate information. It has been employed in a research role that involves acoustic analysis of underwater sound propagation. In this application, data from multi-channel arrays of hydrophones positioned in various oceanic regions are collected in real time by both shipborne and land-based computers. The data are collected and analysed and the basic research results are used to aid in the development of systems to detect and locate submarines.

The Concurrent Model 6450 computer as configured for this project features:

(a) Dual 68030 CPUs, one for data acquisition and transfer, the other for real-time digital signal processing.

(b) Lightning Floating Point Accelerator for performing FFT operations in real time

(c) an AD 12V26 analogue-to-digital converter, two SH 12V26 simultaneous sample and hold modules for setting up data acquisition experiments

(d) a 19 in colour graphics processor for contour plots; transmission loss plots; and colour, two dimensional plots of signal versus transmission loss relative to distance

(e) 64 channels (expandable to 1024) with a data transfer rate up to 30 kHz per channel

(f) a plotter for hard copy printouts.

Operational Status

In operational service with the US Naval Ocean Research and Development Activity (NORDA).

Contractor

Concurrent Computer Corp, Trenton Falls, New Jersey.

DSP SONA-GRAPH ANALYSER

Type

Analysis station for acoustic signals.

Description

The DSP Sona-Graph is a complete signal analysis workstation for the acquisition, analysis and display of acoustic events. It combines the features of a real-time spectograph, a computer-based acquisition system (with PROM-based software) and a high speed, dual-channel FFT analyser to provide a complete system approach. It uses a combination of general microprocessors and special digital signal processing chips.

Any signal in the frequency range from DC to 32 kHz can be analysed in real time, which means that the signal selected by the operator is displayed instantaneously on a high resolution monitor during input. The DSP Sona-Graph offers a wide selection of analysis techniques. Included are waveform, LOFAR, Sonagram, amplitude, power spectrum and waterfall displays.

Features of the programmable and expandable DSP Sona-Graph include:

(a) real-time operation with simultaneous acquisition, analysis and display
(b) time varying signal analysis
(c) high speed digital signal processing
(d) dual channel analysis
(e) complete signal conditioning including digital anti-aliasing filters
(f) high resolution graphics system
(g) sampling rates to 81 Hz per channel
(h) 2 or 8 Mbytes of signal storage
(i) over 1 Mbyte of PROM memory with signal processing programmes
(j) high speed FFT analysis
(k) fast SCSI interface to other computers
(l) interactive cursors for quick, precise measurements.

Contractor

Kay Elemetrics Corporation, Pine Brook, New Jersey.

SUBMARINE SONAR SYSTEMS

FRANCE

DUUA-2A ACTIVE/PASSIVE SONAR

Type
Active and passive sonar for search and attack.

Description
The DUUA-2A sonar is fitted to modernised 'Daphne' class submarines and 1200 tonne submarines, and provides for simultaneous surveillance and attack. It can be used:
(a) for active detection in single (FP) or frequency modulation (FM) modes
(b) for passive detection
(c) as an interceptor (for location of a sonar source).
    The DUUA-2A may also be used for ultrasonic

communications purposes and for depth sounding in very deep water.

Operational Status
In service in the French 'Daphne', and 'Narval' classes of submarine. A number of other navies equipped with Type 209 submarines are also fitted with this sonar. These include Ecuador, Columbia, Greece, Peru and Venezuela. No longer in production.

Specifications
Transmitter: Frequency 8.4 kHz
Power: 20 kW in nominal operation; and 1 kW in reduced operation, SFG and IC. Each emission is manually triggered. Emission duration: 30 ms in SFG; 300 ms in FP; 500 ms in FM

Receiver
Pass band: ± 170 Hz in FP; ± 350 Hz in FM in the neighbourhood of 8.4 kHz; ± 500 Hz in EBM in the neighbourhood of a selected frequency between 2.5 and 15 kHz
Range scale in active mode: 3, 6, 12 and 24 km
Transducer: Directional, Type B 88T 8.5, driven by servo mechanism MSS 5 directivity on site and bearing 2 Θ^3 = 10° at 8.5 kHz. Manual orientation by crank on azimuth or automatic orientation at 4 or 8°/s from −175 to + 175° on bearing. Reset between stops at 60°/s. Site positioning from + 15 to −30°

Contractor
Thomson Sintra Activités Sous-Marines, Valbonne Cedex.

DUUA-2B SONAR

Description
The DUUA-2B sonar can be fitted in all types of submarine and provides for simultaneous search, attack, sonar interception and depth sounding in deep water.
    Two versions exist:
        Model I: active sonar
        Model II: passive sonar.
    Either one or two sonars can be fitted in each submarine, but if two active sonars are fitted only one transmitter need be employed.

Operational Status
In service with 'Agosta' and 'Rubis' class submarines. No longer in production.

Specifications
Transducer: Directional. Manual or automatic search in continuous rotation
Transmitter
Frequency: 8 kHz
Power: 3 kW
Pulse: FP or FM mode
Receiver: In passive mode, from 2.5 to 15 kHz

Contractor
Thomson Sintra Activités Sous-Marines, Valbonne Cedex.

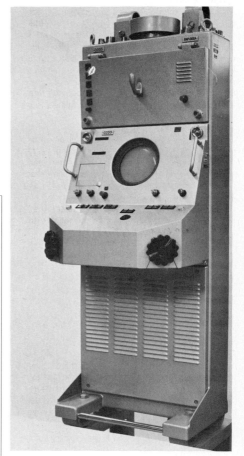

Control console of DUUA-2B active/passive sonar

DUUX-5 (FENELON) SONAR

Type
Passive panoramic surveillance sonar.

Description
Fenelon is the name given to a panoramic sonar capable of automatic and simultaneous tracking of three targets. The equipment incorporates a passive acoustic rangefinder for measuring the range of three targets within 120° sectors, and a panoramic sonar interceptor measuring the true bearing of all sonar transmissions received within the 2 to 15 kHz band.
    The DUUX-5 equipment was developed by Alcatel as a successor to the DUUX-2 series, of which at least 120 equipments were built and supplied to 14 navies for fitting in eight types of submarine. The Fenelon equipment enables range and bearing information on targets to be obtained by the submarine without the need for any transmissions and with minimum delay. Speed of operation permits target course and speed to be computed rapidly, also allowing any changes in either speed or course to be detected without delay. Four targets can be tracked simultaneously (three on radiated self-noise, one on sonar pulses), and there is a continuous panoramic bearing display over 360°. Range information is provided over arcs of 120° on each side of the submarine. There are facilities for transmission of target data automatically to the ship's weapon control system and plotting table.
    Performance characteristics include high accuracy and discrimination, immunity against sonar pulse interference, simplified calibration, and integrated test facilities. Under normal conditions, results in the

Fenelon: DUUX-5 control and display console

middle sector are 0.3° for bearing accuracy, 2° discrimination accuracy between two targets and 5% of range on radiated noise for a target at 10 km distance. Two types of hydrophones are available. The hydrophonic unit is composed of two bases with three hydrophones each on the starboard and port sides of

the submarine. All other technical details remain classified.

Operational Status
In series production. Replaces DUUX-2 passive acoustic rangefinder. Fitted to 'L'Inflexible' and 'Agosta' class submarines.

Contractor
Thomson Sintra Activités Sous-Marines, Valbonne Cedex.

ELEDONE SONAR SYSTEMS

Type
Passive and active sonar systems for submarines.

Description
Eledone is a family of modular integrated sonar systems for use in submarines. These systems are designed to fit any size of submarine and to provide for any type of operational requirement. As an

example of this the family includes the DSUV 22 and Amethyste systems for the French Navy, Argonaute for the Royal Navy and Scylla for the Royal Australian Navy. The new generation Eledone, designated TSM 2233, has the same operational functions as the

Eledone Sonar System, but is enhanced by new generation components and increased modularity. This increased modularity and open architecture make the TSM version particularly well suited to retrofit programmes and offers substantial growth potential for new systems.

Simultaneous handling of various functions is achieved by the use of multi-display workstations, with high definition and colour displays, which integrate all relevant data. Built-in redundancy, performance monitoring and fault localisation facilities ensure a high level of availability.

Eledone systems provide a set of the following functions:

Passive detection
This facility gives a high sensitivity, fine bearing accuracy and high flexibility of installations in all types of submarines. The number of hydrophone staves can be selected; either 32 staves for small diameters down to 1.2 m, 64 for large diameters up to 3.5 m, or 96 staves for the largest.

Automatic anti-jamming
The anti-jamming feature automatically rejects narrowband jammers, thereby hardening trackers and enhancing CMA performance (range, accuracy and convergence delay).

Passive adaptive processing
The adaptive processing function is based on optimal array processing theory which minimises the effect of jamming by strong signals when listening to low level signals. It is a most useful facility in the discrimination between two targets when both are within a common limited sector.

Automatic detection and tracking
By automatically initiating the trackers, this feature allows the operator to avoid repetitive tasks and cope with more complex situations.

Interception
This function allows interception of all active sonar pulses from low frequency surface ship sonars to high frequency torpedo acoustic heads. Very early warning against enemy sonars, with a low false alarm rate is provided. Interception warnings are integrated on the passive listening scope with specific colour (visual correlation). Accurate parameters of pulses are presented on digital readouts and systematically transmitted to the tactical system.

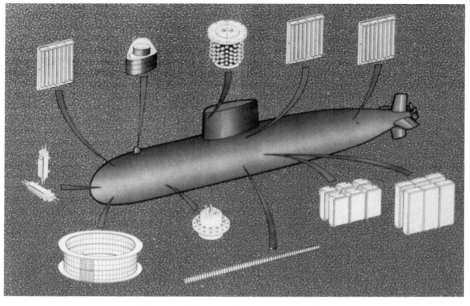

Diagram of Eledone submarine sonar system

Active capability
Although initial bearings are provided by the passive detection function, the operator can also initiate a low frequency wide sector transmission on the same bearing to determine accurate range. In this mode, reception is performed by the passive listening array, and associated with high detection sensitivity.

Classification
The spectrum analysis facility complements the audio function by providing the operator with specific data in target identification. This analysis can be performed in real time or played back from memory store. Paper recorders, tape recorders, cavitation recorders, additional mass memory data and other facilities are provided.

Hostile weapon alarm
A specific algorithm sorts the contacts by taking into account azimuth and level variations, and warns the operator in the event of a hostile weapon.

Contact motion analysis
Contact localisation and motion analysis of contacts are performed automatically on all tracks.

Operational Status
The Eledone series of submarine sonars is in widespread worldwide service with a number of navies. The French Navy operates the DSUV 22 and Amethyste versions and the Argonaute version is fitted to the first batch of the Royal Navy 'Upholder' class submarines under the RN designation of Type 2040. The Scylla variant has been ordered by the Royal Australian Navy for its new Type 471 submarines, and the Octopus version for the Royal Netherlands Navy 'Walrus' class.

Contractor
Thomson Sintra Activités Sous-Marines, Valbonne Cedex.

VELOX M7

Type
Sonar intercept receiver.

Description
The Velox M7 sonar intercept receiver is designed to intercept signals in the frequency range 2.5 to 100 kHz. Detected signals are automatically measured to provide direction, level, frequency and pulse length and to identify recurring identical pulses. The system is also capable of monitoring the noise radiated by ships.

The receiver is designed for integration into a federated combat system, or can be used in a stand-alone configuration.

The receiver uses multi-band processing based on a constant false alarm rate operation and provides a permanent digital display of the parameters of the last two intercepts; synthetic analogue display of the threat direction; extensive audio facilities, including direct, compressed or heterodyned, and anti-saturation features.

Velox M7 comprises a small receiving array and an electronics cabinet or rack drawer. A second identical array can be installed in another location for manual selection to improve the horizontal coverage.

Operational Status
In service with the French Navy.

Contractor
Safare Crouzet, Nice.

GERMANY

ATLAS ELEKTRONIK SONARS

Description
These sonars are designed for installation on various types and sizes of submarines. The sonars are built to a modular concept and depending on the particular application and size and task of the submarine, a wide range of versions can be used, either as stand-alone systems or in combinations. The main equipments include: passive sonar (PSU); active sonar (ASU); intercept sonar (ISK); passive ranging sonar (PRS); flank array sonar (FAS) and towed array sonar (TAS).

These sonars provide some or all of the following functions, depending on the individual type of sonar: detection and bearing of target noise in various frequency bands; automatic target tracking; analysis of intercept target noise; passive ranging; target motion analysis; detection, measurement and analysis of pulses from other sonars; active ranging in special tactical situations; data transfer to and from weapon control equipment, including navigational and status data; recording and storage of target data for post mission analysis.

Contractor
Atlas Elektronik GmbH, Bremen.

CSU 90 SONAR SYSTEM

Type
Attack, intercept and flank array sonar.

Description
The basic version of the CSU 90 submarine sonar involves three main sonars; an attack panoramic passive sonar, an intercept passive sonar and a low frequency flank array sonar. The attack sonar has a 2.8 m cylindrical hydrophone array and full azimuth coverage, a DEMON detection facility uses the array of the attack sonar, and the flank array is used for long-range detection. Any target detected by one of these sonars initiates an automatic process of data collection.

Automatic Target Tracking (ATT)
CSU 90 is equipped with eight independent compound tracks. Each provides the warfare system with a complete set of tactical data. Tracks are automatically incorporated in a compound track and each sensor can initiate a compound track. The total set of compound tracks comprises:
(a) eight ATT channels for the attack sonar

(b) eight ATT channels for the DEMON path, each with eight ALTs
(c) the Target Motion Analyser (TMA) which controls eight compound tracks and generates a sonar tactical display
(d) TMA controls providing further manual tracks
(e) flank array sonar with eight ATT channels and eight MTs with eight ALTs each.

Optional extensions include a passive ranging sonar, an own noise analyser, acoustic passive classification, automatic warning channel, target motion analyser, a disk memory unit, plotting devices and simulation facilities. The disk memory unit allows all displays and sonar data to be recorded for subsequent analysis and for training. It also handles the data for comparison purposes of the sonar information processor and the classification file for the intercept sonar.

Operational Status

The CSU 90 has been ordered as part of the modernisation programme of four Greek Type 209 submarines and the two Chilean 'Oberon' type submarines as well as for the new A19 type submarines of the Swedish Navy.

Contractor

Atlas Elektronik GmbH, Bremen.

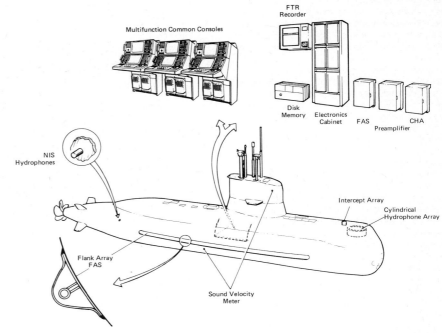

System layout of the CSU 90 submarine sonar

CSU 83 SONAR SYSTEM

Type

Passive/active sonar system.

Description

The CSU 83 (DBQS-21) submarine sonar is intended for boats of 400 tons and upwards. It is the primary sensor of the submarine fire control and command system of the German Navy. It comprises a passive bow array operating in the 0.3 to 12 kHz band, a passive ranging array and a sonar intercept array. In addition, two sensors to detect the submarine's own noises are included.

The packaging of the electronics is based on a modular design and provides easy access to all components for servicing, as well as low size and weight. The basic features of the CSU 83 are:
(a) high probability of detection over 360°
(b) computer-aided processing and evaluation
(c) discrimination between data on integrated displays
(d) simple operation by two operators via interactive, colour-coded control/displays
(e) sequential digital signal processing
(f) automatic fault detection and localisation.

Operational Status

The CSU 83 has been selected for modernisation of 12 Type 206 SSKs of the German Navy and is being installed in new-build Brazilian, Norwegian and Swedish boats.

Contractor

Atlas Elektronik GmbH, Bremen.

PSU 83 SONAR

Type

Passive surveillance and detection sonar.

Description

The PSU 83 is designed for new submarines of small and medium sizes, and can also be retrofitted to older submarines to improve their detection facilities. This is achieved by the use of a newly developed programmable and north-stabilised beam former which can handle all types of hydrophone arrays. High technology pre-processing units with newly designed pre-amplifiers and filter units ensure sensitive and undisturbed reception.

The very good detection performance is achieved in two independent detection channels: one for surveillance tasks with short integration times, and the other for long-range detection using longer integration times. Both detection results are displayed simultaneously to provide a comparison facility. The sonar is equipped with eight independent automatic target tracking channels.

For redundancy purposes and to improve performance, the pre-processing unit and the beam former are split into two independent electronics units. Additionally, a redundant beam former and a separate direct signal path to the operator's console are provided. A built-in target data analyser and a bearing prediction facility improve the operational value of the sonar.

All the signal processing electronics of the PSU 83 are combined in one operator's console and two pre-processing units. The operator's console consists of software-driven buttons, trackerball for accurate cursor movement, high resolution raster scan colour display and a video output for slave display. For extended functions the operator's console is available with two displays.

The PSU 83 can be extended with a number of options, including:
(a) intercept sonar
(b) flank array sonar including LOFAR detection, tracking and analysis
(c) DEMON detection, tracking and analysis
(d) mass storage unit for target data and display recording
(e) sonar graphic recorder
(f) onboard training simulator
(g) own noise analysis.

When combined with the FSU fire control system, the PSU 83 forms the complete submarine warfare system OSID 83.

Operational Status

Entering service or on order for Danish, German and Norwegian submarines (forming part of the CSU 3-4

PSU 83 operator console

outfit on the latter) and Egyptian 'Romeo' class submarines.

Contractor

Atlas Elektronik GmbH, Bremen.

FAS 3-1 SONAR

Type

Flank array sonar.

Description

The FAS 3-1 flank array sonar is designed to perform long-range early detection and classification of enemy surface ships and submarines by picking up their low frequency noise emissions, and carrying out classification on the basis of evaluated LOFAR/DEMON/SPECTRUM data. For directional long-range detection the FAS 3-1 has two flank arrays (linear arrays mounted at the sides of the pressure hull). Because of the direct coupling to the submarine, the flank arrays are specially designed for maximum decoupling from the unwanted noise.

The system consists of the two flank arrays, electronics cabinet, operator's console and recording systems.

The system covers the frequency band from 10 Hz

to 2.5 kHz. The arrays are 20 to 48 m long, depending on the length of the submarine, and each has 96 staves with a total of 288 hydrophones. Bearing accuracy is 1° on a port/starboard sector of 90° between relative bearings port/starboard 45-135°. This can be extended with reduced performance to port/starboard 10-170°.

Operational Status
In service with the Swedish Navy.

Contractor
Atlas Elektronik GmbH, Bremen.

TAS 83 TOWED ARRAY SONAR

Type
Very low frequency towed array sonar system.

Description
Atlas Elektronik has developed a towed array sonar that fulfils all the requirements of submarine sonar for long-range detection, target classification and target motion analysis.

The main features of the sonar are:
(a) early detection of targets
(b) independence of bathythermal conditions in that the towed array can be operated in the optimum layer
(c) detection of the acoustic spectral lines of quiet submarines
(d) detection of transients produced by submarines
(e) detection of low frequency broadcast noise by cavitating surface vessels

(f) semi-automatic classification by acoustical analysis of spectral lines, transients and broadband noise
(g) target motion analysis computed by algorithms specially adapted to the towed array sonar
(h) ability to be fitted on existing hulls using the 'clip-on/clip-off' technique

To perform the above tasks the TAS 83 can be fitted with two different 'wet ends', and task-oriented electronics. Signals received are fed to a north-stabilised beam former which creates 128 different beams over the azimuth. Separate paths are used for signal enhancement for the detection display and audio channel. Up to eight targets can be tracked automatically and a separate signal path is used for frequency analysis using DEMON and LOFAR techniques. Automatic multi-line trackers can be set on characteristic frequency lines of the target's signature. Together, the automatic target tracker and multi-line tracker can follow a target even at long ranges or

under extremely severe conditions such as target crossing or convoy acoustic environment. The target's shaft revolutions per minute and the number of propeller blades are computed in the data enhancement section. The built-in multi-point divider, cursor and other support measures assist the operator in defining and classifying the type of engines, gearbox ratio and other information in order to classify the target. This is achieved using LOFAR, DEMON and SPECTRUM survey analyses, the comparisons of LOFAR and DEMON grams, transient noise detection, audio analysis/classification and bearing/time record.

Operational Status
Fully developed.

Contractor
Atlas Elektronik GmbH, Bremen.

CSU 3-4 SONAR

Type
Active and passive search and attack sonar.

Description
The CSU 3-4 is a conventional submarine sonar designed for a single operator. It is a fast-scanning medium-range active sonar which can be used as a long-range passive sonar, an intercept sonar and as an underwater telephone. In its active mode, the sonar operates using a single beam. The equipment has a cylindrical transducer array with electronic beam steering. The sonar is equipped with automatic interference suppression against acoustic countermeasures.

The CSU 3-4 is intended primarily for shallow water use, and is able to track four targets simultaneously. The basic functions of the CSU 3-4 are described below.

Passive mode
(a) simultaneous azimuth coverage over 360°
(b) panorama presentation of target information on a CRT display with true or ship-relative bearing
(c) bearing/time recorder for long-term plotting of all contacts
(d) high bearing accuracy and angular resolution
(e) automatic target tracking of up to four targets simultaneously.

Active mode
(a) single ping operation
(b) presentation of a 30° sector (waterfall) on the CRT, with continuous memory-refreshed display of target echoes

(c) reverberation threshold for high detection performance
(d) magnified CRT display
(e) cylindrical transducer array with electronic beam steering.

Intercept mode
(a) presentation of up to eight external pulses at the centre of the CRT
(b) additionally, a numerical indication of bearing, frequency, signal level and pulse duration of one of the eight pulses is provided
(c) determination of target elevation angle for estimation of distance and optimum sound ray path
(d) presentation of the received pulses on the audio channel.

Data received by the CSU 3-4 are processed and analysed by the SIP 3 sonar information processor (see separate entry).

The system is capable of evaluating sonar data in the following ways:
(a) analysis of noise spectra in various frequency bands
(b) automatic line tracking
(c) automatic manoeuvre detection
(d) high resolution of target bearings
(e) automatic target tracking of targets set by the sonar, including target crossing situations
(f) measurement of range, bearing and Doppler in the active mode.

Using these features the system can determine certain target parameters such as speed, propeller shaft revolution rate and number of blades or a characteristic low frequency spectrum. These data can be stored in an optional database for later recall for comparison with current target data.

Control/display unit of the CSU 3-4 sonar

Operational Status
In operational service or entering service with the Australian, Argentine, Chilean, Danish, Egyptian, German, Indian and Norwegian navies.

Contractor
Atlas Elektronik GmbH, Bremen.

PSU 1-2 SONAR

Type
Passive sonar for surveillance and detection.

Description
The PSU 1-2 is a passive sonar for use in small conventional submarines. The equipment has the capability for detection, classification and bearing determination of noise radiating targets with

simultaneous azimuth coverage over 360°. The target noise information received by the cylindrical hydrophone array (CHA) is processed into 32 preformed beams and presented on a CRT display in true or ship's relative bearing. Four targets can be automatically tracked using the ATT equipment. Bearing angles of targets appear alphanumerically on the CRT display, either north-relative or ship-relative, as desired. Two targets can be transferred to the fire

control system and one target can be transferred to the plotting table.

Operational Status
In service with the South Korean, Yugoslav, Pakistani and Argentine navies.

Contractor
Atlas Elektronik GmbH, Bremen.

PRS 3-15 SONAR

Type
Passive ranging sonar.

Description
The PRS 3-15 is a submarine passive ranging sonar

designed to be operated in conjunction with the CSU 3-4 (see separate entry), or as an independent system. It can be used for the passive detection of targets and for passive rangefinding on submarines. The 15-stave array with a high directivity index provides a good signal-to-noise ratio, and suppression of adjacent targets that are not of interest. Six individual

arrays are mounted along the hull, three on the starboard side and three on the port side.

The PRS 3-15 performs the following functions:
(a) panoramic detection with high array gain
(b) measurement of target bearing
(c) automatic tracking of target bearing on up to four targets simultaneously

(d) range measurement on up to four targets
(e) calculation of target course and speed without the need for manoeuvres by own submarine
(f) narrowband audio channel via PRS array or CSU cylindrical array
(g) automatic calculation of a confidence factor for the range data.

The signals received by the arrays are correlated with each other, the sampling values of the correlation function are integrated according to range and target dynamics, and are then analysed by the processor in order to estimate the time delays of the signals. From the time delay measurements the range is calculated. In addition, the variance of the time delay estimations is determined and converted to a confidence factor of the range value.

Operational Status
In operational service.

Contractor
Atlas Elektronik GmbH, Bremen.

OSID SONAR

Type
Integrated passive panoramic sonar.

Description
OSID is an integrated sonar system developed for use in small submarines. It comprises a passive sonar, data distribution with a tactical display, and a torpedo guidance system. The sonar has a broadband channel for bearing determination and classification, and utilises minimum correlation techniques for the location of weak targets. It features a 360° azimuth panoramic presentation of targets on a CRT display. The processing computer calculates the position and vectors of the target, own submarine and the torpedo, and the information is presented on a tactical display.

Contractor
Atlas Elektronik GmbH, Bremen.

SIP 3 SIGNAL PROCESSOR

Type
Sonar information signal processor.

Description
The SIP 3 is a sonar information processor developed for use with fast-scanning multi-target tracking sonars of the CSU series. The prime function of the equipment is to provide passive classification of targets using low frequency detection and analysis, spectrum survey and comparison, high target resolution and DEMON analysis. Doppler analysis of target echoes received by the active sonar of the own submarine is used to determine the target's speed. Sound raypath analysis produces an estimate of the conditions for sound propagation. The system has the capability to determine and present specific sonar signal characteristics, comparing them to sound signatures of selected targets held in a data bank.

Operational Status
In operational service with the Argentine Navy.

Contractor
Atlas Elektronik GmbH, Bremen.

ITALY

IPD70/S INTEGRATED SONAR

Type
Passive and active surveillance and attack sonar.

Description
The IPD70/S integrated sonar suite is an improved version of the IPD70 currently installed on board 'Sauro' class submarines, and is the result of a joint development by ELSAG and USEA. The system features a passive sonar, and an active sonar for search and attack, and various optional equipments such as:
(a) IN-100A high precision LF interceptor
(b) ISO-100 panoramic HF interceptor
(c) TS-100 long-range directional/omnidirectional underwater telephone
(d) MD-100 passive rangefinding sonar (see separate entry).

A digital console and processor with two high resolution raster scan monitors with image memories are used to control operations, display processed information and interface the command and control subsystem.

The IPD70/S sensor configuration includes:
(a) a large passive, bow-mounted conformal array for long-range, high accuracy surveillance, detection and tracking
(b) an integrated cylindrical array to perform the functions of passive sensing, passive interception, active sonar and underwater telephone
(c) the MD-100 linear array equipment for passive rangefinding
(d) a multi-mode sensor for panoramic broadband interception.

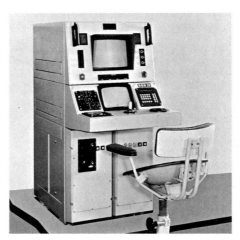

Control console for IPD70 integrated sonar

The passive sonar component provides for detection and tracking of targets (up to four per subsystem) over a wide frequency coverage, divided in adjacent bands. Two separate beam formers and a high resolution compensator allow long detection range with high bearing accuracy.

In the active mode, a panoramic search high-directivity sector coverage can be selected.

The IN-100A LF interceptor provides for detection of active sonar emissions at ranges of several tens of kilometres, and measurement of bearing, pulse length, frequency and repetition rate to classify enemy sonars.

The ISO-100 panoramic HF interceptor has the capability to detect sonar emissions in the medium and high frequency ranges, evaluating bearing, frequency, amplitude, length and repetition rates of intercepted pulses.

The TS-100 underwater telephone is used for long-range directional or omnidirectional communications. Operating frequencies and modulation techniques comply with NATO standards.

The receiving subsystem includes the CM10 set, a figure-of-merit evaluation equipment. Based on an estimation of the most significant acoustic parameters which affect the performance of both the active and passive systems, the CM10 provides synthetic data displayed at the operator's request.

The display console, based on two high performance ESA-24 digital processors, features an improved design man/machine interface, with two high resolution raster scan displays with memories. These provide presentation of easy-to-read graphic and alphanumeric data, and a variety of ergonomic controls (keyboards, trackerballs, etc) for single operator monitoring and management of all system components.

An extensive built-in self-test and diagnostic feature is provided.

Operational Status
In operational service with the Italian Navy.

Contractor
Elettronica San Giorgio – ELSAG SpA, Genoa.

MD-100 Mk 1 RANGEFINDER SET

Type
Passive rangefinding sonar system.

Description
The MD-100 Mk 1 is an upgraded version of the MD-100 and is the result of co-operation between the companies USEA and ELSAG. The system has been designed for installation on small or medium size submarines and can be supplied either as a stand-alone device or as an optional component of the integrated sonar set IPD70/S. In the latter case, all necessary space allocation, as well as command and display facilities, are shared on a common basis with the host system.

The operational capability of the equipment includes:
(a) passive panoramic, high resolution surveillance
(b) automatic passive tracking of targets in range and bearing.

The MD-100 Mk 1 sensor configuration consists of an independent set of acoustic sensors arranged as two linear arrays, each composed of three high directivity streamlined transducer arrays.

The signals collected by the transducers, after appropriate conditioning, are fed to the main unit for all the relevant stages of signal processing. Handling of signals entails the evaluation of correlation and coherence relationships existing among the various couples of waveforms involved, so as to obtain bearing and distance estimates of a noise source with high accuracy.

The MD-100 features simultaneous tracking in bearing and range of up to four targets.

The display unit, in both the stand-alone and in the integrated versions, is based on a high resolution raster scan type screen, which allows constant monitoring of trajectories and all relevant data (both in graphic and alphanumeric form) pertaining to the targets being tracked. In addition, continuous indication is given of special parameters, such as reliability coefficients of processed data, tracking losses and so on, which could be valuable in deciding the operator's course of action.

Operational Status
The integrated version is installed in the Sauro III submarines.

Contractor
Elettronica San Giorgio – ELSAG SpA, Genoa.

SARA SYSTEM

Type
Spectral analysis and classification system.

Description
SARA is a spectral analysis and classification equipment developed for Italian Navy submarines. The system can be supplied as an add-on to an integrated sonar set. The operational capabilities of the system are to perform narrowband analysis of acoustic signatures to identify and classify targets.

The processor, after initial conditioning, performs automatic and/or manual detection and extraction of spectral lines, via LOFAR and DEMON analysis, and automatic comparison, based on inference rules, of signatures stored in appropriate databases which are also generated during the process.

A high resolution colour display and associated processor is used to present the output of the analysis (LOFARGRAMS, DEMON, VERNIER) in operator-selectable formats. These can then be stored on magnetic tape, if required.

The total system configuration is composed of an onboard equipment and a land-based system for

further processing, archiving and updating of the database.

Operational Status
In development for the Italian Navy.

Contractor
Elettronica San Giorgio – ELSAG SpA, Genoa.

IP64/MD64 SONARS

Type
Passive and active search and attack sonar.

Description
The USEA IP64 sonar set is designed for installation aboard small or medium size submarines, and it combines passive sonar equipment operating in the band 175 to 5000 Hz, and an active echo-ranging sonar operating at a frequency of 4000 Hz. It is also designed for integration with the MD64 sonar set for range measurement of noise sources by passive means. The receiving transducer elements, which are common to the passive and active mode, are arranged in the bow of the vessel in a long conformal array to enhance the directional capabilities. The transmitting transducer elements are arranged in a circular array located on deck.

Two linear arrays are provided for range measurement by passive means. Each array has a total length of 30 m and is formed by three transducers. The arrays are located on deck, port and starboard.

Search and detection are carried out primarily by means of the passive listening set, which incorporates advanced correlation techniques and records the

received signals on a graph plotter. After target detection, a single transmission pulse in the direction of the target by the echo-ranging sonar is sufficient to provide range and relative speed target data.

The MD64 equipment measures the range of a detected noise source and direction is determined by the measurement of the relative time delay of sound waves detected by three hydrophones. Two groups of three hydrophones each are provided for port and starboard measurements. Each hydrophone uses a number of elements arranged in five vertical staves and phased in the horizontal plane to produce a reception beam, directed towards the source. Main lobe beamwidth is ±15°.

The audio frequency band employed is 1000 Hz wide, centred on any frequency between 4 and 20 kHz. Operation is automatic, and after acquisition of the noise source the equipment locks on and continuous bearing data are produced. Range is measured continuously and is recorded graphically.

Operational Status
No longer in production.

Contractor
USEA, La Spezia.

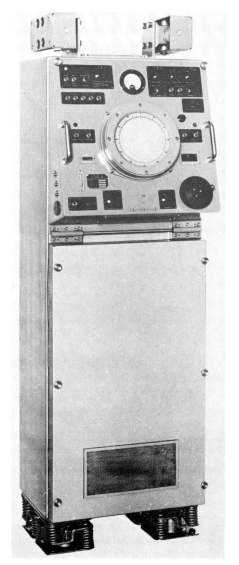

IP64/MD64 sonar display console

NETHERLANDS

SIASS-1 AND -2 SONARS

Type
Passive and active attack and surveillance sonar.

Description
The Submarine Integrated Attack and Surveillance Sonar (SIASS), developed by Signaal, is available in two versions: SIASS-1 and SIASS-2. The former is a stand-alone system with its own sonar consoles interfacing with a submarine data handling and fire control system. SIASS-2 is an integrated part of the Signaal Submarine Sensor Weapon and Command System (SEWACO).

Both SIASS-1 and SIASS-2 perform and integrate the following tasks:
(a) low (long-range) and medium (medium-range) surveillance of noise radiating targets through the use of broadband and multi-channel narrowband signal processing for surveillance, analysis and tracking (LOFARGRAMS, ZOOM-FFT, DEMON, ALI displays)
(b) passive range detection
(c) automatic target tracking for a multitude of targets, including automatic line tracking
(d) contact motion analysis (position determination on bearings-only information)
(e) active sonar
(f) LF, MF and HF intercept
(g) classification
(h) noise level monitoring and cavitation indication.

The sonar suite is divided into three groups of equipment, each with its own processing cabinet:
(a) a cylindrical passive array performing the functions of narrowband and broadband operation in the medium frequency band, low frequency acoustic intercept, and active operation by means of an associated active transmit array
(b) port and starboard flank arrays to carry out narrowband and broadband passive operations in the low frequency band
(c) port and starboard passive ranging arrays, together with a high frequency intercept sonar.

Each sonar can operate autonomously, in combination with one or more display consoles, if other parts of the system are unavailable. This allows, for example, the passive ranger to operate as a backup for the cylindrical array sonar. A flank array sonar which performs low frequency coverage and full primary sonar operations is available.

SIASS can store 35 tracks and performs automatic contact motion analysis on each sonar target being tracked, up to the maximum that can be handled by the system. The broadband subsystems of the cylindrical array and flank array sets each have four operating channels with eight preselected frequencies. The high frequency and low frequency acoustic-intercept sonars each have 16 channels, and the active sonar has four; in these cases tracking is based on the use of associated algorithms. The electronic classification for each sonar can hold information on up to 300 specific platforms.

The cylindrical array sonar acts as the primary attack sensor. A combination of broadband and narrowband processing is used to extend detection ranges, to enhance tracking, and to provide quick and accurate classification of contacts. Simultaneous panoramic coverage is achieved by the use of pre-formed beams.

The flank array sonar is a low frequency passive equipment that can act as the primary attack sonar within its frequency range. The port and starboard line arrays, used in conjunction with real-time delay beam forming, provide coverage of large bearing sectors on each side of the submarine. The processing chain is similar to that of the cylindrical array sonar and can be matched to interface with a towed array if required.

Each of the pre-formed beams produced by the cylindrical array sonar and the 64 beams produced from the flank array undergoes on-line spectral analysis in parallel, a complex fast Fourier transform being used to give reliable detection and tracking of noise emitted by a contact. The frequency-analysed pre-formed beam signals from the cylindrical array processor are used as a low frequency intercept sonar that can associate a large number of different transmissions simultaneously, and incorporates extensive algorithms to reduce false alarms.

The cylindrical array and flank array sonars also have enhanced classification channels for high resolution vernier analysis, and to enable frequencies below 100 Hz to be measured. They can also provide multi-channel automatic tracking.

The active sonar can be operated in directional single-ping, and omnidirectional single-ping or multi-ping modes. Return echoes are detected and processed by the cylindrical array sonar, and when operating in the multi-ping mode the active sonar can automatically track multiple targets.

The passive ranging sonar operates in the lower

SIASS display and computer consoles

part of the frequency band and is normally slaved to the cylindrical array. In this slaved operating mode, range is automatically calculated once a broadband contact that is being tracked on the cylindrical array has entered the ranging sector. The passive ranger's signal processing uses a method of frequency-domain correlation to measure wavefront curvatures and thus provide range information.

The high frequency intercept sonar provides coverage up to 100 kHz and employs a fully digital compressor using fast Fourier transform techniques to detect, correlate and classify intercepts of torpedoes or sonars, and to warn the operator automatically.

Control is exercised from multi-purpose consoles which may be integrated parts of a total SEWACO system. These consoles are multi-function operator workstations equipped with a high resolution raster scan colour display. As an optional extension, signal processing for a towed array can be offered.

Operational Status
In production and in operational service.

Contractor
Hollandse Signaalapparaten GD, Hengelo.

UNION OF SOVEREIGN STATES

SUBMARINE AND SUBMERSIBLE SONAR SYSTEMS

Many submarines of the former Soviet fleet feature prominent, large bow-mounted sonar arrays and it has been estimated that over 60 nuclear attack submarines have been commissioned, supported by as many as 93 diesel attack submarines. These are in addition to the large classes of nuclear and diesel ballistic and cruise missile submarines. *Jane's Fighting Ships 1992-93* lists a total of 265 submarines of various types, with a further 25 in reserve and 14 under construction. This is a truly formidable force, and all of these have some type of sonar installation.

Early sonars appeared during the Second World War, and by 1958 two sonar systems appeared that became standard on all following classes of submarines for nearly two decades. These were the Feniks and Herkules, and a 'top-hat' looking underwater telephone, code-named Fez. The wraparound Feniks HF bow array would require at least a crude compensator switch to steer listening beams. The medium-sized Herkules dome on deck was a combination active/passive system, and was virtually identical to that fitted in surface ships.

Four submarine classes had a larger size topside sonar dome and, since these classes all carried new 16 in ASW torpedo tubes aft, this dome was probably related to the fire control computers. Since passive

fire control solutions require triangulation it is quite likely that the vertical fin aft on many later submarine classes may include sonar hydrophones.

The November class SSNs were the first with a rakishly streamlined bow array, although this was probably of a medium frequency type. The first low frequency sonar appeared in the early 1960s on the 'Echo II' and 'Juliet' classes, and by 1967 streamlined bow shapes with sharply sloped low frequency bow array windows had appeared.

The 'Victor III', 'Sierra' and 'Akula' have a teardrop shaped pod on top of the aft rudder fin. Most references state that it houses either a towed array indicating a high level of signal processing power and beam steering, or perhaps a wire communications antenna. The later 'Oscar' class also has a vertical aft fin that has towed array capability.

Very few details of submarine sonar systems have emerged and the list below gives only the types of sonar that are understood to be fitted to various classes of submarines. All systems are hull-mounted passive/attack sonars, the principal difference between classes being the frequency. No information is available on the use of towed arrays by USS submarines. There is also a large number of smaller submersible vessels, both manned and unmanned, and some of these undoubtedly have sonar systems, in particular the unmanned types for mine countermeasures purposes.

'Foxtrot' class submarine bow sonars

Submarine class	Sonar frequency
'Typhoon' SSBN	Low/medium
'Delta IV' SSBN	Low/medium
'Delta III' SSBN	Low/medium
'Delta II' SSBN	Low/medium

Bow sonar on 'Kilo' class submarine

'Golf' class submarine bow sonars

'Foxtrot' class submarine bow sonars seen from abeam

'Sierra' class SSN bow with missing plates behind bow sonar belt

Large sonar arrays feature prominently on bows of Soviet submarines

'Juliet' class submarine bow sonars

Submarine class	Sonar frequency
'Delta' I SSBN	Low/medium
'Oscar' SSGN	Low/medium
'Papa' SSGN	Low/medium
'Charlie II'	Low/medium
'Charlie I'	Low/medium
'Echo II'	Medium
'Yankee' SSGN	Low/medium
'Juliet' SSG	Medium/high
'Akula'	Low/medium
'Sierra'	Low/medium
'Alfa'	Low/medium
'Victor III'	Low/medium
'Victor II'	Low/medium
'Victor I'	Low/medium
'Echo I'	Medium
'November'	Medium
'Kilo'	Medium
'Tango'	Medium. There is also a large array mounted above the torpedo tubes, as well as a bow-mounted dome
'Foxtrot'	High
'Zulu IV'	High
'Whiskey'	High
'Lima'	High
'India'	High

UNITED KINGDOM

TYPE 2007 SONAR

Type
Passive search and detection sonar.

Description
The Type 2007 is a conventional submarine sonar which uses steerable beams, enabling the submarine to maintain a straight course while the beams are steered to search the waters for detection and tracking of hostile submarines. The Type 2007 search beams listen at a frequency of between 1 and approximately 3 kHz and are used for initial long-range target detection and for gaining a bearing resolution. The sonar listens for propeller cavitation, nuclear cooling pumps and turbine reduction gears for initial target detection.

The Type 2007 is a more modern version of the original Type 186 passive conformal hydrophone system. This latter equipment consisted of a series of hydrophones evenly spaced along both sides of the hull.

Operational Status
The Type 2007 was installed in 'Oberon' class submarines operated by a number of countries, including Australia, Brazil, Canada and Chile, as well as the Royal Navy. It has also been fitted to the Royal Navy's 'Trafalgar' class submarines. Still in operational use.

Contractor
Graseby Marine (Graseby Dynamics Ltd), Watford.

TYPE 2020 SUBMARINE SONAR

Type
Passive and active bow-mounted sonar.

Description
The Type 2020 is one of Marconi's range of sonar equipments. Fitted in the Royal Navy's hunter-killer nuclear submarines, it provides both passive and active detection capabilities from its bow-mounted array. Although considerable emphasis has been placed on passive detection, the equipment incorporates a highly effective computer-aided active capability.

Passive detection is subdivided into low frequency and high frequency bands with information presented on dual speed multi-pen recorders. The system is intended primarily for operation in a computer-aided mode but can be automatically reconfigured to a fall-back mode of operation, with target information being passed to the action information organiser in both modes. Facilities for expansion are included in the design to allow interfacing with future sonar systems. Solid-state switching is used for the transmit/receive function and electronic beam steering techniques are employed. The design provides high detection sonar performance with a full 360° coverage within a frequency range reported to be within 2 to 16 kHz for passive operation. The active transmitter uses an automatic beam steering unit with manual override capability. It transmits in the low/medium frequency bands, reportedly between 3 and 13 kHz.

A computer-controlled automatic monitoring sub-system is incorporated in the equipment, with a

Display/console of Type 2020 submarine sonar

central monitoring unit driving local monitoring units within each equipment cabinet. Diagnosis of computer faults and programme loading is accelerated by the use of a floppy disk.

Operational Status
The first production model was accepted by the Royal Navy in October 1982 and production deliveries are now complete.

Contractor
Marconi Underwater Systems Ltd, Templecombe.

TYPE 2026 SONAR

Type
Passive towed array sonar.

Description
The Type 2026 sonar is a submarine passive towed array sonar and consists of a multi-octave towed array, a tow cable and an inboard processor and display system. The system provides surveillance and classification on data from both towed and flank arrays. It contains tracking, combat information system interface, data recording, DEMON, aural and hard copy facilities. The signal processor may also be used to provide a signal analysis and display function for data from other sonars.

The sonar may be operated by one or two operators and is controlled using a touch-screen, joystick and switches. It includes BITE and may be reconfigured in the event of a fault to minimise its effects.

Type 2026 submarine sonar

Data are displayed on three acoustic data displays, with comprehensive cursor and annotation facilities. Other facilities for driving remote displays (such as in the submarine's control room) or external video recorders are provided. The signal processor is cooled by an onboard chilled water supply, with an optional heat exchanger. The operator's console may be fitted in the sound room, with the signal processor in a remote cabinet space.

The Type 2026 sonar was developed in conjunction with the Admiralty Research Establishment by GEC Avionics and Plessey Naval Systems, both now part of Marconi Underwater Systems. GEC Avionics manufactured the signal processor and display system, STC the tow cable and Plessey the array and terminal unit.

Operational Status
Systems are operational on a number of Royal Navy SSNs and a variant has been supplied to the Royal Netherlands Navy.

Contractors
Marconi Underwater Systems Ltd, Templecombe.
STC, Cable Systems Division, Newport, Gwent.

TYPE 2032 PROCESSOR

Type
Electronics updating package.

Description
The Type 2032 is an electronics processing package designed to be integrated with the bow-mounted active array to give a limited narrowband capability. As well as certain electronic processing equipment, the Type 2032 add-on package includes a number of special transducer elements, a beam former cabinet and control facilities. It is believed to be complementary to the existing Type 2020 array, and is understood to be intended for retrofit and new build applications in the Royal Navy's 'Trafalgar' and 'Swiftsure' class SSNs.

Operational Status
Delivery of the initial systems began in March 1989.

Contractor
Ferranti-Thomson Sonar Systems UK Ltd, Aldershot.

TYPE 2046 SUBMARINE SONAR

Type
Passive hull-mounted sonar.

Description
The Type 2046 sonar is an integrated, passive system designed for the Royal Navy nuclear-powered attack submarines, and is also being fitted into the Type 2400 'Upholder' class of conventional submarines.

Developed from advanced 'Curtis' digital signal processing technology, and including FMS 12 display processing with Ferranti M700 Argus processors, the Type 2046 provides long-range target detection, classification and tracking information.

The system provides 360° bearing cover, multiple beam resolution and broadband surveillance with a high resolution narrowband frequency expansion facility. The operator interface consists of a four-screen, two-operator console suite with touch sensitive plasma panels.

Like all the Ferranti sonar systems the Type 2046 is built to a modular design to make maximum use of the latest technology, to save space and cost, and to assist through-life enhancement of the system. It is also equipped with fully integrated built-in test equipment for ease of maintenance.

Operational Status
A total of 27 systems have been ordered for Royal Navy submarines together with a follow-on order for a further two systems for training purposes. All systems have now been delivered.

Contractor
Ferranti-Thomson Sonar Systems UK Ltd, Stockport.

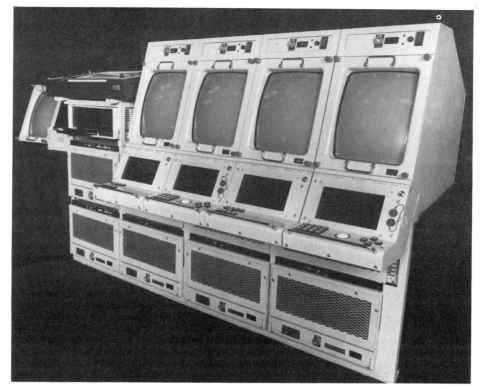

Operators' interface console of Type 2046 sonar

TYPE 2051 SONAR

Type
Passive and active sonar system.

Description
The Type 2051 sonar system is being provided for the update programme of the Royal Navy 'Oberon' class submarines. The prime contractor for the complete sonar programme, known as Triton, is Marconi Underwater Systems with Ferranti-Thomson Sonar Systems as co-contractor. The Triton fit is designed to bring the 'Oberon' class submarines up to date with the latest advances in sonar technology, including computer-assisted detection and classification.

The main elements of the system are a compact narrowband sonar which is linked to the submarine's existing flank array or towed array to provide all-round surveillance and classification, broadband passive sonars for both bow and flank arrays, an active sonar with variable power and sector capabilities, and an intercept sonar. A newly designed bow sonar dome designed to improve the hydrodynamic performance of the submarine and reduce water flow noise has been fitted.

Operational Status
Nine systems have now been ordered under a contract worth about £25 million. Deliveries began in 1986 when systems were fitted to HMS *Otter* and HMS *Ocelot*. In May 1989, Marconi received an order for the urgent supply of three Triton sonar systems for the Canadian Navy 'Oberon' class submarines.

Contractors
Marconi Underwater Systems Ltd, Templecombe (prime contractor).

Ferranti-Thomson Sonar Systems UK Ltd, Stockport.

TYPE 2052 SUBMARINE SONAR

Type
Passive broadband towed array sonar.

Description
The Type 2052 is a passive broadband towed array sonar system designed specifically for submarine use. This sonar has been retrofitted to some of the earlier nuclear-powered fleet submarines of the Royal Navy, such as the 'Valiant' and 'Churchill' classes which were laid down in the 1960s.

Operational Status
In operational service with the Royal Navy.

Contractor
Marconi Underwater Systems Ltd, Templecombe.

TYPE 2054 SONAR

Type
Passive/active intercept and towed array sonar suite.

Description
Marconi Underwater Systems' contract for Sonar 2054 for the new Royal Navy 'Vanguard' class Trident ballistic missile-carrying submarines is well advanced. The complete equipment will consist of a suite incorporating passive, active, intercept and towed array sonars.

Marconi, as the prime contractor and system design authority, will supply the bow-mounted hydrophone array together with the necessary beam forming and signal processing equipment (both narrowband and broadband), the intercept sonar equipment and the towed array. Major subcontractors are: Ferranti Computers for the consoles, data processing and displays; Marconi Underwater Systems Ltd for the intercept sonar equipment and STC for the handling system for the towed array.

The signal processing in the Type 2054 is almost entirely digital, the arrays being sampled by the pre-processor and immediately digitised before being passed to the beam former. Signals are then stabilised for both ship's position and geographical location before passing down for further processing in FM, CW and passive modes. These provide bearing and range, target motion and torpedo detection capabilities.

Subsequently the control and monitoring processing adds the ship's data to the detector signals and allows the operator to select particular modes of operation. This part of the system consists of two cabinets, one for target classification and the other for data display handling, which are interlinked using a Eurobus system and connected to the outside world by a 1553B standard data highway.

Operational Status
In production. Since May 1986 Marconi has received orders of around £200 million from the UK MoD for the Type 2054 project.

Contractors
Marconi Underwater Systems Ltd, Templecombe (prime contractor).

Ferranti-Thomson Sonar Systems UK Ltd, Stockport.

STC, Cable Systems Division, Newport, Gwent.

TYPE 2074 SONAR SYSTEM

Type
Passive and active hull-mounted sonar system.

Description
The Type 2074 passive and active hull-mounted sonar is being fitted to six of the Royal Navy's nuclear-powered hunter-killer submarines. It will replace the Types 2001 and 2020 in these vessels.

The new sonar makes full use of the advantages of latest technology and will be incorporated into a well-established equipment practice developed by Marconi Underwater Systems and used in all major new Royal Navy sonar development programmes.

The Type 2074 will provide improved sonar performance and is significantly smaller (by a ratio of 5:1) than the sonar it replaces. It is also being produced at a cost-effective price.

Due to the modular flexible nature of its design, the sonar can be readily configured to suit a customer's

particular requirements and offers good stretch potential for future enhancements.

Operational Status
In production for the Royal Navy. The modular nature of the design is being used as the basis for Marconi's current export activity.

Contractor
Marconi Underwater Systems Ltd, Templecombe.

TYPE 2075 SONAR

Type
Passive/active hull-mounted sonar.

Description
The Type 2075 is a fully integrated sonar system for the Royal Navy 'Upholder' class submarines, to detect, classify, localise and track targets. Principal functions of the equipment consist of passive surveillance (with a towed array being produced by Ferranti under a separate contract), passive ranging, passive/active search, tracking, intercept and attack. Additional subsystems also include an underwater telephone and an acoustic simulator to provide onboard training.

The system encompasses a complete range of outboard transducer arrays, six of which are new design, and inboard electronics for signal and data processing, plus all related displays. Outboard elements include a new design cylindrical bow array for passive/active reception, together with a transmit array; a new passive ranging array system; and new intercept array modules around the fin in which a new high frequency scanning array for mine detection and a stern arc array are also sited. The flank arrays are standard and a clip-on towed array will also be used with the system.

The inboard electronics are based on a distributed network of powerful 32-bit general processors supported by dedicated 'transputer' modules for specific signal processing tasks. The programming language is Ada to be compatible with the submarine command system. The modular design concept, plus ample

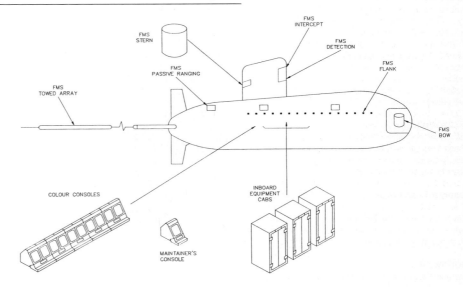

Diagram of the Type 2075 layout

processing capacity, provides scope for further enhancements.

Display facilities will be provided by a suite of five operators' consoles in the control room, each console having two colour display screens, trackerball, and touch-sensitive plasma panel controls. Connected to the signal and display processing units via a fibre optic local area network, each console is capable of displaying and controlling any sonar function.

Operational Status
Ferranti received a contract in mid-1989 from the UK MoD to design, develop and produce the Type 2075 sonar for the Royal Navy 'Upholder' class submarines.

Contractor
Ferranti-Thomson Sonar Systems UK Ltd, Stockport.

TYPE 2076 SONAR

Type
Hull-mounted passive/active sonar.

Description
The Type 2076 is a hull-mounted passive/active search and attack sonar which is intended as a mid-life update for the Royal Navy 'Swiftcare' and 'Trafalgar' class submarines. It will replace the existing 2001 and 2020 sets.

Operational Status
In May 1991 Marconi Underwater Systems and Ferranti-Thomson Sonar Systems were awarded parallel definition contracts for this sonar suite.

TYPE 2077 SONAR

Type
HF sonar system.

Description
The Type 2077 is a highly classified HF sonar project for nuclear submarines (probably an obstacle avoidance sonar). No technical details are available.

Operational Status
In development.

Contractor
Marconi Underwater Systems Ltd, Templecombe.

TYPE 2078 SONAR

Type
Integrated sonar system.

Description
The Type 2078 is the designation for the integrated sonar system for the next class of fleet submarines (SSN 20). Since the SSN 20 project has been can-

celled and replaced by an updated version of the 'Trafalgar' class, there is no longer a requirement for this system.

MARCONI FLANK ARRAY SYSTEM

Type
Submarine flank array sonar system.

Description
The Marconi submarine flank array sonar is part of the Hydra family of systems and offers the submarine operator an alternative form of linear array sonar. The flank array sonar provides a large acoustic aperture without directional ambiguity.

This flank array is a modular system enabling

acoustic performance to be tailored to the requirements of a particular platform. Standard hydrophone modules which are 2 m long contain Marconi-patented large area hydrophones embedded in a special polyurethane resin providing optimum decoupling from the flow noise. The spacing of the hydrophones can be varied during manufacture, depending on customer requirements, to a maximum of 12 hydrophones per module. The number of modules that can be fitted is dependent on the length of the parallel section of the submarine hull.

A special end fairing module provides a smooth contour at the end of the array to minimise flow noise.

The array modules are fitted to prepared sites on the submarine pressure hull comprising studs or tapped bosses welded to the hull. Each hydrophone incorporates its own pre-amplifier which enables long cable lengths to be driven from the array without impairing performance. The pre-amplifiers are housed in the fairings of the module, enabling easy access for maintenance in dry dock without removing the module.

The flank array can be interfaced to existing inboard narrowband and broadband processors, or can be supplied with a Marconi processor as a complete system.

Operational Status
In production. A system has been delivered to the Royal Swedish Navy and is in service on the 'Sea Serpent' class submarines.

Contractor
Marconi Underwater Systems Ltd, Templecombe.

In-water testing of a submarine flank array

FIS³ SONAR SUITE

Type
Integrated submarine sonar suite.

Description
The FIS³ (Ferranti Integrated Submarine Sonar Suite) has been derived from the Types 2046, 2050 and 2075 to provide a complete submarine sonar capability. It integrates the sonar subsystems with the display consoles and, as a logical progression, makes the array data available to all sonar subsystems. It uses a modular design philosophy which gives the system the flexibility necessary to be configured to fit the submarine concerned.

The use of a totally integrated approach also leads to a reduction in manning levels. In normal operation the suite requires two operators, and ease of use is enhanced by the latest in man/machine interfaces. Extensive use of touch-sensitive plasma display screens, trackerballs and nudge keys has also been made.

Operational Status
Fully developed.

Contractor
Ferranti-Thomson Sonar Systems UK Ltd, Stockport.

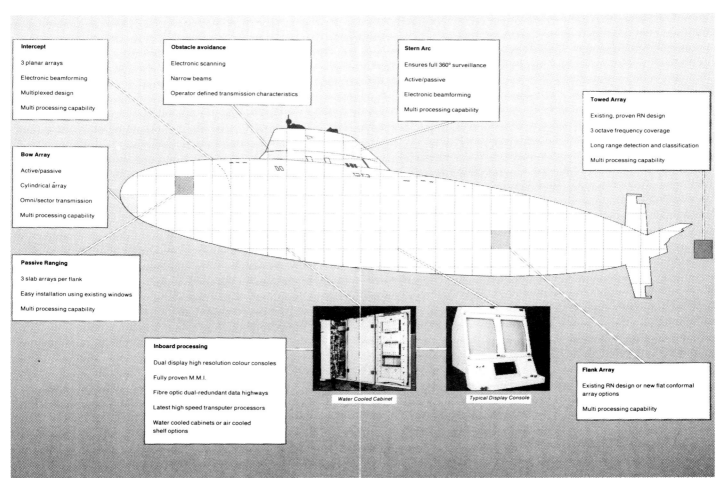

Intercept

3 planar arrays

Electronic beamforming

Multiplexed design

Multi processing capability

Obstacle avoidance

Electronic scanning

Narrow beams

Operator defined transmission characteristics

Stern Arc

Ensures full 360° surveillance

Active/passive

Electronic beamforming

Multi processing capability

Towed Array

Existing, proven RN design

3 octave frequency coverage

Long range detection and classification

Multi processing capability

Bow Array

Active/passive

Cylindrical array

Omni/sector transmission

Multi processing capability

Passive Ranging

3 slab arrays per flank

Easy installation using existing windows

Multi processing capability

Inboard processing

Dual display high resolution colour consoles

Fully proven M.M.I.

Fibre optic dual-redundant data highways

Latest high speed transputer processors

Water cooled cabinets or air cooled shelf options

Water Cooled Cabinet

Typical Display Console

Flank Array

Existing RN design or new flat conformal array options

Multi processing capability

Ferranti Integrated Submarine Sonar Suite

INTERCEPT SONAR

Type
Submarine and submersible intercept sonar.

Description
The intercept sonar captures and analyses active transmissions from sonar platforms. The sonar automatically determines positional information and pulse characteristics and provides the operator with analysis tools for the extraction of subtle target details.

BAeSema, in conjuction with Drumgrange Ltd, has developed a multi-channel intercept sonar processor, SPRINT, which has been successfully trialled with the Royal Navy at sea using an existing Sonar 2019 array and is a contender to meet the Sonar 2082 programme which is planned to replace the Sonar 2019.

The small-sized, modular SPRINT uses high performance algorithms in a multi-channel implementation and versatile array interfacing, with standard interfacing to boat services.

Contractors
BAeSema Marine Division, Bristol.
Drumgrange Ltd, Hampton.

UNITED STATES OF AMERICA

AN/BQR-21 DIMUS SONAR

Type
Passive hull-mounted search and detection sonar.

Description
The AN/BQR-21 DIMUS sonar is a hull-mounted pre-formed beam passive sonar with improved digital signal processing for the detection and tracking of quiet nuclear submarines. It is intended for fitting on all pre-Trident SSBNs and selected SSNs, and will be a direct replacement for the AN/BQR-2. The AN/BQR-21 will incorporate the same array configuration as the earlier equipment but will embody significantly enhanced signal processing capabilities. In addition to improvements in the electronics, performance will be improved through platform noise reductions which are being made in the bow dome of the submarine. These include the addition of an array baffle.

Operational Status
A total of 56 units of this equipment, for fitting in fleet ballistic missile submarines and certain attack submarines, has now been procured. System delivery began in August 1977 and was completed in December 1979. Two engineering models have been refurbished and delivered for use as trainers. In November 1982 Honeywell was awarded a $5 million contract for engineering support. The system is no longer in series production.

Contractor
Honeywell Inc, Training and Control Systems Operations, West Covina, California.

AN/BQR-21 DIMUS sonar display console

MULTI-FREQUENCY SONAR SYSTEM

Type
Scanning sonar system for various applications.

Description
This is a multi-frequency scanning sonar system which can be used in a variety of applications, including minehunting, obstacle avoidance, diver and ROV tracking, and navigation. Up to four different transducers at up to four different frequencies can be used. It is particularly applicable to ROV applications.

The processing is a 36-channel within-pulse sector scanner using a digitally controlled modulation method. The system can be expanded to 72 channels. The beam former is contained within the 9 in diameter × 18 in long electronics pod. The arrays are connected via a multi-way connector system using short umbilicals and Y-connectors as required. The arrays can point in different directions, for example a horizontally mounted transducer can be combined with a vertically mounted transducer to give simultaneous range, bearing and height above the seabed on selected targets.

The sonar information is presented on long persistence displays together with range, bearing and cursor markers. The control functions to the electronics pod are sent via a serial link which may be part of a multiplexed system, as on ROVs. Each sonar picture is sent over a video channel, and is preceded by a short digital message indicating which configuration is being used by the system.

Operational Status
In production.

Contractor
Sonar Research and Development Ltd, Beverley.

MODEL 1110 SONAR

Type
Passive flank array sonar.

Description
The Model 1110 uses a passive flank array to enhance the overall sensor capability of submarines. It is designed to be an adjunct to existing sonar, combat and fire control systems already planned or installed. The Model 1110 provides for long-range detection, tracking and classification of surface and submarine targets, while the existing hull sonar can be optimised for close-in detection.

The Model 1110 is a single operator system built on engineering principles well proven in the naval sphere. It consists of an operator console, processor cabinet, signal conditioning electronics and acoustic sensor arrays.

The system provides capabilities for expanded target detection and analysis through use of low frequency beam forming, split-beam correlation, bearing determination techniques and frequency analysis. It has the capability to track more than 15 targets simultaneously. Target tracking is accomplished by broadband and narrowband signal processing interpolation. Track processing allows the Model 1110 to automatically track many targets with minimal supervision. Selected targets can be assigned to 'high quality' track processors for improved position solution accuracy. The trackers can maintain track even in a crossing target environment without operator intervention. Tracking can be initiated by the operator using the cursor and soft-key controls.

The Model 1110 system single operator console has two high resolution colour displays. In existing submarines the control unit can be located adjacent to the host submarine sonar console and together will create an integrated sonar system. The microprocessor-based host submarine interface allows the Model 1110 to be integrated easily into the combat system of any new or existing submarine.

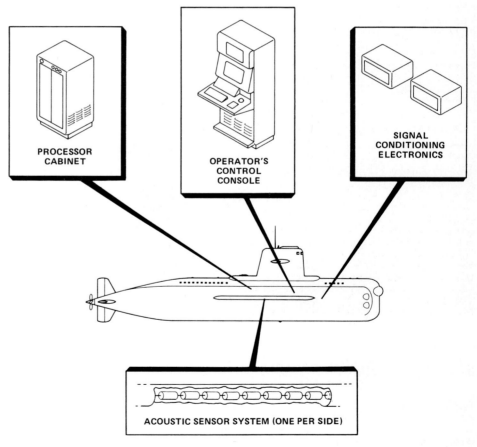

Model 1110 sonar layout

Operational Status
In production.

Contractor
EDO Corporation, Government Systems Division, College Point, New York.

MICROPUFFS PASSIVE RANGING SONAR

Type
Passive search and detection sonar.

Description
The MicroPUFFS system is a high accuracy, long-ranging passive sonar designed for installation aboard submarines. The name suggests that this equipment is a derivative of the original US Navy PUFFS (passive underwater fire control feasibility study) hardware; the PUFFS designation was also applied to USN AN/BQG-1/2/3/4 anti-submarine warfare systems. It was developed for the detection and tracking of submarine and surface targets, and provides automatic target tracking of up to four targets simultaneously, instantaneous and continuous ranging solution, fire control range and bearing output, and secure operation. The basic technique is signal enhancement by cross correlation of the signals received at the arrays with conversion of two measured time differentials into bearing and range data.

The system consists of three hydrophonic baffled arrays mounted on each side of the submarine. The arrays are connected to a display control console, which utilises an analogue processor, digital processor and AN/UYK-26 computer. The console also includes built-in fault localiser equipment. The system is built to US military specifications and uses standard electronic modules throughout.

Operational Status
MicroPUFFS systems are currently installed aboard six 'Oberon' class submarines of the Royal Australian Navy. A subsequent model has been built for the US Navy for test installation on the USS *Barb*.

Contractor
Unisys Corporation, Defense Systems, Great Neck, New York.

AN/BQQ-5 SONAR

Type
Hull-mounted and towed array.

Description
The AN/BQQ-5 sonar system is the principal sensor system of the US Navy's nuclear attack submarines, the SSN-688 class, all of which have been fitted in addition to retrofitting all 'Permit' and 'Sturgeon' class SSNs during regular refits. The AN/BQQ-5 is a digital, multi-beam system employing both hull-mounted and towed acoustic hydrophone arrays. The latter of the two arrays was produced in two types; initially a manually detached and attached receiver array connecting to a fixed towing point was employed, replaced later by a towed array that is retractable and housed along the submarine's hull. Both fixed tow and retractable array versions have provision for mechanically cutting the array adrift in the event of an emergency, except for the first seven equipped with the fixed tow point. The polyethylene-covered tow cable, which has a maximum length of about 800 m, is 9.5 mm in diameter, and the array at the end of the cable where the hydrophones and electronics are located is 82.5 mm in diameter. The array is tapered fore and aft to minimise flow noise. Drag is stated to account for a maximum reduction in speed of 0.5 kts, with no serious inhibition on submerged manoeuvres and little adverse effect on surface manoeuvres with the one exception of those entailing going about.

During the late 1970s the US Navy became concerned that the AN/BQQ-5 would not be able to handle the developing threats and embarked on a programme to update the capabilities of the system. An improved control display console has been developed and sonars that have been upgraded with the new consoles are known as the AN/BQQ-5B. The other major piece of equipment being developed under this programme is the AN/BQQ-5C(V) Expanded Directional Frequency Analysis and Recording System (DIFAR), which replaces the standard signal processors. The AN/BQQ-5D uses a thin-line array and utilises some of the technology associated with the AN/BSY-1-system. It became operational in 1988. A contract for the AN/BQQ-5E was awarded in December 1988. A new thin-line array for passive ranging is to be integrated with the AN/BQQ-5E.

The US Navy has also been developing a thin-line tow array and handling system and other modifications for the AN//BQQ-5 system (see Operational Status). Although most of this programme is classified, it is understood that the development is complete and entered service in 1989. Systems upgraded will then be known as the AN/BQQ-5D. The thin-line towed array, known as the TB-23, forms part of the AN/BSY-1(V) combat suite. It is a lightweight system which can be reeled into the vessel's main ballast tank, instead of being housed on the side of the submarine, and will replace the TB-16 'thick-line' array currently employed in the AN/BQQ-5.

The latest planned product improvement consists of interfacing to a further improved towed array, the TB-29; and the new fire control system CCS Mk 2.

The combining of the AN/BQQ-5E, TB-29, and CCS Mk 2 will result in a system referred to as QE2 which is scheduled for Technical Evaluation in 1992.

The AN/WLR-9A acoustic intercept receiver, produced by Norden, has been successfully evaluated and forms an integral subsystem of the AN/BQQ-5 and AN/BQQ-6 on new attack and Trident missile submarines. The subsystem features a CRT display, a digital readout and a remote unit for the submarine commander. A 'sensitivity improvement' kit for the AN/WLR-9A has been developed by Norden and deliveries to the US Navy are in progress.

Operational Status
The AN/BQQ-5 underwent extensive developmental testing during 1972 and 1973, and was approved for production later in 1973.

Over the period 1973-1986 a number of companies, including IBM, Raytheon, Gould, Bendix, Norden and Tracor, have been awarded contracts for systems, subassemblies and technical support. In December 1986, IBM was awarded a $54 million contract for upgrade kits to bring the AN/BQQ-5 to its latest configuration. Martin Marietta has received a $32 million contract for thin-line passive towed arrays for the AN/BQQ-5 and AN/BQQ-6 systems.

Under a contract issued to IBM in July 1989, more than 15 AN/BQQ-5D conversion kits were due to have been delivered by the end of 1990.

Contractors
IBM, Federal Systems Division, Bethesda, Maryland.
Raytheon Company, Rhode Island.

AN/BQQ-6 SONAR

Type
Passive and active sonar.

Description
The AN/BQQ-6 is an advanced active/passive sonar set developed for the Trident submarine forming part of the Trident command and control system, which is an integrated complex of command, control, communications and ship defence equipment.

The AN/BQQ-6 is based on the AN/BQQ-5 and has many of the same parts. The primary detection group is a digital integrated system employing spherical array, hull-mounted line array and towed array sensors, with an active emission acoustic intercept receiver and high frequency active short-range sonar. In the passive mode the system used is identical to that of the AN/BQQ-5. Support equipment has been added to provide for underwater communications, environmental sensing, magnetic recording and acoustic emergency devices. All other technical details are classified.

Support equipment has been added to provide for underwater communications, environmental sensing, recording and acoustic emergency devices. A version of the AN/BQQ-5E system is currently being developed for inclusion in the Trident submarines. This subsystem will represent a major upgrade to the present AN/BQQ-6 sonar. By utilising to a great extent the current AN/BQQ-5 hardware, a large improvement in logistics (provisioning, spares, training, and so on) is envisioned.

Operational Status
Installed on 'Ohio' class Trident submarines. In December 1986, IBM was awarded a $42 million contract for production of AN/BQQ-6 and AN/BQQ-5 systems, and a further $75 million in July 1989. Various modification and upgrading programmes are continuing in the fields of improved signal processing and towed array technology. Martin Marietta has received a $32 million contract for thin-line towed arrays for the AN/BQQ-5 and AN/BQQ-6 systems.

Contractors
IBM, Federal Systems Division, Bethesda, Maryland.
Martin Marietta, Defense Electronics Division, Glen Burnie, Maryland.

AN/BQR-15 SONAR

Type
Passive towed array sonar.

Development
In May 1972 contracts were awarded for the design and development of SPAD (signal processing and display) systems to exploit more fully the signal gathering potential of the AN/BQR-15 towed array sonar, but difficulties and delays led to the cancellation of this effort in March 1974. Since then funds have been allocated for procurement of a multiple channel signal processing and display equipment for use on FBM submarines and at FBM training facilities with the AN/BQR-15 sonar system.

Description
The AN/BQR-15 is a towed array sonar used by US Navy fleet ballistic missile submarines (SSBNs). Because the array is towed behind the submarine, these sonars provide the first long-range passive detection capability astern. In addition, since the sensors are removed from the submarine hull and internal ship's noise, detection of other submarines is possible at greatly increased ranges. As installed, the AN/BQR-15 includes specially designed hydraulically operated handling equipment that permits the cable array to be partially or completely streamed, retrieved, and adjusted automatically while the SSBN is submerged. The equipment forms part of the SSBN Unique sonar system which also includes the AN/BQR-19 receiving set and the AN/BQR-21.

Operational Status
In operational service. A modified array entered production during 1984.

Contractor
Western Electric Company, Winston-Salem, North Carolina.

AN/BQR-19 SONAR

Type
Passive search and detection sonar.

Description
The AN/BQR-19 was developed in the late 1960s for the detection of surface vessels in the vicinity of surfacing submarines. It was originally designed as a collision avoidance equipment to listen for and detect any surface ships in the vicinity of a surfacing ballistic missile submarine.

Operational Status
Although no longer in production the AN/BQR-19 is still in service with the US Navy aboard SSN-608, SSBN-616 and SSBN-726 class submarines.

Contractor
Raytheon Company, Submarine Signal Division, Portsmouth, Rhode Island.

AN/BQS-13 SONAR

Type
Passive/active sonar system.

Development
Acceptance testing of the multi-purpose subsystem began in late 1973 and technical and operational testing were conducted aboard the *Archerfish* during 1974. The AN/BQS-13 multi-purpose subsystem (BQR-24) was service approved in late 1974. The extended operational test programme delayed procurement to FY75-76, when 11 systems plus support were procured for $25.3 million. These subsystems provide interim improvement in sonar and fire control capability for those submarines not scheduled to receive the AN/BQQ-5 sonar until late in that programme. The BQR-24 has been installed only in submarines where it will have a useful life of at least three years prior to receiving an AN/BQQ-5 sonar system.

Description
The AN/BQS-13 sonar system is the primary sonar in USN 'Permit' and 'Sturgeon' SSN class nuclear attack submarines. The system was service approved in December 1971 and is in operation. A system addition, providing new functional capability, was service approved in late 1974. This addition, the AN/BQS-13 multi-purpose subsystem (BQR-24), is being procured for a limited number of SSNs.

Operational Status
Commencing in October 1979, Raytheon was awarded a series of US Naval Sea Systems Command contracts calling for refurbishment and conversion of AN/BQS-11, 12, and 13 sonar transmission subsystems to AN/BQQ-5 transmission subsystem configuration. These subsystems are part of the AN/BQQ-5 system.

Contractor
Raytheon Company, Submarine Signal Division, Portsmouth, Rhode Island.

AN/BQS-14A SUBMARINE SONAR

Type
Mine detection sonar.

Description
The AN/BQS-14A is an upgraded development of the AN/BQS-14 ice detection sonar developed during the late 1960s. In 1979 the US Navy initiated a programme to upgrade the system to provide an improved mine detection and avoidance capability to the 'Sturgeon' class of nuclear-powered attack submarines. The improved system has been introduced into the fleet as an upgrade and is the primary sonar used for under-ice navigation. Later versions of the sonar array have been integrated as part of the AN/BSY-1 and AN/BSY-2.

Contractors
Hazeltine Corporation, Greenlawn, New York.
IBM, Federal Systems Division, Manassas, Virginia.

AN/BQS-15 SUBMARINE SONAR

Type
Passive/active mine detection system.

Description
The AN/BQS-15 is an under-ice, close contact avoidance, submarine sonar system designed to locate and track a target in the active or passive mode. It is intended primarily for operation in heavily mined areas to provide improved mine detection and avoidance capability to the SSN 637/688 class submarines of the US Navy. Using hardware elements in common with the submarine active detection sonar and the AN/BQQ-5, the AN/BQS-15 is being introduced into the submarine fleet as an upgrade, integrated with the AN/BQQ-5, to become part of a submarine advanced combat system.

Operational Status
In service with the US Navy.

Contractor
EDO Corporation, Electro-Acoustic Division, Salt Lake City, Utah.

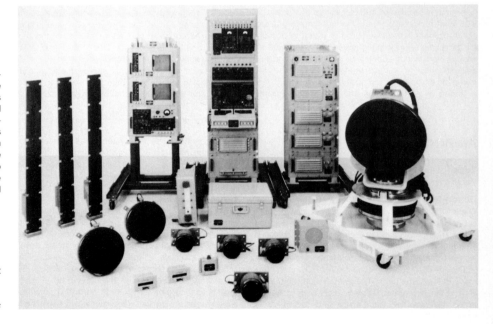

AN/BQS-15 submarine sonar elements

TB-16 TOWED ARRAY

Type
Towed array component of the AN/BQQ-5 sonar.

Description
The TB-16 is the towed array component of the AN/BQQ-5 submarine sonar system. It is a 'thick-line' system which is normally stowed in a tube running outside the submarine pressure hull, with the winch, cable and reel located in the forward main ballast tank, although some earlier installations used a permanent 'clip-on' array. The array cable is 82.5 mm in diameter and is towed on a 800 m long, 9.5 mm diameter cable. The acoustic module is approximately 75 m long.

Operational Status
In operational service.

Contractor
Martin Marietta Electronics and Missiles Group, Aero and Naval Systems, Baltimore, Maryland.

TB-23 TOWED ARRAY

Type
Thin-line towed array for the AN/BQQ-5 submarine sonar.

Description
The TB-23 is a thin-line towed array which is intended for the latest types of the AN/BQQ-5 sonar system. It is a lightweight system which can be reeled into the vessel's main ballast tank instead of being housed on the side of the submarine pressure hull. It will eventually replace the TB-16 in most classes of submarines, and forms part of the AN/BSY-1(V) combat suite (see separate entries).

Operational Status
In production.

Contractor
Martin Marietta Electronics and Missile Group, Aero and Naval Systems, Baltimore, Maryland.

MODEL 1550 CTFM SONAR

Type
Obstacle avoidance sonar system.

Description
The Model 1550 obstacle avoidance sonar provides rapid scanning, high resolution display of subsea objects at ranges from 3 to 1500 m, and a full 360° area coverage. It was designed for installation on conventional submarines and manned submersibles. The Model 1550 features a TV formatted colour display with memory circuits that provide a continuous presentation of the sector scan selected until updated by the next scan. The Model 1550 is based on the design of the 500A and uses the same outboard unit. Its size and connections are compatible with that system.

The Model 1550 features automatic sector modes and manual scan selection. Forward sectors of 60° and 120° are offset while a 240° sector is centred on the display. In addition, a 60° sector can be positioned anywhere in 360° for higher information rate operation. Incorporated in the equipment is the latest digital analyser with 128 channels, providing 3.2 lines higher resolution than the 500A.

The system has a transponder mode for operation with CTFM transponders. The mode can be used alone or simultaneously with the regular sonar mode at the operator's choice. Transponders are used primarily as a navigation aid, but can also be dropped from submarines in a submerged condition to mark positions of interest.

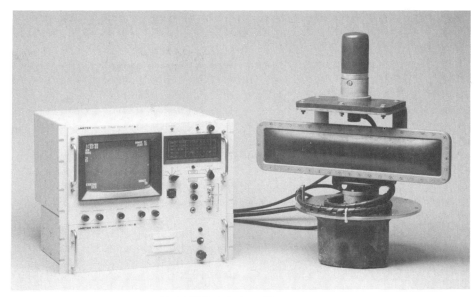

Model 1550 obstacle avoidance sonar

Operational Status
In service with the US Navy and Japan.

Contractor
EDO Corporation, Electro-Acoustic Division, Salt Lake City, Utah.

AIRBORNE DIPPING SONARS

FRANCE

HS 12 HELICOPTER SONAR

Type
Passive and active dipping sonar system.

Description
The HS 12 is an active/passive panoramic helicopter version of the SS 12 small ship sonar and uses the same electronics as the surface vessels version. It has similar capabilities for operation in shallow/noisy waters, and has a system weight of less than 240 kg, thus suiting it for installation on light naval helicopters such as the Lynx. The HS 12 transducer is raised and lowered by a hydraulic winch at a high speed.

Operation in CW and FM modes is possible and digital signal processing is employed by the system's microprocessor. Automatic tracking of two targets and transmission of elements to an external equipment, such as a plotting table, are provided.

The system operates on 13 kHz in the active mode and in the 7 to 20 kHz range passively. A total of 12 pre-formed beams are employed giving 30° sectors, with a maximum range of about 10 km. The display consists of four quadrants, these being obtained by processing adjacent beams. The operator can select a CW mode which provides target range and Doppler. He can also select FM processing, and a sector mode is also provided.

Operational Status
In series production. It is believed that three Chinese SA-321G Super Frelon helicopters have been

Typical helicopter installation of the HS 12 sonar on Dauphin naval helicopter

equipped with HS 12 systems. It is also in operation in Saudi Arabia.

Specifications
Frequencies: Medium frequency
Transmission level: 212 dB/μPa/m
Passive search: Noise or pinger. Display of 12 pre-formed beams, each subject to adaptive signal

processing. Other characteristics as for the SS 12
Total weight: 230 kg

Contractor
Thomson Sintra Activités Sous-Marines, Valbonne Cedex.

FLASH DIPPING SONAR

Type
Helicopter dunking sonar system.

Description
FLASH (Folding Light Acoustic System for Helicopters) is a new low frequency dipping sonar being developed by Thomson Sintra for a number of applications. It is based on the HS 12 system, and operates at three frequencies below 5 kHz, with 24 pre-formed beams. The processing system is based on Motorola 68000 microprocessors in a VME bus environment and is programmed in the Ada high-order language.

Operational Status
Thomson Sintra ASM is teamed with prime contractor Ferranti-Thomson Sonar Systems UK Ltd (a joint venture company established by Thomson-CSF and Ferranti International) to supply FLASH for the long-range dipping sonar requirement for the Royal Navy's Merlin helicopter. This system will combine the Thomson Sintra ASM low frequency array and winch

with the Ferranti-Thomson signal and data processing expertise. Thomson Sintra has also teamed with Hughes Aircraft Company to bid FLASH for the US Navy's ALFS requirement. FLASH has already been evaluated in trials for the US Navy.

Contractor
Thomson Sintra Activités Sous-Marines, Valbonne Cedex.

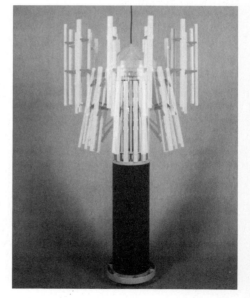

Wet end of the FLASH dipping sonar

DUAV-4 HELICOPTER SONAR

Type
Passive and active sonar system.

Description
The DUAV-4 is an active/passive directive sonar designed for submarine surveillance and location (azimuth, distance and radial speed). It is specially designed for use on board light, versatile, ship-based helicopters of which the Lynx is a typical example. It may also be fitted aboard small surface vessels as either a hull sonar or as a VDS system.

The DUAV-4 differs from conventional sonar in its signal processing system, which is designed to give improved detection, especially in severe reverberation conditions such as shallow waters. The sonar can be operated in either the active or passive mode: true

bearing, range, and radial speed are measured in the former mode; true bearing only in the passive mode.

A combined display unit permits surveillance display (initial detection) or plotting display (precise azimuth determination). Total weight including electronic rack, cable and dome is 250 kg.

Operational Status
The DUAV-4 sonar is in active service with the French Navy, the Royal Netherlands Navy and several other navies. Over 90 have been produced. The French Navy has opted to replace it in some of the Lynx helicopters with the HS 12 (see separate entry).

Contractor
Thomson Sintra Activités Sous-Marines, Valbonne Cedex.

DUAV-4 helicopter sonar installation in Lynx

UNION OF SOVEREIGN STATES

Virtually no information on Soviet airborne submarine warfare equipment has been made public. There are at least two types of dipping sonar, and Magnetic Anomaly Detectors (MAD) are carried by helicopters. The photograph shows an Mi-14PW 'Haze-A' helicopter which is equipped with both MAD and dipping sonar. The variable depth dipping sonar used in a number of Soviet corvettes, and NATO code named 'Rat Tail', is fitted to the 'Hormone-A' helicopter. It apparently operates at a frequency of about 15 kHz and has a maximum range of about 5000 m. The dipping sonar operated from the Mi-14 is almost certainly a medium frequency type.

The Soviet Union undoubtedly uses large numbers of sonobuoys, and a certain amount of information on these has recently emerged (see next section). Airborne processing and analysis/display equipment must therefore be fitted to both helicopters and fixed-wing ASW aircraft, but no information is available.

Mi-14PW of the Polish Air Force showing the towed MAD 'bird' stowed against the rear of the fuselage. The Mi-14PW is also fitted with a dunking type sonar

UNITED KINGDOM

TYPE 195/TYPE 2069 SONAR

Type
Helicopter dunking sonar.

Description
The Type 195 sonar is a helicopter dunking system installed in Sea King helicopters. It provides full 360° coverage and is understood to be effective at ranges up to 8000 yards (7300 m). Full azimuth coverage is provided in stepped fashion over 90° arcs progressively, but manual control allows the operator to concentrate on any particular sector. The system is programmed to undertake an automatic search. The operator is provided with audio, visual Doppler and visual sector sonar information, and close contact maintenance is provided for the tracking of nearby targets and those at greater depths.

The Type 195 may be employed in either surveillance or attack control, or both simultaneously. Pulse length and detection range settings are operator-selectable to optimise the system according to sea conditions and the tactical situation.

The current model is the Sonar Type 195M, which incorporates many improvements over the original equipment. The Type 2069 sonar is the latest upgrade of the 195M. It employs new solid-state transmitters and has a longer cable. The sonar transducer has been re-engineered to provide a greater operating depth and is now integrated with the AQS-902 G/DS acoustic system.

The total system is the first in-service sonar with simultaneous control, processing and display of sonobuoys and sonar. It is installed in the Royal Navy Sea King Mk 6.

Operational Status
More than 250 Marconi Type 195 systems are in service with the Royal Navy and several other navies. In June 1989, Ferranti Computer Systems announced a UK MoD contract for 44 sets of submersible units and suspension cables for the Type 2069.

Contractors
Marconi Underwater Systems Ltd, Templecombe.
Ferranti-Thomson Sonar Systems UK Ltd, Stockport.

Type 195/2069 dunking sonar sensor

CORMORANT SONAR

Type
Active and passive helicopter dunking sonar.

Description
The Cormorant lightweight dipping sonar is a major advance in helicopter sonar, developed by Marconi Underwater Systems to provide a unique, low frequency active and passive acoustic sensor capable of high performance levels typically twice that of existing systems. When combined with a suitable acoustic processor, Cormorant provides an integrated sonics system which complements the area surveillance of modern sonobuoys to provide the maximum tactical freedom for the ASW helicopter.

The advanced design of Cormorant overcomes the size and weight constraints that have limited the operational performance of previous generations of dipping sonar. Its all-up weight of 220 kg has extended the sonar option to smaller helicopters down to 4000 kg in overall weight.

The Cormorant design is based on an expanding array, provided by a simple folding construction, which achieves a large acoustic receive aperture with a volumetric configuration. The hydrophone array is housed within the protection of the submersible unit for stowage in the aircraft and during deployment and retrieval operations. Once the submersible unit has reached its operating depth, the hydrophone array is powered out to its operating position.

The projector array, hydrophone array, interface electronics and auxiliary sensors are carried in the submersible unit body in a modular configuration.

The compact transducer body design, constructed in high strength, corrosion-resistant titanium alloys, achieves a hydrodynamic profile ensuring low drag and stable deployment and recovery. Aerodynamic stability enables 'flight at the trail' at maximum cruising speed, and a minimum winding load gives short recovery and deployment times and long cable life.

Operational Status
Cormorant has successfully completed a full technology demonstrator programme for the Royal Navy, as well as Foreign Weapons Evaluation and Airborne Low Frequency (ALFS) demonstrations for the US Navy. It has been integrated with the UYS-503 processor as part of the AN/AQS-503 HAPS system and delivered to the Canadian Armed Forces.

Specifications
Sonar: Omnidirectional, 360° field-of-view. CW and LFM, multi-frequency active acoustics, broadband passive acoustics
Depth: 15-457 m
Hoist speed: 3.7 m/s (deployment); 5.4 m/s (recovery)
Auxiliary sensors: −10 to + 30°C (bathythermal); 0-670 psi (pressure depth sensors); ± 45° (transducer tilt); array heading compass; >5 *g* (body impact); submersed indicator (wet/dry)

Cormorant sensor array

Underwater communications: AN/UQC-1 and AN/WQC-501 compatible
Dimensions: 1165 × 261 mm (diameter) (submersible unit); 600 × 640 × 1060 mm (winch unit); 1030 × 483 mm (diameter) (housing assembly); 490 m × 8.5 mm (diameter) (cable); 194 × 256 × 492 mm

(transmitter); 194 × 256 × 320 mm (sonar interface unit)
Weights: 60 kg (submersible unit in air); 40 kg

(submersible unit in water); 70 kg (winch unit); 16 kg (typical) (housing assembly); 39 kg (cable); 18.7 kg (transmitter); 16 kg (sonar interface unit)

Contractor
Marconi Underwater Systems Ltd, Templecombe.

AS380 SONAR

Type
Active and passive dunking sonar system.

Description
The AS380 is a combined active and passive dunking sonar providing comprehensive detection and classification capabilities in a single underwater package. It can be deployed from both ships and helicopters. Multiple operating modes allow the system to be used for either ASW or MCM scenarios.

The AS380 has two main active sonar operating modes: detection and classification.

Detection
In the detection mode the AS380 provides a rapid search facility to a range of 3 km whereby a 60° sector coverage, using electronic beam forming, can be steered in any direction. The operator has the choice of selecting either chirp or Doppler transmission, providing alternative target detection processes.

Chirp: frequency modulated transmission and coherent processing ensures, with optimum range capability, improved range resolution and high interference immunity. The PPI sonar display is presented with target intensities represented by a 16-level grey scale or colour. A 60° sector is updated at every sonar pulse. Adjacent 60° sectors may be scanned in rotation on demand. Different frequency coded sonars may be used for more than one sonar system hunting the same target.

Doppler: detects the presence of a moving target within the isonified 60° sector, providing the operator

with an instantaneous measurement of closing or opening velocity, together with its bearing direction. Selectable Doppler filter eliminates spurious Doppler returns created by sensor movement, surface waves and so on. The auxiliary display produces a scan graph of targets representing the velocity (colour) and relative signal strength (amplitude) of the targets.

Classification
In the classification mode a detected target may be examined further for a representation of its shape and size. Short transmission pulses and narrow horizontal beam pattern (3°) necessary for reproducing clearly defined sonar images can be further enhanced using a zoom facility.

Display
The single operator display console consists of two 14 in high resolution display monitors, a keyboard and a trackerball with three push-button controls. The main display shows the current sonar picture and provides the main system control by means of a set of screen icons. The upper display is reversed as an auxiliary for retaining reference sonargrams, bathymetry graphs, gain laws and so on. Complete transfer between the screens is provided in terms of both display data and system control.

Specifications
Nominal frequency: 50 kHz
Beamwidth: 60° (horizontal); 10° (vertical)
Displayed resolution: 5°
Bearing accuracy: 1°
Modulation: FM (chirp); CW (Doppler)
Ranges: 1/2/3 km
Maximum depth: 60 m

AS380 dunking sonar

Contractor
Marconi udi (a subsidiary of Marconi Underwater Systems Ltd), Aberdeen.

UNITED STATES OF AMERICA

AN/AQS-13 HELICOPTER SONAR

Type
Active and passive helicopter dunking sonar.

Description
The AN/AQS-13 is a helicopter dunking system, the AN/AQS-13B and 13F models being in production. It is one of a series of such equipment which began with the Bendix AN/AQS-10 in 1955, the latest variant being the AN/AQS-13F.

The B system is a long-range, active scanning sonar which detects and maintains contact with underwater targets through a transducer lowered into the water from a hovering helicopter. Opening or closing rates of moving targets can be accurately determined and the system also provides target classification clues.

The AN/AQS-13B has significant advantages in operation and maintenance over earlier systems. To aid the operator, some electronic functions were automated to eliminate several controls. Maintenance was simplified by eliminating all internal adjustments and adding built-in-test (BITE) circuits.

These advantages were brought about by the use of the latest electronic circuits and packaging techniques, which also reduced system size and weight.

To enhance detection capability in shallow water and reverberation-limited conditions, while essentially eliminating false alarms from the video display, Bendix developed an adaptive processor sonar (APS) for the system.

The APS is a completely digital processor employing fast Fourier transform (FFT) techniques to provide narrowband analysis of the uniquely shaped CW pulse transmitted in the APS mode. The display retains the familiar PPI readout of target range and bearing but APS adds precise digital readout of the radial component of target Doppler.

With APS, processing gains of greater than 20 dB with zero false alarm rates have been measured for target Dopplers under 0.5 knot.

The AN/AQS-13E was the first system to integrate

sonar (APS) and sonobuoy processing in a common processor, the Sonar Data Computer (SDC). Improvements of this system led to the AN/AQS-13F.

The higher energy transmitted with the longer pulse APS mode, combined with the narrowband analysis, also substantially improves the figure of merit (FOM) in the non-reverberant conditions more typical of deep water operation. Measured processing gains for APS under ambient, wideband noise limited conditions exceeded 7 dB.

The AN/AQS-13F has been designed to provide rapid tactical response against the most advanced submarine threats. It is a sister equipment to the AN/AQS-18 and is identical in many respects. A new state-of-the-art transducer, when lowered to depths of up to 1450 feet, permits instantaneous range improvements of over 100% compared to previous systems. Very high speed reeling allows a dip, cycled to maximum depth, to be completed in approximately three minutes. The powerful omnidirectional transducer providing 216 ± 1 dB source level is integrated with a sensitive directional receiver array providing azimuthal resolution in a small rugged unit. The sonar data computer offers digital matched filter processing for 200 ms and 700 ms sonar pulses, as well as sonobuoy control and processing. The azimuth and range indicator and receiver provides a video display for the operator.

Operational Status
The AN/AQS-13 is widely used by American forces and over 1000 sets have been supplied or ordered for naval helicopters of 15 foreign navies in Europe, Asia, the Middle East and South America. The AQS-13F was selected by the US Navy and is now in operation on the new SH-60F carrier-based ASW helicopters. The AN/AQS-18 is a similar system to the AN/AQS-13F sonar featuring smaller size, lighter weight and increased performance. It was developed for and is operated primarily on the Westland Lynx helicopter.

AN/AQS-13F transducer and winch (Raymond Cheung)

Specifications
Frequencies: 9.25, 10, 10.75 kHz
Sound pressure level: 113 dB ref 1 microbar (13B); 216 dB (13F)
Range scales: 1, 3, 5, 8, 12, 20 yds (0.9, 2.7, 4.6, 7.3, 11, 18.3 km)
Operational modes: Active 3.5 or 35 ms, MTI, APS,

passive, voice communications, key communications (13A only). The 13F has the following operational modes: active–3.5 m/s and 35 ms rectangular pulse, 200 ms and 700 ms shaped pulse; MTI; passive–500 Hz bandwidth 9 to 11 kHz; communicate SSB at 8 kHz (voice)

Visual outputs: (13A) range, bearing; (13B) range, range rate, bearing, operator verification

Audio output: (13A) single channel with gain control; (13B) dual channel with gain control plus constant level to aircraft intercom

Recorder operation: Bathythermograph, range, aspect, MAD self-test

System weight: (13A) 373 kg; (13B) 282 kg; (13F) 280 kg

Operating depth: 1450 ft at 50 ft hover (13F)

Contractor
Bendix Oceanics Inc (a unit of Allied Signal Inc), Sylmar, California.

AN/AQS-18 HELICOPTER SONAR

Type
Active and passive helicopter dunking sonar.

Description
The AN/AQS-18 is a helicopter-borne, long-range active scanning sonar. The system detects and maintains contact with underwater targets through a transducer lowered into the water from a hovering helicopter. Active echo-ranging determines a target's range and bearing, and opening or closing rate, relative to the aircraft. Target identification clues are also provided.

The AN/AQS-18 is an advanced version of earlier dunking sonars made by Bendix and includes digital technology, improved signal processing and improved operator displays. The system consists of a small high density transducer with a high sink and retrieval rate, a built-in multiplex system to permit use of a single conductor cable, a 330 m cable and compatible reeling machine and a lightweight transmitter built into the transducer package. The Adaptive Processor Sonar (APS), which provides enhanced performance in shallow water areas, is an integral part of the system.

The AN/AQS-18 offers a number of improvements over earlier dipped sonars. These include increased transmitter power output giving longer range, high speed dip cycle time and reductions in weight of all units.

The APS increases detection capability in shallow water and reverberation-limited conditions while essentially eliminating false alarms from the video display. The APS is a digital processor which uses fast Fourier transform (FFT) techniques to provide narrowband analysis of the uniquely shaped CW

pulse transmitted in the APS mode. The PPI display retains the normal readout of target range and bearing.

The APS processing gain improvement over the normal AN/AQS-18 analogue processing is 20 dB for a two knot target and 15 dB for a five knot Doppler target. The higher energy transmitted with the longer pulse APS mode, combined with the narrowband analysis, also improves operation in the non-reverberant conditions more typical of deep water. The gain improvement outside the high reverberation zone (10 knots or greater) under wideband noise limited conditions exceeds 7 dB. The latest version is the AN/AQS-18(V) which is available with both 300 and 450 m length cables.

Operational Status
In production and/or operational use with the German, Japanese, Italian, Taiwanese and Spanish Navies, and in use by the US Navy in a similar version, the AN/AQS-13F, on the new SH-60F carrier-based ASW helicopter. The Portuguese Navy is also believed to have specified the system for its helicopters. The AN/AQS-18 is in production for and operates in the Lynx, SH-3, SH-60J and S70C(M)-1 helicopters.

Specifications
Operating depth: 330 m
Operating frequencies: 9.23, 10, 10.77 kHz
Sound pressure level: 217 dB/μPa/yd (0.9 m)
Range scales: 1000, 3000, 5000, 8000, 12 000, 20 000 yd
Operational modes: 3.5 or 35 ms pulse (energy detection) and 200 or 700 ms pulse (narrowband analysis)
Visual outputs: Range, range rate, bearing, operator verification

AN/AQS-18 sonar deployed from German Navy Westland Lynx Mk 88 helicopter

Audio output: Dual channel with gain control, plus constant level to aircraft intercom system
Recorder operation: Bathythermograph, range, ASPECT, MAD, BITE
System weight: 252 kg plus 13.3 kg for APS

Contractor
Bendix Oceanics Inc (a unit of Allied Signal Inc), Sylmar, California.

ALFS – AIRBORNE LOW FREQUENCY SONAR

Type
Next generation low frequency dipping sonar.

Description
ALFS is a US Naval Air Systems Command project for a new generation dipping sonar which will largely replace the existing AN/AQS-13F, and is scheduled to be fitted to the SH-60F helicopters. No technical details have been announced, except that the associated processor will be the UYS-2 manufactured by

AT&T. A number of companies/consortia have submitted proposals for this project. As of February 1991 these were understood to be as follows:
(a) Allied Signal (Bendix) teamed with British Aerospace and FIAR to develop the Helras sonar
(b) Martin Marietta with Dowty to develop the HIPAS system
(c) Marconi Underwater Systems with IBM with a version of the Marconi Underwater Systems Cormorant system
(d) Sanders and GEC Avionics with Osprey
(e) Thomson Sintra and Hughes with a version of the HS 12, known as FLASH (Folding Light Acoustic System for Helicopters).

Operational Status
As this edition went to press it was announced that the consortium of Hughes Aircraft Co and Thomson Sintra had been awarded a $31.3 million contract to develop the ALFS system. The five year contract provides for full-scale development and a follow-on production option of up to 50 systems.

Contractors
Hughes Aircraft Company, Fullerton, California.
Thomson Sintra Activités Sous-Marines, Valbonne Cedex, France.

HIPAS SONAR SYSTEM

Type
Active and passive helicopter dipping sonar.

Description
HIPAS is an acronym for High Performance Active Sonar system being developed under the leadership of Martin Marietta. This system is designed to the requirements of the US Navy's Airborne Low Frequency Sonar (ALFS) and is an active and passive system that provides detection and tracking of the most technologically advanced threat submarines. It is intended to provide increased performance with longer range, greater depth and modern signal processing.

Operational Status
In development. The system is aimed at installations on the SH-60F carrier-based and SH-60B LAMPS Mk III helicopters, as well as the V-22 Osprey tilt-rotor aircraft. The development team consists of Martin Marietta, DRS and Breeze-Eastern. Martin Marietta is responsible for systems integration as well as the underwater array, beam forming and display subsystems. DRS is supplying the signal and video processing capability, and Breeze-Eastern the reeling machine and control. Dowty Maritime Systems, UK, has teamed with Martin Marietta to bid the system for the US Navy ALFS programme.

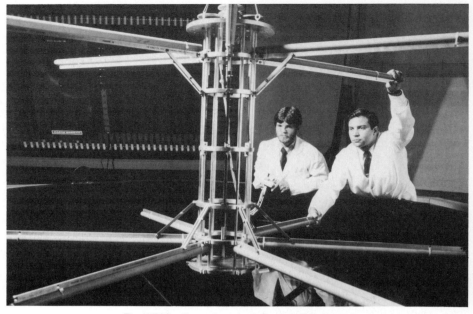

The HIPAS active and passive helicopter dipping sonar

Contractor
Martin Marietta, Electronics & Missiles Group, Bethesda, Maryland.

AN/AQH-9 SYSTEM

Type
Helicopter mission recording and playback system.

Description
The AN/AQH-9 Mission Recorder System captures critical anti-submarine warfare data for contact validation, post-mission analysis and crew training. The equipment can be readily reconfigured for the missions of rotary-wing, fighter/attack and reconnaissance aircraft, and for UAVs.

The system consists of a mission tape recorder interface unit, a VHS video cassette data recorder, a remote-control unit and a video cassette. The video cassette links the airborne equipment to a ground station playback system.

The AN/AQH-9 is capable of recording four analogue channels of any mix of LOFAR, DIFAR, DICASS and BT sonobuoy signals. It records a single channel of frame-sampled, stroke-type CRT display video, six channels of analogue audio, including voice communications from radio and internal systems, and tactical/navigation data. The interface unit contains an IRIG 'B' time-code generator for recording mission time. The remote-control unit contains all the operating controls and indicators required to provide proper system support.

To keep weight and volume to a minimum (37 lb and 0.65 cu ft) for critical applications, the equipment was configured to 'record only' using standard VHS tape cassettes. Other configurations are available to record and provide in-flight playback facilities. The unit is shock-mounted near the sonar operator's console.

The ground element of the system is provided by the AN/AQH-9 Mission Data Playback System which is capable of reproducing all the data from the

AN/AQH-9 mission tape recorder system (left to right) interface unit, remote-control unit, data recorder

mission recorder system. It recalls acoustic, tactical, communications and avionics data from a standard 1553B databus. It contains an IRIG 'B' time-code generator which allows the playback system to provide real-time, post-mission analysis reconstruction of mission data, as well as high speed search for specific information.

The acoustic data are recorded from four sonobuoy receiver channels, maintaining precise phase information, and presented in analogue form for analysis by the AN/SQR-17A. Intercom and UHF radio are reconstructed and made available both as analogue electrical outputs and in audible form through a front panel loudspeaker.

The man/machine interface consists of a high resolution, full colour display, presented to the operator as three interactive windows. The operator interface is accomplished primarily via a trackerball located next to the keyboard. The real-time graphics display

window provides symbolic reconstruction of the mission's tactical scenario. Two supplementary windows are provided for additional data relevant to the current status of the aircraft and the significant events logged during the mission. Interactive data filtering is provided to control the type and extent of the data displayed in these windows.

Operational Status
The recorder system has been manufactured for operational use on board the US Navy SH-60F CV carrier-based helicopters. Two playback systems have been provided to the US Naval Air Development Center.

Contractor
Diagnostic/Retrieval Systems Inc, Oakland, New Jersey.

SONOBUOYS

AUSTRALIA

PROJECT BARRA

Type
Submarine detection system.

Description
Project Barra concerns the development and production of an advanced submarine detection system for use by long-range maritime patrol aircraft.

The project evolved from successful development work under Project Nangana, a research and development programme, undertaken by the Weapons Research Establishment in close collaboration with the RAN and RAAF for the development of improved methods of submarine detection.

Following successful completion of an engineering study, initial development and the establishment of production facilities, together with the first production contract, Amalgamated Wireless (Australasia) Limited was appointed prime contractor, and completed the production of this initial Barra sonobuoy contract by the end of 1981. A further A$55 million contract was let in early 1982 for full production against RAF and RAAF requirements. This contract has been fulfilled and further orders have been received.

The project is a collaborative programme with the UK; Australia developed the SSQ-801 Barra sonobuoy and the airborne processing equipment was developed in the UK under the direction of the UK Ministry of Defence (see AQS-901 acoustic data processing and display system in the *Airborne Acoustic Processing Systems* section). The platform equipment is suitable for installation in a range of aircraft, and has been fitted in RAF Nimrod MR2 and RAAF P-3C Orion maritime aircraft.

RAAF P-3C Orion maritime patrol aircraft equipped with Barra ASW system

Development of the Barra system generated the need for a ground support facility for the P-3C Orion aircraft. AWA designed and built for the RAAF a Compilation, Mission Support, Integration and training facility (CMI). This self-contained modular cabin complex can be reconfigured to suit various RAAF requirements. The CMI is a computer-based system providing training, mission briefing/debriefing and analysis facilities.

Contractor
Sonobuoys Australia (a partnership of AWA Defence Industries Pty Ltd and GEC-Marconi Systems Pty Ltd), Meadowbank, NSW.

Advanced sonobuoy communication link receiver for Barra sonobuoy data exchange

---

SSQ-801 BARRA SONOBUOY

Type
Passive directional sonobuoy.

Development
Development of the SSQ-801 sonobuoy was undertaken as part of the Australian Barra project (see above). A derivative, the SSQ-981, has been developed by Marconi Underwater Systems Ltd (see separate entry). Sonobuoys Australia is also developing a Barra variant which incorporates the in-buoy signal processing to make the transmitter signal compatible with standard airborne receiver and sonic processors such as the AQA-7 and UYS-1. This variant known as EBarra, uses VLSI technology to produce LOFAR compatible data, containing spectral, broadband and accurate directional information.

Description
The SSQ-801 sonobuoy was developed specifically for the Australian Project Barra anti-submarine programme and is a passive directional sensor which may be used either singly or deployed in patterns to detect and locate submarine targets. It provides both range and directional information. Within a standard 'A' size buoy, two parts separate on reaching the water; the upper portion containing radio transmission equipment rises to the surface while the other part housing the acoustic sensor array sinks to a predetermined depth. A cable links the two parts.

The lower portion contains a compass and arms fitted with 25 specially developed hydrophones mounted in a horizontal planar array configuration. These hydrophones are sequentially switched to give the required directional beaming. The beams can be steered, permitting selective target signal enhancement and reduction of noise interference. Multiplexed signals from the lower section are fed to the floating portion of the buoy for transmission by radio to patrol aircraft, where data are analysed by the Barra system AQS-901 sonics processor. Each sonobuoy transmits information on one of 50 preselected VHF radio channels so that the aircraft crew can select a particular buoy's data for analysis. Digital techniques are used in the SSQ-801 to convert the combined sonar and compass information into signals to modulate the buoy's VHF transmissions. Complete azimuth coverage of both broadband and spectral data is transmitted to the airborne acoustic processor.

The SSQ-801 is suitable for release from fixed-wing aircraft such as the P-3C Orion or Nimrod, and from helicopters of various types.

Operational Status
Adopted by both the Royal Air Force and the Royal Australian Air Force. A number of other countries are reported to be interested in the Barra system, including France, Japan, the Netherlands and the USA.

Since the original contract in 1982, additional contracts have been placed for the continuing production of the Barra sonobuoy. A new company, Sonobuoys Australia, has been formed to manufacture the Barra sonobuoy.

Specifications
Type: Passive directional array
Size: A
Length: 914 mm
Diameter: 124 mm
Weight: 12.7 kg
RF transmitter frequencies: 136-173.5 MHz
RF modulation: Narrowband FSK
Channel spacing: 0.75 MHz
Power output: 1 W (min)

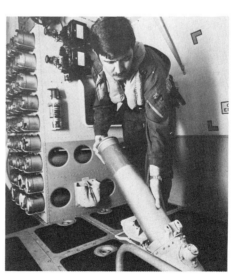

SSQ-801 Barra sonobuoy about to be launched from an RAAF Orion ASW aircraft

Operating life: 0.5, 1, 2, or 4.5 h
Array depth: 21.5 or 128.5 m
Activation time: 60 s at 21.5 m; 180 s at 128.5 m
Acoustic bandwidth: 10-2000 Hz
Compass accuracy: 1° RMS for dip angles less than 65°

Contractor
Sonobuoys Australia (a partnership of AWA Defence Industries Pty Ltd and GEC-Marconi Systems Pty Ltd), Meadowbank, NSW.

FTAAS

Type
Fast time acoustic analysis system (FTAAS).

Description
The FTAAS is a high speed digital processing system that reprocesses acoustic data recorded during airborne maritime surveillance missions on up to 16 parallel channels at speeds of up to eight times real time for post mission analysis.

Data reprocessing provides incident reconstruction, missed detection facilities, signature analysis and crew training.

The system is housed in a portable shelter that can be divided into two units for road/rail transportation. FTAAS contains a deployable subsystem which can be relocated at forward operating bases. This feature

provides local support for aircraft operating away from their main base. The deployable system can process eight channels simultaneously at eight times real time.

Both DICASS and RANGER active buoys can be handled by the system which also has the growth potential to process HARP and VLA buoys when these become available.

The system is controlled by a single operator using either one of two identical workstations. Control functions include tape control, processing modes, processing parameters and display format. Each of the 16 channels can be set up to apply a unique process and display format to the data. These can be reset at any time during replay without affecting the other channels.

Any tape channel may be assigned to one or more of the 16 processing channels. A single channel of data may be simultaneously processed and displayed

using different processes to optimise data extraction. Multiple tape channels may likewise be processed simultaneously with a single process.

Processed video may be displayed at the workstation or on one of eight hard copy units and a colour screen printer. Each hard copy unit can record up to four channels of processed data.

Audio reproduction, including intercom, is available through loudspeakers and headphones to augment the display during real-time processing.

Operational Status
Entered service with the Royal Australian Air Force in 1990.

Contractor
AWA Defence Industries Pty Ltd, Elizabeth, South Australia.

CANADA

AN/SSQ-53D SONOBUOY

Type
Passive directional sonobuoy.

Description
The AN/SSQ-53D DIFAR (DIrectional Frequency Analysis and Recording) sonobuoy is the US Navy's primary sonobuoy sensor. Compared with earlier omnidirectional buoys, the SSQ-53D provides target bearing as well as improved acoustical sensitivity, particularly in the low frequency ranges.

The SSQ-53D is the result of a US Navy sponsored development programme to improve the operational capabilities of the SSQ-53. Changes include extension of the lower limit by one octave to 5 Hz. In addition, omni-acoustic performance has been improved by 10 dB across the frequency range, and the directional hydrophones have been improved by 16 dB at the 10 Hz point.

The sonobuoy may be dropped from an aircraft at indicated airspeeds of 30 to 425 kts, and from altitudes of 30 to 12 200 m. Descent of the buoy is stabilised and slowed by a parachute assembly.

Immediately after water entry, the seawater-activated battery system is energised. Buoy separation

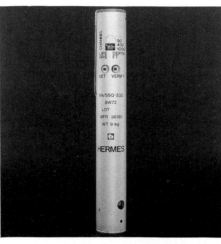

AN/SSQ-53D directional sonobuoy

occurs, jettisoning the parachute assembly and erecting the VHF transmitting antenna. This allows the surface assembly to rise and separate from the sonobuoy housing. The housing serves as a descent

vehicle and separates from the sub-surface assembly at the operating depth. Prior to launch the buoy is programmed via push-buttons to one of 99 channels, life and depth. Readout of these is on an LED display within the buoy.

Operational Status
In production. Deliveries began in mid-1989. Hermes Electronics was the first company to produce and deliver the SSQ-53D. To date Hermes has delivered, or has contracts for, a total of nearly 200 000 SSQ-53D sonobuoys.

Specifications
Frequency: 136-173.5 kHz
RF power output: 1 W (min)
Dimensions: 91.44 × 12.4 cm (diameter)
Weight: 9.1 kg
Operating life: 0.5, 1, 2, 4, 8 h (preselectable)

Contractors
Hermes Electronics Ltd, Dartmouth, Nova Scotia, Canada.
Sparton Corporation, Jackson, Michigan, USA.
Magnavox Corporation, Fort Wayne, Indiana, USA.

INFRASONIC SONOBUOY

Type
Passive omnidirectional LOFAR sonobuoy.

Development
The infrasonic sonobuoy is being developed by Sparton of Canada under its own research and development funding for future use by the Canadian Department of National Defence. Several prototype models have been successfully deployed. The three preselectable hydrophone depths and life can be altered, at no extra cost, to suit the operational requirements of any user.

Description
The infrasonic sonobuoy is an omnidirectional, passive sonobuoy which enables fixed-wing maritime patrol aircraft, helicopters and surface ships to conduct underwater search and surveillance using low frequency analysis and recording (LOFAR) techniques.

The infrasonic sonobuoy is similar in function to the US AN/SSQ-57B and the Canadian AN/SSQ-527B but incorporates features which improve detectability of low level, low frequency signals in high sea state environments (up to state 5). The Sparton VLF sonobuoy is ambient noise limited, even in sea state 0. The sonobuoy features include a sub-surface suspension assembly system which maintains hydrophone depth within specified limits in shear currents up to 2.5 kts.

The sonobuoy can be launched at airspeeds between 45 and 370 kts from altitudes of 45 to 9144 m. Descent is controlled and stabilised by a parachute. Immediately after entry the seawater-activated battery energises a mechanism which deploys the sonobuoy. A flotation bag is inflated and carries the upper electronics assembly, containing a VHF transmitter, to the surface while the hydrophone descends to a preselected operating depth. Three depths are available: 30, 100 or 300 m. The sonobuoy becomes operative within 30 seconds of impact at the minimum depth and within 200 seconds at the maximum depth. It remains in operation for a minimum of one, three or eight hours, according to the preselected operating life.

The sonobuoy can be launched by hand from a ship moving at up to 30 kts without it being necessary to activate mechanical or electrical devices before launch. The electrically activated deployment mechanism functions regardless of attitude on entry since it is not dependent on impact forces acting on the bottom plate.

The sonobuoy is designed to operate in sea conditions up to and including sea state 5. In more extreme conditions the sonobuoy will survive the combination of sea state 6 and twice the ocean shear current profile of 2.5 kts without permanent damage.

Operational Status
In development.

Specifications
Description: LOFAR, passive, omnidirectional
Function: Search, surveillance
Average weight: 16.6 lb (7.53 kg)
Activation time (after splash): 10 s (average), 30 s (max)
Transmitter: 31 channels
RF Power: 1 W
Operating depth: 30, 100 or 300 m (preselectable)
Descent time (in water): 30 s, 65 s, 200 s
Directivity: Omni in both horizontal and vertical plane, ±1 dB
Power source: Seawater-activated battery
Operating life: (1, 3 or 8 h preselectable)
Scuttling: Not longer than 16 h
Launch envelope: CAN STD, NAC 'F-1'
Processor and display: Radio receiving sets: AN/ARR-52A, -72, -75, -76
Aircraft data processor: AN/AQA-4, -5, -7, -7(V); OL-82/AYS; AN/ASA-20, -26; OL-5004/AYS
Operating environmental conditions
Launch air temperature: −20°C to +55°C
Seawater temperature: 0°C to +35°C
Wind velocity: 30 kts
Sea state: 5 (survives in sea state 6)

Contractor
Sparton of Canada Limited, London, Ontario.

MICROBUOY AND MTTF SYSTEM

Type
Miniature sonobuoy and data processing system.

Development
The microbuoy and MTTF system were developed under Canadian government contract by Sparton of Canada Limited in close co-operation with the Defence Research Establishment Atlantic (DREA). The small, lightweight buoy introduces high technology electronics and versatility to the sonobuoy world. MTTF components were developed as a stand-alone system which is easily transferred to other vehicles, yet the output display is as normally seen by ASW operators. The advances achieved by microbuoy developments enable the use of sensors in large quantities for cost-effective search operations.

Description
A small, lightweight sonobuoy has been developed for use by light aircraft or vehicles constrained by storage space. Although the microbuoy is one sixteenth the size of a NATO standard 'A' type sonobuoy, it is able to closely duplicate the performance of current passive omnidirectional sonobuoys. A/20 sized microbuoys are also in development. Microbuoys are packaged in 'A' sized sonobuoy launch containers (SLCs).

Sonobuoys are no longer limited to use by large aircraft. Furthermore, helicopters and light aircraft need not be operationally limited by sonobuoy stores. Weight savings can be converted into longer time on station, and space allocation will now be capable of satisfying several more mission requirements. Nowhere is this new capability more evident than in the fact that sonobuoys are now a viable option for use by UAVs. ASW pods can carry these mission sonobuoys without the need to pierce the skin of the launching aircraft, thus turning any light aircraft or helicopter with external hard points into a potential ASW platform.

A total of 16 buoys is packaged in an SLC in four columns of four buoys. Dividers separate the columns of buoys and provide electrical connections. When

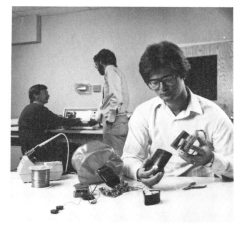

Microbuoy

MTTF

deployed, all sixteen buoys transmit digitised acoustic data on a single RF channel using a Time Division Multiplexed (TDM) scheme. Data are received by the launching vehicle using the MTTF (Microbuoy Transportable Test Facility) system.

The MTTF consists of a ruggedised computer system, keyboard, trackerball, VHF receiver, high resolution colour monitor and graphics printer. It is contained in a standard 19 in equipment rack. It controls the initialisation and launch sequence of the microbuoys and also demodulates and decodes the acoustic data transmitted from the microbuoys. The data from each buoy are stored on computer disks and can be displayed as a colour LOFARgram on the monitor.

Operational Status
Advanced development models tested successfully.

Specifications
Description: Aerially deployed, passive, omnidirectional sonobuoy
Function: Search, surveillance
Size/Weight: A/16 size 0.91 kg, A/20 size 0.85 kg
Activation time: Immediate after launch

Transmitter: 99 channels (programmable before launch)
Channel multiplexing: Shares one VHF channel with up to a maximum of 20 microbuoys
RF power: 0.25 W (min)
Sensor type: Double bender element
Acoustic range: 30-5000 Hz
Operating depth: 40 m (min)
Descent time: 60 s (max)
Operating environmental conditions
Launch air temperature: −20°C to +55°C
Seawater temperature: 0°C to +35°C
Wind velocity: 20 kts (max)
Sea state: 3.5 (survives in sea state 5)
Current profile: 0-90%
Power source: Lithium/Thionyl-chloride batteries
Shelf life: 7 years (min)
Operating life: 4 h (min)
Scuttling: Not longer than 24 h
Launch envelope: NADC Class 1
Processor and display: MTTF

Contractor
Sparton of Canada Limited, London, Ontario.

FRANCE

TSM 8030 (DSTV-7) SONOBUOY

Type
Passive omnidirectional sonobuoy.

Description
The A/3 size TSM 8030 (DSTV-7) is an omnidirectional passive buoy which provides a number of additional facilities. These include the selection of the VHF frequency channel from the 31 or 34 available before the buoy is dropped, choice of 31 or 34 channels from those available between 136 and 173.5 MHz at 365 or 375 kHz spacing, a three position selector to give a choice of hydrophone depth immersion and an increased maximum depth. Other improvements are reduction in overall weight and

volume, and the use of a small balloon to act as a brake during the trajectory of the buoy in the air, and then as a floater on the water surface.

Three variants are available: TSM 8030A, B and C. All are airdropped from 150 to 10 000 ft within a speed range of 80 to 250 kts. They are NATO interoperable.

Operational Status
In production.

Specifications
Frequency coverage: 10 Hz to 20 kHz (optionally 5 Hz to 20 kHz)
VHF emission: TSM 8030A — selectable among 31 channels spaced 375 kHz apart from 162.25 to 173.5 MHz; TSM 8030B — selectable among 34 channels

spaced 375 kHz apart from 136 to 158.375 MHz; TSM 3080C — selectable among 34 channels spaced 365 kHz apart from 148.75 to 161.125 MHz
RF power: 1 W
Deployment: Airdrop from 150 to 10 000 ft within the speed range 80 to 250 kts
Hydrophone immersion depth: 20, 100 or 300 m (selectable)
Power supply: Magnesium/lead chloride battery
Operational life: 1, 3 or 8 h (selectable)
Size: A/3 ± 123.82 mm (diameter) × 304 mm (high)
Weight: 4.5 kg

Contractor
Thomson Sintra Activités Sous-Marines, Valbonne Cedex.

TSM 8050 SONOBUOY

Type
Active omnidirectional sonobuoy.

Description
The TSM 8050 is an omnidirectional active sonobuoy designed for submarine localisation from aircraft. The buoy can be dropped from fixed- or rotary-wing aircraft from altitudes of 150 to 10 000 ft at speeds between 80 and 250 kts. Two versions are available, TSM 8050A and B, which have different hydrophone immersion depths. Onboard selection of one from 12 possible channels spaced 750 kHz apart from 162.25 to 170.5 MHz is available.

Operational Status
In production and qualified for use in the ATL-2 maritime patrol aircraft.

Specifications
Acoustic frequencies: 6, 6.71, 7.5, 8.4, 9.4 or 10.5 kHz
CW pulses: 4 pulses at the selected frequency duration 100 ms
FM pulses: 1 pulse at the selected frequency duration 320 ms
Recurrence period: 11 s
Deployment: Air drop from 150 to 10 000 ft at speeds between 80 and 250 kts.
Hydrophone immersion depth: TSM 8050A — 20 or 150 m; TSM 8050B — 20 or 450 m

VHF emission: On one of 12 possible channels spaced 750 kHz apart from 162.25 to 170.5 MHz
Emission power: 0.25 W
Power supply: Lithium cell; magnesium/lead chloride battery
Operational life: 30 mins
Size: 123.82 mm (diameter) × 914 mm (height)
Weight: 17.69 kg

Contractor
Thomson Sintra Activités Sous-Marines, Valbonne Cedex.

ALKAN LAUNCHERS

Type
Sonobuoy launchers for aircraft.

Description
Alkan has designed a number of sonobuoy launchers for use in both fixed- and rotary-wing aircraft. They meet a range of operational requirements, such as inflight reloading and release or ejection from pressurised or unpressurised cabins. Various launcher sizes and capacities are available to allow for easy installation in the cabin or equipment bay.

Container-launcher
Each buoy is packed in a container-launcher which provides for its storage and ejection, and is equipped with a pressurised vessel at the top and a releasable cap at the bottom. The power for ejection is provided by compressed gas stored in the vessel at the top of the buoy.

Type 8030/8031
These launchers have been designed specifically for the Dassault-Breguet Atlantic Mk 1 and Atlantique Mk 2 aircraft which can house up to four launchers. The capacity of each launcher is 18 size 'A' and 18 size 'F' buoys respectively.

Type 8050
This is a new type of launcher designed for the launch of 'F' size sonobuoys from naval Lynx helicopters. It has a capacity of 18 'F' size buoys.

Operational Status
The Type 8030 is in production. The Type 8050 is in final development and trials.

Sonobuoy launch pod for the Lynx

Specifications (Type 8050)
Capacity: 18 'F' sized buoys
Ejection speed: 6 m/s
Installation: 14 in (standard lugs)
Dimensions: 2500 × 576 × 412 mm
Weight: 80 kg (empty)

Contractor
R Alkan & Cie, Valenton.

ITALY

BI SERIES SONOBUOYS

Type
Passive omnidirectional sonobuoys.

Description
The Servomeccanismi organisation produces three types of passive sonobuoy:
BIT-3: this is an 'A' size omnidirectional passive sonobuoy for use at depths between 20 and 100 m, with a selectable life of one to three hours, and transmitting data over a 31-channel IW link.
BIT-8: this is similar to the BIT-3 but has different frequency/sensitivity characteristics.
BIR: this is a miniature (500 mm long × 100 mm diameter, 3 kg) sonobuoy for use with helicopters. It is omnidirectional and can be deployed at depths to 20 m. Life is one hour.
Receivers for use with these sonobuoys are the REA-16 and REA-31, having 16 and 31 reception channels respectively. The BI series buoys have been designed primarily for use in the Mediterranean, under conditions normally found in that sea area.

Operational Status
In production for the Italian Navy.

Contractor
Servomeccanismi, Pomezia.

MSR-810 SONOBUOY

Type
Passive omnidirectional sonobuoy.

Description
The MSR-810 passive sonobuoy, for detecting and locating submarines from aircraft, is manufactured by Whitehead under Thomson Sintra licence for the Italian market. The sonobuoy transmits, by VHF/FM radio, the underwater acoustic AF signals received from the hydrophone. The MSR-810 complies with NATO and Italian Navy specifications.

Operational Status
Production and in service.

Specifications
Type: Passive
Frequency: Low
Weight: 8 kg
Operating life: 1.3 or 8 h
RF channels: 31
RF output: 500 mW (at antenna)
Antenna: ¼ wavelength

Contractor
Whitehead, Montichiari (Brescia).

UNION OF SOVEREIGN STATES

Virtually no information had ever been made public on Russian sonobuoys until the US Department of Defense released the photograph shown here and published it in *Soviet Military Power*. Early sonobuoys were the RGB-56/64, the latter being a smaller version of the former. In the late 1960s a passive omnidirectional buoy, the BM-1, was produced. It apparently has a 29-channel FM data link at about 171 MHz.
The buoy shown here is the Type 75 which is similar to the US AN/SSQ-41B LOFAR sonobuoy, although somewhat larger.

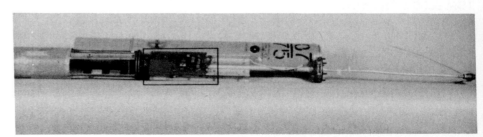
Type 75 LOFAR sonobuoy (Soviet Military Power)

UNITED KINGDOM

SSQ-906/906A/907 LOFAR SONOBUOYS

Type
Miniature passive omnidirectional sonobuoys.

Description
These 'F' size sonobuoys are designed to be launched from aircraft or helicopters at airspeeds between 60 and 300 kts and at altitudes between about 50 and 10 000 m. Once deployed, they detect and amplify underwater sounds to modulate the self-contained FM transmitter. The buoy is capable of operating with a range of airborne signal processors, including the GEC Avionics AQS-901, AQS-902 and AQS-903.

A significant feature is the fact that these buoys have an exceptionally low frequency acoustic performance which results in a significant improvement in operational performance over previous LOFAR sonobuoys. The 907 is a calibrated version of the 906.

The output from the hydrophone is amplified in a pre-amplifier to establish a satisfactory signal-to-noise ratio. The output from the pre-amplifier is further amplified in the sonic amplifier which consists of three IC active filter stages. The first two provide virtually all

SSQ-906 sonobuoy

the shaping with the third stage giving HF roll-off, gain and symmetrical limiting. The amplifier is directly coupled through a buffer to the modulator, to ensure good low frequency performance.

The VHF transmitter consists of a crystal controlled oscillator with a variable capacity diode in series with the crystal. The output from the sonic amplifier is

applied across the variable capacitance diode and results in frequency modulation of the crystal oscillator. The second harmonic of the oscillator drives a doubler stage, the output of which drives a second doubler stage. Finally a power amplifier is used to produce an output of 1 W. The radiation pattern is omnidirectional using a vertical quarter wave antenna and ground plane.

Operational Status
In service with the Royal Navy, Royal Air Force and the French Navy.

Specifications
Type: Miniature passive omnidirectional
Frequency coverage: 4 Hz to 3 kHz
Hydrophone depth: Selectable for either 60 or 300 ft (18 or 91 m); or 60 or 450 ft (18 or 137 m)
Operating life: 1, 4 or 6 h
RF channels: 99
RF output: 1 W (min)
Dimensions: 30.5 × 12.5 cm (diameter)
Weight: 4.5 kg

Contractors
Marconi Underwater Systems Ltd, Templecombe.
Dowty Maritime Sonar and Communication Systems, Greenford.

SSQ-954B DIFAR SONOBUOY

Type
Miniature passive directional sonobuoy.

Description
The SSQ-954B DIFAR (DIrectional Frequency Analysis Recording) sonobuoy is a fully automatic directional passive acoustic sensor which has been developed to meet specific needs of Royal Navy and Royal Air Force ASW operations. The sonobuoy design employs a high performance DIFAR hydrophone system, packaged and developed into a 'G' size sonobuoy configuration. The DIFAR sonobuoy design is fully compliant with UK MoD specifications and also achieves the full performance capabilities of the AN/SSQ-53B.

The SSQ-954B may be launched from a helicopter or fixed-wing aircraft using standard sonobuoy launch systems, at speeds of between 50 and 300 kts and at an altitude of between 45 and 9200 m. After launch a parachute is automatically deployed to retard descent to approximately 30 m/s and stabilise the sonobuoy trajectory. Upon water entry the pre-inflation system (provided for helicopter safety in the event of ditching) prevents deployment until the buoy reaches a depth of 12 ft; at this point the parachute is jettisoned and separation is activated. The surface unit float is automatically inflated to deploy the RF antenna while the hydrophone assembly descends to its selected operating depth. Acoustic signals received by the omni and directional hydrophones are multiplexed, together with azimuth bearing reference data, onto the VHF carrier as a frequency modulated signal for transmis-

SSQ-954B DIFAR sonobuoy

sion to the aircraft. At the end of its selected life the unit automatically scuttles.

The SSQ-954B includes a user-friendly autonomous function selection system to select, set and verify electronically the sonobuoy RF transmitter frequency and its operating life and depth.

The directional hydrophone provides sine and cosine signals which are multiplexed with the omnidirectional hydrophone and compass signal. This complex signal is then used to modulate the VHF transmitter in the surface unit.

Buoys manufactured by Marconi Underwater Systems use the Magnavox SSQ-53B low profile blade array sensor.

Operational Status
In service with the Royal Air Force and the Royal Navy. An anchored variant is in service with Sweden. Orders for the D variant are expected to be placed in the Spring of 1992.

Specifications
Length: 419 mm
Diameter: 124 mm
Weight: 6.8 kg
Transducer depth: 30/140/300 m (selectable)
RF channels: 99
RF power: 1 W (min)
Frequency: 136-173.5 MHz (synthesised and programmable)
Life: 1, 4 or 6 h (selectable)

Contractors
Marconi Underwater Systems Ltd, Templecombe.
Dowty Maritime Sonar and Communication Systems, Greenford.

SSQ-963B SONOBUOY

Type
Command active multi-beam sonobuoy.

Description
The SSQ-963A command active multi-beam sonobuoy (CAMBS) is an 'A' size sonobuoy which was developed in 1987 by Dowty under contract to the British Ministry of Defence. The latest version of CAMBS, the SSQ-963B, entered service with the Royal Air Force at the end of 1991 and offers further significant operational enhancements.

CAMBS is an advanced active directional sonobuoy which features a high data rate, and directional information is obtained by the use of a transmitting acoustic projector combined with a directional receiving hydrophone array. Hydrophone depth is

adjustable by radio command and there is a 31-channel radio telemetry link operating on standard NATO frequencies for communication with the parent aircraft. The antenna is a monopole with ground plane, erected by and integral with the buoy flotation bag. The design incorporates a flux-gate compass with very rapid stabilisation.

While in the passive mode the buoy relays all signals detected by the receive array hydrophones together with magnetic compass information to the controlling platform via the telemetry link. In the active mode acoustic pulses are emitted from the omnidirectional projector and the returned echoes are detected by the receive array and, along with compass information, are amplified and fed to a multiplexer. There they are combined to provide a multiplexed signal which is transmitted to the airborne platform where it is processed to derive a 'fix' of detected targets.

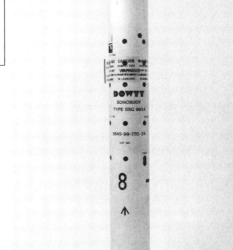

SSQ-963A CAMBS sonobuoy

The power output of the transmitter is 0.5 W and this is fed to the antenna whose radiation pattern is omnidirectional in the horizontal plane with a single lobe typically between 5° and 25° in the vertical plane.

Operational Status

The SSQ-963B sonobuoy is part of the equipment developed for the Mk II version of the Nimrod ASW

SSQ-981A BARRA SONOBUOY

Type

Passive broadband directional sonobuoy.

Description

The SSQ-981A is the third generation Barra sonobuoy to be developed by Marconi Underwater Systems for the UK MoD. Derived from the Australian SSQ-801, the SSQ-981A has been designed to reduce manufacturing costs while maintaining the complete performance capability of the Barra concept.

SSQ-981A is an advanced passive broadband sonobuoy sensor; 25 hydrophones mounted on five arms and arranged over a large acoustic planar array achieve an enhanced detection gain over conventional buoys. Spectral and broadband energy is digitally telemetred to the ASW aircraft. A Barra compatible acoustic processor forms 60 beams providing 360° cover to give excellent signal-to-noise discrimination resulting in good azimuth resolution and extremely accurate contact bearings.

SSQ-981A is configured as a conventional 'A' size sonobuoy which may be launched from helicopter or fixed-wing aircraft using existing stowage and launching systems. It operates with preset VHF channels and allows operator selection of life and depth prior to launch.

aircraft. In production. Dowty received additional orders for CAMBS in 1990 and 1991.

Specifications

Length: 914 mm
Diameter: 124 mm
Weight: 11.3 kg
Transducer depth: 3 selectable depths

Operational Status

In production.

Specifications

Length: 914 mm
Diameter: 124 mm
Weight: 8.16 kg
Transducer depth: 21.5 or 121.5 m (selectable prior to launch)
RF channels: 50
RF power: 1 W (min)
Frequency: 136-173.5 MHz
Life: 1, 2, 3 or 4 h

Contractor

Marconi Underwater Systems Ltd, Templecombe.
  Dowty Maritime Sonar and Communication Systems, Greenford.

RF channels: 31
RF power: 0.5 W (min)
Frequency: 162-173.5 MHz
Life: 1 h, automatic or command setting

Contractor

Dowty Maritime Sonar and Communication Systems, Greenford.

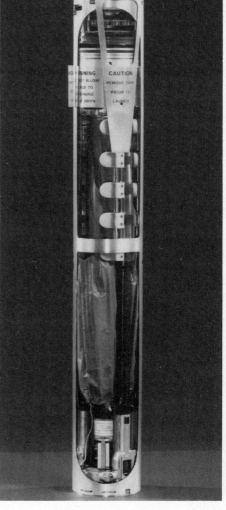

SSQ-981A Barra sonobuoy

SSQ-991 SONAR TRANSPONDER

Type

Helicopter-launched expendable transponder.

Description

The SSQ-991 air-launched sonar transponder is a first generation, helicopter-launched, expendable sonar equipment. It has been developed as a combined radar/dipping sonar alignment aid for Royal Navy ASW helicopters. Based on established sonobuoy techniques, it is housed in an 'A' size configuration. It may be launched from the helicopter using existing sonobuoy stowage and launch systems.

Although programmed to operate with the Type 195M dipping sonar, the SSQ-991 can be adapted to operate with other sonar equipments in the medium ASW acoustic frequency band. The transponder is capable of many pulse transmission modes, including CW, shaped CW, Doppler simulation frequency shifted pulses and FM pulse forms. The octahedral geometry of the radar reflector is optimised around I-band radar frequencies to provide a radar cross-section varying between 3 and 12 m. With an

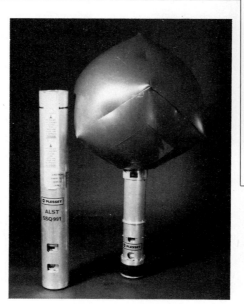

Air-launched sonar transponder

operating range of around 8000 m, the SSQ-991 provides an effective test capability for the ASW system.

Contractor

Marconi Underwater Systems Ltd, Templecombe.

AD 130 SONAR RECEIVER

Type

Sonobuoy homing and DF receiver.

Description

The AD 130 provides sonobuoy homing and DF facilities on 99 channels over the frequency range 136 to 173.5 MHz at a channel spacing of 375 kHz. It is intended for use on both fixed- and rotary-wing aircraft. The system consists of a receiver with computer interface, a remote-control unit and antenna switching controls. Additional units provided include a DF antenna, a left/right and fore/aft antenna for sonobuoy

'on-top' indications, and cross-pointer and radio magnetic indicators. Dual receivers, with independent channel selection, may be operated from a single controller.

The system can be operated in either automatic or manual channel selection modes. In automatic mode the computer carries out selection by decoding the serial bit stream, while manual selection is accomplished by means of pairs of thumbwheel switches. An LED readout indicates the selected channel. The homing system provides left/right and fore/aft indications on the meter and the DF section detects and processes AM signals from the DF antenna for azimuth display. Built-in test circuitry continuously

monitors all modes of operation. The system is stated to be compatible with all known and planned NATO and other allied services sonobuoys.

Operational Status

In production and in service.

Specifications

Frequency range: 136-173.5 MHz
Channel spacing: 375 kHz
No of channels: 99
Frequency accuracy: 5 parts per million
Selectivity: 3 dB bandwidth 250 kHz (min), 60 dB bandwidth 750 kHz (max)

Dimensions
Receiver: 197 × 125 × 324 mm (0.5 ATR short)
Controller: 152 × 102 × 302 mm
Switched antennas: Each 50 × 80 × 193 mm
Weights: 8 kg (receiver); 2.3 kg (controller); 0.8 kg

(each switched antenna)

Contractor
GEC Sensors Ltd, Basildon.

AD 130 sonar homing and DF receiver

TYPE 843 SONAR RECEIVER

Type
Sonobuoy signal receiver for helicopters.

Description
The Type 843 is a multi-channel sonar receiver of modular construction for use in ASW helicopters. It simultaneously receives four VHF channels which are selected by serialised digital commands from a processor or control unit. Advanced filter technology, extensive use of hybrid circuits and wideband techniques ensure high availability. An antenna pre-amplifier unit gives a low noise figure and the

system's high dynamic range and wide bandwidth enable high speed telemetry data to be received.

The receiver provides its full RF performance when installed at up to 50 m from the pre-amplifier. All interfaces are to ARINC specification 404A. Outputs are compatible with existing aircraft data processing systems and with more advanced systems under development. A built-in test signal generator provides extensive check facilities for the receiver and associated processing equipment, and is designed to facilitate servicing by ATE systems.

Specifications
Frequency: 134.5-174.625 MHz

Frequency stability: ±3 PPM (at −26°C to +45°C ambient temperature)
No of channels: 4
Channel coverage: 99 RF channels plus 2 test channels
Channel spacing: 375 kHz
Input impedance: 50 Ω
Modulation: FM/FSK
Weights: 1.3 kg (pre-amplifier); 2.5 kg (control unit); 11.4 kg (receiver)

Contractor
Marconi Underwater Systems Ltd, Addlestone.

ARR 901 SONAR RECEIVER

Type
Multi-channel sonar receiver.

Description
The ARR 901 is a multi-channel sonar receiver of modular construction for use in fixed-wing maritime patrol aircraft. It simultaneously receives eight VHF channels which are selected by serialised digital commands from a processor or control unit. Advanced filter technology, extensive use of hybrid circuits and wideband techniques ensure high availability. An antenna pre-amplifier unit gives a low noise figure

and the system's high dynamic range and wide bandwidth enable high speed telemetry data to be received.

The receiver provides its full RF performance when installed at up to 50 m from the pre-amplifier. All interfaces are to ARINC specification 404A. Outputs are compatible with existing aircraft data processing systems and with more advanced systems under development. A built-in test signal generator provides extensive check facilities for the receiver and associated processing equipment, and is designed to facilitate servicing by automatic test equipment.

Specifications
Frequency band: 134.5-174.625 MHz

Frequency stability: ±3 PPM (at −26°C to +45°C ambient temperature)
No of channels: 8
Channel coverage: 101 RF channels plus 2 test channels
Channel spacing: 375 kHz
Input impedance: 50 Ω
Modulation: FM/FSK
Weights: 1.3 kg (pre-amplifier); 23.6 kg (receiver)

Contractor
Marconi Underwater Systems Ltd, Addlestone.

TYPE R605 RECEIVER

Type
Sonobuoy data receiver.

Description
The R605 receives, demodulates and amplifies FM sonobuoy signals in the frequency range from 136 to 173.5 MHz over 99 channels. Simultaneous operation can be carried out on four channels and a demodulated output for analysis and display is provided. The equipment is DIFAR and DICASS compatible.

The R605 is a much improved version of the AN/APR-75 and uses many of its mechanical and electrical design features to simplify retrofit applications.

Control options include MIL-STD-1553, RS 422, ARINC 429 or a control box. MTBF is quoted as better than 2000 hours, and the life of the system is better than 20 000 hours of continuous operation.

An optional radio set provides independent selection and meter monitoring of any of the 99 channels for each of the receiver modules, and an optional remote pre-amplifier allows receiving sets to be mounted at a distance from the system antenna.

Operational Status
In production and in operational service. It is understood to have been selected for the Royal Navy's Merlin helicopter.

Specifications
Noise figure: 5 dB at 50 Ω
Stability: ± 3 PPM
Weight: 10 kg

Contractor
Dowty Maritime Sonar and Communication Systems, Greenford.

R605 sonobuoy receiving set

TYPE AN/RR-XX

Type
Sonobuoy data receiver.

Description
This set is similar to the R605 (see above) with the

added ability that it can interface with the R612 (see next entry). Extended control is provided from a control box or via ARINC 429 and the receiver also features a wide dynamic range and high sensitivity over a broadbase bandwidth.

Operational Status
Selected for use by the US Navy.

Contractor
Dowty Maritime Sonar and Communication Systems, Greenford.

TYPE R612 RECEIVER

Type
On-top position indicator sonobuoy receiver.

Description
The Type R612 is a modular 99-channel, on-top position indicator which covers the frequency range 136 to 173.5 MHz. It is suitable for use on both fixed-wing aircraft and helicopters. The equipment is fully compatible with all known sonobuoys and a variety of antenna systems. When used in conjunction with an ADF system it enables the operator to home on and locate sonobuoys on any of the 99 channels. The R612 is interchangeable with the R-1651/ARA without changes to existing wiring. Control options include RS 422, MIL-STD-1553B (via R605), or a manual control box.

The R612 functions as an amplitude demodulator for the direction finding system, and produces an indication that an adequate signal exists when the RF signal input is sufficient for reliable direction-finder operation. The presence of the adequate signal is indicated in three ways; a 400 Hz tone to the operator's headset, a light on either side of the control box, or as a discrete signal to the R605 sonobuoy receiver.

Operational Status
In production and in operational service.

Specifications
Sensitivity: 10 dB S/N ratio at 2.5 μV input
Dimensions: 126 × 135 × 78 mm
Weight: 2.16 kg
MTBF: 3000 h

Contractor
Dowty Maritime Sonar and Communication Systems, Greenford.

TYPE T613

Type
Sonobuoy data receiver.

Description
The T613 is an improved version of the T843 receiver which it replaces. The lightweight system is encased in an aluminium chassis, offering higher stability and a lower noise signature. The receiver operates on four simultaneous channels and has a full 99-channel capability but without the high level of sophisticated BITE provided with the other receivers. A dedicated control unit and separate RF amplifier for increased signal-to-noise performance is also incorporated.

Contractor
Dowty Maritime Sonar and Communication Systems, Greenford.

AA34030

Type
Sonobuoy command transmitter.

Description
The AA34030 is a UHF transmitter developed specifically to function as the sonobuoy command transmitter for the RF downlink of a sonics data processing system, allowing selection of the sonobuoy mode, uplink channel, ping and so on. The transmitter provides AM and FM, analogue and digital communications over the frequency range 282 MHz to 292 MHz with 50 kHz channel spacing throughout the band.

The design of the transmitter is related to the

AA34030 sonobuoy command transmitter

AD3400 multi-mode VHF/UHF airborne communications transmitter/receiver in using the same 'sliced' modular construction, with four of the six modules making up the transmitter being common to both units.

Specifications
Frequency range: 282-292 MHz
Channel spacing: 50 kHz
No of channels: 200
Frequency accuracy: 2 parts per million
Transmitter power: 50 W FM, 130 W AM
Weight: 6.5 kg

Contractor
GEC Sensors Ltd, Basildon.

UNITED STATES OF AMERICA

AN/SSQ-41B SONOBUOY

Type
Passive omnidirectional sonobuoy.

Description
The AN/SSQ-41B is an omnidirectional, passive sonobuoy with a multiple life and depth capability. In volume production since 1963, the series has been the mainstay of US Navy sonobuoy operations but is gradually yielding this role to the newer directional AN/SSQ-53B.

The AN/SSQ-41B, extensively redesigned from the earlier AN/SSQ-41, provides an audio bandwidth expanded to 10 kHz and improved dynamic range capabilities. The depth selection is 60 or 1000 ft and life selection is one, three or eight hours.

The AN/SSQ-41B may be launched from an aircraft at airspeeds between 30 and 425 kts and from altitudes of 100 to 40 000 ft (30.5 to 12 200 m).

After a controlled and stabilised descent and upon impact with the water, the termination mass and the hydrophone are released and descend to a preselected operating depth of either 60 or 1000 ft (18.3 or 305 m). On contact with seawater, the battery is activated and the sonobuoy becomes fully operative within 60 seconds.

Modulation is accomplished in the crystal oscillator stage by a variable-capacity diode. Two frequency-doubling amplifiers multiply the RF oscillator-doubler frequency to the desired operating frequency of the sonobuoy, which is preset to one of 31 channel frequencies within the 162.25-173.50 MHz band. Available in both standard and low noise versions.

Operational Status
Although large scale production was contracted until 1982, the US Navy has discontinued further development. This unit will continue to be available.

Specifications
Type: Passive
Audio frequencies: 10-10 000 Hz
Depth: 18.3 or 305 m
Size: A
Weight: 9 kg
Operating life: 1, 3 or 8 h
Launch altitude: 30.5-12 200 m
Launch speed: 30-425 kts
RF channels: 31
RF power: 1 W

Contractors
Sparton Corporation, Electronics Division, Jackson, Michigan.

AN/SSQ-47B SONOBUOY

Type

Active omnidirectional sonobuoy.

Description

The AN/SSQ-47B is an active, non-directional, 'A' size sonobuoy. The operational function is that of detection, classification, tracking and location of submarines.

Sonobuoy AN/SSQ-47B provides an active sonar capability for fixed-wing aircraft. It can be operated from a minimum range of zero to 10 nautical miles at an altitude of between 500 and 10 000 ft and in sea conditions up to sea state 5. The AN/SSQ-47B permits launching and operation of up to six sonobuoys, either individually or simultaneously, without encountering RF or sonar interference.

Operational Status

In operational service. The AN/SSQ-47B, while no longer carried in the US Navy inventory, is in widespread use with other navies and is still in production for this purpose. Licensing arrangements with Canada, the UK, Norway and Japan are in existence. It is produced by Sparton of Canada as the AN/SSQ-502.

Specifications

Type: Active
Sonar modes: Automatic keyed CW
Depth: 18.3 or 244 m
Size: A
Sonar channels: 6 (HF)
RF channels: 12
RF power: 0.25 W
Weight: 14.5 kg

Contractor

Sparton Corporation, Electronics Division, Jackson, Michigan.

AN/SSQ-53B DIFAR SONOBUOY

Type

Passive directional sonobuoy.

Description

The AN/SSQ-53B DIFAR (DIrectional Frequency Analysis and Recording) sonobuoy is the US Navy's primary sonobuoy sensor. As compared to earlier passive buoy types, the SSQ-53B provides target bearing information as well as improved acoustical sensitivity, particularly in the low frequency ranges.

The SSQ-53B is the result of a US Navy-sponsored development programme to improve the operational capabilities of the earlier SSQ-53. Changes include a triple depth capability (100, 400 or 1000 ft (30, 120 or 300 m)), an extended launch envelope and improved low frequency performance. The sonobuoy's electrical and mechanical design was extensively revised to improve reliability and to lower production costs.

The AN/SSQ-53B may be dropped from an aircraft at indicated airspeeds of 30 to 425 kts and from altitudes of 100 to 40 000 ft (30 to 12 200 m). Descent of the sonobuoy is stabilised and slowed by a parachute assembly.

Immediately after water entry, the seawater-activated battery system is energised. Buoy separation occurs, jettisoning the parachute assembly and erecting the VHF transmitting antenna. This allows the surface assembly to rise and separate from the sonobuoy housing. The housing serves as a descent vehicle and separates from the sub-surface assembly at the operating depth. Prior to launch the buoy is programmed via push-buttons to one of 99 channels, life and depth. Readout is on an LED display contained within the buoy.

Operational Status

In production. Deliveries began in early 1984 and to date Magnavox has delivered, or has contracts for, a total of nearly one million SSQ-53B sonobuoys.

Specifications

Type: Directional LOFAR
Size: A
Weight: 8.4 kg
Operating life: 1, 4 or 8 h
Launch altitude: 100-40 000 ft
Launch speed: 30-425 kts
RF channels: 99
RF power: 1 W

Contractors

Sparton Corporation, Electronics Division, Jackson, Michigan.
Magnavox, Fort Wayne, Indiana.
Dowty Maritime Sonar and Communication Systems, UK (licence).
Marconi Underwater Systems Ltd, UK (licence).
Hermes Electronics, Canada (licence).

AN/SSQ-53B sonobuoy

AN/SSQ-53C DIFAR SONOBUOY

Type
Dwarf passive directional sonobuoy.

Description
The Sparton Dwarf DIFAR sonobuoy is the electrical and acoustical equivalent of a standard AN/SSQ-53B sonobuoy, but occupies only a third of the volume. The Dwarf DIFAR has the 4.87 in (124 mm) diameter of a conventional 'A' sized sonobuoy but an overall length of only 12 in (305 mm). The small size of the Dwarf DIFAR sonobuoy enables a threefold increase in the number of units an ASW aircraft can carry and is particularly valuable with small fixed-wing ASW aircraft and helicopters. Dwarf DIFAR sonobuoys are compatible with existing shipping containers and handling equipment and the cost saving in shipping and handling is substantial. They can be incorporated into existing fleet operations with minimal changes to associated equipment and deployment philosophy.

The Dwarf DIFAR sonobuoy can be deployed from fixed-wing or rotary-wing aircraft at indicated airspeeds of 0 to 370 kts and from altitudes of 40 to 30 000 ft (12 to 9144 m). Air descent is stabilised and slowed by a parachute assembly.

The major components of the Dwarf DIFAR sonobuoy are the external sonobuoy housing, the antenna/VHF assembly, the suspension assembly (which consists of the cable pack, the stabilising kite, the damper disc, and the compliant cable), and the integrated hydrophone and electronics assembly.

Upon entry to the water, the Dwarf DIFAR lithium-sulphur dioxide battery pack is activated by a seawater switch that detonates a squib, causing gas-filled cylinders to inflate the sonobuoy flotation bag, jettisoning the parachute assembly and erecting the VHF transmitting antenna. Inflation of the flotation bag provides the pressure force that activates the release plate in the top of the sonobuoy. This causes the parachute to be jettisoned and permits the surface assembly to separate from the sonobuoy housing and the surface housing. The sonobuoy housing serves as a descent vehicle and falls away from the sub-surface assembly at the operating depth, either 30 or 300 m.

Operational Status
The Dwarf DIFAR buoy was developed in the mid-1980s but no information has been released on any later funding, or any orders for the buoy. It is understood that this buoy has now been cancelled.

Specifications
Type: Passive
Depth: 30 or 300 m
Size: 'A' diameter × 305 mm
Weight: 4.5 kg
Operating life: 1 or 4 h
Launch altitude: 12-9144 m
RF channels: Multi-channel

Contractor
Sparton Corporation, Electronics Division, Jackson, Michigan.

AN/SSQ-57A SONOBUOY

Type
Passive intelligence gathering sonobuoy.

Description
The AN/SSQ-57A sound reference sonobuoy features a hydrophone that affords exceptional performance stability under temperature and pressure extremes. The hydrophone is a piezo-electric ceramic segmented cylinder with a smooth frequency response to 20 kHz. The sonobuoy is calibrated to allow determination of underwater acoustic sound pressure levels over this wide frequency range.

The AN/SSQ-57A can be air-launched from altitudes between 150 and 10 000 ft (46 and 3048 m) into sea conditions of up to sea state 5 and is effective at ranges up to 10 nautical miles.

Immediately after entering the water the seawater-activated battery system is energised. Gas-filled cylinders inflate the sonobuoy float, jettisoning the parachute assembly and erecting the VHF transmitting antenna. This permits the surface assembly to rise and separate from the sonobuoy housing. The sonobuoy housing serves as a descent vehicle and separates from the sub-surface assembly at the operating depth.

Operating life is one, three or eight hours, selected before launch. At the end of the operating life, the transmitter is shut off. The watertight sonobuoy housing is equipped with a seawater-soluble plug that dissolves to effect scuttling. The plug's dissolution rate varies as a function of water temperature but is never less than eight hours (maximum selectable operating life) or more than 20 hours.

The AN/SSQ-57A (XN-5) is a version of the AN/SSQ-57A that has a fixed operating depth of 1000 ft (305 m) and is available in both standard and low noise versions.

The AN/SSQ-57B, now in the development stage, provides similar performance to the AN/SSQ-41, but with a slightly greater bandwidth.

Operational Status
The AN/SSQ-57A has been in production since 1968. It is in widespread operational use in the US Navy and a number of other navies. Magnavox has a contract to produce over 40 000 AN/SSQ-57B units.

Specifications
Type: Passive, calibrated audio
Audio frequencies: 10-20 000 Hz
Depth: 18-91 m
Size: A
Weight: 8.6 kg
Operating life: 1, 3 or 8 h
Launch altitude: 30-12 200 m
Launch speed: 30-425 kts
RF channels: 31
RF power: 1 W

Contractors
Sparton Corporation, Electronics Division, Jackson, Michigan.
Magnavox, Fort Wayne, Indiana.

AN/SSQ-58A MOORED SENSOR BUOY

Type
Passive/active sensor buoy.

Description
Sparton Electronics has developed the Moored Sensor Buoy, a multi-purpose building-block system for 'at-sea' measurements. Two separate versions of the buoy are available: a passive system for harbour surveillance, and an active system for use in test ranges and so on. Other applications of the buoy include a variety of oceanographic and environmental requirements. The buoy is retrievable and reusable and can also be configured in a free-floating version. Powered by commercially available batteries, or extended life alternative power sources, it has a long operational life. It is available with 99 VHF selectable frequency channels and is compatible with all sonobuoy receivers and processors.

The buoy consists of a foam-filled glass fibre float, from which is suspended its mooring and hydrophone array. It is normally deployed from a small vessel, usually in depths between about 12 to 35 m. A deep mooring version is also available. A VHF antenna and a radar reflector are mounted on top of the float.

Operational Status
In production for the US Navy.

Specifications
Frequency range: 50 Hz to 10 kHz
Min hydrophone depth: 6 m
Operational life: 100 h (between battery recharging)
Dimensions of float: 1 m (diameter) × 60 cm (deep)

Contractor
Sparton Corporation, Electronics Division, Jackson, Michigan.

AN/SSQ-62B DICASS SONOBUOY

Type
Active directional sonobuoy.

Description
The AN/SSQ-62B is the sonobuoy component of the DIrectional Command Active Sonobuoy System (DICASS). A high performance sonobuoy, it detects the presence of submarines using sonar techniques under direct command from ASW aircraft. The AN/SSQ-62B can also determine the range and bearing of the target relative to the sonobuoy's position.

The DICASS sonobuoy is composed of three main sections: an air descent retarder, a surface unit and a sub-surface unit. The air descent retarder consists of a parachute release assembly and a parachute to retard descent immediately after launch. The surface unit receives commands from the controlling aircraft, via a UHF receiver, and sends target information to the aircraft, via a VHF transmitter. The sub-surface unit transmits sonar pulses in the ocean upon command from the aircraft and receives sonar target echoes for transmission to the aircraft.

Command signals are received by the sonobuoy and are accepted if the correct address code is identified by a decoder. The command capability includes depth selection, scuttle and selection of transducer (sonar) transmission signals. The echoes from the selected activating signal are multiplexed in the sub-surface unit before being transmitted to the receiving station or aircraft.

The AN/SSQ-62B may be launched from fixed- or rotary-wing aircraft within the parameters specified in the launch envelope.

Upon impact with the water, the transducer is released for descent to its shallow operating depth. Immediately after entry, the float is inflated and the VHF-transmitter/UHF-receiver antenna is erected. When the float surfaces, the VHF transmitter begins emitting a continuous FM carrier signal. The sonobuoy is now operating in the passive mode and is ready to receive commands.

The main power source for the sonobuoy comes from a lithium battery pack instead of the more costly silver chloride batteries commonly used in sonobuoys. The sonobuoy is designed for economical volume production without compromising performance or reliability. Electronic design is exclusively solid-state with maximum use of multi-function integrated circuits.

Operational Status
In full production. In service with the US Navy, and exported to Australia, Norway, the Netherlands and Canada.

Specifications
Type: Commandable, omnidirectional active, directional receive
Sonar modes: Pulse CW or linear FM
Depth: Commandable, 27, 119 and 457 m (89, 396 and 1518 ft)
Size: A

Life: 30 mins
Weight: 15 kg
Sonar channels: 4
RF channels: 31
RF power: 0.25 W
Launch envelope: Altitudes up to 10 000 ft and air-speeds up to 300 kts.

Contractors
Sparton Corporation, Electronics Division, Jackson, Michigan.
 Magnavox, Fort Wayne, Indiana.

An AN/SSQ-62B sonobuoy being made ready for launch in US Navy SH-3H helicopter (W Donko)

AN/SSQ-75 ERAPS SONOBUOY

Type
Long-range active sonobuoy.

Development
Initial development commenced in the late 1970s, under a US Navy programme, when feasibility trials were carried out. In 1980 Bunker Ramo received a $12.5 million contract for design and development. In 1985 Bunker Ramo was taken over by the Allied Bendix Oceanics Division and a $4 million contract was awarded to carry on with design and development. The programme was restructured in 1988 and development and initial operating testing was completed in 1990-91. In 1988, Magnavox and Sparton announced that they had formed a joint venture partnership (ERAPSCo) to develop an ERAPS buoy, and had received a $2.5 million initial contract.

Description
The Expendable Reliable Acoustic Path Sonobuoy (ERAPS) is a command-activated long-range (in the region of 15 nm) active search sensor designed to exploit the long-range direct propagation mode known as the reliable acoustic path. It provides a capability

to search actively for a submarine that is virtually undetectable by passive sonar methods, and can rapidly localise the target for an attack. The detection range is gained by using low frequency, high power low frequency transmitted pulse and a volumetric receiving array of hydrophones. Range, bearing and Doppler are provided.

 ERAPS weighs approximately 100 kg so it is carried on aircraft weapon racks rather than being launched as a conventional sonobuoy. The sensor array resembles a dipping sonar, and descends to a maximum of nearly 5000 m to find a reliable acoustic path. The array can pause on command at intermediate levels during descent.

Operational Status
ERAPSCo is scheduled to deliver examples in 1992-93 for technical and operational evaluation.

Contractor
ERAPSCo (a Magnavox/Sparton joint venture partnership).

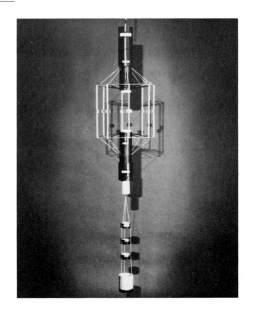

AN/SSQ-75 ERAPS sonobuoy

AN/SSQ-77A (VLAD) SONOBUOY

Type
Passive search and surveillance sonobuoy.

Description
The AN/SSQ-77A VLAD (vertical line array DIFAR) is a passive, tactical search and surveillance sonobuoy designed to improve detection and tracking capability for the DIFAR system in a noisy, high traffic environment.

 The AN/SSQ-77A concept utilises a vertical line array of 11 omnidirectional hydrophones in place of the single omnidirectional hydrophone used in the standard DIFAR unit. A directional DIFAR hydrophone mounted at the array phase centre provides target bearing data.

 The signal format for the AN/SSQ-77A sonobuoy is identical to that for the AN/SSQ-53 and is compatible with the AN/AQA-7 processor and the Sparton TD-1135/A demultiplexer/processor/display. All beam forming functions are accomplished within the sonobuoy, with provisions for sea noise equalisation and omnidirectional phase tracking.

 The AN/SSQ-77A may be dropped from an aircraft at indicated airspeeds of 45 to 380 kts and from altitudes of 100 to 30 000 ft (30 to 9144 m). Descent is stabilised and slowed by a parachute assembly.

 An internal microprocessor electronically selects the desired RF channel and sonobuoy life.

Verification of these settings is provided by an LED display, readable through the SLC.

 Lithium batteries are activated upon entry to the water. These detonate a squib, causing gas-filled cylinders to inflate the sonobuoy flotation bag and initiate deployment. Inflation of the flotation bag provides a pressure force, releasing the release plate in the top of the sonobuoy and causing the parachute to be jettisoned. This permits the surface assembly to rise and separate from the sonobuoy housing. The sonobuoy housing serves as a descent vehicle and separates from the sub-surface assembly at the operating depth.

Operational Status
In quantity production.

Specifications
Type: DIFAR with VLA omni
Size: A
Weight: 10.4 kg
Operating life: 1, 4 or 8 h (selectable)
Launch altitude: 30-9144 m
Launch speed: 45-380 kts
RF channels: 99, electronically selected from 136 to 173.5 MHz
RF power: 1 W

Contractors
Sparton Corporation, Electronics Division, Jackson, Michigan.
 Magnavox, Fort Wayne, Indiana.

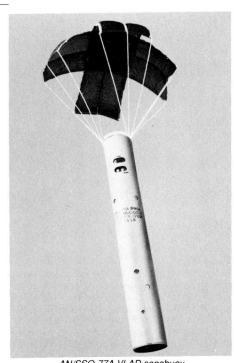

AN/SSQ-77A VLAD sonobuoy

AN/SSQ-77B (VLAD) SONOBUOY

Type
Passive directional search and surveillance sonobuoy.

Description
The SSQ-77B is the result of a US Navy sponsored development programme to improve the operational capabilities of the AN/SSQ-77A. Improvements include multiple depth settings, horizontal or vertical selectable beam patterns and an increased number of omnidirectional hydrophones in the vertical line array.

Operational Status
In development.

Contractors
Sparton Corporation, Electronics Division, Jackson, Michigan.
Magnavox Corporation, Fort Wayne, Indiana.
Hermes Electronics Ltd, Dartmouth, Nova Scotia, Canada.

AN/SSQ-102 TACTICAL SURVEILLANCE SONOBUOY

Type
Long-term surveillance sonobuoy.

Development
The AN/SSQ-102 Tactical Surveillance Sonobuoy (TSS) is in full-scale engineering for the US Navy. It is intended to complement existing buoys by providing long-term surveillance over several days. This will enable a number of maritime patrol aircraft to monitor a larger area of ocean and, providing that the sonobuoy field is replaced every few days, give a continuous record of submarine activity in the area.

Description
Digital processors within the TSS evaluate incoming signals, select information that relates to possible targets, and record it within solid-state memories. When the sonobuoy field is interrogated by an overflying ASW aircraft, which may be hours or days later, only the buoys with potential targets will respond.

Magnavox, and a joint venture comprising Sippican and Hazeltine, are developing designs independently under a full-scale engineering design contract worth approximately $12 million each. The US Navy intended to order about 625 buoys from each of the two contenders in early 1990 for technical and operational evaluation. The winning design is due to be selected in 1992.

Contractors
Magnavox and Sippican/Hazeltine are competing for the programme.

AN/SSQ-103 SONOBUOY

Type
Low-cost passive sonobuoy.

Description
The AN/SSQ-103 is a Low-Cost Sonobuoy (LCS) which is one-sixth the size of the standard 'A' size configuration and is designed for deployment in dense fields covering large areas. It differs from conventional sonobuoys in that it does not transmit acoustic signals for processing and analysis in the aircraft. It uses in-buoy signal processing to alert ASW aircraft to the probable presence of a target. More sophisticated sonobuoys may then be deployed in the contact area once an LCS buoy has made a detection.
The A/6 LCS is less than 14 cm long, is 11.43 cm in diameter and weighs 0.64 kg. It is packaged in a six-buoy sonobuoy launcher, also designed by Sippican, which will deploy a single LCS buoy on command. A passive multi-element array is deployed to a 300 ft depth. Each LCS transmits on one of 330 different RF channels to prevent mutual interference and to provide locating information upon target detection.

Operational Status
Large scale production of the AN/SSQ-103 was expected in the 1989/90 time scale. The buoy was reported to be cancelled in early 1989 but recent reports, allied with the P-3C Update IV programme, indicate that it may yet survive.

Contractor
Sippican Inc, Marion, Massachusetts.

AN/SSQ-103 low-cost sonobuoy

STRAP PROGRAMME

Type
Beam forming sonobuoy system.

Description
STRAP (Sonobuoy Thinned Random Array Project) is a programme initiated by the US Navy to emulate the collective processing beam forming techniques of normal sonar systems, but using sonobuoys. The technique involves the use of large numbers of sonobuoys, probably around 15 to 20, which are deployed randomly over a given area. Since it is essential to know the exact position of each sonobuoy, four of the buoys would be fitted with a low power transmitter and deployed in different areas of the pattern. Each sonobuoy would then receive two sets of signals, one from the water and the other from the four emitting sonobuoys. The acoustic processor in the aircraft would then correlate these positions and use the information for beam forming purposes.

Operational Status
The programme is being managed by the US Naval Ocean Systems Command with Lockheed using the AN/UYS-1 Proteus processor for adaptive processing techniques, and Magnavox modifying the sonobuoys. STRAP forms part of a continuing acoustic search sensors programme. Sea trials took place in 1986 but no further information has emerged on the state of this programme.

AN/ARR-78(V) SONOBUOY RECEIVING SYSTEM

Type
Sonobuoy receiver for operation and management of buoys.

Description
The AN/ARR-78(V) radio receiving set, also known as ASCL (Advanced Sonobuoy Communication Link), is for the operation and management of anti-submarine sonobuoys by the crew of ASW aircraft such as the P-3C Orion (Update III) maritime patrol aircraft and S-3B Viking. The equipment comprises five main items (described below) and is employed with an associated onboard ASW signal processor. The main units of the AN/ARR-78(V) are:

(a) the AM-6875 RF pre-amplifier, an optimised high performance, low noise VHF pre-amplifier which provides amplification and pre-filtering of received RF signals
(b) the R-2033 radio receiver, which contains 20 fully synthesised high performance receiver modules (16 acoustic and 4 auxiliary), and one each of the following modules: RF/ADF amplifier multi-coupler, reference oscillator, Proteus digital channel, I/O processor, clock generator, BITE, and DC power supply module. Each single conversion receiver module includes mixer conversion, frequency synthesised LO, demodulator and output interface circuits. Each of the acoustic receiver modules processes FM/analogue or FSK digital signals at any of the 99 channels in the extended VHF band. Each of the four auxiliary receiver modules processes signals at any of the 99 channels and provides: one channel for selection and processing of the on-top position indicator (OTPI) signals; two channels for the operator to monitor acoustic information; and one channel to monitor the RF signal level in any of the RF channels. Common receiver modules are interchangeable.
BITE circuits provide comprehensive end-to-end evaluation of each receiver from the VHF pre-amplifier to the receiver output interface circuits. BITE is initiated automatically by the computer (such as Proteus) and/or by the operator through the indicator control unit (ICU). Performance status is displayed on the ICU and routed to the computer
(c) the C-10126 indicator control unit (ICU), which

provides the operator with a means for manual control of each receiver channel frequency assignment, receiving mode and self-test. It also displays status of the receiving set and operator entry information

(d) the ID-2086 receiver status indicator, which continuously displays the control mode setting, the RF channel number and the received signal level for each receiver

(e) the C-10127 receiver control unit, which provides the operator with control over the OTPI channel frequency.

Surface acoustic wave (SAW) IF filters are incorporated to provide high selectivity, coupled with good linear phase characteristics which yield low levels of distortion. Microprocessors are used in the ICU and I/O module to process the commands to, and display status data from, the receivers. Versatile programs accommodate changes without requiring aircraft modification.

The auxiliary receivers permit the operator to aurally monitor signals in any of the 99 RF channels without interrupting data processing in that channel.

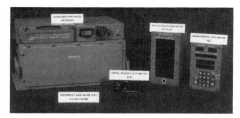

AN/ARR-78(V) advanced sonobuoy communications link equipment (P-3C configuration)

Operational Status
Full-scale development was completed in 1980 under US Navy contracts. The first production contract for P-3C systems and S-3B systems was awarded in December 1982 and further contracts have been awarded for deliveries through to 1993. As of August 1991, 408 systems had been delivered. In December 1991 Hazeltine was awarded a fixed-price contract for the manufacture of 79 AN/ARR-78(V)1/1 advanced communications links for P-3 Orion and S-3 Viking

ASW aircraft for the US Navy and one replacement set for Norway.

Specifications
Frequency: Extended VHF
Receivers: 20 (16 acoustic/4 auxiliary)
Channels: 99 per receiver
Audio output
Analogue: 2 V or 4% V RMS (balanced)
Power input: 115 V AC ± 10%, 380-440 Hz, 3 phase 450 W; 18-32 V DC, 7 W; 26.5 V AC, ± 10%, 400 Hz, 50 W
Dimensions
AM-6875: 7.6 × 14.6 × 10.8 cm; 0.95 kg (weight)
R-2033: 30.9 × 54.1 × 38.9 cm; 45.9 kg (weight)
C-10126: 22.9 × 14.6 × 16.5 cm; 3 kg (weight)
ID-2086: 26.7 × 14.6 × 12.7 cm; 1.9 kg (weight)
C-10127: 5.7 × 14.6 × 8.3 cm; 0.6 kg (weight)

Contractor
Hazeltine Corporation, Greenlawn, New York.

AN/ARR-72 RECEIVER

Type
Sonobuoy receiving system.

Description
The AN/ARR-72 sonobuoy receiver system is currently used on the P-3C patrol aircraft in conjunction with acoustic signal processors and a digital computer. The system receives, amplifies and demodulates frequency-modulated radio signals in the 162.25 to 173.5 MHz VHF band transmitted by deployed sonobuoys. These radio signals contain information related to the frequency and directional pattern of the sound spectrum received by each sonobuoy hydrophone array. The baseband information from the receiver is demultiplexed and submitted to a special analysis and/or time correlation and Doppler analysis by the acoustic signal processing equipment. Bathythermograph buoys and equipment enable determination of the seawater temperature for acoustic range measurement.

Monitoring of the RF level signals from the 31 on-line receivers enables the digital computer to perform many tactical functions such as the determination of the channel usage or presence of jamming prior to sonobuoy launching, connections of the proper channel RF receiver to each processor, and a determination of remaining sonobuoy life. These same RF

level monitoring and channel selection functions may also be accomplished manually through the use of the receiver control-indicator boxes located at the individual sensor stations.

The AN/ARR-72 system is compatible with LOFAR, CODAR, BT, RO, CASS, and DICASS equipment currently being deployed. Specific sonobuoy types include: AN/SSQ-36, AN/SSQ-41, AN/SSQ-47, AN/SSQ-50, AN/SSQ-53 and AN/SSQ-62.

The AN/ARR-72 sonobuoy receiver system is a dual conversion superheterodyne VHF receiver system designed for receiving FM signals from deployed sonobuoys. Radio frequency signals are received at the dual aircraft VHF blade antennas, amplified by the system's AM-4966 dual RF pre-amplifiers, and passed to the CH-619 31-channel receiver assembly. Within this assembly a multi-coupler distributes the pre-amplified RF to the 31 fixed tuned receivers, where it is further amplified and demodulated to provide baseband audio and RF level signals.

Each of the 31 receiver channels contains a channelised first converter, an IF filter, a second converter, and a discriminator/amplifier. The first converter contains a crystal-controlled local oscillator and mixer/IF circuit. The plug-in discriminator/amplifier provides RF level and FM signal detection. The optional phaselock discriminator is directly interchangeable with the discriminator/amplifier module. The first converter,

second converter and discriminator/amplifier assemblies are plug-in units, and, excepting the local oscillator crystals, are identical for the 31 RF channels.

The SA-1605 audio assembly includes 19 audio switching and amplifier cards and two audio power supply regulators. The audio assembly accepts the baseband and RF level outputs of the 31 receivers and outputs them to the computer and processing equipment. Each of the 19 audio channels contains a 31 × 1 switching matrix to select the output of a given receiver. This receiver signal is then amplified and provided through two individually buffered outputs to the processing equipment. Selection of a particular receiver channel may be accomplished by the digital computer or the C-7617 dual channel control indicator. An individual 31 × 1 switching matrix is provided within the audio assembly to service the RF level requirements of the digital computer. This switching assembly is controlled by the computer and provides an integrated RF level for the selected receiver in a digital format upon computer interrogation.

Operational Status
In production.

Contractor
Flightline Electronics Inc, Fishers, New York.

AN/ARR-75 RECEIVER

Type
Sonobuoy receiving system.

Description
The AN/ARR-75 sonobuoy receiving set is a 31 RF channel FM receiver designed for use on ASW fixed-wing aircraft as well as shipboard applications. Independent receiver modules provide four simultaneous demodulated audio outputs each capable of selecting one of 31 RF input channels. The AN/ARR-75 is used on the LAMPS Mk III, LAMPS Mk I and SH-3H, as well as various other surface combatant platforms.

The AN/ARR-75 receiving set is composed of two units. The first is the OR-69/ARR-75 receiver group assembly, which in turn consists of a power supply (PP-6551/ARR-75), four receiver modules (R-1717/ARR-75) and the chassis (CH-670/ARR-75). The receiver group assembly contains the majority of the system's electronics.

The second unit is the radio set control (control box), C-8658/ARR-75 or C-10429/ARR-75. This control unit provides independent RF channel selection and signal strength monitoring of any of 31 RF channels for each of the four receiver modules.

Flightline Electronics also manufactures a maintenance kit for the AN/ARR-75. This optional equipment is designated MK-1634/ARM maintenance kit (also

known as the suitcase tester) and is designed to facilitate interim maintenance of the AN/ARR-75 at the intermediate level. The MK-1634 contains module extenders, adaptors, cables and a test bench panel for interconnecting with and testing the AN/ARR-75.

Operational Status
In production. Fitted to US Navy SH-2, SH-3 and SH-60 helicopters.

Contractor
Flightline Electronics Inc, Fishers, New York.

AN/ARR-84 SONOBUOY RECEIVER

Type
Airborne and surface vessel sonobuoy receiver.

Description
The AN/ARR-84 represents the latest generation of 99-channel sonobuoy receivers. It is designed to receive signals from all current and planned future US and allied sonobuoys. This receiver features four acoustic channels capable of receiving signals from up to four deployed sonobuoys simultaneously on any of 99 RF channels. It is suitable for installation on a variety of rotary- and fixed-wing aircraft, as well as

large surface vessels and fast patrol craft. It is a form and fit replacement for its predecessor, the AN/ARR-75 (see previous entry).

Additional features of the AN/ARR-84 include dual selectable IF bandwidths for optimum performance with a wide variety of sonobuoy types. It provides high AM rejection and mechanical vibration immunity severely reduces interference due to propeller/rotor multi-path and platform vibration, and has a very low susceptibility to conducted and radiated energy allowing the set to be used near strong onboard emitters and shipboard search radars.

An optional radio set control, intended for use when on-line computer control is not available, provides power control and BIT operation for the receiver

group. RF channel selection, sonobuoy type selection and RF level readout is provided independently for each of the four receivers in the group. Operator input is through a multi-function keypad with signal strength provided as a histogram display and entry readback/failure data provided by a 16-character message display.

Operational Status
In production and in use on board US Navy SH-60B, SH-60F, LAMPS Mk III, Royal Australian Navy S-70B and Taiwanese S-2T.

Contractor
Flightline Electronics Inc, Fishers, New York.

AN/ARR-146 SONOBUOY RECEIVER

Type
Sonobuoy receiving system.

Description
The AN/ARR-146 radio receiving set is a 99 RF channel on-top position indicator used on board ASW rotary- and fixed-wing aircraft to provide bearing and on-top position indication of deployed sonobuoys. When used in conjunction with a suitable DF system, the equipment enables the operator to locate and verify the position of deployed sonobuoys operating on any of 99 channels.

The AN/ARR-146 has a modular, solid-state design and it is designed to be used with computer control, either via RS 422 directly, or when connected with the AN/ARR-84 sonobuoy receiver. The AN/ARR-146 is form and fit interchangeable with the R-1651/ARA and R-1047 A/A OTPI receivers. It is compatible with ARA-25, ARA-50 and OA-8697/ARD (DF-301E) DF antenna systems.

The AN/ARR-146 set consists of the R-2330/ARR-146 radio receiver and the C-11699/ARR-146 radio set control. The R-2330/ARR-146 is a VHF, AM receiver, which receives signals from sonobuoys operating on any of 99 RF channels, in the frequency range from 136 to 174 MHz. This receiver houses all of the electronics, including internal transfer circuits for the switching of RF, baseband, power and phase compensation circuits for sharing the DF system between the OTPI and an associated UHF receiver system for the DF. The R-2330/ARR-146 also contains a PLL synthesised local oscillator, digital address decoder, voltage tuned BP filters, AGC and BIT circuitry.

The C-11699/ARR-146 radio set control is an optional manual control box which is used in place of the RS 422 bus or the MIL-STD-1553B dual bus of the AN/ARR-84. This control box provides the capability to apply power to the AN/ARR-146 system, to select any one of 99 RF channels, to activate the BIT circuitry and to display adequate signal strength and results of the BIT via the adequate signal strength indicator.

The equipment is compatible with all known or planned sonobuoys, including DIFAR, LOFAR, RANGER, BT, CASS, DICASS, VLAD, CAMBS, Barra, ERAPS, HLA, ATAC and SAR.

Operational Status
In production for the US Navy.

Contractor
Flightline Electronics Inc, Fishers, New York.

AN/ARR-502 SONOBUOY RECEIVER

Type
Sonobuoy receiving set.

Description
The AN/ARR-502 is a 99 RF channel sonobuoy with 16 acoustic channels and an on-top position indicator channel. It is capable of receiving all types of sonobuoys, including LOFAR, DIFAR, BT, VLAD, Ambient South, Range Only, and DICASS. Additionally, capability is provided to operate with CAMBS, Barra and ATAC buoys, as well as future developmental sonobuoys such as LCS, ELCS, TSS and ERAPS.

The AN/ARR-502 can be controlled either by a MIL-STD-1553B bus or by a manual control unit. Multiple bandwidths and channelisations are provided for compatibility with all present and future analogue, digital and low-cost sonobuoys.

Operational Status
The AN/ARR-502 is being developed under contract from the Canadian DND. The equipment has been delivered to Canada.

Contractor
Flightline Electronics Inc, Fishers, New York.

R-1651/ARA RECEIVER

Type
Sonobuoy radio receiver.

Description
The R-1651/ARA radio receiving set is a 31 RF channel on-top position indicator (OTPI) used aboard rotary- or fixed-wing aircraft to provide bearing and on-top position indication of deployed sonobuoys. When used in conjunction with a suitable DF system, it enables the operator to locate and verify the position of deployed sonobuoys operating on any of 31 channels.

The R-1651/ARA is form and fit interchangeable with the R-1047 A/A OTPI. This radio receiving set is compatible with ARA-25 and ARA-50 antennas and it is also available in a modified configuration which is compatible with the DF-301E antenna system. The R-1651/ARA is controlled by the optional C-3840/A control box.

The R-1651/ARA is a single-conversion, 31 channel, superheterodyne VHF receiver which is designed to receive signals from deployed sonobuoys over the frequency range of 162.25 to 173.50 MHz.

The antenna input is switched by a relay to the receiver or the UHF receiver, on external command. The RF signals (R-1651/ARA) pass through band-pass filters, are amplified, and converted to an IF of 25 MHz. After IF amplification and detection, the resultant audio signal is amplified to the level required by the ADF system. The AGC level is developed at the detector, amplified and used for gain control of the IF and RF stages and to operate the 'adequate signal strength' circuits.

The local oscillator provides one of 31 possible LO signals. Thirty-one crystals are appropriately selected using external controlled code lines containing a 6-bit binary code. Line receivers, decoding gates and matrix drivers connect one of 31 crystals to the local oscillator circuits.

The optional C-3840/A radio receiver control is used to switch the antenna system, by energising an external coaxial relay, from the UHF/ADF mode of operation to the sonobuoy-OTPI mode of operation, to select any one of the 31 RF channels and to indicate adequate signal strength when the R-1651/ARA receives an RF signal of -86 dBm or greater.

Operational Status
In production for the US Navy.

R-1651/ARA sonobuoy OTPI radio receiver

Contractor
Flightline Electronics Inc, Fishers, New York.

AN/AKT-22(V)4 TELEMETRY SET

Type
Telemetry transmitter/receiver for sonobuoy control.

Description
The AN/AKT-22(V)4 telemetry data transmitting set forms an essential element of an airborne ASW installation for the operation of sonobuoys and the relaying of sonobuoy data to co-operating ASW surface vessels.

Two AN/ARR-75 radio receiving sets are included in the airborne element, providing eight VHF receiving channels. This configuration permits relay of up to eight passive sonobuoy transmissions, up to six channels of R−Θ³, and passive data of which up to four channels may be R−Θ³, or two channels of DIFAR sonobuoy data. An air-to-ship voice channel is available except when the DIFAR mode is used. The AN/AKT-22(V)4 produced by EDMAC consists of the following major units:
(a) transmitter-multiplexer T-1220B
(b) control indicator C-8988A
(c) antenna AS-3033

Two main elements of the AN/AKT-22(V)4 telemetry data transmitting set: transmitter-multiplexer (T-1220B) left, and control indicator (C-8988A) right

(d) actuator TG-229.

The multiplexer section of transmitter-multiplexer T-1220B/AKT-22(V)4 uses low-pass filters and VCOs to provide the eight-channel capability. The trigger filter assembly for this set is located in control indicator C-8988A/AKT-22(V)4. Setting any of the trigger-select switches on the control indicator to the 'on' position disables data channels 3C and 4C.

Composite trigger tones are brought into the multiplexer on a separate line and combined with the sonic and voice channels at the mixer-amplifier to form the composite FM modulation signal. This composite FM signal modulates the transmitter.

In the DIFAR operating mode, the composite FM modulation signal is disconnected from the transmitter input. Two DIFAR sonobuoy transmissions, received on dedicated receiver channels, are demodulated in the receivers and are provided to the transmitter-multiplexer. The receiver output is conditioned in an amplifier-adaptor and the DIFAR-A sonic signal is conditioned in a low-pass filter. The DIFAR-B sonic signal is used to drive a VCO centred at 70 kHz and then passes through a bandpass filter. The two filter outputs are linearly combined, and the composite modulation signal is coupled to the transmitter input when the control indicator mode select switch is in DIFAR position. Composite sonic or composite DIFAR input to the transmitter is selected by a relay in the DIFAR filter controlled by a switch on the control indicator.

The shipboard receiving equipment required for the eight-channel data link is telemetric data receiving set

AN/SKR-( ). The receiver demodulates and demultiplexes the composite sonic or DIFAR transmission, and routes the separate outputs to appropriate processing equipment.

Operational Status
In production.

Contractor
Flightline Electronics Inc, Fishers, New York.

AN/AQR-185 99-CHANNEL SONOBUOY RECEIVER

Type
Sonobuoy receiving system for the P-3C Update IV.

Description
The AN/AQR-185 99-channel sonobuoy receiver performs the functions of analogue and digital sonobuoy reception. Each receiver will tune the 99 RF channels between 136 and 173 MHz. The system is designed for applications where weight and size is important and the sonobuoy field area is limited.

The receiver system is divided into a set of modular units, a dual RF pre-amplifier, a control unit, and from one to five receiver units. Each receiver unit will accommodate up to 20 interchangeable receiver modules, providing a growth potential to monitor 100 different sonobuoys simultaneously.

The receiver system accepts optional plug-in assemblies that perform the functions of RF spectrum scanning to determine channel activity, on-top position indication in conjunction with a direction finding antenna and a heading indicator, phase angle measurement equipment to support sonobuoy location determination in conjunction with sonobuoy reference system antennas and software, acoustic test signal generation, provision of acoustic audio signals to operators' headsets and display.

Operational Status
In development for the US Navy P-3C Update IV.

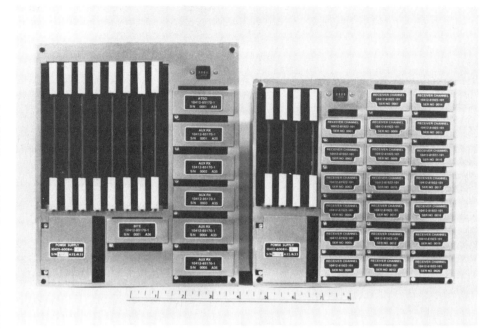

Sonobuoy receiver for Update IV Electronics suite to be fitted to P-3C ASW aircraft

Contractor
Dowty Avionics, Arcadia, California.

AN/ARR-76 SONOBUOY RECEIVER

Type
Fixed frequency sonobuoy receiver system.

Description
The AN/ARR-76 sonobuoy system receives up to 31 VHF/FM channels in the frequency range of 162 to 174 MHz, and provides 16 separate video output channels for monitoring. A general-purpose digital computer or a manual computer simulator is used to control the unit relative to assignment of receiver channels to the video output channels at any one time. Communications between the AN/ARR-76 and

the computer are carried out by Manchester Code at a 6 MHz rate.

Operational Status
In production for US Navy S-3A and Canadian CP-140 aircraft.

Contractor
Dowty Avionics, Arcadia, California.

AN/ARR-76 sonobuoy receiver

OL-320/AYS

Type
Acoustic processing system.

Description
The system incorporates the CP-1584/AYS post-

processing and display elements and the C-11262/AYS monitoring control unit. The latter provides switching between sonobuoy audio processing and the integrated communications system on the S-3B Viking ASW aircraft.

Operational Status
The computer is operational in all S-3B Viking aircraft.

Contractor
Lockheed Sanders Inc, Manchester, New Hampshire.

STATIC DETECTION SYSTEMS

CANADA

CSS-80AS STATIC SONAR

Type

Harbour surveillance sonar system.

Description

The CSS-80AS is an active sonar for underwater surveillance. It consists of a control indicator with a CRT display, an interface unit, and an underwater unit consisting of transducer and transmitter/receiver sub-assemblies. The underwater unit can be bottom-mounted or suspended from a ship. An underwater cable connects the interface unit and the underwater unit.

The equipment uses the omni-sonar concept to provide simultaneous 360° coverage of a 3° conical field which can be stepped through a vertical sector of up to ±15° with respect to the horizontal plane. Adaptive processing techniques are employed to reduce noise and to provide a clear image of targets in the operating area.

Information from each observation station may be simultaneously viewed at central command via a data transmission mode. This is designed to provide a total detailed picture of a developing threat so that effective action can be taken.

The CRT presentation is non-fading in eight colours, each colour representing a certain echo strength. The range of colours represents a total range of approximately 20 dB in echo strength. There is also an alphanumeric display of key sonar operating data as well as operating time and date.

A joystick-positioned cursor provides digital readout of target position (bearing, range and height) on the screen. An operator selectable static suppressed mode allows suppression of all stable static background targets allowing changing target echoes to be easily identified.

A target detect function allows new and moving targets at varying depths to be detected and highlighted on the video display for classification and tracking. Either a built-in audible alarm or a remote alarm can be operator programmed in conjunction with the selection of the target detect function to provide an automatic interfusion detection system.

Contractor

C-Tech Ltd, Cornwall, Ontario.

CSAS-80

Type

Harbour surveillance sonar.

Description

The CSAS-80 harbour surveillance sonar is a third generation static harbour suveillance system developed from the CSS-80AS static sonar. The system comprises a control indicator and CRT display, an interface unit and an underwater unit which can be bottom-mounted on the seabed. The underwater sensor is connected to the interface unit by a cable. The sonar is an omnidirectional, pre-formed beam, active/passive sonar providing an instantaneous search over a 6°, 12°, or 24° conical field which can be automatically tiled 24° in the vertical plane. Adaptive processing techniques eliminate stationary targets and reduce noise interference enabling new and moving targets to be more easily detected.

The system can be expanded using a number of observation stations which can be simultaneously viewed in a central command headquarters via a data transmission mode. This provides a detailed picture of the developing threat and allows effective counter action to be taken. A permanent record can be made of each station using a video tape interface option.

Targets are automatically detected and tracked and presented within range scales at 250 m, 500 m, 1000 m, 1500 m and 2000 m. All echo ranging data in all modes is displayed on a video monitor in the control indicator unit. The presentation is non-fading in 16 colours, each colour representing a certain echo strength. The range of colours represents a total range echo strength. An alphanumeric display presents key sonar operating data and target data as well as operating time and date. A cursor is positioned by trackerball to give a digital readout of target position (bearing, range and height) on the CRT screen. Operating control settings are displayed on the control indicator in the icon format. As a further

aid to detection and classification the system incorporates a zoom facility.

In addition to the normal real-time PPI display, a static suppressed image mode is operator selectable. In this mode echoes from stable background targets such as sea walls, moorings, seabed and bottom projections are suppressed, providing a video display of any new or moving targets. The mode also adjusts to accommodate slowly changing environmental conditions.

With the selection of the 'target detect' function, new and moving targets at varying depths are detected and highlighted on the video display for classification and tracking. A built-in audio alarm or remote alarm provides an automatic intrusion detection system with low false alarm rate.

Contractor

C-Tech Ltd, Cornwall, Ontario.

FINLAND

FHS-900

Type
Fixed passive underwater surveillance system.

Description
The Elesco Hydroacoustic Surveillance System FHS is a fixed and passive system which is undetectable to intruders. The FHS is optimised for the difficult acoustic conditions prevailing in shallow coastal waters typically ranging from 20 m down to 300 m.

The system architecture allows for economical distribution of the sensor network over a large surveillance area. This approach permits the user to deal with the special problems occurring in shallow waters where sound propagation is more variable and unpredictable than in deep waters.

The FHS is configured around sensors placed on the seabed and connected to an onshore station (operational control centre) via submerged cables, which can be up to 30 km long without any buffer amplifiers. The sensors are arranged in groups, each of which is comprised of three hydrophones with integral amplifiers and a common microprocessor for control and communications.

Using a self-optimising sensor configuration steered by independently operating correlators, the hydrophones of the FHS utilise the entire acoustic spectrum from 0.5 Hz to 100 kHz, thus covering audible sounds, ultrasounds and infrasounds as well as impulses. The omnidirectional pattern of the microphones and triangular configuration of the group provides constant detection sensitivity and resolution of signals coming from any direction.

In the operation control centre the front-end configuration includes a multi-processor system performing single-target as well as multi-target detection using patented Fiskars multi-target detection methods; a maximum of 32 interface modules per cabinet, allowing for expansion by up to 96 hydrophones (32 groups of 3 hydrophones) per each additional cabinet.

The operator facilities include a Unix-based 32-bit main computer in the operator's control station interfaced to a high resolution colour graphic display (resolution 1280 × 1024 and up), complemented with a dedicated FFT analyser for real-time noise signature analysis, and a high capacity multi-channel recorder with waveform display for the archiving of acoustic signals.

Operator's console for the FHS-900 system

In addition to the above, the operator interface features simultaneous real-time display of continuous tracking of target movements against a digitised background map (with an optional indication of the 90 per cent location probability circle), alphanumeric data on target location, course and speed, as well as such additional information as signal level of the acoustic signature, seawater temperature and different alarms.

Contractor
Elesco Oy Ab, Espoo.

GERMANY

ATLAS HARBOUR PROTECTION SYSTEMS

Type
Static detection systems.

Description
These systems incorporate modular detection and classification sonars designed to protect vital installations such as commerical ports, naval bases, oil platforms and other offshore equipment from sabotage and terrorist attack.

Hostile intruders such as small submarines, mini-submarines, swimmers and swimmer delivery vehicles are automatically detected, tracked and classified and alarms are triggered. The systems offer high probability of detection with low false alarm rates.

Contractor
Atlas Elektronik GmbH, Bremen.

APSS (ATLAS PORT SURVEILLANCE SYSTEM)

Type
Active static surveillance system.

Description
APSS is a modular surveillance system designed for the permanent monitoring of the underwater scene of harbour areas and to protect jetty and harbour areas against the risk of covert attack, especially by swimmers and swimmer delivery vehicles within a predefined surveillance sector. The system is particularly designed for shallow waters with dense shipping traffic.

APSS is specified to detect swimmers and mini-submersibles at ranges of more than 1000 m. The system consists of an operator's console for the presentation of the tactical and the sonar display. The advanced signal processing includes computer-aided detection and classification, automatic tracking of targets and automatic alarm generation. The underwater components are designed for permanent installation and have no moving parts, which improves the system availability and reduces underwater maintenance tasks to a minimum.

Operational Status
The system is in its final design phase and is based upon proven components which are derived from other applications with permanent underwater installations.

Contractor
Atlas Elektronik GmbH, Bremen.

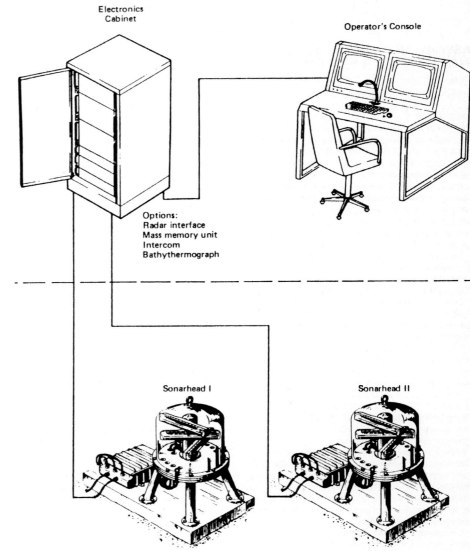

Two sonar head configuration of APSS

ASES

Type
Active Sonar for Entrance Surveillance (ASES).

Description
ASES is an omnidirectional sonar developed for the detection of submarines, mini-submarines and underwater swimmers. The system indicates the operating depth of the target through volume electronic scanning which is performed in a number of different depth layers sequentially using narrow beams. To increase performance the vertical angle can be tilted electronically to match the environmental conditions.

Switching between local and remote-control is available and more than one sonar can be used in the same area without creating mutual interference. The system comprises a seabed sensor unit and a shore-based control room. The sensor unit comprises a sonar head with embedded electronics boards and cylindrical transmitting and receiving arrays.

The shore-based control room houses the operator's console and electronics cabinet. The console is fitted with up to two 19 in high resolution colour raster scan monitors for the presentation of the tactical and sonar analysis displays. Control is exericsed through an alphanumeric keyboard and trackerball with control keys. In addition the console incorporates a numeric keypad and function keys. Data are transferred from the sonar head to the control room via a fibre optic cable.

The system can operate in one of two modes – manual or automatic. In the automatic mode the system carries out continuous monitoring of ambient noise and reverberation within a specified area, providing automatic detection, tracking and classification of targets together with track history and the assigning of track numbers to moving targets and

classification symbols to detected intruders, and triggering video and audible alarms. The display presents the contours of a simplified sea chart of the area with colour coded guard zones. Additional information such as parameters of selected sonar contact (course, bearing, speed, distance, depth, track number) and threshold settings, can be displayed at the operator's request.

In the manual mode the operator is provided with highly sophisticated tools for target analysis. Parameters such as transmission power, detection threshold, transmission mode (FM/CW, pulse length, ping period), multi-ping integration, classification data,

scanning mode (sector or full picture), and magnification can be varied. The operator can also zoom in and display the superimposed data from more than one sonar head or prescribe data on a single head or array. Access to raw sonar data is available so that manual detection and classification can be performed. An A-scan waterfall window and an audio channel are available for analysis and evaluation of critical target data.

Contractor
Atlas Elektronik GmbH, Bremen.

Specifications

	ASES	APSS
Antenna system coverage/horizontal	360°	70° per array
vertical	24°/12°/6° (steerable ± 15°)	15°
Display ranges	100/200/400/600/800/ 1200/2000 m	200/400/600/800/ 1200 m
Update rate (800 m)	1.5 s	1.5 seconds per array
Max operating depth	100 m	40 m
Transmitting signal	CW/FW selectable	CW/FW selectable
Range accuracy	1%	1.5%
Bearing accuracy	2°	1°
Number of targets tracked simultaneously	8	16
Track history	15 min	15 min
Sector mode sector	60°	-
direction	freely selectable	-
Audio channel coverage	360°	70°
receiving beam pattern	up to 10°	up to 10°
frequency range	1-3 kHz	1-3 kHz

ITALY

ASWAS (ANTI-SUBMARINE WARFARE AREA SYSTEM)

Type
Underwater area surveillance system.

Description
ASWAS is an area defence system designed to integrate many different types of sensors for the detection, tracking and classification of underwater targets such as submarines, submersibles, swimmer delivery vehicles, mine delivery vehicles, propelled mines and swimmers. The system is able to monitor a wide area using an integrated combination of subsystems; each subsystem relying on magnetic, geophonic and active/passive acoustic sensors (LF/MF/HF) deployed on the seabed.

ASWAS guarantees a very high degree of threat detection in large areas near naval bases, coastal installations, harbours, offshore platforms, straits and 'chokepoints'.

The system relies on a number of multi-functional consoles for a real-time tactical display of raw and processed data. A surveillance radar and/or optronic system can be connected to the control station to monitor surface targets even at long distances.

ASWAS acts as an early warning detector and as an underwater gate giving accurate information about incoming intruders. The operators have a complete panorama of the tactical situation and, if required, they can concentrate their activities on dedicated sectors of the area. Immediately a preset level of warning is achieved it is possible to alert a 'fast reaction force'.

ASWAS can also be integrated with other similar systems for the widespread defence of an area.

Operational Status
In advanced development.

Contractor
Societa Sistemi Subacquei WELSE SpA Consortile, Genoa.

Control console of the ASWAS system

UNION OF SOVEREIGN STATES

FIXED SONAR SYSTEMS

Type
Static sonar surveillance systems.

Description
There is little doubt that the former Soviet Union has an extensive fixed and moored sound surveillance system and it is almost certainly deployed at a number of locations around the world. Virtually no information is available on this sphere of operations.

It is known that the former USSR was continuing to deploy Cluster Lance planar acoustic arrays in Pacific waters near the Russian land mass where broad area surveillance is required. Barrier arrays have been deployed at points of egress and ingress from SSBN operating areas, such as in the Barents, Greenland and Kara Seas. These are placed in or near trenches at 'chokepoints' which would serve as 'tripwires' more than long-range surveillance systems.

The USA considers that by the early to mid-1990s acoustic sensors will have considerably increased their detection ranges and will include improved hydrophone arrays and acoustic processing capability. There is even a suggestion that long-range, low frequency arrays could be mounted on the permanent Arctic ice pack.

The former Soviet Union is also believed to have been developing non-acoustic surface signature detection systems which sense the disturbance on the surface caused by a passing submarine. Wake detection systems that sense turbulent wake, internal wave wake or contaminant (chemical or radioactive) wake could also be in development.

UNITED KINGDOM

AS370 SEABED SURVEILLANCE SONAR

Type
Active static surveillance sonar system.

Description
The AS370 is a seabed surveillance sonar designed with long-term immersion capabilities for coastal waters and harbour surveillance installations. It is designed to detect underwater swimmers out to 500 m in sea state 3 and small submersibles out to its maximum range of 1 km. It is intended for permanent installation on the seabed and can be deployed with single or with multiple underwater sensors combined to provide total coverage of an area.

The transducer array rotates within an acoustically transparent GRP dome which prevents marine growth forming on sensitive array surfaces. All components exposed to seawater are manufactured in GRP, PVC or stainless steel material. As an added precaution the dome can be filled with clear, fresh water prior to final seabed installation. The interconnecting single line cable length to the surface control unit can be up to 6 km, although this can be increased by means of in-line signal repeaters.

There are three sonar modes employing different transmission and signal processing techniques. These are:
(a) PCM – for optimum range resolution
(b) chirp – for maximum interference immunity
(c) Doppler – for target relocating analysis.
Alarm facilities enable the detection and tracking of sonar targets which satisfy selectable predetermined parameters such as amplitude/threshold, target size and Doppler content. By selection the operator may

AS370 showing underwater unit on tripod

be presented with either alarm contacts only or sonar plus alarm contacts. Nuisance alarm triggering can be eliminated while using high sensitivities by means of a paintbox facility.

The surface control unit offers various display and processing features, including six page storage and recall, zoom ×5 magnification, origin shift ×2 magnification, normal PPI display and recordable video output. The keyboard allows further facilities on the screen.

Operational Status

The system is currently undergoing long-term evaluation trials with the Swedish Defence Materiel Administration.

Specifications

Beam pattern: 4° × 6°
Range: 50, 100, 200, 500, 1000 m
Transmission power: Variable
Scan: Sector or continuous
Transmission pulse: Variable
Gain law: Variable or auto
Beam tilt: ± 15°
Sonar transmission: PCM, chirp, Doppler
Underwater unit
Depth rating: 60 m
Dimensions: 1900 mm (high) × 1049 mm (diameter)

Weight: 155 kg
Display unit
Dimensions: 575 mm (wide) × 475 mm (high) × 515 mm (deep)
Weight: 56 kg
Cable: Combined electrical/fibre optic with Kevlar stress member
Cable weight: 40 kg/100 m

Contractor

Marconi udi (a subsidiary of Marconi Underwater Systems Ltd), Aberdeen.

SWIMMER DETECTION SONAR

Type

Static or mobile surveillance sonar system.

Description

The Marconi Swimmer Detection Sonar is designed to counter the threat of terrorism and sabotage, offering comprehensive protection to naval bases, commercial ports and other economically vital assets such as offshore oil platforms, oil refineries and nuclear power stations. It can be supplied either as a fixed system or in a mobile form.

The system will continuously detect, classify and track intruders at ranges that allow successful interception. It is fully automatic. When a target is detected a series of detection, classification and tracking algorithms are initiated prior to an alarm being activated. This level of interaction with the possible intruder provides very low false alarm rates. The auto-detection process is programmable to meet the requirements of target threat versus environmental conditions. The special characteristics, or signatures, of swimmers and small vehicles ensure that they can be accurately tracked while similar-sized harmless debris is eliminated electronically.

Having confirmed the presence of an intruder, the system will predict its track and estimate the time of interception. Using the sonar beam, a police launch or similar craft fitted with a transponder can be directed to a position immediately above the intruder.

Due to the high degree of automation, it is possible for an unskilled person to interrogate the system and, by means of the tactical display, to identify quickly the type of intrusion, its range, speed and bearing.

The Swimmer Detection Sonar uses a mirror array as the sensor. The array provides near-perfect beam forming with a minimal number of components, in a robust, low-cost configuration. The fixed installation consists of a 120° sonar unit, giving an azimuth resolution of 1.3° over the total field-of-view. The trainable mobile installation has a 40° azimuthal field-of-view. The electronically scanned sonar heads have no moving parts, thus ensuring high reliability and ease of maintenance.

Marconi also offers complete intruder detection system packages comprising radar, IR sensors, acoustic fences and physical barriers. Used in

Marconi Swimmer Detection Sonar and deployment system

combination, these systems offer protection against intruders from up to 30 miles away right up to the inner harbour zone.

A flexible communications system with facilities for point-to-point, broadcast and group communications is available at all displays. Interfaced external communications provide up-to-date data for naval forces and harbour authorities. Full command and control facilities allow optimum co-ordination and prevent any possibility of undetected escape or reinforcement.

Operational Status

The Swimmer Detection Sonar system has successfully undergone extensive proving trials in a variety of demanding environmental conditions worldwide.

Contractor

Marconi Underwater Systems Ltd, Templecombe.

TG-1 HF SONAR SYSTEM

Type

HF static surveillance sonar system.

Description

Ferranti has designed, developed and produced a new high frequency sonar system which offers electronic scanning at a cost which, the company states, makes it competitive with mechanically scanned equivalents. It has been given a provisional nomenclature of TG-1. The equipment, which is understood to have a range of some 700 m, is built to withstand continuous operation down to depths of 500 m. Targets are identified in both azimuth and depth.

The modularity of the system makes it suitable for a diversity of applications. These include harbour surveillance, offshore platform defence, mine identification and classification, ROV navigation, and oil/gas pipeline profiling.

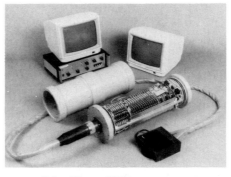
Units of Ferranti TG-1 sonar system

Typically a single unit comprises two 64-element transducer arrays each connected to its own underwater pod, and using a digital beam forming technique. An umbilical cable provides both power and

data link to the pods from the surface control unit. A standard VDU presents high resolution data which include the bearing, range and depth of sonar contacts. An auto-alarm system causes an audible alarm to sound when a possible sonar return is received.

In more general surveillance applications up to 255 pods can be linked to form a system. The control system then acts as a master and accesses each of the slave pods in turn.

Operational Status

The equipment has already been tested successfully. The basic system was developed by Universal Sonar Limited, Bridlington, which was acquired by Ferranti plc in December 1986.

Contractor

Ferranti-Thomson Sonar Systems UK Ltd, Stockport.

UNITED STATES OF AMERICA

FIXED DISTRIBUTED SYSTEM

Type
New static underwater sonar surveillance programme.

Description
The Fixed Distributed System (FDS) is a highly classified US Navy undersea sensor system programme that will listen for submarines at specific locations or 'chokepoints'. It is almost certainly part of the overall Underseas Target Surveillance Program which examines the application of varied technologies to the detection, localisation, classification and tracking of underwater targets, including submarines, small submersibles, ROVs, ordnance and swimmers.

During the past few years new fibre optic trunk cables have been designed, as part of the Fiber Optic Distributed System's programme, and tests have been carried out. In addition processing algorithms have been developed to perform detection and classification for the Fiber Optic Distributed System.

The FDS is scheduled to supplement the existing SOSUS system (see separate entry), which operates at different ranges.

Operational Status
The US Navy's Space and Naval Warfare Systems Command is receiving proposals from four industrial teams for design of the shore signal information and processing segment of the FDS programme. The four teams competing are AT&T Technologies, IBM Federal Systems Division, General Electric and Hughes Ground Systems. The US Navy plans call for the award of two $10 million design contracts to two teams for a 20-month long design phase. One team will then be selected to proceed to full-scale engineering development of the data processing segment. AT&T already has a $100 million contract for continued research that will lead to manufacture of the sensors and other underwater portions of the system.

Contractors
To be selected.

PILOT FISH

Type
Static seabed sonar system.

Description
Pilot Fish is understood to be an underwater detection and position locating system in development, which involves the use of a transmitter placed on the ocean bottom to send data to submarines. The transmitter would have sufficient self-contained power to transmit sonar signals. The echoes would be reflected from a hostile submarine, and would reach a listening submarine which would be far enough away so that echoes rebounding off it would not in turn be strong enough to disclose anything to the enemy.

Operational Status
The programme is highly classified and no official information has been released. It is understood, however, that the programme remains a high priority item for the US Navy.

RDSS

Type
Seabed moored passive surveillance system.

Description
RDSS (Rapidly Deployable Surveillance System) is designed to use air-deployable, bottom-moored, long-life passive acoustic buoys as the main sensor in a new redeployable monitoring system. It will be emplaced by aircraft to provide undersea surveillance coverage on an urgent basis if required in areas of special interest. It could be used to augment or replace existing systems for long periods. RDSS buoys are capable of use in shallow water and areas unsuitable for either SOSUS or SURTASS techniques.

The basic design of RDSS allows for transmission from the buoy to existing maritime patrol aircraft for either real-time readout or recording for later analysis and replay. Two versions of RDSS buoys have been identified as Mod 0 and Mod 1 respectively. The former transmits its data to P-3 or S-3 maritime aircraft for analysis and presentation on the aircraft. The data are also recorded for later processing. The Mod 1 design is intended to send information direct to an ASW processing centre which will utilise the data without delay. The Mod 0 buoy is regarded as a 'near-term' item and the Mod 1 is thought likely to take longer to reach operation.

Operational Status
The approval for full-scale development of the system was given in August 1981 after a lengthy study and research programme. In December 1984 the US Navy cancelled the programme for a variety of reasons but 16 days later it was resurrected and re-examined, with a view to restructuring the whole concept. However, development funds for 1987 were deleted by the US Congress and the programme was shelved. Although the programme may be re-activated at some stage it appears unlikely to be in this configuration.

SOSUS

Type
Fixed sound surveillance system.

Development
The current US philosophy regarding defence against the principal ASW threat (submarines of the USS) can be summarised as relying on engaging these submarines in forward areas and at barriers before they can get within attacking range of US forces. To do this, reliance rests primarily on American attack submarines and long-range P-3 patrol aircraft supported by underwater systems.

The ability to locate enemy submarines within broad ocean areas is essential to the task of containing the large USS submarine force, and fixed undersea surveillance systems have played a major role in this for some years. Consolidation and expansion of such systems continues, and funding of the Sound Surveillance System (SOSUS) is sustained at a consistently significant level. To improve US submarine surveillance capabilities, the 1986-90 programme continues funding for the mobile SURTASS system, described in the *Surface Ship ASW Systems* section. Development of the rapidly deployable air-dropped system (RDSS) has been halted.

Research in the undersea surveillance systems programme, which funds future upgrades, is concentrating on the development of integrated acoustic display and wideband acoustic recall software packages. The US Navy is also conducting experiments with an acoustic survey system in shallow and moderate depth waters.

Description
SOSUS (Sound Surveillance System) consists of fixed undersea acoustic networks of passive hydrophone detector arrays deployed in the Atlantic and Pacific Oceans, and this is stated to provide significant detection capabilities. However, over the years (SOSUS was conceived in the 1950s) it was inevitable that the network gradually expanded as increments were added to enlarge the coverage. Most of these additions were the subject of various classified programmes, of which most details remain secret apart from isolated programme code names. Among the latter are 'Caesar' and 'Collosus', referring to the US Atlantic and Pacific segments of the SOSUS network. Hydrophones designated AN/FQQ-10(V) are located at intervals of 5 to 15 miles along a linking cable connected to the shore station(s). Similar shore stations are understood to exist beyond the USA (for example, the Aleutian Islands, Canada, Denmark, Iceland, Italy, Japan, South Korea, the Philippines, Spain, Turkey and the UK), these being included in programmes code named 'Barrier' and 'Bronco'. Constantly changing threat characteristics, particularly those associated with much quieter running submarines and the use of acoustic cladding on the hull, call for significant system improvements and the USN embarked on a two-phase SOSUS improvement programme. The specific aims of this programme remain classified in detail but are known to include improvements to the communications elements of the system and complementary improvements to the signal and data processing segments, thereby facilitating greater operational utilisation of the data obtained by the undersea sensor arrays. In June 1985 a contract awarded to AT&T referred to the 'FY85 SOSUS Backfit Program'. No details of this have been disclosed.

In the SOSUS system the raw data by LOFAR techniques (detection by signature) is passed to local centres for initial processing, and thence to regional centres. From the regional centres the processed data are passed to main evaluation centres which combine the data with information from other sources.

Funding of the SOSUS programme remains fairly constant, but security classification prevents further information. Considerable funding for both development and procurement is likely for at least 10 years. It is understood that the Canadian government is considering the installation of a similar system to SOSUS to detect submarines beneath the ice cap and in Canadian territorial waters.

The ultimate objective for the SOSUS system is the complete integration of all undersea surveillance sensor systems into a fully co-ordinated and centrally controlled network, IUSS (Integrated Underwater Surveillance System).

Operational Status
In operational service on a worldwide basis.

Contractors
AT&T, Greensboro, North Carolina (research and engineering).
Ernesco, Springfield, Virginia (processors).
TRW, McLean, Virginia (engineering/integration).
Simplex Wire & Cable, Portsmouth, New Hampshire (coaxial cable).

COUNTERMEASURES

This section covers those countermeasures which are used to protect the vessel. This includes towed and expendable acoustic anti-torpedo systems and submarine-based electronic warfare systems. Both submarine and surface ship EW systems are covered in detail in *Jane's Radar and Electronic Warfare Systems.*

ACOUSTIC DECOYS

INTERNATIONAL

SURFACE SHIP TORPEDO DEFENCE (SSTD)

Type
Development programme for anti-torpedo defence.

Description
For some years the US Navy and the Royal Navy have been studying various methods of anti-torpedo capability, under the Surface Ship Torpedo Defense (SSTD) and Talisman programmes respectively. These studies have culminated in a memorandum of understanding (MoU) signed in October 1988 to collaborate in the development of an anti-torpedo defence system to protect surface ships against attack from the increasingly sophisticated wake-homing torpedoes. The project will see the SSTD programme merged with the Talisman project, with the US Navy taking the lead.

An Initial Operational Capability is expected for the late 1990s. The US Navy envisages a requirement for about 400 systems while the Royal Navy is believed to want 80.

The SSTD programme will be designed to counter any projected torpedo threats. The priority is to provide a vessel with defence against 70 km range, wake-homing torpedoes which lock onto the turbulence created by a passing warship. Detection of such a threat is difficult since the attacking weapon needs to be picked up within the turbulence created by the ship – an area which represents a poor detection environment for sonar. To overcome this problem a specialised variable depth sonar or towed array may be employed.

Following detection the torpedo threat will be categorised and countermeasures selected to provide the most appropriate defence. Various 'hard kill' and 'soft kill' options are being examined.

Hard kill options include the use of a modified lightweight torpedo, such as the UK Mk 46 or the UK Sting Ray, depth charges which are dropped or fired from the rear arc of a vessel, and mortars such as the present Royal Navy 1 km range Mk 10.

Soft kill options under consideration include the use of nets and anti-torpedo decoy countermeasures. A combination of both hard and soft kill options could eventually form a complete anti-torpedo package. Since signing the MoU, the USA has veered towards developing a solely hard kill solution, but the UK favours a combined soft/hard kill.

Operational Status
The US Department of Defense and the UK Ministry of Defence are conducting the concept evaluation phase of the project. Three bi-national teams have received parallel contracts for demonstration and validation, following which they will compete for a single full-scale development contract. The expected in-service date is 1997-98.

One team includes General Electric and Honeywell Underseas System in the USA with Marconi Underwater Systems and Plessey Naval Systems in the UK. Under this arrangement Plessey had responsibility for the bow array of the torpedo detection, classification and localisation system, together with the command and control system and Marconi, with Honeywell, is working on soft and hard kill countermeasures, although with the merger of Marconi and Plessey this may change.

The second team is led by Westinghouse in the USA, with Librascope (USA) and Dowty Maritime Systems (UK) for the soft kill, and Ferranti (UK) for command and control. A third team has now been nominated, consisting of Martin Marietta, British Aerospace, Hughes Aircraft Company and Frequency Engineering Laboratory.

It has been reported that the French Ministry of Defence is interested in joining the programme and is currently examining the US/UK MoU. If France joins, then the French Navy would have a requirement for up to 100 systems.

ISRAEL

ATC-1 TORPEDO DECOY

Type
Acoustic torpedo decoy.

Description
ATC-1 (Acoustic Torpedo Countermeasures-1) is a towed decoy, developed as a solution to the threat of modern, high speed acoustic-homing active/passive torpedoes. Simple and effective, it generates and transmits strong acoustic signals which are accepted as a legitimate target by attacking torpedoes. Consequently, the torpedo acquires the decoy and attacks and re-attacks until its power source is exhausted and it sinks.

The ATC-1 does not restrict the speed or manoeuvrability of the towing ship. Lightweight, compact and reliable, it is simple to install and requires a minimum of deck and below-deck space.

Compatible with a wide range of both naval and merchant vessels, from fast missile boats, corvettes, frigates and destroyers to cargo ships and tankers, the ATC-1 has been designed, built and tested to MIL-E-16400. The decoy does not interfere with other systems on board, and is not affected by them. Optimal vessel protection is ensured by a frequency range that covers all existing homing heads.

The ATC-1 features a towed body 120 cm long and 30 cm in diameter weighing 25 kg, together with winch, launch and recovery units weighing a total of 1325 kg. Also on board the ship are an electronics cabinet weighing 190 kg and a remote-control unit weighing 10 kg, neither of which require much space. The manufacturer claims that a single operator is required for the system.

Operational Status
In production.

Contractor
Rafael, Haifa.

Rafael ATC-1 towed body

SCUTTER

Type
Expendable torpedo decoy.

Description
Scutter is a self-propelled expendable torpedo decoy providing omnidirectional protection for submarines. On receipt of a torpedo alert, Scutter is launched from any standard submarine signal ejector without requiring any special tools. No pre-launch tests or presets are required. System operation is automatically initiated after launch, the decoy propelling itself autonomously to the operating position. Torpedo transmissions detected are analysed and the decoy selects the appropriate deception signals for emission, allowing the submarine to perform evasive manoeuvres.

The decoy can operate for up to 10 minutes at depths down to 300 m. The decoy self-destructs and sinks on completion of the mission, thus preventing retrieval by hostile forces.

Scutter can be deployed by all types of submarine and has no special installation requirements. Due to its light weight and compact size it takes up very little space.

Specifications
Diameter: 100 mm
Length: 1000 mm
Weight: 7.8 kg

Contractor
Rafael, Haifa.

ITALY

C303 TORPEDO COUNTERMEASURES

Type
Submarine anti-torpedo countermeasures system.

Description
The Whitehead C303 submarine anti-torpedo countermeasures system has been designed to counter attacking acoustic-homing, active/passive torpedoes by means of expendable 'soft' devices (decoying and jamming actions). These actions give a consistent escape probability to the submarine.

The C303 countermeasures system consists of:
(a) effectors (jammers and decoys)
(b) launching system, either on the pressure hull or inside a torpedo tube
(c) control panel, which can be integrated with the torpedo detection system.

Operational Status
In development.

Components of the C303 decoy systems

Contractor
Whitehead, Livorno.

UNITED KINGDOM

BANDFISH

Type
Expendable torpedo countermeasures system.

Description
An expendable countermeasures system for use against acoustic torpedoes, Bandfish has been developed by Dowty Maritime Sonar and Communication Systems. It is designed to be launched from a submarine's submerged signal ejector or from a surface ship launcher. Once launched, the device operates independently of the launch platform.

In operation Bandfish hovers in mid-water and transmits a high intensity acoustic signal. The operating depth and acoustic pattern can be designed to meet the customers' requirements. It has a minimum shelf life of five years and requires no user or depot maintenance. Simple go/no go tests are provided to ensure that it can be launched with maximum confidence.

Operational Status
In full production.

Specifications
Length: 995 mm
Diameter: 102 mm
Weight: <12 kg

Contractor
Dowty Maritime Sonar and Communication Systems, Greenford.

Bandfish expendable torpedo countermeasure

G 738 DECOY

Type
Towed torpedo decoy equipment.

Description
The G 738 is an improved and re-engineered version of the Royal Navy Type 182 towed torpedo decoy, now being manufactured for several other navies. Known features of the equipment are modern solid-state signal generation equipment and compact, lightweight deck machinery.

The equipment is designed to decoy both active and passive homing torpedoes from a ship fitted with the G 738/Type 182. Decoy signals are electronically generated within the ship and fed via the towing cable to electro-acoustic transducers within the towed body; the signals thus produced divert passive homing torpedoes astern and confuse or jam the steering mechanisms of active homing torpedoes. It can also be used as a high frequency noise source for trials purposes.

The G 738 simultaneously produces two independently controllable output signals which are radiated from the towed decoy. An amplitude modulated noise signal is provided, to simulate the ship's propeller noise and to cause passive homing torpedoes to be seduced away from the ship towards the towed decoy.

A swept CW signal is provided to simulate sonar echoes so as to confuse or jam the guidance system of active homing torpedoes, and cause them to miss the ship.

The frequency bands covered by both types of output are switchable, as required for tactical scenarios and to enable concurrent operation of other sonar and communication equipments.

A variety of deck-handling arrangements is produced, and dual systems may be fitted. Remote-control of selection of signal characteristics provides

Specifications

Dimensions and Weights	Height	Depth	Width	Weight
Electronics cabinet	1725 mm	640 mm	585 mm	375 kg
Remote-control unit	152 mm	190 mm	305 mm	2.3 kg
Deck connection box	203 mm	115 mm	265 mm	13 kg
Resistance box	444 mm	115 mm	381 mm	14.3 kg
Towed decoy	533 mm	2005 mm	457 mm	74 kg
Cable reel	1245 mm	2340 mm	1390 mm	2540 kg
Gantry	203 mm	5918 mm (retracted)	440 mm	810kg
Davit	3500 mm (Jib raised)	3400 mm (max)	1270 mm	818 kg
Triple control unit	357 mm	240 mm	440 mm	188 kg

Towing cable
Length: 411 m
Diameter: 14.48 m
Weight: 269 kg

for instantaneous changes from operations room or bridge. The equipment can incorporate handling systems for open deck on extended flight deck ships.

The operating characteristics remain confidential between the manufacturer and the purchaser.

Operational Status
In production and in operational service with the Royal Navy, and a number of other navies.

Contractor
Graseby Dynamics Ltd, Watford.

SUBMARINE SIGNAL AND DECOY EJECTORS

Type
Reloadable submarine discharge systems.

Description
Submarine signal and decoy ejectors provide a small reloadable discharge system, capable of launching cylindrical stores approximately 100 mm in diameter and 1250 mm in length, over the complete submarine operating envelope. These ejectors can also be designed and manufactured to meet particular requirements.

Existing ejectors, currently in service, are used to deploy communications buoys, bathythermal probes, exercise and emergency signalling stores, decoys and countermeasures, including programmable stores and stores requiring data links. The ejectors can be configured for purely manual operation, or they can be provided with an automatic operating capability, enabling discharge to be initiated either locally or from a remote position.

The submarine signal and decoy ejector contains two main assemblies built into a common hull insert. The launch tube assembly, into which the store is loaded prior to launch, and a ball valve closure outboard. Mounted alongside is the water/air ram assembly. The hull insert is normally positioned radially in the pressure hull, discharging across the ship's flow-field. It can be positioned with its exit flush with the pressure hull, or it can be moved radially outboard if required to bridge the space between the pressure hull and the free flooding casing.

The stores are loaded into the ejector which is then

Submerged signal ejector trainer

automatically flooded and equalised at depth pressure. Launch is effected by admitting high pressure air to the air ram causing it to move and discharge water into the tube, thereby ejecting the store. Ejectors of this type combine a virtually instantaneous discharge capability with a very rapid cycle time, and have a proven record of high reliability and low maintenance.

Operational Status
In operational service. Strachan & Henshaw are currently the design authority for the Royal Navy.

Contractor
Strachan & Henshaw, Bristol.

SEA SIREN DECOY SYSTEM

Type
Towed anti-torpedo decoy system.

Description
The Sea Siren is a new, microprocessor-based towed torpedo decoy system which is claimed to provide significant improvements in performance and cost over existing systems. It is a logical follow-on to the G 738, but much more sophisticated, and is suitable for all classes of ship over 400 tons or 50 m in length.

Sea Siren produces a realistic noise signature of the ship to decoy passive homing torpedoes from the ship, and generates sonar-like echoes to confuse active-homing torpedoes. User-programmable

parameters are employed to ensure total confidentiality and the decoy can be easily updated to counter the latest threats. Easy deployment and handling is ensured using a davit or gantry system. One or two decoys can be used and can be controlled remotely from any position in the ship over a serial data link.

Operational Status
In advanced stage of development.

Specifications
Acoustic/operational data
Active decoy signal (CW)
Sweep frequency bands: Up to five programmable sweep frequency bands are provided to cover a wide range of torpedo sonar frequencies
Sweep rate: User-programmable

Chopped CW: On/off chopping periods programmable by the user
Output modes: CW only; noise only; CW and noise simultaneously; CW and noise alternately
Passive decoy signal
Noise bandwidth: Pseudo-random noise spectrum band to cover a wide range of passive torpedo sonar frequencies
Modulation: Choice of modulation envelope shape and programmable modulation depth
Modulation frequency: Dependent on rpm and number of propeller blades
Chopped noise: On/off chopping period programmable by the user

Contractor
Graseby Dynamics Ltd, Watford.

MARCONI ACOUSTIC COUNTER-MEASURES

Type
Range of acoustic countermeasures devices and launcher units.

Description
Marconi Underwater Systems has developed a family of acoustic countermeasures which can be used to

protect both submarines and surface ships against homing torpedoes. Both passive and active threats are affected. These devices are value engineered as a self-contained and expendable store, and various parameters, such as frequency bands, acoustic power levels and jamming signals can be designed to individual requirements.

The launch system consists of a cradle, launch canisters and a launch control unit. The canisters, containing the countermeasures, are supplied as environmentally sealed units ready for immediate use.

Once discharged the canisters can be retained for refurbishment. The launch cradle supports up to six canisters and can be mounted in any deck position. The launch control unit is mounted in the operations room and deploys selected countermeasures in response to a torpedo alert.

Contractor
Marconi Underwater Systems Ltd, Waterlooville.

FERRANTI ACOUSTIC COUNTERMEASURES

Type
Range of acoustic countermeasures units.

Description
Ferranti has designed and produced a range of acoustic countermeasures units. These are high power, high efficiency devices for deployment against acoustically guided weapons to protect both surface vessels and submarines. Both spatial targets and jammers have been designed and can be tailored to individual requirements in submarine or surface (air-launched) configurations. These devices are available in formats which can be deployed by most standard launch systems. Special requirements, such as overall frequency bands, source levels and noise structures can be made available as required.

An anti-torpedo acoustic countermeasure known as ATAAC is a noisemaker launched from a standard chaff launcher. There are both 3 and 4 in versions and the rounds are programmed before launch to produce broadband noise. An 8 in version, for discharge from a submarine countermeasures tube, is also available.

Operational Status
In production.

Contractor
Ferranti-Thomson Sonar Systems UK Ltd, Aldershot.

UNDERWATER NOISE PROJECTORS

Type
Low frequency acoustic projectors.

Description
Gearing & Watson has been involved with the production of depth compensated, low frequency acoustic projectors (hydrosounders) for a number of years. These consist of a user-programmable digital signal generator, a power amplifier, hydrosounder with pressure compensator and a test unit.

Systems which use the UW300S hydrosounder have been installed on Royal Navy SSN and SSK submarines. These systems are known as Sonar Type 2071. A system has also been installed on an 'Agusta' class submarine.

Using the UW300F hydrosounder housed within a glass fibre hydrodynamic body a towed version has been produced. This is known as Sonar 2058 and has been produced for the Royal Navy to be deployed from various classes of surface vessels.

By producing complex signals consisting of tones, noise bands or pulses the underwater noise projector systems can provide:
(a) noise countermeasures against sonars and torpedo homing heads
(b) realistic target signatures for operator training
(c) improved performance assessment for all types of passive sonar and sonobuoy systems
(d) a low frequency sound source for noise ranges and experimental work
(e) acoustic adjunct to mine countermeasures systems.

The complex low level signals required are generated by a digital signal generator (DSG) which is capable of simultaneously generating up to 11 different signals. These signals may comprise single tones, swept tones, bands of random noise, modulated noise, or pulses. The type, frequency, amplitude and duration of these signals is programmed into the DSG using a keypad and displayed for the operator via a VDU. The DSG is able to store in memory up to eight sets of programmes, thus eliminating the need to reprogramme each time the system is used. Pre-recorded data may also be routed through the DSG.

Operational Status
In operational service with the Royal Navy.

Contractor
Gearing & Watson (Electronics) Limited, Hailsham.

UNITED STATES OF AMERICA

Note: For security reasons detailed information regarding US acoustic countermeasures is rarely available. However, various items mentioned in the FY90 long-range estimates document put out in early 1990 by the US Navy Sea Systems Command are given below for the interest of readers. All are submarine devices, except for the AN/SLQ-25 decoy. Some of the devices mentioned are also described later in this section.

Countermeasures Command and Control Unit – an advanced submarine countermeasures controller unit with an expandable capability for countermeasures device inventory management, processing tactical solutions, target data management, and launch sequencing of all externally configured countermeasures launchers. The equipment is in early design with engineering development models due in 1993-94. It is believed that about 50 sets are likely to be procured by 1998.

ADC Mk 1 Mod 0 – 5 in expendable acoustic countermeasures device produced by Conax Corporation. Procurement quantities for 1990-94 exceed 2000 units.

ADC Mk 1 Mod 1 – 6 in expendable acoustic countermeasures device.

ADC Mk 2 – 3 in expendable acoustic countermeasures device produced by Hazeltine (Emerson Electric).

ADC Mk 3 – 6 in submarine-launched acoustic expendable countermeasures device produced by Bendix Oceanics.

ADC Mk 4 – submarine-launched acoustic countermeasures device produced by Hazeltine (Emerson Electric).

ADC Mk 5 – expendable acoustic countermeasures device in early development by Bendix Oceanics.

NAE Beacon Mk 3 – 5 in sonar countermeasures device produced by Pique Engineering. Procurement quantities scheduled for 1990-94 exceed 3000 units.

Mobile Multi-function Device – 6 in expendable submarine-launched advanced countermeasure device with mobile capabilities for confusing adversary sonar and torpedo threats. In early design with 20 engineering development models due to be delivered in 1994-95. This may be the ADC Mk 5 (see above).

CSA Mk 2 Mod 0/Mod 1 – 6 in launching system for acoustic countermeasures stores.

AN/SLQ-25 – Nixie towed surface ship torpedo decoy in full production by Frequency Engineering Laboratories.

AN/SLQ-25 COUNTERMEASURES

Type
Torpedo countermeasures transmitter (Nixie).

Development
Development of the AN/SLQ-25 Nixie was accomplished under a US Navy project in the mid-1970s. This project was involved with the development of systems to enable ASW and non-ASW ships to counter attack. The first production contract was awarded in 1975.

Description
The AN/SLQ-25, also known as Nixie, is a towed electro-acoustic device that provides surface ships with an effective countermeasure against homing torpedoes. Together with a command/display and information processing system, the AN/SLQ-25 comprises the Ship Acoustic and Torpedo Countermeasures (SATC) system. In addition to detecting and recognising threats posed by hostile acoustic sensors and acoustic homing torpedoes, the Nixie device attempts to jam the acoustic sensor and transcribe necessary data so that the ships can take the appropriate action.

Nixie consists of two small lightweight towed bodies which are streamed and recovered by two winches. Below decks equipment consists of a number of electronics cabinets and a control console sited either in the command centre or near the sonar consoles. The towed bodies transmit acoustic signals intended to complement the ship's acoustic signature, and decoy the torpedo away from the ship.

The towed bodies are streamed in tandem in case one should be hit by a decoyed torpedo, and thus leave the ship defenceless against a salvo attack. Among the decoy's features are those of simulating specific noises created by the towing vessel, such as machinery and propellers, or generating noises of a specific nature at different frequencies, depending on the nature of the anticipated threat and its mode of operation.

Operational Status
The AN/SLQ-25 is still in production. The device is in operational use in the US Navy and a number of other navies, including those of Australia, Italy, Japan, South Korea, Portugal and Spain.

Contractor
Frequency Engineering Laboratories, Azusa, California.

ADC Mk 2 COUNTERMEASURE

Type
Expendable acoustic torpedo countermeasure.

Description
The ADC (Acoustic Device Countermeasure) Mk 2 is a submarine-launched decoy designed to counter acoustic torpedoes. It is an 8 cm diameter device which hovers vertically at a pre-selected depth emitting an acoustic signal.

The uppermost section consists of ceramic transducers and impedance-matching networks for the acoustic transmitter. Below this is the electronics section in which the signals are generated and amplified. At the bottom of this section is a seawater-activated battery which powers a pressure-controlled motor driving a small, shrouded propeller which keeps the decoy hovering.

Operational Status
In production since 1978 and more than 3000 units have been produced. A follow-on contract was awarded to Emerson Electric Co in 1988 for modifications to the ADC Mk 2, followed by a production award for additional decoys in 1989.

Contractor
Hazeltine Corporation, Greenlawn, New York.

ADC Mk 3 COUNTERMEASURE

Type
Submarine countermeasures equipment.

Description
ADC Mk 3 is a submarine-launched countermeasures equipment. The device is a wideband 6 in (15 cm) unpowered jammer ejected from submarines, designed primarily to jam torpedo sonar homing heads. Information on this device is restricted and no other technical details have been released.

Operational Status
In production. Bendix Oceanics was awarded a contract worth over $13 million in mid-1989 for 367 systems.

Contractor
Bendix Oceanics Inc, Sylmar, California.

ADC Mk 4 COUNTERMEASURE

Type
Expendable acoustic countermeasure.

Description
The ADC (Acoustic Device Countermeasure) Mk 4 is a submarine-launched acoustic countermeasure. No other details are available.

Operational Status
In development since 1988.

Contractor
Hazeltine Corporation, Greenlawn, New York.

ADC Mk 5 COUNTERMEASURES

Type
Expendable acoustic countermeasures.

Description
The ADC Mk 5 is a wideband expendable acoustic countermeasures device in development. No details have been released but it is probably a mobile device for use against sonar and torpedoes.

Operational Status
In early development.

Contractor
Bendix Oceanics Inc, Sylmar, California.

US NAVY TORPEDO DECOYS

Type
Development of new anti-torpedo decoys.

Description
The curren US Navy decoy system for use against acoustic torpedoes is the AN/SLQ-25 Nixie, but this is now regarded as inadequate against the present generation of threats. Two new programmes, Anti-Ship Torpedo Defense and Surface Ship Torpedo Defense, were inaugurated. The former was cancelled but the latter forms part of a joint programme with the United Kingdom.

In October 1988, a memorandum of understanding was signed between the United States and the United Kingdom for the anti-torpedo programme. This is discussed in more detail in the International section.

ELECTRONIC WARFARE

CHINA, PEOPLE'S REPUBLIC

921-A ESM EQUIPMENT

Type
Shipborne radar ESM receiver.

Description
The 921-A is a wideband pulse radar direction finding receiver and is installed on submarines to detect emitters of airborne, shipborne and shore-based radars. It provides coarse measurement of azimuth, frequency band and operational state of the hostile emitters.

Operational Status
In service with the Chinese Navy.

Specifications
Frequency: 2-18 GHz in four bands
Sensitivity: 1.5×10^{-3} to 10^{-4} W/M²
Bearing accuracy: Better than ±30°
Dimensions and weights
Antenna unit: 560 mm (diameter) × 515 mm (high); 80 kg
Receiver and display: 450 × 468 × 124 mm; 40 kg
Distribution unit: 145 × 214 × 291 mm; 6 kg

Contractor
China National Electronics Import and Export Corporation, Beijing.

921-A ESM equipment

FRANCE

DR 2000U (TMV 434) ESM RECEIVER

Type
Submarine radar threat warner.

Description
The DR 2000U is the submarine variant of the DR 2000 series of ESM systems. The system incorporates a crystal video receiver and has some analysis capability. When coupled with the Dalia analyser it can be converted into a basic ESM suite, known as the TMV 434.

The receiver is built up of six DF antennas and one omni-antenna and a processing control/display console. It provides a virtual 100 per cent probability of detection over the complete 360° of azimuth. The ESM system carries out passive search, detection of all pulse and CW signals and gives an instantaneous audible and visual alert. The DR 2000U Mk 3 is an improved version with better sensitivity.

The Dalia analyser provides alarm analysis and identification facilities through a library of 1000 radar modes and parameters and is easily reprogrammable.

Operational Status
In production.

Contractor
Thomson-CSF, RCM, Malakof.

DR 3000U ESM SYSTEM

Type
Submarine ESM system.

Description
The DR 3000 series is the latest family of ESM system from Thomson-CSF. Available in a number of variants (the submarine version is the DR 3000U), the surface ship configuration has been selected by the French Navy to equip the 'La Fayette' class frigates.

The DR 3000 is a fully automatic system providing 360° of coverage over the C to J frequency bands, and uses GaAs technology and radar identification by an expert system, supplemented with a high sensitivity IFM receiver. The lightweight (150 kg) DR 3000 uses autonomous air cooling of low power consumption units. It features an instantaneous high probability of interception of modern radars, high sensitivity providing range advantage compared with other systems, accurate direction finding, and reliable instantaneous identification of radars and platforms.

DR 3000 missions include radar threat warning, tactical situation buildup and updating, target designation, ECM control and ELINT.

The DR 3000 is offered in a variety of configurations. Four different types of antennas are available, including lightweight periscope aerial, small standard DF antennas, and a very high accuracy DF unit. Associated consoles are either a high definition colour TV display or a simplified lower weight LCD display unit. A bus interface is provided for integration of the DR 3000U unit with multi-purpose consoles in different submarine combat systems.

Operational Status
In production.

Contractor
Thomson-CSF, RCM, Malakof.

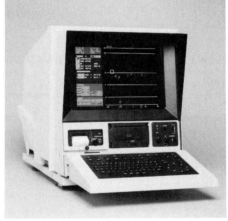

DR 3000U submarine ESM system

DR 4000U ESM SYSTEM

Type
Submarine ESM surveillance system.

Description
The DR 4000U is a variant of the DR 4000 series of surface ship and submarine ESM systems. It is a radar detection, analysis and identification system which has an extremely fast reaction time and a high instantaneous data collection capability. High sensitivity IFM receivers allow 100 per cent intercept probability, even on a single pulse. The highly automated system has an instantaneous 360° DF capability with bearing accuracies of about 5°. Frequency coverage is over the C- to J-bands with the possibility of extension to the range as an option.

The highly sensitive antenna system includes an omnidirectional unit for frequency measurements and two concentric six-port assemblies, one for C- to G-bands and the other for H- to J-bands for DF. The information is presented in a variety of forms to the operator on advanced colour consoles.

Operational Status
In production.

Contractor
Thomson-CSF, RCM, Malakof.

DR 4000U IFM ESM system console

ISRAEL

NS-9034 ESM/ELINT SYSTEM

Type
Submarine radar warning and ESM system.

Description
The NS-9034 is a compact and sophisticated ESM/ELINT radar warning system for submarines. It features instantaneous high sensitivity frequency measurement combined with accurate direction finding. It will carry out accurate and rapid detection and analysis in dense electromagnetic environments. The system uses an ESM mast incorporating six spiral antennas, covering the 2 to 18 GHz frequency range, plus an optional fine DF array incorporating an RKR lens array (7.5 to 18 GHz coverage). An omnidirectional array could also be mounted on the periscopic mast. Coarse DF accuracy is 6°, while fine DF measurements to within 1° are claimed to be possible. Elevation coverage is from −10° to +30°. A variety of data display facilities is available.

The NS-9034 can be installed in new submarines or retrofitted to existing vessels.

Operational Status
In development.

Contractor
Elisra Electronic Systems, Bene Beraq.

ELBIT SUBMARINE ANTENNAS

Type
Range of submarine antennas for ESM purposes.

Description
Elbit Computers manufactures a range of submarine antennas, designed primarily for ELINT purposes. Descriptions of the main systems are given below. Antennas meet the environmental requirements of MIL-E-16400 and are tested to withstand high pressure.

Model EA-0240/P
This is a wideband omnidirectional antenna operating over the frequency range from 2 to 40 GHz. The antenna is small and is designed to be installed on top of the submarine search periscope.

The antenna is built as a stack of three vertically polarised biconical antennas covered by a slant 45° polariser and sealed by a radome. The lower deck covers the 2 to 8 GHz band, the middle section operates over 8 to 18 GHz, and the upper one covers the 18 to 40 GHz range. The gain of the antenna varies smoothly with frequency and no 'holes' are detected in swept gain measurement curves, in vertical or horizontal polarisations. Dimensions of the antenna are 120 mm diameter × 230 mm high, with a weight of 7 kg.

Model EA-2818/SM
This is a wideband omnidirectional antenna, designed for submarines and covering the 2 to 18 GHz frequency range. The antenna consists of a stack of two biconical antennas, covered by a polariser and sealed by a radome. The lower deck covers the 2 to 8 GHz band with linear slant 45° polarisation, and the upper one operates in the 8 to 18 GHz band with circular polarisation.

Model EA-0818/SMH
Model EA-0818/SMH is a horn antenna operating from 8 to 18 GHz, and was developed for a submarine IDF array. The wide bandwidth and relatively constant beamwidth are achieved by the unique shape of the horn's flare, and by the specially designed feed section. The horn is covered by a shaped thin radome and is sealed to withstand high hydrostatic pressure.

Model EA-0208/SMS
This is a cavity backed spiral antenna designed for a submarine IDF antenna array. It is circularly polarised, and has a fairly constant beamwidth over the operating frequency range of 2 to 8 GHz. The wide bandwidth and constant beam behaviour are the results of the special layout of the antenna and the Balun feed. The antenna is covered by a hemispherically shaped radome and is completely sealed to withstand high hydrostatic pressure.

Model EA-0040/SM
Model EA-0040/SM is an ultra broadband omnidirectional antenna system operating over the frequency range 100 kHz to 40 GHz. The antenna system is small and is designed to be installed on top of a submarine search periscope.

The antenna system consists of three basic sections:
(a) microwave antenna – built as a stack of three vertically polarised biconical antennas covered by a slant 45° polariser and sealed by a radome. The lower deck covers the 2 to 8 GHz band, the middle section operates over 8 to 18 GHz, and the upper one covers from 18 to 40 GHz
(b) sub-microwave active stubs – this consists of two short monopoles encircling the microwave antenna radome in slant 45° helix shape to minimise interference, with appropriate active matching circuits

(c) RF module – contains limiters, amplifiers and detectors for the different receiving channels of the antenna system.

The gain of this antenna varies smoothly with frequency, and no 'holes' are detected in swept gain measurement curves, in vertical or horizontal polarisation. Dimensions of the antenna system are 350 mm height × 120 mm radome diameter and 250 mm base diameter. Weight is 15 kg.

Contractor
Elbit Computers Ltd, Haifa.

Model EA-0040/SM broadband omnidirectional antenna

TIMNEX-4CH ELINT/ESM SYSTEM

Type
Radar detection and analysis system.

Description
The Timnex-4CH system is an ELINT/ESM receiver intended for intercepting radar signals and for operating in land, sea or airborne environments. It covers the frequency range from 2 to 18 GHz and is designed to intercept simple or complex transmissions and to sort and compare them with a pre-programmed radar library.

The system features a channelised IFM receiver, a sophisticated IDF subsystem and real-time computer processing. Tabular display of the signal parameters with instantaneous spectra and gonio displays are presented to the EW operator. Hard copy printout and data logging on magnetic media are additional features.

Timnex-4CH has been designed to intercept radars at a range of several hundred kilometres. By making several measurements during a known path, computations of the radar location are carried out to provide a full tactical picture of the area. The data acquired may be either used at the platform level or transmitted to ground/shore control stations by data link. Processed information is displayed on a 19 in (483 mm) colour raster display in several desired cross-sections.

Operational Status
In production for land and sea applications, and in service with the Israeli armed forces.

Contractor
Elbit Computers Ltd, Haifa.

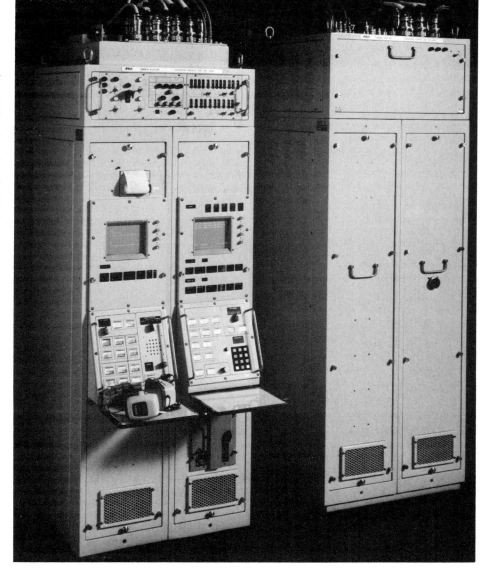

Timnex-4CH-V2 automatic ELINT/ESM system

ITALY

THETIS ESM SYSTEM

Type
Submarine ESM surveillance system.

Description
Thetis is a modular ESM system designed to fulfil the EW requirements of a submarine. The system is produced in two configurations. The first and simpler of these, the ELT/124-S, performs high sensitivity instantaneous threat warning and DF functions, while the second, the ELT/224-S, extends these functions to those of a full ESM system or tactical ELINT. Both variants use the same DF antenna as the sensor, and both can be fitted with an additional and separate RF pre-amplified omnidirectional antenna as an option. This latter antenna is very small and can be installed on top of the submarine's attack or search periscope for primary warning.

The ELT/124-S variant consists of a threat warning and DF receiver system. The hardware comprises a DF antenna and receiver, a processor unit and a warning display. It provides instantaneous, high

sensitivity, automatic warning of threats programmed into the processor and continuous surveillance of the radar environment with 100 per cent intercept probability. A dedicated operator is not required.

The ELT/224-S consists of the units of the ELT/124-S plus an IFM receiver and an ESM display. It provides instantaneous DF with 100 per cent intercept probability of pulse and CW emissions, display of all intercepted emissions, IFM, automatic technical analysis and identification of emitters and their platforms. It will also provide warning that the vessel is entering an area under hostile radar surveillance. A single operator is required for this system.

The antenna assembly consists of one conical spiral omnidirectional antenna plus eight plane spiral DF antennas assembled in a single casing. It has a very low radar cross-section and is also covered with microwave absorbing material. It is light and strong enough to be mounted on the search periscope. The small, lightweight optional antenna can be mounted on the attack periscope and will provide the submarine commander with a first alarm facility but without DF facilities.

Thetis submarine ESM system (display/control console and two antennas)

Operational Status
In production and operational service.

Contractor
Elettronica SpA, Rome.

RQN-5 ESM/ELINT SYSTEM

Type
Integrated naval ESM/ELINT system.

Description
The RQN-5 ESM/ELINT equipment has been developed with almost complete system automation which allows a single operator to perform the ESM and ELINT functions with ease. The modularity of the

system enables it to be installed on board a variety of platforms, including submarines. It can be provided with an additional small omnidirectional antenna, installed on top of the search or attack periscope, for early warning.

The system is characterised by a number of features:

(a) high probability of detection, complete real-time analysis and tracking of ships or helicopters in a completely automatic operation mode, over a wide frequency band and in a very dense electromagnetic environment

(b) automatic warning capability for threat emitters whose parameters may be stored easily in a dedicated memory, and for any locked-on signals, even of an unknown type

(c) processing and analysis of the radar signals, and storing the results for post-mission data collection

(d) consistent range advance factor versus all the radars present in the scenario

(e) ability to drive active ECM automatically, for example the Selenia TQN family designed to operate against search, tracking and homing radars through sophisticated noise, jamming and deception techniques

(f) ability to interface a passive ECM system to initiate chaff and/or infra-red decoys launch

(g) availability of several alternative standard interfaces for integration with a very large variety of peripheral sensor weapon systems, and command and control systems

(h) high system modularity allows it to be changed or upgraded at any time by addition of replacement modules.

The RQN-5 has complete commonality with the Selenia RQH-5 airborne ESM/ELINT system.

RQN-5 ESM/ELINT system

Operational Status
In service with several navies.

Contractor
Selenia Industrie Elettroniche Associate SpA, Defence Systems Group, Rome.

UNITED KINGDOM

MANTA ESM SYSTEM

Type
Submarine radar detection and analysis system.

Description
Manta is a range of advanced ESM systems optimised for submarine applications. Each of the Manta systems intercepts, analyses, classifies and identifies enemy radars operating in the 2 to 18 GHz frequency band.

Instantaneous detection of radar threats at long range provides submarine commanders with maximum warning of the presence of enemy surface ship or airborne ASW systems. In addition, Manta provides information for tactical battle management, targeting and information gathering. Instantaneous warning enables the submarine commander to maintain a covert posture.

Manta is designed as a modular equipment that enables the most cost-effective system to be built up to match the operational requirement for a particular class of submarine.

Different antenna configurations provide options for extended frequency coverage and high resolution targeting. The ESM can be interfaced with the central command system and the output data may be recorded on either magnetic tape and/or hard copied.

Using wide open ESM techniques, the systems provide 360° coverage and 100 per cent probability of intercept. Intercepted signals are automatically analysed and identified by reference to a comprehensive library of known radar types. Processed data are immediately displayed to the operator in both cartesian and alphanumeric formats on two interchangeable colour displays. Manta uses a central management computer controlling subsystems, each containing advanced microprocessors carrying out local processing.

A version of Manta with DF accuracy down to 2° is available for use in long-range weapon targeting. Manta variants are available for ocean-going and

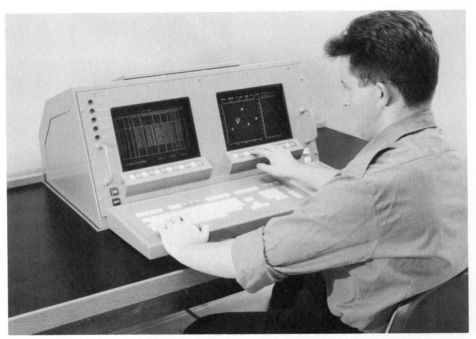

Operator's console for Manta ESM system

patrol submarines, and the Manta antennas can be mounted on a dedicated EW mast, or elsewhere as requested.

Operational Status
In production. Sea trials of a variant of Manta, on an 'Oberon' class submarine, were completed in August 1988 as part of a programme to update the Royal Navy's submarine ESM system. Two Manta variants, known as Outfits UAH and UAL, are being supplied to the Royal Navy to retrofit seven 'Oberon' class submarines and five 'Churchill' class nuclear attack submarines respectively. Another variant is being supplied to Inisel to retrofit the 'Agosta' class submarines of the Royal Spanish Navy. Manta outfits were also ordered in 1990 for installation in the Kockums Type 19 submarine programme for the Royal Swedish Navy, and it is also the preferred fit for export orders.

Contractor
THORN EMI Electronics Ltd, Crawley.

SEALION ESM SYSTEM

Type
Submarine ESM search and warning system.

Description
Sealion is an ESM system for submarines and is designed to provide radar warning and a high bearing search capability. It uses an omnidirectional antenna array and a rotating antenna. The antennas are separate units and may be mounted on separate masts.

The omnidirectional antennas provide signals for measurement of amplitude, pulsewidth, frequency and time of arrival of received radar pulses. Processing of these data, together with identification of the radar through the system library, provides the warning function of the ESM system. The rotating dish provides signals for direction finding for passive targeting and surveillance.

Sealion DF antenna with radome removed

Data are presented on a large format colour display controlled through a normal keyboard. The out-of-hull items are designed to minimise the use of

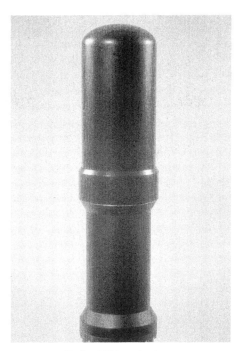

Sealion ESM omni-antenna

active electronics and consequently assure maximum reliability in an area which is non-maintainable at sea. The number of cables has been minimised because of the restricted space within the hull penetrating masts.

Operational Status
In production for submarines of the Royal Danish Navy.

Contractor
Racal Radar Defence Systems Ltd, Chessington.

Operator's console of the Sealion ESM system

PORPOISE ESM EQUIPMENT

Type
Submarine radar surveillance system.

Description
A submarine version of the Cutlass ESM equipment, Porpoise is a fully automatic ESM system operating throughout 360° of azimuth. The equipment receives signals in the 2 to 18 GHz frequency range, measures their parameters and compares these with those contained in a pre-programmed radar threat library. Processing of the radar emitter signals is carried out against the library to give the operator an alphanumeric or graphic display of identification in threat significance order. Up to 2000 radar modes (equal to about 400 radars) are held in the library and the operator can feed in data on up to 100 other emitters.

Intercepted signals are pre-amplified in the mast unit before being passed to the processing and analysis equipment inside the hull. Bearing data are extracted using amplitude comparison. Amplitude, pulsewidth, frequency and time are combined in a single digital word before being passed to the processor unit in the operator's console. There the pulse trains of the different radars are de-interleaved and identified from the library information.

The system is capable of integration with the vessel's fire control and communications systems and may also be integrated with periscope-mounted radar warning equipments. Porpoise also has the ability to give an alert warning when prime threats, such as helicopter or maritime surveillance radars, reach a pre-programmed danger level.

The Porpoise antenna is a compact six-port system giving adequate bearing accuracy and may be mounted on either hull penetrating or non-hull penetrating masts. It is built of titanium to reduce weight

Porpoise console unit

and overcome corrosion, and is pressure resistant to 60 bar. With the exception of the antenna system, the primary subassemblies of Porpoise are fully compatible with those of Cutlass and so enable common logistic facilities.

Porpoise antenna unit

Porpoise 2 is a development with improved DF accuracy (3° instead of 5°) and about 5 dB greater sensitivity. It is designed for fitment on even very small submarines, and studies are being carried out on the possibilities of mounting some of the electronics outside the pressure hull.

Operational Status

In production and in service with several navies. A recent contract is for Turkey. A variant has been supplied to the Royal Navy (as Outfit UAJ) to equip some 'Oberon' class SSKs.

Contractor

Racal Radar Defence Systems Ltd, Chessington.

UAP FAMILY OF ESM SYSTEMS

Type

ESM systems for submarines of the Royal Navy.

Description

Racal Radar Defence Systems supplies the major part of the ESM equipment for the submarines of the Royal Navy. Outfits UAB and UAP(2) have been supplied for some of the earlier UK nuclear-powered submarines to replace the original UA-11 and UA-12 equipments, also supplied by Racal. Both UAB and UAP(2) use a suite of antennas feeding a flexible arrangement of receivers and processors/displays. Outfit UAP(2), which has been installed in the 'Trafalgar', 'Swiftsure' and 'Upholder' classes, both as new and retrofit, has been found to have higher accuracies than predicted.

UAP(2) has now been followed into service by the UAP(1) ESM Outfit. This equipment will also be fitted in the SSNs and new design SSKs, additionally a variant UAP(3) will be fitted into the four 'Vanguard' class SSBNs. Outfit UAP(1) has a frequency range covering all the required radar bands, and a frequency extension capability is already in place, in anticipation of future operational scenarios.

The UAP equipment fulfils the four principal submarine ESM requirements of self-protection, surveillance, data gathering and over-the-horizon targeting for weapon release. It also incorporates Sadie, the latest Racal high performance signal analysis processor. UAP(1) also benefits from a number of operational and technical enhancements as follows:

(a) automatic alarm levels – the calculation of the alarm level which indicates that the submarine is at risk of detection by radar has been automated. Factors considered in this calculation include the number of masts raised, the sea state, command risk factor, frequency band, and library identity of the intercepted radar

(b) disposition of antennas – the minimising of mast exposure is a critical requirement of submarine ESM. The disposition of the UAP(1) antenna elements allows the command to carry out in excess of 95 per cent of his radar ESM role using the search periscope only. The main ESM mast is used only for high accuracy direction-finder targeting, or as a reversionary mode in the event of search periscope failure

(c) Man/Machine Interface (MMI) – UAP(1) is a single operator system employing a number of special function keypads, which together with a rollerball allow the operator to achieve the majority of his interaction with the system by using a single key stroke. The MMI features twin high resolution colour displays which present real-time and processed data to the operator

(d) onboard training – UAP(1) will be the first submarine ESM system to have an integral onboard trainer. This will allow pre-recorded scenarios to be played into the system, while the submarine is deep. The onboard trainer is capable of simulating the most complex scenario with full environmental representation. Realistic onboard training

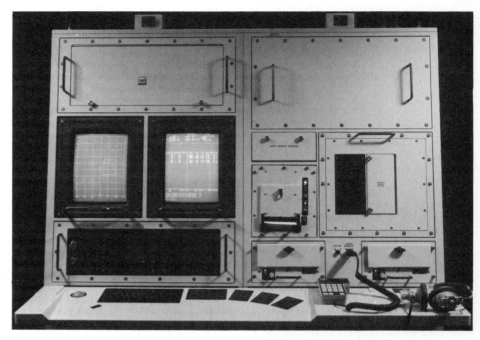

UAP(1) operator's console unit showing twin colour displays, advanced MMI, and an onboard training unit

UAP(1) ESM mast antenna

is a critical asset in maintaining an operator's professional skill both at sea and in harbour.

(e) data gathering – UAP(1) allows the operator to log to disc both parametric and pulse-by-pulse data of any selected radar intercept. The recordings can be reviewed on board prior to submission for high level analysis on shore.

Racal has also supplied a variant of its Porpoise ESM system (see separate entry) as Outfit UAJ to equip some of the Royal Navy's 'Oberon' class submarines.

UAP(1) search periscope top assembly

Operational Status

In production.

Contractor

Racal Radar Defence Systems Ltd, Chessington.

UNITED STATES OF AMERICA

AN/BLD-1 DF SYSTEM

Type

Submarine ESM direction finding system.

Description

The AN/BLD-1 is a submarine-based precision ESM system, installed on the US Navy's 'Los Angeles' class of attack submarines, which passively detects and tracks airborne, sea surface and land-based

radar threats. It employs a mast-mounted antenna that is raised just above the sea surface for operation, and delivers precise threat bearing information that is integrated with other sensor data for tactical surveillance and over-the-horizon targeting.

No technical details are available, although it is understood that Litton Systems has developed a number of complex algorithms required to solve the multi-path problems caused by the close presence of saltwater.

Operational Status

In service with the US Navy. Some 38 systems have been delivered or are on order. Two systems have been ordered for the Seawolf SSN where they form part of the AN/WLQ-4(V)1 EW system. They are due to be completed in September 1993.

Contractor

Litton Systems Inc, College Park, Maryland.

AN/WLQ-4 SIGINT SYSTEM

Type
Submarine SIGINT detection and analysis system.

Description
The AN/WLQ-4 performs SIGINT missions aboard 'Sturgeon' class submarines of the US Navy. It is an automated, modular signal collection system which allows for the identification of the nature and sources of unknown radar emitter and communications signals. It incorporates a network of minicomputers and microprocessors, and data from these computers are correlated with information received from satellite sensors. The system is part of Sea Nymph, a highly classified US Navy NAVELEX programme.

The AN/WLQ-4 system has a number of key features, including:
(a) automatic search, acquisition and signal processing
(b) automatic logging, book-keeping and reporting
(c) semi-automatic correlation of real-time measured data with input from an external system
(d) 400 000 lines of AN/UYK-44 source code
(e) 50 000 lines of executable code in 40 microprocessors
(f) a significant growth capability to handle new threats.

Operational Status
In full-scale production for the US Navy submarines. In November 1987, GTE received a $10 million contract for repackaging the AN/WLQ-4 for the Seawolf submarine. A further $30 million was awarded in 1988 for product improvement to the AN/WLQ-4(V)1.

Contractor
GTE Electronic Defense Systems Sector, Mountain View, California.

AN/WLQ-4(V) skeletal system in a US Navy SSN-637 submarine

AN/WLR-8 EW RECEIVER SYSTEM

Type
Submarine tactical radar detection and analysis system.

Description
The AN/WLR-8 is a tactical electronic warfare and surveillance receiver designed for fitting in both surface ships and submarines of the US Navy. The system is of modular construction and provisions are made for operation in conjunction with numerous types of direction finding or omni-antennas, and a wide range of optional peripheral equipment, to provide comprehensive ESM (electronic support measures) facilities. The WLR-8 is compatible with NTDS (navy tactical data system) and similar action information automation systems. The system can be expanded in frequency or signal handling capability by means of simple additions and/or software changes. Four versions are available: V(1) for submarines, V(2) for 'Los Angeles' class submarines, V(4) for large surface ships and V(5) for Trident submarines.

Two digital computers are incorporated: a Sylvania PSP-300 for system control, automatic signal acquisition and analysis, and file processing; and a GTE PSP-200 microcomputer for hardware level control functions. Digital techniques are employed throughout the WLR-8 system, which is all solid-state.

Operational facilities provided include:

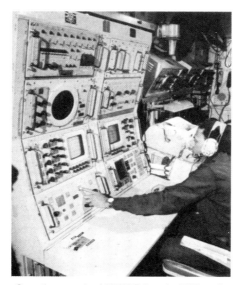

Operating console of AN/WLR-8 tactical EW receiver

(a) automatic measurement of signal direction of arrival
(b) signal classification and recognition
(c) sequential or simultaneous scanning over a wide frequency range
(d) signal activity detection for threat warning
(e) analysis of signal parameters such as frequency, PRF, modulation, pulsewidth, amplitude and scan rate
(f) logging of signal parameters for display to operator(s), and printout of hard copy to teletype or printer
(g) extensive built-in test equipment
(h) directed priority searches of specific frequency segments.

Direct reporting to onboard computers, such as NTDS, permits response times in the millisecond range with minimal operator involvement. A two-trace CRT is provided for display purposes, and this can be supplemented by an optional five-trace panoramic display for presentation of signal activity data. Another CRT display is incorporated if the WLR-8 is used with automatic or manual DF antenna systems.

Operational Status
Operational with USN and possibly other navies. More than 95 sets have been produced. The system is in use on board SSN-688 class submarines, and the V(5) version is fitted to Trident fleet ballistic missile submarines.

Contractor
GTE Electronic Defense Systems Sector, Mountain View, California.

PHOENIX ESM SYSTEM

Type
Submarine radar ESM system.

Description
Phoenix is a range of ESM systems which have been designed for use in submarines to provide automatic identification and bearing of intercepted radar emissions in the frequency range 2 to 18 GHz. The system comprises an antenna array and a highly sensitive IFM receiver, which provides precise measurement of the threat radar operating frequency. A detailed display is provided to the operator, giving all significant parameters of the incoming signals. The system is modular in concept and can be configured to meet individual customer requirements. Both manual and fully automatic systems and analysis capabilities are available. The fully automatic version is known as Phoenix IV.

Operational Status
In production. It is understood that systems have been supplied to India and Sweden.

Contractor
ARGO Systems Inc, Sunnyvale, California.

S-2150 ESM SYSTEMS

Type
Submarine radar intercept, analysis and warning systems.

Description
The S-2150 ESM systems are shipborne threat warning equipments normally fitted to smaller ships and submarines. They are automatic in operation and provide warning and direction finding facilities.

The S-2150-01 provides instantaneous direction finding cover over the frequency range 2 to 18 GHz, with an accuracy of 10° RMS. The system consists of a masthead unit containing the antenna array and receiver, and a processor/display and control panel. An emitter library is referenced with each signal intercepted and an audio alarm is activated if the signal is matched with a possible threat. Information is also displayed in alphanumeric and graphic form on the display panel.

The S-2150-03 is also an automatic surveillance system that detects and analyses signals in the 2 to 18 GHz band and provides direction finding with a 5° accuracy. The equipment operates independently or as part of an overall EW suite. An operator display provides threat symbology, bearing and approximate range as well as other data.

Operational Status
In service.

Contractor
EM Systems Inc, Fremont, California.

S-3000 SURVEILLANCE SYSTEM

Type
Submarine intercept and warning system.

Description
The S-3000 operates over the frequency range 0.5 to 18 GHz and is designed as a radar surveillance system for rapid, high sensitivity signal interception, and to provide an audio warning of threats. It uses a narrowbeam, high gain scanning antenna in conjunction with an omnidirectional antenna and an IFM receiver. The system contains a comprehensive emitter library which is used to compare parameters of the intercepted signal. These parameters are also displayed on an operator's CRT. The operator controls the DF antenna, choosing between scan, sector and point modes. The system is designed to operate in dense signal environments and will identify priority signals as defined by the user.

Operational Status
EM Systems is currently working on a $20 million contract to supply antennas and receivers to THORN EMI Electronics (UK) for retrofitting in the RN 'Oberon' class submarines and in nuclear-powered SSNs.

Contractor
EM Systems Inc, Fremont, California.

SEA SENTRY

Type
Submarine ESM system.

Description
Sea Sentry is a range of passive ESM systems which have been designed to detect, track and analyse radar emissions. The Sea Sentry III is a completely automatic broadband radar threat detection and analysis system for integration into a submarine surveillance platform. The system consists of four elements: a multiple antenna array and microwave receiver, a display and control unit, a processor, and an optional low frequency receiver.

Sea Sentry is available in four different configurations for submarine applications. The equipment covers the 1 to 18 GHz frequency range, with instantaneous detection of both pulsed and CW signals. Bearing accuracies of better than 10° RMS are claimed. Advanced automatic data handling and signal analysis techniques are used to process all threats in a dense environment.

Operational Status
In service and available for submarines.

Contractor
Kollmorgen Corporation, Northampton, Massachusetts.

GUARDIAN STAR SHIPBORNE EW SYSTEM

Type
Shipborne and submarine radar detection and surveillance systems.

Description
Guardian Star is a family of EW systems designed to cover requirements from threat detection to electronic intelligence (ELINT). Various configurations allow the system to be configured to the specific requirements of surface ships and submarines.

The Mk 1 system offers early warning to small surface craft or patrol boats, with an average bearing accuracy of ±10° provided by octave frequency measurements. The Mk 2 system is designed to meet the basic ESM needs of surface ships and submarines. It provides YIG tuned frequency measurement and emitter average bearing accuracy of ±5°. The Mk 3 is an ELINT system for surface ships and submarines. It carries out instant threat warning and accurate frequency measurement by IFM devices in the receiver front end. Average bearing accuracy of less than ±5° is achieved.

The systems consist basically of an antenna assembly, an RF/digital interface unit (RFDIU) and a display/controller unit (DCU). The antenna assembly comprises an omnidirectional and six spiral DF antennas covering the frequency range from 2 to 18 GHz. All the necessary pre-amplifiers and pre-processing components are included in the assembly. The RFDIU processes all signals before transfer to the main digital processor in the DCU. All signal conversion electronics, auxiliary outputs and power supplies are contained in this RFDIU. The DCU includes the input/output section, main processor, magnetic tape unit, keypad and all required operator interfaces. The operational program and library file data are loaded into the processor memory via the magnetic tape unit.

In the surveillance mode the system operation is broadband from 2 to 18 GHz. This mode is entirely automatic and the operator has only to view the situation summary display. The display will provide data on up to 50 emitters (10 per page). The displayed emitter characteristics are frequency, PRF/PRI, pulse width, pulse amplitude, true bearing (instantaneous and averaged), scan rate, emitter name (if matched in the library), and threat level (if matched in the library). The library file handles up to 2000 sets of emitter parameters. The operator can store previously known emitter data together with pre-assigned threat levels in the file. The system matches incoming emitter data with the file and issues an immediate alert on high interest threats. Library data can be changed or entered via the display/controller unit.

Operational Status
In production.

Specifications
Frequency range: 2-18 GHz (optional expansion down to 0.5 GHz and up to 40 GHz)
Library: 2000 emitter parameters
Bearing accuracy: 2-8 GHz <10° RMS; 8-18 GHz <5° RMS
Dimensions: Antenna assembly 360 mm (high) × 15 cm (diameter) (including radome); RFDIU 430 × 360 × 350 mm; DCU 250 × 430 × 530 mm
Weights: Antenna 5 kg; RFDIU (Mk 3) 33.5 kg; DCU 27 kg

Contractor
Sperry Corporation, Arlington, Virginia.

UNDERWATER COMMUNICATIONS

Radio communications between a submerged submarine and ships, aircraft or land-based systems pose many problems because most radio signals are unable to penetrate water to any extent. The exception to this is extremely low frequency (ELF) signals which will penetrate to a considerable depth. Very low and low frequencies (VLF and LF) will also penetrate to a few feet. For all other frequencies the submarine must either come to the surface or have an antenna which is on or above the sea surface. This can be achieved by having an antenna mounted on the hull or periscope, or by having a towed antenna or one mounted on a buoy. Other methods involve the use of small expendable transmitters or receivers which are ejected from the submarine whilst it is at operating depth. These rise to the surface and, after their useful life is exhausted, are scuttled by a timing system.

This section, therefore, covers those systems that are primarily devoted to underwater communications. Information on these systems is frequently sparse because of security restrictions. Equipment covered includes communications between surface ship/submarine and vice versa, aircraft/submarine and vice versa, and submarine/submarine. Research is still being carried out on a viable satellite/submarine communications system but this is still not in a sufficiently advanced stage to describe in any detail.

It should also be remembered that many surface and submarine sonar systems incorporate an underwater telephone facility as part of the overall system. Details of these can be found in the appropriate sonar section.

SUBMARINE COMMUNICATIONS SYSTEMS

FRANCE

MCA 30/MCA 45 CABLE ANTENNA

Type
Buoyant cable antenna for submarine VLF/LF reception and navigation.

Description
The buoyant cable antenna is a dispensing and retrieval system for submarine VLF/LF/HF reception and navigation. The cable acting as an antenna depth is paid out swiftly, according to the speed and submersion of the submarine, in such a way that the end of the antenna rises to the surface of the sea, close enough to receive electromagnetic signals.   The MCA 30 (300 m version) or MCA 45 (450 m version) is a compact system consisting of a buoyant cable dispensed by a winch equipped with a cable pusher, and a pressure hull throughway.

The buoyant cable is an insulated conductor terminated with a short electrode in direct contact with the seawater. For HF communications the cable is fitted with an on-line HF pre-amplifer.

The cable pusher unit drives the cable in and out and is fitted with a cable length detector which will automatically dispense the cable to the required length. The pressure hull throughway is a complete subassembly equipped with a series of pneumatic and manual redundant safety devices. This sub-assembly ensures watertightness in the different phases of the operation, which are, cable in, cable out and cable in motion.

Operational Status
The buoyant cable is a standard equipment on the French Navy's submarines and has been installed on submarines of a number of other countries.

Specifications
Deployment speed: 0.5 m/s
Total weight: 500 kg
Electrical power: 2.75 kW
Inflated seal pressure: 42 bars
Radio channels: VLF 14-23 kHz; LF 60-70 kHz; OMEGA 10.2-13.6 kHz; DECCA 100 kHz; MF/HF 500 kHz–30 MHz

Contractor
Société Nereides, Les Ulis.

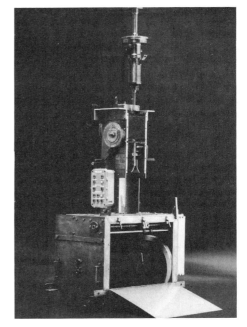

MCA 30/MCA 45 buoyant cable antenna

ISRAEL

SACU 2000 COMMUNICATION UNIT

Type
Stand-alone communications unit for submarine use.

Description
The SACU 2000 stand-alone communication unit is a flexible and rugged digital communication unit for use on both surface ships and submarines. It provides two-way data transmission in standard formats or free text through three parallel channels, and is compatible with all types of standard communication equipment. The display leads the operator through a user-friendly message entry sequence. Entries are made through the keyboard, such as free text or coded messages, according to a pre-stored list. Incoming messages are stored in the memory and displayed upon operator request.

The main system features are:

SACU 2000 stand-alone communication unit

(a) automatic and/or manual acknowledgements
(b) interfaces with standard communication equipment – FM or AM, UHF, VHF or HF, and two-wire telephone equipment

(c) up to three parallel communication channels
(d) multi-network relay capability
(e) single or dual language capabilities
(f) interface capability with C³I systems with central computer.

Specifications
Display: Rugged electroluminescent graphic display; 256 × 512 pixels, 2000 characters in a 5 × 7 dot format
Keyboard: Sealed QWERTY type, illuminated for night operation
Processing: Multi-microprocessor-controlled operation and data processing, memory up to 1 Mbyte of mixed EPROM/RAM for the main processor. CMOS static RAM with battery backup.

Contractor
Elbit Computers Ltd, Haifa.

ITALY

ELMER SUBMARINE COMMUNICATIONS

Type
Integrated radio communication system for submarines.

Description
Main features of Elmer's integrated communication system for submarines include a complement of wideband and tunable antennas for transmission and reception over the LF to UHF frequency bands, the use of MF/UHF multi-couplers and antenna filters, the assembly of equipment in pre-configured racks, centralised system control and supervision, and the use of serialised data transfer.

Communication facilities available include those for ship-to-ship, ship-to-shore and ship-to-air.

Operational Status
Manufactured in a number of variants, the systems have been installed on a large number of naval units belonging to Italy and other countries.

Specifications
Frequency range: LF to UHF
Power output: 30 or 400 W

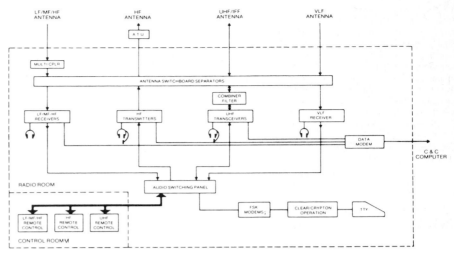

Typical integrated communication system for submarines

Types of service: Analogue and digital voice, data
Antenna types: Magnetic loop, whip. UHF section of UHF/IFF antenna
Number of users: Up to 3

Contractor
Elmer, Pomezia.

UNITED KINGDOM

SEAFOX INTEGRATED NAVAL COMMUNICATIONS SYSTEM

Type
Submarine integrated C³ system.

Description
Seafox is an integrated C³ system designed for submarines and light naval forces. An alternative range of equipment is available, designed to suit vessels ranging from patrol boats, mine countermeasures vessels and offshore patrol vessels, up to large corvettes and light frigates.

Each system provides for intercom for combat systems co-ordination and control, plus ship management, tactical radio communication with other ships and aircraft and strategic radio communication with the command on shore.

The Seafox range of equipment includes the H1073 amplifier, H4640 transceiver, H6701 control outfit, H6702 internal communication system and the H1473 automatic tuning unit.

H1073 250/400 W MF/HF Amplifier
When used with the H1473 automatic antenna matching unit and a suitable drive (for example, ICS3 or H4640), the H1073 provides a 250/400 W transmission system. Tuning is fully automatic over the 1.5-30 MHz frequency range. The frequency range can be extended down to 240 kHz as an option. Tuning time is typically 1.5 seconds.

H4640 Transceiver
The H4640 is a fully synthesised drive/receiver designed for naval shipborne use. The basic equipment comprises four units mounted in a single cabinet. These are the drive, receiver, synthesiser and a common power supply unit. This basic configuration allows for simplex or two-frequency working; other configurations provide for duplex working and for the use of pre-/post-selectors where improved out-of-band performance is required.

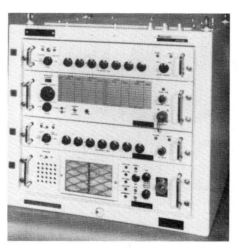

H4640 transceiver

H6701 Control Outfit
The H6701 control outfit has been designed to provide remote-control of radio communications equipment. The system consists basically of a central Control Selection Unit (CSU) and a number of remote-control units situated at remote operating positions, together with radioteletype and data terminals and loudspeakers as required.

Time division multiplex techniques are employed to achieve switching between the various user positions and the radio equipment. The controls necessary for selecting the user equipment combinations, HF transmitter power level and HF transmit/receive muting are centralised. Frequency channel and audio characteristics are selected remotely by the operator.

The standard system has a capacity for a maximum of 16 user lines and 16 equipment lines but can be sub-equipped for a smaller number of lines. For larger systems two CSUs can be interlinked by cross-connecting two or more user and equipment lines according to system requirements.

H6702 Tactical Internal Communication System
The H6702 system is designed to provide tactical internal communication (voice) between operational users. Forms of communications available are: group, in which a number of users are connected in conference on a single speech channel; interphone, for single user-to-user connection; and broadcast, for a single user transmitting to multiple users on receive only. The basic system is designed around a central internal communication exchange (ICE) and provides for up to 16 user positions. There are 16 separate speech channels available for voice communication between any combination of user-to-user, conference groups and general broadcast. Each user position can have up to 16 keys available (depending on the number of internal CSUs fitted) which can select particular connections to other users or combinations of users via the switching exchange.

The system can be expanded to accommodate up to 32 user positions by adding a second ICE. Facilities are provided for interconnection to the external communication circuits.

H1473 Automatic Antenna Tuning Unit
The H1473 antenna tuning unit automatically matches the HF output from an H1073 amplifier (or the 500 W combined output from a pair of such amplifiers) to whip antennas 7-12 m long, or equivalent wire antennas, over the 1.6-30 MHz frequency range.

Operational Status
The Seafox equipment was introduced in 1983. The Royal Navy has ordered 32 systems. Also sold to Greece and Indonesia.

Specifications
H1073 Amplifier
Frequency range: Ed 01 – 1.5-30 MHz; Ed 02 – 1.5-30 MHz and 240-1496 kHz
Power output: 6-400 W mean or PEP into 50 ohm load with level adjustable in 7 steps

Contractor
Marconi Communication Systems Ltd, Chelmsford.

GS7110 RECEIVER

Type
VLF/LF multi-channel receiver.

Description
Designed for use in submarines, surface vessels and shore stations, the GS7110 VLF/LF receiver interfaces with a variety of antenna systems and decryption equipment, and complies with NATO STANAG 5030 requirements.

Operational over a frequency range of 10 to 200 kHz, the microprocessor-controlled receiver provides local/remote-control of frequency, modulation mode and receiver status, and has a 99-channel non-volatile memory with a BITE facility. The incorporation

GS7110 multi-channel VLF/LF receiver

of a software modulator gives optimum performance for two or four channel MSK, FSK (50 or 75 baud), CW and MCW. AM speech is also provided.

Software options available for the equipment include various spread spectrum modes, high data rate single-channel MSK and other modulation modes. The receiver measures 133 × 483 × 519 mm and weighs 32 kg. It meets both Royal Navy and German Navy environmental specifications.

Operational Status
The GS7110 receiver has been accepted as the NATO standard and has been sold to the Netherlands, Canada, Denmark, Norway, Turkey, Australia and Spain. It has also completed a successful evaluation by Germany.

Contractor
Marconi Underwater Systems Ltd, Addlestone.

MARCONI VLF SHIFT KEYING SYSTEM

Type
Communications and management control of submarine forces.

Description
The Marconi VLF control system has been designed for management and control of submarine forces. Sophisticated broadcast preparation increases the capacity, efficiency and security of processing command traffic, while networked communications to the broadcast transmitters provide high interconnectivity and information exchange across the entire

theatre of operations, thus permitting effective use to be made of the new minimum shift keying (MSK) technology.

This system, developed for NATO, and complying with the latest NATO standards, equips two broadcast preparation stations and one broadcast transmitter in each of three countries to cover the Atlantic and Mediterranean NATO fleets.

Each equipped broadcast preparation centre can accept up to 24 simultaneous incoming circuits from anywhere in NATO, and provides electronic editing and collation of messages for the generation of up to eight broadcasts. These are then re-encrypted and sent via a complex line switching matrix which permits full interconnectivity between all sites to the broadcast transmitters, where they are transmitted using the

Marconi MSK equipment.

Computing and processing and software controlled automatic switching are essential features of the Marconi VLF system which employs Ada software throughout. To complete the upgrading of the broadcast stations' off-the-air monitoring, error detection and correction systems are provided, while the associated electronics automatically tune the broadcast variometers to ensure optimum broadcast quality.

Operational Status
In operational use.

Contractor
Marconi Secure Radio (a division of Marconi Defence Systems), Portsmouth.

SUBMARINE COMMUNICATION MANAGEMENT SYSTEM (SCMS)

Type
Message handling and management system.

Description
The Submarine Communication Management System (SCMS) is a high integrity message handling and management system, providing fully automated message and broadcast handling capabilities for the Royal Navy. The system incorporates advanced, in-house designed and built, hardware and software modules.

The system includes a message and broadcast schedule database handling messages in ACP 127 format. The routeing of messages both locally and externally is directly under the control of a system supervisor. Once the system routeing information has been installed by the supervisor the system will automatically route messages through the system to their destination.

Local message processing is made possible by the system's advanced full screen message editing facility; message distribution can be achieved electronically or by the use of paper tape.

The system provides a means by which broadcasts can be automatically issued at pre-defined times. Broadcasts may be of short or long duration and will normally be directed through VLF assets.

SCMS has the capability to handle multiple broadcasts with each broadcast being monitored for the occurrence of broadcast corruption. If corruptions occur the system will inform the system supervisor who will be able to identify the particular corrupted broadcast and will have the option to cancel the outgoing broadcast.

Utilising a modular architecture incorporating dual redundancy facilities, the system employs state-of-the-art technology and provides a flexible distributed processing capability, including highly advanced processing units designed to provide a high degree of availability.

The system comprises two separable items, a workstation and the main SCMS cabinet. The workstation provides a WIMP interface to the user, direct access to its own paper-tape reader and writer and direct access to the full SCMS facilities via a VT100 window. The SCMS main cabinet contains main and standby store and forward units with an automatic change-over facility in case of equipment failure. The cabinet houses two disk storage units. These units

act as main and replicated standby to prevent the loss of messages in the event of disk failure; each disk being accessible by each of the store and forward units. The disk capacity will be adequate to maintain full system information and message data for an extensive period.

The main system unit will provide ports for connection to operator consoles, paper-tape readers and writers, telegraph type printers, external devices and external systems.

The system also employs a KW46 Crypto Remote-Control Unit. The KW46 CRCU is one of the Siemens Plessey Defence Systems MRS range of intelligent communications products. The basic unit provides for the remote-control and status monitoring of up to 32 KW46 devices, reducing the need for error-prone and repetitive manual interventions.

Operational Status
In development. Scheduled to be delivered in the Autumn of 1991.

Contractor
Siemens Plessey Defence Systems, Christchurch, Dorset.

SUBMARINE COMMUNICATIONS MAST

Type
Integrated non-hull-penetrating communications mast for submarines.

Description
The mast forms part of an integrated submarine communications system comprising HF, VHF and UHF aerials which are supported on a non-hull-penetrating low radar cross-section GRP mast.

The integrated communications system performs multiple functions, from MF, HF, VHF and UHF communications through JTIDS and IFF, to SATCOM and SATNAV (both Transit and NAVSTAR), using only one mast. The system is capable of simultaneous transmission and reception for most functions, and reception only for others. Spatial coverage is excellent, from low angle line-of-sight communications to high angle SATCOM.

The hydrodynamic shape of the mast enables deployment whilst submerged, at high submarine speeds and in severely adverse weather conditions to provide high operational availability for radio communications.

The low drag, streamlined mast profile exhibits excellent vortex-shedding characteristics and provides extreme resistance to vibration whilst extended, thus conferring stability to the masthead payload at high speeds. With hydraulically actuated extension and retraction, the mast is reliable in operation under all conditions. Being non-hull-penetrating, the mast allows flexibility in submarine design and saves valuable space inside the pressure hull. The mast can be supplied as a single drop-in module.

To minimise radar signature the upper portion of the mast incorporates multi-band RAM in the GRP laminate.

The Submarine Communications Mast meets all the requirements of the Royal Navy AJU communications mast.

Marconi AJU integrated communications mast fitted in HMS Upholder

Operational Status
In production. In April 1990, Marconi announced a multi-million pound contract from the Australian Submarine Command to design a new lightweight multi-function non-penetrating mast for the new submarine fleet of the Royal Australian Navy.

Contractor
Marconi Underwater Systems Ltd, Templecombe.

SUBMARINE COMMUNICATIONS MULTI-FUNCTION ANTENNA

Type
Broadband submarine mast-mounted multi-function communications antenna.

Description
The Submarine Communications Multi-function Antenna consists of a pressure-tight broadband antenna system containing a head amplifier unit, together with an inboard control unit.

This compact, unobtrusive antenna affords excellent spatial coverage, from low angle line-of-sight (LOS) to high angle SATCOM. The single antenna element is capable of performing many communications functions, covering VHF, UHF, JTIDS, IFF, SATCOM and SATNAV, with simultaneous transmission and reception for most functions.

Two separate antenna elements meet the need for low angle line-of-sight (LOS) communications over the VHF and UHF bands and satellite communications from UHF to D-band. A conical log spiral antenna is used for transmission and reception of circularly polarised 'top cover' radiation patterns. The low angle vertically polarised LOS coverage is achieved from a dipole assembly within the log spiral antenna structure. An electronics package mounted in the lower half of the dipole assembly provides the switching, filtering and amplifiers functions for the selected mode of operation.

The antenna outfit has been fully tested under extreme depth conditions, and meets all the requirements of the Royal Navy AVD1 communications antenna. Dimensions are 580 mm high with a maximum overall diameter of 356 mm, and a weight of less than 20 kg for the complete outfit of antenna, flange and electronics package.

Operational Status
In production.

Contractor
Marconi Underwater Systems Ltd, Templecombe.

Submarine communications multi-function antenna

MARCONI LOW PROFILE ANTENNA

Type
Submarine periscope-mounted communications broadband antenna system.

Description
The Marconi Submarine Communications Low Profile Antenna consists of a pressure-tight broadband antenna system containing a head amplifier together with an inboard control unit. The base of the antenna provides a general-purpose mounting arrangement suitable for fitting on a wide range of submarine masts and periscopes.

The single element antenna is capable of receiving communication and navigation signals in the VLF, LF, MF, HF, VHF and UHF bands. Transmit/receive operation for V/UHF line-of-sight and SATCOM is also provided.

The antenna presents a very low profile and its special water-shedding coating ensures quick response communication. It has been fully tested under extreme depth conditions and meets all the requirements of the Royal Navy AVS antenna. Height of the antenna is 323 mm, with a maximum overall diameter of 210 mm. Weight, including periscope-mounting adaptor, is 9.9 kg.

Operational Status
It is widely fitted to submarines of the Royal Navy and of other navies.

Contractor
Marconi Underwater Systems Ltd, Templecombe.

Marconi low profile communications antenna

MARCONI COMMUNICATIONS AND NAVSTAR ANTENNA

Type
Submarine broadband periscope-mounted antenna for communications and navigation.

Description
The Submarine Communications and NAVSTAR Antenna consists of a pressure-tight broadband antenna system containing a head amplifier, together with an inboard control unit. This is a compact antenna which affords excellent spatial coverage. The antenna/head amplifier combination gives a signal-to-noise ratio at the receiver which meets the NAVSTAR/GPS requirements for worldwide coverage.

The single element antenna is capable of performing communication functions covering VLF, LF, MF, HF, VHF, UHF and D-band GPS reception (with simultaneous capability), and V/UHF line-of-sight transmission.

The antenna presents a very low visual profile, and its special water-shedding coating ensures quick-response communications. It has been designed specifically for operation in highly reflective environments, thus ensuring full operational low angle NAVSTAR coverage. It has been fully tested under extreme conditions.

A universal flange fitted to the base of the antenna provides a general-purpose mounting suitable for a wide range of submarine periscopes. Size of the antenna is 323 mm high by 197 mm maximum overall diameter. Weight of the complete outfit, including the electronics package, is less than 6 kg.

Contractor
Marconi Underwater Systems Ltd, Templecombe.

Communications and NAVSTAR antenna

R800 RECEIVER

Type
VLF/LF receiver system.

Description
The R800 receiver offers full coverage of the VLF and LF frequency bands, and is designed for fully automatic reception on MSK, FSK, LW and FDM. Full operating parameters can be programmed for up to 63 channels, 10 of which can be protected from accidental reprogramming.

The R800 is suitable for use in submarines, surface ships and shore stations, and incorporates BITE. It can be used with normal Crypto equipment, and a number of spread spectrum and frequency-hopping techniques can be included.

Operational Status
The R800 is the specified receiver of the Royal Navy and is in service with the Royal and Portuguese Navies.

Contractor
Redifon Ltd, Crawley.

R800 VLF/LF receiver

UNITED STATES OF AMERICA

EXTREMELY LOW FREQUENCY (ELF) COMMUNICATIONS PROGRAMME

The concept of using extremely low frequency radio signals to communicate with submerged submarines was first suggested over 30 years ago. However, the extremely large antenna size and environmental worries prevented its introduction until recently.

ELF signals can travel great distances with low loss and can penetrate seawater to considerable depths. In practice, an ELF consists of one or more shore-based transmitters, operating at around 40-80 Hz, connected to long horizontal wire antennas (either just above ground or buried for additional security) that are earthed at each end. Orthogonal antennas are used to provide omnidirectional radiation patterns. The transmitted signals are sensed by an antenna on the submarine and decoded by a sophisticated, computer-based receiver. Because the bandwidth is low at ELF, the message transmission rate is necessarily very slow but, even by employing a simple three-letter code system, any of a great number of messages can be transmitted in reasonable time.

The most favourable site in the US for the installation of the system is in northern Michigan or northwestern Wisconsin where the low-conductivity bedrock formations in the Laurentian Shield greatly enhance propagation.

Development

In 1969, the US Navy constructed an experimental transmitter in Wisconsin to study propagation and environmental effects of ELF. The antenna consisted of two 22.5 km long pole-mounted lines at right angles. In 1976, a message-handling capability was added and a small quantity of shipboard receivers was built and installed to prove conclusively that an ELF system would perform as anticipated.

Encouraged by this trial, the Navy planned to construct an operational ELF system with a completely buried antenna and redundant transmitters. Code named Sanguine, this vast system would have been highly resistant to blast overpressure and could have absorbed a moderate number of direct nuclear hits. However, in 1975 a defence analysis group reached the conclusion that the increasing accuracy and number of Soviet nuclear warheads could neutralise the system. Sanguine was, accordingly, downgraded and eventually cancelled.

The Navy, undeterred by this setback, persisted and developed a more modest system with aboveground transmitters and a pole-mounted antenna 45 km long. This system, called Seafarer and installed at Clam Lake, Wisconsin, immediately ran into opposition from the residents of Wisconsin and the environmentalist lobby, who were concerned about the environmental impact of the system and health issues. Despite numerous Navy-sponsored biological studies showing no adverse environmental or ecological effects from ELF, and a 1977 study carried out by the National Academy of Sciences giving ELF a clean bill of health, the project was cancelled in 1978.

In 1981, President Reagan ordered that the Wisconsin transmitter be reactivated and upgraded to operational status and, at the same time, ordered the Department of Defense to conduct a study of ELF requirements. That study resulted in the conclusion that ELF would enhance the US strategic C³ posture. It recommended that in addition to the upgrade of the Wisconsin system, a supplementary 90 km system should be constructed at the nearby KI Sawyer Air Force Base on the Upper Peninsular of Michigan. Congress approved funds for the project in 1982 and research and development resumed. Construction commenced at both sites in 1983 but, in January

1984 a court injunction halted further work pending the preparation of a supplemental environmental impact statement (EIS) to evaluate health effect studies of ELF fields performed since the original EIS was filed in 1977. A study was carried out by the American Institute of Biological Sciences which reaffirmed the results of previous investigations and the EIS was filed in 1985. Construction restarted following the cancellation of the injunction by the US Court of Appeal. The Supreme Court subsequently upheld that decision.

By the end of 1986, both stations were completed. Prototype submarine receivers were delivered in April 1985. The first test in May 1985 aboard a submarine of the Pacific Fleet was successful. In subsequent tests, submarines in the Mediterranean, the western Pacific and on patrol under the North Polar ice-cap have successfully received signals from the Wisconsin station.

Description

There are four segments to the Wisconsin/Michigan ELF communications system: the broadcast control segment (BCS), the message input segment (MIS), the transmitter segment (TS) and the receiver segment (RS).

The main input port is the BCS, which is controlled by the Commander, Submarine Forces Atlantic (COMSUBLANT) in Norfolk, Virginia. The MIS, the secondary input port for ELF messages, is located at KI Sawyer AFB and can take over from the BCS should that become disabled. It can if necessary pre-empt the BCS.

The TS comprises the ELF stations at Wisconsin and Michigan which normally operate simultaneously but can operate independently when required. Two frequency bands are used, 40-50 Hz and 70-80 Hz and each transmitter facility uses commercial prime power from its local utilities companies. In the event of power failure, each site has backup diesel generators. Each transmitter facility has installed two spare backup power amplifiers and an uninterruptive power system to ensure necessary reliability to continue transmitting in case mechanical or power failures occur. The TS is a soft, surface deployed subsystem with ECCM and electromagnetic pulse protection. However, it is not expected to withstand a hostile physical attack.

The RS is located on SSBN and SSN submarines, although the BCS and MIS have receivers for monitoring functions. Signals are normally sent to the BCS, but the MIS can also input messages, and then to the TS via dedicated communication lines. Both the BCS and MIS include a message entry element consisting of a data terminal set (teletype Model 40), a link encryption device (KG-84) and a data link selector panel. This arrangement allows the operator to

enter messages into the message queue at the master transmitter facility via telephone company lines. The BCS and MIS also include an order-wire for operator-to-operator communications between facilities. The master transmitter facility is in contact with the BCS and maintains the message queue and automatically updates the backup queue at the slave facility.

At each transmitter facility a processor element converts the encrypted ELF message from the message processor into drive signals for the power amplifiers, monitors antenna current and controls the transmitter facility master/slave protocol.

The antenna arrays at each transmitter facility consist of antennas oriented north-south and east-west. Two pairs of power amplifiers exist at each facility, one pair for each antenna configuration. Only two power amplifiers are used at any one time, one for each antenna, and each is rated at an output level of 660 kW.

Signals which are picked up via the OE-315 towed antenna are then fed to the ELF receiver terminal group via an in-line amplifier. The receiver performs analogue and digital signal processing to detect any message that may be present. The receiver terminal group (OR-279(XN-1)/BRR) consists of five main units: the pre-amplifier, receiver timing and interface unit, the combined processor and key generator unit, time and frequency junction box and navigational interface junction box. To extract the ELF message, the receiver has first to filter out atmospheric and ocean noise and eliminate interference caused by the submarine's onboard power systems.

The ELF system is synchronous requiring that the receiver know accurate time information relative to the transmission, and time compensation has to be allowed for the ELF signal propagation delay. A band spreading key stream is generated and removed from the message, which is then decrypted. By using an embedded AN/UYK-44 militarised reconfigurable processor, many of these functions are performed digitally.

Operational Status

Full operational capability is anticipated in the mid-1990s.

Contractors

GTE Government Systems Corp, Needham, Massachusetts (prime).
Illinois Institute of Technology, Chicago, Illinois.
Computer Sciences Corp, Falls Church, California.
Mitre Corp, McLean, Virginia.
R M Vredenburg Company, McLean, Virginia.
Spears Associates, Norwood, Maryland.
Booz-Allen and Hamilton Company, Washington DC.

AN/WQC-2A COMMUNICATION SET

Type

Sonar underwater communications set.

Description

The AN/WQC-2A is a sonar underwater communication set for surface ships, submarines and shore installations. The system provides an SSB general voice and CW communication set consisting of a control station, remote-control station, receiver/transmitter and LF and HF transducers.

The system employs voice, audio and low speed telegraphy in two frequency bands: high (8.3 to 11.1 kHz) for close range, and low (1.45 to 3.1 kHz) for long range. In addition the system facilitates the use, through separate input and output connectors, of a final transmitter amplifier in a frequency range of 100 Hz to 13 kHz.

The receiver/transmitter contains the main electronic assemblies of the sonar communication system. It consists of a rack-type cabinet with removable drawer assemblies containing the power supply and test panel assembly, the receiver/transmitter assembly and the final amplifier assembly. The

Control and remote-control stations for AN/WQC-2A system

primary functions of the receiver/transmitter are to develop the high powered single sideband transmission for voice, audio and CW signals to drive the LF or HF transducers and to receive and demodulate signals received from the transducers. The receiver and transmitter outputs are made available to external tape recordings and monitoring equipment. The circuits are also capable of muting, or being muted, by other external equipment.

The control station contains the required controls, indicators, microphones and so on. The remote-control station is a secondary operating position, if required.

The two transducers (LF and HF) have a horizontal (radial) omnidirectional beam pattern. The electro-mechanical energy conversion is accomplished by piezoelectric ceramic elements which are totally encapsulated and covered by an acoustically transparent neoprene boot.

Operational Status

In production for US Navy and overseas navies.

Specifications

Frequency bands: 1.45-3.10 kHz (LF); 8.3-11.1 kHz (HF); 100 Hz to 13.0 kHz (auxiliary mode)
Output power: 600 W (LF); 450 W (HF); 1000 VA (auxiliary mode)
Weights
Receiver/transmitter: 223 kg
Control station: 6.8 kg
Remote-control: 4.1 kg
Transducers: (with 175 ft of cable) 200 kg (LF); 5.5 kg (HF)

Contractor

General Instrument Corporation, Undersea Systems Division, Westwood, Massachusetts.

MCDONNELL DOUGLAS LASER COMMUNICATIONS

Type
Air-to-submarine communications.

Description
Laser links which provide secure communications between aircraft/satellite/submarines are currently in the development stage. McDonnell Douglas has carried out a number of trials with data rates ranging from low kilobits to hundreds of megabits for voice, data transfer and co-operative avionics.

Models for propagation of laser signals through both ocean water and clouds have been developed and these models have been verified through a rigorous series of at-sea tests. Sophisticated signal processing algorithms have been developed to convert and decode the received optical signals into usable information streams.

In 1986 McDonnell Douglas installed the first laser receiver on an operational submarine, which participated in a series of exercises to demonstrate the operational utility of a submarine laser communications system.

Operational Status
Design and development of operational laser communications systems is continuing.

Contractor
McDonnell Douglas Astronautics Company, Defense Electronics, St. Louis, Missouri.

GTE SUBTACS

Type
Submarine tactical communications system.

Description
The GTE Submarine Tactical Communication System (SUBTACS) is designed for land-based transmission, as well as tactical battle group communications, to submarines with the installation of shipboard transmitters. With a data rate 300 times faster than the original GTE extremely low frequency (ELF) systems, SUBTACS functions at ranges in excess of 1000 nm, and at submarine operating depths and speeds.

The system is more jam-resistant than a VLF system, is less expensive to install, has a greater range and far greater water penetration. The ELF system operates at frequencies between 30 and 300 Hz. (See also the Extremely Low Frequency Communications Programme entry.)

Contractor
GTE Government Systems Corporation, Needham, Massachusetts.

SPEARS SUBMARINE ANTENNA SYSTEMS

Type
Range of antenna systems for submarine use.

Description
Spears Associates designs and manufactures a wide range of antenna systems for use on submarines. This range includes hull-mounted and periscope-mounted antennas, buoyant cable antenna systems, and towed buoys. Examples of some of these systems are given below.

BCA-100/-200/-300/-400 Buoyant cable antenna systems
These systems provide a submarine with the ability to receive radio transmission within the range of VLF (10 kHz) to VHF (200 kHz), and to transmit in the HF band, while allowing the submarine to remain at operating depth. This broad operating frequency range allows reception of both communications and Omega and Loran C navigation signals.

A system consists of three major assemblies; the buoyant cable antenna, the deployment mechanism or reeling machine, and the antenna coupler.

The antenna is 610 m in length with a diameter of 16.5 mm, and is fully recoverable when not in use. Shorter lengths are also available. The antenna incorporates an in-line amplifier which operates in the frequency range of 2 to 30 MHz (and beyond in some versions).

The antenna coupler is required since, within the VLF/LF frequency range, the antenna exhibits a resonance resulting in widely varying signal levels and impedances, as a function of frequency. This makes the antenna unsuitable for direct connection to the receiver. In the MF/HF frequency range response is not flat, and is further degraded by attenuation in the cable assembly. The antenna coupler corrects these anomalies and provides more sensitive amplifiers than most receivers for improved overall system sensitivity. Selection of the proper gain and response is automatically made by switching the antenna coupler mode for the antenna in use.

VLF/LF/GPS communications and navigation antenna systems
These are submarine communications and navigation antennas that allow the user to select from a VLF/LF (communications and Omega) receiving antenna to one with both VLF/LF and GPS satellite reception capability, combined in one unit.

Each system consists of an antenna/pre-amplifier, outboard transmission line, hull penetrator, inboard transmission line and an amplifier/combiner. All antennas are submersible to at least 600 m.

The antenna/pre-amplifier covers the range from 10 to 160 kHz, with extended coverage up to 300 kHz, and is usable for both communications and Omega navigation. The antenna/pre-amplifier can be easily upgraded to GPS receiver capability by means of a modification kit.

Towed buoys
These free flooding buoys were designed for the US Navy as submarine towed communications platforms. At shallow depths the buoys become sea following which avoids surface broaching, and are extremely stable over a wide range of operating speeds. The buoys can accommodate a wide range of antennas depending on the communications requirements. The onboard electronics are housed in pressure-proof, non-corrosive titanium canisters. The buoys are tethered to the submarine by a multiconductor tow cable, and are deployed and retrieved using either an inboard hydraulic winch, or an outboard palletised submersible electric winch.

Operational Status
In operational service with the US Navy.

Contractor
Spears Associates Inc, Norwood, Massachusetts.

UNDERWATER TELEPHONES

FRANCE

ERUS-3

Type
Underwater emergency telephone.

Description
The ERUS-3 ultrasonic telecommunications equipment has been designed for telephone and telegraphic communication between a surface ship and one or several submerged craft, diving bells or submarines, or between several submerged craft.

The ERUS-3 operates with the ship's mains power supply or with an autonomous battery power supply. A selector switch enables the operator to choose the required power supply. The equipment can operate with batteries only.

The ERUS-3 incorporates an automatic transmission device, for use in an emergency, and a responder, enabling ships to estimate the distance between them (this device is also used for standby operation). The range is greater than 10 km under normal propagation conditions with the ships running at silent speed.

The ultrasonic signals are transmitted and received by a type 2Z9A omnidirectional transducer. Signals are transmitted with SSB AM.

In the passive mode, when a signal with a particular call frequency is received, the receiver is unlocked to enable it to receive telephone signals. In the responder mode, reception of a signal which has a particular call frequency initiates retransmission of a signal of the same frequency.

The power supply batteries are placed in special battery containers. The following operating times are quoted per container and should, therefore, be doubled or trebled if two or three battery containers are installed.

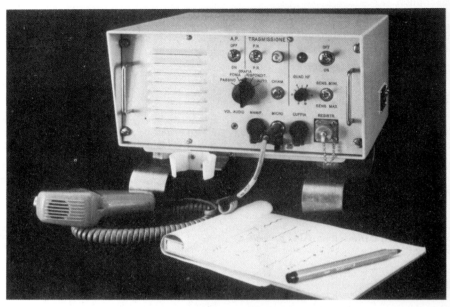

ERUS-3 transmitter/receiver unit

Autonomous operating time on transmission is 3½ hours, intermittently transmitting for 3 minutes per hour for 72 hours, for example. Autonomous operating time on reception is 72 hours for continuous operation.

The equipment can operate either on a classified NATO frequency for military use or on 10.5 kHz for commercial use. For particular needs the two frequencies can be used on the same equipment with a change-over switch.

Operational Status
In production for the French Navy.

Contractor
Safare Crouzet, Nice.

TSM 5152A, 5152B

Type
Transmitter/receivers for underwater telephony.

Description
The TSM 5152 transmitter/receivers (TSM 5152A for submarines and TSM 5152B for surface vessels) allow two-way telephony and telegraphy between submerged submarines, between surface ships and submarines, and between surface ships.

The TSM 5152 has a range of operating modes. These include: a watch mode for silent survey when the receiver is normally silent but is activated by reception of a call; a phone mode for communication; a CW transmission mode controlled with the signalling key; a CW and reception mode with the equipment automatically switched to reception during breaks of transmission; an automatic transmission mode used to obtain a first contact; a transponder mode in which a pulse is automatically transmitted on receipt of a CW transmission; and a test mode to allow the operator to test the equipment automatically.

Operational Status
In production for the French and other navies.

Specifications
Mode: USB transmitted-call FM
Carrier frequency: in accordance with NATO or other specifications
Power output: 400 W for 0.33 form factor with AC supply

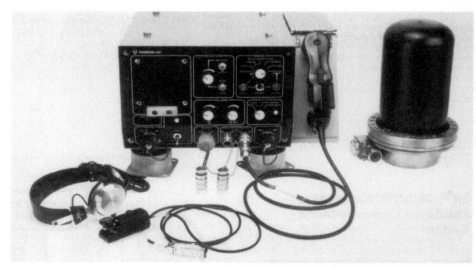

TSM 5152 transmitter/receiver and associated projector

Power supply: 115 V AC ±5%, 48-63 Hz, 3-phase
Consumption: 800 VA (max)
Weight: 33 kg

Projector
Type A
Each 1 of 3 groups includes 2 independent and identical hydrophones
Bearing aperture: 120° (3 dB)
Elevation aperture: 50° (3 dB)
Operating depth: 300 m (max)

Type B
Omnidirectional in bearing
Elevation aperture: 50° (3 dB)
Range: 20 km (approx)

Contractor
Thomson Sintra Activités Sous-Marines, Valbonne Cedex.

TUUM-1E/F UNDERWATER TELEPHONE

Type
Short-range acoustic underwater communications system.

Description
The TUUM-1E/F is an underwater telephone operating in the HF frequency band and, because of its short range, is intended for discrete communications between submerged submarines, or between submarines and surface ships. It provides for both voice communications and transmission and reception of telegraphic signals.

The submarine equipment consists of a transceiver and four directional transducers installed forward, aft, starboard and port on the vessel. That of the surface ship comprises a transceiver and an omnidirectional transducer. Transmission and reception from the submarine can be directional or omnidirectional, using one transducer or four simultaneously. Output power is 6 or 60 W from both surface ship and submarine.

Operational Status
In service with the French Navy.

Contractor
Safare Crouzet, Nice.

TUUM-2C/D

Type
Underwater wireless telephone.

Description
The TUUM-2C/D is a miniaturised equipment intended for underwater telephone and telegraph communication.

Because of its size, this equipment can be used on standard ships (surface craft, submarines) as well as on midget submarines, diving bells and so on. It consists of a transceiver unit with one or several transducers. The omnidirectional range is greater than 10 km with normal propagation. It can be extended to over 20 km by addition of a 1 kW amplifier.

A telemetry device allows localisation of submerged transmitting buoys, and the transmitter can be modulated with an external audio generator.

Connection with the active sonar of the ship allows reciprocal operation of the receivers during transmission by one or other equipment.

Operational Status
In service with the French Navy.

Specifications
Transducers
Surface craft: 1 omnidirectional transducer

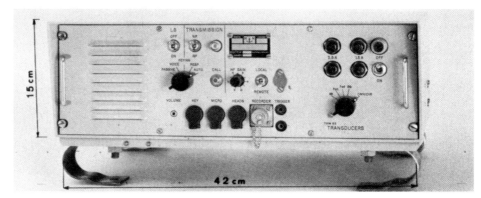

TUUM-2C/D underwater telephone

Submarine: 4 directional transducers (90°) which can be switched for omnidirectional operation
Electronic unit: single model for both surface and diving crafts
Modes
Passive: equipment is silent and switches to receive operation on reception of signal
Telephone
Telegraph
Transponder: transmission on reception of call signal, allowing distance measurements between 2 equipments
Automatic transmission with reception between dashes
Socket for tape recording.

Contractor
Safare Crouzet, Nice.

TUUM-4A

Type
Underwater multi-channel telephone.

Description
The TUUM-4A is a miniaturised system intended for underwater telephone and telegraphic communication. Its small size and weight allow the system to be fitted on surface craft and submarines as well as on midget submarines, diving bells and so on. It comprises a transceiver unit, with one or several transducers. The omnidirectional range is in excess of 14 km with normal propagation at NATO standard frequency.

A telemetry device allows localisation of submerged transmitting buoys, and the transmitter can be modulated with an external audio generator. Connection with the ship's active sonar allows reciprocal operation of the receivers during transmission by one or other equipment.

Operational Status
In service with the French Navy.

Specifications
Transducers
Surface craft: 1 omnidirectional transducer per frequency range
Submarine: 4 directional transducers (90°) per frequency range which can be switched for omnidirectional operation
Electronic unit: single model for both surface and submarine

Contractor
Safare Crouzet, Nice.

AN/WQC-501(V) UNDERWATER TELEPHONE

Type
Acoustic underwater communications system.

Description
The AN/WQC-501(V) is an underwater telephone system operating on NATO frequencies and designed for communications between surface ship and submarine. Normal operation is in the passive mode, the receiver remaining silent until a call signal is received. An automatic encoder input and an audio output can be used for transmission and reception of a slow speed teleprinter transmission.

The surface equipment consists of a transceiver, a remote-control unit, a power amplifier and an omni-directional transducer. That on board the submarine comprises a transceiver, power amplifier, four directional transducers and a transducer switching unit. The directional transducers are installed foward, aft, starboard and port.

Transmission and reception from the surface vessel is omnidirectional. The submarine transmission and reception can be either directional or omnidirectional as selected by the operator, using one transducer or four simultaneously.

Operational Status
In operational service with the Canadian Navy.

Specifications
Operating modes
Voice: Voice transmission and reception
Keying: Transmission and reception of telegraphic signals
Responder: Automatic transmission on reception of call signal
Surface ship output power: 5.5 45, 55 or 450 W as selected by the operator
Submarine output power: 5.5 or 55 W (directional); omnidirectional transmission is 5.5, 22, 55 or 220 W (1.4, 5.5, 14 or 55 W per transducer respectively)

Contractor
Safare Crouzet, Nice.

ITALY

TS-200 SYSTEM

Type
Underwater telephone system.

Description
The TS-200 underwater telephone set has been designed, following Italian Navy specifications, to be installed on surface vessels and submarines. By means of the TS-200, both phonic and telegraphic underwater communications between surface and underwater vessels are available. This equipment can transmit and receive in either omnidirectional or directional (with three selectable bearings) modes following NATO standards concerning signal modulation and operating frequency.

Transmitted power is such as to allow communications at distances of several tens of kilometres in optimum sound propagation conditions. The emitted power can be reduced to half or quarter of maximum for shorter range communications.

The equipment consists of a set of piezoelectric transducers, a cabinet containing all the electronics required, and a main and auxiliary control panel from which the various operating modes can be selected by the operator.

In addition to standard IFF applications, the equipment can also integrate on-line an automatic communications coding system by means of an optional interface.

Operational Status
Currently installed on the *Giuseppe Garibaldi* helicopter carrier, 'Maestrale' class frigates and 'Minerva' class corvettes.

Specifications
Frequency range: 8.3-11.1 kHz
Modulation: SSB/SC and CW
Carrier frequency: NATO standard
Horizontal beam width: 360° (omni); 30° (directional)
Vertical beam width: 90° (omni) (approx); 28° (directional) (approx)

Contractor
USEA SpA, La Spezia.

TS-200 auxiliary control panel

TS-300 UNDERWATER TELEPHONE

Type
Emergency telephone for underwater vessels.

Description
The TS-300 is an emergency telephone designed for installation on underwater vessels. It provides voice and telegraphic communications, and coded communications by means of an optional interface, as well as selectable pinger and transponder functions. The tele-

phone can operate on 115 V AC or on backup floating batteries.
Autonomous operating time on batteries is 10 h on transmission and 200 h on reception, for continuous operation.
The equipment can operate in omnidirectional receiving and transmitting modes, and on either the NATO standard frequency or such frequency as the customer requires. The transmitting section provides communication over a range of about 10 km under optimum conditions of propagation.
The TS-300 consists of a cylindrical piezoelectric transducer, a transmission and control unit, loud-

speaker, microphone and headphone, battery and battery charger unit.

Specifications
Modulation: SSC/SC and CW
Frequency range: To customer requirements
Carrier frequency: To customer requirements
Power supply: 115 V AC or backup battery

Contractor
USEA SpA, La Spezia.

TS-400 UNDERWATER TELEPHONE

Type
Modular underwater telephone.

Description
The TS-400 is a modular underwater telephone designed for use by both surface and underwater vessels. It provides voice and telegraphic communications, and coded communications by means of an

optional interface, as well as selectable pinger and transponder functions. Omnidirectional and directional transmitter/receiver modes are available.
The main components of the TS-400 are a transceiver and control unit including the loudspeaker, microphone, headphone, and an omnidirectional transducer for the base version. Optional equipment includes an additional power amplifier unit, and three linear arrays (each 100° horizontal beamwidth) for the directional mode, and/or high emission omnidirectional transducer.

There are two versions of the equipment:
(a) a basic version with source level of 190 dB/μPa/m
(b) an extended version with source level of 200 dB/μPa/m, by the addition of a power amplifier. Four emitted power values are selectable for this version.

Contractor
USEA SpA, La Spezia.

TS-500 UNDERWATER TELEPHONE

Type
Standard and emergency underwater telephone.

Description
The TS-500 was designed to Italian Navy specifications for installation on surface ships and submarines. The telephone can operate on 115 V AC or on a backup battery. It provides voice and coded communication facilities between surface ships and submarines. Omnidirectional receiving and transmitting modes are possible within the frequency band

according to the modulation types specified by NATO standards.
The transmitting section provides communication over a range of about 10 km under optimum conditions.
The TS-500 is composed of a cylindrical piezoelectric transducer and a small sprayproof cabinet containing all the transmission and reception electronics.

Specifications
Modes: SSB/SC and CW
Frequency range: 8.3-11.1 kHz
Carrier frequency: NATO standard
Power supply: 115 V AC, SP

Backup battery operating life: 70 h at rate of 1.5 min of broadcast per hour

Electronics Cabinet
Height: 280 mm
Width: 360 mm
Depth: 330 mm
Weight: 40 kg

Transducer
Overall diameter: 170 mm
Height: 90 mm
Weight: 5 kg

Contractor
USEA SpA, La Spezia.

UNITED KINGDOM

TYPE 3200 COMMUNICATION SYSTEM

Type
Throughwater acoustic underwater telephone.

Description
The submarine communications system Type 3200 is

a throughwater acoustic telephone permitting high quality speech communication between submersibles, or between submersibles and surface vessels. The system uses single sideband suppressed carrier modulation to ensure optimum transmitting efficiency. The 3200 offers dual frequency operation with selectable up/down transducer selection for the submersible installation.

A 37.5 kHz distress frequency pinger beacon is included for emergency purposes. Both surface and submersible electronic packages are identical, minimising the necessity for holding operational spares.
The transducer units are fully encapsulated assemblies and are supplied with integral cable for overside requirements, or with certified through hull penetrators to customers' requirements.

Specifications
Frequency: 10 and 27 kHz
Modulation: Upper sideband suppressed carrier

Transmit power: 50 W peak
Audio output: 1 W

Contractor
Marconi udi, (a subsidiary of Marconi Underwater Systems Ltd), Aberdeen.

G732 Mk II UNDERWATER TELEPHONE

Type
Acoustic throughwater communications system.

Description
The G732 Mk II is an improved underwater telephone system for use in submarines and surface vessels. It provides communications facilities for voice, Morse, digital and coded messages, plus an emergency pinger mode of operation.

The system operates on the standard NATO frequency of 8 to 11 kHz using upper single sideband transmission. Alternative operational frequency bands can also be supplied, and a choice of transducer installations is available to suit customer requirements. The transmit direction, which can be omni, port, starboard, ahead and above, is user selectable, as are also the transmit source levels to maintain security. A narrow band is also available via a suitable transducer.

The receiver/transmitter electronics are housed in a ruggedised bulkhead-mounted unit which connects to one or more transducers to produce directional transmissions as required by the user. The system normally operates from the main power supply but a battery-powered unit can be supplied to maintain communication in the event of complete failure of the vessel's own electrical power supplies.

A portable variant of the G732 Mk II is also available for shore-based and helicopter operation.

Operational Status
In production.

Specifications
Operating frequency: 8.4-11.3 kHz (nominal)
Mode: Upper SSB (NATO compatible)

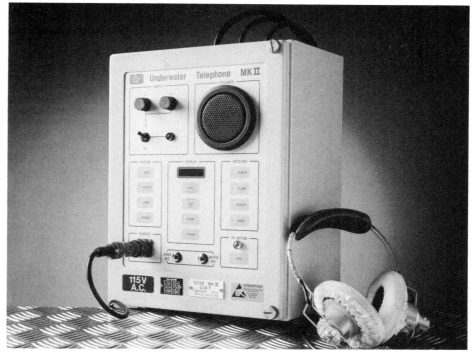

G732 Mk II underwater telephone

Transmitter source level: 186 ±2 dB/μPa/m (max) (port, starboard, ahead, upper); 183 ± 2 dB/μPa/m (max) (omni)
Dimensions: 400 × 300 × 200 mm
Weight: 18 kg (max)

Contractor
Graseby Dynamics Ltd, Watford.

TYPE 2073 UNDERWATER TELEPHONE

Type
Emergency underwater telephone.

Description
The Type 2073 is an underwater telephone/pinger which has been designed to assist in locating and communicating with a submarine or submersible in distress. In an emergency the equipment can operate as a pinger producing low frequency (10 kHz) tone pulses, with the receiver operational at the same time. As a rescue vehicle approaches, the pinger can be changed to a higher frequency (37.5 kHz) to enable the vessel to home in on the submarine or submersible escape hatch. For naval use a 43.5 kHz frequency is included for communication with divers.

The equipment can also operate as a throughwater communication system, on any of three standard channels to communicate with rescue vessels, divers and other ships.

Operational Status
In production.

Specifications
Underwater telephone
Frequency: 10, 27 and 43 kHz
Range (min): 4000 yd (low frequency); 2500 yd (mid frequency); 750 yd (high frequency)
Pinger
Frequency: 10 and 37.5 kHz
Range (to a sonar): 8000 yd (10 kHz); 2000 yd (37.5 kHz)

Contractor
AB Precision (Poole) Ltd, Poole.

UNITED STATES OF AMERICA

MODEL 5400 UNDERWATER TELEPHONE

Type
Series of acoustic underwater telephones.

Description
The Model 5400 underwater telephone is a compact AN/UQC-compatible system. Primarily intended for voice communication, the equipment is based on a synthesised transceiver which covers the frequency band from 5 to 45 kHz in 1 Hz steps plus the AN/WQC band. Up to 50 frequency combinations can be stored in memory, for easy recall at any time. Upper and lower sideband modulation is operator selectable, as is output power in five steps; 2, 10, 50, 100 and 200 W.

The Model 5400 is equipped with TIPE, an acronym for Transponder, Interrogator, Pinger/Echosounder. The transponder/interrogator functions provide automatic ranging between two 5400s (or any

Model 5400 underwater telephone

other compatible device, such as ATM-504A equipped with TIPE). The Pinger mode is primarily intended to be used as a tracking pinger during exercises, but can also be used in emergency situations. Used with a vertically directed transducer, the Model 5400 can provide digital readout of depth below the surface or distance to the seabed. Transducers are available for operation to full ocean depth. The operating frequencies for the TIPE mode are operator selectable within the system's entire frequency band.

Up to four remote-control units can be connected, or three remotes and one data communications equipment. The equipment can be provided with an optional modem for serial data communication via the water column. Typical applications are telex, computer-to-computer communications, and slow-scan TV.

The Model 5400 is form-fit compatible with the ATM-504A underwater telephone in service with US and overseas users.

Operational Status

In operational service.

Contractor

EDO Electro-Acoustic Division, Salt Lake City, Utah.

COMMUNICATIONS BUOYS

UNITED KINGDOM

ECB-680(1) BUOY

Type
Expendable communications buoy.

Description
The Expendable Communications Buoy ECB-680(1) system has been developed by Marconi for Royal Navy operations to provide a one-way line-of-sight VHF/UHF radio communications relay between a submerged submarine and surface receivers.

ECB-680(1) is designed for launching from either forward or aft submerged signal ejector (SSE) tubes at all operational depths up to 15 kts. Prior to launch, the ECB is programmed by means of an inboard control interface unit, with its operating parameters and plain or encoded messages as dictated by the tactical situation. Alternatively, in an emergency, ECB-680(1) can be launched with no programme input; in this mode of operation a pre-programmed distress SARBE signal is transmitted.

On release, the buoy ascends to the surface after a preset transmission delay has elapsed, allowing the submarine to leave the area discreetly. The aerial unit

ECB-680(1) expendable communications buoy

with its flotation collar is then deployed and message transmission is initiated. Transmissions are made in the VHF/UHF band between 168 MHz and 310 MHz; modulation may be programmed for either AM or FM transmissions.

The message cycle allows up to three minutes of message with a one-minute gap and may be repeated up to 60 times, allowing the buoy four hours transmission time. For the emergency role, an unlimited

number of cycles may be selected, in which case the buoy will transmit in excess of eight hours. On completion of the preset number of message cycles, the buoy scuttles automatically.

ECB-680(1) is a highly reliable assembly with an unattended shelf life of five years. It incorporates stabilisation components which ensure excellent transmission continuity. The buoy's aluminium hull is capable of withstanding the extreme water pressures experienced during normal submarine operations. The hull is reinforced to permit launch at high speed for suitably equipped submarines.

ECB-680(1), operating in conjunction with the submarine's communication system, provides a comprehensive capability for emergency distress and tactical communications. By employing existing submarine SSE launch facilities, ECB-680(1) is easy to install, requiring minimum maintenance and support.

Operational Status
Widely fitted to submarines of the Royal Navy and other navies.

Contractor
Marconi Underwater Systems Ltd, Templecombe.

ECB-699 COMMUNICATION BUOY

Type
Expendable buoy for satellite communications from submarines.

Description
The ECB-699 UHF Satellite Communications System provides a digital communications link from submarines to shore.

A Control Interface Unit (CIU) is used to load messages, frequency and timing information for subsequent transmission. The CIU also provides built-in test facilities.

The ECB-699 buoy is launched from the submarine's Submerged Signal Ejector (SSE) at normal operating depths and speeds. After launch, the buoy rises to the ocean surface. Once the programmed delay has elapsed, the buoy automatically erects its antenna, and transmits, via satellite, to shore. If the submarine is in distress, the buoy can be used to

transmit a distress signal on 406 MHz via the COSPAS/SARSAT system, which provides reliable, rapid, worldwide indication of the buoy's identity and location.

Operational Status
In production for the Royal Navy.

Specifications
Buoy
Weight: 4.4 kg
Dimensions
Length: 502 mm
Diameter: 101 mm
Communications mode: Digital data preset for 2400 bits/s. Selectable for 75, 300, 600, 4800 and 9600 bits/s
SSE holding and transmission delay: Selectable up to 8 h
Message cycles: 1 to 15 programmable messages each repeated once

Buoy scuttle: After message cycle
Transmit power output: 100 W ERP
RF transmission modes: Dual phase frequency shift keying (FSK)
Operating frequency range: 290-320 MHz
Channel Selection/Spacing: Multiple programmable frequencies/25 kHz channel increments
Alert mode: COSPAS/SARSAT tone at 406 MHz
CIU
Weight: <10kg
Dimensions
Width: 482 mm
Height: 889 mm
Depth: 350 mm

Contractor
Dowty Maritime Sonar and Communication Systems, Greenford.

UNITED STATES OF AMERICA

SUS Mk 84 SOUND SIGNAL

Type
Air-to-submarine communications device.

Description
The SUS Mk 84 Underwater Sound Signal is an expendable electro-acoustic device which provides one way air-to-submarine communications. It may be dropped or deployed from fixed-wing aircraft or helicopters, and may also be used to simulate the drop of an ASW weapon during a tactical exercise. It may be launched from altitudes up to 10 000 ft and speeds up to 380 kts. It can also be deployed from surface ships.

The SUS Mk 84 is able to transmit five different acoustic signals, each of which may convey a predetermined message to the submarine. The device

transmits two acoustic tonals, one at 3.5 kHz and the other at 2.95 kHz, each of which may be pulsed for either 0.5 or 1.5 s. Four of the five coded SUS signals are produced by generating these two tonals alternately and varying the length of each pulse. This results in four two-tone combinations. The fifth is a steady transmission of the 3.5 kHz tone. The appropriate signal is selected immediately prior to launch. The submerged submarine may use either passive sonar or underwater telephone to receive the SUS output.

The SUS is a compact device, measuring 38 cm in length by 7.6 cm diameter and weighing 3 kg. A heavy zinc nose houses a seawater battery, a hermetically sealed inlet port to the battery compartment and most of the electronics.

Operational Status
In production and in service.

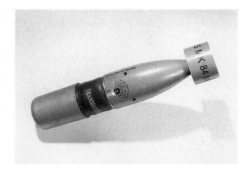

SUS Mk 84

Contractor
Sippican Inc, Marion, Massachusetts.

AN/BRT-1 COMMUNICATIONS BUOY

Type
Submarine-launched communications buoy.

Description
The AN/BRT-1 is an expendable one-way transmitter which is launched through the submarine's signal ejector. It carries a taped message (maximum four minutes) which can be either voice or CW, and a small transmitter. The buoy can function in sea conditions up to sea state 5.

Transmission from the buoy can be delayed for either five minutes or one hour after launch to give the submarine time to get well away from the transmission point. Transmission is by VHF on a sonobuoy frequency, and can be turned off by an aircraft or ship by means of a UHF signal. Selectable life time of the buoy is 1, 3 or 8 hours, after which time it will be sunk by its scuttle timing unit. Weight of the buoy is 3.3 kg.

Operational Status
In service.

Contractor
Sippican Inc, Marion, Massachusetts.

BRC-6 EXPENDABLE TRANSCEIVER

Type
Submarine-launched expendable communications buoy.

Description
The BRC-6, also known as XSTAT (an acronym for Expendable Submarine Tactical Transceiver) is a two-way communications buoy which is launched via the submarine's signal ejector. After launch, the buoy rises to the surface but remains connected to the submarine by a very fine wire (approximately 0.2 mm thick), up to 3000 m long. When the buoy reaches the surface it deploys an antenna for two-way voice communication in the UHF frequency band.

Power is by a battery which has approximately 45 minutes of working life. However, since the submarine is leaving the area at some speed the actual life depends on the time it takes to exhaust the length of wire. At a speed of 10 kts this is approximately eight minutes.

Operational Status
In operational service. Sippican is understood to be developing a version of this buoy which uses a 20 km fibre optic link in place of the fine wire.

Contractor
Sippican Inc, Marion, Massachusetts.

AN/SSQ-86(XN-1) DOWNLINK COMMUNICATION SONOBUOY

Type
One-way surface-to-submarine communication equipment.

Description
The AN/SSQ-86(XN-1) downlink communication (DLC) sonobuoy is a one-way communication device designed to transmit a pre-programmed message to a submerged submarine. Developed to meet the need for a reliable means of transmitting a message without revealing the receiving submarine's position, the DLC is compatible with existing sonobuoy launch platforms. The unit is packaged in a standard 'A' sized sonobuoy envelope and can be launched from any properly equipped fixed-wing aircraft or helicopter, or from a surface ship.

The desired message is programmed into the DLC using a single push-button switch and four seven-segment LED displays. Four groups of three digits make up the message. The message can be verified and corrected if necessary with a second push-button, provided to ensure proper data insertion. Once programmed, all remaining operational functions are automatic. The buoy is tunable to one of 99 RF channels.

When air-launched, the DLC is slowed and

AN/SSQ-86(XN-1) downlink communication sonobuoy

stabilised by a small parachute. For launches from surface vessels, it is merely thrown overboard. Upon water entry, the DLC immediately deploys. As soon as the subsurface unit reaches the shallow operating depth, the message, coded into appropriate tones, is acoustically transmitted. The first transmission is followed by a five-minute pause while the subsurface unit deploys to the deep depth. The message is repeated at the deep depth; after a second five-minute pause, the message is transmitted a third time. At the end of the final transmission, the DLC automatically scuttles. The nominal life of the DLC from water entry to scuttle is 17 minutes.

For maximum reliability and minimum development time, extensive use has been made of existing technology and hardware from other Sparton sonobuoy designs. Examples include the lithium battery power supply, the air descent and float inflation hardware, and much of the circuit design. The omnidirectional projector is a simple resonant bender element designed to operate efficiently at the desired bandwidth.

Operational Status
No longer in production but still in service.

Specifications
Depth: Shallow and deep
Size: A (123 × 910 mm)
Weight: 11.4 kg
Operating life: 17 mins (nominal)

Contractor
Sparton Corporation, Electronics Division, Jackson, Michigan.

AN/SSQ-71 ATAC BUOY

Type
Expendable acoustic communications device.

Description
The AN/SSQ-71 air-transportable acoustic communication (ATAC) buoy is an expendable two-way sonic RF receive-transmit communications device tunable to one of three RF channels. It is launched from surface vessels or either fixed-wing or rotary-wing aircraft. Incoming UHF signals are encoded into a multiple-tone format for sonic transmission; incoming sonic signals are likewise processed for VHF transmission. The AN/SSQ-71 borrows heavily from designs developed and refined in other Sparton sonobuoy programmes. The antenna system, for instance, which receives UHF and transmits VHF signals, is derived from the Sparton AN/SSQ-62 sonobuoy. The entire mechanical layout is based upon the AN/SSQ-53A, designed and produced in volume by Sparton.

Operational Status
No longer in production.

Contractor
Sparton Corporation, Electronics Division, Jackson, Michigan.

AN/BRT-6 COMMUNICATIONS SYSTEM

Type
UHF satellite communications system.

Description
The AN/BRT-6 UHF satellite communications system was developed by Hazeltine Corporation, in conjunction with the US Naval Underwater Systems Center and the Space and Naval Warfare Systems Command (formerly NAVELEX), as a digital communications link from submarines at operating depth to shore communications stations and operating naval forces. The UHF SATCOM buoy provides reliable, high fidelity, one-way communications using a uniquely designed Hazeltine right circular polarised floating antenna. The SATCOM buoy is manually or automatically launched from standard 3 in signal ejectors on fleet submarines and can be operated in conditions up to sea state 5.

The buoy operates on a satellite frequency between 290 MHz and 315 MHz, which is selected prior to launch in 25 MHz increments.

Operational Status
In production. Hazeltine was awarded a follow-on competitive production contract for up to 4500 buoys (including options) in September 1989.

Contractor
Hazeltine Corporation, Greenlawn, New York.

ELECTRO-OPTICAL SENSORS

PERISCOPES

FRANCE

SOPELEM PERISCOPES

Type
Submarine search and attack periscopes.

Description
For more than 70 years SOPELEM has been successfully involved in the design, development and production of submarine periscopes. The new generations of submarines, including the 'Agosta' class and SSBN nuclear boats, as well as nuclear hunter-killers, have resulted in more demanding requirements for periscope operational capabilities in more severe tactical environments.

SOPELEM produces optical transfer reference tubes of extremely high accuracy, necessary for the built-in sextant.

The company also designs and manufactures advanced surveillance periscopes in co-operation (see Pivair entry). These incorporate both infra-red and visual daylight channels. They have a two-axis stabilisation system.

The Type K periscope incorporates a light intensifier for night use. This provides a magnification of ×5 with a 10° field-of-view and an elevation arc of ±10°. In daylight, magnification is ×1.5 and ×6 with 36° and 9° fields-of-view respectively, and an elevation arc of −10 to +80°.

ST 5 attack periscope head
This is a newly developed periscope head stabilised by a rate gyroscope with an image-intensified TV micro-camera for night vision. The design is so compact that it is fitted in the small ST 5 periscope head, which itself has been specially shaped and covered by RAM to reduce its radar cross-section. It uses a fixed eyepiece with magnifications of ×1.5 and ×6 giving fields-of-view of 36° and 7° over an elevation arc of −10 to +30°.

Contractor
SOPELEM, Paris Cedex 18.

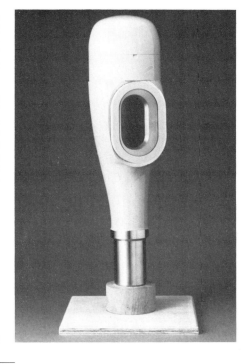

ST 5 attack periscope head

PIVAIR SEARCH PERISCOPES

Type
Submarine visual and infra-red periscopes.

Description
The Pivair optronic search periscope is built around three fundamental characteristics: two-axis gyro-stabilisation of the line-of-sight, built-in vertical reference, and integration of infra-red vision and an optical channel.

A two-axis gyroscope located in the head overcomes problems from mast vibration and improves visual and IR observations by gyro-stabilising the line-of-sight in elevation and bearing.

Associated with the built-in vertical reference, very accurate sextant measurements can be obtained, providing a reliable position fix.

The use of 8 to 12μm band infra-red extends the operational capabilities, both at night and in poor meteorological conditions. The periscope can be operated in a real panoramic IR search mode providing long-range airborne and surface detection.

A number of options are also available, including:
(a) choice of magnification (optical channel) ×1.5 to ×12
(b) gyroscope rangefinder
(c) built-in ESM antenna on top of the mast
(d) built-in Navstar receiver antenna on top of the mast
(e) RAM coatings
(f) low light-level TV camera
(g) display and recording of video and infra-red images on remote console.

Operational Status
Operational on board all French Navy nuclear submarines (SSBN and SSN).

Contractor
SAGEM, Paris.

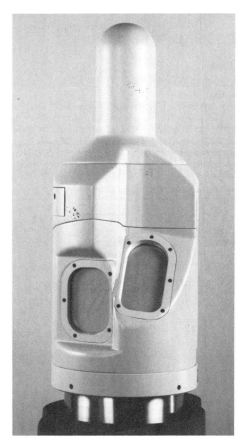

Search head

SAGEM ATTACK PERISCOPES

Type
Small diameter attack periscopes.

Description
The small diameter attack head includes a single axis stabilised line-of-sight for both optical and low light level TV channels. Two magnifications (×1.5 and ×6) are available for the optical channel. The LLL CCD TV is mounted in the head.

Optional features include a laser rangefinder (replacing) the TV, a built-in ESM antenna on top of the mast, RAM coatings and display and recording of TV pictures on a remote console.

Operational Status
Developed from attack periscopes operational on board French submarines.

Contractor
SAGEM, Paris.

Attack head

SAGEM OPTRONIC MAST

Type
Submarine optronic non-penetrating mast.

Description
The optronic mast combines the advantages of SAGEM's optronic search periscopes with the advanced safety of non-penetrating masts. With the mast fully located outside the pressure hull, control room space is increased and more comfortable

operating conditions provided. The optronic non-penetrating mast is thus more easily accommodated in submarine designs and facilitates retrofit upgrades.
The SAGEM optronic mast includes:
(a) thermal imaging system (conventional imagery mode – optional panoramic search mode)
(b) high resolution ×4 magnification TV system
(c) two axis gyro-stabilisation of the line-of-sight
(d) interface for ESM antenna and/or GPS antenna
(e) electrical rotation
(f) RAM coatings.

SAGEM optronic masts can be fitted on any type of non-penetrating hoisting device.
A remote-control and display console includes a 19 in CRT display, VCR, joystick and keyboard as well as electronic PCBs and power supply.

Operational Status
A demonstrator is operational.

Contractor
SAGEM, Paris.

SAGEM OPTORADAR MAST

Type
Submarine optoradar non-penetrating mast.

Description
The optoradar mast combines the capabilities of SAGEM's optronic mast while including integration of a navigation radar. The optoradar mast allows elimination of separate masts for these functions, thus improving the discretion of the submarine. The non-hull penetrating system facilitates submarine retrofit upgrades.

The SAGEM optoradar mast includes:
(a) thermal imaging system (conventional imagery mode – optional panoramic search mode) or alternatively low light level ×2 magnification TV system
(b) high resolution ×2 magnification TV system
(c) one axis gyro-stabilisation of the line-of-sight
(d) X-band navigation radar
(e) electrical rotation
(f) interface for ESM antenna and/or GPS antenna
(g) RAM coatings.
The SAGEM optoradar mast can be fitted on any type of non-penetrating mast hoisting system.

Separate control and display consoles are provided to operate the radar and the optronic system.
The radar console includes PPI display, software keys and a trackerball. The optronics console includes a 19 in CRT display, VCR, joystick and keyboard as well as electronic PCBs and power supply.

Operational Status
In development. Scheduled to be operational on board French Navy SSBNs in 1993.

Contractor
SAGEM, Paris.

ITALY

RIVA CALZONI NON-PENETRATING MASTS

Type
Range of optronic, ESM and radar masts for submarines.

Description
Riva Calzoni produces a range of submarine masts for various applications. All masts are non-penetrating into the control room; there are no translating parts and sliding seals through the hull boundary. Only a short fixed head is sometimes fitted inside for hydraulic connections. This completely frees the control room and provides the crew with an unobstructed view and access.

Each hoist consists of three main parts:
(a) guide fixed to the main sail structure
(b) streamlined mast which slides, by means of sliding shoes, inside the guide
(c) hydraulic cylinder, inside the mast and fixed to the hull.
Electrical connections to the antennas on the mast top are generally of flexible cables located inside the mast and forming a loop housed in a side recess, making the hoist units compact and self-contained.

The integration of hydraulic, electrical and mechanical parts ensures reduced size and weight, and also fast and easy installation. The masts are streamlined in order to reduce wake and vibration. They are fully free-flooding except for the hydraulic cylinder, where the actuating rod has double seals for water and oil tightness.

Masts are applicable to a number of configurations, including optronic, radar, radio, ESM/ECM and other special purposes. These configurations refer to single-stage and double-stage masts.

Available are compact solutions of two or more masts integrated within one modular structure.

Operational Status
Riva Calzoni has supplied or is supplying non-penetrating masts to the Italian Navy and a number of other countries' navies, including Argentina, Holland, Norway, Germany and Australia.

An integrated solution of three masts in one modular structure has been installed in the new Norwegian 'Ula' class submarines. In late 1989, Kollmorgen Corporation, USA announced the award

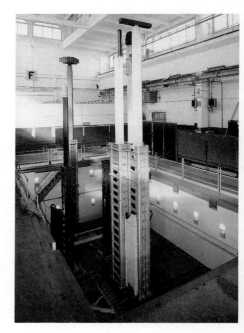

Riva Calzoni non-penetrating masts

of a contract to Riva Calzoni, funded by the Defence Advanced Research Project Agency (DARPA), to develop a non-hull penetrating hoisting mast, to interface a new range of optronic sensors for future and existing SSN-type submarine platforms.

In 1991 it was announced that following successful full acceptance trials carried out in Italy, the advanced mast would be transferred to the US for SIT testing, and final installation on board later the same year.

Contractor
Riva Calzoni, Bologna.

UNITED KINGDOM

PILKINGTON SEARCH PERISCOPES

Type
Submarine search and surveillance periscopes.

Description
Pilkington Optronics produces periscope and mast systems for all types of submarines. Periscope main tube diameters vary between 127 mm, 190 mm and 254 mm, depending on submarine size and the required performance specification.

The 127 mm compact submarine periscopes (CK32, CK37, CK39 and CK41) offer high performance optics coupled with image intensification for night vision, stabilisation, electronic weapons system interface, still camera facility, elevation of line-of-sight (−15° to +60°). ESM, GPS and TV camera upgrades can be offered. The periscopes are the optimum solution for small submarines without compromising performance.

The search version of the 190 mm periscope (CK038) incorporates an image intensifier (II) at the top of the optical system. This uses a high reliability tube which provides the periscope with night vision capability. All functions on the periscope are by electronic control, with additional facilities such as azimuth torque drive and electronic weapons system interface offered as standard equipment.

The CK043 search periscope is 254 mm in diameter and equips the Royal Australian Navy 'Collins' class submarines. The CK043 is the most advanced of Pilkington Optronics' search periscopes and features implementation of full remote-control from a multi-function console. It offers four powers of magnification together with the following features:
(a) thermal imaging (TI)
(b) 35 mm still camera
(c) LLTV camera
(d) stabilisation of line-of-sight
(e) heated top window
(f) digital display of data
(g) range transmission and display
(h) azimuth torque drive
(i) remote-control operation
(j) sextant
(k) ESM.

Operational Status
All 127 mm and 190 mm periscopes currently in service. The CK043 is currently in production. All other 254 mm search periscopes are currently in service.

Contractor
Pilkington Optronics, Glasgow.

CK043 search periscope

PILKINGTON ATTACK PERISCOPES

Type
Submarine attack periscopes.

Description
Pilkington Optronics designs and manufactures a range of periscope and mast systems for all classes of submarines. This includes both search and surveillance periscopes (see separate entry), and attack periscopes.

The CH088 190 mm attack periscope incorporates a high degree of design commonality with the proven CK038 search periscope. It is designed to operate as a stand-alone instrument or as part of a 190 mm periscope pair. It also features azimuth torque drive and an electronic weapon system interface. TV camera with video recording, ESM and GPS can be offered as an upgrade.

CH085 Pilkington attack periscope

The CH085 and CH093 periscopes are 254 mm in diameter and offer greatly enhanced optical performance over the 190 mm periscope designs. The CH085 equips the Royal Navy 'Upholder' class submarines while the CH093 will equip the Royal

Australian Navy 'Collins' class submarine and together with the CK043 search periscope will offer the most advanced periscope system currently available. The CH085 incorporates a Pilkington thermal imager whilst the CH093 is fitted with an image intensifier and LLTV camera. The II and optical channels in the CH093 are both stabilised. Both periscopes have a weapons system interface and can be fully remote-controlled from a separate console. Video recording of either the TV or TI data is offered as standard. ESM upgrades are also available.

Operational Status
CH085 periscope is currently in service. CH093 is in production.

Contractor
Pilkington Optronics, Glasgow.

PILKINGTON OPTRONIC MAST

Type
Electro-optical submarine mast.

Description
The Pilkington non-hull penetrating optronic mast uses a family of hoisting gear compatible with all submarine above water sensor systems, giving total commonality.

The system comprises a multi-function remote-control and display console with plasma panels. The non-penetrating mast comprises four main elements: the optronics pod which contains all the E/O and RF sensors, azimuth drive module and mast hoisting gear, and the mast processing unit. Powerful image processing facilities support the E/O sensors offering facilities such as contrast enhancement, image capture, zoom and noise suppression. The mast flexibility allows a variety of mission configurable pods to be fitted with an optimum sensor mix. The mast has been designed to provide an ESM capability, an integrated day/night thermal imaging and TV facility, and radio communications. The mast features multiple pre-programmable modes which enable it to carry out automatic search with the thermal picture recorded for future analysis.

Principal features of the system are:
(a) thermal imager
(b) combination of TV and NIR sensors
(c) stabilised line-of-sight (two-axis)
(d) programmable control facility with full data recording
(e) integration with submarine weapons system, or stand-alone console
(f) radar absorbent material
(g) streamlined fairing
(h) artificial horizon sextant
(i) electronic support measures
(j) communications transmit/receive antenna.

Operational Status
Pilkington has received a £60 million contract from the UK MoD to supply integrated electro-optical periscope systems for use in the Royal Navy 'Vanguard' class submarines.

Contractor
Pilkington Optronics, Glasgow.

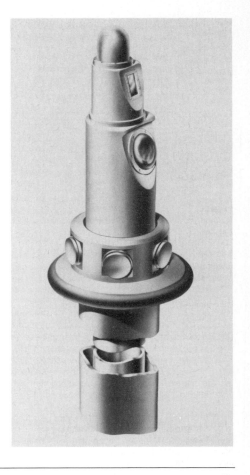

Pilkington optronic mast

UNITED STATES OF AMERICA

MODEL 76 PERISCOPE

Type
Modular attack and search periscope.

Description
The Kollmorgen Model 76 is a modular periscope system with common components for the attack and search versions. The basic difference is that the attack periscopes have smaller heads, while the search periscopes have larger heads to act as multi-purpose reconnaissance platforms.

The system consists of a mast unit (whose weight depends upon customer requirements) with optical train, a display and control unit including a split beam binocular eyepiece, a 35 mm camera and training handles. In addition to the mast unit there is a hoisting yoke weighing 90 kg, a control unit, and a junction box unit weighing 86 kg.

The control unit includes a control panel, system focus, mode select, stadimeter control and microphone. The attack periscope includes a broadband antenna and crystal video receiver ESM system, together with a display and control panel on the control unit. The search periscope also includes a system focus control.

The periscopes can be fitted with a Galilean telescope to provide ×12 magnification, line-of-sight stabilisation, a power rotation drive, a remote-control unit with 9 in television monitor, an image intensifier, a television camera, thermal imager, video, laser or radar rangefinders.

Operational Status
In production for export customers.

Specifications
Diameter: 190.42 mm
Elevation: −10 to +74° (attack); −10 to +60° (search)
Magnification: × 1.5, × 6
Field of view: 8°, 32°

Contractor
Kollmorgen Corporation, Electro-Optical Division, Northampton, Massachusetts.

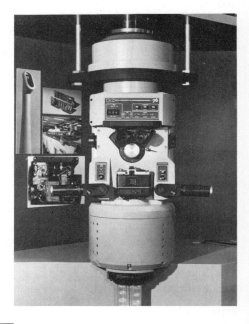

Model 76 periscope

MODEL 86 OPTRONIC MAST SERIES

Type
Submarine non-hull penetrating optronic masts.

Description
Capabilities which previously required a hull penetrating optical periscope can now be included in one of these sensor packages. The Model 86 Optronic Mast Series includes a sensor unit, a hydraulically operated streamlined mast including connection cabling external to the hull, and electronic interface unit and a control/display console internal to the hull. Electronic processing and data transmission permit all sensor information to be processed and displayed in a

dedicated operating console, or incorporated into the main combat consoles to further reduce space requirements and enhance operation.

The Model 86 Optronic Mast Series has been developed in three basic configurations: the Tactical Optronic Mast (TOM) which combines visual and infra-red sensors as well as electromagnetic sensors on a single mast. The TOM is suitable for all missions on all submarines; the Optronic Mast Sensor (OMS) which utilises a visual sensor on a multi-purpose pressure-compensated rotational pod. The OMS can be utilised on all submarines; and the Compact Optical Mast (COM) which provides a small size television sensor in an enclosed pressurised housing which allows hemispheric search. The COM is best suited for special applications, refits and midget submarines.

Features of the Tactical Optronic Mast include:
(a) thermal imaging for day/night/adverse weather viewing without major hull penetration
(b) television for daylight, low light-level and 'quick-look' viewing
(c) two-axis line-of-sight stabilisation to eliminate ship's motion and mast vibrations
(d) ESM warning to detect radar threats
(e) unique, rotating sensor package (sealed statically) with quick response and low power consumption
(f) manual or automatic mast control with a 'quick-look' mode.

Features of the Optronic Mast Sensor include:
(a) television for daylight, low light-level and 'quick-look' viewing
(b) elevation stabilisation

(c) ESM warning system (optional)
(d) unique, rotating sensor package (sealed statically) with quick response and low power consumption
(e) thermal imaging systems upgrade (optional).
Features of the Compact Optical Mast include:
(a) low light-level TV or high resolution monochrome TV
(b) one-axis stabilisation
(c) hemispherical field-of-view
(d) low weight, small size
(e) ESM/GPS (optional).

The TOM and OMS are configured in a similar way. The sensor unit is divided into three main sections, with the upper rotating section containing the appropriate cameras, ESM early warning antenna, line-of-sight control elevation system, and packaging space for additional capabilities. The middle section contains the rotating motor, a unique oil-filled system which is self-pressure compensated to maintain seal integrity. The third and lower section contains support electronics and interface connections.

The COM is configured in a unique, small size and

low weight, completely sealed housing with a hemispheric optical dome on top, allowing for a television sensor to be installed on a dedicated mast or other existing submarine mast.

Operational Status

The basic optronic configurations are directly derived from the Model 90 Optronic Periscope System (see separate entry which is currently in production for export and will be operational in 1991). A more advanced version of the TOM is being developed under a contract from the US Defence Advanced Research Projects Agency. A demonstration unit is currently being evaluated on board the USS *Memphis*.

Contractor

Kollmorgen Corporation, Electro-Optical Division, Northampton, Massachusetts.

Tactical optronic mast

MODEL 90 OPTRONIC PERISCOPE SYSTEM

Type
Submarine optronic periscope system.

Description
The Optronic Periscope System has been developed to allow the operator inside the submarine to search the sea surface during day and night utilising a thermal imaging subsystem and, at the same time, to supply a direct viewing visual channel. The periscope system combines a wide range of sensors in one periscope; a thermal imaging camera, CCD TV camera, 35 mm photographic camera, laser rangefinder as well as passive TV and visual stadimeter, omni radar early warning antenna, a radar direction finding antenna and GPS. The periscope provides high optronic performance by utilising accurate stabilisation to compensate for induced vibrations and platform motion to the visual and thermal lines-of-sight. Additionally, in combination with this function, the operator is provided with a periscope rotation and line-of-sight elevation rate control which allows fast direction and target tracking. The operator has a direct view of the scene in addition to a video display and eyepiece data display of target range, target bearing and line-of-sight elevation angle.

A remote-control station is supplied as part of the Optronic Periscope System, in addition to a complete control data link to the submarine fire control system.

Model 90 optronic masthead unit

Operational Status
In production for export customers.

Specifications
Periscope tube diameter: 190.42 mm
Line-of-sight elevation: −10° to +74° (visual and

Model 90 optronic mast eyepiece unit

TV); −10° to +55° (thermal imaging)
Magnifications: ×1.5, ×6, ×12, ×18

Contractor
Kollmorgen Corporation, Electro-Optical Division, Northampton, Massachusetts.

MISCELLANEOUS ELECTRO-OPTICAL SENSORS AND SYSTEMS

ITALY

GALIFLIR EXTRA

Type
Extended range thermal imaging system.

Description
The Galiflir Extra thermal imaging system is specifically intended to provide maritime patrol aircraft with the ability to detect and identify targets such as submarines at long range by day or night.

The system comprises a high accuracy stabilised platform housing the optomechanical head of the imager. The platform can be manually controlled by joystick, slaved to the radar set, or set to scan along programmed patterns within predefined sectors.

Images are displayed on a standard TV monitor together with indications on aiming direction and system operating status.

Two different fields-of-view can be selected, with medium or high magnification. The imager uses a series parallel scanner in combination with the SPRITE detector. The imager can be integrated with other onboard systems such as radar and automatic TV tracker and other equipment through suitable interfaces.

Operational Status
In production and in service on Italian AB 212 helicopters.

Specifications
Infra-red operating wave band: 8-12 μm
Wide field-of-view (WFOV): 6.7° × 4.6°
WFOV magnification: ×6
Narrow field-of-view (NFOV): 2.5° × 1.7°
NFOV magnification: ×16
Detector: 8 elements CMT SPRITE
Scanning: Series parallel
Display: CCIR 625/50
Infra-red lines: 512
Platform field of regard: -75° to +45° (elevation); -170° to +170° (azimuth)
Stabilisation accuracy: better than 40 μrad RMS

Contractor
Officine Galileo, Florence.

Galiflir Extra imager

NETHERLANDS

UA 9053

Type
Thermal imaging camera.

Description
A passive sensor for day and night tracking in fire control systems. Operating in the 8-12 μm window, the UA 9053 has a TV-compatible output signal for remote viewing and recording on standard equipment. In operation, it is undetectable by electronic support measures and immune to electronic countermeasures.

Operational Status
In production and in use.

Specifications
Total field-of-view: 42 × 23 mrad
Instantaneous field-of-view: 0.15 × 0.15 mrad
Objective lens: 300 mm, f/2
Time to reach operational status: <10 min
Camera output: CCIR-compatible video signal
Power consumption: 250 W

Contractor
Signaal-USFA, Eindhoven.

UA 9053 thermal imaging camera

UP 1043

Type
Thermal imaging camera.

Description
A passive thermal imaging camera which operates in the 8-12 μm window, and is designed for tracking and observation purposes. Capable of tracking at distances required in fire control systems, the lightweight camera is compact and weighs only 18 kg.

It has a 35 × 18 mrad field-of-view for long-range tracking and identification and an optional 260 × 130 mrad field-of-view, ideal for naval navigation and surveillance purposes. The camera employs digital signal handling and has a CCIR video signal output which can be displayed and recorded on standard video equipment. The UP 1043 is able to withstand NBC contamination and, unlike radar, is insensitive to electronic countermeasures.

Operational Status
In production.

Specifications
Spectral range: 8-12 μm
Field-of-view: 35 × 18 mrad
Optional field-of-view: 260 × 130 mrad
Focusing distance: 100 m to infinity
Video output: CCIR-compatible video signal
Cooling: closed-cycle stirling cooler
Controls: ON/OFF switch
    Focus
    Brightness
    Contrast
    Polarity
Power consumption: 50 W
Supply voltage: 18 to 31 V DC
Weight: 18 kg
Operating temperature: −35 to +63°C

UP 1043 thermal imaging camera

Contractor
Signaal-USFA, Eindhoven.

MODEL 1242/01

Type
Biocular night sight.

Description
A biocular night sight with a second generation image intensifier for passive observation and target identification in light conditions approaching total darkness. Its low weight and compact construction ensure comfortable operation even from moving or vibrating craft. Automatic Brightness Control and Bright Source Protection have been incorporated. The sight is completely sealed to withstand all types of battle conditions and has been designed for operation in all climatic conditions and extreme temperatures.

Powered by two standard R14 1.5V dry cells, the

sight requires no maintenance except for changing the batteries and cleaning the lenses. Also available is the UA1242/02, a CCD version which converts the image to a CCIR video signal so that the scene viewed directly by the user can be displayed simultaneously on a TV monitor for remote viewing or recording.

Operational Status
In production and in use.

Specifications
Magnification: ×5
Field-of-view: 7.5° circular
Resolution (contrast 30%): 1 mrad at 1 millilux (starlight); 0.5 mrad at 100 millilux (moonlight)
Focus range: 25 m to infinity
Objective: f/1.6, catadioptric
Eyepiece adjustment: -5 to +3 diopters

Interocular adjustment: 58 mm to 72 mm
Image intensifier tube
XX 1380 series: 20/30 micro-channel plate tube
Automatic brightness control: Stabilises image brightness
Bright source protection: Limits excessive bright spots in image
Power source: Two R14 (C Type) standard dry cells
Operating voltage: 3 V DC nominal
Power consumption: 120 mW maximum
Weight excluding batteries: 1.9 kg
UA 1242/02 CCD version
Field-of-view: 6.0° (horizontal) × 4.5° (vertical)
Video output: composite video, 1 Vp-p (75Ω) CCIR/625 lines
Resolution: 450 TV lines horizontally
Pixels: 604 (horizontal) × 588 (vertical)
Gain control: automatic
Operational temperature range: −10 to +55°C

Biocular night sight

Contractor
Signaal-USFA, Eindhoven.

UNITED KINGDOM

FD 5000 PERISCOPE CAMERA

Type
Periscope-mounted video camera.

Description
This lightweight (135 g) monochrome video camera has been designed to be mounted within a periscope head to assist in periscope observation. It measures 37 × 52 × 55 mm and features a 24 mm lens and offers a resolution of 400 lines.

Submarines equipped with the system would come to periscope depth long enough for a scan of the horizon. The submarine can then dive while the images are examined on a TV monitor linked to the camera by a screened cable.

The camera has no moving parts, giving high reliability.

Contractor
GEC-Ferranti Defence Systems Ltd, Edinburgh.

FD 5000 periscope video camera

GEC SENSORS OPTRONIC SENSORS

Type
Submarine periscope optronic sensors.

Description
The increasing need to supply accurate information day and night in all weather conditions requires high performance sensors and information processing. The GEC Sensors naval-based experience in visible and thermal systems, expertise in image stabilisation and signal processing have been brought together to fulfil current contracts to provide submarines with high performance optronic mast and periscope sensors.

The system for periscopes and non-hull penetrating masts incorporates fully stabilised, high performance daylight, low light, colour and thermal imaging sensors. Processing includes automatic target detection, image enhancement, and other image analysis. The data from the multiple sensors may then be correlated to provide optimum information. Fibre optics and signal multiplexers are used to minimise the system complexity.

In the visible spectrum use is made of high resolution CCD cameras for colour, monochrome or low light. In the thermal spectrum the most advanced Staring Array camera technology is employed to provide a miniature IR sensor in the 3-5 μm range. The TICM II mini-scanner and electronic modules are utilised, operating in a single or dual mode configuration over the 8-13 μm and 3-5 μm wavelengths.

Operational Status
In production.

Optronic sensors equip submarine periscopes

Contractor
GEC Sensors Ltd, Electro-Optical Guidance Division, Basildon.

MODEL OE 0285

Type
Submarine ice operation camera.

Description
Developed for use aboard submarines for operating in ice conditions, the OE 0285 is an ISIT camera capable of viewing objects in ultra low light level conditions in the region of overcast starlight, enabling thinner patches of ice to be identified. This provides an important aid to submarines surfacing through ice when engaged on Arctic operations. The cameras, which are the first to be developed specifically for under-ice operations where low-light levels present a major obstacle to obtaining well-defined images, are connected via suitable cabling and hull gland penetrators to the control equipment inside the submarine's hull. The control equipment comprises a dual redundant camera power supply, data overlay generator, display monitor and high quality video monitor. The

Model OE 0285 ice cameras in bronze housings

Model OE 0285 combined display monitor, video monitor, power supply and control unit

cameras are housed in nickel-aluminium bronze castings and are positioned on the submarine's fin, allowing both forward and overhead images to be obtained.

Operational Status

In production and operational aboard Royal Navy submarines.

Contractor

Osprey Electronics, Aberdeen.

MAGNETIC ANOMALY DETECTION SYSTEMS

CANADA

AN/ASQ-504(V) AIMS SYSTEM

Type
Magnetic anomaly detection equipment.

Description
The AN/ASQ-504(V) Advanced Integrated MAD System (AIMS) is designed to provide optimum performance when used on helicopters, fixed-wing aircraft and lighter-than-air platforms. AIMS is a fully automatic magnetic anomaly detection (MAD) system which improves detection efficiency while significantly reducing operator workload. For helicopter installation, the detecting head is fixed to the aircraft body. This provides true 'on-top' contact when over a target by eliminating the delay inherent in a towed detecting system.

The AN/ASQ-504(V) system combines sensitivity and accuracy with ease of operation, eliminates aircraft-generated interference fields and delivers continuous, automated feature recognition. Contact alert, both visual and audible, is provided. Slant range via an LED display allows the operator to determine if he is within target acquisition range.

AIMS eliminates hazards associated with towed systems and permits high speed surveillance and manoeuvrability, thereby increasing patrol range, improving detectability and substantially reducing false alarms. When used with dipping sonar, transition from one system to the other can be performed quickly and effectively.

The AIMS package weighs 49 lb (21.7 kg) and consists of a 0.005 nano tesla optically pumped magnetometer, a vector magnetometer, an amplifier computer and a control indicator. The system operates from a single phase, 115 V, 400 Hz power source and requires less than 200 VA power. AIMS has the flexibility either to perform independently or to accept and execute commands from common control/display units via a 1553 databus.

Operational Status
Production quantities are currently being delivered for requirements on the RAN Sea Hawk, Grumman Turbo-Tracker, RAF Nimrod and RN Lynx and Sea King. Total contracts to date exceed 340 systems.

A CAE MAD detector head being inserted into the tail of a CP-140 Aurora maritime aircraft

Specifications
Detecting head: Caesium, optically pumped, oriented
Sensitivity: 0.01 γ (in-flight)
Features: Automatic target detection. Operator alert, visual and audible. Slant range estimate to 2800 ft (854 m)
Outputs
Visual: Contact alert and estimated slant range
Analogue: Adaptive MAD signal
Digital: Control and display interface to MIL-STD-1553B
Audio: ICS alert
Dimensions: 178 mm (diameter) × 813 mm (long) (detecting head); 193 × 257 × 559 mm (amplifier/computer); 152 × 152 × 152 mm (vector magnetometer); 190 × 145 × 145 mm (control indicator)
Weights: 6.1 kg (detecting head); 17.7 kg (amplifier/computer); 0.7 kg (vector magnetometer); 1.8 kg (control indicator)

Contractor
CAE Electronics Ltd, Saint Laurent, Montreal, Quebec.

FRANCE

MAD Mk III DETECTOR

Type
Magnetic anomaly detection system.

Description
The MAD Mk III is derived from several years' experience with the Mk I and II systems and is specifically designed for inboard use on fixed-wing aircraft and helicopters to detect the presence of a submersible by measuring the disturbance to the earth's magnetic field. It is an integral, digital, solid-state and highly reliable airborne system which becomes operational at switch-on without any warm up time. Target parameters are automatically delivered in real time on a CRT control and display unit.

The basic equipment can be broken down into three separate units: a detection unit, a computer and a control/display unit. A 1553B databus interface card is included in the system. A graphic recorder is available as an option.

The sensor operates on the nuclear magnetic resonance principle and uses the precession of protons in a liquid, the frequency of which, measured by the pick-up coils, is proportional to the magnetic field to be measured. Compensation is employed to eliminate from the received signal all disturbances created by the magnetic element in the aircraft and their movement in the earth's field. The compensator uses a 16-terms model, representing the magnetic components of the aircraft. The MAD Mk III includes rapid aircraft identification modes.

The target is detected and located automatically. This is a fundamental role, allowing the MAD system to give a high detection probability with a very low false alarm rate, in extracting the target signal from the background noise. The Sextant Avionique MAD system is based on a mathematical comparison between the current MAD signal and an analytical model of the target signals, as opposed to the more conventional method using threshold detection in several frequency bands. All computing tasks are performed by the ALPHA 732 proprietary 1 MOPS digital computer operating in Pascal.

Operational Status
Sextant Avionique has sold MAD systems in many parts of the world since 1972. This includes the French Navy Alouette III and more recently the Saudi Arabian Dauphin helicopters. The MAD Mk III is now operational with the French Navy Atlantique 2 maritime patrol aircraft.

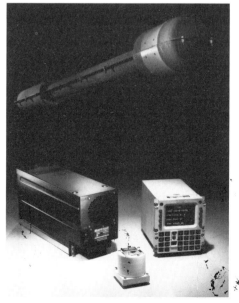

'Inboard' MAD Mk III

Specifications
Dimensions: 1250 mm (long) × 125 mm (diameter) (detection element); 1/2 ATR (long) (computer); 190 mm (long) × 146 mm (wide) × 162 mm (high)(control/display unit)

Weights: 5.5 kg (detection element); 10 kg (computer); 3.5 kg (control/display unit)

Contractor
Sextant Avionique, Meudon-la-Forêt.

MAGNETOMETER Mk 3

Type
Towed magnetometer for helicopters.

Description
This is a version of the Sextant Avionique MAD equipment intended for helicopters. It operates in exactly the same way as the MAD Mk III system described earlier, and in order to eliminate disturbances by the helicopter carrier, the detector probe is placed in a streamlined 'bird' which is towed at the end of a 70 m cable. The digital computer measures the signals from the sensor and transforms them into suitable formats for the graphic recorder at the operator's station. Maintainability is assisted by in-flight checking of the 'bird' (the towed magnetometer), computer, control unit and recorder. The ground test points are easily accessible and the subassemblies are plug-in units.

The 'bird' assembly includes the detection probe and an electronic unit, combining to produce a nuclear oscillator whose Larmor frequency is a function of the magnetic field exerted on the probe. The geometry of the probe is chosen so that its position

MAD sensor mounted on a P-3C aircraft (W Donko)

with respect to the magnetic field vector can be ignored.

Operational Status
In production and service.

Specifications
Background noise: Typical deviation 0.006 γ
Measurement range: 25 000-70 000 γ
Sensitivity for relative field output: 1, 2, 5 or 10 γ

for 100 mm stylus deviation
Paper speed: 6, 75 or 300 mm/min
Streamlined body
Dimensions: 1300 mm (long) × 160 mm (diameter)
Weight: 16 kg
Computer
Dimensions: 346 × 124 × 194 mm
Weight: 7.5 kg
Power: 100 W at 200 V AC, 400 Hz, 3-phase, plus 5 W at 28 V DC
Control unit
Dimensions: 165 × 150 × 57 mm
Weight: 1 kg
Recorder
Dimensions: 190 × 150 × 190 mm
Weight: 4 kg
Power: 50 W at 115 V AC, 400 Hz; plus 20 A at 28 V DC
Winch and cradle
Dimensions: 1330 × 350 × 755 mm
Weight: 44 kg
Power: 40 A at 27 V DC

Contractor
Sextant Avionique, Meudon-la-Forêt.

UNITED STATES OF AMERICA

AN/ASQ-81(V) ASW MAGNETOMETER

Type
Magnetic anomaly detection system.

Description
The AN/ASQ-81(V) magnetic anomaly detector (MAD) system was developed for the USN for use in the detection of submarines from an airborne platform. The system operates on the atomic properties of optically pumped metastable helium atoms to detect variations in total magnetic field intensity. Changes in the Larmor frequency of the sensing

elements are converted to an analogue voltage which is processed by bandpass filters before it is displayed to the operator.

Two configurations of the AN/ASQ-81(V) are available: one for installation within an airframe and one for towing behind an aircraft. The USN uses the AN/ASQ-81(V)-1 inboard installation with carrier-based S-3A/B aircraft as well as with the land-based P-3C ASW aircraft. For towing, the configuration is the AN/ASQ-81(V)-2, which is employed by USN SH-3H and SH-2D helicopters. The towed version is also fitted in the USN's latest helicopter, the SH-60B Seahawk.

The equipment has been upgraded with a form, fit and function replaceable digital implementation

version which incorporates microprocessor technology to achieve automatic aircraft compensation and perform all signal processing and built-in test. The upgraded version features improved performance and can eliminate up to nine Weapons Replaceable Assemblies (WRAs).

Operational Status
The AN/ASQ-81(V) is in production for the USN and a variety of international customers on helicopters such as the WG-13, HSS-2 and 500D.

Contractor
Texas Instruments Inc, Dallas, Texas.

AN/ASQ-208(V) MAD SYSTEM

Type
Digital magnetic anomaly detection system.

Description
The AN/ASQ-208(V) is a derivative of the AN/ASQ-81 magnetometer that operates on the same principle of the atomic properties of optically pumped metastable helium atoms to detect variations in total magnetic field intensity. The AN/ASQ-208(V) is a digital

implementation which incorporates microprocessor technology to achieve aircraft compensation, multiple channel filtered display and threshold processing. Two system configurations are available; one for inboard installations and one for towed installations. The system has been configured for ease of installation on either existing or new aircraft and uses existing AN/ASQ-81(V) wiring where applicable.

This system is form, fit and function replaceable with the AN/ASQ-81(V) and eliminates up to nine WRAs. Improved signal processing yields a 25 per

cent improvement in open water detection range and a 50 per cent improvement in shallow water detection range compared to the AN/ASQ-81(V).

Operational Status
In production.

Contractor
Texas Instruments Inc, Dallas, Texas.

Underwater Weapons

Torpedoes

Guided Weapons

Rockets

Mines

Depth Charges

UNDERWATER WEAPONS

This section contains information on those weapons concerned primarily with underwater targets (submarines) or which operate in the underwater environment for the major portion of their functional life (such as torpedoes).

Related entries will also be found under the Anti-submarine Warfare section.

Torpedoes

The torpedo tube has now become a general delivery system for missiles and mines as well as the torpedo, and most boats today ship a mix of missiles and torpedoes. As a general rule two or three mines can be substituted for each torpedo and submarine mines are available from numerous international sources. A feature now available is an external add-on section for mine stowage so that there is no reduction in the number of missiles/torpedoes carried.

Owing to the increased space demanded by bow sonars, there is a tendency to re-site torpedo tubes from their traditional bow position to a position slightly farther aft, and angled out from the centre line.

The type of torpedo carried depends, to some extent, on the boat's mission. Torpedoes are optimised for anti-submarine or anti-surface vessel missions. However, it is both a restriction on mission capability and a waste of valuable space for a submarine to carry two types of torpedoes. Modern torpedoes are thus being optimised for use against both surface and submerged targets.

Torpedo design has undergone very considerable changes since the Second World War. Both range and speed have been improved and wire guidance has given the weapon a very high hit probability. With a wire-guided torpedo the salvo and high expenditure of weapons to secure a hit against a single target are no longer necessary – a single wire-guided torpedo, properly handled, should be sufficient to ensure a hit. Tactics involving the use of two wire-guided weapons aimed at the same target from different directions give almost complete assurance of a hit – although it should always be remembered that in warfare nothing is certain!

For long-range stand-off anti-ship missions against prime targets, enabling the submarine to keep well out of the way of hostile ASW forces, the submarine-launched anti-ship missile has been developed.

Weapons stowage is an area of considerable importance and shock requirements, position of tubes, shape of compartment, size of weapon and handling arrangements all dictate the number and availability of weapon reloads that can be carried.

Two prime methods of discharge have been developed: positive discharge and swim-out. In the swim-out tube the weapon is discharged using its own propulsive system. In the positive discharge system energy is imparted to the weapon by an external source to push it out of the tube either by pneumatic water pulse or mechanical means.

While the swim-out tube requires less space it does require large volumes of water to be transferred during weapon loading, which requires large capacity tanks. The swim-out system is of low weight and does not require great energy. It does suffer from a disadvantage that the torpedo propulsion makes a small noise, but with modern torpedoes this is fairly insignificant.

Four types of positive discharge are available. The compressed air type can be independent of depth, but at greater depths requires greater energy to accomplish discharge. The piston pump, turbine driven rotary pump and telescopic ram methods are all independent of depth.

The turbine rotary driven pump system is perhaps the most effective for it requires less space than a piston pump, weighs less and the discharge cycle is quicker. The system can also be adapted to suit various types of weapon (for example, mines, torpedoes, missiles).

One of the most important aspects of weapon discharge is the safety interlocks which must be incorporated to ensure that tubes are closed and drained of water before loading occurs.

Drones and Missiles

Drone and missile torpedo launchers will become increasingly important in the years to come as surface platforms seek to increase the range at which they engage underwater targets. Coupled with the ability to detect targets at extreme ranges using towed array sonars, the one method of rapidly deploying an anti-submarine torpedo in the vicinity of the target is to launch it from a supersonic missile vehicle or drone, which itself can be launched from a surface platform and remain undetected by the intended target until the moment the torpedo is dropped into the water from the missile.

First generation systems such as Ikara and ASROC have been in service for some time, but only a few navies have so far embarked on the acquisition of such systems, notably the American, Australian, Brazilian, British, French and New Zealand Navies. The American ASROC system has a dual role – as a carrier of either an A/S torpedo or a depth charge. The submarine-launched missile SUBROC is not at present a torpedo carrier. Implementation of the US Sea Lance (combined anti-submarine warfare stand-off weapon programme) should remedy this deficiency as the Mk 50 lightweight homing torpedo is under consideration as one of the probable payloads planned for this new weapon.

Details of USS weapons, because of the usual lack of direct evidence, should be regarded as provisional, although the information given is considered to be accurate within these constraints.

Depth Charge Mortars

Towards the end of the Second World War a more streamlined type of depth charge with a higher rate of sinking was introduced by the RN and subsequently by other navies. The intention was to project charges ahead of an attacking ship so that they could be fired more accurately while still in sonar contact, which was otherwise lost when the attacking ship ran over the submarine to deliver conventional depth charges. Such mortars have been developed in several countries, Squid and Limbo in the UK being two examples and the Italian Menon being a third. The Menon single-barrel launcher is notable for the length of its tube and is said to have a range of about 1000 m and a rate of fire of 15 DCs/min.

Another weapon in this category is the French four-barrelled mortar. This is mounted in a turret and is automatically loaded. It fires a heavier projectile and has a longer range (about 2750 m) than any of the others. All these mortars fire 12 in (305 mm) depth charges.

It is believed that the USS Navy has not adopted the streamlined form of depth charge. They do, however, use depth charge mortars, but it is thought possible that they use compressed air to propel the charge whereas other countries use an explosive cartridge.

Medium-Range Rocket Launchers

One of the most widely adopted developments of the years since the Second World War has been the medium-range anti-submarine rocket launcher. Brief details of the French system, and Swedish system on which the French system is based, are given. Although it differs in some important respects from these systems, the Norwegian Terne system is also in this general weapon category. The USN has also developed systems of this kind, the current one being known as Weapon Alfa, successor to Weapon Able. Many such developments have also taken place in recent years in the USS.

Other entries give details of the widely used Swedish Ordnance four-tube launcher and the more recent two-tube launcher.

Underwater Guided Weapons

The latest weapon to be developed for subsurface deployment is the anti-ship guided missile. In the Western world such weapons tend to be variants of existing surface-launched anti-ship missiles such as Exocet and Harpoon.

Mines and Depth Charges

One of the most cost-effective forms of naval warfare is the mine. Mines are small, easily concealed, cheap to acquire, require virtually no maintenance, have a long shelf life, are easy to store in considerable numbers, and can be laid easily and simply from almost any type of platform, which need not necessarily be a military platform. They can be used both strategically and tactically to deny waters to hostile forces and to defend high value targets such as ports, anchorages and offshore structures from amphibious or seaborne attack, and can very quickly wipe out or very seriously impair the effectiveness of surface forces. To counter and neutralise the mine requires an effort out of all proportion to its size. In short, the mine is probably one of the most deadly weapons that any navy can deploy in its armoury.

Moored Mines

The moored mine has changed little from its forbears of the First and Second World Wars. Although a relatively unsophisticated weapon, it remains, nevertheless, an extremely effective one. Being simple to manufacture and relatively easy to lay (all that is necessary is a set of rails and a reasonably accurate navigation system which can be quickly installed on a wide variety of available ships), an extensive minefield will require a large force of minesweepers to sweep the field, an exercise which may take considerable time. The moored mine is therefore an ideal weapon for defensive purposes. A moored minefield can be used to protect extensive areas of vulnerable coastline and important port areas and anchorages. It can be used to create specified navigation channels leading to important areas such as ports; channels which can be subsequently rapidly closed using ground mines in face of impending attack.

Because of the nature of the weapon the modern moored mine can be laid in much deeper waters than the ground mine (that is, beyond the limits of the continental shelf). The moored mine is therefore eminently suitable for use in a barrier system, denying passage in open waters. Although post Second World War developments concentrated on the ground or influence mine, the moored mine has again come into prominence in new forms for use in submarine barriers in deep waters and to control choke points.

While many moored mines are of the older horned variety which are detonated by physical contact, the new moored mines for deep water barriers are detonated by magnetic or acoustic influence. These new types of moored mine can be countered using deep sweeps towed between two minesweepers. Naturally the mines incorporate anti-sweep devices such as time delays and snag lines.

The normal contact mine is usually triggered in one of three ways:
Contact – hydrostatic
Contact – mechanical
Contact – chemical.

The moored mine comprises the mine itself coupled to an anchor box for laying. Mine and anchor are laid together by the minelaying ship. On reaching the seabed the mine is released from the anchor mechanism and rises on a cable attached to the anchor to a preselected height, determined in advance from operational requirements. The body of the mine is made of steel or GRP and contains the explosive charge, fuzes, detonators and other elements (such as influence devices, batteries and security devices) required for the correct functioning of the weapon.

Ground Mines

Ground mines are detonated using one of three influences created by a target: acoustic, magnetic or pressure signature, or a combination of these influences. Influence mines are technologically the most sophisticated of all types of mine, and their reactivity to the specific type of influence on which they operate is infinitely variable. So sophisticated is the technology associated with these types of mine that they can literally be programmed to react not only to a specific class of ship, but, where sufficient data are available, even to a named ship within that specific class.

Unlike the moored mine, the ground mine houses all its operating equipment within the steel or GRP casing which is usually cylindrical in shape. A cylindrical shaped mine has obvious advantages in that it is compatible for laying from a wide variety of platforms ranging from submarine torpedo tubes to surface vessels of all types and sizes to aircraft. However, cylindrical shaped objects reflect fairly definite sonar patterns or shadows which can be identified by a minehunter. To compound the difficulties of the minehunter, therefore, certain types of ground mine have been developed of non-standard shape which, when laid, cast very little sonar shadow or reflect an indefinite pattern. Sonar patterns are further confused by the use of anechoic coatings.

For use on soft or sandy beds, some modern ground mines have the capability to bury themselves in the seabed, thus becoming virtually undetectable to modern minehunting methods. Such mines, however, are small in size, with a small explosive charge and are better suited to very shallow water operation such as in estuarial areas or in areas where hostile submarine activity is anticipated fairly close to the coast or in specific navigable channels.

The two principal types of influence are the acoustic and magnetic. The acoustic influence can be programmed to react to specific types of ship's signature such as propeller cavitation, engine noise, hull cavitation or shaft revolutions.

Magnetic influence mines react to changes in the ambient magnetic field and can be fuzed to react to magnetic polarity in either a vertical plane or fore and aft or athwartships in a horizontal plane on a target, and to specific fields and strengths, reacting to specific parts of a ship such as machinery area or stern area.

In addition to the multitude of fuzing refinements available, triggering mechanisms can also be controlled with activate/deactivate devices, and various types of delay, counters and time responses.

The other principal type of influence used is the pressure influence. As a ship moves through the water an area of low pressure is created beneath the hull which varies according to the speed and draught of the vessel. This pressure change can be used in a number of ways to operate the pressure sensing device in the mine, the decrease in pressure opening and completing an electrical circuit to detonate the mine. Because of the method of operation, mines incorporating this type of influence certainly cannot be swept using normal methods (as can some magnetic and acoustic influence mines) and have to be 'hunted' using sonar and neutralised using either clearance divers or remote-control submersibles.

Tethered Mines

Tethered mines are specific types of mine which actively react in a particular way to the approach of the target, which is principally the submarine, as distinct from the normal type of moored mine which remains completely passive until struck by the victim.

The most common form of tethered mine is known as the 'rising mine'. Both the US and the USS have developed such weapons. Most tethered systems rely on passive detection of an approaching target to activate a powered, mobile homing system carrying the explosive charge.

Tethered mines are very much a barrier weapon to prevent passage of hostile shipping and to control major choke points.

The other major type of tethered system comprises a mobile homing system such as an acoustic torpedo tethered to the seabed in depths below 3000 m. The system is linked to a passive acoustic sensing system laid on the seabed (such as the US SOSUS system) which detects, analyses and classifies underwater sounds, and in response to pre-programmed target signatures activates the release of the homing torpedo which then reacts in its normal way to the approach of the target.

Controlled Mines

These were, in fact, the very earliest type of mine ever devised. They simply consisted of a series of large explosive charges laid on the seabed and connected by electric cables to a position on the shore overlooking the minefield. As the target was observed passing over the position of the mines they were detonated by an observer in the shore station.

Such systems using modern explosives and technology are still valid today and indeed some navies do operate such controlled minefields.

However, more sophisticated means are being developed to provide the remote activation of minefields. Experiments in this area have been geared towards the control of deep water minefields controlling principally acoustic influence mines. This is necessary due to the fact that while a mine remains in water its casing acts as home to a variety of marine life which affects the sensitivity of the acoustic transducers. Consequently the longer a mine remains in water, the less sensitive are the transducers.

Furthermore there is a finite limit to the life of the batteries which power the mine's various electronic systems. If the mine can be remotely activated at intervals well beyond the capacity of any in-built timing device in the mine, then it could remain dormant in the water for considerable lengths of time, its onboard sensors all being activated by a simple remote command. Naturally any such remote activating system must be proved to be almost infallible before any system relying on this method is deployed. This is essential if any such system is to be deployed prior to active hostilities.

Remote-control of such minefields will, in all probability, rely on coded VLF transmissions which are the only satisfactory means of long-range underwater communications at depth. Such a communications system, however, requires very high power and would most probably be shore-based.

Depth Charges

Relatively simple cylindrical depth charges that can be rolled or catapulted into the sea or dropped from aircraft are the longest-established anti-submarine weapons; they were first used by the RN in the First World War. They are generally depth-fuzed and have a low sinking rate to give the launching vessel time to get clear. They are still used extensively by many navies: a typical weight is 150 kg and a launcher can project the charge up to about 150 m.

Nuclear Depth Charges

Also introduced by the USN is the nuclear depth charge. To take this clear of the launch vessel a rocket is required, and these depth charges have so far been associated only with ASROC and SUBROC. In the UK, it is understood that Royal Navy Sea King helicopters can be equipped with 'Bomb 600 lb special ordnance'. Based on the US B-57 depth charge, this is intended for the attack of deep-diving submarines which are running too deep or too fast for homing torpedoes. These nuclear devices are of UK manufacture.

TORPEDOES

FRANCE

L3 TORPEDO

Type

Submarine/surface ship-launched acoustic torpedo.

Description

This is a conventionally shaped, ship-launched or submarine-launched, anti-submarine torpedo with a strong body in light alloy and a laminated nose cone with the following five compartments: AS-3 active acoustic self-guidance and electromagnetic firing device; explosive charge and impact fuze; secondary battery; air tank and automatic pilot; electric motor for propulsion. Location in bearing and elevation is used to guide the torpedo along a pursuit curve. As the torpedo nears the target, the pulse rate increases to improve the accuracy of pursuit and of the acoustic detonator. If after a computed time the target has not been detected at the previous position the torpedo commences a circular search (in shallow water) or a helical search (deep water). The torpedo can attack vessels proceeding at speeds up to 20 kts and down to a maximum of 300 m depths.

L3 torpedoes are currently being upgraded to E15 Mod 2 standard.

Operational Status

Quantity production. The 550 mm diameter version of the weapon is in service with the French forces. A 21 in (533 mm) version has also been designed and is

L3 acoustic torpedo launch

available for manufacture but not in production. Its performance is the same as that of the 550 mm version, but its length is 4318 mm (170 in), its weight 900 kg, and its diameter 21 in. Approximately 600 have been manufactured.

Specifications

Length: 4300 mm
Diameter: 550 mm
Weight: 910 kg
Warhead: 200 kg Tolite A1
Propulsion: Electric

Speed: 25 kts
Range: 5500 m
Guidance: Acoustic, active, range approx 600 m with favourable inclination of the target submarine. Type AS3T
Fuze: Impact/acoustic

Contractors

Direction des Constructions Navales (DCN), Paris (programme direction).

Etablissement des Constructions et Armes Navales (ECAN), Saint Tropez (manufacturer).

L4 TORPEDO

Type
Air-launched acoustic torpedo.

Development
Developed by Direction des Constructions Navales (DCN) for anti-submarine warfare, the L4 torpedo was designed to be launched from aircraft or the ASW missile Malafon and to attack submarines travelling underwater at speeds below 20 kts.

Description
The L4 is a conventionally shaped torpedo having a body made of removable sections of moulded magnesium alloy and comprising the following main compartments: head section containing the guidance system, acoustic firing circuits and the warhead (contained in removable canister) with its inertial percussive firing system; centre section with battery and priming elements; tail section with the air reservoir, propulsion unit (electric motor driving two contra-rotating propellers through a reduction gear) and the

L4 air-launched torpedo

steering mechanism. There are also the launching devices to ensure that the torpedo makes a smooth entry into the water, comprising a parachute stabiliser and release mechanism aft and an ejection cap forward.

In operation, the torpedo describes a circular path until its detection system locates the target. It then changes course and homes on the target, when the warhead detonates either by the acoustic proximity mechanism or by impact fuze.

The torpedo is capable of operating in shallow water and attacking submarines from periscope depth down to 300 m. Delivery platforms include the Super

Frelon helicopter, the ASW aircraft Atlantique as well as the Malafon ship-launched missile.

With minor modifications the weapon can be used against surface vessels.

Operational Status
Currently in service with the French Navy.

Specifications
Length: 3130 mm including parachute stabiliser
Diameter: 533 mm
Weight: 525 kg
Warhead: 100 kg HE (HBX)
Propulsion: Electric
Speed: 30 kts
Range: 6000 m (approx)
Guidance: Active acoustic homing
Fuze: Impact/acoustic

Contractors
Direction des Constructions Navales (DCN), Paris (programme direction).

Etablissement des Constructions et Armes Navales (ECAN), Saint Tropez (manufacturer).

L5 TORPEDO

Type
Surface/submerged launch dual role torpedo.

Description
Most recent of the 'L' series of torpedoes, the L5 is powered by silver-zinc batteries which are activated at launch and provide the power for the motor, which drives twin contra-rotating propellers.

There are three models: L5 Mod 1; L5 Mod 3; L5 Mod 4.

The Mod 1 and Mod 3 have been in operational service with the French Navy for a number of years. The Mod 4, derived from the Mod 1, is carried by surface ships of the French Navy and several other navies. All models are designed to attack both underwater or surface vessels. The Mod 3 is submarine-launched and the Mod 1 and Mod 4 are launched by

surface ships and submarines. All models are fitted with a passive/active homing head which has various operating modes, including direct attack or programmed search. Maximum operating depth is 555 m. The compact size of this weapon makes it ideal for operation from small displacement submarines.

Operational Status
Operational with the French Navy and other navies. The Spanish Ministry of Defence purchased nine L5 Mod 4 torpedoes during 1987.

Specifications (L5 Mod 4)
Length: 4400 mm
Diameter: 533.4 mm (can be increased to 550 mm if required)
Weight: 935 kg
Warhead: 150 kg HE
Propulsion: Electric
Speed: 35 kts

L5 torpedo about to enter the water (DCN photo)

Range: 9.5 km
Guidance: Passive/active homing
Fuze: Impact/magnetic

Contractors
Direction des Constructions Navales (DCN), Paris (programme direction).

Etablissement des Constructions et Armes Navales (ECAN), Saint Tropez (manufacturer).

E14 TORPEDO

Type
Submarine/surface ship-launched acoustic anti-ship torpedo.

Description
This is a conventionally shaped, submarine- or surface ship-launched, anti-ship (or anti-submarine against noisy targets close to the surface) torpedo with a strong body in light alloy, and a laminated nose cone with the following five compartments: acoustic/passive self-guidance and electromagnetic firing device; explosive charge and impact fuze; secondary battery; air tank and automatic pilot; and

electric motor for propulsion. As many parts as possible have been commonalised with those in the L3 torpedo. The torpedo can attack surface vessels proceeding at speeds up to 20 kts and submarines at shallow depth. The depth setting is variable between 6–18 m. The E14 has the same geometry and mechanical features as the L3. E14 torpedoes are currently being upgraded to E15 Mod 2 standard.

Operational Status
Quantity production. The equipment is in service with the French forces and supplied for export. About 100 are believed to have been manufactured.

Specifications
Length: 4279 mm

Diameter: 550 mm
Weight: 927 kg
Warhead: 200 kg Tolite
Propulsion: Electric
Speed: 25 kts
Range: 5500 m
Guidance: Acoustic/passive, average range 500 m
Fuze: Impact/magnetic

Contractors
Direction des Constructions Navales (DCN), Paris (programme direction).

Etablissement des Constructions et Armes Navales (ECAN), Saint Tropez (manufacturer).

E15 TORPEDO

Type
Submarine/surface ship-launched acoustic anti-ship torpedo.

Description
A conventionally shaped, submarine- or surface ship-launched, anti-ship (or anti-submarine against noisy targets close to the surface) torpedo with a strong body in light alloy and a laminated nose cone, the E15 is a lengthened version of the E14 model 1 torpedo. It has the following five compartments: acoustic/passive self-guidance and electromagnetic firing device; explosive charge and impact fuze; secondary battery; air tank and automatic pilot; electric motor for propulsion. As many parts as possible have been commonalised with those in the L3 torpedo. The

torpedo can attack surface vessels proceeding at speeds up to 20 kts and submarines at shallow depth. Depth setting is variable between 6–18 m. The self-guidance system is the same as the E14, but geometry, range and explosive charge are different.

The E15 is currently being upgraded to Mod 2 standard. This constitutes fitting a new housing head of the AH-8 type, and replacing the batteries with Ag/Zn primary batteries.

Operational Status
Quantity production. The equipment is in service with the French forces and supplied for export.

Specifications (Mod 2)
Length: 5900 mm
Diameter: 550 mm
Weight: 1387 kg

Warhead: 300 kg Tolite A1
Propulsion: Electric
Speed: 25 kts
Range: 12 000 m
Detection range: 2000 m
Search depth: 25 m
Attack depth: 6-18 m presettable
Max depth rating: 300 m
Guidance: Passive, medium range
Fuze: Impact/magnetic

Contractors
Direction des Constructions Navales (DCN), Paris (programme direction).

Etablissement des Constructions et Armes Navales (ECAN), Saint Tropez (manufacturer).

F-17 TORPEDO

Type

Submarine/surface ship-launched wire-guided torpedo.

Description

The F-17 torpedo family comprises the following models: F-17 S1, F-17P, F-17 S2 and F-17 Mod 2.

These torpedoes are operable from surface vessels or submarines and are able to destroy surface ships and submarines (except for the F-17 S1 which can only be used against surface ships).

Launching by submarine may be performed in one of the three modes:

(a) pneumatic rammer
(b) water ram
(c) swim-out.

The F-17 torpedo carries an extensive warhead charge which enables it to operate successfully against all naval vessels. The sonar homing system can operate either in passive or active mode (except F-17 S1) giving a high detection performance. An internal countermeasures complex enables the F-17 Mod 2 and F-17 S2 to strike their targets under the most adverse conditions. These performance features, associated with a long range, high speed, great operating depth, plus the ability to operate with considerable stealth give the F-17 torpedoes a formidable capability.

The F-17 Mod 1 and F-17P can be updated to the Mod 2 or S2 standard.

Moreover, the maintenance concept of these torpedoes has been optimised according to the reliability of the former F-17 Mod 1 which has been in service with the 'Marine Nationale' since the 1970s.

As a result the time interval between two consecutive maintenance operations has been increased to 18 months, and also simplified.

Physical characteristics of the torpedo and its tube container for 'Agosta' class submarines

	Diameter[1]	Length	Weight
F-17P	533 mm	5914 mm	1410 kg
F-17 S1	533 mm	5914 mm	1428 kg
F-17 S2	533 mm	6300 mm	1428 kg
F-17 Mod 2	533 mm	5406 mm	1402 kg

Common characteristics

Launcher : surface ship or submarine
Warhead : 250 kg of HBX3 (equivalent 400 kg TNT)
Guidance : wire guidance with self guidance in terminal phase

Notes:

1 can be increased to 550 mm if required
2 ss = surface ship
   sub = submarine

F-17 Mod 2 torpedo

Operational characteristics

	Target[2]	Speed	Range	Counter-countermeasures capabilities	Maximum navigation depth	Search depths[3]
F-17P	ss/sub	35 kts	18 500 m	low	>500 m	Za or 30 m / Za or 60 m
F-17 S1	ss	35 kts	18 500 m	low	>500 m	Za or 30 m / 100 m or 200 m
F-17 S2	ss/sub	35 kts	18 500 m	high	>600 m	Za or 30 m / 100 m or 200 m
F-17 Mod 2	ss/sub	40 kts	20 000 m	very high	>600 m	Za or 30 m / 100 m or 200 m

3 Operational depth against surface ship. Za = 6 m or 20 m

The F-17 Mod 2 is the latest version of the F-17 torpedo family which is in current operational service (F-17 Mod 1 and F-17P). The Mod 2 has been upgraded as a result of experience gained and in the light of recent technological advances in all technical spheres. The F-17 Mod 2 is a multi-mode wire-guided torpedo, capable of submarine or surface-ship launch, and is intended for attack against both underwater and surface vessels. The torpedo is electrically powered by silver-zinc batteries which are activated automatically at launch.

The F-17 Mod 2 carries an extensive warhead charge which enables it to operate successfully against all naval vessels. The multi-mode homing system can operate in either passive or active mode to give high detection performance. An internal counter-countermeasures complex makes the F-17 Mod 2 able to attain its target in the most adverse conditions. Maximum operating depth is 600 m. These performance features, associated with a long range, high speed and deep operating depth, plus the ability to operate with considerable stealth, places the F-17 Mod 2 in the premier rank of heavyweight torpedoes.

The F-17 Mod 1 and F-17P can be updated to the Mod 2 standard.

CNIM has proposed an update for the F-17 Mod 1 incorporating a new AH8 homing head and primary silver-zinc battery in place of the nickel cadmium battery. Performance objectives for the uprated torpedo include a detection range of 2000 m, a search depth of 25 m, a presettable attack depth variable between 6-18 m, a maximum operating depth of 300 m, a speed in excess of 31 kts, and a range of 12 000 m. The warhead would be 300 kg of aluminium Tolite.

Operational Status

F-17 Mod 1 and F-17P are operational with the 'Marine Nationale' and other navies. The F-17 Mod 2 is in final integration. The first batch entered operational service in 1989.

The F-17 S2 torpedo, whose operational characteristics are very similar to the F-17 Mod 2, is offered for export, with the same time delivery conditions as F-17 Mod 2.

Specifications (Mod 2)

Length: 5406 mm
Diameter: 533.4 mm (can be increased to 550 mm if required)
Weight: 1402 kg
Warhead: 250 kg HBX3
Propulsion: Electric
Speed: 40 kts
Range: 20 km
Guidance: Wire and automatic homing

Contractors

Direction des Constructions Navales (DCN), Paris.
   Etablissement des Constructions et Armes Navales (ECAN), Saint Tropez (manufacturer).

MURÈNE TORPEDO

Type

Lightweight anti-submarine torpedo.

Development

Development of the Murène, intended as a replacement for the L5 and Mk 46 torpedoes in French service, commenced in 1982, with an in-service date scheduled for the beginning of the 1990s. The French requirements for the new torpedo were that it should be capable of operating worldwide irrespective of temperature and salinity; should be NATO interoperable and easily adaptable to all types of foreign launcher; should be capable of operating in both deep and shallow water; and should be able to be deployed from all types of platform, including missiles. The torpedo has been developed by DCN and ECAN with the principal objective of destroying the latest generation of nuclear submarines, although it is equally effective against all other types of submarine. Development trials were completed in 1988 with the accompanying software completed by mid-1989.

Description

The Murène is electrically driven with a motor powered by SAFT silver-oxide/aluminium battery activated by seawater. Later versions of the weapon might use a lithium/thionyl/chloride battery with potassium hydroxide electrolyte. To overcome the slight delay experienced before the saltwater activates the batteries, two auxiliary batteries provide power for the electrolyte pump, the four cruciform rudders and the guidance unit in the interim period before full power from the saltwater batteries is available. The batteries drive a DC electric motor to power a pump jet with fixed stator and seven-blade propeller. It can operate at two speeds: 40 kts during the search and detection phase and higher than 55 kts for tracking and attack. Endurance is about six minutes at maximum speed, or about 12 minutes when the two-speed requirement is necessary. Its manoeuvrability allows it to operate in all attitudes including the vertical, and it can attack submarines at all depths from shallow water down to 1000 m. It has a maximum range of 14 km and can operate worldwide in all waters.

The torpedo has an extremely large data processing capacity giving it a very high degree of analysing power. It has an excellent signal processing capability with a rapid detection facility at considerable ranges. These features combine to offer high resistance to countermeasures. The SFIM-developed central command and control unit, known as Capitole, is located amidships and is responsible for guiding and controlling the torpedo. It incorporates four Motorola 68000 16-bit microprocessors and integrates the functions of attitude determination, automatic direc-tional control and stabilisation, optimum autonomous guidance, target tracking and automatic programming of the operating modes. The Thomson Sintra acoustic homing head incorporates three separate sonars which use four transducer arrays, two lateral and one downward pointing (used as a depth finder in shallow water), each of four transducers, and one forward array with 30 transducers. The two lateral and forward looking sonars used for target detection can operate in a combination of modulated frequency and pure frequencies, with or without amplitude modulation. It can also operate in multi-configuration modes with continual commutation between numerous possible arrangements of preformed beams in transmission and reception. These confer on the weapon high ECCM and anti-reverberation capabilities. Central to the homing head's operational capability is the 50 million operations per second Mangouste computer used for signal processing and which itself incorporates another three Motorola 68000 16-bit microprocessors. All these systems combine to give Murène the capability to track up to 10 targets (including real targets, echoes, decoys etc and distinguish between them) simultaneously with complete autonomy.

The torpedo has a shaped charge warhead which is guided to the most vulnerable section of the target, attacking it at an optimum impact angle. The warhead is designed to penetrate both the outer hull and inner pressure hull, allowing water pressure to complete the

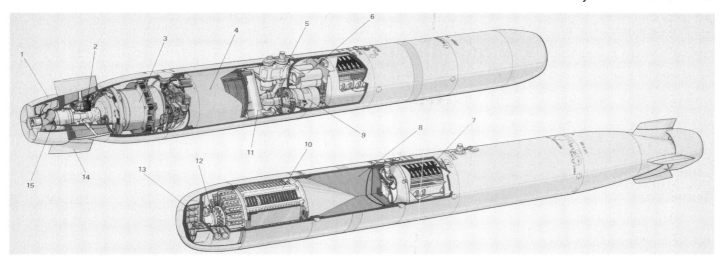

(1) *Stator,* **(2)** *Rudder actuator,* **(3)** *Motor,* **(4)** *Electrochemical unit,* **(5)** *Priming valve,* **(6)** *Auxiliary batteries,* **(7)** *Command central unit,* **(8)** *Warhead,* **(9)** *Motor-pump set,* **(10)** *Acoustic head,* **(11)** *Degausser,* **(12)** *Side antenna,* **(13)** *Plane antenna,* **(14)** *Control surfaces,* **(15)** *Rotor*

submarine destruction. The torpedo can be launched from aircraft, helicopters, missiles or ships.

Operational Status

Sea trials with prototype torpedoes are almost complete. Estimated in-service date with mass production weapons was scheduled for 1992. A report from the US states that Chile has ordered the Murène for its naval helicopters. The French Navy indicated that it would order in the region of 25 weapons per year to start with. A new weapon combining features of Murène and the Italian A290 is now to be developed.

Specifications

Length: 2.96 m
Diameter: 324 mm
Weight: 295 kg without 15 kg parachute pack (approx)
Warhead: 59 kg shaped charge HE
Propulsion: Electric using AgO-Al battery with KOH activated by seawater
Speed: >50 kts
Range: 14 km

Contractor

GIE, La Londe les Maures.

Murène lightweight torpedo

GERMANY

SUT TORPEDO

Type

Anti-surface vessel and anti-submarine torpedo.

Description

The SUT (Surface and Underwater Target) torpedo is the latest and most versatile member of the Seal, Seeschlange and SST 4 family of torpedoes. It is a dual purpose wire-guided torpedo for engaging both surface and submarine targets. The SUT can be launched from submarines and surface vessels, from fixed locations or mobile shore stations. Its electrical propulsion permits variable speed in accordance with tactical requirements, silent running and wakelessness. The wire guidance gives immunity to interference with a two-way data link between vessel and torpedo. The acoustic homing head has long acquisition ranges and a wide search sector for active and passive operation. After termination of guidance wire, SUT continues operation as a highly intelligent homing torpedo, with internal guidance programmes for target search, target loss and so on. The large payload with combined fuze systems ensures the optimum effect of explosive power. The SUT operates

Surface and Underwater Target (SUT) torpedo

at great depths as well as in very shallow waters. Consort operation permits exploitation of the full over-the-horizon range of the SUT. The body is made of reinforced glass fibre or aluminium. SUT at present exists in three different versions with slight differences in the internal guidance programs and the extent of data transferred via the guidance wire. The latest version is SUT Mod 2 with its special feature of additional data from the homing head being signalled back to the vessel including an 'Audio Channel'.

Operational Status

In service and in production for several navies (NATO, South America, Asia). In the early 1980s Indonesia signed a contract for indigenous manufacture of the SUT within a long-term programme still running.

Specifications

Length: 6150 mm (6620 mm with guidance wire casket)

Diameter: 533 mm
Weight: 1420 kg (without casket)
Warhead: 260 kg
Fuze: Magnetic proximity and impact

Contractor

STN Systemtechnik Nord GmbH, Bremen/Hamburg.

SEAL (DM2/DM2A1) AND SEESCHLANGE (DM1) TORPEDOES

Type

Anti-surface vessel and anti-submarine torpedo respectively.

Description

These two wire-guided, heavyweight, 21 in (533 mm) torpedoes were purpose-developed for use in the German Navy: Seal for use against surface ships, and the smaller Seeschlange against submarines. There is a high degree of equipment commonality between the two weapons. Major differences are that the anti-submarine model has half the propulsion battery capacity of the Seal, but is fitted with three-dimensional sonar. The following main features apply to both types.

All essential data are sent to the torpedo throughout its run via a dual core guidance wire. Similarly, actual torpedo running data are simultaneously transmitted to the ship. An active/passive homing

head is fitted with a steerable transducer array. Attack options following acquisition are either by manual or computer control from the launch ship, or by self-homing by the torpedo.

Provision is made for a programmed run after guidance wire pay out or loss of signals from the onboard fire control system. Different programmes adapted to various tactical situations are available and can be selected via the guidance wire.

Launch arrangements include compressed-air firing from surface ships and swim-out from submarines. There are no limitations on ship movements during the launch and guidance phase. The electric propulsion system employed provides long running distances, permitting launch from beyond target defence area. Torpedo speed is selectable.

Other features are: combined impact and proximity fuze; full performance in shallow or deep water; three-dimensional internal stabilisation; identification of different targets and high hit probability; automatic system check before firing.

Operational Status

Both torpedoes are in service with the German Navy

aboard '206' class submarines. Seal is also deployed on surface ships, in particular on the Type 142 and 143 fast patrol boats. Seal has been modified to form the special surface target torpedo SST 4 and the dual purpose surface and underwater target torpedo SUT.

Specifications

Seal
Length: 608 cm (655 cm with guidance wire casket)
Diameter: 533 mm
Weight: 1370 kg
Warhead: 260 kg
Fuze: Magnetic proximity or impact
Seeschlange
Length: 415 cm (462 cm with guidance wire casket)
Diameter: 533 mm
Warhead: 100 kg
Fuze: Magnetic proximity or impact

Contractor

STN Systemtechnik Nord GmbH, Bremen/Hamburg.

SEEHECHT (DM2A3) TORPEDO

Type

Anti-surface and anti-submarine torpedo.

Development

This wire-guided, heavyweight, 21 in (533 mm) acoustic homing torpedo has been developed to replace the Seal and Seeschlange torpedoes in service with the German Navy and NATO countries.

In a further development it is planned to modify the DM2A3 into the DM2A4 configuration incorporating a new propulsion system leading to increased speed and range.

Based on the DM2A3 a new generation of torpedoes for a specific export market designated Seahake is under development in accordance with the regulations of the German MoD. There will be some differences in hard- and software in order to meet the export regulations and the special requirements for worldwide application. However, the main characteristics of the DM2A3 will be incorporated in the new design.

Description

Seehecht is a dual-purpose torpedo which can engage surface and submarine targets and can be launched from both surface vessels and submarines. The main features of the weapon are: extremely long guidance distance; silent running resulting from the greatly improved propeller design and other special measures; impact and improved magnetic proximity fuze with high ECCM capability; improved communication system for two-way transmission of extended volume of data, including the complete acoustic panorama and wideband noise samples; stabilisation system with highly accurate sensors and regulator systems optimised for multiple operational requirements; multi-stage guidance concept including highly intelligent internal guidance programs; homing head using PFB technology and covering a wide panorama; operation in active and passive modes in several frequency bands; ability to handle multiple target situations with numerous features capable of identifying or suppressing jammers and decoys; the latest technology using microprocessors connected

Seehecht torpedo on test-stand

via a MIL-BUS system and fibre optic conductors; and extensive BITE (Built-In Test Equipment).

Operational Status

DM2A3 – First-series production is under way for the

German Navy and another NATO navy.
DM2A4 – Concept phase.

Contractor

STN Systemtechnik Nord GmbH, Bremen/Hamburg.

SST 4 TORPEDO

Type

Special anti-surface vessel torpedo.

Description

The SST 4 (Special Surface Target) torpedo is comparable in its dimensions, construction and capabilities to the Seal weapon, except for those features which can only be applied within the operational area of the German Navy. In this respect the SST 4 has been adapted to a standard international version. It has an operating depth down to 100 m.

The basic design has been continually improved over the years, especially in the homing head functions and in the stabilisation and related control system.

Further improvements covering additional return signals (actual course, speed and depth) and magnetic proximity fuze are incorporated in the SST 4 Mod 1 configuration. A modification kit is available for conversion of SST 4 weapons to SST 4 Mod 1 configuration.

Operational Status

In service on '209' class submarines, fast patrol boats of the 'Combattante I', 'Combattante II' and 'Jaguar II' classes; introduced into various NATO and South American navies.

Specifications

Length: 608 cm (655 cm with guidance wire casket)
Diameter: 533 mm
Warhead: 260 kg
Propulsion: Electric
Guidance: Wire-guided, dual core with two dispensing systems; active/passive homing sonar
Fuze: Impact (SST 4 Mod 1 proximity also)

Contractor

STN Systemtechnik Nord GmbH, Bremen/Hamburg.

ITALY

A 184 TORPEDO

Type

Submarine- or surface-launched torpedo.

Description

The A 184 is a compact, dual-purpose, wire-guided, electrically propelled torpedo equipped with an AG 67 panoramic homing head, controlling both course and depth. It is suitable for use against both submarines and surface vessels. It is a dual speed weapon, carrying a high explosive charge and capable of operating to considerable depths. Dedicated fire control systems exist in different versions for submarines and surface vessels. On board a number of non-Italian built submarines, the A 184 is controlled by existing fire control systems. The launchers are Whitehead B 512 and B 516 for submarines and surface ships, respectively, although the A 184 may also be used with a large variety of 533 mm discharge systems using both swim-out and pulse techniques.

Commands carried by the guidance wires include: course, depth, acoustic mode (active, passive and combined), enabling range, stratum allowed, speed,

Type A 184 dual-purpose torpedo

impact and influence fuze setting, torpedo stop. Replies from the weapon include: course, distance, depth, acoustic mode, speed, other data on interrogation. The fire control system displays the tactical scenario, acquiring data from onboard sensors and allows the underwater weapon selection, check-presetting, start and guidance against the designated targets. The fire control system is modular, each module being capable of being used independently in case of failure of the others.

Operational Status

In production. The A 184 is in service with the Italian Navy and is due to enter service with several foreign countries.

Whitehead was one of four companies bidding for a US contract for 2000 torpedoes. Only the Whitehead contender reached the initial evaluation sea trials phase, which has been successfully completed. However, the US Navy has not requested funding for the Anti-Surface Warfare (ASUW) torpedo in FY89 and FY90.

Specifications

Length: 6 m
Diameter: 533 mm
Weight: 1300 kg
Warhead: HBX HE
Propulsion: Electric, silver-zinc battery
Range: 20 km (under wire guidance)

Contractor

Whitehead, Livorno.

A 244/S MOD 1 TORPEDO

Type

Lightweight torpedo.

Development

Developed for the Italian Navy by Whitehead Moto Fides, the A 244 is a lightweight torpedo designed to replace the older US Mk 44 type. The A 224 Mod 1 was designed to provide surface units (including small ships), fixed-wing aircraft and helicopters with an ASW weapon system which could meet operational requirements facing submarine technical and tactical evaluation.

Description

The A 244/S Mod 1 is a lightweight torpedo based on an advanced computerised homing system with re-programmable software. This means that its search and/or homing behaviour can be adjusted without any hardware changes. The acoustic homing head

is capable of active, passive or mixed modes for closing on to its target. It can discriminate between decoys and real targets in the presence of heavy reverberations by specially emitted pulses and signal processing. The head has a large search volume covered by multiple preformed beams following a number of self-adaptive search patterns. The computerised homing system also provides for presettable combinations of signal processing, spatial filtering and tactical torpedo manoeuvring to match the torpedo's performance in the ever changing operational situation as dictated by the threat and ASW tactics.

Operational Status

In production. In service with several foreign navies. The Swedish FMV (Defence Material Administration) has placed an order worth SEK70 million for Italian torpedoes understood to be the A 244/S Mod 1. The weapons will equip the 'Stockholm' and 'Goteborg' class corvettes pending availability of the Swedish Type 45.

Type A 244 lightweight torpedo on helicopter

Specifications

Length: 2700 mm
Diameter: 324 mm
Weight: 221 kg (approx) (warshot version)
Warhead: Shaped charge HBX-3 HE
Propulsion: Electric
Guidance: Active/passive sonar, self-adaptive programmed patterns

Contractor

Whitehead, Livorno (prime contractor).

A290 TORPEDO

A290 lightweight, self-homing torpedo

Type
Lightweight anti-submarine homing torpedo.

Development
The requirements for the characteristics of new torpedoes designed to combat future deep diving, high speed submarines are exceedingly demanding compared to current weapons operational and under development. The size and weight of the A290 had to be such that it would be capable of being deployed from smaller helicopters such as the Agusta-Bell 212, Westland Lynx, and Aerospatiale Dauphin. Operating depth had to be increased to over 1000 m, which in turn demanded improved materials and more sophisticated structural design in order not to increase the total weight of the weapon beyond the established limits.

Warhead capability had to be increased in order to ensure a 90 per cent hit probability against the largest and most sophisticated targets with double hull and interhull spacing filled with water.

The homing system has to be a more sophisticated system than that already developed for other modern weapons such as the A244/S Mod 1. The requirement for the homing system includes: higher detection range in passive mode; higher detection range in active mode; increased detection capability against anechoically and stealth coated submarines; full protection against anti-torpedo devices using built-in ECCM measures; and full operating capability in both deep and extremely shallow water. The torpedo features very low self-radiated noise making it difficult to detect except at short range.

The weapon's speed and endurance are increased in order to engage and hit the fastest of targets. A speed almost 5 kts above the maximum target speed was required together with an endurance enabling medium-range attacks to be carried out. The weapon is powered by a high efficiency, low noise electric motor/pump jet propulsion system.

Operational Status
Under development and was due to become operational at the end of 1992 beginning of 1993.

Development work is now being co-ordinated with development of the French Murène (qv) for a new lightweight torpedo for both countries.

Specifications
Length: 2750 mm (max)
Weight: 300 kg (max)

Contractor
Whitehead, Livorno.

B 515

B 515 torpedo launcher

Type
Surface vessel torpedo launching system.

Description
The B 515 torpedo launching system is designed for the operation of lightweight anti-submarine torpedoes such as the A 244/S, A 244/S Mod 1, Mk 44, Mk 46 and A290.

The torpedo launcher is designed so as to be installed on the ship's deck. Normally the launchers are mounted in pairs, one at the starboard and the other one at port side of the ship, both being arranged with the muzzle facing forward.

The barrels are mounted on a basement which can be manually trained by means of a suitable retractable handle. The torpedo is fired after the tube has been trained into the allowed sector. The barrels are provided with an electrical de-icing system, meant to assure suitable temperature for the torpedo loaded into the barrel. Similarly heaters are installed on the rotating base. Moreover, the launcher is equipped with an alarm circuit which signals overheating in the tubes.

Each barrel is also provided with manual safety cock of the pneumatic circuit and a series of electric interlocks to prevent anomalous and/or unwanted firing operations.

Firing is realised by means of compressed air stored in air bottles secured to the barrel's outer surface and refilled via the air charging station (also made by Whitehead).

The torpedo presetting is carried out electrically via a connecting wire (snap connector) connected to a plug mounted on each tube. Presetting is normally remote-controlled but in case of a failure of the launching network or/and of the remote-control panel

the presetting and firing sequence can be performed by means of a portable presetter which has to be directly connected to the aforementioned plug. The torpedo firing can be controlled either electrically, by means of the Shipborne Remote-Control Panel (Whitehead production), or manually, acting on a special push-button located on each barrel.

Operational Status
In production. In service with the Italian and several other navies.

Specifications
Length: 1200 mm
Depth: 3400 mm
Height: 1285 mm
Weight: 1050 kg

Contractor
Whitehead, Livorno.

JAPAN

GRX-3

Type
Anti-submarine warfare torpedo.

Development
Full development of the GRX-3 was funded under the FY88 budget. Work on the GRX-3 has progressed slowly, test and design problems associated with the weapon having slowed development. In the mid-1980s, 20.45 million yen was allocated for 'First Phase' fabrication. Under the FY86 budget, 3090 million yen was set aside to implement 'Second Phase' development.

Design
The GRX-3 ASW torpedo is intended to replace three torpedoes currently operational with the JMSDF, namely the Mk 46 Mod 5, the Model 73 and the Model 73 Mod. Prime responsibility for development and Second Phase testing of the GRX-3 lies with Japan's Research and Development Institute. Under FY82 funding a new battery and motor were authorised, with inertial navigation and electric propulsion system in FY83. It is planned that the GRX-3 will be capable of tracking and attacking the latest high speed nuclear submarines. The weapon features higher speed, improved operating depth and target detection capabilities over existing Japanese torpedoes.

GRX-2

Under the Medium Term Defence Programme (FY86-90) another torpedo, the GRX-2, will be developed as a future replacement for some Mk 46 and Model 73 torpedoes.

SWEDEN

TP 43XO TORPEDO

Type
Lightweight ASW torpedo with anti-surface vessel capability.

Description
The Swedish Ordnance TP 43XO is a homing, wire-guided multi-purpose torpedo for launching from submarines, surface ships and helicopters against submarines and surface targets. It is based on the well-proven TP 42 and retains the electrical propulsion system with a silver-zinc oxide primary or secondary battery.

The TP 43XO is designed around a multiprocessor computer system using Pascal language which gives flexibility to meet changing requirements. The homing system is optimised for use against very silent submarines in shallow water because the acoustic detector distinguishes between target and disruptive noise. The combination of homing and wire guidance is claimed to result in very high hit probability even at long firing ranges, the torpedo having three speeds selectable during its run to the target. The warshot torpedo is equipped with both an impact and a computerised proximity fuze. Both fuzes can be controlled from the launcher fire control system. The proximity fuze is tested on every practice run.

The guidance wire communication link permits the launching craft to transmit orders to the torpedo, controlling speed, depth, course, target data and so on. The torpedo reports its position, speed, course and depth, homing systems parameters and target noise.

Homing head	Homing head	Ag-Zn battery	Wire dispenser	Motor	Propellers
Tracking light	Charge	Switching unit	DC power	Gearbox	Rudders
Balloon	Safety device	Computer unit	supply unit	Control	Wire outlet
Recorder	Impact fuze	Control		servomotor	
Electronics for		Signal processing	Tube supplies	Flap	
exercise runs		Proximity fuze	connector		

In the event of communications being interrupted, the torpedo computer calculates the expected target position, guides the torpedo to the predicted point of impact and initiates one of the several possible search patterns.

Modular design makes it easy to configure the TP 43XO for special purposes, for example to accommodate special warheads or two battery sizes. Built-in test equipment is incorporated and computerised test equipment is used in maintenance and preparation. The test equipment is also used for evaluating practice runs. Recorded torpedo data can be presented together with recorded data from the ship fire control system.

In May 1990 the Swedish FMV contracted Swedish Ordnance to continue with the final phase of development of a new version lightweight wire-guided torpedo with new homing head. The new system will replace or upgrade the 43XO.

Operational Status
In operational use with the Royal Swedish Navy. An order worth SEK120 million for the TP 431 torpedo was given by the Swedish Navy in February 1986.

Specifications
Length: 2645 mm
Diameter: 400 mm
Weight: 280/310 kg (min/max) including battery and guidance wire
Warhead: 45 kg shaped charge HE
Propulsion: Secondary Ag/ZnO battery 4.2/5.6 kWh (primary battery optional); 3-speed thyristor battery switching unit; electric DC motor with gearbox

Contractor
Swedish Ordnance, Eskilstuna.

TORPEDO 43X2

Type
Multi-purpose lightweight torpedo.

Description
The 43-series of torpedo has been developed to meet the challenge of ASW in shallow waters. The 43X2 is the fourth generation of Swedish ASW torpedoes. Like previous 43-series torpedoes, the 43X2 is wire-guided, using a hydro-acoustic homing system for the terminal guidance phase. The weapon can be launched from submarines, surface ships and helicopters (the latter flying at up to 70 kts), and can be wire-guided from a flying or hovering helicopter, being deployed without the use of a parachute.

The weapon is fitted with an onboard computer system which monitors the comprehensive volume of data which it both gathers itself and which is fed to it via the wire communication link and which makes highly accurate inertial navigation possible. The wire link enables more than 80 different types of message to be transmitted in both directions. This information controls the weapon's parameters and targeting, and supervises homing procedures and so on. In the event of breakage in the wire communication, the computer takes full control of the weapon using the latest data received as well as computed search patterns incorporating safe/attack zones and so on.

The homing system is of new design (almost identical with that of the Torpedo 2000) with advanced signal processing carried out by the computer housed in the electronics module of the weapon. Three selectable homing modes are possible, active, passive or simultaneous active/passive. The weapon can track several targets simultaneously, classifying target signals and rejecting all false signals from the environment and from acoustic countermeasures. Data generated includes among others: target presence in one or more of the preformed beams; the sign and value of the target's position in both azimuth and depth; plus the data necessary for countermeasures action.

The main computer software is written in Pascal/Ada high level language, which simplifies software maintenance.

The torpedo is fitted with a multi-frequency hydro-acoustic proximity fuze operating on several frequencies, which detects the presence of a surface target above the weapon. An impact fuze is also incorporated. The transducers are positioned in a recess in the front end of the hydrodynamically shaped nose, which is designed for low self-noise at high speed. Advanced analysis of the received echoes enables waves, wakes and countermeasures from the hull of a ship to be discriminated.

Propulsion is by a secondary Ag/Zn 4.2 kWh battery feeding a DC electric motor with gearbox. The torpedo uses three selectable speeds (managed through a thyristor battery switching unit) in order to optimise hit probability and minimise time from launch to strike.

The body of the weapon is made from aluminium alloy castings and comprises interchangeable modules.

Operational Status

The weapon has completed its final development phase for the Swedish Navy and a production contract worth in the region of SEK80 million was awarded to Swedish Ordnance in the Autumn of 1991 for series deliveries to commence in 1993.

Specifications

Diameter: 400 mm
Length: 2800 mm
Range: c. 20 km

Contractor

Swedish Ordnance, Eskilstuna.

TP 617 TORPEDO

Type

Anti-surface vessel torpedo.

Description

The Swedish Ordnance TP 617 is a homing, wire-guided, long-range torpedo for launching from sub-

TP 617, latest version of Swedish Type 61 long-range torpedo

marines and surface ships against surface targets. It is based on the well-proven TP 61 which entered service 20 years ago and retains the thermal propulsion system with hydrogen peroxide/alcohol/water as propellants.

The TP 617 is designed around a programmable digital computer which controls the homing system, communication with the launching fire control system and the torpedo navigation system. The homing system is designed for use against surface targets and has demonstrated excellent acquisition capabilities against a broad range of targets. The combination of homing and wire guidance results in very high hit probability, even at long firing ranges.

The warshot torpedo is equipped with both an impact and a computerised proximity fuze. Both fuzes can be controlled from the launching fire control system. The proximity fuze is tested on every practice run.

The guidance wire communication link allows the ship to transmit orders to the torpedo controlling speed, depth, course, target data and so on. The torpedo reports its position, speed, course and depth, homing system parameters, and target noise. In the event of communications being interrupted, the torpedo computer calculates the expected target position, guides the torpedo to the predicted point and initiates one of several possible search patterns.

Built-in test facilities are incorporated and computerised test equipment facilitates maintenance and preparation. This test equipment is also used for evaluation practice runs. Recorded torpedo data can be presented together with recorded data from the launching fire control system.

Torpedo 2000 is a shorter, lighter, wire-guided heavyweight torpedo offering extended range and speed compared to the TP 617. It will be powered by a thermal engine (see below). The weapon will also be equipped with new electronics and a new homing head.

Operational Status

The TP 617 has been in operational service with the Royal Swedish Navy since early 1984 and is in service with some other Scandinavian navies.

Specifications

Length: 6980 mm
Diameter: 533 mm
Weight: 1860 kg
Warhead: 240 kg (approx)
Propulsion: Energy produced by combusting alcohol with hydrogen peroxide as an oxidiser. Two speed valve steam engine with gearbox

Contractor

Swedish Ordnance, Eskilstuna.

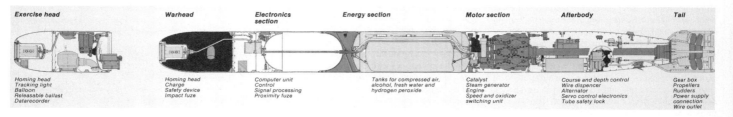

Exercise head	Warhead	Electronics section	Energy section	Motor section	Afterbody	Tail
Homing head Tracking light Balloon Releasable ballast Datarecorder	Homing head Charge Safety device Impact fuze	Computer unit Control Signal processing Proximity fuze	Tanks for compressed air, alcohol, fresh water and hydrogen peroxide	Catalyst Steam generator Engine Speed and oxidizer switching unit	Course and depth control Wire dispencer Alternator Servo control electronics Tube safety lock	Gear box Propellers Rudders Power supply connection Wire outlet

TORPEDO 2000

Type

Heavyweight torpedo.

Description

The 533 mm dual-purpose heavyweight Torpedo 2000 is being developed for the next generation of Swedish submarines and surface vessels for use against both underwater and surface targets. The weapon is powered by a new, high powered thermal pump jet propulsion system with a very high energy content based on Swedish Ordnance's long experience of high test peroxide (HTP) systems. This results in a wakeless system giving the weapon high speed and long range down to considerable depths with low radiated noise level. The motor is an axial two-stroke piston engine with seven cylinders and bore and stroke of 70 mm. Admission temperature is 800°C and the engine uses a two-stage compressor with compression pistons linked to the engine pistons. The propellant comprises 85 per cent HTP and paraffin, the engine developing between 25 and 300 kW of power.

The weapon is equipped with an advanced active/passive homing head and a wire guidance communication link which ensures high hit probability. In the event of loss of communication between platform and weapon, the torpedo's onboard computer takes over full command of the weapon and calculates the target's anticipated position and guides the weapon to the predicted point of impact initiating one of several possible search patterns programmed into the computer.

Operational Status

All parts of the weapon have been developed and tested, and full scale integration has been completed ready for sea trials which are due to commence in 1992. Delivery of the weapon is scheduled for the mid-1990s. A contract worth approximately SEK200 million was awarded for the last development phase in April 1991.

Specifications

Length: 5750 mm
Diameter: 533 mm
Weight: 1249 kg
Speed: <50 kts
Range: <40 km

Contractor

Swedish Ordnance, Eskilstuna.

TORPEDO PROPULSION SYSTEM

Description

Swedish Ordnance has developed a new torpedo propulsion system to meet future torpedo requirements which will demand a minimum speed of not less than 25 kts, a running depth of 500 m, and very low noise levels. A number of propellants has been considered and Swedish Ordnance has elected to develop a system based on a bipropellant using a combination of hydrogen peroxide (HTP) and diesel oil. This combination has been chosen because of its high energy content, high density, the fact that the propellants remain liquid at prevailing temperatures and the combustion reaction products (steam and carbon dioxide) can be fed directly into a heat exchange engine, the exhaust leaving no visible wake.

By using a mixture of HTP and diesel oil a number of working modes can be chosen for the engine. The semi-closed cycle method has important advantages in that engine performance is not significantly influenced by the ambient pressure and no heat exchanger is required. The HTP is decomposed by a catalyst prior to feeding into the steam generator, where the HTP reacts with the diesel oil to generate a large amount of energy. By adding fresh water the temperature of the reaction products is reduced to 700-800°C, which is acceptable to the engine.

The exhaust from the engine mainly consists of steam and carbon dioxide at a temperature of 200-300°C. The steam is condensed in a condenser and most of the condensation is recycled back into the system. The remaining part of the exhaust gases is compressed in a compressor before release into the surrounding sea.

The HTP is stored in a flexible bag located in a special tank which is directly connected to the water and diesel oil tanks, so that all tanks are the same pressure. A water pump supplies pressurised sea-water to the water tank to provide feed pressure. The capacity of this pump is controlled by the engine speed.

The engine is a seven-cylinder axial piston engine (cylinder bore 70 mm, stroke 70 mm), the engine pistons acting against a sinusoidal double rise cam. With the selected shape of the cam curve a complete dynamic balance of the inertial forces is achieved. The engine is designed to run at low speed so that the cam rollers of the pistons are always in contact with the cam.

The admission gas is supplied via a central connection, distribution being controlled by inlet and outlet valves. The engine pistons are linked to the compression pistons with compression being achieved in two stages. The condenser is built around the engine which assists in dampening any engine noise, and the whole unit is mounted in shock absorbing elements in the torpedo shell.

It is estimated that the prototype engine will produce a power of 25-300 kW at 600-1500 rpm at an admission pressure of between MPa 2-8 and with a fuel consumption of between 3.5 and 5 kg/kWh.

UNION OF SOVEREIGN STATES

533 mm TORPEDO

Type
Submarine-launched torpedo.

Description
Certainly until recently, and possibly on a continuing basis, the standard calibre of torpedo in service has been 533 mm. The standard 533 mm torpedo of the Second World War, which is still likely to be in service, was the Type 3, which was 7.27 m long, weighed 1610 kg (dry) and had a maximum range of 12 km. Maximum speed was 46 kts and the torpedo had a 300 kg TNT/tetryl warhead.

It may confidently be assumed that torpedo development in the USS has generally proceeded along lines similar to those of developments in Britain, France and the USA and has probably taken the form of development from wartime German pattern running and homing torpedoes.

It is however worth noting that somewhere around the late 1950s a version of this torpedo with a nuclear warhead was deployed in some submarines, and this was confirmed in 1981 when a USS 'Whiskey' class submarine ran aground in Swedish waters. The former Soviet Union regarded it as normal practice to fit alternative nuclear warheads to their torpedoes. Swedish sources estimate a warhead yield of about 15 kT.

Two distinct stages have been distinguished in the development of surface vessel launchers for torpedoes of this calibre, and the fact that the launchers are of slightly different lengths at these two stages

Stern view of 'Kara' class cruiser Petropavlovsk *shows hangar and alighting area for 'Hormone' ASW helicopter, SA-N-3 missile launcher and associated Headlight radar group, and port and starboard 5-tube 533 mm torpedo launchers* (G Jacobs)

may indicate that the corresponding torpedoes are also different, but not necessarily so. One 533 mm torpedo believed to be the 1957 pattern is said to be 825 cm long – which is as large as torpedoes go. It is carried by submarines and torpedo boats. Both surface target and ASW models exist in the range of 533 mm diameter torpedoes.

The dates of introduction of these two launcher groups appear to have been about 1948/49 and 1957/58. Apart from the difference in length just mentioned the principal change appears to be from manual local training to remote-power training. One-, three- and five-tube launchers are found in the earlier series; two-, three-, four- and five-tube launchers in the later.

Operational Status
All the above types of launcher seem still to be in service, as also, presumably, are the torpedoes. The 533 mm torpedo is also a standard weapon on USS submarines, and the 'Foxtrot' and 'Whiskey' class boats are equipped with four and two 533 mm stern torpedo tubes, respectively.

Other countries operating USS 533 mm torpedoes include: Albania, Bulgaria, China, Egypt, Finland, Germany, India, Indonesia, Iraq, North Korea, Libya, Poland, Romania, Somalia, Syria, Vietnam and Yugoslavia.

AIR-LAUNCHED TORPEDOES

Type
Helicopter/aircraft/missile deployed lightweight anti-submarine torpedoes.

Description
Two lightweight torpedoes have been developed for deployment from aircraft or missiles. The Type 45 weapon is 450 mm in diameter and 3.9 m long and is designed to be launched by helicopters or the SS-N-16 ASW rocket. The torpedo is fitted with an active/passive homing head and a 100 kg charge. Speed is 30 kts and the range 8.1 nm. The Type E53 is a 533 mm diameter torpedo with a length of 4.7 m. It is designed to be launched from fixed-wing aircraft or the SS-N-14 ASW rocket system. Warhead charge is 150 kg and the weapon uses active/passive homing. Range is 8.1 nm and maximum speed 40 kts.

TYPE 65 TORPEDO

Type
Submarine-launched 650 mm torpedo.

Description
Reports have recently indicated that a large diameter anti-ship torpedo is currently entering service with USS submarines. The main data are understood to be a length of some 1000 mm with a diameter of 650 mm. The propulsion is given as a closed cycle thermal system or MHD giving a range of 27 nm at 50 kts, and 54 nm at 30 kts. The weapon appears to use wake homing terminal guidance with wire guidance.

It is stated that an extensive modification programme is under way to provide the larger torpedo tubes needed for the larger diameter weapon.

Warhead charge is said to be about 900 kg.

Operational Status
It has been reported that the torpedo is in the early stages of production and may well equip the latest SSN classes. No other information is available.

TYPE 40 TORPEDO

Type
Surface-launched lightweight torpedo.

Description
Of comparatively recent development, a 406 mm torpedo of some 4.5 m length is now in service with some submarine chasers and light destroyers of the Navy. The active/passive homing torpedo is launched from trainable tubes that are only about 5 m long. These torpedoes are believed to be similar in concept to the American Mk 44 and Mk 46 torpedoes (see later in this section) and are principally anti-submarine weapons with a speed of 40 kts and range of 8.1 nm. They are deployed aboard 'Hotel II' and 'Hotel III' class ballistic missile submarines, 'Echo I' class nuclear attack submarines, 'Echo II' class missile submarines, and 'November' class boats, and are fired from special stern countermeasures tubes. Quintuple mounts are fitted in 'Petya 1' class frigates, and 'Mirka' class frigates. Fixed mounts exist on a number of small patrol classes. Bulgaria is known to operate USS 406 mm torpedoes. The torpedo carries a 100 kg warhead.

HEAVYWEIGHT TORPEDOES

Type
Submarine/surface ship-launched anti-ship torpedoes.

Description
Two types of anti-ship torpedo have been identified, both 533 mm diameter weapons with a length of 7.8 m. One weapon is a passive wake homing torpedo armed with a 300 kg charge and powered by a turbine giving it a speed of 50 kts and a range of 13.8 nm.

The other anti-ship torpedo is armed with a 400 kg charge and is electric powered, with a speed of 40 kts and range of 8.1 nm.

UNITED KINGDOM

Mk 8 TORPEDO

Type
Submarine-launched torpedo.

Operational Status
Obsolescent. Being replaced in RN service by the Tigerfish torpedo.

Quite possibly the longest-lived of all torpedo designs and certainly a remarkable survival in an otherwise reasonably sophisticated navy, the Mk 8

torpedo was designed in the early to mid-1930s and was certainly in service in the RN, including nuclear-powered submarines, until 1973, and is almost certain to be found in some other navies.

Although now being replaced by Tigerfish as the primary armament of the principal attack submarines of the RN, it should perhaps be said that, within its limitations, the Mk 8 performed well and gave service which was honourable as well as long. The latest known versions are the Mods 2, 3 and 4, all for surface ship targets.

Specifications
Length: 670 cm
Diameter: 21 in (533 mm)
Weight: 1521 kg
Warhead: 340 kg HE
Propulsion: Compressed air
Speed: 40-45.5 kts
Range: 4570-6400 m
Guidance: Preset course angle and depth

TIGERFISH Mk 24 TORPEDO

Type
Submarine-launched torpedo.

Development
Development of Tigerfish Mod 2 was completed in 1985. It incorporates important homing and wire guidance improvements which enable targets to be attacked successfully in the most difficult operational scenarios. The homing improvements are based on techniques developed in the Sting Ray lightweight torpedo and the wire guidance system is identical with the new Spearfish heavyweight development. Performance of the Mod 2 was demonstrated successfully in a series of trials which culminated in the sinking of a decommissioned Type 12 frigate. A modification kit has been developed which enables Mod 1 torpedoes to be upgraded to Mod 2 standard.

Description
Tigerfish is the name given to the production weapon derived from a redevelopment of the Mk 24 torpedo. The weapon is a 533 mm wire-guided/acoustic homing torpedo fitted with a dual action inertia-type impact fuze and an all-round proximity fuze designed to operate at the point of closest approach to the target, either submarine or surface vessel. In line with the Royal Navy policy of covert submarine operation, the torpedo is used in the passive mode whenever possible.

Wire guidance is used in the initial stages of an engagement up to the point where the torpedo's automatic three-dimensional passive/active acoustic homing system can control the run into the target. Wire is dispensed from both torpedo and submarine so as to avoid any wire stress due to their relative motion. The torpedo is roll-stabilised by controlling ailerons on retractable mid-body stub wings and is steered by hydraulically powered cruciform control surfaces mounted at the tail.

The torpedo carries its own computer which is connected through the guidance wire to the computer of the submarine's torpedo fire control system. During the wire guidance phase the torpedo's computer responds to the demands of the submarine computer, and during the homing run it interprets the data from the homing system sensors and calculates and commands the appropriate course, subject to a priority overriding steer-off azimuth control from the submarine.

The sonar can operate in either passive or active mode. During the attack the interrogation rate is progressively increased, as the torpedo nears the target,

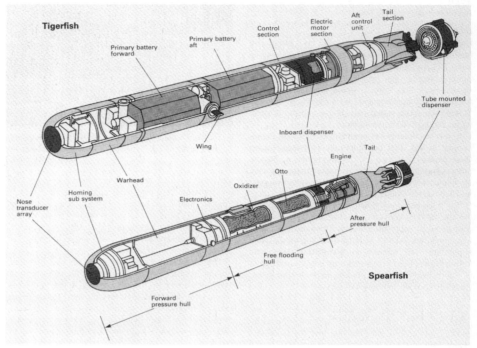

Tigerfish (top) and Spearfish (below) heavyweight torpedoes

so as to improve system accuracy. The interrogation rate is controlled by the onboard computer, which is thus performing several functions during this phase: interrogation control, sonar data computation, torpedo steering control and data transmission to the submarine to update its own computer memory.

Tigerfish is propelled by a powerful two-speed electric motor driving a pair of high efficiency, low noise, contra-rotating propellers. Fast run-up gyros for directional and attitude stability provide a very quick reaction time. High or low speed is selectable at all times.

Operational Status
In production. The Mod 1 version was in service with the Royal Navy for several years but has been entirely replaced by the Mod 2 higher performance model. Data given below relate to the warshot torpedo; there are also exercise and dummy (handling) versions. The exercise version is similar to the warshot but has rechargeable batteries, becomes buoyant at the end of the run, and has an instrumentation pack for data analysis in place of the warhead.

Tigerfish has been released for export to some foreign navies, the first of these to be identified being Brazil, which in 1982 ordered the Mk 24 Tigerfish for its Type 209 submarines. Tigerfish exports to Brazil now exceed £24 million and it has been successfully integrated into several different types of submarine with varying fire control systems.

Specifications
Length: 6.464 m
Diameter: 21 in (533 mm)
Weight: 1550 kg (in air)
Propulsion: Electrically driven contra-rotating propellers; 2 speeds
Speed: Dual high/low selectable at all times, max 50 kts (estimated)
Range: 21 km (estimated)
Fuze: Impact and proximity

Contractor
Marconi Underwater Systems Ltd, Waterlooville.

SPEARFISH TORPEDO

Type
Submarine-launched torpedo.

Development
In 1979-80 there was a competition between the projected Spearfish and the proposed US Mk 48 ADCAP. The UK MoD decided that Spearfish met the design requirements, particularly for covert operation and, taking into account progress made in the Sting Ray programme being managed by Marconi, awarded a fixed price contract for development and procurement understood to be worth about £350 million.

Description
Spearfish is a submarine-launched heavyweight torpedo designed to operate totally autonomously from all Royal Navy submarines and interfacing with a minimum of modification to all weapon handling and fire control systems. The torpedo is about 6 m long with a standard diameter of 533 mm and has an in-air weight of approximately 1850 kg. Spearfish meets the Royal Navy staff requirement to counter faster, deeper diving, quieter and stronger-hulled submarines, as well as dealing with surface targets. It will be phased-in gradually to replace the present Tigerfish Mod 2 in all nuclear-powered attack submarines (SSNs), the 'Trident' class SSBNs and the new diesel 'Upholder' class SSKs.

The variable speed of Spearfish is achieved by a 900 hp turbine thermal engine powered by an advanced liquid fuel and oxidiser contained in separate tanks. The top speed (widely reported at 70 kts), endurance and diving depth are understood to be nearly double that of the Tigerfish torpedo. The propulsor itself is a pump jet consisting of a rotor and stator using design techniques from the successful Sting Ray development. The thermal propulsion has been designed to be very quiet in operation to provide the covert operation needed to protect the firing submarine from detection on torpedo launch and subsequent counterattack. Small control surfaces mounted in the efflux from the pump jet make Spearfish extremely agile.

The torpedo's advanced sonar transducers and electronic system enable it to operate in a primary passive mode. However, when required to operate against a very quiet target, or in the final stages of attack, the active mode is used. In this mode, the powerful transmitters give Spearfish a long detection range and enable it to 'burn through' enemy counter-measures. The detection capabilities are further enhanced by an array offering many times the search volume of Tigerfish, with frequency-agile transducers which enable salvo firing to be carried out.

Communication between submarine and torpedo is via a guidewire link. The torpedo normally operates autonomously, relaying data back to the submarine, but the command team can assume control at any time. The wire is dispensed from two reels, one in the torpedo and one in the submarine launch tube, through a new design of dispenser which allows discharge at high submarine speeds as well as complete freedom of manoeuvre for the submarine after launch.

Spearfish contains a number of homing and tactics computers to control the weapon, enabling it autonomously to select search, detection and attack modes, to classify signal returns, to decide on appropriate tactics including re-attacks on the target if necessary, and to classify, track and overcome countermeasures and decoys. The torpedo's capability can be enhanced by software updates as target and countermeasure characteristics evolve in the future.

The warhead is a combined blast and shaped charge (directed energy) type, designed to pierce the double hulls of submarines entering service. The advanced homing and computer system enables the torpedo to be guided to the optimum point on the target's hull. Spearfish has an equal capability against surface targets where the fuzing system detonates the warhead under the hull of the target. The blast creates a whipping effect which breaks the ship's back.

Data given above relate to the warshot torpedo;

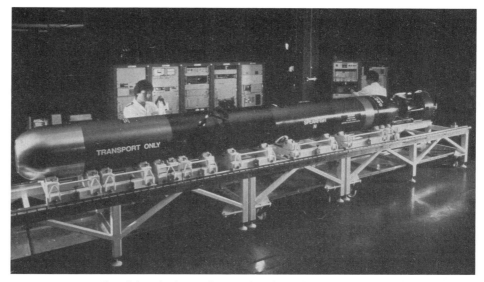

Spearfish production warshot torpedo and associated test equipment

there are also exercise and dummy (handling and discharge) versions. The exercise version is identical to the warshot except that the warhead is replaced by a novel design of recovery and instrumentation section. This 'R and I' section carries a tracking system and a very advanced data recorder with a large signal capacity, together with a recovery system. At the end of run the torpedo, which is heavier than water, is floated to the surface by means of a toroidal bag contained in the R and I section. Developed from the Sting Ray system, it is however an advance in that it uses cold gas (CO_2) to inflate the bag rather than the Sting Ray hot gas system. The system is clean, effective and reliable.

Operational Status
Marconi states that development has been to programme and within cost. Contract Acceptance Trials are complete. Spearfish is now in production, using additions to the company's manufacturing facility initially purpose-built for Sting Ray, and the first production torpedoes have been delivered to the RN Armament Depot at Beith. Further tenders for Spearfish production are expected in 1992.

Contractor
Marconi Underwater Systems Ltd, Waterlooville.

STING RAY TORPEDO

Type
Lightweight torpedo.

Development
Development was started in 1977 by Marconi Underwater Systems and the first fixed price production contract was placed in November 1979. First production weapon was delivered in August 1981 and acceptance trials began April 1982, resulting in the issue of the design certificate in December 1982. Fleet Weapon Acceptance trials were successfully completed in 1985 when the weapon's complete performance envelope was checked out. Subsequent to these trials an RAF Nimrod successfully launched a Sting Ray and actually sunk a submarine target in the open sea. The torpedo will remain capable of countering submarine threats effectively until well into the next century.

Description
Sting Ray is an advanced lightweight torpedo designed for launching from helicopters and fixed-wing aircraft as well as from surface ships. It has a multi-mode, multi-beam sonar, quiet high speed propulsion system and a fully programmable onboard digital computer, which together give the weapon a high performance not only in deep water, but also in shallow water where sonar conditions are difficult. A very considerable effort was put into gathering acoustic data in shallow water with many hundreds of trials in order to optimise the torpedo's tactics and software algorithms under such conditions. Its computer system enables it to make tactical decisions during an engagement to optimise the various homing modes to suit the environment and target behaviour. The computer can be reprogrammed 'through the skin' whenever updated software, or programmes tailored to a particular customer's operational requirements are required. These programmes then adapt themselves to the particular tactical and environmental conditions in which the torpedo operates. The high speed performance of Sting Ray comes from its pump-jet propulsion system. Vehicle hydrodynamics are

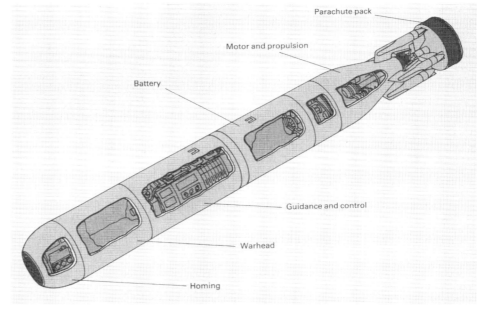

Sting Ray outline

accomplished by an electrohydraulically driven proportional control system developed by BAe Dynamics Group. The four control surfaces are mounted aft of the propulsor. The new seawater battery gives extended endurance and no performance degradation with depth.

Sting Ray carries a directed energy warhead required to counter modern submarine design. This type of warhead demands high accuracy guidance to ensure the torpedo strikes the most vulnerable part of the target.

In operation, when the torpedo enters the water and frees itself from its parachute, it sets up a preprogrammed search pattern designed to maximise the chances of target detection. On acquiring the target, the computer and active/sonar sensors make it almost impossible for the target to evade attack. The

advanced software of the guidance/homing system allows it to filter out background noise and decoys and to make an interception rather than perform a conventional tailchase attack. It is believed to have a maximum speed of about 45 kts with an estimated endurance of some eight minutes at that speed. Maximum operating depth is believed to be about 1000 m.

Data given below relate to the warshot torpedo; there are also exercise and dummy (handling, carriage and release) versions. The exercise version is identical to the warshot except that the warhead is replaced by a novel design of recovery and instrumentation section. This 'R and I' section carries a tracking system, a data recorder and a recovery system. At the end of the run the torpedo, which is heavier than water, is floated to the surface by means of a toroidal bag contained in the R and I section which is inflated by a solid propellant hot gas generator. The system is reported as effective and reliable.

Over 1000 in-water runs have now been completed both in development and operational usage, mainly against targets, not only making it one of the most tested torpedoes in the world but also ensuring an ever expanding bank of in-water and environmental data to enable continuous upgrading of its software and algorithms to be made.

Operational Status

In full production. In service with RN and RAF and the Thai and Egyptian navies. A £400 million production order for 2000 plus torpedoes was signed in January 1986 by UK MoD and well over 1000 are now in service. In 1988 Thailand placed a follow-on order for 12 Sting Ray torpedoes worth an estimated 202.6 million baht.

On December 15, 1989 Marconi signed a contract with the Norwegian Navy and Air Force for a substantial number of Sting Ray torpedoes with a contract value in excess of £25 million. Delivery commenced early in 1991 together with the work necessary to modify the Mk 32 launch tubes to take Sting Ray. Trials with Norwegian Air Force P-3 Orions are in hand and it was anticipated that an air clearance certificate would be issued during 1991.

Specifications

Length: 2597 mm

Sting Ray torpedoes on a Sea King helicopter of the Royal Navy

Diameter: 324 mm
Weight: 265 kg
Warhead: Shaped Charge HE
Propulsion: Details not yet revealed

Contractor

Marconi Underwater Systems, Waterlooville (prime contractor).

TORPEDO GUIDANCE SYSTEMS

Type

Integrated guidance systems for torpedoes.

Description

Typically the source of power for a long-endurance control system necessary in torpedoes such as Spearfish is a hydraulic pump with a return oil system. To cover essential control activity in the immediate post-launch period and prior to the run-up speed of the propulsion engine, a stored-pressure system gives a smooth hand-over to pump-driven operation.

Fairey hydraulic systems are specifically configured to suit the space envelope constraints in a torpedo tail cone. A ring main system integrates the following essential components in an electrohydraulic system: pump, hoses (pressure and return) with quick-disconnect couplings, pressure and return ring-main pipes, relief valve, accumulator with gas cartridge release valve and filter, pressurised reservoir, electrohydraulic servo valves (four), rotary actuator with position feedback (four), control system's electronic package and control system test set.

In electrohydraulic systems a pump, usually near the torpedo motor remote from the ring main, is connected via hoses and quick-disconnect couplings. Particular attention is paid to pump noise levels and to mounting structure stiffness for maximum noise attenuation.

Quick-disconnect couplings enable a modularised torpedo to be broken down easily for servicing without draining oil from the system.

A ring main of concentric pressure and return rigid pipes links mounting pads to which are attached accumulator; reservoir; relief valve; and four electrohydraulic servo valves. These components are arranged around the propeller shaft giving an optimum use of space. A low hysteresis relief valve limits the pressure peaks in the hydraulic system and discharges from the high to the low pressure ring main. An accumulator provides a backup source during the starting period and when actuator demands momentarily peak above the pump output flow.

On torpedo release an electrical signal actuates a frangible valve in a sealed gas bottle. This gas immediately pressurises accumulator fluid to provide hydraulic power to the actuators via the electrohydraulic valves.

A spring-loaded reservoir maintains a positive return-line pressure at all times and absorbs fluid displacements due to accumulator discharge and fluid expansion.

Four two-stage electrohydraulic servo valves mounted on the ring main control four rotary

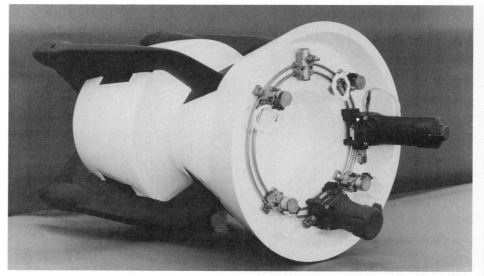

Part of the Fairey Hydraulics' electrically signalled, hydraulically powered control system to the control fins of the Spearfish heavyweight torpedo

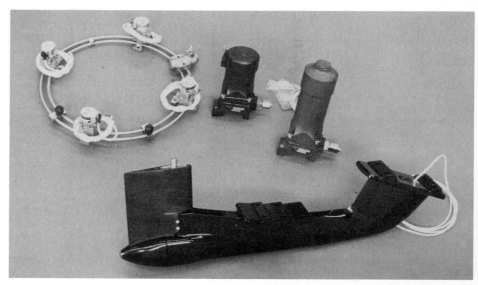

Fairey Hydraulics system components for the Spearfish heavyweight torpedo showing a control fin with control surface

actuators. Each valve receives an electrical signal from the torpedo system via an electronic control loop-closure package.

Four rotary actuators directly coupled to the control rudders are fitted in the fins. The actuator/control rudder position is monitored through a mechanically linked rotary potentiometer which provides the position feedback signal to the electronics package. The fins are individually attached to the torpedo tail cone.

Test sets are provided to put the four-axis control system through its paces by injecting various electrical demand input signals, including steady-state, sinusoidal and other forms. Hydraulic power is used to energise the ring main of the control system under test; and an electronic section simulates the torpedo control system and provides identical output signals.

A microprocessor is installed, programmed to take each control system through the complete test routine. If any test result is not met the fault will be indicated and identified. According to the programming the test will either stop at that point or be continued with the fault recorded.

Contractor
Fairey Hydraulics, Bristol.

PMW 49A

Type
Deck launcher for lightweight torpedoes.

Description
The system comprises two sets of triple GRP torpedo tubes arranged with their axes parallel in a triangular configuration, mounted on a training mechanism which, in turn, is supported by a stationary base bolted to the ship's deck. Muzzle closures and anti-condensation heaters provide an all-weather storage capacity for the torpedoes.

Torpedoes, including the US Mk 46 (all variants), Sting Ray (all variants) and Italian A 244/S, are fired by the release of compressed air stored in the breech assembly air reservoir. Air reservoir charging terminals are installed at each launcher position and connected to the ship's high pressure air supply.

The system is operated through a Dowty ATLAPS torpedo launch controller and presetter. ATLAPS comprises a main processor/command unit housed in the operations room, together with environmentally sealed slave control units, mounted either above or below decks in close proximity to the torpedo launchers. The system provides for automatic presetting of the torpedoes and control of the launch sequence with manual override fallback.

The system is a multi-microprocessor system with built-in reversionary modes in order to maintain operational capability under all conditions.

Various configurations are available through the

PMW 49A triple torpedo launcher

use of modular construction, including the ability to fit small ships which lack a combat information system. Various interfaces can be fitted allowing the system to integrate with the ship's other ASW systems including sonar.

Contractor
Marconi Underwater Systems, Waterlooville.

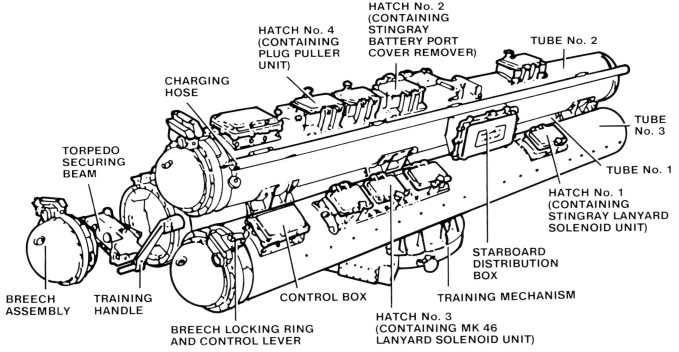

PMW 49A triple torpedo launcher

WEAPON HANDLING AND DISCHARGE SYSTEM

Type
Positive discharge system for submarines.

Description
The design can accommodate various weapon types and sizes including torpedoes, wire-guided missiles and mines, including the following weapons:
(a) Mk 24 'Tigerfish' wire-guided torpedo

(b) Spearfish heavyweight torpedo
(c) Royal Navy anti-ship missile Sub-harpoon
(d) Mk 48 US wire-guided torpedo
(e) Mk 5 Mine
(f) Mk 6 Mine.

The equipment takes the weapon through embarkation, stowage, loading and launching from four, five or six tube boat configurations, incorporating a variety of stowage requirements, including hull-mounted dependent or independently shock-mounted stowages and integral raft structures.

The equipment provides for all the handling systems, the torpedo tubes and electronic firing equipment necessary for a positive discharge launch method, utilising one of the following processes:
(a) air turbine pump (ATP)
(b) water/air ram discharge
(c) high pressure air discharge system.

There is a growing demand for positive discharge systems brought about by the adoption of tube-launched missiles with no 'swim-out' capability and the increasing need for quiet discharge.

Operational Status

Positive discharge systems are in use in overseas navies, and most recently a positive discharge system has been chosen by Kockums of Sweden and the Royal Australian Navy to equip the new 'Collins' class submarine. Systems are operational with the Royal Navy, Royal Australian Navy, Canadian Armed Forces, Brazilian Navy and Chilean Navy aboard the following classes of submarine:

(a) 'Oberon' class SSK
(b) 'Oberon' Type SSK in service with RAN, RCN and Chile
(c) 'Valiant' class SSN
(d) 'Resolution' class SSBN
(e) 'Swiftsure' class SSN
(f) 'Trafalgar' class SSN
(g) 'Upholder' class SSK
(h) 'Vanguard' class SSBN
(i) 'Collins' class SSK.

Contractor

Strachan & Henshaw Ltd, Bristol.

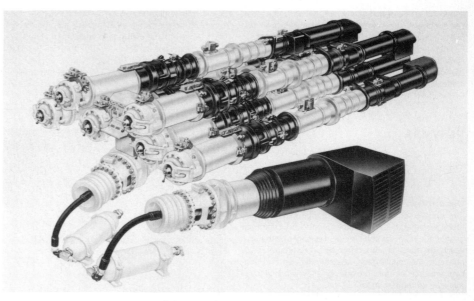

Submarine weapon ejection system

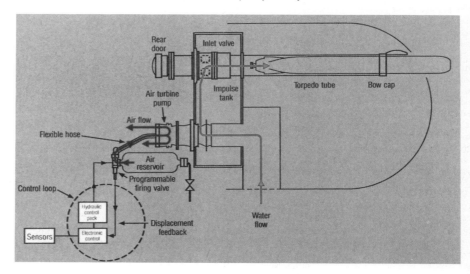

Diagrammatic arrangement of ejection system

LIGHTWEIGHT TORPEDO RECOVERY SYSTEM

Type

Torpedo recovery system.

Development

The system was originally designed for the recovery of the Sting Ray torpedo in a vertical floating recovery position. It has since been adapted to recover both Mk 44 and Mk 46 lightweight torpedoes, including the horizontally floating Mk 46 exercise variant. The system has been engineered for use from Sea King (RN), Lynx (RN), Puma (RAF), Sikorsky SH-3 (USN), Agusta 109, Dauphin and Bell LongRanger helicopters. The system has also been cleared by the Joint Air Transport Establishment in the UK to fly under any suitable UK military aircraft.

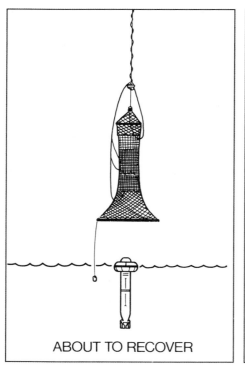

ABOUT TO RECOVER

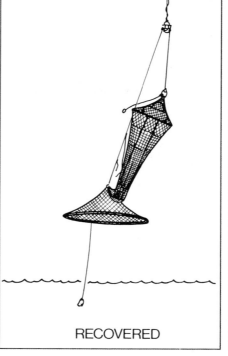

RECOVERED

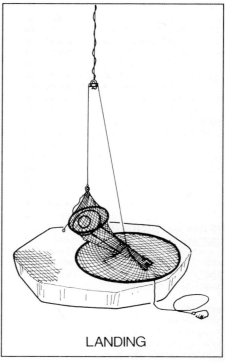

LANDING

Principle of operation of the lightweight torpedo recovery system

Description

The system comprises two elements: a recovery net and a landing platform. The conical net arrangement is suspended by a 9 m cable from the standard cargo hook on a helicopter. A remote SACRU (Semi Automatic Cargo Release Unit) is attached to the end of the cable and is activated from the aircraft cabin by a completely independent power supply. The SACRU supports three colour coded lines whose relative lengths are changed by SACRU release, to effect weapon capture. Auxiliary lines consisting of a rubber bungee and weak link arrangement also assist in this automatic procedure. The polyester braided net is kept in shape by means of two aluminium hoops which are coated to prevent damage to the torpedo.

The transportable landing platform is a sectioned octagonal frame of tubular aluminium across which is stretched a braided net which is kept under tension by means of ratchet tensioners. This strong but soft platform all but eliminates the risk of damage to the weapon. Attached to the inside of the frame is a PVC sump to catch any effluent.

Weapons can be recovered by the system in up to sea state 6. Turn round time on weapons is kept to a minimum and the requirement for swimmers has been eliminated.

Specifications

Length of net: 5.48 m (from apex to bottom hoop)
Bottom hoop diameter: 3 m
Top hoop diameter: 1.2 m
Weight: 87 kg

Operational Status

To capture a floating weapon the aircraft hovers over the weapon until the catchment area of the bottom hoop is approximately centred over the weapon. The aircraft and net are then lowered until the nose of the weapon is seated in the apex of the net. The remotely operated cargo hook is then released allowing the net to drape fully over the weapon. The pilot then slowly increases hover height, allowing the net's trapping lines to operate thus enclosing the weapon with the bottom hoop folded under and to one side of the net. The aircraft then lifts the net complete with the weapon from the water and proceeds to the landing site where the net is lowered onto the landing platform and released from the aircraft.

Contractor

Bridport Aviation Products, Bridport.

UNITED STATES OF AMERICA

Mk 44 TORPEDO

Type

Air/surface/missile-launched anti-submarine torpedo.

Description

This is a lightweight torpedo designed for launching from aircraft or helicopters, from surface vessels (using Mk 32 tubes), or by the ASROC rocket system.

Two models have so far been produced but the differences in dimensions are trivial. Both torpedoes are electrically propelled and their calibre is 12.75 in (324 mm). Approximate length is 2.56 m and the torpedoes weigh about 233 kg with a 34 kg warhead. Active acoustic homing is used. Depth and course settings are entered by umbilical cable. Estimated maximum submersion depth is approximately 300 m. A range of about 5000 m at a speed of 30 kts has been reported. Arming is by seawater scoop.

Operational status

Obsolescent. Licence production was initiated in a number of foreign countries. Replaced by Mk 46 in USA and UK service, and in some other navies. Navies still equipped with the Mk 44 are those of Argentina, Australia, Brazil, Canada, Chile, Columbia, Germany, Greece, Indonesia, Iran, Italy, Japan, Netherlands, New Zealand, Norway, Pakistan, Philippines, Portugal, South Africa, South Korea, Spain, Thailand, Tunisia, Turkey, Uruguay and Venezuela.

Mk 46 TORPEDO

Type

Air-launched lightweight torpedo.

Development

Developed as a successor to the Mk 44 in the early 1960s. Deliveries of first Mod 0 model were made to US Navy in 1965. The Mod 1 was introduced into US Navy in April 1967 followed by Mod 2 in 1972. The latest version is the Mod 5 resulting from the Near-Term Improvement Programme (NEARTIP), which was aimed at improving acoustic performance, countermeasure resistance, guidance and control system and fire control system. This latest version is in full production and was introduced to combat the adoption of anechoic coatings by USS submarines (Codename Clusterguard).

Description

This is a deep diving, high speed torpedo fitted with active/passive acoustic homing and is intended mainly for use against submarines. After water entry it starts a helical search pattern, acquires and attacks its target; if it misses the target it is capable of multiple re-attacks. It has a maximum speed of about 40 kts. The Mk 46 Mod 0 used a solid fuel motor whereas the Mod 1, which is slightly lighter, uses a five cylinder liquid mono-propellant (Otto) motor. This latter propulsion system was introduced because of the maintenance problem with the original solid fuel motor.

The latest version is the Mod 5 which features a new passive/active sonar capable of detecting most types of underwater submarine target including anechoically coated hulls. The sonar offers improved target acquisition capabilities in all types of acoustic environment, including shallow water. The Mod 5 is fitted with the Mk 103 Mod 1 conventional warhead with proximity fuze. Guidance and control systems have been upgraded and feature re-attack logic. The propulsion system features a two-speed capability and range and endurance have been increased.

The Mk 46 can be launched by surface vessels and the ASROC rocket system as well as by aircraft and helicopters.

Mk 46 Mod 5 (NEARTIP) torpedo produced by Honeywell which succeeded the Mk 46 Mods 1 and 2

Operational Status

Mk 46 Mod 0 weapons, which were produced in limited quantities, were in use as air-launched torpedoes only by the US Navy.

By early 1975, most Mod 1 torpedoes were converted to Mod 2 by modification action. Currently, NEARTIP implementation is being applied to all existing Mod 1 and Mod 2 torpedoes in the form of modification kits applied on a retrofit basis.

Since 1965 more than 20 000 Mk 46 torpedoes have been produced for the USN and other navies. The FY84 report by the US Secretary of Defense included funding for both the Mk 46 Mod 5 (NEARTIP) torpedoes and conversion kits to upgrade older Mk 46s. For 1986 and 1987 procurement of 500 torpedoes each year was proposed with annual costs of $132.8 million and $106.8 million respectively. In February 1987, a contract worth $295 million was awarded to Honeywell for the Mk 46 Mod 5 torpedoes for the US Navy, Canada and Turkey.

Users of Mk 46 torpedoes include: Australia, Brazil, Canada, France, Greece, Indonesia, Iran, Israel, Italy, Japan, Morocco, Netherlands, New Zealand, Norway, Pakistan, Portugal, Saudi Arabia, South Korea, Spain, Taiwan, Thailand, Turkey, United Kingdom and USA.

In addition and subject to Congressional approval, Belgium plans to purchase 65 Mod 5 Mk 46 torpedoes from Honeywell Defence Systems. These will replace French-built ECAN L5 torpedoes currently deployed on four 'Wielingen' class frigates.

A total of 331 Mod 5 torpedoes and 93 conversion kits have been ordered by the USN as a combined purchase for Brazil and Spain under the Foreign Military Sales programme.

Specifications

Mk 46 Mod 1 Version
Length: 2590 mm
Diameter: 324 mm
Weight: 230 kg
Warhead: 44 kg HE
Propulsion: Monopropellant
Speed: 40 kts (max)
Range: 11 000 m (max)
Acquisition range: 460 m (estimated)

Contractors

Aerojet Electro Systems Company, Azusa, California (development and early production of Mod 0 and Mod 1) (prime contractor).

Honeywell Incorporated and Gould Ocean Systems Division now manufacture the Mk 46 Mod 5, with Honeywell being the prime contractor.

Mk 50 TORPEDO

Type
Advanced lightweight torpedo.

Development
The Advanced Lightweight Torpedo (ALWT) (now designated the Mk 50) was developed as a successor to the Mk 46 Mod 5 (NEARTIP) torpedo to counter the advancing and sophisticated Soviet submarine threat. It was designed for delivery by surface ships, submarines and aircraft, and to be interoperable with existing Mk 46 launch platforms. Development commenced in the late 1970s and full-scale development is now near completion.

Description
The Mk 50 has similar dimensions to the Mk 46 but has improved speed (in excess of 40 kts), greater endurance, greater operating depth (more than 600 m) plus improved terminal homing, signal processing and greater destructive effect. In addition to the ability to operate against fast, deep diving submarines there is also a requirement for capabilities against shallow, slow moving submarines and surface ships. According to the US DoD, the most difficult object to achieve is the realisation of accurate, reliable terminal homing which is essential for directed energy warhead use.

For maximum effect the directed energy blast must penetrate the submarine hull and not glance off, and at the same time it would be advantageous to hit the target submarine amidships. The US Navy has established a programme to improve the performance of the warhead. A Mod 1 improved warhead is expected to enter production in FY93, and a Mod 2 in FY96.

From the two original competing contractors under US procedure A-109, Honeywell was selected in 1981 as the prime contractor for Mk 50 development and production, with the Garrett Division of Allied Signal Corporation being responsible for propulsion. Propulsion is achieved by a closed cycle steam turbine engine in a stored chemical energy propulsion system (SCEPS) with the energy being supplied by a chemical reaction in which solid lithium is ignited to produce an oxidant with extremely high energy density. This is then injected into the boiler and mixed with lithium to generate the very high temperatures needed to produce superheated steam for driving the turbine. The steam is then converted back to water for recycling through the boiler. The propulsion unit develops full power at all depths and is capable of multi-speed settings as required by the tactical situation.

Development testing of the Mk 50 terminal homing system has been carried out by using seven modified

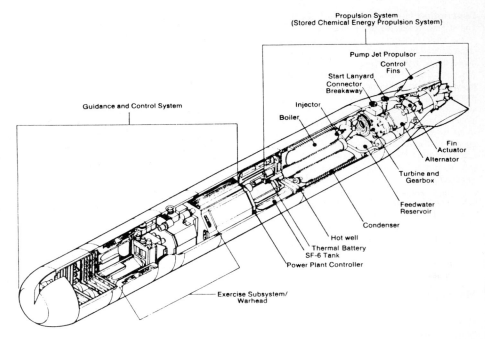

Mk 45 torpedoes as high speed mobile torpedo targets (ADMATT). These run at 41 kts and tow a hydrophone/echo repeater array to simulate the spatial extent of a real submarine. A three-dimensional instrumented shallow water range has also been developed for Mk 50 evaluation.

The sonar comprises a low noise nose array assembly, transmitter unit, receiver and two digital signal processing units.

The active/passive sonar operates with multiple, selectable transmit and receive beams.

Mission control, including commands to the sonar and propulsion subsystems, autopilot navigation and target detection, classification, tracking and countermeasures, is carried out by an AN/AYK-14 standard US Navy computer. Guidance is carried out using two digital signal processors, processing sonar data to generate a range map for the AYK-14 missile computer.

Operational Status
Now in full-scale development, the Mk 50 should enter USN service by late 1992. Low rate initial production commenced in 1987 with a purchase of 340 weapons in a two-part production. September 1990 deliveries were contracted on the basis of 76 torpedoes from Honeywell and 64 from Westinghouse. Deliveries under a second contract for 100 torpedoes

from each company will begin in 1992. Airdrop tests were carried out in February 1987, followed by operational tests in March 1987. In June 1987, Westinghouse Defense and Electronics Center won a $7.5 million initial contract to act as a second manufacturing source. In December 1988 Westinghouse was awarded a $188 million contract for initial production of the Mk 50 torpedoes. Operational Evaluation (OPEVAL) tests commenced in July 1990. Work was expected to be completed in July 1991. Full production contracts were expected to be awarded during FY91.

Specifications
Length: 2896 mm
Diameter: 324 mm
Weight: 363 kg
Warhead: Shaped charge HE
Propulsion: Closed-cycle
Speed: 55-60 kts (approx)
Range: 20 000 m (approx)

Contractors
Honeywell Inc, Defense Systems Division, Hopkins, Minnesota (prime contractor).

Westinghouse Defense and Electronics (second source).

Mk 37 Mods 0 and 3 TORPEDOES

Type
Submarine-launched anti-submarine torpedo.

Description
Designed primarily as a submarine-launched anti-submarine torpedo, but suitable for deck-launching by the Mk 23 and Mk 25 torpedo launchers, the Mk 37 torpedo is a 19 in (482.6 mm) weapon that has been described as the first successful high performance anti-submarine torpedo. The 19 in diameter of all versions of this torpedo was chosen to enable the torpedo to swim-out from a standard 21 in (533 mm)

launch tube, an arrangement with obvious operational advantages.

Mods 0 and 3 are free-running torpedoes. After one has been launched on a target interception course it maintains course until, at a preset range, it initiates a process which arms the warhead and switches in the attack logic circuits. The latter include various preselectable options such as depth limits, search pattern, and type of homing. The final attack is by sonar autohoming which can be active, passive, or both combined.

Mod 3 is an updated version of Mod 0, the updating consisting of the incorporation of a large number of minor modifications based on operational experience.

Operational Status
The Mk 37 has been largely replaced by the Mk 37E and Mk 48, although quantities of the Mk 37 are probably still in service.

Specifications
Length: 352 cm
Diameter: 483 mm with guides to fit 533 mm tubes
Weight: 645 kg (warshot); 540 kg (practice)
Warhead: 150 kg HE
Speed: 24 kts (Mod 3)
Submersion: To 270 m (Mod 3)

Mk 37 Mods 1 and 2 TORPEDOES

Type
Submarine-launched anti-submarine torpedo.

Description
These versions of the Mk 37 torpedo differ from Mods 0 and 3 in their size and method of guidance; in other respects they are substantially similar.

Both versions are wire-guided, whereas the Mk 37 Mods 0 and 3 are free-running torpedoes. Both are 4.09 m long and weigh 766 kg (warshot) or 657 kg (practice).

Just as Mod 3 is an updated version of Mod 0, so is Mod 2 a version of Mod 1, updated by the incorporation of a number of minor modifications.

Operational Status
Operational.

Specifications
Length: 409 cm
Diameter: 483 mm with guides to fit 533 mm tubes
Weight: 766 kg (warshot); 657 kg (practice)
Warhead: 150 kg HE
Speed: 24 kts (Mod 2)
Submersion: To 270 m (Mod 2)

NT 37F™ MODERNISATION

Type
Anti-surface vessel/anti-submarine torpedo.

Development
The original Mk 37 upgraded version was called the NT 37C and included the thermochemical propulsion system and anti-ship kit.

Alliant Techsystems, at the request of principal NATO NT 37C user navies, instituted a development programme to improve the logistic support capability and the acoustic and guidance performance of the torpedo. This effort resulted in the new, supportable, solid-state acoustic system and nose assembly. An upgraded control system was introduced in 1990.

Alliant Techsystems has introduced the NT 37F™ configuration which combines the original thermo-chemical engine upgrade with a new long-range sonar and reliability enhancements to the control system. This new configuration is currently under contract for delivery to customer navies in 1993.

User navy exercises on the NT 37D model have demonstrated long-range active and passive homing attack on acoustic and live targets. The tests also demonstrated the following improvements: increased dynamic control and accuracy, complete wire guidance capability, ease of tactical software, low sonar background noise levels, low acoustic false alarm rate, long-range active and passive acoustic attacks on acoustic targets, increased torpedo reliability, and reduced maintenance. A semi-automatic test set has been designed to replace manual test equipment.

Description
The NT 37F™ torpedo is a modernised, improved version of the standard, battery-electric propulsion Mk 37 heavyweight torpedo. This dual purpose ASW and anti-ship torpedo can be fired from both submarines and surface ships. Interfacing to a standard Mk 37 fire control interface or a standard RS 232 or RS 422 bus digital fire control, the torpedo is compatible with the following platforms and fire control systems: German 205/206/MSI-70U; German 209/HSA-SINBADS (later version); RNLN Swordfish/HSA-M8; or any class equipped to fire Mk 37 torpedoes and with a digital interface.

The NT 37E performance improvement kit and hardware, largely installed in the field, completely

Sea trials of NT 37E heavyweight torpedo

upgrades existing Mk 37 torpedo inventories. It consists of three major subsystems:

(a) an Otto fuelled thermo-chemical propulsion system. When compared to the performance of an unmodified Mk 37 silver/zinc oxide battery and electric motor, the new propulsion system demonstrates a 40 per cent increase in speed, a 150 per cent increase in range, and an 80 per cent increase in endurance. The propulsion system is adapted to the existing Mk 37 tail section and is compatible with existing Mk 46 fuelling and engine maintenance facilities, engine tools, and engine refurbishment kits. A fuel bag replaces the DC electric battery.

(b) a solid-state acoustic system and a noise reduction laminar-flow nose assembly replaces the Mk 37's vacuum-tube acoustic panel and hemispherical nose. The new sonar substantially improves the passive detection range against high speed surface targets and active detection range against small silhouette submarine targets; in most cases, target acquisition range has been doubled. The new self-noise reduction nose assembly increases transducer isolation while reducing flow noise effects, reducing the likelihood of self-decoying at all depths.

(c) a computer-based control system upgrade eliminates the original relay logic, resulting in greater system stability, reliability and accuracy. The new system utilises an Intel microprocessor, allowing

fully programmable tactics. Should future naval tactics require modified torpedo command and control, changes can be implemented by revision of software and replacement of programmable read-only memory units (PROMs).

Operational Status
The US Navy Mk 37 family of torpedoes was purchased by at least 16 navies. Life cycle support is provided for 15 years. The NT 37E version is on order for the Egyptian Navy. Deliveries are due to begin in 1994.

Specifications
Type: NT 37F™, wire-guided, Mod 2; non wire-guided, Mod 3
Length: 4505 mm (Mod 2); 3846 mm (Mod 3)
Diameter: 485 mm
Weight: 750 kg (Mod 2); 642 kg (Mod 3)
Run modes: (1) Straight run/salvo anti-ship; (2) straight run with acoustic miss indicator to initiate acoustic re-attack; (3) active snake and circle – ASW; passive snake and circle – anti-ship
Warhead: 150 kg HE
Propulsion: Thermo-chemical rotary piston cam engine with Otto fuel

Contractor
Alliant Techsystems, Marine Systems West, Mukilteo, Washington.

Mk 32 TORPEDO

Type
Surface-launched anti-submarine torpedo.

Description
The Mk 32 is an acoustic anti-submarine torpedo which is still in use aboard a number of former USN ships now transferred to other navies in various parts of the world, mostly aboard escort types. They are launched by the Mk 4 rack-type launcher unit in most cases, in which a single torpedo is carried in the ready-to-launch position with two more torpedoes stored nearby on deck. The current version is the Mk 32 Mod 2, which is equipped with an active

acoustic homing head. It is also capable of airdrop launch.

Propulsion is by electric motor, powered by a silver/zinc oxide battery which is activated by seawater. After launch, the torpedo descends to a depth of 6 m and the active sonar begins to operate, transmitting 50 kHz pulses while the torpedo follows a spiral search pattern, circling to left or right as selected. The search pattern is maintained until the sonar obtains a target response and the homing sequence commences. In the event of failure to obtain a target contact, the circular search pattern continues until the battery is exhausted, after which the torpedo dives deeper prior to self-destruction.

Operational Status
Operational but obsolescent.

Specifications
Length: 2080 mm
Diameter: 483 mm
Weight: 350 kg
Warhead: 49 kg HE
Propulsion: Electric motor
Range: 8600 yd (7864 m) at 12 kts (approx)
Max acoustic range: 560 m

ADCAP TORPEDO

Type
Submarine-launched anti-submarine torpedo.

Development
Anticipation by the US authorities of impending advances in submarine technology by the USSR noted in the 1960s and 1970s led to studies of the Mk 48's capabilities against likely threats, and in 1975 this resulted in an Operational Requirement issued by the Chief of Naval Operations for a programme to develop appropriate modifications to the Mk 48 torpedo to keep pace with anticipated submarine threat developments. The origins of the Mk 48 ADCAP (advanced capabilities) programme lay in this requirement, but the extent and rate of Soviet submarine technology advance hastened both ADCAP

progress and another Mk 48 improvement programme.

Recognition (by the USA) of the impressive operational characteristics of the Soviet 'Alfa' class submarine in late Spring 1979 resulted in a decision, taken in September 1979, to accelerate the ADCAP programme. It was also responsible for an intensive test and analysis programme to determine the true limits of the then current Mk 48 in terms of depth, speed and acoustic capabilities. This was known as the expanded operating envelope programme, and showed that the Mk 48 was structurally reliable at the depth needed to engage 'Alfa' class submarines, the target speed recognition capability required could be achieved, the vertical coverage was adequate as was the self-noise at higher speeds with the existing nose and array, and additional speed could be achieved. Laboratory modifications were made to a few torpedoes for tests, and these changes were implemented

in the form of a programme to update fleet Mk 48 torpedoes to what is now the Mk 48 Mod 4 standard.

Of the performance requirements demanded by the ADCAP programme, the most important are:
(a) sustained long acquisition range
(b) minimised adverse environment and countermeasure effects
(c) minimised shipboard tactical constraints
(d) enhanced surface target engagement capabilities.

Hardware changes involved in ADCAP entail replacing the entire nose of the weapon housing the acoustics and beam forming circuits, and replacement of the signal processing by the latest electronics. The latter will also incorporate the current command and control electronics. Warhead sensor electronics will be improved.

Application of the expanded operating envelope programme findings to ADCAP has resulted in the upgraded ADCAP, which incorporates: upgraded acoustics and electronics; an expanded operating envelope (depth, target speed, weapon speed options); increased fuel delivery rate and capacity for optimum speed and endurance; improved surface target capabilities.

Description

The torpedo is wire-guided through a two-way communications link in the current Mod 3 version.

The Mk 48 torpedo is propelled with an axial flow pump-jet propulsor driven by an external combustion gas piston engine. This engine, like that in the Mk 46 torpedo, is a Gould design. The fuel for the engine is a monopropellant: Otto fuel II.

The Mk 48 torpedo is capable of operation in wire-guided active or passive, acoustic and non-acoustic modes. The acoustic modes of operation allow active or passive target detection capabilities. A complete description of the torpedo will be found in *Jane's Weapon Systems 1987-88*.

Operational Status

The Mk 48 has been in production since 1972 and is used aboard USN attack submarines and strategic submarines for self-defence. By early 1980, more than 1900 torpedoes of this type had been delivered to the USN and an estimated 800 plus were in the production and procurement line. It was then estimated that another 1050 might be required to meet inventory objectives and to allow for peacetime training and testing.

In August 1979, Hughes Aircraft Company received a contract for development of digital guidance and control electronics for the Mk 48 ADCAP programme.

The first test run was carried out by the USN at Nonoose Bay in early 1982, using the inertial guidance system developed for the ADCAP programme.

Data processor cards which are part of digital guidance and control system being developed by Hughes for Mk 48 torpedo ADCAP programme

About 240 more runs were programmed before completion of this phase of the programme, after which it was expected that entry into service would take place in 1983/84. However, the 1985 report by the US Secretary of Defense amended this to indicate anticipated deployment of the system in the mid- to late-1980s.

In the US Secretary of Defense's Annual Report to Congress in February 1985, it was stated that following completion of a successful test programme in 1984, it had been decided to accelerate production of the ADCAP torpedo. The five-year programme called for production of 1890 ADCAP units, and included 123 in 1986 at a cost of $433 million and 280 units in 1987 costing $671 million. Hughes began production of the ADCAP version in 1985. In 1986 Gould (subsequently taken over by Westinghouse) was named the second ADCAP source.

Hughes Aircraft Company has been awarded contracts for 370 plus Mk 48 ADCAP torpedoes and related test equipment, and the Westinghouse Electric Corporation a contract for 96 Mk 48 ADCAP. In June 1989 the US House Armed Services Seapower Subcommittee cut $331 million from the US Navy's $493 million procurement request for Mk 48 ADCAP for 1990, reducing the proposed buy from 320 to 140 for 1990. At the beginning of 1989 Pentagon had approved ADCAP for full rate production. The USN wants to purchase a total of 3401 Mk 48 ADCAP torpedoes.

The only known foreign users are Australia and the Netherlands.

Specifications

Length: 5.8 m
Diameter: 21 in (533 mm)
Weight: 1600 kg (approx)
Max speed: 55 kts
Max range: 38 km
Max depth: 500 fathoms (914 m)

Contractors

Hughes Aircraft Company, Ground Systems Group, Fullerton, California (Mk 48 ADCAP version).

Westinghouse Electric Corporation, Baltimore, Maryland.

GUIDED WEAPONS

AUSTRALIA

IKARA

Type
Surface/subsurface ASW torpedo missile system.

Development
Initial design was undertaken by the Australian Government. The version to meet RN requirements was subsequently developed in a joint Australian/British programme.

Description
The concept of Ikara is the employment of a guided missile to deliver a lightweight anti-submarine homing torpedo to the target submarine. The missile is launched from a surface ship which uses a computer to calculate the torpedo dropping position. Target information from the firing ship's own sonar, or via a remote radio linked source, together with information regarding the in-flight position of the missile and other data such as ship's position, wind, and so on, are processed by the computer.

The guidance system ensures that the missile flies to the continuously updated optimum dropping position to maximise kill potential.

After release from the Ikara vehicle the torpedo descends by parachute, which is discarded when the torpedo reaches the sea. The torpedo then carries out a homing attack on the target submarine.

In 1982 a number of additional optional torpedo payloads were announced; these included the British Sting Ray, Swedish TP42, Italian A244/S and Japanese Type 73.

The main structural strength is provided by the two-stage Murawa combined boost and sustainer rocket motor that powers it. In previous versions of the system Ikara is launched at a fixed elevation of 55°, generally from a trainable launcher, but in 1982 a 'boxed' version of the system was revealed in which Ikara vehicles are carried in individual container/launchers which are bolted to the ship's deck.

Ikara is capable of attacking submarines out to the maximum range of the ship's sonar regardless of the weather conditions.

Target information from the ship's long-range sonar is fed into a computer which calculates the dropping position, taking into account such factors as ship's own course and speed, wind effect and target movement during time of flight. The outputs from the computer are passed to the missile via the guidance

Ikara being launched from Royal Australian Navy destroyer. This system is in service with the RAN and RN and a variant (BRANIK) has been installed in four Vosper Mk 10 frigates of the Brazilian Navy

system, a ship-mounted radio/radar system enabling the missile to be tracked and guided accurately to the drop zone where command signals initiate the torpedo release sequence.

The Ikara launcher ensures that the missile takes up its correct flight path as quickly as possible, while the automatic handling system ensures rapid reloading from the magazine, where the missiles are stowed with their torpedoes attached.

The layout of the magazine and handling area varies considerably between the classes of ships already fitted with Ikara, and designs are available to cater for the differing requirements of ships ranging from 1500 tons upwards.

BRANIK
The requirement to fit Ikara to the Brazilian 'Niteroi' class Mk 10 frigates gave rise to the development of a third version of the system. Known as BRANIK, this version again differs from the Australian and RN versions in the way in which the launcher and missile obtain computer service. In the 'Niteroi' class weapon control system two fire control computers (Ferranti FM 1600B) are used to control all the ship's weapons. The BRANIK system employs a special-purpose missile tracking and guidance system which is fully integrated with one of these computers. A lightweight semi-automated missile handling outfit is also incorporated in this version.

BRANIK was the subject of a joint development programme by the Australian Government, Vosper

Thornycroft, British Aerospace Dynamics and Ferranti.

In April 1982 the Australian Minister of Defence announced a programme to improve the Ikara system and extend its operational life by reducing size and weight and improving reliability and performance. Current Australian plans relate to development of a system in which Ikara may be used in a containerised form (possibly modified to have folding wings) together with SSMs such as Otomat to comprise a ship's weapons fit.

A modified Ikara could carry as its payload any one of the US Mk 44 or 46, Swedish Type 42, Italian A244/S or UK Sting Ray lightweight torpedoes.

Operational Status
Australia has now withdrawn the last of its Ikaras from service, but the system is still in service with the Brazilian Navy, supported by BAeSema and ASTA.

Specifications
Length: 3.42 m
Span: 1.52 m
Height: 1.57 m
Propulsion: 2-stage Murawa rocket motor
Speed: High subsonic
Range: 20 km (approx)

Contractors
BAeSema Marine Division, Bristol.
    Office of Defence Production, Canberra.

SUPER IKARA

Type
ASW torpedo-carrying missile system.

Development
Super Ikara is an anti-submarine weapon system being developed by British Aerospace in collaboration with the Government of Australia. The new shipborne system will complement the helicopter and significantly contribute to the philosophy of layered defence.

Description
In operation, a box-launched air flight vehicle accurately delivers a torpedo to a predicted target position. The vehicle can carry any one of a range of lightweight torpedoes and engage a submarine over 60 miles (96 km) distant from the point of launch. After the torpedo is released the air flight vehicle continues for some distance beyond the seduction range of the torpedo before diving into the sea. Folding wings enable the vehicle to be launched from compact deck-mounted box launchers that can be

readily fitted to existing ships down to the size of fast patrol boats.

A unique feature of Super Ikara is the ability to loiter in the vicinity of the target, and mid-course corrections can be provided if the target changes course. It will be possible to hand over control to a helicopter or other forward platform for extended range engagements, enabling maximum use to be made of the latest target data.

With emphasis on a compact, versatile, reliable system, which can also be used on oil rigs or shore-based platforms, Super Ikara is designed to be an effective counter to the modern submarine threat.

Operational Status
Although development work has currently stopped, BAeSema and ASTA will be presenting a Super Ikara concept to the US Navy.

Contractors
Office of Defence Production, Canberra.
    BAeSema Marine Division, Bristol.

Trials launch of box-launched, folded-wing Ikara missile as part of the Super Ikara development programme

CHINA, PEOPLE'S REPUBLIC

CY-1

Type
ASW surface-launched missile system.

Description
The CY-1 single-stage anti-submarine ballistic missile was first revealed in 1987. It is designed to be used by surface ships from multi-cell launchers. No further details are available.

Contractor
China Precision Machinery Import and Export Corporation, Beijing.

FRANCE

MALAFON

Type
Surface-to-subsurface torpedo missile system.

Description
Malafon is a shipborne weapon consisting of a radio command guided winged vehicle carrying a homing acoustic torpedo. It is intended primarily for use from surface vessels against submarines, but may also be used to attack surface targets.

Malafon has the appearance of a small conventional aircraft with short, unswept tapered wings and a tailplane fitted with endplate fins.

The missile is ramp-launched and propelled by two solid-fuel boosters for the first few seconds of flight. Subsequent flight is unpowered. A radio altimeter is fitted to the missile to maintain a flat trajectory at low level. On reaching the target area, approximately 800 m from the target's estimated position, a tail parachute is deployed to decelerate the missile. The homing torpedo is ejected from the remainder of the vehicle and enters the water to complete the terminal guidance phase of the attack by acoustic homing.

Target detection and designation in the case of submerged targets is by means of sonar, and by radar in the case of surface targets. These sensors, as appropriate, are used during the flight of the missile to provide data on the target for the generation of command guidance signals, which are sent via radio command link to guide the missile. Missile tracking is aided by flares attached to the wingtips.

Operational Status
Malafon was deployed in an interim form on French

Malafon anti-submarine missile on its mount on the French destroyer Aconit *(Stefan Terzibaschitsch)*

Navy vessels while full development trials were still in progress, these installations being updated as development continued. These fittings have been referred to as Malafon Mk 1 systems. Deployment of the system includes installations in the destroyer *Aconit*, two 'Suffren' class destroyers, three 'Tourville' class destroyers and the destroyer *La Galissonnière*. Malafon is now obsolescent and will be gradually replaced by Milas beginning in 1993.

Specifications
Length: 6.15 m
Diameter: 0.65 m
Wing span: 3.3 m
Weight: 1500 kg (launch)
Range: 13 km (max, approx)

Contractor
Société Industrielle d'Aviation Latecoere, Paris.

SM POLYPHEM

Type
Submarine air defence system.

Description
SM Polyphem is a missile system designed for the self-defence of submarines against ASW aircraft and helicopters. The system will have a range of 10 km and is designed to achieve a kill before the air platform has a chance to launch its ASW weapons.

The munition, which can be launched at depths of several hundreds of metres, will be carried in a launch tube designed to fit the submarine's torpedo tubes. The launch tube (which will resemble the tube used to deploy the submarine-launched Exocet), will propel the missile payload up to the surface. On breaking the surface the launch tube will separate into two halves, allowing the missile's motor to ignite the moment it emerges into the atmosphere. The missile will be linked to the submarine by a broad bandwidth fibre optic two-way link (in much the same way as a wire-guided torpedo), through which the missile will communicate with the fire control system in the submarine. Using the two-way link and the one or more sensors with which the missile will be fitted, the system will carry out automatic target recognition and identification, automatic tracking and engagement of selected targets, image correlation, and hit confirmation. The missile will have a high ECCM capability against probably both IR and jamming. It is anticipated that the missile will achieve a speed of between 150 and 300 m/s with a manoeuvrability of 10 g. The sensor package will include off-the-shelf sensors operating in the visible or IR spectrum in the 3-5 micron band.

Contractor
EUROMISSILE, Fontenay-aux-Roses.

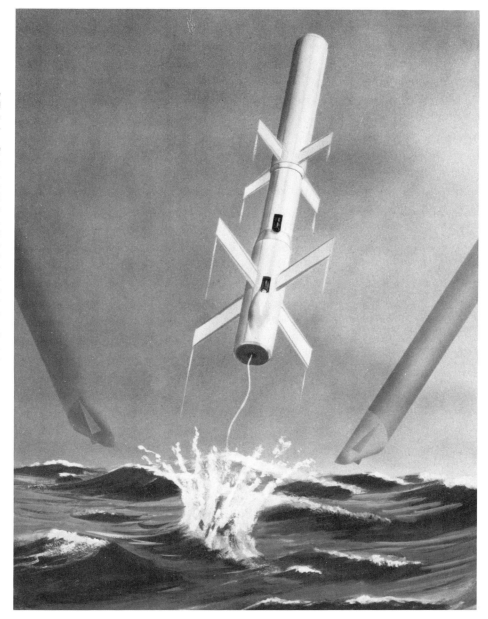

SM Polyphem – an underwater launched anti-air missile for submarine defence

SM 39 EXOCET

Type
Subsurface-to-surface anti-ship missile.

Description
The French Navy employs a submarine-launched version of Exocet, known as the SM 39.

Like the other Exocet missiles, with which it has many components in common, the SM 39 is an all-weather, fire-and-forget, sea-skimming weapon. Target acquisition and tracking is carried out by an active electromagnetic homing head.

The SM 39 missile consists of a watertight, highly resistant, powered and controlled underwater capsule, housing an aerial missile before the latter's ejection following surface broach.

Known as the VSM (*véhicule sous-marin*), the capsule can be discharged from a submarine's 533 mm standard torpedo tube at any speed. At a safe distance from the submarine, the solid propellant motor of the VSM is ignited and the VSM is piloted up to ejection of the aerial missile by jet deviation of gases; this deviation is performed by four interceptors actuated by electromagnets placed to the rear of the nozzle outlet section.

The guidance orders for pitch and yaw, which slave the VSM to a preset bearing and attitude depending on the launch depth and offset angle, are elaborated by the inertial system and guidance computer in the aerial missile.

The missile breaks the surface at an angle of 45° in the direction of the target, whatever the sea state. The nose cone is jettisoned. Jet interceptors enable the VSM attitude to be stepped down to limit the

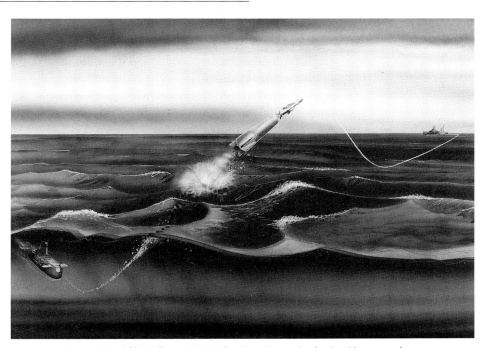

Impression of SM 39 Exocet ejecting from launch capsule after breaking sea surface

culmination altitude. After unlatching, the aerial missile is ejected, its wings and control surfaces are deployed, the boost motor is ignited and then it flies towards impact with the target like all other Exocets. All VSM's components sink rapidly to avoid detection.

Operational Status
Development of the SM 39 was completed in 1984. Successful launches have been carried out by the diesel-electric submarine *La Praya*, the nuclear-powered submarine *Saphir* and the nuclear-powered

ballistic missile submarine *Inflexible* from periscope to maximum launching depth at the Centres d'Essais de la Mediterranée (CEM) or des Landes (CEL). SM 39 is now operational in 15 French submarines.

Specifications
Length (launch vehicle): 5.8 m
Weight: 1345 kg (approx)
Range: 50 km

Contractor
Aerospatiale, Division Engins Tactiques, Châtillon.

INTERNATIONAL

MILAS

Type
Surface-launched ASW torpedo-carrying missile system.

Development
Milas was first revealed at the Le Bourget Naval exhibition in 1986 and is being developed to meet a joint French and Italian naval requirement for an air flight vehicle which will be carried by frigates and be capable of carrying a lightweight anti-submarine torpedo.

The first two development firing trials of Milas were conducted at the San Lorenzo range in Sardinia on 21 and 27 June 1989. The successful firings validated the separation of the torpedo from the carrier.

Description
The air vehicle is derived from the Otomat missile and will have a range of more than 28 nm. Milas will be a quick reaction system and target data will be updated in flight. The Milas vehicle will be 6 m long and will weigh 800 kg. French versions will carry the Murène torpedo while Italian versions will carry the Whitehead A290 or any other lightweight torpedo.

Operational Status
Under development with first production delivery

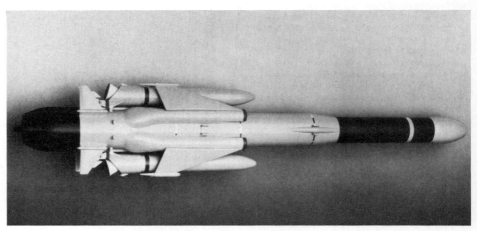

The Milas ASW missile under development by Matra and OTO Melara

schedules for 1993. Beginning in 1993 Milas will arm French Navy vessels currently equipped with Malafon.

Specifications
Length: 6 m
Diameter: 0.46 m

Span: 1.06 m
Weight: 800 kg
Speed: 1080 m/s
Range: 55 km

Contractors
Gie Milas formed by: OTO Melara, La Spezia.
    Engins Matra, 78-Vélizy.

UNION OF SOVEREIGN STATES

FRAS-1

Type
Anti-submarine rocket.

Description
FRAS-1 (Free-Rocket Anti-Submarine) is the designation that has been assigned to the anti-submarine missile associated with the forward twin launcher

(SUW-N-1) in 'Moskva' class helicopter carriers and 'Kiev' class aircraft carriers. FRAS-1 is reported to have a maximum range of up to 30 km and to consist of an ASROC-type weapon in which a rocket is used to project a nuclear depth charge to the neighbourhood of the suspected target, as determined by sonar and other means, after which the rocket separates from the charge which sinks to a suitable depth before being detonated by either a straightforward pressure/depth fuze or some type of proximity fuze. It

has been conjectured that the probable yield of the nuclear payload is in the region of 5 kT.

Specifications
Provisional dimensions
Length: 6.2 m
Diameter: 55 cm

SS-N-14 'SILEX'

Type
Missile-carrying ASW torpedo system.

Description
The SS-N-14 'Silex' is thought to be somewhat similar in concept to the French Malafon or the Australian Ikara anti-submarine weapons in which a subsonic winged vehicle carries a homing torpedo to the position of the target submarine and can have its course corrected during flight so as to allow for submarine movements and make an accurate drop. In this weapon, however, it is believed that a nuclear warhead with a weight of about 150 kg and a yield in

the low kiloton range can be carried instead of the torpedo. No confirmed photographs of the SS-N-14 have been released for general publication, and indeed, for many years, it was thought to be an anti-ship missile system rather than an anti-submarine system. The evidence which has caused this change of assessment has not been released but is believed to be conclusive.

It is also believed possible that the SS-N-14 has an anti-ship capability and may be able to drop a homing torpedo outside a ship's normal defensive cover. No confirmation of this possibility has ever been seen.

Operational Status
Operational since 1968 and fitted in 'Kirov', 'Krivak I',

'Kresta II' and 'Kara' class ships. Later USS classes noted with this weapon system include the 'Udaloy' class of anti-submarine destroyers which are fitted with two quadruple launchers.

Specifications
Length: 7.6 m
Max range: 55 km
Payload: Nuclear warhead or homing torpedo
Trajectory: Programmed/radio command flight to target area under autopilot control with command override capability. Speed is about Mach 0.95 at 750 m above sea

SS-N-15 'STARFISH'/16 'STALLION'

Type
Subsurface-launched anti-submarine rocket.

Description
The SS-N-15 is reported to be an anti-submarine weapon of the same general type as the American SUBROC, a system in which a submarine-launched missile follows a short underwater path before

broaching the surface to follow an airborne trajectory for the major part of the distance to the target area. On reaching the latter, a depth charge bomb is released to continue on a ballistic trajectory until it enters the water near the target, sinking to the optimum depth before detonation. This type of weapon relies upon accurate localisation of the target in the first instance, followed by rapid launch and flight to the target area before the target has time to travel far from its last known position. Maximum range was originally estimated as about 35 km, but later official US figures suggest between 45 and 50 km. The same source also confirms the torpedo-tipped version

of this weapon, which carries a homing torpedo payload instead of the depth charge. In the homing torpedo-carrying version the weapon is designated SS-N-16.

Operational Status
The SS-N-15 is probably in 'Papa', 'Alfa', 'Victor III', and 'Tango' class USS submarines. Other classes of submarine which may be equipped to operate this weapon include some 'Charlie II' class and 'Victor I' or 'Victor II' boats.

UNITED STATES OF AMERICA

ASROC

Type
Anti-submarine missile system.

Development
ASROC entered service with the USN in 1961, becoming widely fitted in anti-submarine vessels of various classes. Plans to develop the ASROC system to utilise a vertical launch system (VLS), initially for some surface ships but later in attack submarines, were at first suspended on grounds of cost and/or a need to merge the project with the ASW-SOW stand-off weapon programme. However, changing circumstances in the USN's overall ASW requirement and its inventory of weapons in this category led to a reversal of the original cancellation of vertical launch ASROC (VLA), and it is intended that the new version will ultimately replace the original ASROC deployed on surface ships.

Description
ASROC (RUR-5A) is an all-weather, day or night, ship-launched ballistic missile carried as the primary anti-submarine warfare (ASW) weapon aboard USN destroyers as well as some cruisers and frigates. The weapon consists of a Honeywell Mk 46 acoustic homing torpedo or a nuclear depth charge (estimated yield 1 kT) attached to a solid-propellant rocket motor. It can be fired from an eight-cell launcher Mk 112, the Mk 26 launching system, or from the Mk 10 Terrier missile launcher (ASROC/Terrier system).

Other major components include a fire control computer and an underwater sonar detector.

After launch the weapon follows a ballistic trajectory. The booster rocket motor is jettisoned at a predetermined point and the payload follows a ballistic flight to the point of airframe separation. At the point of airframe separation a parachute is deployed to gently lower the torpedo payload to the water. On entering the water the protective nose cap shatters, the parachute is detached and the torpedo's motor starts, the weapon commencing its search function. Depth charge payloads sink to a predetermined depth before detonating.

Range of the weapon is classified but has been estimated to be from 2 to 10 km. The missile is 4.6 m long, has a diameter of 32.5 cm, and a span of 84.5 cm. Launch weight is about 435 kg.

Operational Status
Users of the ASROC system include: USA (27 cruisers, 87 destroyers, 65 frigates), Japan (15 destroyers, 11 frigates), Spain (five each destroyers and frigates), Greece and Taiwan (four destroyers each), Canada (four frigates), Germany (three destroyers), Turkey (three destroyers), Brazil, South Korea and Pakistan (two destroyers each), and Italy (one missile cruiser). In the FY85 report by the US Secretary of Defense, funding for ASROC (VLA) was included for development with $30.8m assigned for 1984, $26.7m in 1985, $37.4m in 1986 and $38.1 m in 1987. In FY87 the US Navy was granted $39.2 million for continued VLA development and a further $39.8 million for production, the latter figure including the introduction of a second source manufacturer. Having won a competition to become the VLA prime contractor in November 1983, Loral Systems Group (formerly Goodyear Aerospace Corporation) conducted a source selection

Eight-round ASROC launcher aboard USS David R Ray *(W Donko)*

and selected Martin Marietta as the VLA second source supplier. ASROC (VLA) will be first installed on 'Spruance' class destroyers and 'Ticonderoga' class cruisers, and subsequently on the new 'Arleigh Burke' class destroyers.

The conventional launch ASROC will continue to be supported by Honeywell as prime contractor to the US Naval Ocean Systems Centre, San Diego.

Contractors
Honeywell Inc, Training and Control Systems Operations, West Covina, California (prime contractor).

Loral Systems Group, Akron, Ohio (prime contractor for VLA development).

SUBROC (UUM-44A)

Type
Submarine-launched anti-submarine missile.

Development
Development of the UUM-44A began in June 1958 at the US Naval Ordnance Laboratory, White Oak, Maryland, under the management direction of the Bureau of Naval Weapons, with Loral Systems Group (formerly Goodyear Aerospace Corporation) as the prime contractor. Technical evaluation was completed in 1964, and production and operational deployment began in 1965.

Description
This is a submarine-launched, rocket-propelled missile which follows a short underwater path before transferring to an air trajectory for the major portion of its journey to the target area. At a predetermined point during boost the nuclear depth bomb is separated from the remainder of the missile and then follows a semi-ballistic trajectory to the point where it re-enters the water. The W55 nuclear charge then sinks to a predetermined depth before detonation. Estimated yield of the warhead is about 1 kT.

Principal characteristics of SUBROC are: length 625 cm, diameter 53.3 cm, weight 1853 kg (approximately), range 56 km (approximately), and speed is supersonic. It is propelled by a Thiokol Corporation solid fuel motor.

The UUM-44A missile forms part of an advanced anti-submarine system designed for deployment in nuclear-powered attack submarines operating against submerged vessels armed with strategic missiles. This system includes the AN/BQQ-2 Raytheon integrated sonar system and the Mk 113 SUBROC fire control system produced by the Librascope Corporation.

After detection and location of a target submarine, co-ordinate data are fed into the attack submarine weapon system which programmes an optimum mission profile for the SUBROC missile. It is assumed that this can be accomplished in virtually real time.

The SUBROC missile is launched horizontally from a standard 21 in (53.3 cm) torpedo tube by conventional means. At a safe distance from the launch vessel (which need not be directed towards the target area for firing) the solid fuel missile motor is ignited and the missile follows a short level path before being directed upward and clear of the water. Missile stability and steering is effected by four jet deflectors, which function in both water- and airborne sectors of the trajectory. Guidance is by means of an inertial system (SD-510) produced by Kearfott.

When free of the water, SUBROC is accelerated to a supersonic speed and guided towards the target area. At a predetermined point, separation of the nuclear depth bomb is initiated by explosive bolts and a thrust-reversal deceleration system, which enables the warhead to continue to the re-entry point on a trajectory controlled by vanes on the depth bomb.

Impact with the water is cushioned to protect the arming and detonation devices. A preset depth sensor detonates the nuclear charge when the bomb is in the vicinity of the target. It is probable that the target position will need to be established with an accuracy that will permit the warhead to detonate within its estimated lethal radius of 5 to 8 km from the target.

SUBROC missiles can be carried in torpedo tubes without attention for long periods, and launched with minimal preparation time.

Operational Status
SUBROC is operational on nuclear-powered attack submarines. Each ship carries four to six SUBROC missiles. An improvement programme to sustain the

SUBROC anti-submarine missile after leaving launch submarine and taking to the air, but before separation of its nuclear depth charge and rocket motor

effectiveness of this weapon against new threats was initiated in 1976-77.

This would have entailed replacement of the SUBROC analogue guidance subsystem by a digital one and regraining of the weapon's rocket motors, and it was estimated that to extend the operational life of the system by five years would have cost about $41 million. A Congressional decision was taken in 1979 not to update the system along these lines, and as a consequence the USN turned to the ASW-SOW (Sea Lance) programme to meet existing and anticipated operational requirements. The SUBROC system is now considered obsolescent and it is proposed to withdraw the system from the USN inventory during the 1990s. It will be replaced by the Sea Lance weapon.

Contractors
Loral Systems Group, Akron, Ohio (prime contractor).

Librascope Corporation, Glendale, California (Mk 113 fire control system).

The Singer Company, Kearfott Division, Little Falls, New Jersey (SD-510 inertial guidance system).

Sandia Corporation, Livermore, California (nuclear warhead).

Morton Thiokol Inc, Elkton, Maryland (solid fuel motor).

SEA LANCE

Type
Submarine-launched ASW stand-off missile.

Development
The USN ASW-SOW programme was approved by the US Secretary of Defense in 1980 to develop a

long-range anti-submarine weapon by the late 1980s. The weapon has now been given the name of Sea Lance and is required to replace the technologically obsolete SUBROC, which the USN plans to withdraw from service during the 1990s. The new weapon is also intended to meet the challenge of new threats such as the latest USS 'Alfa' class of submarine, and it will be designed for launching from US nuclear

attack submarines, and surface fighting ship vertical launch systems. Specific platform applications of Sea Lance have not been defined more precisely to date.

In USN attack submarines the weapon will be used with existing torpedo tubes and the Mk 117 digital fire control system. (The latter is planned to be the first type of existing shipboard hardware to incorporate Outlaw Shark capabilities, now in research and

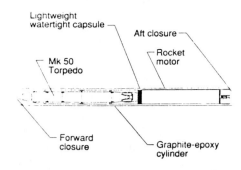

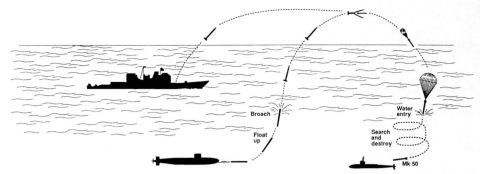

development to provide a correlated, computer-formatted, all-source data handling facility for forces at sea. A major objective of this project is over-the-horizon targeting.) US attack submarines cannot at present launch a weapon against a target unless it closes to within reach of the Mk 48 torpedo. Sea Lance will extend this range appreciably, as well as permitting further attacks on submarines which successfully evade a torpedo attack.

Four separate designs for Sea Lance were proposed but they have in common a combined underwater/air-flight/underwater path from submarine launch platform to target, similar to that of the SUBROC predecessor system, but variations include the method(s) of propulsion employed. Originally, alternative payloads consisting of either a nuclear depth bomb or a Mk 50 homing torpedo were considered, but only the latter is now under development.

As already stated, the proposals made by the various contractors selected to submit concept formulation plans in response to the operational requirement differed in a number of respects and all four were described in *Jane's Weapon Systems 1981-82*. The team selected to continue development of the project was that linking Boeing, Westinghouse and Hercules Aerospace: Boeing as the leader with responsibility for the vehicle; Westinghouse the aft body and capsule components; and Hercules for the single solid-propellant rocket motor.

Description

The solid-propellant rocket motor is apparently designed to propel the weapon during its flight and trajectory. On clearing the surface of the water, four small wraparound fins at the rear of the rocket motor casing deploy automatically to aid stability. After booster burnout, the vehicle follows a ballistic path to the target area, where after deceleration the payload re-enters the water for the payload/warhead section (nuclear or torpedo) to come into action. The missile, including the payload, is 20 ft long.

It will be possible to fire the Sea Lance from surface ships or submarines and rounds will be stored in canisters ready to fire from the Mk 41 vertical launch systems of destroyers ('DD 963' class), guided missile cruisers ('CG 47' class), and guided missile destroyers ('DDG 51' class). From SSN '688' class attack submarines, the missile will be carried by capsule to the surface before the solid-propellant rocket motor ignites to propel it to the target area.

Both the surface-launched and submarine Sea Lance missiles will be able to carry the advanced Mk 50 lightweight torpedo.

Operational Status

The Sea Lance programme was initiated in 1980 and it was then estimated that the research, development, test and engineering cost would amount to a total of about $550 million. Funding is under Programme Element 63367N of the US DoD budget and the actual funding for FY82 was $35.4 million, with $20.2 million in 1983, $27.4 million in 1984, $51.3 million in 1985 and $75.3 million in 1986. In July 1986 Boeing was awarded a contract valued at $380 million to

Launch of the Sea Lance missile from a US Navy facility on San Clemente Island off San Diego. The complete system included full missile, avionics, flight software and all flight controls. The dynamic test launch was a significant milestone in the full-scale engineering development phase of Sea Lance for the US Naval Sea Systems Command

carry out full-scale development. The FY86 report by the US Secretary of Defense stated that Sea Lance will be developed to replace SUBROC aboard USN attack submarines, with provisional entry into service during the 1990s. The programme contract was scheduled to be finally terminated on 15 September 1990, but has now been extended. Sea Lance has been restored to the FY91 budget, but its future depends on the outcome of budget conference committee and the reaction of Congress. The first production buy may take place in the mid-1990s and the

total requirement is for 1052 units at a cost of $1.9 billion. One of the first classes to be fitted will be the '688' class of nuclear attack submarines, although the weapon will be compatible with surface vessels equipped with vertical launch facilities.

Contractors

The Boeing Company, Seattle, Washington; with Westinghouse and Hercules Aerospace.

ROCKETS

FRANCE

ANTI-SUBMARINE ROCKET

Type
Surface-to-subsurface anti-submarine rocket.

Description
This system comprises an anti-submarine rocket launcher associated with a sonar and a computer. The rocket launcher is remotely controlled, aiming and rocket fuzing being determined by the computer, which in turn receives input data from the sonar.

The launcher is made by Mécanique Creusot-Loire under licence from Swedish Ordnance (formerly Bofors) and is a six-tube device with automatic reloading from a magazine. It fires single rockets or salvoes as required, and will accept any of the range of Swedish Ordnance 375 mm rockets, thereby giving a choice of ranges from about 655 m to about 3625 m. Rate of fire can be up to one round per second.

The computer calculates ballistic data for initial velocities of 100, 130, 165 and 205 m/s for the different rockets that may be used with the systems.

Input data from the sonar comprise the location and rate of change of position of the target.

Mécanique Creusot-Loire is also licensed for the manufacture of the twin-tube rocket launcher.

Operational Status
The most recent version is that equipping the destroyer *Du Chayla* and 17 'Type A69' frigates of the French Navy and four 'E71' class escorts of the Belgian Navy.

The French Bofors ASW rocket launcher on the frigate Lieutenant de Vaisseau le Hénaff
(Stefan Terzibaschitsch)

Contractors
Mécanique Creusot-Loire, Immeuble Ile de France, Paris La Défense (launcher).
Thomson-CSF (computer and sonar for French Navy).

NORWAY

TERNE III

Type
Anti-submarine rocket-propelled depth charge.

Description
Terne III is a short- to medium-range anti-submarine weapon system ripple-firing rocket-propelled depth charges, in salvoes of six, against submarine targets. It is designed to be the principal anti-submarine weapon for ships up to frigate size, and a secondary weapon for ships carrying long-range anti-submarine armament. Salvo patterns can be formed according to the tactical situation up to the moment of firing. Recoverable practice missiles are available. The remote power-controlled six-tube launcher is automatically loaded from a manually loaded magazine

hopper. The launcher is specially protected against arctic conditions.

The main features of Terne III are:
(a) a unified and balanced design
(b) low weapons weight and small space requirements. A complete installation weighs less than 10 tonnes
(c) the weapon part can be interfaced with existing sonars and other target data sources. Norwegian frigates have a new modular fire control unit to which the Terne III system is interfaced via databus STANAG 4156
(d) freedom of choice in the position on board
(e) remote-control firing at elevation angles between 47° and 77°, giving variable ranges throughout 360° of bearing
(f) depth charge is fitted with proximity/time/percussion fuze
(g) high rate of fire.

The system is scheduled to be upgraded to Mk 10 version between 1991 and 1993.

Operational Status
In production since 1960. Fitted in Royal Norwegian Navy ships.

Specifications
Length: 1.97 m
Diameter: 20 cm
Weight: 120 kg
Warhead: 70 kg TNT equivalent
Fuze: proximity, time and impact
Range: 400-1600 m (Mk 8 version)

Contractor
NFT, Kongsberg.

SWEDEN

TYPE 375

Type
Shipborne surface-to-subsurface short- and medium-range ASW rocket system.

Description
The ship's sonar provides submarine position for prediction of Swedish Ordnance 375 launcher elevation and bearing data. The launcher has either two or four

tubes and can fire single- or multiple-shot salvoes. A special design ensures a predictable and accurate underwater trajectory. The launcher is reloaded by automatic means from the magazine, which is disposed of directly below the launcher.

Rockets have two types of motor giving differing range brackets. The rocket trajectory is flat, thus giving a short time of flight to minimise target evasive action. Fuzes are fitted with proximity, time, and DA devices.

The 80 mm MI type is launched from a special holder that is handled as a live rocket in the operating room, hoist and launcher.

Operational Status
The twin-tube launcher and anti-submarine rockets are in production for a number of navies.

Contractors
Swedish Ordnance, Eskilstuna.

Specifications

375 mm ASW	Weight with fuze	w/out fuze	Length time fuze	proximity fuze	Charge	Max velocity 1 motor	2 motors	Final sinking speed	Range to 30 m depth
Erika	250 kg	236 kg	2000 mm	2050 mm	107 kg hexotonal	100 m/s	130 m/s	10.7 m/s	655-1635 m
Nelli	230 kg	216 kg	2000 mm	2050 mm	80 kg hexotonal	165 m/s	205 m/s	9.2 m/s	1580-3625 m
Practice									
Type H	100 kg	86.5 kg	—	—	—	135 m/s	—	—	980-1570 m
80 mm MI 15.5	—	—	1120 mm	—	—	130 m/s	—	—	980-1600 m

SHIPBORNE FOUR-TUBE ANTI-SUBMARINE ROCKET LAUNCHER

Description
The Swedish Ordnance 375 mm four-tube anti-submarine rocket launcher is electrohydraulically driven, has an integral fixed-structure loading hoist with a rotatable rocket table, and is remotely controlled. For further details see previous editions of *Jane's Weapon Systems*.

Operational Status
Design was started in 1948 and the launcher first went into service with the Royal Netherlands Navy, which had initiated the work, in 1954. It was subsequently supplied to Argentina, Colombia, France, the then West Germany, Japan, Portugal and Sweden.

No longer in production but believed to be still in service with a number of navies; it has been superseded by the two-tube launcher.

Contractor
Swedish Ordnance, Eskilstuna.

SHIPBORNE TWO-TUBE ANTI-SUBMARINE ROCKET LAUNCHER

Description
The Swedish Ordnance ASW rocket launcher is used for firing anti-submarine rockets at close and medium ranges. The launcher has an integral motor-driven twin hoist, for loading both tubes at once, and a rotating loading table. The laying mechanism is electro-hydraulic with a choice of local or remote operation. Fuze setting is with the rocket in the tube. The rocket table holds four rounds and the total number of rockets in the operating room is 24. The shortest time between successive firings is one second; the time for firing six ready rockets (four on the table and two in the tubes) is one minute, and the firing rate for continuous fire is two rockets every 45 seconds. The manning complement during continuous firing is three men.

Operational Status
In production.

Specifications
Weight, excluding rockets but including flame guard and deck plate: 3.8 t
Traversing speed: 30°/s
Elevating speed: 27°/s
Traverse limits: Unlimited
Elevation limits: 0 to +90° (mechanical); 0 to +60° (for firing)
Power supplies: 440 V, 60 Hz, 3-phase
Power consumption: Tracking 6 kW (mean)

Bofors twin-tube ASW rocket launcher

Contractor
Swedish Ordnance, Eskilstuna.

ELMA/SAAB ASW-600

Type
Lightweight anti-submarine weapon system.

Development
Elma was developed to meet the threat of small and medium sized submarines operating in coastal waters in the Baltic. Development was undertaken as a collaborative venture between the Swedish Navy, Swedish Ordnance and Saab Missiles. The system's commercial designation is Saab ASW-600.

Description
Elma has been designed to achieve a direct kill. The system comprises four launchers, each containing nine non-magnetic stainless steel tubes for firing the projectiles. These are mounted on a support structure at an angle of 30°.

The projectile is fitted with a shaped charge specially designed to achieve good penetration of submarine hulls. A jet beam punches a hole in the submarine's pressure hull after passing through the water-filled outer casing. The projectile features a device which prevents it from bouncing or tumbling when it strikes a moving target.

The projectiles are loaded down the muzzle of the tube, while the propulsive charge is loaded from the base of the tube. The weapon is fired from the ship's bridge in a programmable pattern.

Grenade firing is initiated from the Combat Information Centre in a programmable pattern when a submarine target has been identified and located. A carpet of 36 grenades can be laid over a pinpointed submarine within a range of 350 m.

Operational Status
Elma is operational on all corvettes, FAC and MCMVs

Elma ASW grenade system launchers aboard a Swedish FPB

of the Swedish Navy and will equip all new ASW vessels. The system is also operational on board Finnish patrol boats.

Specifications
System
Number of grenades: 9, 18, 27 or 36 per salvo
Range: 300-400 m (approx)
Dispersal: Adjustable

Grenade
Calibre: 100 mm
Length: 477 mm
Weight: 5.5 kg

Equipment weights and dimensions
Launcher: 103 kg, 460 × 920 × 620 mm
Relay unit: 7 kg, 330 × 230 × 110 mm
Firing unit: 6 kg, 330 × 230 × 148 mm
Rectifier: 34 kg, 610 × 315 × 210 mm

Contractors
Saab Missiles, Linkoping.

UNION OF SOVEREIGN STATES

ASW ROCKET LAUNCHERS

There is a wide variety of anti-submarine rocket launchers in the USS fleet, although all appear to operate on much the same principle of firing charges in a pattern ahead of an attacking ship from a mounting which can be trained and elevated under remote-control. Two different calibres of 250 mm and 300 mm have been reported and the weight of the bombs in the 250 mm versions has been estimated as 180 to 200 kg. Few details are available of the many different systems but such as are known are given below. The systems may be designated as either RBU or MBU and the associated number is understood to mean the maximum range possible in metres.
MBU 1800: A five-barrelled 250 mm calibre system fitted in some of the smaller and older ships and now probably obsolescent. The launching tubes are in two horizontal rows with three in the top row and two in the bottom row.
MBU 2500: A 16-barrelled 250 mm calibre system of much the same vintage and design as the MBU 1800 with two horizontal rows each of eight launching tubes about 1.6 m long. Depth bombs are loaded by hand. The system is fitted in the older cruisers and destroyers although also in the more modern but smaller 'Petya I'.
MBU 4500A: A six-barrelled version hitherto seen only in the one major fleet support ship *Berezina*. It is thought to be of 300 mm calibre with a launching tube 1.5 m long and to use automatic reloading.
MBU 6000: The most modern system, probably of 300 mm calibre, and fitted very widely. Launch tubes are arranged in a circular fashion and it is believed that they are reloaded automatically by bringing them to the vertical and then indexing one by one while depth bombs are loaded from below. Range is 6000 m and the launch barrel length has been extended to about 1.8 m.
MBU 1200: A six-barrelled system which is fitted normally on the quarters of larger ships such as the *Kirov*, *Kara* and *Kresta*. Almost certainly it has a range of no more than 1200 m and is probably intended to deal with torpedoes if they can be detected in the closing stages of an attacking run.

UNITED KINGDOM

LIMBO AS Mk 10

Type
Surface-launched anti-submarine mortar.

Description
This is a shipborne surface-to-subsurface medium-range anti-submarine mortar system. Mortars are stabilised in pitch and roll by a metadyne system referenced to the ship's stable platform. For details of this operation see *Jane's Weapon Systems 1987-88*.

Weight of the projectile is about 175 kg with a 92 kg HE warhead and range is about 900 m.

Operational Status
Only a few systems remain in service: in the UK frigate HMS *Ariadne*, on the three ex-UK 'Tribal' class frigates in the Indonesian Navy, and two UK-built frigates of the Malaysian Navy.

Contractors
Manufactured to MoD (Navy) designs by several contractors.

MINES

BRAZIL

MCF-100

Type
Moored contact mine.

Description
The MCF-100 moored contact mine is designed for use against submarines and surface ships in depths between 10 m and 100 m in strong currents. The mine can be programmed to remain inert on the bottom for a fixed period of time and then to self-release and anchor itself at the desired depth.

The newly developed installation procedure allows the mine to be moored with an accuracy in anchoring level of 0.50 m independently of currents and surface waves.

During 1992 Consub will complete development of an influence (magnetic and acoustic) sensor which will increase the efficiency of the MCF-100. This sensor could also be used in combination with up to three explosive modules, a priming charge and fuze, resulting in a powerful ground influence mine which could also be deployed from submarine torpedo tubes.

Operational Status
A contract for 100 mines was placed by the Brazilian Navy in August 1991.

Specifications
Length: 1400 mm
Width: 1020 mm
Height: 1500 mm
Weight: 770 kg
Charge: 160 kg Trotyl
Operating depth: 10 to 100 m

MCF-100 mine on launching rail

Contractor
Consub Equipamentos e Servicos Ltda, Rio de Janeiro.

CHILE

MS-L

Type
Ground mine.

Description
The MS-L magnetic ground mine is designed for operations against both surface ships and submarines. The mine can be laid from 533 mm submarine torpedo tubes, by surface vessels and from cargo aircraft when the mine is fitted with a special parachute.

The mine incorporates the following subassemblies: fuze, safety and arming unit; body and chamber with explosive charge; intermediate body and electronic unit; stern and ogive.

The microprocessor control unit is activated by a master switch (manual- or remote-operated) and becomes operable when the hydrostatic pressure exceeds one bar (corresponding to a depth of about 10 m). At this point the weapon's clock begins to run and after a predetermined period of time the fuze unit is armed. Ninety seconds later the microprocessor logic becomes active.

Signals from the magnetic sensors are amplified and analysed and, according to the setting of one of five parameters, either activate the detonator or inhibit the acceptance of signals for a predetermined period of time. The mine can also be de-activated electronically by putting very low resistors across the batteries when a predetermined period of time has elapsed or the batteries have reached a predetermined level below their nominal voltage.

The magnetic sensor generates a signal of double polarity which is filtered to reject either extremely slow or very fast varying magnetic fields. The change in polarity of the signal and the fact that in both polarities the signal amplitude must exceed a prefixed value are essential requirements for the signal to be further analysed.

The time which elapses between the receipt of signals which exceed the prefixed amplitude values must fall within a fixed time window. This window is determined by a minimum and maximum time measured in seconds (it is used to determine specific targets travelling at specific speeds). If this condition is satisfied, the ship counter advances one step. When the number of steps in the ship counter corresponds to its prefixed number, the mine is detonated.

To counter minesweeping operations, irrespective of whether the sequence of operations outlined above has been completed or not, the circuit inhibits itself for a prefixed period of time.

The various programmable parameters include magnetometer sensitivity (2, 4, 8, 16 and 32 Mega Gauss); fuze arming times (2, 4, 16, 32, 64, 128, 256 and 512 hours); inhibition time (32, 256 and 512 seconds); lower limit window time (2 and 4 seconds); upper limit window time (16, 32, 64 seconds); ship counter 1 to 9; sterilisation time (2, 4, 8, 16, 32 and 64 weeks).

Contractor
Fabricas & Maestranzas de Ejerato (FAMAE), Santiago.

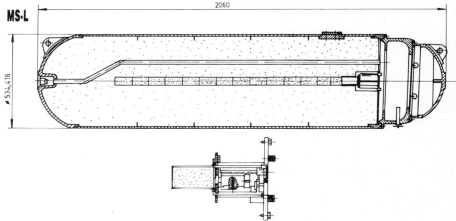

MS·L
2060
⌀534,416

MS-L

MS-C

Type
Ground mine.

Description
This magnetic influence ground mine is designed for anti-invasion duties against landing craft. It is easily carried and laid from aircraft, enabling a threat area to be rapidly mined. It operates in a similar way to the MS-L mine.

Contractor
Fabricas & Maestranzas de Ejerato (FAMAE), Santiago.

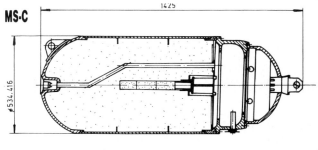
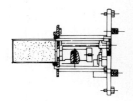

MS·C
1425
⌀534,416

MS-C

CHINA, PEOPLE'S REPUBLIC

CHINESE NAVAL MINE UPGRADING SYSTEM

Description

The Chinese Navy, which largely relies upon USS-designed naval mines, has developed an 'intelligent' sea mine actuation system for what are described as 'large' and 'medium' moored mines.

The system features a programmable central processor which can accept inputs from acoustic and magnetic sensors and, optionally, pressure sensors as well. It incorporates a ship counter system which can permit up to 15 actuations before detonation, a delay mechanism of up to 250 days before arming and a self-destruction timer for up to 500 days. There are eight operating modes which are believed to be mixtures of fuze and logic settings to meet different operational or environmental conditions.

Contractor

Dalian Warship Institute, Dalian.

DENMARK

MTP-19

Type

Cable controlled mine.

Description

The MTP-19 is a fully remote-controlled mine designed to block channels in uncontrolled minefields, the entrances to harbours and protected anchorages, and so on. The mine system comprises a portable mains-/battery-powered weapon control unit, distribution box and the mine itself which consists of two sections – the flotation unit and the weapon which incorporates a microcomputer, which enables settings to be changed to meet particular needs and situations, and the explosive charge. A number of mines can be linked together via the distribution box and each can be independently controlled from the control unit. The weapon can be laid from minelayers, ferries or any other suitable vessel. It is mounted on a trolley which is designed to fit the transport rails in Danish ferries. After launching the mine is turned to the correct attitude within the first 3 m of water and the weapon's electronics actuated. After laying the two parts of the mine separate and the buoyancy unit is retrieved enabling the cable connecting the buoyancy unit to the mine to be hoisted and reconnected to another cable or the distribution box. The control unit retains in its memory the exact location of every mine it controls and enables the acoustic and magnetic

sensors to be set in four steps of sensitivity. The control unit also indicates the complete status of each mine. Mines can be armed/detonated manually from the control unit ashore, which can be linked to an automatic sensor system indicating the approach of targets. In the auto-alarm mode the mine uses its own sensors to actuate the detonator. The final arming mode can be set in the event that should the cables be cut, damaged or disconnected, the mine automatically activates itself, after which it remains fully active all the time. In view of the range of automatic modes available, the mine incorporates a very high degree of safety. In an emergency the system can be sterilised by short circuiting the batteries and disconnecting the detonator, which can be done from the shore. The mine can also initiate self-sterilisation under certain conditions, for example, when battery power drops below a certain sensitivity. The normal operating depth of the mine is 20 m, but it can operate at greater depths. Weapon control can be exercised at distances up to 12 km, but this too can be extended if required. The charge is 300 kg of high explosive.

Specifications

Height: 1128 mm (max)
Length: 1000 mm
Width: 1090 mm
Weight: 800 kg
Charge: 300 kg
Operating depth: 3-20 m

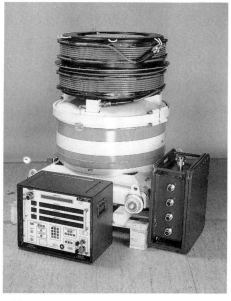

MTP-19 mine with sinker unit, drum of cable, portable weapon control unit and distribution box

Contractor

Nea-Lindberg A/S, Ballerup.

FRANCE

TSM 3510 (MCC 23)

Type

Submarine-launched ground influence mine.

Development

A sophisticated cylindrical mine developed by the French Navy Mine Warfare Laboratory (GESMA) and Thomson-CSF.

Description

The mine is constructed in two sections, one housing the charge and the other the electronics. The mine is equipped with magnetic and acoustic sensors and a safety device for minelaying. The influence sensors, clock settings for arming delay and inhibition time,

TSM 3510 (MCC 23) submarine-launched sea mine

sensor sensitivity thresholds, shape and duration of influence signals, event-time sequence and ship counter are preset according to operational requirements. The mine is fitted with protective devices to prevent accidental detonation in the event of

magnetic storms and sympathetic detonation due to explosion from other causes in the immediate area. The mine has a very low power consumption and is claimed to have an active seabed life of up to two years. It has also proved to be fairly difficult to sweep using classical methods.

Specifications

Length: 2368 mm
Diameter: 530 mm
Weight: 850 kg (loaded)
Charge: 530 kg
Service depth: 150 m (max)

Contractor

Thomson Sintra Activités Sous-Marines, Brest Cedex.

TSM 3530 (MCT 15)

Type

Surface vessel-laid ground influence mine.

Description

This hemispherical ground influence mine is a seabed defensive mine designed to be launched from rails on surface ships. It is fitted with a parachute to retard its

descent which ensures that it is properly positioned on the seabed as it is laid. The mine is fitted either with a magnetic sensor or a combined magnetic/acoustic sensor. All parameters can be preset. The TSM 3530 is armed by clockwork activated time delay, the elapsed time between sowing and arming being reset.

Operational Status

In service with French and other navies.

Specifications

Length: 1100 mm
Diameter: 1200 mm
Weight: 1500 kg (loaded)
Charge: 1000 kg
Service depth: 100 m (max)

Contractor

Thomson Sintra Activités Sous-Marines, Brest Cedex.

GERMANY

FG 1 SEA MINE

Type
Surface or submarine-launched ground influence mine.

Description
The FG 1 is designed to damage or destroy either surface shipping or enemy submarines and warships.

The mine casing is of cylindrical configuration, constructed mostly of non-magnetic material, and the mine itself consists of two sections which are combined by means of a clamping ring. The first (smaller) section is the control section which contains the firing mechanism and the safety and arming unit with its ancillary items. The other section contains the explosive charge.

An interface section, located at the bottom of the control section, houses the firing system, which is an electro-pyrotechnic link between the fuze system and the charge. The FG 1 mine is equipped with a proximity fuze. Sensors in the fuze system detect physical fields generated by the hull of a vessel which are then evaluated by the fuze system. If the vessel is within the effective range of the mine, (that is, if the signals reach a predetermined level) the charge is detonated.

Depending upon the version, the mine is equipped with various sensors and fuzes. These differences can be established from the external modifications to the mine and the different equipment and sensors attached to it.

Prior to laying, the power circuit is connected by inserting a plug into the carrier system. In a live launch, arming of the safety and arming section and water pressure switch is by means of a rope attachment. For surface vessel laying, this function can be performed manually.

Operational Status
In production. For training, a recoverable mine, FG 1-Ex, is available.

Specifications
Length: 2310 mm; 710 mm (control section); 1600 mm (charge section)
Diameter: 534 mm
Weight: 770.5 kg; 143.5 kg (control); 627 kg (charge)
Charge: 535 kg HE
Service depth: 60 m (max)
Negative buoyancy in water: 291.5 kg

Contractor
Faun-Hag, Lauf ad Pegnitz.

SM G2

Type
Ground influence mine.

Description
The SM G2 heavyweight, non-magnetic ground mine is designed for blocking shipping lanes, and for laying defensive minefields in coastal waters and sea areas.

The mine comprises an equipment section incorporating acoustic, magnetic and pressure influence sensors and signal processing electronics and an explosive charge section with detonator and safety devices.

The sensors detect target signatures which are then analysed by the signal processor and compared with pre-programmed data held in the computer's programmable memory. When certain preset parameters are recognised the detonator is actuated and the mine explodes. The analysis programs and parameters are fed into the onboard computer via the testing and programming units; particular tactical parameters can still be programmed into the weapon just before laying.

Operational Status
Serial production has been authorised under a MoU between Denmark and Germany and assembly will be carried out in German and Danish naval arsenals.

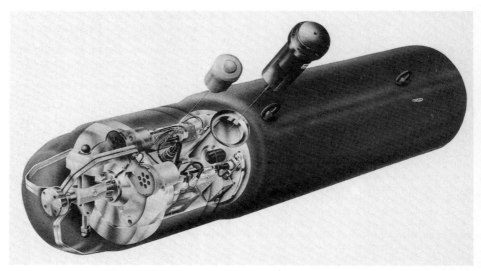

For training purposes an exercise mine is available, and for data acquisition of ships' signatures a special version is also available.

Specifications
Length: 2 m (approx)
Diameter: 600 mm (approx)
Weight: 750 kg (approx)
Operating temperature range: −2.5 to +38°C

Contractor
Atlas Elektronik, Bremen (prime contractor).

SAI/AIM

Type
Anti-invasion mine.

Development
STN Systemtechnik Nord (formerly DMT Marinetechnik) has developed the SAI (Seemine Anti-Invasion) mine under a bilateral programme in conjunction with the German MoD and another NATO country.

Operational Status
SAI has been produced for the German Navy and another NATO country, and is generally available for NATO navies. STN Systemtechnik Nord is also studying the market for a slightly modified version of the SAI mine for non-NATO countries. This version will be designated AIM (Anti-Invasion Mine) in accordance with the regulations of the governments co-operating in the bilateral development and production programme.

Contractor
STN Systemtechnik Nord, Bremen/Hamburg.

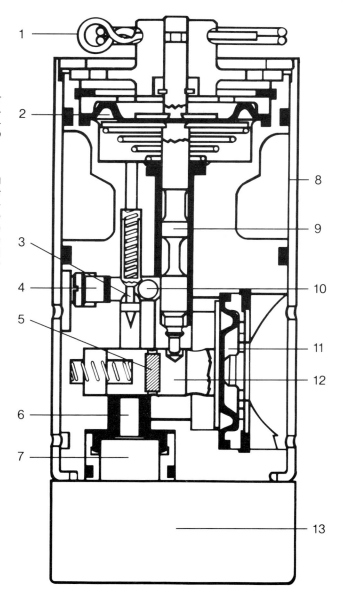

1 Safety pin
2 Diaphragm
3 Firing pin
4 Calibrated leak
5 Detonator
6 Lead Cup
7 Booster
8 Case
9 Plunger
10 Steel ball
11 Diaphragm
12 Slider
13 Explosive Charge

DM 211 and DM 221

Type
Anti-frogman depth charge and underwater signal charge.

Development
The anti-frogman depth charge is designed to provide an effective weapon for the protection of ships and harbour installations against frogmen. The underwater signal charge is used for encoded submarine-to-surface ship communications. Both charges incorporate safety features providing for safe handling and use by untrained personnel.

Operational Status
In production and used by the German Navy.

Specifications
DM 211
Length: 268 mm
Diameter: 60 mm
Weight: 1400 g
Charge: 500 g
Operating depth: 6 m
Throwing distance: 15 m
DM 221
Length: 145 mm
Diameter: 60 mm
Weight: 800 g
Charge: 50 g
Operating depth: 6 m
Throwing distance: 20 m

Contractor
Rheinmetall GmbH, Düsseldorf.

IRAQ

SIGEEL/400

Type
Ground mine.

Description
This ground mine is designed for laying in both deep and shallow water for use against medium and large sized surface targets. It can be launched from ships and helicopters. The mine is supplied with both safety and sterilising devices.

Specifications
Height: 850 mm
Diameter: 700 mm (upper); 980 mm (lower)
Weight: 535 kg
Charge: 400 kg

Sigeel/400 mine

AL KAAKAA/16

Type
Floodable submersible mine.

Description
The Al Kaakaa is possibly the largest mine in the world. It is a floodable submersible mine designed to destroy large offshore structures such as drilling platforms, bridges and under-seabed pipes. The shape and structure of the charge has been designed in such a way that detonation achieves its maximum effect, even in very deep water.

Specifications
Dimensions: 3.4 × 3.4 × 3 m
Weight: 16.1 t
Charge: 9 t (equivalent to 13 t of TNT)
Detonation: By timer or remote-control

Model of Al Kaakaa/16 mine

ITALY

MR-80 SEA MINE

Type
General-purpose ground influence mine.

Description
The MR-80 cylindrical seabed mine, which can be laid by any type of platform, is actuated by a combination of influences (magnetic, pressure, low-frequency acoustic, audio frequency acoustic) from the target and is able to damage or sink surface vessels and submarines of all types, either conventional or nuclear-powered. The body is of epoxy resin and glass fibre, which renders the mine very resistant to sea corrosion and makes it lighter than previous generation mines. The fore part contains the explosive charge, while the tail section contains all the actuation, priming and operating devices, and it is closed by means of a cover which houses the acoustic, magnetic and pressure sensors together with the safety and arming device.

All influence devices are connected to a central unit (the 'Final Logic'), where all delay functions, safety intervals, sterilisation, influence combinations, ship counting, anti-removal and firing control are located.

The sole plate supporting all the electronic modules and the magnetic sensor is linked to the cover by slides that run inside the tail section. By unscrewing the bolts of the crown, the tail cover is easily removed with all the devices linked to it; in this way every part of the mine is accessible for assembly and adjustment operations.

The MR-80/1 is an exercise version of the warshot mine incorporating a data transmission device or smoke signals. As such it can be used for training, research, analysis or evaluation.

Operational Status
In production and in service.

Specifications
Length: 2750 mm (Mod A); 2096 mm (Mod B); 1646 mm (Mod C)
Diameter: 533 mm
Weight: 565-1035 kg (according to model and type of explosive)
Charge: TNT, HBX-3 or similar HE; 380-865 kg (according to model)
Service depth: 5-300 m (for surface targets and for submerged submarines)
Power supply: Dry lithium or alkaline-manganese batteries
Arming and sterilisation delay: Variable up to 999 days. Electronic (quartz) clock
Working temperature: −2.5 to +35°C

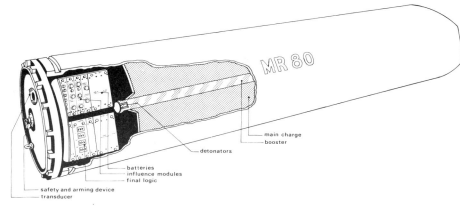

Diagram of MR-80 general-purpose ground influence mine

Storage life: 30 years with a maintenance mean time of 5 years
Operational life: 500-1000 days

Contractor
Whitehead, Montichiari (Brescia).

MP-80

Type
General-purpose ground mine.

Development
Designed by MISAR in co-operation with the Italian Navy to meet a requirement for a general-purpose ground mine capable of meeting the latest concepts on mining doctrine.

Description
The MP-80 uses the same general mechanical structure as the previous MR-80, but with an updated firing mechanism and the addition of sophisticated data processing to improve performance. The target detection device uses four microprocessors to provide digital processing of target signatures and comparison with an onboard threat library. The system can discriminate between targets and countermeasures and when presented with multiple choice targets selects the most suitable.

Activation is by combined magnetic, acoustic and pressure sensors controlled by the programmable microprocessor. Arming and neutralisation delay time is also programmable.

Operational Status
In service with the Italian and other navies.

Specifications
Length: 2096 mm
Diameter: 533 mm
Weight: 780 kg
Service depth: 5-300 m

Contractor
Whitehead, Montichiari (Brescia).

MRP SEA MINE

Type
General-purpose ground influence mine.

Description
The MRP uses a microprocessor firing device that increases the capability of computation of the system, so achieving the maximum effectiveness and discrimination of the target countermeasures. Its mechanical, structural and explosive characteristics are the same as those used in the MR-80 general-purpose ground mine.

The central programming unit (CPU) is used to compile the operator settings, mine programming and testing.

Detonation is microprocessor-controlled by pressure, acoustic and magnetic influences with a wide range of selectivity on the threshold, times, combination logistics and localisation logic.

MRP advanced general-purpose sea mine

Operational Status
In production and in service. MRP customers include India.

Specifications
Length: 2096 mm
Diameter: 533 mm
Weight: 680-780 kg

Operator with MRP and CPU central programming unit

Charge: 520 kg TNT; 620 kg HBX-3
Service depth: 6-300 m

Contractor
Whitehead, Montichiari (Brescia).

MANTA MINE

Type
Shallow water ground influence anti-invasion mine.

Description
The Manta mine can damage, sink or destroy amphibious and landing craft, small and medium-sized surface vessels, and submarines. It operates at depths of between 2.5 and 100 m, and is a dual influence type (acoustic and magnetic). The mine remains effective underwater for more than one year. Its weight and shape are such that it rests firmly on the sea bottom even in running or tidal waters. The mine is 470 mm high and has a diameter of 980 mm. Total weight is about 220 or 240 kg depending on the explosive. The mine consists of two units: the body

Manta shallow water ground influence mine. Key to diagram: (1) main charge (2) booster (3) detonator (4) priming device (5) firing mechanism (6) battery (7) acoustic transducer (8) magnetic sensor (9) calibrated holes

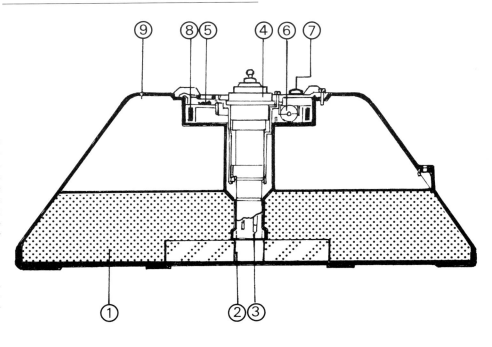

containing the explosive, and the igniter which comprises all the safety, target detection and firing devices. The mine is equipped with all the safety devices needed for handling and transport. The priming device keeps the detonator away from the explosive until the operating depth is reached so as to prevent any explosion due to a casual ignition.

Electronic circuits include:

(a) Delay circuit: started by the priming device, this enables an actuation delay from 0 to 127 days adjustable by steps of half a day to be set

(b) Pre-alarm device: this detects the noises caused by the passage of the targets and energises the sensing circuit

(c) Sensing circuit: activated by the pre-alarm circuit, this actuates the firing circuit

(d) Sterilising circuit: this sterilises the mine after an adjustable preset time.

The mine is laid by surface vessels or by frogmen.

Operational Status
In production and in service.

Specifications
Diameter: 980 mm
Height: 470 mm
Weight: 220 kg (approx)
Charge: 150 kg TNT or 180 kg HBX-3
Service depth: 2.5-100 m
Firing system: Pre-alarm and influence firing circuit

Contractor
Whitehead, Montichiari (Brescia).

MISAR-designed Manta mine

SEPPIA

Type
Moored influence mine.

Description
Seppia is a programmable moored influence mine which can be laid by submarines, surface vessels equipped with rails (manual or automatic in operation) or by aircraft. It is designed to operate on any type of bottom against all types of target, both surface and submerged. As such it complements the range of MISAR-designed ground and shallow water mines to provide a complete range of mines.

Using its built-in microprocessor the mine is fully programmable, enabling it to select specific targets and discriminate against countermeasures. Before laying, the mine's various parameters are programmed according to operational requirements to set depth of operation, sensitivity of influence firing mechanism, arming delay and life time, and ship counter.

Specifications
Length: 1560 mm

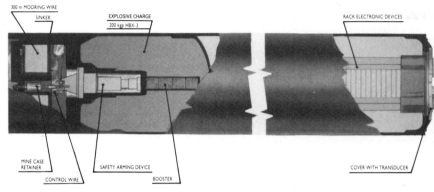

Diagram of Seppia moored influence mine

Diameter: 660 mm
Weight: 870 kg (approx)
Charge: 200 kg HBX-3
Storage life: 30 years

Contractor
Whitehead, Montichiari (Brescia).

EXERCISE MINES

Type
Exercise ground mines.

Description
Exercise variants of all MISAR-designed mines are available, designed to support all mining activities both in workshops and at sea for training and trials in minelaying and minesweeping/hunting. The exercise mines can be used for minehunting/sweeping and mine laying and counter mining exercises, maintenance training, collecting data on the influence signatures of surface vessels and submarines, simulating hostile mines, and the study of new MCM techniques. The exercise variants are almost identical to the warshot versions, which enables operators to become thoroughly familiar with all aspects of the weapon and

to provide a very effective, realistic and accurate method of obtaining data.

All exercise mines are remotely controlled via cable and/or acoustic link. Using these links various functions relating to data collection, hardware and software tests and programming can be performed according to the type of mine being simulated.

Exercise variants available include Seppia/I, which incorporates a microprocessor-controlled target detecting device. The MRP/I exercise mine is supported by a complete range of optional equipments which allow operators to carry out programming exercises on the mine both in the workshops ashore and at sea, testing of hardware and software and the gathering of influence data as well as simulation activities. The MR-80/I differs from the warshot version only in that it contains no explosive, and is thus a very realistic training aid. Manta/I is a shallow water exercise mine which, like the other exercise mines, differs

only in that it does not contain any explosive. For exercises and trials, actuation is indicated either by cable or by a pyrotechnic signal. Use of the cable enables a more accurate study of the firing mechanism in different environmental conditions and against various types of target to be undertaken. From this it is possible to determine the most effective mechanism settings according to the type of target. Using the cable and the RC-Mk 4 data console, pre-alarm (acoustic) device actuation and detection (magnetic) device actuation can be studied. The pyrotechnic signal GP-Mk 4 is more suitable for use on exercises as it enables the mine to simulate an explosion by means of light and smoke produced by the signal.

Contractor
Whitehead, Montichiari (Brescia).

UPDATING MINES

Type
Influence mines.

Description
Large quantities of old sea mines, manufactured shortly after the end of the Second World War, still remain in many inventories around the world. Both technically and operationally these mines are now considered obsolete and are unable to comply with the modern requirements of naval mine warfare. In addition their firing devices, designed and using technology of 40 years ago, are no longer reliable and are thus often unusable. Modern targets present operational, constructional and passive countermeasure techniques completely different from those extant when these mines were designed. To be effective such mines require a modern, sophisticated and intelligent firing mechanism.

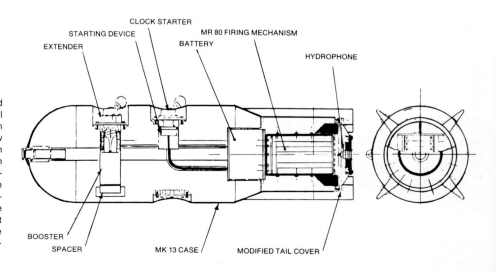

To meet these requirements MISAR designed and developed a refit package based on the Target Detection Device (TDD) which can be fitted inside the original mine casing and allowing the mine to retain its main charge. The TDD can operate over a wide range of pressure and magnetic sensors as well as different acoustic frequencies. TDD can be used to update any type of mine, both ground and moored. The refit TDD package is also available in exercise versions.

Contractor
Whitehead, Montichiari (Brescia).

EPR/2.5

Type
Small anti-frogmen/limpet mine.

Description
The EPR/2.5 can be deployed from surface vessel or helicopter and is designed principally for harbour protection. It can operate at depths down to 40 m and is an effective counter against frogmen and other underwater threats. It can easily be converted into a limpet mine by the addition of a simple magnetic or mechanical coupling device. The mine is designed to detonate after a preset time variable between 1 to 16 hours in one hour steps. The mine is equipped with a safety pin to prevent accidental detonation. To arm the mine the safety pin has to be removed before deployment or after attachment to the ships hull. This allows the detonator box to move and sets the detonators in the firing position, at the same time allowing the electric circuit to be closed through the microswitch. The detonation delay is selected by removing the bottom cap and selecting the time by a thumbwheel switch.

Specifications
Diameter: 260 mm
Height: 90 mm
Weight: 5 kg
Charge: 2.5 kg HE
Service depth: 40 m

Contractor
Tecnovar Italiana, Bari.

MAS/22

Type
Ground anti-invasion mine.

Description
The hemispherical MAS/22 is designed for use as a shallow water anti-invasion mine. Three horns are fitted on top of the hemisphere each of which carries a 250 g primer of exogene and a type M41 detonator. The mine sensor is fitted with three safety devices with the detonators moving to the operative position when they are deflected by a target. The mine can be secured to the bottom by a mooring body, three spikes or by three spring catches fixing the mine to a preset net.

Specifications
Diameter: 380 mm
Weight: 22 kg (approx)
Charge: 17 kg TNT

Contractor
Tecnovar Italiana, Bari.

MAL/17

Type
Moored anti-invasion mine

Description
The MAL/17 is a spherical anti-invasion mine fitted with three horns at 45° from the top and at 120° from each other on the upper hemisphere of the mine's body. Each horn has 250 g of exogene primer and a Type M41 detonator. Each sensor is equipped with three safety devices, the detonator of each sensor only sliding into position when the sensor is deflected by a target. From the bottom of the lower hemisphere protrudes a pipe with four fins, which act as a stabiliser. At the end of the stabiliser is a stainless steel mooring chain which can be adjusted in length according to the water bottom depth. This chain moors the mine to a 50 kg sinker which can be made of various materials.

Specifications
Length: 470 mm; 630 mm (stabiliser)
Diameter: 380 mm
Weight: 22 kg (without mooring chain and sinker)
Charge: 17 kg TNT

Contractor
Tecnovar Italiana, Bari.

SWEDEN

GMI 100 ROCKAN

Type
Anti-invasion ground influence mine.

Description
This ground influence mine is mainly intended for use as an anti-invasion deterrent against small and medium tonnage vessels in coastal waters, inlets, and confined waters such as harbours. It may also be deployed in deeper waters against submarines to protect friendly routes and bases. A feature of the design is a shape which enables minelaying over a wide area while the minelayer covers the minimum distance. This is achieved by shaping the mine in a casing that enables it to plane or glide in the water for a distance of up to twice the depth of the water it is being laid in. This also enables mines to be sown in pairs from each side of the ship, or for mines to be spread out directly from the quay.

Another advantage of this shape is its low profile, which makes it difficult to detect by minehunting sonar and other anti-mine techniques. The mine is constructed of an outer shell of reinforced plastic containing an explosive charge, a fuze with arming unit, and a sensor/electronic unit. The outer shell has sliding runners to enable the mine to be moved on a mine rail, orientated either along or across the rails. The shell can be opened easily for maintenance or for adjustments and so on.

After resting on the seabed for a time the Rockan ground mine becomes very difficult to detect both visually and by minehunting sonar

Operational Status
In production.

Specifications
Length: 1015 mm
Width: 800 mm
Height: 385 mm
Weight: 190 kg
Charge: 105 kg
Service depth: 5-100 m
Glide coefficient: 2 (approx)
Glide speed: 2 m/s (approx)
Minimum distance between mines: 25 m

Contractor
NobelTech Systems AB, Järfalla.

MMI 80 MOORED MINE

Type
Moored influence mine.

Description
This influence mine is designed for use against small and medium-tonnage surface ships and submarines. It is suitable for laying in water 20 to 200 m deep. The mine itself consists of a buoyancy section of cellular plastic, which also contains the explosive charge, the fuze and the sensor/electronics package (which is similar to that of the Rockan, but adapted to the moored mine role).

The MMI 80 is programmed to anchor itself automatically at the desired depth, which is set on a pressure sensitive device before sinking. A Manostat 80 sinker unit is employed with the MMI 80, and the mine can be laid from rails.

Operational Status
In production.

Specifications
Length: 1125 mm
Width: 660 mm
Height: 1125 mm
Total weight: 450 kg

Sinker weight: 240 kg
Charge: 80 kg
Bottom depth range: 20-200 m
Mine depth range: 8-75 m
Minimum distance between mines: 25 m

Contractor
NobelTech Systems AB, Järfalla.

BUNNY

Type
Ground influence mine.

Development
Development ended in May 1990 and was followed by a production contract.

Description
The Bunny ground influence mine is being developed as part of a new submarine weapon which will be carried in a mine girdle attached to the outer hull of submarines. The mine will also be able to be launched from any type of surface ship.

The mine incorporates a buoyant upper section which gives the weapon an underwater weight of 200 kg. Against surface targets the mine will have an operating range of 40 to 50 m but can be laid to deeper water down to 150 m for deployment against submarines.

The mine is fitted with the NobelTech 9SP 180 programmable naval mine sensor unit which incorporates multiple sensors using sophisticated logic, making it resistant to countermeasures. It will be able to discriminate specific types of target and initiate detonation at the lethal range for that particular target. The sensor unit is powered by a lithium battery pack and/or external power source. The sensor unit can remain in water for up to 12 months. The PCHS-1 combined hydro-sensor unit senses depth, dynamic pressure and sound in separate channels, while the three-axis magnetometer measures variations from the locally determined magnetic field. The measured signals are filtered and evaluated against pre-programmed criteria by the microprocessor. The logics unit reports the evaluation result through a standard electrical interface to the firing device or corresponding unit in non-mine applications. The sensor unit can also be retrofitted to existing mines.

The volume of explosive carried is 400 litres.

A moored version of the mine has been successfully tested for submarine deployment.

Operational Status
In production for the Swedish Navy. The mine is also being offered as an option for the new submarines for Australia.

Contractors
NobelTech Systems AB, Järfalla.

TAIWAN

M89A2

Type
Portable magnetic limpet mine.

Development
The M89A2 magnetic mine has been developed by Taiwanese engineers to penetrate up to 15 mm of strengthened steel. The mine comprises three sub-units: a moulded high density 'ABS' waterproof computer timer, a cone-shaped percussion cap and a detonator. The mine is simple to operate, all that is required is to plug in the detonator and pull off the safety pin, at which point the timer is automatically switched on.

Specifications
Diameter: 270 mm
Depth: 140 mm

Weight: 6 kg
Time delay: 15-360 min
Normal operating depth: 6 m
Maximum operating depth: 25 m

Contractor
Military Technology Consulting, Taipei.

UNION OF SOVEREIGN STATES

In 1987 the US DoD estimated that the former Soviet Union possessed an inventory of 300 000 naval mines. According to US intelligence the inventory includes about 100 000 moored contact mines. In addition, mines have been provided to other Warsaw Pact navies and also exported to many other countries. Among countries outside the Warsaw Pact that are believed to have received stocks of USS mines are China, Egypt, Finland, Iran, Iraq, Libya, North Korea and Syria. It is also likely that other Third World navies possess mines of USS design. At least some of these countries have set up production lines to manufacture mines based on USS designs.

Moored Mines
The USS is known to have at least 11 different types of contact mine. Some of these are no longer in production, but many of them could be encountered.

The main type is the basic contact moored mine using an inertia-type firing mechanism which can be either galvano-contact, contact-mechanical or contact-electrical.

The two smallest mines are designated MYaRM and MYaM. Both have a conventional spherical shape with horns and small sinker units. The MYaRM has an explosive charge of 3 kg and is used in rivers and lakes. The MYaM has an explosive charge of 20 kg and is used in lakes and shallow coastal areas to protect them from small boats and landing craft. The MYaM is thought to have entered service immediately after the Second World War and was first encountered in 1952 by the US Navy. It has been used by both the Iranians and Iraqis in the Persian Gulf.

The medium-sized moored contact mine is largely confined to the extensive stocks of the M-08 series used for coastal defence barriers. The M-08 was developed in 1908 and remains in widespread service in various navies. The robust, reliable, spherical mine is filled with 115 kg of explosive and the firing mechanism consists of five Hertz horns. Intended for use against surface ships, the M-08 can be laid in up to 110 m of water. The M-08 mines used in the Persian Gulf were, according to US intelligence reports, assembled by the Iranians, probably from components

M-AG antenna mine

supplied by the North Koreans, who apparently manufacture the M-08 themselves.

There are two improved versions of the M-08: the M-12 and M-16, which are identical to the M-08 except that they have longer cables connecting the mine case to the anchor, enabling them to be laid in deeper waters. In addition, the USS has devised a version of the M-16 in which two of them are linked together in tandem, with only one mine case being released at a time. If the first mine is swept or detonates, the second mine is released.

The M-26 is totally different from the M-08, using a PLT inertial firing mechanism. When armed, a shock causes the trigger shaft to move out of position, thus allowing the trigger to strike percussion caps that detonate the main charge. The entire mine comes to rest on the bottom before the mine case is released from the anchor. Although this makes it possible to use a simple anchor, it also makes it impossible to use the M-26 in deep water where the case might be crushed by high pressures. The M-26 has been used by the North Koreans.

The M-KB is a large mine with a 230 kg explosive charge intended for use in deep waters down to 300 m. It operates in much the same way as the M-08 and even uses some of the M-08's components. The M-KB is known to have been supplied to a number of countries including Egypt and North Korea.

Firing mechanism for KMD mine

Several variants of the M-KB mine have appeared. The case of the M-KB-3 is 6 in shorter with a corresponding reduction in the size of its explosive charge. The M-AG is almost identical to the M-KB except that it also has an antenna firing device and can be laid in considerably deeper water. The mine has a 35 m long upper antenna and a 24 m long lower antenna. The antenna operates by generating an electric current when the copper wire of the antenna comes into contact with the steel hull of a surface ship or submarine. The mine is intended primarily for use as an ASW weapon as it increases the vertical area covered by the mine. Antenna mines tend to be unreliable, however, and most countries now use influence mines for ASW.

The AMG-1 is a version of the M-KB adapted for air delivery. Not being fitted with a parachute, it must be laid at low altitudes and slow speeds. However, because of its high impact velocity it can penetrate several inches of ice. Because of this feature it is possible that this otherwise obsolescent mine might still be used operationally by the USS.

Two contact mines exist which can be laid by submarines, the PLT and PLT-3. The large PLT can only be laid by specially designed submarines fitted with an internal minelaying mechanism. As there are no such submarines operational, the PLT is likely to be encountered only if laid by surface ships. The newer

PLT-3, however, can be laid through submarine torpedo tubes.

It is likely that the USS has additional types of moored mines in its inventory. According to one report there is an acoustic moored mine, and it would seem probable that a magnetic influence moored mine has also been developed. These could be versions of existing moored mines fitted with influence sensors. It is also probable that the USS has developed a new aircraft-delivered moored mine. The AMG-1 is a primitive weapon not suited for deployment from modern high speed aircraft. Hence, the appearance of a new moored mine fitted with either a contact or an influence firing mechanism would not be surprising.

Mine Protectors

Because of the ease with which moored mines can be swept, the USS has developed the MZ-26 mine protector. Attached by a cable to an anchor is a magazine with a buoyancy chamber and four floats. The magazine is normally set to float at a depth of 18 m, but it can be up to 46 m deep. Each of the floats is attached to the chamber by a cable usually 12 m long. The floats contain a small explosive charge which is detonated when a sweep cable comes into contact with it, cutting the sweep cable, and halting minesweeping operations. Although the float is destroyed, a replacement float is automatically released when one is destroyed. Thus each MZ-26 has the potential for cutting up to four sweep wires. The protectors can be used in conjunction with M-16 tandem mines.

Ground Mines

Relatively little is known about USS influence mines. The first ground mine available in any quantity was the KMD developed in the late 1940s. The AMD is an aircraft-delivered version of the KMD. It was first identified in the late 1950s but may have entered service earlier in the decade. The mines can be fitted with one of four different types of sensor: acoustic (using either or both low frequency and high frequency noise generated by the target), magnetic (relying on either the intensity of the horizontal or vertical component of the target's magnetic field or the rate of change of its field), pressure, or a combination firing mechanism using two or all three different influences in conjunction. The KMD mine is fitted with a timer that can be set to activate the mine after a delay of up to 10 days, and it can also be fitted with a ship counter that requires up to 11 activations before the mine actually detonates. Both mines can carry either 500 kg or 1000 kg of explosives, and both have been exported to other former Warsaw Pact navies and to countries in the Third World.

The USS is known to have developed additional influence mines in the past 30 years. The first of these was identified in 1985 after being used by the Libyans in the Red Sea. According to published reports the mine was 533 mm in diameter, suggesting it was designed for submarine deployment. The mine is believed to have been of modular design with replaceable circuit boards, which enabled the sensitivity to be altered according to the type of target to be countered. The mine carried an explosive charge of 680 kg.

It is probable that several different types of ground mine are available, including an air-deployed version of the submarine-launched mine. As that particular mine was large, it is also probable that smaller anti-invasion ground mines also exist. The latest types of ground mine probably follow Western practise in that they are programmable microprocessor controlled types, whose mechanisms can also be retrofitted into existing warstocks.

While the USS has exported its newer mines to countries as diverse as Libya and Finland, it is unlikely that they will incorporate firing devices as sophisticated as those fitted in the USS Navy's inventory.

Specialised Mines

Several special-purpose mines intended for a deep water ASW role have been developed. Two such weapons are the torpedo-shaped, rocket-propelled, tethered rising mines, NATO designation 'Cluster Bay' (operating in water 80 to 200 m deep) and 'Cluster Gulf' (which can be laid in 2000 m of water).

Egyptian KMD mines on display in 1977

Iranian M-08 mines captured by the US Navy in the Persian Gulf

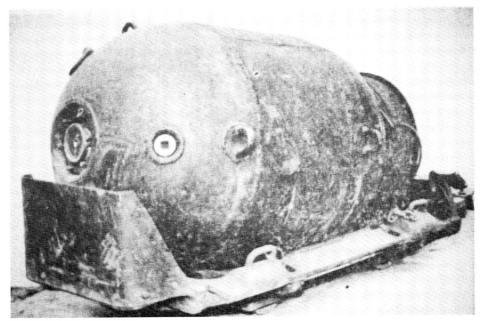

M-26 mine

Targets are initially detected by a passive acoustic sensor and located by transmissions from an active acoustic sensor. If the target is confirmed as being within the vertical attack zone, the tether is cut and the rocket ignited. The very fast upward speed allows very little time for the target to evade the device if its launch has been detected.

The USS is also believed to have developed an Underwater Electric Potential (UEP) mine. Nothing is known about this mine, except that the firing mechanism operates by detecting the electrical field generated by a target.

It is possible that the USS Navy has also developed a mobile mine similar to those used by the USN. Mobile mines are attached to torpedoes, making it possible for a submarine to lay a mine at a distance from the location where the mine is released into the water.

Nuclear Mines

The USS is believed to have a small stockpile of nuclear mines with yields ranging between 5 and 20 kT for use against high value surface units and base targets. Laying of these mines is almost certainly assigned to specially selected SSK/SSN units.

Type	Mine designation	Firing mechanism	Overall weight	Explosive charge	Min mine-laying depth	Max mine-laying depth	Min case depth	Max case depth
Bottom	AMD-1000	Influence	987 kg	782 kg	4 m	200 m	0 m	54.9 m
	AMD-500	Influence	–	299 kg	4 m	70 m	0 m	24.4 m
	KMD-1000	Influence	978 kg	782 kg	4 m	200 m	0 m	54.9 m
	KMD-500	Influence	500 kg	300 kg	4 m	70 m	0 m	24.4 m
	Mirab	Influence – magnetic	279 kg	64 kg	2 m	–	0 m	9.1 m
Mobile	?	Influence	–	–	40 m	70 m	0 m	–
Moored	AMG-1	Contact – chemical horn	1034 kg	262 kg	13 m	100 m	2 m	9 m
	M-08	Contact – chemical horn	–	115 kg	6 m	110 m	0 m	6 m
	M-12	Contact – chemical horn	–	115 kg	6 m	147 m	0 m	6 m
	M-16	Contact – chemical horn	–	116 kg	6 m	366 m	0 m	6 m
	M-26	Contact – inertial	–	240 kg	6 m	139 m	1 m	6 m
	M-AG	Antenna	1089 kg	230 kg	80 m	454 m	0 m	91 m
	M-KB	Contact – chemical horn	1089 kg	230 kg	0 m	300 m	6 m	9 m
	M-KB-3	Contact – chemical horn	1061 kg	200 kg	0 m	273 m	0 m	9 m
	MYaM	Contact – chemical horn	175 kg	20 kg	3 m	50 m	1 m	3 m
	MYaRM	Contact – chemical horn	–	3 kg	3/4 m	50 m	–	–
	PLT	Contact – impact-inertial	839 kg	230 kg	9 m	139 m	0 m	9 m
	PLT-3	Contact – chemical horn	998 kg	100 kg	0 m	128 m	0 m	9 m
	UEP ?	Influence – electrical	–	227 kg	0 m	490 m	0 m	–
Obstructor	MZ-26		413 kg	1 kg	24 m	46 m	0 m	34 m
Rising	'Cluster Bay'	Influence – acoustic	–	230 kg	80 m	200 m	0 m	609.6 m
	'Cluster Gulf'	Influence – acoustic	–	230 kg	80 m	2000 m	0 m	–

UNITED KINGDOM

SEA URCHIN

Type
Programmable influence mine.

Description
Sea Urchin is an intelligent ground mine which utilises the mine modernisation components developed by BAeSema for the Royal Navy. It can be programmed to detonate on a range of influence characteristics either singly or in combination, including a ship's acoustic or magnetic influences. Its advanced microprocessor control ensures that detonation occurs at the closest approach point of the target within the damage radius of the mine.

Mine setting and self-testing procedures are simple and secure, being achieved by means of a small plug-in setting box weighing less than 5 kg.

Equipped with a 600 kg warhead, Sea Urchin can be deployed in water depths from 5 to 200 m; various launch attachments are available for the nose and tail of the mine to permit deployment by surface vessels, submarines or aircraft.

Operational Status
Development has been completed but no information is available on production or orders.

Contractor
BAeSema Marine Division, Bristol.

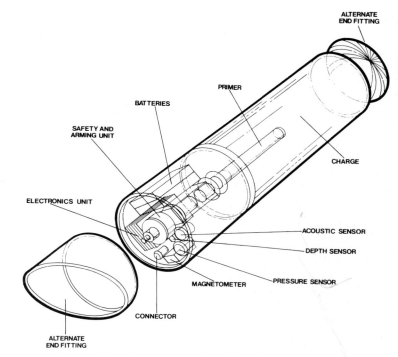

Diagram of internal arrangement of Sea Urchin mine's component parts

Sea Urchin naval ground mine, 600 kg variant

STONEFISH MINES

Type

Surface-, air-, submarine-launched ground influence mine.

Development

Marconi Underwater Systems Ltd continues to develop its family of advanced mines comprising warstock, exercise, training and assessment versions. Stonefish, as this family is known, is designed on the modular concept enabling mines to be made up to meet any tactical, exercise or training requirement.

Description

The warstock variants are provided with an advanced design plastic bonded explosive warhead. The warhead, together with the appropriate launch kit and a standard tail unit containing the sensors, signal conditioning and processor package, can be configured to cover the range of launch vehicles, depths and targets. The basic sensor pack comprises acoustic, magnetic and pressure sensors. The signal processing includes signal conditioning and information processing, the setting of thresholds, mine logic and sterilisation delay times. The mine programme can be modified by means of a portable presetter, changing threat and operational conditions before the mine is deployed.

The Stonefish warstock versions have a storage life of over 20 years and an in-water or laid life of over 700 days. They are intended to be laid at depths of between 10 and 90 m for surface targets and down to depths of 200 m for submarine targets. The refurbishment life for stored warstock weapons is nominally six years. To provide training in mine countermeasures, Stonefish is available as an exercise variant.

The exercise variant is capable of assessing mine-hunting and minesweeping effectiveness in its true environment. When laid it is available for data gathering relevant for target selection, together with a facility to emulate known mine characteristics. The main features are:

(a) command activated recovery system
(b) acoustic command system, enabling mine states to be interrogated or changed
(c) digital signal processing and large solid-state data store
(d) it is available as cable controlled or a seabed test platform for sensor/processing development.

Using the command link the exercise variant can accurately measure target range, change its operational status and recover itself, even if buried in deep mud on the seabed by means of an inflatable flotation collar.

It can be refurbished at sea and recharged within four hours ready for relaunch.

The assessment variant is used to gather mine and target data for use by the tactical planner so that the optimum minefield can be planned. The variant uses a modified version of the warstock version's electronics system and is connected by underwater cable to a shore-based computer.

The training variant is inert but can accept all the available equipment options to facilitate classroom instruction, handling, programming and test trials.

Mines of the Stonefish family are up to 2 m in length, with a diameter of 533 mm and all-up weight varying up to 770 kg depending on configuration.

The Stonefish presetter, designed as a self-

Stonefish mine

contained unit, employs a microprocessor and incorporates built-in data checking during input and automatic test facilities.

Operational Status

In production. Marconi has received orders for Exercise Stonefish from the Royal Australian Navy and has sold warstock units to Finland and Pakistan.

Contractor

Marconi Underwater Systems Ltd, Waterlooville.

Specifications

	Mk II Warstock Mine	Exercise Mine	Assessment Mine
Length	1812 mm	1912 mm	
Diameter	533 mm	520 mm	533 mm
Weight	770 kg		
Warhead	500 kg PBX\		

MINE MODERNISATION

Description

BAeSema has designed and developed a new target information module (TIM) which can be quickly and easily installed in older mines to convert them cost-effectively into modern weapons, thus considerably extending their useful service life.

The TIM enables modernised mines to be configured to meet a user's operational requirements. Maximum use may be made of existing internal components. For example subsystems such as safety and arming units could be retained. The TIM houses acoustic and magnetic influence sensors and an electronics unit employing the latest in digital microprocessing technology.

If required the flexibility exists to add other types of sensors. A new safety and arming system is available, designed to cater for all types of ground mine, including submarine-launched ground mines. High energy batteries ensure the modernised mines have a long storage and in-water life.

Mine preparation is simple and rapid using a hand-held plug-in setting box. The minefield planning officer programs the units with the appropriate mine mission parameters, which are then downloaded into the mines immediately prior to laying. Before transferring

TIM installed in updated Royal Navy mine

the mission parameters, the setting box automatically tests the mine's electronics, and only if they are functional will the information be downloaded. Mine testing and setting takes approximately 30 seconds. The safety and arming unit, TIM and battery module have been installed in mines of the Royal Navy and are suitable for installation in most mine types.

Operational Status

In service with the Royal Navy.

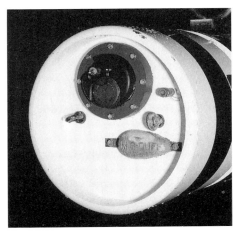

TIM showing installation in Sea Urchin exercise mine

Contractor

BAeSema Marine Division, Bristol.

VERSATILE EXERCISE MINE SYSTEM (VEMS)

Type

Exercise ground mine.

Description

The Versatile Exercise Mine System (VEMS) is designed to assess the effectiveness of mine countermeasures equipment and tactics. Using advanced microprocessor technology it accurately simulates sensor characteristics and firing actuation systems of any known ground mines and is capable of expansion to cover future trends.

Typically VEMS can be programmed to simulate either magnetic, acoustic, pressure or combined influence mines, to enable realistic mine hunting and

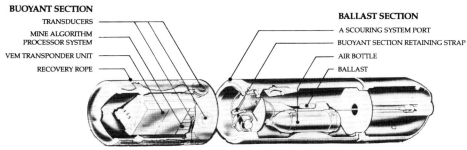

BUOYANT SECTION
TRANSDUCERS
MINE ALGORITHM PROCESSOR SYSTEM
VEM TRANSPONDER UNIT
RECOVERY ROPE

BALLAST SECTION
A SCOURING SYSTEM PORT
BUOYANT SECTION RETAINING STRAP
AIR BOTTLE
BALLAST

Diagram of the Versatile Exercise Mine System (VEMS)

sweeping exercises to be carried out. After recovery and extraction of recorded exercise data, which can be accomplished up to six months after laying, the VEM can be re-programmed as an entirely different type of mine.

Mine emulation programs from a standard library of mine types can be used, or those of particular mines may be developed, according to specific customer requirements, enabling tactical evaluation of MCM forces against a wide range of mines.

In addition to mine emulation, VEMS may be programmed to record sensor influence levels in order to check ship and sweep signatures.

VEMS provides the facility to create a library of mine warfare tactical planning and operational data which can be expanded as new mining technologies are developed.

An overside transducer is required – in conjunction with the mine transponder unit – for two-way communications between the ship and the exercise mines. When a mine is activated, coded signals are sent to the overside transponder which then transmits the information to the Mine Actuation Indicator in the ship's operations room to provide the mine actuation signal. It also transmits coded signals from the mine actuation indicator to the mines for recovery or ranging purposes.

BAeSema are funding improvements to the system which will greatly enhance the capability of the VEM. These will include sonar insonification, greater operating depth, improved communication systems and a land line.

Operational Status
VEMS is currently in production and in service with the Royal Navy, the US Navy and the Royal Thai Navy. Current orders will make the US Navy the largest operator of the VEM.

Contractor
BAeSema Marine Division, Bristol.

EXERCISE LIMPET MINE

Type
Training device for divers.

Description
The Exercise Limpet Mine is designed to realistically train divers in placing, locating and removing limpet mines underwater. The mine is attached to its target either by four powerful magnets on its base or by its carrying handles. The mine is realistic in shape and size and has a zenon flash tube which simulates actuation and which can be seen through a perspex window in the dome shaped cover.

The components and fixing magnets are mounted on a cast aluminium base plate, and the mine contains no explosive or pyrotechnics. The controls, which are situated in the base, comprise a variable firing timer ranging from 2 to 120 minutes and an arming switch which provides an initial fixed time delay of 20 minutes. There is also an anti-lift switch which prevents removal of the mine without flash tube activation.

The unit is powered by standard 9V batteries housed in a watertight compartment in the base.

Operational Status
In use by naval diving schools in several countries.

Contractor
BAeSema Marine Division, Bristol.

LNR3-3 MAGNETIC SENSOR

Type
Three-axis magnetic anomaly detector.

Description
The LNR3-3 is a new three-axis magnetic sensor which detects small magnetic anomalies or measures low frequency magnetic fields. Although designed primarily for use in sea mines, other applications include geophysical surveys, surveillance, security screening, and monitoring for unwanted ferrous matter. The salient feature of the LNR3-3 is that it provides a lower power DC output.

LNR3-3 can detect magnetic anomalies as low as 2.4 nT in the earth's static field and requires less than 5 mW of power from a ± 5 V DC supply. Two low impedance outputs are provided for each axis; one is AC coupled and the other is DC coupled. Both outputs are suitable for interfacing directly to a multiplexer and an analogue to digital converter. The device is compact (80 mm diameter × 40 mm), robust, totally self-contained and encapsulated. Reliability is quoted as 300 000 h MTBF.

Operational Status
In service.

Contractor
THORN EMI Electronics Limited, Naval Systems Division, Rugeley.

LNR3-3 three-axis magnetic sensor

UNITED STATES OF AMERICA

MINE WARFARE PROGRAMMES

The current stock of sea mines consists of the Mks 52, 55, 56 and 57 types.

Specifications
Type: Mk 52, submarine/aircraft-laid bottom mine
Length: 2.25 m
Diameter: 844 mm
Weight: 542 kg (Mod 1); 567 kg (Mod 2); 572 kg (Mod 3); 570 kg (Mod 5); 563 kg (Mod 6)
Charge: 300/350 kg HBX-1
Service depth: 45.7 m (Mod 2, 183 m)
Actuation: Acoustic (Mod 1); magnetic (Mod 2); pressure/magnetic (Mod 3); acoustic/magnetic (Mod 5); pressure/acoustic/magnetic (Mod 6)

Type: Mk 55, submarine/aircraft-laid bottom mine
Length: 2.89 m
Diameter: 1.03 m
Weight: 580 kg (Mod 2); 992 kg (Mod 3); 992 kg (Mod 5); 996 kg (Mod 6); 995 kg (Mod 7)
Charge: 576 kg HBX-1
Service depth: 45.7 m (Mods 2/7, 183 m)
Actuation: magnetic (Mod 2); pressure/magnetic (Mod 3); acoustic/magnetic (Mod 5); pressure/acoustic/magnetic (Mod 6); dual channel magnetic (Mod 7)

Type: Mk 56 Mod 0, aircraft-laid moored mine
Length: 3.5 m
Diameter: 1.06 m
Weight: 1010 kg
Charge: 500 kg HBX-3
Service depth: 366 m
Actuation: Total field, magnetic dual channel, acoustic/magnetic

Type: Mk 57 Mod 0, submarine or ship-laid moored mine
Length: 3 m
Diameter: 510 mm
Weight: 934 kg
Charge: 154 kg HBX-3
Service depth: 350 m
Actuation: Total field, magnetic dual channel, acoustic/magnetic

Also employed by the USN is a range of air-deployed munitions based on modified general-purpose low drag bombs and which can be released without requiring a parachute. The modification involves the use of a Mk 75 Mod 0 Destructor Modification Kit which can be added to 500 lb, 1000 lb and 2000 lb Mk 80 series bombs to form the Service Destructors (DST) Mks 36, 40 and 41, respectively. These are mostly intended for use in shallow waters such as estuaries, against typical coastal targets. There is also the DST 115A, which can be employed with either aircraft or surface craft for use against surface targets.

Specifications
Type: Mk 36, aircraft-laid bottom mine
Length: 2.25 m
Diameter: 400 mm
Weight: 240 kg (with fixed conical fin); 261 kg (with tail retarding device)
Charge: 87 kg H-6
Service depth: 91.4 m
Actuation: magnetometer (Mods 0/3); magnetic/seismic (Mods 4/5), magnetic dual channel, acoustic

Type: Mk 40, aircraft-laid bottom mine
Length: 2.86 m
Diameter: 570 mm
Weight: 447 kg (with fixed conical fin); 481 kg (with tail retarding device)
Charge: 204 kg H-6
Service depth: 91.4 m
Actuation: magnetometer (Mods 0/3); magnetic/seismic, magnetic dual channel, acoustic
Type: Mk 41, aircraft-laid bottom mine
Length: 3.83 m
Diameter: 630 mm
Weight: 926 kg (Mods 0/3); 921 kg (Mods 4/5)
Charge: H-6
Service depth: 90 m
Actuation: magnetometer (Mods 0/3); magnetic/seismic (Mods 4/5), magnetic dual channel, acoustic

Type: 115A, aircraft-laid surface mine
Length: 0.45 m
Diameter: 620 mm
Weight: 61 kg
Charge: 24 kg HBX-3
Service depth: 45 m
Actuation: Magnetic/seismic

Quickstrike
The Quickstrike bottom mine development programme embraced a family of mines using different size cases but with common target detection and classification mechanisms. The four members of the Quickstrike family are the Mks 62, 63, 64 and 65. The last of these (Mk 65 Mod 0) is in the 2000 lb (900 kg) class and is in full production by Aerojet TechSystems in Sacramento, California. The Mk 64 will probably be the next to enter production and this also is in the 2000 lb (900 kg) class, based on a Mk 84 2000 lb bomb and measuring 3.8 m long with a 633 mm diameter.

Quickstrike mines are for shallow water deployment (to approximately 100 m) and targets will have to approach to within a few hundred feet for it to act. It will use existing Mk 80 series GP bomb cases as well as a new mine case. Quickstrike mines are deployed by aircraft, surface ships or submarines, but principally from the former.

This family of mines is based primarily on conversion of existing ordnance (bombs and torpedoes). An exception is the Mk 65 mine which is not a bomb conversion. It has a thinner case than the equivalent bomb and contains the effective underwater PBX explosive. It is 3.25 m long and 533 mm in diameter. The Mk 65 has now been deployed with the US Navy and is being evaluated by the Italian Navy.

Type: Mk 62, aircraft-laid bottom mine
Weight: 227 kg
Charge: 87 kg
Actuation: Magnetic and pressure

Type: Mk 63, aircraft-laid bottom mine
Weight: 454 kg
Charge: 202 kg
Actuation: Magnetic and pressure
Type: Mk 64, aircraft-laid bottom mine

Weight: 908 kg
Actuation: Magnetic and pressure
Type: Mk 65, submarine/aircraft-laid bottom mine
Weight: 908 kg
Charge: PBX
Actuation: Magnetic and pressure and combined magnetic/pressure

SLMM

The submarine-launched mobile mine (SLMM) Mk 67 is intended to provide the US fleet with a capability for planting mines in shallow water (to approximately 100 m) by submarine, using a self-propelled mine to reach water inaccessible to other vehicles. It is also meant for use in locations where covert mining would be particularly desirable from a tactical standpoint. It measures 4.09 m long × 485 mm diameter and weighs 754 kg.

The Mk 67 SLMM consists essentially of a modified Mk 37 torpedo; alterations involved include some reworking of the Mk 37 torpedo bodies and replacement of the torpedo warhead with the applicable mine components. Tooling and other plant facilities were installed in FY78 for production of Mk 67 submarine-launched mobile mines in 1979.

Procurement plans for the Mk 67 SLMM for 1987 was 273, but it appears that this has now been cancelled.

CAPTOR – ENCAPSULATED TORPEDO

Type
Encapsulated torpedo.

Description
Captor, a contraction of 'encapsulated torpedo', is the name given to an anti-submarine system comprising a Mk 46 torpedo inserted into a mine casing.

Deployment is in deep water, generally in the vicinity of strategic routes travelled by enemy submarines, and submarines are the intended targets. US officials have stated that Captor has the ability to detect and classify submarine targets while surface ships are able to pass over a Captor field without triggering the Mk 46 Mod 4 torpedo which carries the warhead (43.5 kg PBXN103 explosive). This capability is reported to have been tested.

The detection and control unit (DCU) that performs these functions is the most costly subassembly of the complete Captor weapon and accounts for about 45 per cent of the total unit cost – approximately $130 000. DoD statements imply that the DCU incorporates facilities for turning itself on and off, in addition to its principal operational functions of detecting possible targets, classifying them by their sound signatures, and initiating release of the Mk 46 torpedo when the target is within range of its homing head. It is probable that the turn-on/turn-off system is quite sophisticated in the interests of power conservation to ensure the maximum operational life for deployed Captor mines. Factors which are likely to be taken into account include the levels of traffic (surface and submarine), ambient conditions, sea-state and so on.

Both active and passive sensor modes are employed and the system first operates in a listening (passive) mode, which continues for a certain length of time sufficient to identify the target as a submarine and not a surface vessel. The system then switches to an active mode during which it is assumed target ranging is carried out to determine the optimum release time for the homing torpedo.

The detection and control unit is gated to ignore surface traffic and has an estimated range of about 1000 m on submarine targets. There is no IFF capability and friendly units must be warned of Captor minelaying and positions of deployed mines.

There are presumably provisions for some method of self-deactivation or self-destruct for those Captors which are life-expired and which are not capable of retrieval, and other measures to prevent unauthorised salvage or interference may be expected.

The current deployed life of a Captor mine is thought to be in the region of six months, but it is not known if this is the USN's target, although it is a fact that considerable effort has been placed on ensuring maximum shelf and operational life.

Captor mines can be sown by surface ships, submarines and by aircraft. In the former delivery mode, mine rails or other delivery systems are not required, the chosen technique being by means of an over-the-side boom (or yard and stay) with a capacity of 1045 kg. The Captor is brought to a point about 10 m above the surface before release, which has to be at an angle to ensure proper entry into the water. Any submarine equipped with standard 21 in (533 mm) torpedo tubes can lay Captor mines, these having the advantage of being capable of covert minelaying. Aircraft employ a parachute technique for delivery of Captor.

According to USN statements, there are two main options for Captor delivery: one utilises P-3 maritime aircraft, nuclear strike submarines and surface ships; the other envisages a combination of USAF B-52 bombers and P-3 aircraft, plus surface ships and nuclear strike submarines. Captor has been tested for delivery on P-3C, A-6, A-7 and B-52 aircraft, LKA and other cargo ships, and aboard 10 different classes of submarine, including conventionally powered boats.

Most Captor minefields will be barriers located at some distance from possible enemy defences, and in the more highly defended areas Captor mines would be delivered by aircraft or submarines. The unique capability of submarines to plant mines covertly and under ice would be employed selectively.

Operational Status
Initial production was started in March 1976, although the US Secretary of Defense's report of January 1981 stated the procurement had been at a low level while development and testing were conducted to correct performance deficiencies. In December 1978, reduced performance against shallow water targets had been indicated. Nevertheless, initial operational capability was achieved in September 1979. Further procurement was cancelled after FY80 by a decision taken in December 1980, and in January 1981 a Captor improvement programme was approved. The weapon was granted approval for service use in February 1981, but no procurement funds were sought in the FY81 budget 'because Captor failed to provide the high level of effectiveness we had sought'. Subsequent testing has demonstrated that recent modifications have corrected its performance deficiencies.

Despite this somewhat chequered history, it was disclosed that about 630 Captors had been procured.

Specifications
Length: 3.7 m (overall)
Diameter: 533 mm
Weight: 908 kg (with torpedo and mooring)

Contractor
Loral Systems Group, Akron, Ohio.

ADVANCED SEA MINE

Description
A sophisticated new naval mine was to have been developed jointly by the United States and the United Kingdom. It was designed to meet the need for a potent and cost-effective weapon for use against surface ships and modern submarines in medium depth water. The origins of the project for the Advanced Sea Mine (ASM) lay in a requirement by the Royal Navy for a defensive system to be laid along the Continental Shelf. The US Navy reviewed its requirements and agreed on a joint programme for both development and production.

The first phase of collaborative work included parallel studies of a joint performance requirement by two transatlantic consortia, one led by British Aerospace Dynamics Division with Honeywell and Plessey Marine, and the other led by Marconi Underwater Systems teamed with Loral.

The ASM was to feature advanced sensors and processors to identify the target, which would then be destroyed by a propelled homing device.

Operational Status
Tenders for project definition were submitted at the end of April 1987; contracts worth $3.4 million were awarded in September 1987. The British Aerospace/Honeywell/Plessey contender was named Crusader, while the Marconi/Loral contender was Hammerhead.

The two consortia undertook competitive project definition, but the British Government subsequently withdrew from the project late in 1988. The future status of the project is not at present known.

YUGOSLAVIA

M66

Type
Diversionary underwater mine.

Description
This practically non-magnetic mine is designed for the destruction of vessels, harbour installations and other fixed offshore installations and in rivers and lakes. The weapon incorporates a pyrotechnical safety device which enables full safety in handling and preparation of the mine. Various clockwork fuze settings from a minimum of 20 minutes up to a maximum of 10 hours are possible. Fuze setting is set during weapon preparation. Once the mine has been attached to its target and the mine safety and pyrotechnical safety elements removed, the weapon cannot be removed from the target without risking detonation of the weapon.

Specifications
Weight: 50 kg
Charge: approx 27 kg TNT
Length: 670 mm
Diameter: 320 mm
Width: 430 mm
Operating depth: 30 m

Contractor
Federal Directorate of Supply and Procurement (SDPR), Belgrade.

M70

Type
Acoustic influence ground mine.

Description
The M70 acoustic influence ground mine is designed to destroy or seriously damage warships up to 5000 tonnes and merchant ships up to 20 000 tonnes and over. Highly sensitive sensors and a large explosive charge make the weapon extremely effective and applicable for either offensive or defensive roles.

The mine can be laid from submarine torpedo tubes in depths suitable for anti-submarine (150 m) and anti-surface vessel (50 m) targets. Targets and laying depths can be preselected.

The detonating system comprises both mechanical and fully transistorised electrical devices using state-of-the-art printed board circuit technology. The detonation system is compact, robust and resistant to vibration/shock and stable under any climatic conditions for long periods of time.

Specifications
Weight: 1000 kg

Length: 2823 mm
Diameter: 534.4 mm
Charge: 700 kg
Operating depth: 50 to 150 m

Contractor
Federal Directorate of Supply and Procurement (SDPR), Belgrade.

M71

Type
Limpet mine.

Description
The M71 limpet mine is designed for use against both ships and submarines. Detonation of the weapon is designed to rupture the underwater part of the hull plating. It is attached to the target by means of magnets, and to wooden structures by special screws fitted on the mine.

The weapon is fitted with a time fuze and clockwork mechanism with time settings ranging from 0.5 to 10 hours and with anti-removal devices. Fuze activation is performed after the mine has been fixed to the target.

Specifications
Diameter: 345 mm
Height: 245 mm

Weight: 14 kg
Charge: 3 kg pressed TNT
Operating depth: 30 m

Contractor
Federal Directorate of Supply and Procurement (SDPR), Belgrade.

DEPTH CHARGES

CHILE

AS-228 DEPTH CHARGE

Type
Air/surface-launched depth charge.

Description
The AS-228 depth charge is an anti-submarine weapon with a hydrostatic pressure activated fuze that permits its use against targets at depths from 100 to 1600 ft (30 to 490 m) and the detonation depth can be preset to any one of 19 depths between these limits. The detonator also incorporates three safety measures for handling and transportation, inertia, and submarine action. The charge itself can be launched by conventional methods from naval vessels or from aircraft, including helicopters. The fuze, manufactured by Industrias Cardoen, is also supplied as a separate unit as a replacement for outdated fuzes in depth charges and bombs of other manufacture.

This company also manufactures underwater hand grenades for use as an anti-diver weapon for the

Cardoen AS-228 anti-submarine depth charge

Cardoen anti-frogman underwater grenades

protection of ships moored or anchored in insecure waters. These are operated by hydrostatic fuzes and can be set to explode at depths between 4 and 12 m.

Thrown overboard at intervals alongside warships they offer a defence against the attentions of frogmen.

Operational Status
In production.

Contractor
Industrias Cardoen Ltda, Santiago.

SWEDEN

SAM 204

Type
Air-launched depth charge.

Description
The SAM 204 depth charge is designed for use against submarines operating in shallow waters or at periscope depth. The weapon can be deployed in patterns, weapons being set to detonate at different depths to achieve greatest shock and damage effect against submarines.

The charge comprises a steel case with fixings for standard NATO helicopter bomb launchers. The fuze Type SAM 104 is of unique design and is a self-contained unit operating on the hydrostatic principle for depth control. It is fitted with devices which eliminate the effect of shock waves and also make it insensitive to inertia forces in any direction. Because the weapon is not subject to sympathetic detonation from nearby exploding charges it can be deployed in patterns.

Four variants of the weapon are available in different weight categories and with different sinking speeds ranging between 5.2 and 6.8 m/s.

A training version of the weapon is available.

Operational Status
The depth charge entered service with helicopters of the Royal Swedish Navy in 1985.

Specifications
Length: 988-1420 mm
Diameter: 240-345 mm
Weight: 61-205 kg
Charge: 50-140 kg

Contractor
SA Marine, Landskrona.

UNITED KINGDOM

Mk II DEPTH CHARGE

Type
Air-launched depth charge.

Description
The Mk II depth charge has been updated to tolerate the harsh vibration levels associated with helicopter operations. The charge case and nose section have been strengthened to withstand entry into the water at high velocities without distortion. The charge is fitted with a modern fuze comprising pistol unit, detonator and primer assembly which can withstand heavy vibration and shock to ensure accurate detonation at the set depth. On impact with the sea the tail section breaks away and the hydro-pneumatic arming system is activated.

The weapon is fully compatible with the carriage and release systems of helicopters such as the Lynx, Sea King and Wasp and it is also cleared for use by other aircraft.

The depth charge is particularly suited for coastal defence operations and is effective against submarines on the surface or at periscope depth.

The depth charge is offered with a Torpex filling. However, development of a Polymer Bonded Explosive (PBX) filling is in progress.

The air-launched (Mod 3) version of the Mk II depth charge

Operational Status
The Mk II has been operational with the Royal Navy for a number of years. A contract to supply a number of Mk II depth charges to the French Navy was signed in 1988.

Specifications
Length: 139 cm
Diameter: 27.9 cm
Weight: 145 kg
Warhead: 80 kg HE

Contractor
BAeSema Marine Division, Bristol.

Mine Warfare

Command and Control and Weapon Control Systems
Combat Information Systems
Positioning and Tracking Systems

Sonar Systems
Hull-Mounted Minehunting Sonars
Variable Depth Minehunting Sonars
Side-Scan Minehunting Route Survey Sonars
ROV Sonars

Mine Disposal Vehicles
Remote Operating Vehicles
ROV Ancillary Equipment

Minesweeping Systems

Divers' Systems
Diving Equipment
Diver Vehicles

MINE WARFARE

The key to mine warfare is the ability to define the threat. The more that is known about the threat, the easier it is to counter. There is little that is not already known about the present day threat covering the moored mine, various types of ground influence mine and the volume threat posed by devices such as rising mines and so on.

Mine countermeasures techniques fall into two main areas: active and passive.

Active MCM

Active countermeasures can also be divided into two main spheres of action: minehunting and minesweeping.

Minehunting has proved to be the only relatively safe and effective method of dealing with modern sophisticated influence mines. If the minehunter is to be effective in its task then it must be equipped with highly sophisticated equipment. In order to be able to pinpoint the exact location of a mine and to record its position precisely requires accurate navigation systems of the highest order. Next, in order to detect the mine and carry out accurate classification, the vessel must be equipped with an efficient sonar system able to detect the smallest of targets under the most adverse of conditions.

Having detected and classified contacts, the next task is to neutralise those classified as mines. The main task of mine destruction is now carried out by small submersibles called Remotely Operated Vehicles (ROVs) deployed from the minehunter. These either position a remotely detonated counter-mining charge next to the mine, or else they are equipped with powerful cutters which sever the tethering wire of moored mines so that they float to the surface where they can be destroyed.

With the advent of the Continental Shelf Mine, which is extremely difficult to deal with and laid in much deeper waters, minesweeping has taken on a new lease of life. The new role of the minesweeper is to deploy the deep armed sweep designed to operate at much greater depths than the normal Oropesa sweep. To achieve this the minesweepers have to operate at least in pairs. This operation also requires extremely accurate navigation and station keeping, and such vessels must be equipped with the most up-to-date navigational aids available if they are to perform their task effectively.

Precise Positioning

The key to MCM is accurate navigation and precise positioning. It is often necessary for vessels to subsequently return to the position of previously defined contacts to re-examine them, or in the case of confirmed mines to carry out countermining. Precise positioning and plotting are absolutely essential elements if an MCMV is to carry out this task effectively. With such a system a vessel can return to the precise position where a contact was previously found with considerable saving in time and with the sure knowledge that the original contact will be found again.

Route Surveying

The most effective form of MCM is that of route surveying. In peacetime this is a primary task not only of the MCM force but also of the hydrographic service. To be effective, route surveying requires that the whole of a proposed wartime shipping route be extremely carefully surveyed to produce bottom contour charts which show in the minutest detail the composition of the seabed and precise data on all objects on it. By regularly surveying and updating the charts, accurate pictures of proposed routes can be maintained which will enable safe ones to be selected in wartime. Furthermore, these routes can be checked by the minehunters much more rapidly and accurately as the vessel then only has to check objects not already marked on the chart.

Command and Control

With such volumes of data now being made available to the command, a co-ordinated, effective means of correlating and presenting this wealth of information is essential if the MCMV is to carry out its task effectively and within a reasonable time span. Modern tactical plan displays frequently using colour graphics indicate to the command the route to be swept or surveyed, the ship's position, all objects on the seabed, the format of the seabed with integrated contour lines, the direction of the sonar beam, danger and specified circles around objects, map overlay, and alphanumeric tote display. As they are of modular construction and use distributed processing and pre-processing techniques, these systems can integrate with a wide variety of minehunting sonars now available and allow the command to accept only valid data required for the task in hand.

Among features now being incorporated into the latest command systems are window techniques which allow the display of information such as raw sonar data, contact data and track data and the possibility of correlating and extracting radar positional data, ESM data and so on. Other features under study include the capability to automatically control the ship in various attitudes such as hovering and heading using autopilot and propulsion control. The problem with propulsion control is integrating the thrust developed by the various manoeuvring control systems. Development of such an automatic control system will greatly ease the strain on the crew during minehunting operations.

Finally, the man/machine interface (MMI) has to be of the highest order. The presentation of the vast amounts of data required in minehunting demand displays of high resolution and clear definition, the careful use of colour techniques, and the possibility of presenting more specific data from various sensors using window techniques.

Passive MCM

Passive MCM involves such techniques as noise reduction (reducing cavitation to a minimum, reducing machinery noise and so on), degaussing (to reduce a ship's magnetic signature to a minimum) and optimising the hydrodynamic shape of the hull with a minimum displacement (to reduce pressure signature to a minimum). Alternatively, it may be possible to modify a ship's signature so that it no longer appears to be what it actually is. Such techniques, however, require that one can precisely define the threat. Mine avoidance is another passive technique which can be employed, but again requires prior knowledge of the precise nature of the threat.

The Future

MCM will continue to lag behind the development of the mine. The gap is narrowing, however, and there are limits to the extent to which mines can be developed. The two main developments in the future will be the self-propelled mine and mines which can be laid in much deeper waters; both of these will pose enormous problems for MCM forces. Other likely developments include increasing the mine's effectiveness against specific targets, in particular minehunters and submarines, and seeking to capitalise on influences so far not developed.

For example, it may be that some mines will be influenced to react to a minehunter's sonar, or for tethering cables of moored mines to react to the operations of an ROV and then to set off a chain reaction of other mines in an attempt to destroy the minehunter.

On the MCM side, more efficient mine detecting sonars, command and control systems will be developed to deal with the mass of data accumulated. Efforts will be made to reduce reliance on human operators, for minehunting can be an extremely tedious task and when operators become bored or tired errors can be made that prove fatal. Greater efforts will be made to make MCMVs more immune to mines, and new generations of submersibles will be developed with much greater capacities for dealing with more sophisticated mines, longer mission times, increased speed and range and deeper diving capabilities. Also, smaller ROVs will become attractive as complementary systems to the larger vehicles.

The use of data links to relay information between vessels on task and the shore-based MCM headquarters are now becoming essential to the speeding up of MCM operations and ensuring that shipping is allowed to use cleared channels at the earliest opportunity.

Sonar prediction techniques, already widely used in ASW operations, will also become more important as a means of improving effectiveness and safety with the ability to define safety circles more carefully.

Finally, the requirement to keep the MCM platform as far removed as possible from danger zones, particularly in areas which have not previously been route surveyed, may well lead to the further development of remote-controlled, autonomous surface and subsurface platforms for minehunting and disposal.

COMMAND AND CONTROL AND WEAPON CONTROL SYSTEMS

COMBAT INFORMATION SYSTEMS

FRANCE

EVEC 20

Type
Data handling system.

Description
The data handling system developed for the Tripartite minehunter is the Thomson-CSF EVEC 20. The EVEC 20 samples and processes data from the sonar, radio navigation systems, navigation radar, gyro-compass and Doppler log to present the operator with a continuously updated display of the minehunting area showing position of own ship, located targets, radar contacts, tracks and lines representing

specified zones. All or part of the displayed images can be recorded on cassette for subsequent recall, updating and maintaining an up-to-date permanent record of operations relating to a specified zone. A repeater on the bridge displays point co-ordinates from the minehunting sonar derived by the EVEC 20 and transmitted to the repeater.

The computer in the EVEC 20 also assists the automatic pilot, providing command inputs and error generation, and helps to control the automatic radar navigation systems providing location computations, error corrections, location of tracking windows and so on.

The automatic pilot is used to keep the ship on a pre-determined track during minehunting operations,

receiving necessary data from the gyro-compass and Doppler log.

The navigation radar and automatic tracking system integrate with the EVEC 20 to provide a buoy location system.

The horizontal plotting table uses an 80cm² screen on which up to 200 sonar contacts can be displayed. The table receives data from the central processing unit which has a 20 Kbyte random access memory.

Operational Status
Operational in French 'Eridan' class MCMVs.

Contractor
Thomson-CSF, Division SDC, Meudon-la-Forêt.

IBIS SYSTEMS

Type
Mine countermeasures data handling system.

Description
IBIS III integrates the Thomson TSM 2021B (DUBM 21B) dual antenna sonar with a Thomson TSM 2060 NAVIPLOT tactical plotter for navigation, plotting the precise location of detected targets and the recording of all relevant data. The NAVIPLOT uses the 15M 05 computer.

Data are displayed on a large four colour CRT which is mounted at an angle to enable the operator to work in a sitting position, giving him greater safety in the event of a mine exploding close to the ship. For operations in the vicinity of complex geographical formations (such as archipelagos and fiords) the raw radar video can be overlaid on the graphic symbology. In all cases, coastlines are presented in the synthetic mode.

The IBIS III system provides a continuous simultaneous display of the planned and ship's actual track, fully labelled display of up to 256 contacts, complete recording of all operational data and storage in the computer's memory. This enables a continuous comparison to be made between previously recorded data and the current situation with regard to the area being surveyed. New underwater contacts are thus instantaneously identified and located, providing increased safety for the ship and resulting in considerable saving in time on minehunting operations.

Associated subsystems of IBIS III are the ship's Doppler sonar log (Thomson-CSF 5730), radio navigation system (for example Thomson-CSF TRIDENT III), navigation radar (for example Racal/Decca 1229), compass, log, SATNAV, echo sounder, radio data link and SWEEPNAV, an acoustic location system for mechanical sweeps. As optional extras the IBIS III can be integrated with the ship's autopilot for automatic track following course to a point and hovering.

The IBIS V is a lightweight system suitable for retrofit as well as new construction. The sonar associated with the IBIS V is the Thomson TSM 2022 single array system, which takes up much less space in the hull than the TSM 2021. The TSM 2026 NAVIPLOT is the second main component of the IBIS V. Whereas the 2021 can provide simultaneous detection and classification in different directions, the 2022 performs detection and classification in sequence and requires only a single operator.

IBIS V weighs just 1.5 tonnes, the array having a span of about 1.5 m but pivoting to retract vertically into the sonar well, which is only 75 cm in diameter. IBIS V Mk II is an upgraded version of IBIS which comprises the TSM 2022 Mk II sonar and the TSM 2061 tactical system. The TSM 2022 Mk II uses a new 19 in high resolution colour console with sonar processing such as CAD (computer aided detection), CAC (computer aided classification) and PI (performance indicator).

The latest system to be developed in the IBIS series is the IBIS 43. Using sophisticated computing equipment and the TSM 2054 side-scan sonar, with antenna incorporated in a towfish, IBIS 43 produces

high quality images of the seabed with a high resolution, enabling operators to identify underwater mines. The computer system ensures real-time optimisation of sonar operating parameters depending on the speed of the towed body, its height above the seabed and the automatic compensation for roll, pitch and yaw. Computerised aids are used in conjunction with an image management system to ensure real-time detection and classification of unknown objects and to compare images obtained in successive missions in a given sector.

The towfish can be navigated manually or automatically at constant depth between 6 m and 200 m with an altitude in relation to the seabed from 4 m to 15 m, programmable from the sonar console keyboard. It can automatically follow seabed terrains with gradients of more than 15 per cent with high precision. It also has an obstacle avoidance capability for obstacles up to 10 m high at maximum speed.

Operational Status
The IBIS V is in service on board Malaysian and Nigerian 'Lerici' class minehunters, Indonesian 'Eridan' class minehunters and the Swedish 'Landsort' class minehunters.

The IBIS V Mk II has been selected to equip the four minehunters on order for the Singapore Navy.

The IBIS 43 is in service on board the STANFLEX 300 multi-role vessels of the Royal Danish Navy.

Contractor
Thomson Sintra Activités Sous-Marines, Brest Cedex.

GERMANY

MWS 80

Type
Minehunting weapons system.

Development
MWS 80 was developed for the German Navy's new 'MJ 332' class minehunters by Atlas Elektronik.

Description
Atlas Elektronik has developed the MWS 80 as an integrated multi-role MCM system which performs search, detection, classification, identification and neutralisation of ground and moored mines, precision navigation and vessel control, full control and co-ordination of all MCM operations, and maintains an accurate record of all areas searched, targets located and their geographic position and details of classification results.

MWS 80 integrates the DSQS-11M minehunting sonar, the NBD (navigation and vessel control equipment), the TCD tactical command system for co-ordinating and documenting MCM operations and the mine disposal system.

The DSQS-11M provides independent detection and classification modes through 360° azimuth. The stabilised sonar beams simultaneously cover a 90° horizontal sector and a 60° vertical sector for three-dimensional target location. Functions include performance prediction, computer aided detection classification and tracking with full colour display to improve detection, performance and discrimination between targets. The man/machine interface features function keys and interactive, colour-coded control displays.

The TCD assists the operator in controlling all phases of an MCM operation by providing means for display, plotting, storage and retrieval of tactical data. The equipment employs a comprehensive database

which is continuously updated during operation.

The navigation sensors such as GPS, Radio Location Systems, Gyro or Inertial Platform and so on, and the integrated Doppler Log DLO 3-2 enable the NBD to compute with a very high degree of accuracy the various navigational parameters such as ship's geographical position, ground speed, speed through water, course, heading and drift.

All data handled are recorded on hard disk, tape and printer to enable missions to be stopped and started at will, provide recall of data for comparison and evaluation, maintenance of a permanent record of the seabed for MCM survey and for training purposes ashore.

Operational Status
A total of 16 MWS 80 systems has been sold, including 10 for the German Navy's 'MJ 332' minehunters and three for the first batch of SM 343s.

Contractor
Atlas Elektronik GmbH, Bremen.

MWS 80 Mine Countermeasures System

NCE (NAVIGATION AND COMMAND EQUIPMENT)

Type
Action information system.

Description
The Navigation and Command Equipment (NCE) is a modular system designed to control minehunting and minesweeping tasks. In order to cope with the requirements of leading navies, the NCE integrates the functions of tactical command, precision navigation and vessel control.

Covering all phases of a mine countermeasure mission, the NCE provides means for preparation, control and evaluation of the MCM mission. For the control of the MCM mission NCE presents tactical information to the operator, which is continuously updated by sonar, radar and other sensor data. The information is stored in a database which contains all tactical and operational data and utilises typically 2 × 190 Mb storage capacity. For vessel guidance the NCE provides the integrated navigation and vessel control package. The ship's position is determined by the advanced integrated navigation which allows the processing of data from navigation sensors like GPS/DGPS, Radio Location System, Gyro or Inertial Platform, Doppler Log, Radar and so on.

The integrated track pilot allows the automatic steering of the vessel along a defined set of tracks which typically cover the operational area. For mission evaluation NCE documents all relevant data on plotter and printer.

Operational Status
The new NCE is in production for the German Navy's 'MJ 332' class vessels. In total 13 systems have been sold.

Contractor
Atlas Elektronik GmbH, Bremen.

ITALY

MACTIS MM/SSN-714(V)2

Type
Minehunting data processing system.

Development
Under the designation MM/SSN-714(V)2, SMA and Datamat have developed a digital navigation and plotting system for the Italian Navy's new 'Lerici' class of mine countermeasures ships.

Description
The main functions of the MM/SSN-714(V)2 MACTIS (minehunting action information subsystem) are:
(a) operations planning
(b) automatic computation and presentation of the ship's current position; navigation control (through autopilot interface)
(c) display of the tactical situation
(d) location of surface and underwater targets
(e) analysis and presentation of target characteristics
(f) guidance of surface and underwater craft
(g) event recording.

The system is based on a computer of the same or similar type to that employed in the SACTIS submarine combat information system, namely a Rolm MSE 14 machine. In the MM/SSN-714(V)2 system this is interfaced with recording units, display units, controls, printer/plotter and ship's sensors. The principal items in the last of these categories are radar, sonar, compass, log and various navigation aids. The operator's display has a vertical screen CRT, an alphanumeric display, a keyboard for communications with the system, supplementary data readout display units and associated input controls on the bridge and in the operations room.

The following units comprise MACTIS: an operator station with one 16 in graphic video screen, a functional keyboard and a trackerball, two display units (bridge and CIC/ops room) for data display and navigation, processing signal distribution and power supply units.

Analysis and presentation of target characteristics is achieved using the FIAR SQQ-14 fully solid-state mine detection and classification sonar. The sonar is a dual frequency variable depth, beam steering sonar enabling simultaneous detection and classification of targets. Separate search, classification and memory display consoles are provided. Comprehensive navigation equipment, in addition to the MM/SPN-703 navigation radar, is provided to ensure precise positioning of the ships and accurate plotting of underwater objects encountered.

Operational Status
In service with Italian Navy 'Lerici' class minehunters.

Contractors
SMA SpA, Florence.
  Datamat, Rome.

Graphic display for MM/SSN-714(V)2 system

NORWAY

MICOS

Type
Mine countermeasures command system.

Description
The MICOS command and control system has been developed for the Norwegian mine countermeasures programme. MICOS is an integrated mission control system which can be configured for either minehunting or minesweeping missions. The system comprises four main subsystems: the navigation and dynamic positioning system; a surface surveillance system; the mine warfare system; and the underwater/sonar system. These are all integrated via a dual redundant Ethernet data highway. The main elements of the subsystems are: the Simrad ADP 701 navigation and dynamic positioning console; the Norcontrol DB 2000 Officer-of-the-Watch console; the Simrad ATC 900 tactical console; the Thomson/Simrad TSM 2023 minehunting sonar; and the Simrad SA 950 mine avoidance sonar.

Operational Status
In total nine systems have been ordered for the Royal Norwegian Navy's surface effect ship 'Oksoy' class minehunters and 'Alta' class minesweepers, with delivery taking place from 1992 to 1995.

Contractor
Simrad Subsea A/S, Horten.

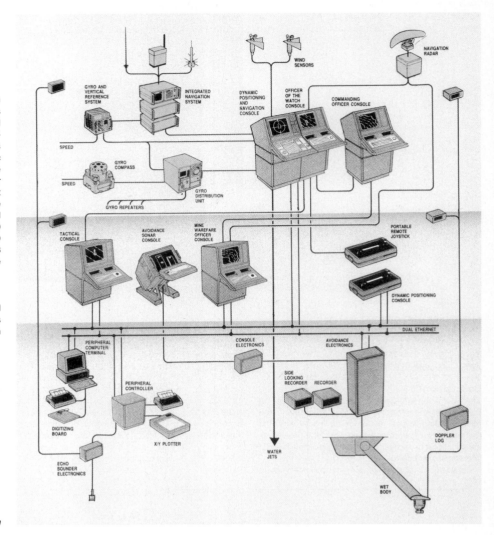

MICOS minesweeper configuration

SWEDEN

9 MJ 400

Type
Integrated navigation and combat information minehunting system.

Development
Developed as a joint project between NobelTech and Racal (UK) the 9 MJ 400 integrated navigation and combat information system provides a wide range of navigation and MCM functions together with sensor interfaces.

Description
The 9 MJ 400 enables the operator to plan in minute detail all manner of MCM tasks. The system computes and displays the navigation plan to and from a search area, and during the MCM task displays computerised search tracks, taking into account the type of sonar fitted, and environmental conditions.

The sonar search is planned using two types of search plan displayed on the conference type combat information console. The operator then defines an area or route to be searched and enters the anticipated sonar coverage and required percentage overlap. The system then computes and displays a search plan, the sonar being controlled in one of two remote-control modes (search or target indication). The search plan and actual tracks, together with the position of located targets, are automatically drawn on the X-Y plotter.

During a search the operator can track other ships and aircraft operating in the area, the system

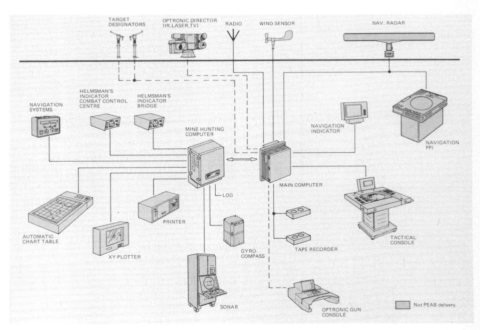

NobelTech Type 9 MJ 400 minehunting system

tracking 16 targets automatically or 30 targets semi-automatically.

Data on mine contacts are stored in the direct access memory of the computer (up to 100) and 'dumped' on to magnetic tape, as well as being fed to the plotter and/or the printer. The magnetic tape can

store data on 1000 mine contacts, which can be read into the direct access memory when planning the operation. All data recorded on tape or printer are available as a permanent record for subsequent comparison when carrying out searches of specific channels.

Own ship position and true speed are continuously calculated from each sensor input which includes Decca Navigator, transponders, navigation radar and dead reckoning. The operator selects the most reliable source for input to the system. The computer can store up to four range-measuring beacons when using a microwave transponder system.

In addition to its MCM function the 9 MJ 400 system provides an ASW capability and weapon control function for depth charges and mortars. When linked with the NobelTech 9 LV 100 FCS, the 9 MJ 400 offers an upgraded self-defence capability for the ship as well as improving MCM tracking functions.

The 9 MJ 400 also features an integrated data link for the onward transmission to shore-based control centres of important MCM data, as well as to other units in the operating area.

Operational Status
In service with the Swedish 'Landsort' class MCMV.

Contractor
NobelTech Systems AB, Järfalla.

UNITED KINGDOM

MC500

Type
Mine countermeasures command and control system.

Development
An up-to-date MCM command system currently under development is the Ferranti MC500, which is on offer to a number of prospective customers. MC500 is a variant of the Ferranti System 500 which features industry standard hardware and software written in Ada.

Description
MC500 is a fully integrated mine warfare and ship control system designed to control and display all MCM operations of mine hunting, sweeping and laying. The system comprises facilities in the categories of mission planning, mission data management, navigation and safety, ship control, onboard training and data recording.

MC500 is based on a stand-alone multi-function console. The system is designed to handle the vast masses of data (static, geographic, environmental, existing route survey data and so on) which can now be made available to the command from all sources. Although designed as a single console system, additional consoles can be networked to provide a second operator position, or a command overview on the bridge. A separate display is provided for the helmsman. The console interfaces with the minehunting sonar, all navigation sensors, mission recording systems (including disc, hard copy printout, X-Y plotter), and the ship's propulsion and steering systems.

Navigation data are automatically processed to provide an extremely accurate assessment of the ship's position, speed and attitude. The processed data are then used to carry out manoeuvring control of the ship, and to control precisely any outboard equipment, sensors and ROVs.

The MC500 operator can exercise complete control of the whole mine warfare operation using either fully automatic ship control or manual control for track keeping and hovering. Using data gathered from previous route survey operations, and with depth contour lines, ground conditions, known mine danger areas, danger depth warning, swept path warnings, track error warnings, and so on and, if required, superimposed radar data, the operator can pre-plan a complete minehunting operation. Results of this pre-planning are programmed into the computer, which will then automatically guide the ship along the track programmed by the operator, leaving him free to concentrate on the minehunting and providing automatic warning when any of the preset parameters are reached.

During the operation, MC500 monitors the ship's position and carries out any necessary corrections to maintain track. It displays all sonar contacts, investigates and classifies selected contacts, calculates the swept path and records all the data for post mission analysis.

The heart of MC500 is an information management system which controls the real-time database. MC500 uses an advanced man/machine interface (MMI) with full colour raster display in which window techniques are used to provide additional displays for specific aspects of the MCM operation. Thus, own ship's data, date/time, position of marker and key, track keeping, threat list and so on can all be displayed in windows of the display. The main display presents the tactical

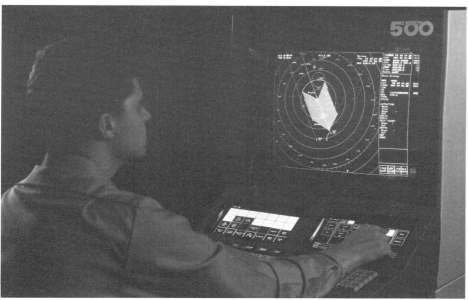

Console of the MC500 system

Typical display of MC500 system

picture compiled from the pre-mission plan, sonar inputs and synthetic data generated during the mission. Ship control can be exercised through a joystick, and many of the advanced MMI facilities of the System 500 are available, including finger touch control of the display screen and electro-luminescent displays for display and task management. Important prompt pages can be directly accessed using special function keys.

The software follows the philosophy of the System 500, being layered and modular in architecture to provide an extremely flexible system.

Operational Status
Private venture development.

Contractor
Ferranti Computer Systems Limited, Naval Command and Control Division, Bracknell.

FIMS

Type
Integrated mine countermeasures system.

Development
FIMS (Ferranti Integrated Mine Countermeasures System) has been developed offering proven systems software and hardware using the latest technology. The flexible design concept will allow users to select their own modules specific to their requirements, and the system is compactly packaged for ease of installation and transportation.

Description
The FIMS is based around the Manta ROV (qv) and combines the capabilities for full route surveying with those of a total mine countermeasures system which can carry out all phases of an MCM operation from detection through to classification and inspection to final disposal.

The Manta ROV can be integrated to any one of a variety of mine warfare command and control systems and underwater tracking, positioning and navigation systems. This flexibility allows the customer to select those elements best suited to his mine countermeasures requirements. By integrating the latest technology systems, FIMS forms an advanced mine countermeasures solution suitable for purpose-built MCMVs or COOP (Craft Of Opportunity).

The system is compact and modular in design enabling it to be housed in a single standard M1 ISO container. This containerisation allows the system to be fitted quickly and efficiently and easily moved from craft to craft ensuring maximum ship flexibility and utilisation.

FIMS has been designed to allow simple, low-cost, low risk interfacing of payload options that make up the total solution. Through-life costs are minimised by the interchangability of the payload elements and the lack of expensive mission-expendable fittings.

Operational Status
The system is currently under evaluation with the Royal Canadian Navy.

Contractor
Ferranti-Thomson Sonar Systems UK Ltd, Stockport.

SYSTEM 880

Type
Minesweeping/minehunting control system.

Description
The Racal System 880 minesweeping package is available for precise navigation in minesweeping and co-ordination between ships and has been fitted to the 'River' class minesweepers of the Royal Navy. The system has also been sold to the USN and is also recommended for Craft Of Opportunity (COOP). A minehunting version of the System 880 is also available using a comprehensive tactical display, in addition to the equipment available on the minesweeping package. The system can integrate with a variety of sonars (an early version in the RN 'Ton' class minehunters integrated with the Plessey 193M sonar) and has provision for control of ROVs. Based on the modular concept and using distributed processing, the system can be easily reconfigured to meet different requirements and to interface with different equipments. A concept currently under study is the possibility of incorporating an automatic hovering capability using the autopilot and propulsion control. Other possibilities include feeding the command and control system with extracted positional radar data, and the display of extracted ESM data.

System 880 is a third generation, low-cost MCM system comprising two major subsystems: the Racal Integrated Navigation System (RINS) and Racal Action Display System (RADS).

RINS is designed for minesweeping vessels, while for minehunters the addition of RADS provides a full MCM capability. System 880 automatically gathers information from sensors and navigational aids and presents it to the command on a ground-stabilised 19 in colour CRT display. The display uses colour graphics to identify route, ship's position, objects, direction of sonar beam, alphanumeric tote display, and danger and safety circles prescribed around the object under examination, all presented on a plan display. Associated with the main display is a control panel, alphanumeric keyboard and smaller alphanumeric display for on-line communication with the computer system.

RINS provides a minesweeper with mission planning capability, accurate navigation display data for the helmsman to aid accurate track keeping, hard copy printout on a plotter or automatic chart table, and autopilot control. For team sweeping a special-purpose version of RINS is available called RAFTS (Racal Aid For Team Sweeping). This provides all the facilities noted above with, in addition, a communications link between the RINS systems on the ships involved in the team sweep to ensure that the wing ship is automatically maintained on station with respect to the lead ship.

RADS can be added to RINS to provide integrated sonar, radar, MCM database and a tactical display for minehunting. RADS also provides an acoustic navigation capability enabling inputs from an underwater acoustic positioning system such as Racal's Aquafix 4 to navigate a remotely controlled submersible.

Operational Status
In service on board Royal Navy 'River' class minesweepers.

Contractor
Racal Marine Systems, New Malden.

System 880 MCM system

MAINS

Type
Minehunting Action Information and Navigation System.

Description
MAINS provides the following main functions:
(a) accurate navigation
(b) ship control guidance
(c) integration of minehunting sonar
(d) combined surface/sub-surface tactical display
(e) automatic hard copy tactical plot with detailed data printout
(f) patrol navigation/combat information operating characteristics as an optional second role.

The main features of the equipment are: simplified operating procedures; flexible characteristics; compact equipment; low cost; and reliable and easy maintenance. Being of modular construction MAINS is capable of interfacing with a wide variety of equipments suitable to meet individual requirements of any navy.

Own ship's position is fixed using one or more high accuracy radio and microwave ranging systems such as the Racal Hi-Fix 6 (large area coverage extending to 150 km) or Hyperfix, or the portable line-of-sight Trisponder with a range of 80 km. Other navaids include the statutory navigational radar which has sufficient resolution to track marker buoys, SATNAV, Decca, Loran C, the Doppler sonar, gyro-compass, and conventional EM or Doppler/acoustic log.

The MAINS computer can also be used for ship control and steering guidance.

An automatic plotter and printer provides a hard copy record of operations necessary for subsequent operations covering the same area.

The interactive PPI display is the Racal ED1202 which combines both radar surface situation and own vessel position relative to sonar search plan together with underwater contacts as a single presentation. The displayed picture has multi-level brilliance and correlated data and is presented in alphanumeric form.

The computer software is modular and is capable of accepting a variety of languages.

One of the major features of the MAINS is the compact size of the equipment, the control and display unit being tabletop-mounted. Being modular and compact means that such a system can be rapidly fitted to suitable vessels to build up a mine warfare capability.

Racal Marine Systems action information system

Operational Status
In service on board the Swedish 'Landsort' class MCMVs where it integrates with the 9 MJ 400 system.

Contractor
Racal Marine Systems, New Malden.

NAUTIS-M

Type
Third generation integrated command, control and navigation system for shipborne mine countermeasures.

Development
NAUTIS is a modular family of naval command and control systems for air, surface and underwater requirements.

NAUTIS-M was originally developed for the Royal Navy's new single role minehunters, the 'Sandown' class. Requirements included the integration of the new Type 2093 variable depth sonar and remotely controlled mine disposal system, multiple navigation and environmental sensors, radar and ship control as an integrated combat system. Functional requirements for NAUTIS-M cover all phases of minehunting operation including mission planning, route surveys, minehunting, classification, disposal and mission reporting.

Since the award of the development contract to Plessey Naval Systems (now Marconi Underwater Systems) in 1984, NAUTIS-M systems have successfully undergone rigorous environmental testing and software proving by the UK MoD (Navy). The first production system successfully completed Royal Navy acceptance trials on HMS *Sandown*, and further systems are in production for follow-on ships of the class.

The US Navy has successfully tested a variant of NAUTIS-M for their 'Avenger' class MCM ships. A three-year programme covering evaluation of hardware and software designs and environmental characteristics against US military specification and functional performance culminated in at-sea testing on a US Navy MCM ship. Final evaluation trials as part of the AN/SSN-2 Phase 3 combat system update programme were due to be completed during 1991.

A 1988 contract from the UK MoD (Navy) has produced NAUTIS consoles with high resolution shock-hardened colour raster displays with radar video superimposed from a scan calculator within each console. The US Navy has ordered a number of these consoles as part of the AN/SSN-2 Phase 3 evaluation programme. Under this programme NAUTIS has been given the US Navy nomenclature of AN/SYQ-15.

NAUTIS-M is included in the minehunter procurement programme for the Royal Saudi Naval Forces.

Variants are currently proposed worldwide for a range of other MCM combat systems and ship types.

Description

The basis of each NAUTIS system is a new technology autonomous intelligent console. The console includes interfaces for sensors, weapons and, if required, a data link, an integral radar auto-tracker, processors and memory to handle an extensive command system database. High level system application software and a selection of interfaces determine the functionality of the console for air/surface/underwater warfare roles. A high-definition colour raster display presents a labelled radar picture with tactical graphics superimposed.

A NAUTIS system comprises a number of consoles networked via a dual-redundant digital highway. The highway is used to automatically maintain a replica of the command system database within each console. Each console user has independent access to the database and to interfaced sensors, weapons and peripherals according to his task requirements. User facilities enable tasks to be readily changed to give functional interchangeability between consoles of a system. Should any console be out of use then the other consoles remain fully operational as a system, using duplicated interfaces, and can take up the additional tasks.

NAUTIS-M systems can be configured with one or two networked consoles in a ship's operations room (CIC) and, if required, with a further console on the bridge (as in the 'Sandown' class). The number of consoles depends on operational requirements, the performance of the overall combat system and the number of operators required to handle the system data.

The NAUTIS-M on-line database covers an area of up to 2000 nm². The data include route plans, above and below water geographic and environmental data, known seabed contacts and other supporting data that will enable a mission to be planned and carried out with best use made of the ship and its combat system. Examples of database content are: at least 200 radar and sonar tracks, threat evaluations and weapon assignments, route/search plans and navigation status, six user-designated tactical maps, 15 synthetic charts, 200 labelled reference points, 5000 fully detailed sonar contacts for MCM, 32 labelled bearing lines and operational warning messages. This database is on-line for access by the operator: radar, labelled graphics, totes and an interactive main area

NAUTIS-M command and control system as fitted to HMS Sandown

The latest technology NAUTIS console with colour raster display

clearly displayed in one viewing area; two different display compilations maintained for alternate selection by single-key action; display scales typically from 0.125 nm to the limits of the database; off-centring anywhere within the database area; true/relative motion stabilisation; labelled electronic range and bearing lines which can be hooked onto fixed points and moving tracks. The database is retained in each console, with the system operating program, in non-volatile memory. It can be loaded and retrieved using a magnetic tape cartridge. The man/machine interface also includes a typewriter keyboard, up to 32 assignable special function keys, a trackerball and electroluminescent panel. There is a comprehensive library of characters, symbols and lines. Size and brilliance can be selected by the operator.

Operational effectiveness of the ship's systems is maximised by full integration of sonar, navigation, collision avoidance and ship control systems through NAUTIS-M together with the on-line presentation of the MCM database.

Advanced display presentations, simplified user controls and operating procedures enable NAUTIS-M to be fully utilised with minimum training.

No specialised technical skills or test equipment are required for onboard maintenance.

Comprehensive built-in tests are performed automatically at start-up and during run-time and any faults are indicated at plug-in module level.

The modular architecture of NAUTIS-M enables it to be readily adapted for MCM combat system applications with various forms of variable depth and hull-mounted sonars, navigation systems, radars, mine neutralisation systems and ship control systems. NAUTIS-M typically comprises one, two or three consoles.

Operational Status

In service with the Royal Navy and chosen by the United States Navy, the Royal Saudi Naval Forces and other navies.

Contractor

Marconi Underwater Systems Ltd, Addlestone.

QMSS

Type

Mine surveillance system.

Description

A portable system for COOP and STUFT as well as enhancing the mine surveillance and analysis capability of dedicated MCMVs, is the Qubit Mine Surveillance System (QMSS), which provides real-time targeting and analysis of data logged during Q-route surveys, search and salvage and similar operations.

The complete package has been based on easily installed, portable commercial survey systems, and includes a high resolution side-scan sonar, the Qubit TRAC integrated navigation/positioning system, and the Qubit QASAR sonar analysis and reduction system.

TRAC is an essential element of the QMSS package and provides repeatability of vessel position, underwater ROV and bottom target, together with

absolute co-ordination, error modelling and data-basing.

As the sonar sweeps the area, the QASAR system helps the sonar operator to look more closely at the targets, to accurately pinpoint their positions and to make comparisons with previously noted seabed objects and data stored on the TRAC database.

It provides full real-time integration of sonar and position, high resolution sonar logging to optical disc, and incorporates a range of digital sonar enhancement techniques.

Built as a compact, transportable unit, it can link any side-scan sonar to any integrated navigation equipment.

Operation is simple and rapid. Sonar information is displayed on QASAR's VDU to give a continuous real-time picture of the seabed beneath the survey vessel. The user merely points to any object of interest on the touch-sensitive screen. The computer's cursor will automatically follow his finger to give an initial position which he can fine-tune using the system's rollerball. He can then simply press a button to categorise the target.

Video enhancement techniques, including contrast stretching, allow the operator to look more closely at his target.

QASAR logs the sonar data to optical disc and integrates it with data from the positioning system.

Later the operator is able to fast-scan the record to review omissions, freeze, zoom and enlarge targets, overrule earlier interpretations, and add and remove contacts. He can plot a complete overview of the mission or isolate specific track lines and groups of targets.

QMSS enables navies to chart precisely the seabed from any vessel, however small.

It provides a powerful tool not only for the experienced MCM officer but has been designed so that non-specialist reservists can be rapidly trained in its use.

Extensive trials have shown positional relocatability over multiple missions of better than 5 m.

Contractor

Qubit UK Ltd, Passfield.

POSITIONING AND TRACKING SYSTEMS

AUSTRALIA

ACOUSTIC TRACKING SYSTEM

Type
Underwater acoustic tracking and positioning system.

Description
This series of Acoustic Tracking Systems (ATS) is particularly relevant to minehunting and mine clearance operations. It is also used for underwater vehicle tracking, equipment handling and for tracking towed objects. By using a sophisticated eight chirp signal in the 15 to 18 kHz frequency range, ATS offers high accuracy, for example better than 0.25 per cent of slant range in X and Y co-ordinates depending on the model.

ATS also offers solutions to difficulties in harsh acoustic environments where other systems will not function, such as in shallow water, highly reflective conditions around underwater structures, and for operations near the water surface. It minimises problems such as background noise, reverberation and multi-pathing effects.

The ATS series is rack-mounted and easy to operate, and needs only a minimum set-up time with maximum flexibility. It is available in four very short baseline models, and a top of the range LR08 model, incorporating very short and long baseline tracking in the same unit. All models use the same hydrophone and beacons.

The hydrophone incorporates all receiving and transmitting acoustic elements in a compact self-contained unit. The ATS operates in a full hemisphere below the hydrophone and tracks up to eight fixed or moving beacons simultaneously, depending on the model.

Operational Status
In operational service. Recent sales include Martin Marietta for the MUST vehicle programme, the Applied Physics Laboratory at the University of Washington, and to Harvey-Lynch Inc of Texas.

Contractor
Nautronix Limited, North Freemantle, Western Australia.

GERMANY

POLARTRACK

Type
Tracking and positioning system.

Description
Polartrack is an all-purpose high precision three-dimensional laser tracking and dynamic position fixing system. Suitable for automatic continuous horizontal and vertical tracking, and for positioning of moving targets equipped with a set of optical prisms for return

of signal, the system can be operated variously as a total station, theodolite, or as an automatic or remote-controlled tracking unit.

Operational range is approximately 1.5 times line-of-sight in adverse weather conditions and extends from 20 m to upwards of 10 km under favourable conditions. Dynamic accuracy is of the order of ±5 cm per km of measured distance with a positional update rate of five or ten times per second. Dead reckoning and pattern search functions for automatic re-acquisition of target without operator intervention are also provided.

Applications of the system include automatic tracking of buoys, land vehicles and helicopters, as well as positioning and tracking of naval vessels for degaussing, antenna calibration and mooring.

Operational Status
Fully developed.

Contractor
Atlas Elektronik GmbH, Bremen.

NORWAY

HPR-300 POSITIONING SYSTEM

Type
Underwater hydro-acoustic positioning system.

Description
The HPR-300 is a low-cost, portable hydro-acoustic positioning reference system, based on the super-short baseline system, with just one onboard transducer necessary. The system consists of a small

transceiver unit in a portable, splashproof cabinet with connectors for transducer, display, joystick, gyro and data output. Normally the system is delivered with mini transponders specifically designed to give position to small ROVs.

From a standard RGB monitor the operation is carried out by joystick control of functions and parameters. Operating instructions are built into the system and this appears on a self-explanatory display menu.

The HPR-300 system has a built-in roll and pitch

sensor in the transducer which enables an easy and quick installation. The transducer may be mounted over the side of a craft of opportunity. The system is able to position five transponders simultaneously, and can be upgraded to position 14 transponders.

Operational Status
In service.

Contractor
Simrad Subsea A/S, Horten.

UNITED KINGDOM

TRAC and CHART SYSTEMS

Type
Integrated navigation, hydrographic survey and data logging system.

Description
Qubit TRAC and CHART systems are a family of automated data acquisition and processing systems used extensively by navies and commercial organisations for a wide variety of operational tasks such as mine warfare, maritime patrol, hydrography, trials and evaluations and offshore engineering.

TRAC provides integrated navigation, data logging, and precise and dynamic positioning for vessel control. It can form an important part of a naval action information organisation (AIO) by providing real-time data in a variety of formats direct from the vessel's sensors.

TRAC is a self-contained system which uses an advanced position computation algorithm to handle up to 20 position lines which may be any combination of hyperbolae, ranges, bearings or latitude/longitude lines. The system can operate on any spheroid and

The Qubit TRAC and QASAR systems

projection and takes into account the various geodetic reference systems.

Control of TRAC is through a conventional keyboard and colour or monochrome display. Alternatively in harsh environments, a Qubit remote-control unit can be used. The system supports a wide range of peripherals, including remote colour displays, plotters and printers.

CHART is a complete data processing system which complements TRAC. It may be supplied as a single-user system, in which it runs on hardware similar to TRAC, or it may be configured as a multi-user system supporting a number of terminals. CHART may be used afloat or ashore for data analysis and the production of fair sheets, and contains all the necessary processing facilities to carry out these functions.

Operational Status
Qubit TRAC and CHART systems are in service with the Royal Navy, Royal Australian Navy, Italian Navy and Turkish Navy.

Contractor
Qubit UK Ltd, Passfield.

OE2059

Type
Acoustic tracking system.

Description
The OE2059 acoustic tracking system is specifically designed for military applications including minehunting. It consists of three units – a controller, a transducer assembly, and the underwater transponder (up to five transponders or responders may be tracked by a single control unit, or alternatively a single acoustic pinger).

The transducer is a miniature hydrophone array containing three receiving elements and one transmitting element. The transmitting element generates the interrogation pulse to which the transponder replies. The three receiver elements are arranged in an orthogonal pattern, and the transponder return signal is computed from the phase difference between the three elements. In addition to the hydrophone array the transducer incorporates a replaceable plug-in circuit card.

The underwater transponder unit generates the acoustic signal which the system tracks. One transponder must be fitted to each underwater unit that requires to be tracked. The transponder generates a 30 kHz acoustic pulse in response to the acoustic interrogation from the transducer. The system can also be used with underwater responders, these are very similar to transponders except that they provide an acoustic reply to an electric trigger supplied via an umbilical cable. The system is also able to track a free-running 30 kHz acoustic pinger in place of transponders/responders.

Operational Status
In operational service.

Specifications
Tracking range: 2500 m slant range
Range resolution: 0.1 m
Range accuracy: 1.23% slant range in the hemisphere 10 to 85° depression; 2.5% slant range in the hemisphere 0 to 10° and 85 to 90°
Operating depth: 1000 m maximum

Surface control unit for OE2059

Depth accuracy: ± 1.5 m
Bearing resolution: 0.1°

Contractor
Osprey Electronics, Aberdeen.

UNITED STATES OF AMERICA

PINS

Type
Integrated navigation system for MCMs.

Description
Magnavox has developed the PINS AN/SSN-2(V) (Precise Integrated Navigation System) for MCMs. The heart of the PINS is a powerful military computer interfaced with an array of commercial and military sensors, control units and data display and recording systems. The PINS uses a Kalman Filter routine to smooth received data and apply appropriate corrections for known error factors. The system integrates and compares data from multiple navigation aids and other sensors to perform precise navigation and position determination, mission planning, plotting and data recording, target location and positioning and post-mission analysis. The TRANSIT SATNAV system is used to provide navigation fixes, interfaced via PINS with the ship's Doppler sonar operating in either ground speed or water speed modes. Other interfaced systems include LORAN C and Hyperfix.

All contacts and their positions are recorded and displayed, together with the ship's track, on a vertical plotter in the operations room. The high speed vertical belt-fed plotter provides a hard copy printout using standard navy charts if required.

Operational Status
PINS is now operational aboard the US Navy MCMV *Avenger* (MCM 1).

Contractor
Magnavox Advanced Products and Systems Co, Torrance, California.

MODEL 4068 NAVTRAK TRACKING SYSTEM

Type
Ultra-short baseline tracking system.

Description
NAVTRAK V is an ultra-short baseline tracking system designed for the positioning of towed vehicles, manned submersibles, ROVs, subsea maintenance and construction operations plus a variety of other subsea applications. The system is capable of tracking up to five targets simultaneously in standard configuration at ranges to 10 000 m.

An operated oriented front panel format simplifies operation and data entry. A 12 in CRT display provides easy-to-interpret graphics in either polar or rectangular formats with alphanumeric position data in both formats. New and expanded capabilities include five methods of depth input, depth priority, target relative display, ship's heading and automatic ranging display.

A fully portable system, the NAVTRAK V operates as effectively in over the side operations as in permanent hull-mounted installations. Ease of operation and portability offer a practical approach when fast requirement is needed on leased vessels or craft of opportunity.

Contractor
EDO Corporation, Electro-Acoustic Division, Salt Lake City, Utah.

MODEL 4268 MICRONAVTRAK TRACKING SYSTEM

Type
Ultra-short baseline tracking system.

Description
The Model 4268 MicroNAVTRAK incorporates all the features and proven reliability of the Model 4068 (see previous entry). The MicroNAVTRAK is an ultra-short baseline system which has been designed as an integrated component to a shipboard navigation system. The processor consists of a single rack-mountable unit used to process position information from the hydrophone assembly.

The position information is transmitted to a host computer using either the RS 232 interface or the IEEE-488 interface. The host computer is used to further process the position information and to control the MicroNAVTRAK system configuration parameters.

The optional Model 871 colour display monitor can be used to display target information and control system configuration parameters. The monitor is a 12 in high resolution equipment with a data entry keypad. Target information can be displayed in polar mode, rectangular mode or gridless mode. All targets are displayed with graphics and annotation.

A pitch and roll unit is used to detect these motions for vertical reference data. This information is used to correct errors in the target's calculated position introduced by the ship's motion.

The MicroNAVTRAK uses a 3 in slimline transducer which operates as effectively in over the side operations as in permanent hull-mounted installations. The transducer operates at 30 kHz and offers a maximum range scale of 2500 m. Other frequency versions of the system are also available.

Model 4268 MicroNAVTRAK tracking system

Contractor
EDO Corporation, Electro-Acoustic Division, Salt Lake City, Utah.

SONAR SYSTEMS

Many of the problems associated with sound transmission in ASW apply equally to MCM, although there are considerable differences between the sonars used in the two forms of warfare. While ASW is essentially a long-range detection problem requiring low frequency, MCM is a very short-range problem requiring high frequencies. The use of acoustics in minehunting is further compounded by the problem of multi-path reflections from the bottom, experienced in shallow waters. The reason for using high frequencies in MCM is because of the need to detect very small objects and produce very high definition pictures of them. Low frequency sonar offers extended range but poor definition in detail, while high frequency gives short range but very high definition. The approach to defining requirements for MCM sonars is therefore quite different from that of ASW. This is now becoming even more important as sonars for MCM are required to pick out small mines on the bottom which may be hidden among rocks, buried in mud or sand, protected by anechoic coatings or provided with irregular shapes to reduce the probability of detection and classification.

To provide as much detail and data over a very small insonified area as possible, a high frequency sonar is needed. The higher the frequency, the more detail will be returned to the signal processor. However, the problem with high frequency sound in water is that it is quickly attenuated, and so sonar ranges in MCM are very limited. This means either that an MCMV with an onboard sonar has to approach closely in order to detect, identify and classify objects, with the possibility of detonating a mine, or that the sonar must be displaced from the MCMV in some way so allowing the MCMV to remain well outside the danger radius of any possible mine detonation. Until recently the use of hull-mounted minehunting sonars has required a trade-off between the high frequency required for high definition, and the lower frequency required to ensure that the MCMV

can stand off at a safe distance for any possible mines (the danger radius of ground mines is considered on average to be about 200 m). To overcome this problem most sonars incorporate a dual frequency capability, using a lower frequency to survey a specific area, before moving in closer to use the high definition sonar to classify and identify the target. However, to ensure ship survivability against modern mines most MCMVs now carry an offboard system (the ROV) as well, carrying either a camera or small high definition sonar (or both), or use a diver to positively identify and neutralise the detected target.

Stand-off Sonars for MCM
There are two methods of deploying minehunting sonars in a stand-off form. One is to use a side-scan sonar towed astern of the platform. Alternatively the sonar may be fitted in an ROV, of which numerous examples now abound, or it may be deployed as a variable depth sonar. The latest systems now the subject of intense development and investigation include autonomous offboard ahead swimming vehicles capable of carrying out a fully integrated minehunting operation against both shallow and deep water laid mines.

Route Surveillance
One of the most important factors in mine warfare is route surveillance. This requires that areas susceptible to mining must be constantly surveyed well in advance of hostilities. Detailed records of the seabed and all objects on it must be maintained by constantly surveying the area and this will enable routes to be quickly checked for new targets in time of rising tension and hostilities. This will enable routes to be cleared more quickly, or alternative routes to be assigned. Surveillance such as this requires a high speed of advance and high coverage rate. Route surveying thus needs to be carried out at speeds up to 15 kts with a path cover about 2000 m wide. This

requires a sonar range of about 1600 m. In good propagation conditions a sonar with an azimuth resolution of 3° and a range resolution of 20 cm would be sufficient to carry out such a route survey.

Classification, on the other hand, demands a much slower speed of advance with much greater resolution. Thus speeds of 5 kts are the norm, and to achieve good classification results using acoustic shadow and echo to allow evaluation of the shapes of detected objects at a range of at least 275 m (to ensure platform safety), requires a bearing resolution of 23° with 5 cm resolution. In addition classification is further aided by employing sector-scan techniques. For close range identification an even higher resolution capability will be required.

As targets become smaller and more elusive the ability to detect and classify accurately will demand even higher frequencies and smaller beamwidths, while identification will be aided by optics and laser systems.

The Future
The key to the whole question of sonars, both for ASW and MCM, is technology. New and improved transducer materials are required to increase cavitation limits, synthetic aperture sonar processing would enable the range and coverage rate of sonars for MCM to be increased without degrading resolution and requiring a large underwater body. Parametric sonar techniques would enable sonars to penetrate the seabed to detect buried mines, a feature which is now of paramount importance in the face of the latest mine technology. As with ASW, there is also the need for increased capabilities in the field of real-time data processing and post-processing techniques. Finally the MMI interface has to be improved, together with computer-aided detection and classification, which will greatly assist the operator in his task.

HULL-MOUNTED MINEHUNTING SONARS

FRANCE

TSM 2021B SONAR

Type
Minehunting sonar system.

Description
The Thomson Sintra TSM 2021B (service designation, DUBM 21B) is a modern minehunting sonar, the design of which applies experience with the DUBM 20A, also developed by Thomson Sintra. The hull-mounted sonar is intended for use in conjunction with precision navigation equipment, and a mine countermeasure (destruction) system such as the PAP 104. A former version is the 2021A with different array housings and minor display variations. TSM 2021 functions are:
(a) the detection of mines at distances up to 600 m
(b) the classification of such targets by the study of the shape of their echo and acoustic shadow, at distances up to 250 m.

Each detection and classification sonar chain consists of an acoustic transmitter/receiver transducer, the associated transmission and reception electronics, and a display console. A subassembly performs the stabilisation, steering, and retraction of the detector and classifier arrays. This part of the system is the result of studies and development by ECAN at Ruelle.

TSM 2021B minehunting sonar, showing the main electronics cabinet flanked by the classification and detection display console (left and right), with the transducer arrays in the foreground

The electro-acoustic subassembly (EAA) includes:
(a) the classification sonar transmitter/receiver chain cabinet
(b) a display console which controls the transmission and reception functions and has CRT presentation of mine detection and location by the detector sonar chain
(c) a display console with CRT displays for presentation of mine locations within the classification chain
(d) bridge repeater display, showing range and bearing of designated targets.

Operational Status
The TSM 2021 is in service with the French and other navies. Altogether some 60 systems have been ordered.

Specifications
Coverage: ±175° in sectors of 30, 60 and 90° for the detector chain; 3, 5 or 10° for the classification chain
Elevation: Variable, between −5 and −40°
Stabilisation: ±15° (roll); ±5° (pitch)
Detection sonar
Frequency: 100 kHz (modulated ±10 kHz)
Pulse duration: 0.2 or 0.5 ms
Range scales: 400, 600 or 900 m
Max sound level: 120 dB
Channels: 20
Beam aperture: 1.5°

Display: PPI
Classification sonar
Range scales: 200 or 300 m
Max sound level: 122 dB
Channels: 80
Beam aperture: 0.17°
Display: CRT and storage magnifier tube

Contractor
Thomson Sintra Activités Sous-Marines, Brest Cedex.

TSM 2022 and TSM 2022 Mk II

Type
Hull-mounted minehunting sonar system.

Description
The TSM 2022 is a lightweight minehunting sonar designed specifically for small and medium sized mine countermeasures vessels. The sonar can also be used for mine avoidance on minesweepers, or in civil applications such as bottom profiling and side-scan surveys.

The TSM 2022 is based largely on experience gained with the DUBM 21B in the field of high resolution beam forming techniques. It is a hull-mounted equipment intended also for use in conjunction with precision display and navigation equipment and a mine countermeasures system.

The main feature of the TSM 2022 is its small-size retractable single array assembly, used for both detection and classification, which enables easy installation and maintenance. The high resolution features of the DUBM 21 are maintained and improvements are mainly in the field of digital processing technology.

The main functions of the sonar are: detection and classification of moored and bottom mines, detection at distances up to 600 m (2000 m for submarines), and classification of targets by analysis of the shape of their echo or shadow at distances up to 250 m. In the classification mode, horizontal beamwidth is 7°; in detection mode either 14° or 28° beamwidth can be selected. In both modes the vertical beamwidth is 15°.

The main assemblies are:
(a) the hoisting and stabilisation system together with the retractable array which is installed in the sonar trunk (0.75 m diameter). The total weight is only 900 kg
(b) electronics cabinet
(c) operator's console.

The operator console provides display of sonar images in the various modes, memory and display of former sonar contacts, and includes interfaces with current mine disposal and navigation equipments.

In the Mk II configuration the TSM 2022 is equipped with a 19 in high resolution colour console and specific sonar processing such as computer-aided detection (CAD), computer-aided classification (CAC) and performance indicator (PI).

Operational Status
To date, 31 equipments have been ordered by seven navies. The Swedish Navy has seven TSM 2022s in operation, Malaysia four, Indonesia two, Nigeria two, and Yugoslavia two. The Singapore Navy has also ordered four TSM 2022s for its new MCMVs.

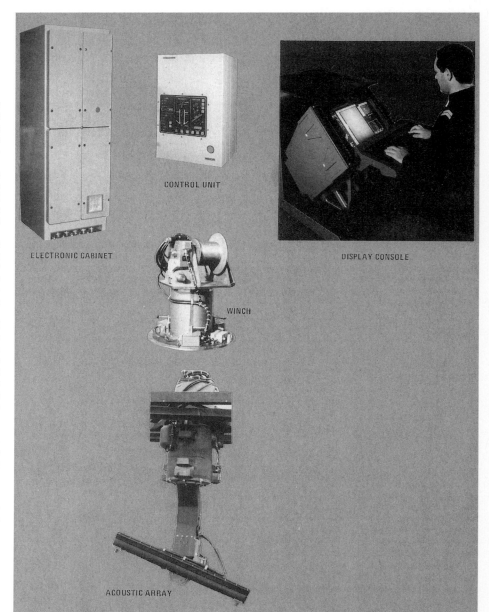

CONTROL UNIT

ELECTRONIC CABINET

WINCH

DISPLAY CONSOLE

ACOUSTIC ARRAY

Elements of the TSM 2022 Mk II minehunting sonar system

Contractor
Thomson Sintra Activités Sous-Marines, Brest Cedex.

TSM 2023 SONAR

Type
Hull-mounted minehunting sonar.

Description
The TSM 2023 has been developed by Thomson Sintra in collaboration with Simrad Subsea, with Thomson Sintra as the main contractor and Simrad Subsea as subcontractor being responsible for the detection sonar.

The TSM 2023 performs mine detection and classification simultaneously, and comprises:
(a) a detection sonar which covers a bearing sector of 90° electronically steerable through 360°
(b) a classification sonar which can perform shadow analysis at three different frequencies. The three different frequencies allow the beam resolution and bearing sector to be adjusted according to the size of the detected object
(c) two sonar consoles, one for detection, one for classification, equipped with CAD and CAC processing and a performance indicator.

Operational Status
The TSM 2023 was ordered by the Norwegian Navy in 1990.

Contractor
Thomson Sintra Activités Sous-Marines, Valbonne Cedex.

IBIS SONAR SYSTEMS

Type
Minehunting integrated sonar systems.

Description
IBIS III, V and VII are fully integrated minehunting systems using a high performance sonar and a tactical display system which integrates and displays all critical information.

The system consists of:
(a) a minehunting sonar, TSM 2021, TSM 2022 or TSM 2023
(b) a tactical display console, TSM 2060 NAVIPLOT
(c) a Doppler sonar log, TSM 5730 or TSM 5722
(d) navigation systems, autopilot, windspeed sensors, and so on.

Mine detection and classification is the function of the sonar (TSM 2021, TSM 2022 or TSM 2023).

Detection is performed at distances up to 600 m (2000 m in the case of submarines). Classification is possible in two modes, echo mode and shadow mode, allowing identification of moored and bottom mines.

The action/information data display and storage are performed by the tactical display system which provide ship's track monitoring, location of contacts and records of operational data. Positions of classified mines, possible contacts and non-mine objects are recorded and displayed simultaneously.

The IBIS V Mark II offers a number of improvements over its predecessor. The TSM 2061 tactical system integrates the digital technology Colibri CRT console with dual screen capability. In addition the Mark II is fitted with a range of detection and classification aids that operate in real time or batch processing modes. By extracting suspicious objects, the classification aids reduce the false alarm rate and offer off-line generation of mine warfare charts for optimised debriefing and mission appraisal.

The IBIS VII Variable Depth Sonar (VDS) is intended for use against advanced mines, such as the mobile types, from the mid-1990s onwards. It will be available in two versions, a towed VDS and a propelled VDS, with the body operating behind and ahead of the ship respectively. The towed VDS can operate at speeds up to 5 kts, can descend to 200 m and trails less than 150 m behind the ship. The propelled version uses a self-propelled vehicle operating at depths to 300 m, distances up to 600 m ahead of the ship, and speeds of up to 6 kts.

Operational Status
See under separate TSM 2021B, TSM 2022 or TSM 2023 entries.

Contractor
Thomson Sintra Activités Sous-Marines, Brest Cedex.

GERMANY

DSQS-11A/H SONARS

Type
Minehunting and mine avoidance sonar systems.

Description
Atlas Elektronik has developed systems for use in mine countermeasures. Few details are available but the following is a brief summary of the equipment.

DSQS-11A
This is a mine avoidance sonar designed for the location of underwater objects, in particular moored mines. It can be used as a navigational aid in mine areas for both surface ships and submarines, and is

operated by a single person. When an object is located the distance, bearing and depth are presented to the operator on a PPI display. The bearing can be referenced to the ship's heading, true north, or shown in a true motion mode. The array is fully retractable.

DSQS-11H
This is a mine countermeasures sonar and is used for the detection and classification of surface, moored and ground mines. The equipment provides independent operation of detection and classification modes through 360° of azimuth, with simultaneous azimuth coverage of 90° in the detection mode. Target images are stored in the system to aid recognition. Electronic beam stabilisation is fitted for both modes, and the

sonar can be operated at speeds up to 10 kts with minimal degradation of performance. The cylindrical search array transmits 45 pre-formed beams, and the steerable flat-plate classification array produces 20 vertically stacked beams. The DSQS-11H is one-man operated.

Operational Status
The DSQS-11M has been adopted for the 'MJ 332' minehunter programme of the German Navy.

Contractor
Atlas Elektronik GmbH, Bremen.

NORWAY

SA950 SONAR

Type
Mine detection and avoidance sonar system.

Description
The Simrad SA950 is an active, high resolution, multibeam sector scanning sonar designed for fast and accurate detection of moored and bottom mines. The design of the sonar ensures simple and accurate operation, a high degree of maintainability, and low size and weight.

The system consists of a hull unit, transducer, transceiver unit, pre-amplifier unit, roll and pitch unit, servo unit, and an operator/display unit. The operating frequency is 95 kHz and the transmission is sectorial, covering a sector of 45 or 60° with 32 beams of 1.7° each. The sector is mechanically trainable and tiltable covering ±190° in the horizontal plane and from +10 to −90° in the vertical plane. The transducer

has 64 individual staves mounted in a plane array. The elements are common for both transmission and reception.

The signal processing is performed in a fast digital system using the full dynamic range of the echo information. The receiver has time varied gain and automatic gain control on the pre-amplifiers, background normalisation after the beam forming, ping-to-ping correlation and target tracking.

The echoes are presented in 64 colour high resolution on a 20 in monitor. There is sector presentation and one beam/range (echogram) presentation. In addition, PPI, relative or true motion is selectable. The system is controlled by a single operator using designated hardware switches on the operator unit front panel and interactive with a menu on the display.

The SA950 can be used as a stand-alone system or form part of the MICOS integrated combat information system.

Operational Status
In January 1989 Simrad received an order from the

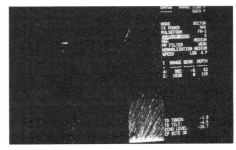

Moored mine at 40 m depth and at a range of 802 m. Own vessel speed 4.9 knots

Turkish MoD for four sonar systems to be fitted to minesweepers.

Contractor
Simrad Subsea A/S, Horten.

UNITED KINGDOM

TYPE 162M SONAR

Type
Side-looking target detection sonar.

Description
Sonar Type 162M, which is fitted in ships of the Royal Navy, detects and classifies both mid-water and seabed targets. It displays port and starboard recordings simultaneously on a single straight-line recorder, which has a maximum range scale of 1200 yd. Operation is simplified by entirely automatic gain control. Reliability is enhanced with solid-state

technology and maintenance is facilitated by ease of access and comprehensive built-in testing and monitoring features.

The three transducers are all similar and employ 49.8 kHz barium titanate elements. Their beam pattern is fan-shaped, about 3° wide and 40° vertical angle; the side-looking elements have their axes 25° below the horizontal.

The recorder design uses a double helix (left and right hand) so that there are two points of contact with the moist electro-sensitive paper. As the helix rotates, the two points of contact move outwards from the centre. Port and starboard signals are fed respectively to the left- and right-hand points of contact so

that they are recorded simultaneously. There are two zero lines near the middle of the paper, and the 18 mm gap between them is used for time marks every five minutes. The paper width is 286 mm, and each trace occupies 133 mm.

An electronic oscillator provides a controlled frequency supply for the driving motor and interval marks. Motor speed changes, for the three range scales, are made by frequency division so as to avoid the use of change speed gearbox, and a stroboscopic arrangement is included so that the helix speed may be easily checked.

A loudspeaker and a socket for headphones enable signals to be monitored aurally if required.

The three range scales are 0 to 300, 0 to 600 and 0 to 1200 yds, and accuracy is better than two per cent assuming a sound velocity of 4920 ft/s. The paper speed changes automatically when the range scale is selected, and the speeds are 6, 3 and 1.5 in/min (152, 76 and 38 mm/min) respectively. At a ship speed of 10.8 kts the display scales are the same across the paper and vertically. A take-up spool is fitted but its use is optional. A fix marker draws a line across the width of the paper when a button is pressed, and it can be operated in conjunction with a Type 778 echo sounder or other equipment.

A new transducer array has now been developed as a direct replacement unit. The new array incorporates redesigned elements and techniques using a low voltage device which virtually eliminates insulation problems. Tuning boxes for the three transducers are no longer required since the transducer tuning coils are incorporated into the individual housing gland.

Operational Status
In service.

Contractor
AB Precision (Poole) Limited, Poole.

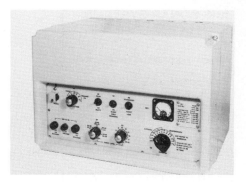

Type 162M sonar transmitter/receiver unit

SONAR 193M/193M MOD 1

Type
Sector scanning minehunting sonar.

Development
Development of the Type 193M was undertaken by Plessey (now Marconi Underwater Systems) in 1968. Since then a continuous programme of development and operational trials in conjunction with the UK MoD has resulted in further system improvements defined as the Type 193M Mod 1.

The system offers greatly improved performance in detection and classification by the use of digitally processed video and display systems, together with computer-aided target classification.

Description
The 193M is a short-range, high definition sector scanning sonar, operating at two frequencies, providing detection and classification of both moored and ground mines. It has been developed from the Type 193 minehunting sonar; the adoption of solid-state electronics and other advances in technology have resulted in a reduction in installed weight to about 860 kg, which compares with a figure of some 2100 kg for the older Type 193. Among the operational improvements are extensive use of digital displays and facilities for interfacing with computer systems.

The basic concept is to employ high definition sonar to locate and classify mines which are then destroyed by explosive charges.

The sonar provides both bearing and range data; the fine range and bearing resolution enables the operator to assess accurately the shape of a target and hence its nature. Since the resolution depends both on operating frequency and pulse length, two selectable frequencies are provided, each with a choice of pulse lengths in the classification mode.

Range and bearing data appear on two displays: in the search mode one display shows the total range covered; in the classification mode, a 27 m (30 yd) section of the search display is expanded to fill the second display screen and permit close examination of the target.

Two frequencies are employed by the 193M sonar: 100 kHz for long-range search and 300 kHz for short-range search and classification. The transducers for these signals are carried beneath the ship on a stabilised, steerable mounting, the whole assembly contained within a dome. A choice of either inflated fabric or GRP material is provided for the dome. The receiver uses a modulation scanning technique with 15 beams, 1° wide in LF and 0.33° wide at HF, giving azimuth coverage of 15° and 6° respectively. The searching of wider areas of the sea bottom is achieved by means of an automatic search sequence selected in accordance with the type of sea bottom. The returned echoes are presented to the operators on separate CRTs at the control console. One of these displays range and bearing of targets within the

Type 193M sonar operator's console

sector being scanned by the search transducer, while the other is used for the presentation of the classification channel data. The controls for adjusting the transducer position, signal parameters such as frequency and pulse length and for co-operating with the rest of the minehunting and destruction team, are also provided at the console. Type 193M data can also be fed into other ships' systems.

The Marconi Speedscan system can be fitted to the Type 193M to allow it to be operated in the side-scan mode and to generate a hard copy print out of the seabed.

Range and bearing data outputs are provided in synchro and digital forms, allowing the sonar to be interfaced with various plotting tables, action information systems and remote indicators. The system is fully integrated with the Marconi NAUTIS-M system as chosen for the Royal Navy single role minehunter.

Operational Status
In service with the Royal Navy and many other navies, the Type 193M/193M Mod 1 is suitable for installation in all current minehunters.

Type 193M is fitted in the Royal Navy's 'Hunt' class MCMVs. In total, 12 navies have now purchased some 50 Marconi minehunting sonar equipments, including both the 193M and the earlier version, 193. A continuing programme of enhancement will ensure that the efficiency of the equipment is kept continually under review.

The Type 193M Mod 1 enhancements have been retrofitted as a modification to RN ships fitted with 193M, while complete Type 193M Mod 1 sets are now being exported.

Recent additions to the system are a surface and near-surface mine detection mode, which also provides a mine avoidance capability while minesweeping, and a new compact sea-chest installation.

Contractor
Marconi Underwater Systems Limited, Templecombe.

PMS 58 operator console

PMS 58 MINEHUNTING SONAR

Type
Hull-mounted minehunting sonar.

Description
The PMS 58 is a new low-cost, lightweight hull-mounted dual frequency minehunting sonar. Use of modern high speed processing modules has created a flexible system architecture which allows the system to be offered in a number of forms:

(a) mine avoidance sonar for minesweepers or non-MCM vessels
(b) standard single-operator minehunting sonar
(c) standard dual-operator minehunting sonar.

All minehunting variants have facilities for search and classification of ground mines, and special array and processing features for detection and classification of moored mines, and surface and near-surface mines. A full route survey capability is incorporated as a standard feature.

A modern lightweight directing gear provides good platform stability characteristics with small hull aperture requirements.

PMS 58 also offers a number of optional facilities.

These include computer-aided detection, and multi-ping processing and computer-aided classification for the minehunting variants.

Operational Status
The system has successfully undergone minehunting trials.

Contractor
Marconi Underwater Systems Limited, Templecombe.

PMS 75 SPEEDSCAN/SONAR 2048

Type
Side-scan modification for minehunting sonars.

Description
The PMS 75 Speedscan is a self-contained equipment for ships already fitted with a compatible sonar system. It allows the sonar system to be operated in a side-scan mode and generates a hard copy printout of the seabed. It can be fitted to all types of minehunting sonar.

When the sonar is operating with Speedscan, its transducer is trained 90° to port or starboard of the ship's track. As the ship proceeds along a predetermined track, an area of seabed parallel to the track is interrogated by the sonar. The received sonar data and reference timing signals are fed to the Speedscan processor which forms a side-scan beam and presents the processed data as hard copy on continuous recording paper.

The Speedscan equipment consists of two main assemblies, each contained in a portable case. In use, the two cases are stacked and locked together with the end covers removed. The upper case houses the processor and the operator's control panel, the lower case the recorder unit.

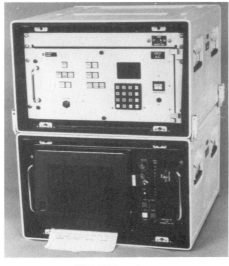

PMS 75 Speedscan

Speedscan presents information on light-sensitive paper exposed in the recorder. On the record, the seabed is represented as a series of parallel narrow strips of selected length, perpendicular to the ship's track. The sonar data from the whole azimuth sector scanned by the sonar is extracted by Speedscan to represent a narrow strip of seabed along the sonar beam axis. This data are extracted by the beam former to generate a parallel beam of sufficient width to overlap the strips covered by the two previous transmissions, so providing the three-ping sample of each point of the seabed along the range axis. The beam forming system is programmed by ship's speed and sonar range to provide a controlled interrogation of the seabed. This information is used to intensify and modulate a light source in the recorder which produces a high definition image of the seabed.

Operational Status
The PMS 75 is designated Royal Navy sonar Type 2048 in mine countermeasures vessels. It is also used by other navies.

Contractor
Marconi Underwater Systems Limited, Templecombe.

UNITED STATES OF AMERICA

SH100 MINEHUNTING SONAR

Type
Mine detection and classification sonar.

Description
The SH100 mine detection and classification sonar is an active, dual frequency, high resolution sonar designed for hull mounting on coastal or inshore minehunting or minesweeping vessels. The operational capability of this very compact sonar is from the amphibious landing zone out to depths of 100 m for detecting and classifying bottom mines and beyond that for detecting moored mines. The SH100 is an extension of the single frequency SA950 mine avoidance sonar.

The transducer arrays are mounted on a fully retractable hull unit which is mechanically stabilised against roll, pitch and yaw. With the stabilised array, reliable detection and classification performance is maintained even in severe conditions.

The SH100 design uses hybrid pre-amplifier circuits to maximise performance and reliability while reducing the size, weight and cost of the total system. The signal processing, including beam forming, is performed in a fast digital signal processing system using the full dynamic range of the signals.

The SH100 operates at 95 kHz (LF detector) and 335 kHz (HF classifier) and has a sectoral transmission and multi-beam reception covering an LF sector of 45°, with 32 beams of 1.6°, and an HF sector of 16° with 64 beams of 0.25°. Vertical coverage is 10°. The sectors, which are aligned along the same axis, are mechanically trainable (±200°) and tiltable (+10° to −90°).

The selection of the two frequencies is optimised for the separate tasks of detection and classification. Extensive test and application results from a range of sea environments have proven this sonar's ability to consistently detect bottom and moored mines at a range of 600 m and classify at a range of 200 m. Mine detection with the SH100 has been demonstrated at ranges greater than 1000 m.

The sonar can operate on both the LF detection and HF classification frequencies simultaneously.

Signal enhancement and display techniques, such as shadow mode normalisation, FM pulse compression, ping-to-ping filtering, echogram detection and zoom, can be applied separately to each frequency. This allows the sonar operator to optimise system performance for the specific mission task. The following operational modes are available:

(a) wide sector search and target acquisition using only the LF mode
(b) high resolution classification and precise localisation using only the HF mode
(c) combination of LF and HF on alternate transmissions
(d) vertical positioning of the transducer for moored mine classification and depth measurement.

The SH100 sonar consists of a hull unit, pre-amplifier, transceiver, servo control unit, servo transformer unit, and a control and display unit. The basic configuration includes software and hardware interfaces to Doppler speed log, gyro-compass, roll and pitch sensor, and surface navigation systems. Simrad can deliver the SH100 as a stand-alone system which includes these additional sensors and power units if required.

The control and display console includes two high resolution 20 in RGB monitors mounted one above the other. The lower monitor displays the sonar echo image while the upper one displays tactical information. The echo display presents the sonar image in B-scan and/or echogram modes. The tactical display presents a map-like image of the sonar situation including the ship symbol and past track, target positions and classifications, LF and HF sector coverage, and tilt and train graphics. Additional tactical data displayed include the ship data such as position, course and speed, and target positions (both relative and geographical). The simple MMI provides access to all the primary controls with the buttons and joysticks on a control panel. Secondary commands are accessed through an on-screen menu.

Simrad's new SH100 tactical extension is an enhancement to the basic configuration which provides data distribution to multiple locations on the vessel, recording of target database, sonar image printing, and more powerful tactical and navigational functions. Interfaces to buyer supplied combat

Control/display unit of the SH100 minehunting sonar

systems for transferring target data or the full sonar image data can be readily adapted.

Operational Status
In service.

Contractor
Simrad Inc, Lynnwood, Washington.

VARIABLE DEPTH MINEHUNTING SONARS

ITALY

AN/SQQ-14IT

Type
Minehunting sonar.

Development
In 1986 FIAR launched a study in reply to a request by the Italian Navy for a deep water minehunting sonar for installation on the new minehunters programmed for 1993. On the basis of experience gained during development of the SQQ-14IT, a proposal has been put forward for this sonar. This constitutes an evolution and improvement of the SQQ-14IT programme and adapts the performance to the Italian Navy's requirements. The system provides the contemporary functions of search and detection of both bottom and close-tethered mines, as well as moored mines situated at considerable depths. Operational tests on a full-scale development model are expected to commence in 1992, followed by installation on the first ship by the end of 1993.

Description
The aim of the SQQ-14IT programme has been to redesign the General Electric AN/SQQ-14 minehunting sonar, and completely new electronics have been developed for both the dry and wet ends. Only the mechanical parts of the old hoist system remain.

In line with the next generation sonars, new functions such as speed scan, memory display and remote display have been included, as well as BITE functions which provide self-diagnosis and the location of malfunctions down to the printed circuit cards.

SQQ-14IT minehunting sonar

The transfer of the transmitter unit into the towed body has provided an increase in the system operational performance, as well as a redundancy of the principal functions, that is, console interchangeability and the possibility of operating with reduced efficiency in the case of malfunction in one or more of the transmitter sections. The adoption of the latest consoles for search and classification operations has greatly improved the man/machine interface, and simplified the use of the system and the training of operators. Recording and mission playback facilities are available, as well as training and simulation programmes.

Operational Status
FIAR has manufactured 15 AN/SQQ-14 minehunting sonars for the Italian and Belgian navies.

Contractor
FIAR — Fabbrica Italiana Apparecchiature Radioelettriche SpA, Underwater Department, Milan.

UNITED KINGDOM

TYPE 2093 SONAR

Type
Variable depth, multi-mode minehunting sonar.

Description
The Sonar 2093 is stated to be the world's first variable depth, multi-mode minehunting sonar. Designed by Marconi, Sonar 2093 is fitted in HMS *Sandown*, the first of the new single role minehunter ships of the Royal Navy. The system is designed to operate in either hull-mount or variable depth mode, deployed through the ship's centre well, and will detect, localise and classify all current and future mine threats. With a total system weight of 13 000 kg, Sonar 2093 is the lightest variable depth minehunting sonar available.

As the culmination of a six-year development programme, the system is designed to carry out its mission at high speeds without compromising the ship's safety. A variable depth, dual frequency search and dual frequency classification capability provides a sonar able to operate under all bottom and sea conditions at a range claimed to be twice that of hull-mounted sonar and with an increase of three times detection depth. The Sonar 2093 is integrated with the ship's command system to allow simultaneous operation in combinations of search, detection, classification and route survey modes. The area coverage rate capability will allow a significant improvement on current route survey operations.

The towed body of the 2093 is cylindrical with hemispherical ends. Stabilising vanes consist of fixed strips mounted a few centimetres away from the shell, largely parallel to its surface and curved so as to be concentric with its axis, and with the centres of the hemispherical ends over an arc of 45°. The sonar arrays within the body consist of two 360° rings for LF and VLF operation, hydrophones for HF and VHF reception, and separate HF and VHF projectors. Below is a cylindrical VLF projector and finally a VLF

Sonar Type 2093 consoles fitted in HMS Sandown

depth sounder. LF and VLF are used for search, survey and moored mine classification; HF and VHF for classification.

Sonar 2093 employs the unique approach of multiple operating modes, which offer the MCM commander the choice of a number of operational configurations to match the prevailing threat and environment. These multiple operating modes comprise

VLF and LF search and moored mine classification, VLF and LF search, search and ground mine classification, and search and route survey. Sonar 2093 can combine these modes to enable concurrent operation.

Operational Status
In production, with 12 systems having been ordered for the Royal Navy and six systems for an export

customer. The contractor's acceptance trials were successfully completed in March 1990.

Contractor
Marconi Underwater Systems Limited, Templecombe.

UNITED STATES OF AMERICA

AN/SQQ-14 SONAR

Type
Variable depth towed mine detection sonar.

Description
The AN/SQQ-14 system is essentially a variable depth, dual frequency sonar for detecting and classifying bottom mines in shallow water. It utilises a towed body in the shape of an elongated sphere towed through a centre well on a US Navy minesweeper.

The AN/SQQ-14 can operate down to 45 m. It transmits at 80 kHz and 350 kHz respectively for search and classification, scanning over azimuths of 100° and 18°, and through an elevation of 10° in each case. Azimuth and range resolutions are 1.5° and 1 m for search, and 0.3° and 8 cm for classification.

A unique aspect of the AN/SQQ-14 design is the towing cable. This consists of discrete, 18 in (457 mm) sections of articulated struts with universal joints at each section, permitting the cable to flex in any vertical plane, but restraining it from torsional motion. This configuration imparts a constant heading to the towed body, thereby eliminating the need for a gyroscopic heading reference system.

A rubber-jacketed electric cable containing 35 shielded, coaxial and individual conductors passes through the centre of the strut sections, terminating in a slip ring in the towed body and in a winch with compound winding drum in lieu of a slip ring installed on the 01 deck of a mine countermeasures ship.

The articulated strut type of tow cable posed unique problems related to winching, storage on the winch drum, and ship's tow point construction. Struts, of which there are 117 per system, are stored in a single layer on a cylindrical drum equipped with specially designed pads serving as contact points for the articulated strut knuckles. Since a conventional level wind mechanism could not be used with the struts, the winch drum itself is designed to traverse axially by means of an Acme lead screw and rotating nut,

AN/SQQ-14 projectors and hydrophone arrays

thereby accomplishing the level wind function and maintaining a fixed fleeting point. All equipment for this application must be non-magnetic. Struts, bearings and knuckles are fabricated of high strength Inconel alloy. The winch drum is cast aluminium; the supporting structure is aluminium and stainless steel.

Operational Status
In operational service. The SQQ-14 sonar is produced under licence in Italy by FIAR SpA, which has

supplied equipment to the Italian Navy. Other users include the Belgian and Spanish Navies. FIAR is offering an improved version, the SQQ-14IT, which uses about 40% of the improvements carried out by GE to upgrade the SQQ-14, to the SQQ-30 (see FIAR entry in the Italian section).

Contractor
General Electric Company, Electronic Systems Division, Syracuse, New York.

AN/SQQ-30 SONAR

Type
Variable depth towed mine detection sonar.

Description
The AN/SQQ-30 is the successor to the AN/SQQ-14 system and is a latest technology minehunting sonar. The system consists of two sonars: a search sonar

for initial detection; and a high frequency, high resolution sonar for classification of targets. The two sonars are separated to give the area coverage needed. They are housed in a hydrodynamically egg-shaped vehicle and towed at various speeds by the minehunter. The towed body is streamed from a well in the forward part of the ship, using a winch driving a 3 m diameter cable drum on the foredeck. Two display consoles are provided for the search and classification sonars.

Operational Status
In production. Fitted to the first few US Navy MCM-1 'Avenger' class minehunters.

Contractor
General Electric Company, Electronic Systems Division, Syracuse, New York.

AN/SQQ-32 SONAR SYSTEM

Type
Advanced minehunting search and classification sonar system.

Development
Technical and operational evaluation of the advanced minehunting sonar system (AN/SQQ-32) for the US Navy, by Raytheon teamed with Thomson Sintra in France, has been completed. The new system is scheduled to be installed in the US Navy's newly constructed minesweepers designated MCM-1 and MHC-51.

Description
The system consists of two separate sonars: a search

sonar for initial detection; and a high frequency, high resolution sonar for classification of the targets. The two sonars are partially housed in a hydrodynamically shaped vehicle and towed at various speeds by the minehunting ship. The latest technologies in beam forming, signal processing, modular packaging and displays are used throughout the system.

The detection sonar is being designed and manufactured by Raytheon and will be able to detect mines over a wide range of distances and bottom conditions. This system incorporates a computer-aided detection facility which is designed to help the sonar operator to discard non-mine objects detected by the sonar. Raytheon is also providing the computer facilities and display consoles and is responsible for overall system integration.

The classification sonar is being designed and

fabricated by Thomson Sintra and will be based on its experience with the DUBM 21 and TSM 2022 minehunting sonars. The classification sonar will provide very high resolution transmission and reception of underwater signals to enable targets to be identified with near-picture quality.

The 'wet end', consisting of a hydrodynamically shaped vehicle, a deployment/retrieval system, a tow cable and a winch, is being designed by Charles Stark Draper Laboratory of Cambridge, Massachusetts. The towed body is housed in a vertical trunk extending from keel to deck, just forward of the bridge. The body is roughly egg-shaped, with a pair of boomerang-shaped vane arms pivoted on either side. When stowed in the trunk, these lie alongside the body itself, once clear of the ship's keel they swing aft to stabilise the body as it is towed along in the direction normal to its axis. The body is slung from pivots on either side to a fork assembly that is connected to the towing cable, so that its axis remains vertical. Within it are the search sonar staves arranged as a 360° ring array belted around the body, together with immediate sonar electronics and associated equipment. Below the body is the rotating scanner for the classification sonar which can operate at three frequencies down to 300 kHz.

The AN/SQQ-32 consists of two identical operator consoles with high resolution displays. Search and classification data can be displayed simultaneously or independently. A search display presents video patterns for the detection operator to examine, and computer-aided detection indicates objects likely to be mines. Search data are presented on six screen displays allowing the operator to determine if a mine-like object is a real target or a natural bottom feature. The system proceeds to long-range classification, with the second operator, by measuring the detected object's height above the seabed in order to assess the probability of it being a moored mine. The classification operator then uses the high resolution classification sonar, which has very narrow beams with dynamic focusing, to examine objects detected on the bottom.

Operational Status

In February 1989, Raytheon received a $46 million contract to build two production systems and refurbish two of the three full-scale development units. In January 1990, Raytheon received a further contract award worth $125 million for 10 systems (seven for the US Navy and three for Japan). In January 1990, Thomson-CSF announced an order from Raytheon for the classification sonar part of the first order. The US MCMV USS *Avenger* (MCM) is fitted with a development model which was used in the Persian Gulf during Desert Storm.

The first MCMV to be fitted with a production version SSQ-32 will be the USS *Warrior* (MCM 10) in 1992. Thereafter units already in service will be back-fitted with the system.

Contractors

Raytheon Company, Submarine Signal Division, Portsmouth, Rhode Island.

Thomson Sintra Activités Sous-Marines, Brest Cedex.

AN/SQQ-32 operators' consoles

Simulation of the AN/SQQ-32 minehunting sonar system under way

SIDE-SCAN MINEHUNTING ROUTE SURVEY SONARS

CANADA

MODEL 972 SONAR

Type
Side-scan sonar system.

Description
The Model 972 system has been designed as a high quality sensor system to gather side-scan sonar information. The 972 supplies this information to larger survey systems which will incorporate navigation systems, tactical displays, mass data storage systems and computerised target recognition systems. Computer equipment and sophisticated software packages will be required, and are available for such systems.

The Model 972 consists of the following major electronic components:
(a) towfish electronics module
(b) left and right side-scan transducers (120 kHz and 330 kHz)
(c) echo sounder transducer (and optional sub-bottom transducers)
(d) two-conductor cable to surface
(e) surface processor module
(f) RGB colour monitor
(g) tape recording system consisting of PCM encoder and a VHS video tape recorder.

An important difference between the 972 and traditional scan systems is the use of a 1280 × 1024 × 128 Levels (42 dB) colour video display instead of the standard paper chart recorder. This video display provides a much greater dynamic range than a paper chart recorder. It is much easier to operate and maintain, particularly for less experienced operators.

The 972 produces undistorted side-scan displays, with the water column removed, and with corrections for slant range and towfish speed. Data may also be displayed in uncorrected format. With additional zoom features added, an off-line true zoom capability can significantly enhance object classification with resolution up to 100 pixels/m. When the sub-bottom profiler option is added, the system allows simultaneous side-scanning and sub-bottom profiling.

A basic system allows display of side-scan and sub-bottom information simultaneously on a single

Model 972 towfish and towfish depressor, display monitor and side-scan processor, remote winch controller and video data recorder/playback

monitor. If an additional processor module is added, side-scan data may be displayed on one monitor, with sub-bottom or zoom data displayed on the other monitor. The side-scan and sub-bottom profile data can be combined and stored on a magnetic media recorder or displayed on a paper chart recorder.

The underwater module is well adapted for incorporation into various kinds of underwater vehicles, including remotely operated vehicles, manned submarines or unmanned autonomous vehicles.

The towfish shell houses the buoyant electronics module in the upper half shell, and the heavier transducers and trim weights in the lower half. The towfish is neutrally buoyant and hydrodynamically shaped by a Kevlar reinforced plastic shell. The fins stabilise the towfish for precise sonar imaging.

Most fish are directly attached or coupled to the armoured tow cable. Direct coupled towfish are susceptible to surface sea-state action which results in significant image distortion. By decoupling the towfish from the armoured cable and using an intermediate length of cable, the surface sea-state can be dampened out without destabilising the towfish, thus resulting in undistorted sonar images.

Operational Status
In full-scale production and in operational service.

Contractor
Simrad Mesotech Systems Limited, Port Coquitlam, British Columbia.

MODEL MS 992

Type
Side-scan sonar system.

Description
The MS 992 is a simultaneous dual or single frequency sonar suitable for both general route survey and minehunting aspects of MCM. A unique aspect of this system is its two wire telemetry which makes the modular system ideal for ROV or sled mounting. A range of transducers is available for the 992, including a long-range system at 120 kHz and a high resolution system (330 kHz) with a range to well over 150 m.

A variety of towfish configurations are offered:
(a) an extremely rugged stainless steel model
(b) neutrally buoyant fish for shallow water operations
(c) compact modular unit for ROV installation.

A safety feature is the integral breakaway cable. Attitude sensing and responder are optional.

The compact surface processor measures only 5¼ in high by 13 in wide. High resolution colour video is standard. Hard copy paper recording and digital audio recording system interface are also available.

A unique feature with the side-scan processor is that it can be configured to operate Simrad Mesotech's fast scanning sonar heads (see Model MS 997). This is possible because the MS 990 processor forms the heart of both the MS 997 Fast Scan and MS 992 Side-Scan Systems. This has tremendous advantages in terms of equipment commonality.

Operational Status
In full-scale production.

Contractor
Simrad Mesotech Systems Limited, Port Coquitlam, British Columbia.

Towfish and cable for the MS 992

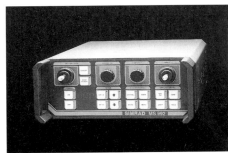

Processing unit for the MS 992

Display monitor for the MS 992

FRANCE

DUBM 41B (IBIS 41) SONAR

Type
Side-scan minehunting sonar system.

Development
Development of the DUBM 41B was conducted in collaboration by the French Navy Mine Warfare DCN (GESMA) and Thomson Sintra.

Description
The Thomson Sintra TSM 2050 (service designation DUBM 41B) system is a high resolution side-scan sonar system designed for the location and classification of objects such as mines, lying on the seabed. The system consists of three towed sonar vehicles, two consoles, the necessary cables, shipboard hoists and handling gear. In normal operation two of the three sonar bodies are used, with the third kept as a ready spare. A permanent record of the sonar data gathered is made on a facsimile type recorder and a tape recorder. The information is also presented simultaneously on two CRT storage tubes, these console-mounted displays providing an image of the seabed. The system is designed for use by low tonnage vessels, and in waters 100 m or deeper. The towed sonars operate at about 5.5 to 7.5 m above the seabed at speeds of 2 to 6 kts. The tow cables are provided with deflector vanes which ensure that the bodies are towed at a distance to left and right of the ship's track, and marker floats are provided to indicate the line of travel for each sonar.

The sonar body is a streamlined vehicle fitted with the side-scan sonars, plus additional sonar transducers for determining height above seabed and for obstacle detection. Servo-controlled fins are provided for depth control, and roll stabilisation is incorporated by means of a separate set of four fins. Dimensions of each sonar vehicle are: length 3.725 m, diameter 36 cm and in-air weight 340 kg. An acoustic pinger is installed as an aid to recovery should the sonar body

Towed sonar DUBM 41B on manoeuvres

break its tow. Manual controls are provided at the operating console to permit manual piloting, either to override automatic control or as a standby mode.

Each of the two sonar bodies operates on their own frequency, the two being separated by 50 kHz, with the area of seabed between them being scanned by both sonars. A typical scanned area covered by the two bodies amounts to a total width of about 200 m, with maximum sonar range selected. There are two other range settings, 50 and 25 m, at the latter setting resolution is stated to be better than 5 × 10 cm in the lateral (scan) direction and in the line of travel.

Operational Status
In service.

Thomson Sintra under contract to GESMA is carrying out a feasibility study of an experimental panoramic sonar for the detection of buried mines. This is part of an international research programme involving France, the UK and Holland.

The sonar and a navigation system are derived from the DUBM 41B towed sonar, and the new towed unit will house non-linear acoustic arrays and part of the processing system. Thomson Sintra is responsible for the towed arrays and transmission circuits, as well as the interface and display systems, and British Aerospace (Dynamics) Limited for the receiver section.

Contractor
Thomson Sintra Activités Sous-Marines, Brest Cedex.

DUBM 42 (IBIS 42) SONAR

Type
Side-scan mine detection sonar system.

Development
Development of the DUBM 42 has been carried out jointly by the French Navy Mine Warfare DCN (GESMA) and Thomson Sintra.

Description
The DUBM 42 is a side-scan sonar able to detect and classify mines at ranges up to 200 m, with an area coverage of 2000 m²/s. The system is intended for fitment to the French Navy 'Narvik' class MCMVs for survey of the Atlantic coasts at an operational speed of 10 kts and a height above the seabed of 30 m.

The basic system consists of one vehicle with three sonars (two lateral and one front-head) on the sonar body, one Colibri sonar display, one Colibri tactical display and one high density magnetic recorder. The data collected by the three sonars are recorded for analysis in a land-based processing centre. In addition a visual display of the seabed is provided on the operator's console. A magnifier provides a more accurate classification. All navigation functions such as piloting and automatic control are integrated on the same console.

The two lateral and single front-head sonars operate on their own individual frequency, the swept channel being 400 m wide.

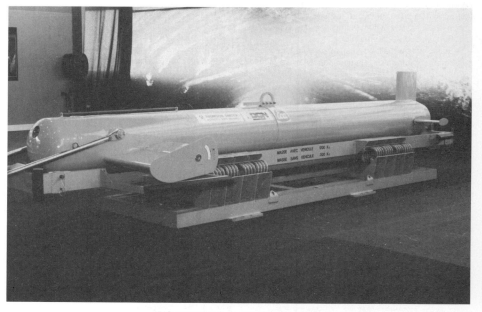
IBIS 42 sonar for the 'BAMO' programme

Operational Status
In production. IBIS 42 is the export nomenclature.

Contractor
Thomson Sintra Activités Sous-Marines, Brest Cedex.

IBIS 43

Type
Mine surveillance system.

Description
IBIS 43 consists of the TSM 2054 multi-beam side-scan sonar and the TSM 2061 tactical system. In base line configuration data collected are analysed off-line in a shore centre. In Mk II configuration data are analysed in real time on board. To carry out data analysing in real time onboard, the IBIS 43 Mk II comprises:
(a) TSM 2054 multi-beam side-scan sonar
(b) TSM 2061 tactical system

(c) detection and classification sonar console
(d) database
(e) CAD and CAC sonar processing.

Contractor
Thomson Sintra Activités Sous-Marines, Brest Cedex.

TSM 2054 SONAR

Type
Side-scan multi-beam minehunting sonar.

Description
The TSM 2054C (IBIS 43) is a high resolution side-scan looking sonar derived from the experience of the DUBM 42. The detection and classification of mine-like objects is performed at a maximum speed of 15 kts, and the swept channel width is 200 m.

The towfish sonar vehicle is 3 m long and its weight is less than 300 kg. The fish operates at depths down to 200 m and also carries an obstacle avoidance sonar that allows it to 'overfly' 10 m high obstacles at its maximum towing speed. It is also equipped with a Doppler sonar for accurate speed measurement. Automatic correction of speed, yaw and pitch, and the precise location of the towed body, is computed aboard the ship to an accuracy of 9 m and less than 1 m in altitude.

Two operating concepts are available:
(a) data collected by the sonar are recorded for analysis in a land-based processing centre
(b) detection and classification of new mine-like contacts is carried out on board ship.

The system is based on a multi-function Colibri display console and is used for tactical detection and classification operations as the IBIS 43.

Operational Status
The Royal Danish Navy has placed an order for two IBIS 43 systems.

Specifications
Manual or automatic towfish navigation at constant depth or altitude
Water depth: 6 to more than 200 m

The TSM 2054 side-scan sonar towfish

Altitude in relation to the seabed: 4-15 m, programmable from the sonar console keyboard.
Automatic, high precision terrain-following of irregular seabeds with gradients of more than 15%
Obstacle avoidance capability: Obstacles up to 10 m high at maximum speed
Speed in relation to water: 4-15 kts
Sonar range: 2×50 m; 2×100 m

Resolution: 2×10 cm or 2×20 cm, according to sonar range
Coverage: Up to 5.4 km²/h

Contractor
Thomson Sintra Activités Sous-Marines, Brest Cedex.

UNITED KINGDOM

SONAR 3000 PLUS SERIES

Type
Hydrographic survey sonar system.

Description
The Sonar 3000 PLUS family of side-scan sonars has been developed from the Type 2034 high definition sonar. The system uses modular construction which allows the assembly of variants to suit a range of applications. The system, originally aimed at the commercial market, is also in use for military applications. One variant is known as the Type 2053 and is in use by the Royal Navy for inshore hydrographic survey work.

The Sonar 3000 PLUS family comprises models ranging from the basic Sonar 3000 to the powerful Sonar 3010. The latter has been developed to provide an affordable solution to the mining threat; it is particularly suited to craft of opportunity and to use by non-specialised personnel. Key features of the Sonar 3010 are:
(a) automatic adaptive gain so that the operator has no sensitivity settings to adjust and the sonar record is consistent on every pass or survey
(b) ranges can be varied from 37.5 to 300 m
(c) video display for classification so that the operator can freeze, zoom, contrast, enhance or colour scale the picture
(d) target marking facility
(e) dual frequency based on the Sonar 2034 design but fitted with four transducer arrays for optimum performance at 100 kHz (long-range) and 300 kHz (high resolution)
(f) high definition data recorder
(g) high resolution thermal linescan hard copy recorder
(h) interface compatibility with integrated navigation processors such as the Qubit TRAC IV or Racal 900 to provide precise geographical co-ordinate data on marked targets.

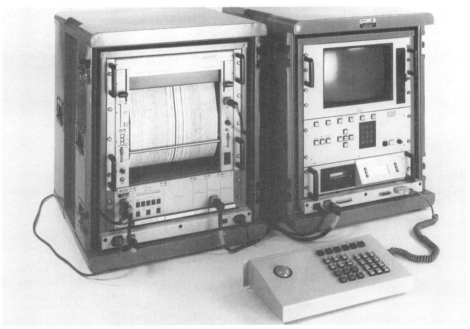

Sonar 3010 processing and display equipment

Operational Status
The Sonar 3000 has been ordered by the Royal Navy Hydrographic Service, under the nomenclature of Type 2053. Dowty received an order for seven 3010/T systems from the US Navy, plus options for up to 52 more, in September 1988. In March 1989 an additional five sets were ordered. The 3010/T system is based on the Sonar 3010 but includes additional items, such as the Ferranti Trackpoint II towfish tracking system, and improvements in sonar range. The system will be used on US Navy craft of opportunity vessels, and for US Coastguard mine classification, route surveillance and mapping programmes. It is described in the next entry.

Contractor
Dowty Maritime Systems Limited, Weymouth.

TYPE 3010/T (MINESCAN) SONAR

Type
Mine detection and route surveillance sonar.

Description
Minescan (3010/T) is a variant of the Dowty 3010 sonar and is a low-cost route surveillance and mine detection side-scan equipment particularly suited to craft of opportunity (COOP) applications, and for use by reserve or regular forces. The 3010/T version has a greater range and includes the Ferranti Trackpoint II towfish tracking system.

Key elements of the system are:
(a) heavy duty, dual frequency towfish, 100/325 kHz, operable to 300 m depth, with optional depressor
(b) dual cabinet transportable system in shock-mounted, splashproof units
(c) single cabinet system for permanent installation applications
(d) colour video display with image enhancement and target marking facilities
(e) range selection – 50, 100, 200 and 400 m port and starboard; total swathe width up to 800 m
(f) image correction – slant range and speed over ground
(g) high density sonar data recorder with random access target image recall
(h) high resolution thermal linescan recorder
(i) interfacing to integrated navigation processors.

The manufacturer states that other elements of the complete COOP mine countermeasures, such as the winch, towfish tracking system, navigation aids, ROV, and shore planning and analysis, can be supplied as a turnkey operation.

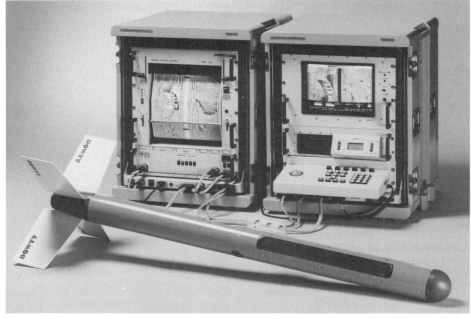

Dowty Minescan sonar in the dual cabinet configuration for the US Navy COOP programme

Operational Status
Dowty has received a number of orders from the US Navy for use in COOP applications. By the end of 1989, 32 3010/T systems had been ordered. Complete minescan-based COOP packages have also been supplied to other navies, including the Singapore Navy for installation on COOP mine countermeasures craft. In June 1989 a 3010 mine-scan system and a Qubit TRAC V were installed on a Royal Navy minehunter to carry out surveys of a practice minefield.

Contractor
Dowty Maritime Systems Limited, Weymouth.

UNITED STATES OF AMERICA

AN/AQS-14 MINEHUNTING SONAR

Type
Helicopter towed side-scan sonar.

Description
This is a helicopter towed side-looking multi-beam sonar, with electronic beam forming, all-range focusing and an adaptive processor, intended for use with the RH-53D Sea Stallion and MH-53E Sea Dragon helicopters used by the US Navy for mine counter-measures. The underwater vehicle is 3 m in length and has an active control system which enables it to be run at a fixed height above the seabed, or at a fixed depth beneath the water surface as chosen by the operator in the helicopter. The tow cable is armoured and is non-magnetic.

Controls in the helicopter include: a TV-type moving window sonar display; underwater vehicle controls and status displays; system status indicators; and a magnetic tape recorder for sonar data recording. The system functions include all that is needed to locate, classify, mark, permanently record and review records of mines, mine-like objects and underwater features in the search area. Because the sonar employs multi-beam techniques, a rapid search speed is possible. At lower tow speeds the towed sonar automatically reduces the number of transmitted beams. The system is adaptable to surface platforms, especially hovercraft and remotely operated drone vessels for route survey missions.

Operational Status
Induction for the US Navy MH-53E and RH-53D helicopters, with over 30 systems delivered.

Contractor
Westinghouse Electric Corporation, Annapolis, Maryland.

5952 MINE COUNTERMEASURES SONAR SYSTEM

Type
Dual beam, high resolution side-scan sonar using simultaneous dual frequency insonification of the target for multi-spectral data collection and processing.

Description
The Type 5952 sonar system uses the latest technology available in lightweight side-scan sonar systems to provide the highest resolution and target discrimination for the detection of current generation bottom mines. The simultaneous dual frequency insonification technique is used for target discrimination because targets reflect various frequencies in different manners, thereby providing the operator with the opportunity to observe the reflectivity of difficult targets with both high and very high frequencies.

The returns of both frequencies are displayed simultaneously on either hard copy, very high resolution thermal paper or on a high resolution video display unit (VDU). The sonar operator has a direct, real-time visual comparison of the target returns for optimum evaluation and target selection.

The sonar system uses a variable depth sonar (VDS) transducer which can be selectively deployed to avoid interference from temperature thermoclines. Combined 100/500 kHz frequency for simultaneous insonification of target areas, and 3.5 kHz for penetration of the sea bottom are available for use in the determination of bottom hardness for use in the assessment of a buried mine threat.

The sonar images are transmitted up the VDS tow cable to a combined sonar transceiver and graphic recorder for processing and display on high resolution thermal graph paper. Alternatively, the combined sonar image can be displayed on a high resolution digital video display unit (VDU). Performance enhancing accessories are available which permit advanced image processing of the sonar data, as well as data reduction. Fully integrated sonar systems, which interface with navigation and shipboard equipment are also available.

The sonar system is lightweight, weighing less than 100 kg (225 lb) and is easily transported and installed aboard vessels of various sizes down to 5 m in length. The VDS transducer weighs less than 25 kg (55 lb) when operated in the simultaneous dual frequency mode and is easily deployed by a single crewman without specialised handling equipment.

A full range of accessories is available, including tow cables, towing winches, as well as various related systems including acoustic positioning, surface navigation, and data management equipment.

Operational Status
The 5952 is currently in production and in service with various navies worldwide for MCM applications.

Contractor
Klein Associates Inc, Undersea Search and Survey, Salem, New Hampshire.

UNDERWATER ORDNANCE LOCATOR MK 24 MOD 0

Type
Dual beam side-scan mine-locating sonar.

Description
The Mk 24 Mod 0 ordnance locator is a dual beam side-scan sonar for use in locating mines and other types of underwater ordnance. It is available with modular accessories which adapt the equipment for use in channel conditioning/route survey mine countermeasures operations, as well as for use in explosive ordnance disposal operations.

The Mk 24 system consists of a side-scan sonar towfish, weighing approximately 22 kg (48 lb), a lightweight Kevlar towcable, and a combined transceiver and graphic recorder weighing 44 kg

(95 lb). The system and its accessories are lightweight and portable, and can be deployed from any available vessel down to 5 m in length. It can also be towed from any available and unmodified helicopter.

The side-scan towfish contains the circuitry to energise the transducers, which project high frequency bursts of acoustic energy in fan-shaped beams, narrow in the horizontal plane and broad in the vertical plane. Echoes from both transponders are processed by the combined transceiver and printed by the graphic recorder. Four types of transducers are available: a 50 kHz long-range unit for covering up to 1.2 km swaths of the bottom; a 100 kHz high resolution version for detection operations covering up to 400 m swaths; a 500 kHz very high resolution unit for target classification applications; and a simultaneous 100 kHz and 500 kHz unit for detection and classification of small targets by comparative techniques through the use of frequency diversity. A 3.5 kHz sub-bottom profiler is available for assessing bottom hardness to determine potential buried mine hazards.

The towfish can be towed at speeds up to 16 kts, but for optimum operation against small targets the recommended speed is 5 to 6 kts. Standard towfish can operate down to depths of 1000 m (3280 ft) with options permitting operation down to 2270 m (7445 ft). Special units are available for operations down to full ocean depth.

Operational Status

All versions of the system are in full production and are in use by the US Navy Explosive Ordnance Disposal units.

Contractor

Klein Associates Inc, Undersea Search and Survey, Salem, New Hampshire.

Components of Mk 24 minehunting sonar

MS 900

Type
ROV minehunting sonar.

Description
The MS 900 compact sonar processor was designed for smaller, lightweight ROVs. It operates with standard Mesotech 971 sonar heads to provide a variety of ROV applications including search and recovery, obstacle avoidance and minehunting. The Model 971 sonar head can be supplied in a full range of transducer frequencies (2 MHz to 120 kHz) and head configurations.

The processor incorporates a built-in long line amplifier which automatically compensates for cable losses. The sonar also features:
(a) built-in 2/4 wire telemetry
(b) built-in conversion from RGB to NTSC or PAL for direct video recording
(c) on-screen date and time
(d) hard copy available in colour via PaintJet printer
(e) on-screen menu.

Operational Status
In full-scale production and in service.

Contractor
Simrad Mesotech Systems Ltd, Port Coquitlam, British Columbia.

Transducer head of MS 900

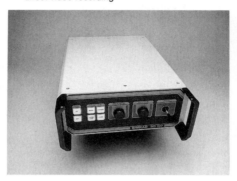

MS 900 processor

PaintJet colour printer for MS 900

MODEL 971/977 COLOUR IMAGING SONARS

Type
ROV sonars for search and minehunting.

Description
The Mesotech Model 971 colour imaging sonar was designed for use in a variety of ROV applications, including search and recovery, obstacle avoidance and minehunting. The standard 971 system consists of a sonar head, processor and display screen, and is lightweight, compact and versatile. A large selection of head and transducer options are available which can be adapted for specific requirements.

The sonar head has two scanning techniques: imaging and profiling. Imaging uses a fan-shaped beam, usually scanning horizontally to provide an overall picture of the underwater situation. The head can be deployed on a long cable with a compass option that shows orientation of the head with respect to the bottom. The profiling technique uses a cone-shaped sonar beam, usually scanning vertically to generate digitised profiles on the bottom.

The display provides a clear detailed image in colour of objects at ranges from 0.5 to hundreds of metres. Display can be in any one of five modes:
(a) Sector: the system scans and displays in a preselected arc

Model 971 sonar system

(b) Polar: 360° continuous scan and a PPI type display
(c) Side-Scan: the sonar head is locked in one position and images are displayed as scanned by vehicle movement
(d) Linear: gives an accentuated view of close objects
(e) Perspective: gives an impression of depth.
A ×2 or ×4 magnification allows detailed inspection of a target.

The Model 977 fast scanning sonar combines the Model 971 high resolution image technology with significantly increased scanning speed. The 330 kHz version was designed for obstacle avoidance or minehunting applications on faster moving vehicles or

Model 977 sonar system

submersibles, while the 100 kHz version is used for stationary deployment in harbour surveillance applications.

Operational Status
In full-scale production and in service.

Contractor
Simrad Mesotech Systems Ltd, Port Coquitlam, British Columbia.

MS 997

Type
Fast scan ROV minehunting sonar.

Description
The MS 997 fast scan sonar system has been introduced to meet the requirement for sonar systems for larger, fast moving ROVs and vehicles and for harbour surveillance applications.

The MS 997 is a multiple beam system combining the MS 990 processor with the MS 997 fast scan head to allow scanning up to 20 times faster than conventional single beam sonars. As well as being ROV/vehicle-mounted, the system can also be deployed over the side of a ship or on a tripod for dockside monitoring and harbour surveillance.

The system offers the following features:
(a) built-in 2-wire telemetry
(b) built-in long line amplifier
(c) adjustable speed for sound
(d) on-screen date and time
(e) menu settings stored and automatically recovered on power-up
(f) user defined text
(g) removable control panel for remote mounting
(h) serial I/O allows remote-control and input of navigational data.

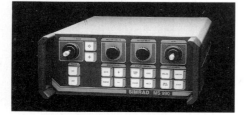

MS 990 processor control panel

MS 990 transducer head

Operational Status
In full-scale production and in service.

Contractor
Simrad Mesotech Systems Ltd, Port Coquitlam, British Columbia.

GERMANY

AIS 11 SONAR

Type
Active identification sonar for minehunting.

Description
The AIS 11 is an active identification sonar intended for use with submersibles. Although the equipment can be used for a variety of underwater tasks, its prime role is that of mine countermeasures, for which purpose it is normally fitted to the PAP 104 or Penguin B3 mine disposal units (see separate entry).

The underwater parts of the AIS 11 consist of an electronics unit and a transducer unit, the latter being fitted in the hull of the submersible. The parent surface ship is equipped with an evaluation unit, a control panel and a display. Weight of the overall system is 160 kg, that of the underwater part being

about 20 kg. Data transfer between the underwater part and the surface takes place along a cable.

During operation, a 60° conical beam is transmitted ahead of the vehicle and the acoustic signals reflected by an object are received by a linear antenna. These signals are then processed in real time, by both analogue and digital methods, and transmitted to the surface vessel where the information is shown on a PPI display in a sector-shaped format with correct angular representation. Fast real-time signal processing enables the underwater scene to be displayed at the frame repetition rate of a normal cine film.

Operational Status
In operational service.

Contractor
Atlas Elektronik GmbH, Bremen.

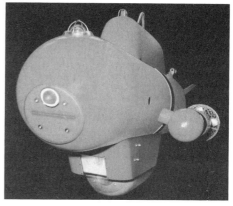

AIS 11 active sonar fitted to an ROV

UNITED KINGDOM

MARCONI ARMS SONAR

Type
ROV-deployed minehunting sonar system.

Description
A new Advanced Remote Minehunting System (ARMS) is being developed by Marconi for use in warships and vessels of opportunity to locate and evade/destroy mines. The system uses a remotely

operated vehicle (ROV) to carry a sonar and a standard design of mine destructor charge. The ROV swims at considerable distances ahead of the parent ship at depths down to 300 m and at speeds up to 6 kts.

The sonar is a development of the Marconi PMS 58 sonar, and provides a full minehunting capability. In addition the vehicle is fitted with two cameras, one looking forward and one in the belly to give the downward-looking capability for mine identification. Sonar and TV data are passed back to the ship by means of a fibre optic cable.

The complete ARMS system is containerised in three units for rapid deployment by air transport.

Operational Status
In development.

Contractor
Marconi Underwater Systems Limited, Templecombe.

HI-SCAN SONARS

Type
Range of surveillance and target location sonars.

Description
The Hi-scan range of sonars is intended for applications such as target location, obstacle avoidance, surveillance and navigation. The equipment is suited for use in surface ships, ROVs and on fixed sites. The system consists of a compact design offering an electronically scanned sonar receiver, and a real-time display.

Hi-scan uses modulation scanning, a highly efficient electronic technique to provide within-pulse sector scanning of an insonified area. This arrangement gathers information from the entire area within a single transmission period. According to the application, this represents a coverage rate around 20 times faster than can be achieved by mechanically scanned sonar.

The sonar electronics are contained in a pressure

housing to which the transducer is coupled via a short multi-way cable and connector. This allows the transducer to be used with an index panning mechanism to increase the coverage up to a full 300°.

The equipment is also available with the transducer mounted directly to the pressure housing for applications where the electronically scanned sector is sufficient.

The sonar electronics communicate via a high speed serial data link with a display system consisting of a scan processing unit, control panel and colour monitor. This provides a high resolution colour or monochrome display which is easy to interpret and comfortable to view.

Hi-scan 600
This version is particularly suited to ROV applications of target location and obstacle avoidance. It provides a 30° electronically scanned sector, the whole of which is refreshed at a rate of 8 Hz on an 80 m range scale, and up to 32 Hz for shorter ranges. This provides TV-like presentation of relative movement, and allows direct sonar guided piloting of the vehicle.

It has also been successfully evaluated during sea trials in both the USA and the UK in connection with long term MCM programmes. These have often involved the sonar fitted to a highly manoeuvrable ROV such as the Phantom HVS4.

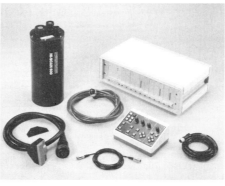

360° display using twelve 30° segments

Specifications
Frequency: 600 kHz
Scanned sector: 30°
Scan rate: 15 kHz
Beam angles: 1.6°
Pulse length: 75 μs
Range resolution: 5 cm
Range scales: 5, 10, 20, 40, 80, 160 m
Optional pan unit: 30°/s (indexed); ±200° (coverage)

Hi-scan 180
This is a proposed derivative based on Hi-scan 600 modules and software. It is intended for hull-mounting on ships, and employs a stabilised pan/tilt mechanism for the transducer and a raise/lower hoist.

Hi-scan 180 is a mine detection and avoidance sonar offering a 30° electronic scan giving 10 cm range resolution and 1.6 angular resolution. It has a detection range of about 600 m, and can be either

Hi-scan fixed head version

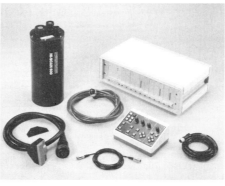

Units of the Hi-scan 600 sonar

Operational Status
Hi-scan 600 is available as a standard production item and has achieved notable sales in the USA in connection with offshore/harbour surveillance, object location and acoustic imaging research.

manually trained in azimuth to the required 30° sector, or automatically indexed in 30° steps for wider coverage.

Hi-scan 100

This variant is intended for fixed site installation and is optimised for the detection of small submersibles and swimmers. It is intended for the protection of harbours, oil/gas rigs, and other sensitive areas. The principle of operation is as for Hi-scan 600, and many modules are common to both systems. Hi-scan 100 is a proposed/experimental system offering the specifications below.

The equipment uses a serial data link from a submersible electronics unit to the display/control unit normally sited at a security post. An alternative arrangement uses several submersible units with their 30° electronically scanned sectors arranged to give effective total coverage.

Specifications

Frequency: 100 kHz
Scanned sector: 30°
Beam angles: 2.2° horizontal; 5° vertical
Pulse length: 200 μs
Range resolution: 15 cm
Range scales: 80, 160, 320, 840, 1280 m
Pan/tilt function: ±200° pan; +20 to −70° tilt

Contractor

Smiths Industries, Cheltenham.

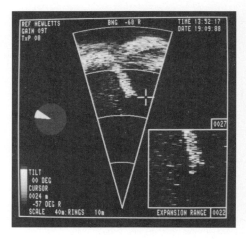

30° electronic sector scan display with expansion zone

SONAVISION 4000

Type

ROV sonar.

Description

The Sonavision 4000 is a new ROV sonar designed specifically for operation in areas of poor visibility where lighting for TV cameras has, for various reasons, to be restricted.

The sonar system uses an entirely new composite transducer element combined with digital signal processing. The new composite array technology developed by Marconi udi results in a wider bandwidth and much greater efficiency in converting electrical energy into mechanical energy which results in superior detection and imaging capabilities.

Transmitter and receiver electronics are fully tunable via software from 150 kHz to 1.5 MHz. This offers various beam angles and frequencies, for example, 1 MHz profile, 200 kHz long-range search.

The system features real-time acoustic zoom for close-up details of the target shape. Using a conventional push-button and joystick the operator can quickly select and fly a zoom box on to any interesting feature on the display and magnify the picture information within the zoom window by a factor of up to five times. This feature provides increased information by faster scanning and sampling of the sonar data within the subsea electronics. The interpretive qualities are defined by target shape, size, colour or shade variations and shadows. This simplifies interpetation by the operator allowing him to make fast decisions.

Other features of the system include an audio output representing range and amplitude which reduces operator workload and fatigue, a built-in joystick for range and bearing cursor control, infinitely variable sector and position overlay enabling rapid update of target data, selection of display modes in recordable S-VHS standard, a choice of telemetry links to suit the umbilical link, and 120 colours for display use. The simulated audio and user friendly features reduce operator fatigue and provide a high degree of confidence in decision making.

Specifications

Transducer: 500 kHz
Beamwidth: Transmit 27° vertical, 2.1° horizontal; receive 27° vertical, 3.0° horizontal
Pulse length: 100 s
Bandwidth: 10 kHz
Scan rates (menu selectable): Slow, normal, fast, super fast
Depth rating: 1000 m

Contractor

Marconi udi, Aberdeen.

AS360 SONAR

Type

Scanning sonar for ROV purposes.

Description

The AS360 is a mechanically scanned sonar of high resolution capability designed specifically for installation on small remotely operated unmanned vehicles. With a range up to 100 m, the search capability is extended beyond that range by television. TV refreshed techniques offer a visual presentation in both monochrome and colour.

The frontal area has been minimised by the use of an independent scanning head and separate external electronics unit. This ensures low drag and allows the sonar and electronics to be placed in the optimum positions on the ROV for both balance and hydrodynamics. A number of versions are available for ROV purposes.

AS360 MS5 ROV sonar equipment

The AS360M and AS360 MS5 are derivatives of the basic AS360 equipment. The AS360M has a range of 100 m and employs a colour TV display. It is designed for use on large manned or unmanned submersibles, and can be used in a number of other applications. The AS360 MS5 is a third generation system, specifically designed for installation of small ROVs.

Operational Status

The Type AS360 MS1/1 has been selected for installation on the ECA PAP Mk 5 ROV to be used by the Royal Navy's single-role minehunters.

Specifications

Frequency: 500 kHz (AS360M and MS1A); 700 kHz (AS360 MS5)
Beam width: 1.4° × 27° (AS360M and MS1A); 1.5° × 25° (AS360 MS5)
Depth rating: 300 m (AS360M); 1500 m (AS360 MS1A); 750 m (AS360 MS5)

Contractor

Marconi udi, Aberdeen.

3000 SERIES HRTS SONAR

Type

Small boat and ROV minehunting sonar.

Description

The 3000 series HRTS is a high scan rate target identification and relocation sonar, developed for short-range minehunting off small boats and remotely operated vehicles (ROVs). The system has a proven performance in locating ground mines, and extensive trials with the equipment attached to the side of a small boat have demonstrated its capability in detecting mid-water targets. This is achieved by the use of beam forming circuitry with a scan rate 10 times higher than single beam sonars. A speed azimuth servo motor scans the beam formed through almost 360°.

There are eight beams, electronically formed on reception, each having a horizontal beam width of 0.625°. The unit transmits a single pulse at 580 kHz, repetition time being dependent on the range selected. The maximum operating range is 125 m. The HRTS is focused in the near field so that targets may be detected at a distance of 0.4 m from the transducer.

Target classification is achieved by the combination of this very narrow beam width and a wide dynamic range, allowing the simultaneous display of target highlights and shadows, where present. The sonar has two subsea units: the scanning transducer head and the remote processing unit. On the surface, the HRTS is controlled by a operator's compact control panel and the associated electronics in the surface control unit, which can be remotely mounted in an equipment bay. The output is displayed on a colour monitor.

Shallow Water Minehunting Sonar

The Shallow Water Minehunting Sonar has been produced by Defence Equipment and Systems Limited, with its associated company Ulvertech. The system consists of the 3020 HRTS for classification, the 3025 search sonar (see separate entry) for collision avoidance and identification, and an SPL combat information system plotting table for navigation and route survey. The system is modular in

3000 series HRTS sonar

concept, is readily fitted to small craft of opportunity, and provides a credible solution for most inshore and estuary minehunting applications.

Specifications

Minimum range: 1 m focused in the near field
Range switch setting: 10, 20, 30, 40, 50, 75, 100, 125 m or yd

Horizontal beam width: 5° or 10°
Vertical beam width: +5° to −30° from horizontal
Scan rate: Varies from 30°/s to 170°/s depending on mode and range selected

Range resolution: Can be programmed by varying pulse length (minimum pulse length 50 μs)
Depth rating: 300 m (options available down to 6000 m)

Contractor
Ulvertech, Ulverston.

MODEL 3025 SONAR

Type
Search sonar for ROVs and harbour surveillance.

D escription
The Model 3025 search sonar is designed for a number of subsea purposes, including ROV application, mine neutralisation vessels, harbour surveillance, and as a defence for surface ships against covert divers. The system has two subsea units: the scanning sonar head and the remote processing unit.

On the surface, the sonar is controlled by a compact operator's panel and the associated electronics.

Electronic beam forming combined with a speed azimuth servo motor gives scan rates much higher than single beam mechanically scanned systems. This is achieved using a 110 kHz transmission covering 25° and beam forming the received echoes into eight 3° beams. The sector is then mechanically scanned over a full 360° arc for full coverage, and provides the operator with a real-time picture of the mid-water and the seabed.

Specifications
Range switch settings: 35, 50, 100, 150, 200, 300, 400, 500 m
Minimum range: 1 m
Horizontal beam width: 25°
Vertical beam width: 12°
Receiver angular resolution: 3°
Scan rate: Varies from 30°/s to 170°/s, dependent on the mode and range selected
Depth rating: 1000 m

Contractor
Ulvertech, Ulverston.

UNITED STATES OF AMERICA

MODEL 4235A SONAR

Type
Doppler sonar for submersible use.

Description
The Model 4235A is a lightweight Doppler sonar designed for use on remotely operated vehicles (ROVs), autonomous underwater vehicles (AUVs) and other submersibles where weight and size are critical. It incorporates a significant feature; an operating frequency of 435/151 kHz which increases the basic resolution capabilities and helps in meeting minimum operating range, 300 to 1000 ft off bottom and water mass tracking depth, and speed requirements.

The transducer assembly consists of four sonar elements aligned at 90° increments around a high pressure-resistant rugged housing. The elements are 60° relative to the horizontal. Four pre-amplifier boards are within the housing which is designed to withstand pressures of 9000 psi (15 kHz to 1000 psi).

The electronic housing is 165 mm in diameter (without the pressure bottle), 660 mm long and weighs 9 kg. The high frequency transducer array with connector is 212 mm in diameter, 220 mm deep and weighs 11.3 kg.

Operational Status
In production.

Model 4235A ROV sonar system

Contractor
EDO Corporation, Electro-Acoustic Division, Salt Lake City, Utah.

MODEL 258 CTFM SONAR

Type
ROV scanning sonar system.

Description
The Model 258 is a compact system designed for use on remotely operated vehicles. It employs CTFM sonar techniques to provide rapid scanning and simultaneous display of multiple subsea objects, including CW markers. The system incorporates dual operating frequencies to achieve optimum performance for both long-range search capability and high resolution display for short-range operations. The long-range search mode operates on a frequency band of 107 to 122 kHz and is capable of detecting objects at ranges up to about 600 m. The high resolution mode uses a narrow beam and 342 to 357 kHz frequency band with a 128 channel digital spectrum analyser to provide a clear, detailed image. The standard version is depth rated to 6000 m.

Contractor
EDO Corporation, Electro-Acoustic Division, Salt Lake City, Utah.

Model 258 CTFM sonar

Q-MIPS PROCESSING SYSTEM

Type
Sonar image processing system.

Description
Q-MIPS is a fully integrated, menu-driven data collection and image processing system which connects to a side-scan sonar. It provides digitising, processing, display, colour hard copy and permanent storage of sonar images. This equipment can be used for a number of applications, including minehunting, Q-route surveillance, range and harbour security and geophysical survey.

Q-MIPS is compatible with any side-scan sonar analogue output and features:

(a) automatic range scale setting, trigger detection and bottom tracking
(b) four simultaneous channels of the highest speed 12-bit A/D
(c) very high sampling rates with the ability to change sampling schemes to user requirements
(d) 256-colour waterfall display with real-time corrections for slant range, water column, radiometric non-linearity and speed
(e) shipboard colour hard copy
(f) permanent safe storage on optical disks for high capacity and instant retrieval in full resolution
(g) automatic navigation interfaces
(h) spatial image processing enhancements including many edge detection and filtering algorithms, plus unlimited zoom and rotation
(i) transform domain image enhancements, including DFT filters.

Operational Status
In operational service.

Contractor
Triton Technology Inc, Watsonville, California.

MINE DISPOSAL VEHICLES

Dealing with the modern influence mine is an extremely complicated affair requiring a very careful approach. Once an object has been detected and classified as a mine by a minehunting system, it has to be neutralised. In the past this highly dangerous task was accomplished either by the ship itself – an operation of considerable danger to the ship concerned – or by clearance divers operating from inflatables carried by the mine countermeasures vessel (MCMV).

The advent of the remotely controlled mine neutralising vehicle (ROV) has given mine countermeasures and vessels used on MCM tasks a much greater flexibility of operation. Using the ROV, MCM operations can now be carried out under much more adverse conditions than would have been the case if only mine clearance divers were available. Moreover, the operation itself is completed more quickly and there is a greater certainty of success without the risk of loss of life, or damage to the ships.

Using technology originally developed for the offshore industry, a range of ROVs has been developed for mine countermeasures tasks. Mine development, however, continues to advance, and in the future new technology will be incorporated into future generations of ROVs making them far more autonomous, enabling the parent vessel to stand off at much greater distances from danger areas, allowing the vehicle to carry out search, location and identification, tasks previously carried out by the MCMV.

While much of the effort has concentrated on MCM, a developing area for ROV operation comprises the use and maintenance of fixed seabed systems such as deep ocean surveillance systems designed to detect and track deep diving nuclear submarines. ROVs are increasingly being used in this area. The ROV also offers an enhanced capability in operating fixed seabed ranges for determining a ship's acoustic and magnetic signatures, degaussing, and in underwater weapon trials and so on. Here too, the ROV can be an extremely valuable tool for weapon recovery. Finally the ROV will become an invaluable tool for the hydrographer, assisting in carrying out seabed surveys, and recovering seabed samples for analysis.

REMOTE OPERATING VEHICLES

CANADA

MANTA

Type
Remotely operated vehicle.

Description
Manta has been developed and operated in Canada over a number of years. Various models of the vehicle have been built for subsea inspection and surveillance at depths of 1000 ft in 4 kt currents. Manta is an extremely quiet vehicle with a very low magnetic signature. As a result Manta is now being developed for mine countermeasures tasks. The vehicle is supplied with its own container to house the vehicle and the control room and is suitable for fitting on to COOP as small as 40 ft long.

Manta is fitted with two manoeuvring thrusters and the version for the Canadian Navy will be self-propelled (previous versions of Manta have been towed vehicles) and capable of operating at depths of 3000 ft ahead of the parent vessel.

Operational Status
On order for the Canadian Forces.

Contractor
Sea Industries Ltd, British Columbia.

Handling Manta ROVs on board

TRAIL BLAZER

Type
Remotely operated vehicle.

Description
Developed in a joint venture by the Canadian firm ISE (International Submarine Engineering) and Fairey Hydraulics of the UK, the Trail Blazer mine disposal vehicle is capable of high underwater speed and deep diving. As with other new vehicles much of the basic technology used in Trail Blazer has already been proven in remotely operated vehicles developed for the offshore industry.

The basic concept of the vehicle is portability and the total MCM system, including displays, control, navigation, vehicle and umbilical cable, can be stowed, transported and operated in a Standard IATA Boeing 747 Combi aircraft load container.

The system is small, compact, lightweight and is easily handled, being capable of deployment from unsophisticated vessels as small as fishing smacks or workboats.

Although it forms part of a family of underwater vehicles, the version developed for MCM duties was designed to meet the following criteria:
(a) high power propulsion system to permit easier and more accurate manoeuvrability in the

Trail Blazer ROV showing rear thrusters

presence of strong tidal currents, and to reduce overall mission time
(b) ability to carry out missions for extended periods unrestricted by the limited availability of onboard vehicle power

(c) low magnetic and acoustic signatures
(d) use of advanced telemetry systems to ensure ease of system operation and ready availability of sensor data
(e) provision of appropriate sonar equipment and

high resolution TV systems to allow rapid location and identification of underwater objects

(f) high payload capacity to allow carriage of a number of countermining charges when on extended missions.

The frame of the vehicle is constructed of thick anodised aluminium girders. A slightly positive buoyancy is maintained by means of high density syntactic foam within the vehicle. Lead ballast is added to the structure to allow for adjustment of buoyancy when heavy payloads are carried. Additional foam may be added to increase vehicle maximum payload if required.

The vehicle is powered by five hydraulic thrusters. Two thrusters are mounted at the after end of the vehicle providing forward and reverse motion. Two amidships thrusters provide lateral movement. A fifth thruster amidships provides vertical motion. Variable thrust and directional control for each thruster are maintained by pressure compensated servo valves mounted on a control valve manifold unit.

Electric power for the propulsion system is fed from the surface ship's supply via a transformer to the vehicle through the umbilical cable. Normally the vehicle's hydraulic powerpack operates at a pressure of 210 bar but provision is made for operation at low pressure where low acoustic noise signatures are required.

The longitudinal and lateral thrusters are controlled from a single joystick on the control console, the stick controlling forward or astern movement, rotation of the stick (creating differential pressure on port and starboard thrusters) allowing the vehicle to be steered, while sideways movement produces lateral movement. A separate control is provided for the vertical thruster.

The use of an umbilical powered electrohydraulic propulsion system offers considerable operational flexibility and provides almost unlimited endurance. The control package uses a half duplex PCM telemetry system for transfer of command signals and system data between vehicle and control console. Video write-on facilities are provided for display of navigation and system function parameters.

The main console provides all the usual vehicle control functions, data on system operation and

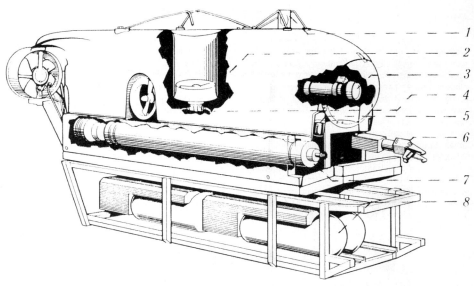

Trail Blazer ROV. Key to diagram: **(1)** *sonar;* **(2)** *vertical thruster;* **(3)** *camera;* **(4)** *side thrusters;* **(5)** *lights;* **(6)** *manipulator;* **(7)** *electrohydraulic powerpack;* **(8)** *mine disposal charges*

displays for TV/sonar. To simplify operation during launch and recovery a portable control unit for use on the deck of the MCMV is provided.

For navigation, Trail Blazer is provided with a flux-gate compass; however, a gyro-compass can be fitted optionally if required and slaved to the flux-gate compass. Vehicle heading and depth is continuously displayed on the console monitor. If required, heading and operating depth can be set and automatically maintained, the data being displayed on the monitor.

Standard sensor fit includes an Osprey OE 1321 SIT TV camera, but colour TV or still cameras can be added if required. A hydraulically operated pan and tilt facility is provided while surveillance is aided by the use of a 500 W variable intensity lighting unit. A wide range of sonar equipment can be fitted to meet varying operational requirements. A range of hydraulically operated manipulators is also available.

The vehicle can carry up to four 50 kg US Navy-type countermining charges or one NATO standard 127 kg charge mounted on a carriage underneath the vehicle.

Specifications
Length: 2.63 m
Width: 0.64 m
Height: 0.86 m
Weight: 772 kg
Charge: 130 kg
Range: 1000 m +
Service depth: 500 m
Speed: 5.75 kts

Contractors
International Submarine Engineering (ISE), Port Coquitlam, British Columbia.

ROVWR

Type
Remotely operated vehicle weapon recovery.

Description
ROVWR is a self-contained weapon recovery system designed to locate and recover weapons weighing up to 3.5 tonnes or more in one diving operation. The system comprises a portable control cabin with fully integrated ROV control, navigation and positioning systems. These integrate with the parent ship's existing navigation equipment or operate as an independent system. The vehicle and its load is recovered by a winch and load-bearing umbilical handling system.

The ROV is powered by a 50 hp motor and the design is based on the ISE Hy-Sub AT. It is fitted with a complete set of equipment including full suite of location equipment and video cameras; two ISE Magnum manipulators – one fitted with five functions for placing recovery grabs and so on, and one with seven functions for tool manipulation; a mud pump that can dig out a completely buried weapon; and a full outfit of accessories for recovering a wide range of weapons. Special equipment is available to suit individual customer needs.

Contractor
International Submarine Engineering (ISE), Port Coquitlam, British Columbia.

DENMARK

FOCUS 400

Type
Towed inspection vehicle.

Development
The Focus 400 towed inspection vehicle is a new generation system based on the Focus 300 fisheries inspection vehicle developed by MacArtney A/S in co-operation with the Danish Maritime Institute. For surveying and inspection duties a highly stable and manoeuvrable platform was required with automatic positioning relative to targets on the seabed. The vehicle's pitch, roll and yaw had to be minimal at speeds up to 5 kts.

Description
The Focus 400 system comprises five elements: the underwater vehicle; control and display console; electrohydraulic winch; fibre optic tow cable; and power supply.

The vehicle consists of a 'box kite' type body with parallel vertical and horizontal control surfaces for manoeuvring and positioning the vehicle mounted forward of the frame and stabilising fins at the rear. The towing point is at the centre of the front end. The low centre of gravity of the vehicle increases stability, but buoyancy and weight can be adjusted according to payload. The body houses a number of standard components including cable termination bottle and electronics bottle housing the power supplies, control electronics, multiplexer, dual axis clinometer, heading sensor and dual depth sensors. The vehicle also incorporates a side-scan sonar (the Klein 590 system dual frequency — 100/500 kHz) and electronics, an obstacle avoidance sonar (the Tritech ST 325) and dual Mesotech Model 808 altimeters. There is also interfacing for optional equipment such as analogue sensors, a pan and tilt unit, SIT TV (Osprey 1323) and still cameras and lights. The vehicle can accommodate a large range of oceanographic sensors without modification (for example, photometers, sound velocity probes, samplers, dissolved oxygen, and so on).

The control and display console houses all the vehicle and winch controls and displays required for controlling the vehicle in either manual or automatic modes, the computer and a slide-out keyboard used to set parameters and control the display graphics. Vehicle control is exercised through a joystick providing for vertical and horizontal movement with a trigger for auto modes. The winch is also controlled by a joystick which controls cable haul and deployment. The console also provides controls for light intensity adjustment (push-button with light bar intensity indicator), switching for cameras and so on.

The display provides a range of status indicators showing altitude and depth; speed through water; flap positions; water temperature; heading, pitch and roll; cable deployed and speed; real time; and user options. The vehicle's position relative to the surface and seabed together with the offset from the target are displayed graphically and can be scaled for accuracy. The strike angle to the target is also displayed. Navigation data from the ship's computer are also displayed. The display screen usually depicts real-time video, but the picture can be frozen and moved to another area of the screen for evaluation and eventual storage on a disk via the computer.

The winch can be controlled either from the winch position or from the control console and can manage

Focus 400 underwater towed vehicle

Focus 400 control and display console

up to 1200 m of fibre optic tow cable. The cable is deployed over a sheave fitted with a line-out indicator which displays data at the control console. The winch drum houses the fibre optic converters and an electrical slip ring.

The fibre optic tow cable is constructed from seven Electro Light copper armoured multi-mode fibres with an overall contrahelical steel armour. The copper armour on the fibres is used to transmit power to the vehicle. The cable is 12.3 mm in diameter and has a breaking strength of 89 kN.

Operational Status
A pre-production model has been demonstrated and a number of navies has shown interest in the system for mine countermeasures applications and volume area search and survey.

Specifications
Depth: 400 m
Speed: 5 kts
Cable length: 1200 m
Power requirements: 220 V AC 50/60 Hz

Contractor
MacArtney A/S Underwater Technology, Esbjerg.

FRANCE

PAP MARK 5

Type
ROV mine disposal system.

Description
The PAP Mark 5, developed and manufactured by Société ECA, is the fifth generation of the PAP 104 family.

The PAP Mark 5 has a capability to simultaneously hunt and destroy ground mines and tethered mines by either placement and remote detonation of a charge or by cutting the mooring rope of tethered mines. To achieve these aims, the PAP Mark 5 embodies both new features and also those maintained from the former PAP versions. Recently two new options have been added to the PAP; an umbilical cable and a small explosive charge (30 kg).

PAP characteristics
The vehicle is to be fitted on a minehunting vessel equipped for locating mines. This can be achieved either by a minehunting sonar or a surface positioning system.

The vehicle is autonomous; it carries its own source of energy – a battery which can be recharged or changed on board between missions.

The vehicle is wire-guided and remote-controlled from a control console situated in the operation room. The wire is consumable and dispensed by the vehicle.

The vehicle is horizontally propelled by means of two side thrusters.

By means of a guide rope, the vehicle can navigate at constant altitude.

The PAP Mark 5 is able to carry the 126 kg charge necessary to ensure the total destruction of the mine. Cutters can be carried with the charge.

New characteristics of PAP Mark 5
The PAP Mark 5 employs modern new electronics, based on microprocessors, providing a reliable, high rate data transmission. It includes an ergonomic control console and provides the operator with different navigation aids.

The front section of the vehicle is devoted to an optional sonar of the customer's choice; it can be either a high resolution nearfield sonar or a mid-range relocation sonar.

Identification is carried out by a tiltable low light

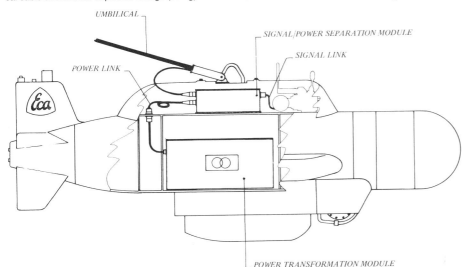

PAP Mark 5 umbilical cable option

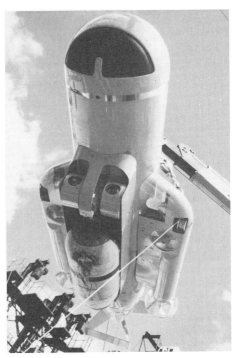

PAP Mark 5 mine disposal system

TV camera and/or by a tiltable colour TV camera.

Vertical thrusters are used which allow mid-water navigation. A new payload can be fitted in place of the charge, consisting of a manipulator with associated camera and projector.

Other features include:

Trim variation to allow faster descent and ascent transits.

Sealed lead acid batteries.

Guide rope length variation to allow fine adjustments of vehicle altitude during a seabed mission.

Radio control for surface recovery in case of cable breakage.

Alternative options include fibre optic wire.

The PAP Mark 5 has a very low acoustic and magnetic signature.

Operational Status

More than 341 PAP vehicles have been sold and 13 navies are equipped with them.

The PAP has an unmatched experience with more than 18 000 runs including many successful combat missions in the Red Sea, the Falklands, the Persian Gulf and so on.

Specifications

Length: 3 m
Diameter: 1.2 m
Height: 1.3 m
Total weight: 850 kg (including 100 kg charge)
Operating range: 2000 m (at depths of 300 m)
Speed: 6 kts (max)

Contractor

Société ECA, Colombes.
   CSI, Colombes (export company).

PAP Mark 5 control console

GERMANY

PINGUIN

Type

ROV mine disposal system.

Development

Under a long-term development programme for the German Navy a series of remotely controlled underwater vehicles for mine countermeasures operations has been developed by STN SystemTechnik Nord. Two of these craft have played a significant part in this programme: the first, known as Pinguin A1, was designed by STN SystemTechnik Nord under a predevelopment contract awarded by the government and led to a second type in the shape of the present Pinguin B3.

Description

The Pinguin B3 is a remotely controlled, unmanned underwater craft which is a very important element in

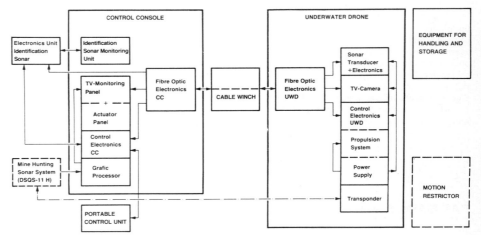

Pinguin B3 system functional block diagram

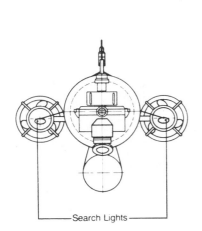

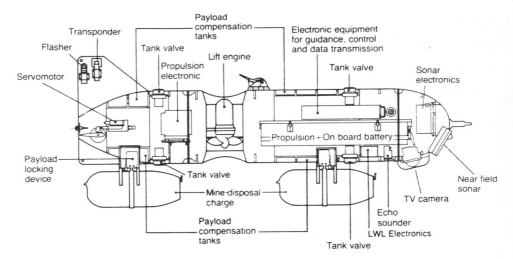

Pinguin B3 component arrangement

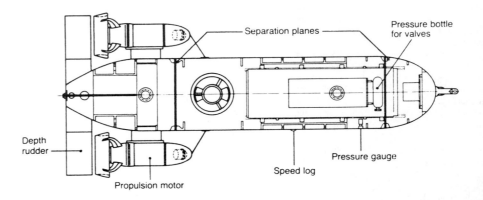

the complete minehunting system. With the input of target information (derived from minehunting sonars) it is able to travel at any selected distance from the seabed and to be guided almost automatically towards a mine-like object. The Pinguin B3 can then identify the mine and, if necessary, release a mine disposal charge to destroy it. A second charge is carried, and the Pinguin craft is intended to continue its mission and proceed to destroy a second mine.

The Pinguin is equipped with a near field, high resolution, electronic scanning sonar featuring a high pulse repetition rate and sector-shaped PPI display with correct angular representation. The Pinguin is fitted with a 1000 m fibre optic cable and in low current speed can operate out to a distance of 900 m at 200 m depth. In currents of 3 kts the vehicle has a range of about 500 m at 200 m depth. The vehicle is also fitted with a light-sensitive CCD underwater camera.

The propulsion system consists of two low magnetic signature, high power horizontal units mounted in pylons welded laterally to the stern section of the vehicle. The motors drive propellers shrouded by Kort nozzles to provide forward and reverse motion. Vertical movement is achieved from a lift motor and propeller fitted at the hydrodynamic neutral point of the vehicle.

The Pinguin vehicle can be operated freely within the whole operating volume and ranges in the minehunting scenario. Equipped with a cutter, it can also counter moored mines. The complete system offers a considerable growth potential for the adoption of alternative sensors, future technology and weapon systems.

Operational Status
In series production for the German Navy to equip the Type 332 minehunters and delivered to a foreign customer.

Specifications
Vehicle: Pinguin B3
Length: 3.5 m
Hull diameter: 70 cm
Wing span: 1.5 m
Height: 1.4 m
Mass: 1350 kg
Propulsion: 2 variable propulsion electric motors, 1 controllable lift motor
Power supply: Internal battery systems
Diving depth: 200 m
Speed: 6 kts (standard), 8 kts (max)
Endurance: 60/120 mins
Payload: 240 kg, 2 mine disposal charges and other payloads

The Pinguin B3 remotely controlled mine disposal vehicle carries two mine destruction charges and sensors for navigation and identification

Sensors: TV camera; different sonars, echo sounders, speed log

Contractors
STN SystemTechnik Nord GmbH, Bremen.

ITALY

MIN

Type
ROV mine disposal system.

Description
The MIN (Mine Identification and Neutralisation) system developed and produced for the Italian Navy by Selenia Elsag Sistemi Navali and Riva Calzoni, has the capability of identifying and neutralising both bottom and moored mines. The system consists of:
(a) one self-contained, hydraulically powered, wire-guided submarine vehicle mounting a TV camera and a high definition sonar transceiver
(b) a main console in the minehunter operations room for vehicle guidance and control
(c) one portable auxiliary console for guidance and visual control of the vehicle during launching and recovery operations
(d) operational accessories, such as auxiliary battery charging station, oleo-pneumatic powerpack recharging, and support system for the vehicle.

The vehicle is guided from the ship via a coaxial cable stored on a reel, through two pairs of horizontal and vertical thrusters, each consisting of a propeller driven by a hydraulic motor, and a main propeller fixed to a shaft orientable in the two planes. The vehicle is powered by a hydraulic system fed by closed-circuit oleo-pneumatic accumulators. Below the powerpack there is a stowage for the bottom mine destruction charge.

When used against moored mines the vehicle must be equipped with a cutting device carrying an explosive cutter. The vehicle is also provided with: a flashing light and an acoustic marker for localisation and recovery; an acoustic transponder to aid manoeuvring of the vehicle from the ship's sonar display; a compass, echo sounder, depth gauge and attitude indication device for supplying information to the main console. This latter is fitted with control levers and a joystick for speed and direction control of the vehicle during the search and approach phase, and for fine positional adjustments of the vehicle when in close proximity to the target.

The TV and the sonar monitors, with their associated control devices on the console, provide the operator with information regarding the underwater scenario. The TV camera is mainly used for ground mine identification under good visibility conditions, and the electronic scanning sonar for operation against moored mines and identification of ground mines in poor visibility.

The MIN Mk 2 features various modifications to the

MIN system Mk 1 underwater vehicle aboard a 'Lerici' class MCMV

Mk 1 version. Vertical manoeuvring to the vehicle's maximum operating depth is achieved through water ballast tanks filled or emptied by means of pressurised air. Hydraulic power is thus used only during transit to/from the parent ship and during the search phase of a minehunting operation.

The TV camera and sonar (a mechanical sector scan, high resolution and frequency model with 40° vertical beam amplitude) are orientable through 150° on the vertical plane. Operational depth has been increased to 350 m from the 150 m of the Mk 1 version.

Operational Status
Series production. Four MIN Mk 1 are operational aboard Italian Navy minehunters. The MIN Mk 2 are in production for the 'Gaeta' class minehunters of the Italian Navy.

Contractors
Consorzio SMIN, Rome.
  Selenia Elsag Sistemi Navali, Rome.
  Riva Calzoni SpA, Milan.

PLUTO

Type
Remotely operated mine disposal vehicle.

Description
Pluto is available in three configurations: battery powered with a 6 mm, 500 m long umbilical cable; battery powered with a 3 mm, 2000 m long fibre optic umbilical cable; or remote-powered of unlimited endurance with 8 mm, 500 m long umbilical cable.

The vehicle is powered by five thrusters: two horizontal for forward/reverse, two vertical for vertical and lateral shift, and a transversal thruster. The thrusters are controlled by two joysticks. The vehicle can maintain automatic depth control within ±10 cm. Forward maximum speed is 4 kts and vertical speed 1 kt.

Sensors are mounted in the forward tiltable section of the vehicle. Optional equipment includes black and white LLTV, colour TV or still camera, search or scanning sonar, acoustic pinger, strobe flash, measuring instruments, manipulators and so on. There are 10 free channels for remote-control and two four-digit telemetry channels for measurements.

The console incorporates a 9 in TV monitor for display of information showing TV image, depth, compass, head tilt angle, elapsed time and sonar diagram. All displayed data can be video recorded.

Operational Status
In service with the navies of South Korea (2 units), Italy (13 units), US (1 unit), Nigeria (4 units), Finland (1 unit), Thailand (4 units), and Spain (1 unit). In addition, Gaymarine operates a prototype unit, Whitehead

Pluto Plus (right) and standard Pluto (left)

uses a vehicle in connection with torpedo trials, ENEA of Italy operates a vehicle, and two others have been supplied to an unknown customer.

Specifications
Length: 1680 mm
Width: 600 mm
Height: 630 mm
Weight: 160 kg
Service depth: 400 m
Payload: 15 kg

Contractor
Gaymarine, Milan.

PLUTO PLUS

Type
Remote-control mine disposal vehicle.

Description
The basic configuration of Pluto Plus is the same as the standard version, but incorporating a number of improvements. Pluto Plus uses the latest fibre optic floating, reusable 2 km long cable and hydrodynamic design to provide improved observation capabilities, extreme stand-off range, an increase in speed to 7 kts, an increase in endurance from 6 to 10 hours (battery capacity has been doubled) and the lowest possible drag factor.

The vehicle is fitted with special sonar sensors for navigation, search, obstacle avoidance and identification. The sensors are all mounted in a single package featuring ±100° tilt and ±80° pan. A single control console monitor gathers and displays the video picture, navigational data, sonar graphics and maps of the investigated area.

The vehicle exhibits low magnetic and acoustic signature and is resistant to shock and vibration to MILSPEC standards for minehunting.

The lightweight, compact Pluto Plus is designed to operate from all kinds of vessels without special or expensive handling equipment.

Operational Status
Three units in service with the Italian Navy.

Specifications
Length: 1950 mm
Width: 510 mm
Height: 550 mm
Weight: 200 kg
Service depth: 400 m
Payload: 15 kg

Contractors
Gaymarine, Milan; and Westinghouse Electric Corporation, Annapolis, Maryland (USA) under licence.

SWEDEN

SEA EAGLE

Type
Remotely operated mine disposal vehicle.

Description
Sea Eagle is a military version of the civil Sea Owl Mk II ROV. The vehicle is driven by seven thrusters and is fitted with heading, depth and attitude sensors and a computerised control system. Sea Eagle has a telescopic arm which carries the mine disposal charge. This is extended in operation to place the charge in the correct position, after which the arm is withdrawn and the vehicle manoeuvres to a safe distance before the charge is detonated. Sea Eagle is fitted with two TV systems as standard and optional systems include forward seeking sonar and altimeter and range warning sonar. The vehicle is handled in a launching cage which is also used to stow the vehicle on board the parent ship.

Contractor
SUTEC, Linkoping.

DOUBLE EAGLE

Type
Remotely operated mine disposal vehicle.

Description
Double Eagle is a comparatively low weight system fitted with powerful high speed thrusters for use in currents up to at least 3 kts. The vehicle is driven by eight thrusters giving over 5 kts forward, 3 kts reverse, 3 kts lateral and 1 kt vertical movement. Double Eagle is fitted with a computerised stabilisation control system and is extremely manoeuvrable, exhibiting unlimited movement in six degrees of freedom (pitch ± 180°, roll ± 180°).

The display comprises two monitors with control keyboard, one monitor displaying the picture from a colour TV camera, the other the picture from a monochrome TV camera mounted at the tip of a telescopic arm. Digital data such as heading, depth, pitch and roll angle, cable twist, leakage warnings, real-time clock with date and time and diagnostics are superimposed on the TV monitor. High quality video is available using fibre optic link in the umbilical cable. The standard umbilical is 600 m long, but as an option a 1000 m cable can be attached. The vehicle is fitted with one colour CCD camera as standard, and tilt arrangement with colour and SIT camera is available as an optional extra.

Optional sensors include echo sounder, and electronic scanning or conventional sonar.

The vehicle uses a unique precision charge placement technique using a small mine disposal charge which also features several advantages. The vehicle can also carry a heavy standard NATO mine disposal charge as well as cable cutters.

Specifications
Length: 2050 mm
Width: 1290 mm
Height: 450 mm
Weight: 250 kg
Service depth: 500 m

Contractor
SUTEC, Linkoping.

SEA OWL Mk II

Type
Tethered remotely controlled submersible.

Description
Sea Owl Mk II comprises four elements: the submersible; an umbilical cable; surface control console; and power unit. The vehicle can be fitted with a variety of equipments including manipulators, grip claws, sonar, stereo TV colour, CP-probe with brush

module, still camera with strobes, altimeters, marking buoys, SIT camera and so on. Sensors include three rategyros, two pendulums, fluxgate compass, depth sensors and leakage warning. A microprocessor controls the automatic depth, roll, pitch, heading (four functions) and fine depth. The surface control console incorporates a TV monitor presentation and digital presentation of heading, depth, pitch angle, cable twist, ROV number, and datalink failure warnings. Optional presentations include a still camera frame counter and flash ready indication together with a real-time clock with date.

Specifications
Length: 1240 mm
Width: 750 mm
Height: 600 mm
Weight: 85 kg (in air)
Maximum depth: 350 m
Maximum speed: 2.5 kts

Contractor
SUTEC, Linkoping.

SEA TWIN

Type
Tethered remotely operated submersible.

Description
The Sea Twin ROV is an extremely stable vehicle offering very high manoeuvrability and unlimited movement in 6° of freedom (pitch ±180°, roll ±180°). With its eight powerful thrusters and high speed capability the vehicle is capable of operating in currents up to at least 3 kts. Various tools and equipments can be fitted to the ROV allowing it to carry out a variety of

missions. Standard equipment fit is a colour CCD camera with optionally a tilt arrangement with colour and SIT cameras. Optional sensors include echo sounder, electronic scanning or conventional sonar and additional cameras. The vehicle can also be optionally fitted with a hydraulic powerpack using standard tooling adapted to various operations. A fibre optic cable in the umbilical provides high quality video to the operator who is provided with TV monitor presentation and digital presentation superimposed on the TV monitor of heading, depth, pitch angle and roll angle, cable twist, leakage warnings, real-time clock with date and time and diagnostics. Optional presentation includes still camera frame counter and

flash ready indication, and presentation of data from optional sensors.

Specifications
Length: 2050 mm
Width: 1290 mm
Height: 450 mm
Weight: 250 kg (in air)
Maximum operating depth: 500 m
Speed: 5 kts (max)

Contractor
SUTEC, Linkoping.

UNITED KINGDOM

TOWTAXI

Type
Towed underwater research vehicle.

Description
Towtaxi is an advanced, highly stable, actively controlled vehicle designed for normal operation at depths down to 100 m, but with a maximum depth capability of 250 m. Experimental payloads and electronics may be housed in the body of the vehicle and an external payload, such as a transducer array, can be mounted underneath. The vehicle is controlled from on board the trials vessel by a command and monitoring system which is linked to the vehicle via the tow cable. It is used in underwater research work, for example in minehunting sonar research.

Operational Status
In use by the UK MoD Admiralty Research Establishment.

Contractor
BAeSema Marine Division, Bristol.

Underwater research vehicle Towtaxi

DRAGONFLY

Type
Remotely operated vehicle.

Description
Dragonfly is a modularly designed vehicle whose payload can be quickly and simply changed. The system comprises a basic vehicle incorporating a hydraulically powered thruster, electrical power system and control telemetry equipment, basic navigation instrumentation and cameras.

Payloads can be rapidly changed by virtue of the electrical power and data handling system which uses easily accessed junction boxes.

Equipment designed to carry out specific functions or tasks is attached to the underneath of the vehicle and 'bolt-on' connections enable a quick changeover between configurations to be made.

Hinged dome ends on the control pod, together with an equipment rack on runners, enables complete withdrawal of relevant units for unobstructed access for maintenance purposes.

A composite electromechanical umbilical cable, incorporating all power and telemetry conductors including fibre optics, provides the connection between the GEC Avionics control console and Dragonfly. A single fibre optic carries three digitally encoded video channels combined with a high speed two-way data channel, and using colour multiplexing techniques, a studio quality, interference-free TV image and error-free data are presented on the displays for the operator. The system can be expanded, allowing Dragonfly to carry up to six cameras from which the surface-controlled video switching system can select any three cameras for display. A second fibre optic is incorporated into the umbilical for redundancy and a coaxial video/data backup system guards against loss of the vehicle in the unlikely event that both fibre optics fail. Reversionary modes are implemented automatically.

The umbilical also carries two other twin axial cables and one coaxial cable to allow sonar or additional payload items to be connected to the system independent of the main data link.

The vehicle is controlled by microcomputers at each end of the umbilical. Together with the flexible

communications system, all designed and manufactured by GEC Avionics, these computers enable the vehicle to carry a wide range of equipment without the need for redesign. The computer in the vehicle also carries out a number of functions locally, which helps to reduce the workload on the data channel.

To ease the workload of the operator, the GEC-designed and built console makes extensive use of modern information display techniques, using technology developed for the aerospace industry. The modular panel arrangement enables controls to be easily reconfigured to suit varying customer requirements. Selectable computer-controlled VDU displays and overlays are central to the design philosophy, and replace many of the discrete panel functions normally encountered.

Contractors
OSEL, Ulvertech.
GEC Avionics Ltd, Rochester.

HYBALL

Type
Remotely operated underwater vehicle.

Description
The Hyball is a miniature vehicle designed principally for inspection tasks of ships' hulls, harbour installations, buoys and moorings, seabed installations and so on. The vehicle's standard equipment fit includes a low light CCD colour TV camera with wide angle auto-iris lens mounted on a chassis which pitches through 360° and allows the camera to view up, down, forward or backwards through the unique patented Meridian viewport. TV pictures are relayed to a 15 in colour monitor in the surface platform. For poor visibility conditions a low light black and white SIT camera may be mounted on the standard camera chassis together with the stills and colour cameras. The vehicle is also fitted with two fixed 75 W quartz halogen lamps pointing forward and two 55 W lights mounted on the camera chassis. These latter follow the camera and provide sufficient light to check the vehicle's umbilical cord for snags.

Vehicle status information is displayed along the bottom edge of the monitor screen and includes text lines for information such as customer name, job number or location which is defined by the operator at the beginning of each dive. The video overlay may be dimmed or switched off by the operator.

The Hyball is powered by four thrusters, two for forward/reverse motion and rotational motion and two for vertical and lateral movement. Power is derived from 0.5 hp 24 V DC brushed motors through 10.5:1 reduction gear boxes driving 5 in diameter propellers mounted in Kort nozzles. Thruster speed and direction are controlled by digital signals proportional to movements of the joystick. Vehicle movement is controlled by the single three-axis joystick and dive and surface buttons on the hand controller. Any combination of vehicle movement control may be held by the use of the trim button for hands-off operation or for re-centering the joystick when operating in a cross current. Vertical thrust power may be preset from the surface unit panel.

Specifications
Length: 460 mm
Width: 510 mm
Height: 470 mm
Weight: 39 kg
Speed: In excess of 2.5 kts
Operating depth: 300 m
Payload: 4.5 kg

MERLIN

Type
Remotely operated underwater vehicle.

Description
Merlin has been designed for a variety of underwater tasks. The system comprises a vehicle with a wide angle, solid-state colour video camera, a surface console with a 10 in colour monitor, a 50 m umbilical cable, and a twin joystick controller.

The vehicle is built of three tubular pressure hulls bonded together. The main hull contains the colour video camera and all the control electronics and has a hemispherical macrolon viewport.

Power is derived from six rechargeable lead acid gel batteries mounted in the two lower pressure hulls. Four 12 V electric motors mounted in an assembly attached to the rear hemisphere power all the systems, including the thrusters and the two halogen 150 W lamps and the video camera. The vehicle is propelled by two thrusters mounted at preset angles at the rear. The thrusters incorporate a 3 in propeller mounted in a Kort nozzle with an external shroud to improve hydrodynamic flow. Average endurance between charges is a minimum of approximately 1-2 hours when all equipment is used in bad visibility and high current areas. Under more normal conditions an endurance of between 3-5 hours can be expected.

Vertical movement of Merlin is controlled by a unique variable buoyancy system. Three copper tanks attached to the main pressure hull are filled with water and air as appropriate through the umbilical from the surface unit. When Merlin reaches its working depth a close-loop depth lock system may be switched in which will then stabilise the vehicle at that depth to ±6 in.

The vehicle control transmitter is fitted with two joysticks which operate all thrust functions, the camera iris and the lights.

Specifications
Length: 840 mm
Width: 390 mm
Height: 330 mm
Weight: 45 lb
Maximum operating depth: 200 ft

Contractor
Bennico, Aberdeen.

V-45

Type
Mine identification and destruction vehicle.

Description
The V-45 ROV is designed for use in support of naval minehunting operations. It can be operated from most existing MCMVs as well as COOP. The vehicle has been designed for a major UK defence contractor for military use by a commercial design team with 15 years' operational, engineering and practical experience of manned and unmanned vehicle designs and operations.

The vehicle can achieve a through water speed of 3 kts in currents up to 1 kt. It can operate down to depths of 100 m (with full integrity to 200 m) at 250 m from the support vessel in tidal streams of 3 kts.

The vehicle carries a payload of two video cameras (one colour, one black and white SIT), variable intensity lights, and a three-function manipulator. The vehicle also carries a scanning sonar, acoustic pingers, flasher and VHF radio beacon, plus an explosive charge as and when required. The vehicle is also fitted with all normal heading, depth and roll/pitch sensors.

The V-45 is powered by thrusters driven by brushless DC motors which are energised by a battery pack.

The ROV is controlled from a console, the pilot retaining full control over all vehicle functions including the manipulator. Depth and heading can be either manually or automatically controlled. In the presence of high tidal currents the pilot is assisted in controlling the vehicle by a software routine which prevents him from over-controlling the yaw angle which can result in loss of orientation. Full automatic modes available include hover, pitch, depth, yaw, and reverse control. Command of the vehicle is exercised through a coaxial cable or fibre optic disposable tether.

The control console, power supply units, spare battery packs, battery charger, tools and spares are all contained in a portable, non-magnetic container. If required the sonar console and video recording system can be installed elsewhere in the ship.

Specifications
Length: 2.86 m
Width: 1.18 m

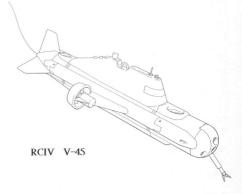

RCIV   V–45

Height: 0.75 m
Weight: 385 kg (in air)
Endurance: 90 mins (in worst tidal conditions)
Depth: 10 m
Speed: 3 kts

Contractor
Winchester Associates Ltd, Aberdeen.

D-001

Type
Disposable mine demolition vehicle.

Development
The D-001 has been designed in response to a request from the UK MoD (N) for ideas on the potential for underwater vehicles. The D-001 is designed for use by dedicated MCMVs, COOP and others where space/weight and cost implications are of importance. The vehicle is deployed and operated in the same way as a conventional ROV, but on the basis that a positive target identification will result in the destruction of the entire vehicle. 'Reload' times are therefore much reduced as no vehicle recovery is involved. The vehicle can also be used in the defence role against swimmers or other threats as well as general subsea investigative operations.

Description
The vehicle is equipped with a CCD black and white, fixed focus, auto-iris/ring light, video camera, a solid-state heading sensor which is set prior to launch, and a quartz resonator transducer as depth sensor. Control is exercised through a console-mounted joystick by which the vehicle is manoeuvred and controlled in speed and direction. The control console houses the video monitor, and indication is provided of heading, depth and battery voltage. The vehicle carries a charge which is armed and triggered from a control console. The vehicle is stored without its disposal charge and battery pack which are only fitted when the system is required for use.

Specifications
Length: 1.15 m
Diameter: 165 mm
Weight: 15.09 kg
Speed: 5 kts
Thrust motor: 0.80 kW
Lateral/Vertical thrusters: 200 W
Endurance: 30 mins

Contractor
Winchester Associates Ltd, Aberdeen.

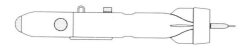

UNITED STATES OF AMERICA

MNS

Type
Remotely operated mine disposal system.

Description
The Alliant Techsystems mine neutralisation system (MNS) developed by the company's Marine Systems and the US Naval Sea Systems Command, is now in production for the US Navy's new MCM and MHC ships. It is the result of more than 10 years' work to provide a system that surpasses the conventional towed mechanical, magnetic and acoustic sweep techniques. It is designed to detect, locate, classify and neutralise moored and bottom mines, using high resolution sonar, low light-level TV, cable cutters and mine destruction charges. A special underwater vehicle carries these sensors and countermeasures and is remotely controlled from the parent vessel.

Initial target detection and vehicle guidance information is provided by the ship's sonar. Initial vehicle navigation is plotted and monitored within the MNS acoustic tracking system. Vehicle sonar is used during the mid-course search and final homing phases, and high resolution enables operations in poor visibility by sonar guidance alone. Low light TV is used in conjunction with sonar during the precision guidance phase near the target. Underwater launch and recovery of the MNS vehicle assists operations in high sea states.

Vehicle power is provided via a neutrally buoyant umbilical cable, which also carries signal and control links between the vessel and the MNS vehicle. Aboard the parent vessel a sonar screen on the control console displays sensor and vehicle status information in digital form. The monitor and control consoles are the focal point for operation and management of the complete system. Display facilities include: vehicle sonar, vehicle TV, deck TV, vehicle control and navigation, cutter and dropped charge and provision for monitoring system status.

Operational Status
In production.

Specifications
Length: 3.8 m

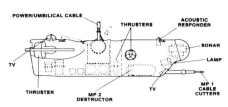

Arrangement of the MNS

Width: 0.9 m
Speed: 6 kts
Weight: 1135 kg (in air)
Propulsion: Hydraulic, twin 15 hp
Thrusters: Horizontal/vertical
Cable length: 1070 m
Power requirements: 60 kW (peak)

Contractor
Alliant Techsystems, Marine Systems West, Mukilteo, Washington.

PHANTOM

Type
Remotely operated vehicle.

Description
The Phantom ROVs have been utilised in a variety of commerical and military applications. Among the tasks performed by the Phantom have been mine countermeasures, ship inspections, patrol tasks in harbours and port areas, bottom surveys, diver backup and assistance, survey and search and recovery operations.

The high speed Phantom HVS4 was first designed and produced for the US Navy as a powerful and precise mine countermeasure/neutralisation vehicle, and formed part of the Phantom Spectrum series of vehicles. The HVS4's cable and console are fully interchangeable with eight other Phantom vehicles. The HVS4 system includes: a hard-wired shock-mounted vehicle; colour CCD TV camera; propulsion units; Spectrum HP control console; high definition scanning sonar; built-in navigation/tracking responder system; two or three function manipulator; payload release mechanism capable of carrying and releasing 20 kg plus payloads such as explosive charges and/or marking beacons.

Among the numerous accessories are special cameras, laser/scaling device, head-up display computer, remote joystick control, fibre optic cable, powerful cable cutter and other versatile tools.

The vehicle is guided from the ship via an umbilical cable connected to the integrated HP control console. It may be driven over umbilical lengths of up to 600 m. The Spectrum series is a hard-wired system which minimises the complexity of subsea electronics and keeps critical functions operational, even if the umbilical sustains damage.

The vehicle has four horizontal thrusters with contra-rotating propellers that deliver over 150 lbs of thrust. Two vertrans thrusters provide precise vertical and lateral thrust.

Phantom HVS4 ROV

With a maximum speed in excess of 4 kts, the HVS4 operates with all onboard systems from the surface to 300 m. The vehicle is able to maintain station in currents over 2 kts, due to its inherent stability and high thrust capability.

The HVS4 is equipped with an electronic depth gauge, auto heading and auto depth to allow the operator to easily position the vehicle.

The control console features a video monitor together with automatic heading and depth control functions combined with digital navigation. It uses an advanced joystick control, with options including advanced head-up display and MILSPEC design. The console also incorporates various controls associated with payload packages. Among the systems which have been proven with the HVS4 are Mesotech 971 sonar, Trackpoint II responder, 20 kg payload release mechanism and cable cutter. The vehicle weighs less than 200 lbs and can deliver more than 150 lbs of thrust, making it possible to sustain a speed of over 4 kts at operating depth.

Operational Status
Over 180 Phantom vehicles have been produced to date, with a number of systems now in service with NATO navies including the Royal Navy, US and Portuguese Navies; and the Australian and Royal New Zealand Navies.

Contractor
Deep Ocean Engineering, San Leandro, California.

RECON IV

Type
Remotely operated vehicle.

Description
The Recon IV is powered by four high efficiency thrusters, each of 80 lb thrust. Automatic depth control is standard. The vehicle uses a small size protected flying tether, allowing it to manoeuvre in confined areas. It is fitted with a minimum of onboard electronics which ensures reliability and ease of maintenance. Each component on the vehicle is hardwire-controlled from the surface for maximum simplicity. The open frame design, 38 spare conductors and positive buoyancy of the vehicle enable a wide range of optional equipment and accessories to be fitted. An optional multiplex system is available to enhance vehicle flexibility. The vehicle has a positive buoyancy of 160 kg with an option for heavier payloads.

Options include black and white or colour video camera, outboard lights, sonar (forward, side-scan or other types), still camera, stereo camera and acoustic tracking system.

Specifications
Length: 2.06 m
Width: 0.9 m
Height: 0.85 m
Speed: 3 kts (forward), 1 kt (lateral/vertical)
Weight: 457 kg
Payload: 160 kg
Service depth: 457 m

Contractor
Perry Tritech Inc, Jupiter, Florida.

VIPER Mk 1

Type
Underwater robotic vehicle system.

Description
The Viper is a high speed robot submarine specifically designed to carry out a variety of waterside security, mine countermeasures and channel conditioning missions. It carries a high resolution sonar system which can acquire targets at a range of several hundred metres, and a low light CCD video camera which enables positive target identification at short range. To meet specific mission profiles, the Viper can carry a variety of payloads including ordnance neutralisation charges and intruder response systems.

Propelled by four high power brushless thrusters, the Viper can travel at speeds exceeding 8 kts and can dive and ascend at over 1.5 m/s. The vehicle is powered by either high energy density batteries or, for longer duration missions, by a solid polymer fuel cell. Control of the vehicle is via a small diameter fibre optic tether drawn from a canister in the rear of the fuselage. This yields an operational radius of 2 km, a sufficient range for both intruder detection and mine countermeasures applications. Conveniently stowed in flyaway containers, Viper installation and operation requires minimal logistic support.

The pilot controls the Viper using a joystick and mouse that are integral with the head-up display, an integrated system enabling ergonomic, precise and rapid vehicle guidance. Control is augmented by the vehicle's auto-heading, depth and altitude functions.

In keeping with its mine countermeasures role, the Viper is designed and constructed so as to minimise acoustic noise and magnetic signature while maximising performance and reliability. Due to low procurement and mission costs, minimal deployment and support logistics and high performance potential, the Viper offers a practical and cost-effective solution to many waterborne requirements.

Specifications
Length: 1.88 m
Width: 0.5 m
Height: 0.53 m
Weight: 46 kg
Service depths: 200 m

Contractor
Allen Osborne Associates, California.

RCV-225

Type
Remotely operated vehicle.

Description
The RCV-225 system is designed to carry out detailed underwater inspections down to 400 m, such as the survey of underwater installations, search and salvage, remote monitoring of the laying of underwater installations, oceanographic surveys, placement of charges and arming. The vehicle is fitted with a low light-level SIT camera and optional colour TV and photographic systems. The SIT camera is equipped with a unique lens assembly that enables the operator to remotely pitch the angle of view ±90° from the horizontal. Two 45 W tungsten halogen lamps provide a viewing range of up to 10 m with no ambient light.

The ROV is driven by four oil-filled electric motors giving a speed of 3 kts in all dimensions. The motors are enclosed in a syntactic foam hull and the camera and electronics are contained in a pressure housing.

Full command of the vehicle is exercised from the control station with desired depth and heading automatically maintained by servo controls. Vehicle depth, heading and lens pitch angle are displayed in the TV picture and continuously recorded on video tape.

The vehicle is deployed using a protective RCV launcher with tether cable and winch, a deck winch with double armoured cable mounted on a skid/A-frame assembly and a hydraulic power supply. The vehicle is carried to the working depth in the launcher, from which it is deployed for the operation, returning to the launcher for transport back to the surface.

Specifications
Length: 5100 mm

CONTROL STATION

CONTROL/DISPLAY   HAND CONTROLLER

DEPLOYMENT UNIT

WINCH/SKID/A-FRAME

VEHICLE

STROBE FLASHER

LIGHT

TV CAMERA DOME

THRUSTERS   TETHER CABLE

ARMORED CABLE

LAUNCHER

WINCH

Width: 6600 mm
Height: 5100 mm
Weight: 82 kg (in air)

Contractor
Sachse Engineering Associates Inc (SEA) Hydro Products, San Diego, California.

SEA SEARCH SYSTEMS

Type
Towed vehicle systems.

Description
These towed vehicle systems are designed for professional search, survey and location operations. The multi-mission Mk I and II systems are intended for use in inland waters and ocean depths to 1000 m. The hydrodynamically stable design permits low altitude flying and precision control of the vehicle. The towed body can carry a variety of sensors/equipments. The Mk I is fitted with a low light CCD TV camera and video control console. The more powerful Mk II system also incorporates a high resolution side-scan sonar with low light TV video camera for confirmation of sonar targets. Both vehicles carry a still camera, underwater light and are controlled from a console fitted with an 8 in colour video monitor. A side-scan recorder is also available with the Mk II. The sonar operates on one of two frequencies, 100 kHz or 500 kHz with a vertical beamwidth of 40°. The compact control console is installed in a special aluminium enclosure, moisture sealed and ruggedly mounted.

The Mk III is a low-cost towed camera system which can operate down to 150 m. The 'fish' can carry any type of underwater camera and the simple desktop control console includes a colour video monitor.

The Mk IV system uses the same 'fish' body as the Mk III but is depth rated to 800 m.

The Mk V system is equipped with a low light, high resolution, black and white SIT camera, and a 100 kHz or 500 kHz side-scan sonar. The 'fish' is controlled from a small control console with special electronics that allow the video and sonar signals to be transmitted over the same cable simultaneously.

Contractor
Sachse Engineering Associates Inc (SEA) Hydro Products, San Diego, California.

LENS

Type
Remotely operated vehicle.

Description
EDO Corporation's Electro-Acoustic Division and Lockheed Missile & Space Company have jointly developed a low-cost, wire-guided mine neutralisation vehicle, designated LENS, which has the capability to be manually steered and by acoustic means to automatically lock on to and home on moored or bottom mines. Its primary function is to permit a vessel or helicopter to deliver a destructor charge to a suspected area which may contain a mine. It is intended to operate in conjunction with more complex auxiliary search and classification sonars which can be used to furnish the approximate location of the mine to be destroyed.

The basic vehicle contains the necessary guidance control circuitry to home in on a mine when used with one or more mine locating sonars. The LENS vehicle was designed to be as low-cost as possible, since it is an expendable device. Additional features can be added when other missions are assigned to the vehicle, for example, automatic navigation capability to manoeuvre in an acoustic transponder nav-system, improved automatic target recognition and classification circuitry to discriminate or enhance bottom mines, and various shaped charges for special missions and so on. At present, however, the LENS vehicle contains only the basic vehicle control circuitry, wire-guided multiplexer control circuitry, altitude sensing and bottom following control sensors, terminal guidance homing sonar, a transponder for tracking the vehicle and necessary vehicle sensors and controls to enable operator involvement in the initial localisation of the target when utilising information from standard minehunting sonars. Finally it includes the necessary safety arming and firing mechanism to activate a suitable neutralisation charge.

Contractor
EDO Corporation, Electro-Acoustic Division, Salt Lake City, Utah.

The LENS vehicle

UUV

Type
Prototype Unmanned Undersea Vehicle (UUV).

Development
The US Defense Advanced Research Projects Agency (DARPA) together with the US Navy and Charles Stark Draper Laboratories has built two prototype UUVs to demonstrate the feasibility of developing a design for an ROV capable of undertaking covert high priority missions. The vehicle is intended for deployment from submarines as well as surface ships. A third prototype is under construction and is due to be delivered next year. The US submarine *Memphis* is currently being adapted as a trials boat for testing the UUV concept.

The UUV will be fitted with a range of advanced acoustic sensors, communications and signal and data processing systems for trials which will demonstrate the vehicle's ability to carry out autonomous operations. Missions envisaged include MCM, surveillance and communications. Three mission systems are being developed for the UUV: a Tactical Acoustic System (TAS), Mine Search System (MSS) and Remote Surveillance System (RSS).

The hull is of titanium and pressure tests on the first hull were carried out in December 1990. An internal pressure hull contains the payload which occupies a section some 5 ft in length. Power for the 12 hp electric motor will be provided by batteries housed in two 4 ft 4 in long sections while the vehicle's control electronics will be contained in another 4 ft 4 in section in the centre section. The electric motor and its control unit will be housed in a 12 ft aft section.

Specifications
Length: 36 ft
Diameter: 44 in

Contractors
DARPA.
 Charles Stark Draper Laboratories.

HYDRA

Type
Cable repair vehicle.

Description
The Hydra-AT 1850 CRS has been designed with an extremely low magnetic signature for the location, tracking, excavation and recovery of buried cables for repair and reburial.

The vehicle is equipped with a powerful jet for blowing away mud and sand covering cables. The vehicle detects and tracks cables using the magnetic field developed around the cable, or by detecting superimposed AC tones or DC current generated through the cable.

The ROV is equipped with a range of tools including two seven-function manipulators, the powerful jet system bolted to a skid on the bottom of the ROV, eight cameras (five colour), a short-range scanning sonar and the usual range of sensors for manoeuvring.

The five video cameras perform navigation, observation and control of jetting operations. Three of the cameras are on pan/tilt mountings for navigation, with another camera aft for observation. The system uses extensive graphic displays which enable the controller to manoeuvre the vehicle in conditions of zero visibility.

Power is provided by two 75 hp hydraulic packs driving seven thrusters; two twin units fore and aft, two lateral units and three vertical units.

Operational Status
Delivered to the USN in 1988, operational at the beginning of 1990. A second system is on order for the USN and a third unit has been ordered by a commercial organisation.

Specifications
Length: 3.5 m
Width: 1.83 m
Height: 1.83 m
Weight: 3775 kg (in air)
Speed: 3 kts (forward)
Depth: 1850 m

Contractor
Ocean Systems Engineering.

ROV ANCILLARY EQUIPMENT

GERMANY

DM 801 B1

Type
Mine disposal charge.

Description
This newly developed and qualified charge consists of an aluminium casing filled with approximately 15 per cent more HE than standard charges, resulting in increased lethality. The mine disposal charge is positioned in the proximity of the ground mine by an ROV, like PAP 104 or Pinguin B3.

It is then remotely detonated with the Rheinmetall remote-controlled fuze system.

Operational Status
In production and used by the German Navy.

Specifications
Length: 932 mm
Diameter: 350 mm
Operating depth: 5-300 m
Weight: 126.5 ± 1.2 kg
Charge: 105 kg HE

Contractor
Diehl, Nuremberg.

DM 59/DM 69

Type
Remote-Controlled Explosive Cutter (RCEC).

Description
Based in part on design features of the Rheinmetall explosive cutter for sweepwire attachment, this cutter is intended for use with underwater ROVs, such as the Société ECA PAP 104 system, STN SystemTechnik Nord GmbH Pinguin or Gaymarine Pluto.

The cutter's remote-control system is functionally identical to that used for detonating the mine disposal charge with the fuze DM 1002 A1. With its hard foam housing, the RCEC is neutrally buoyant in water.

Operational Status
Under development for use in the Royal Navy. Series production commences in 1992.

Specifications
Cutting capacity
Stud link chains: 20 mm
Steel wires: 26 mm
Synthetic ropes: 40 mm
Operating range: 300-2000 m
Operating depth: 5 to <300 m
Weight in air: 6.4 kg
Weight in water: neutral buoyant
Dimensions: 865 × 195 × 121 mm

Contractor
Rheinmetall GmbH, Dusseldorf.

Remote-controlled explosive cutter

DM 1002 A1

Type
Remote-controlled fuze.

Description
The fuze is ignited by a coded acoustic signal only and deactivates automatically if no such signal has been received within a certain time span. The fuze is impervious to shocks from detonations or similar occurrences above or below the water, as well as to any other acoustic noise or signal.

Operational Status
In production and used by the German Navy as well as by the British, Dutch, Belgian, Norwegian, Australian and Indonesian navies.

Specifications
Length: 364.5 mm
Diameter: 200 mm
Operating range: ≤2000 m
Operating depth: 5-300 m
Weight: 8.0 kg
Weight of explosives: 0.92 kg
Self-deactivation: 45 mins (after laying)

Contractor
Rheinmetall GmbH, Dusseldorf.

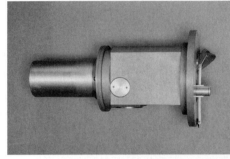

Remotely controlled fuze DM 1002 A1

E 67/E 67 MOD 1

Type
Firing transmitter.

Description
The firing transmitter generates a coded acoustic signal which is led via cable to the transducer in the water. The coded signal sets off the mine disposal charge via its remotely controlled fuze at operating distances of up to 2000 m.

The transmitter exists in two versions. One is supplied in its own carrying case for use on the quarterdeck, the other is for incorporation in a console below deck.

Operational Status
In production and used by the German Navy as well as the British, Dutch, Belgian, Norwegian, Australian and Indonesian navies.

Specifications
Firing Transmitter E 67 capacity
Output: 14 W
Supply voltage for battery charger: 110 V
Batteries: 2 × 12V/500 mAh
Code setting: By two rotary switches

Firing transmitter E67 supplied in its own carrying case for use on the quarterdeck

Weight of transmitter: 6.5 kg
Weight of cable with transducer: 10.8 kg
Length of cable: 33 m
Weight of storage container: 11.7 kg
Dimensions of transmitter: 230 × 200 × 130 mm
Dimensions of transducer: 200 mm (max diameter)
Dimensions of storage container: 440 × 600 × 250 mm
Firing Transmitter E 67 Mod 1 capacity
Output: 14 W
Power supply: 115 V AC

Firing transmitter E67 Mod 1 for incorporation in a console below deck

Code setting: By two rotary switches
Weight transmitter: 3.5 kg
Weight of cable with transducer and cable drum: 15 kg
Length of cable: 100 m
Dimensions of transmitter: 340 × 153 × 153 mm

Contractor
Rheinmetall GmbH, Dusseldorf.

---

158 R

Type
Exercise Fuze 158 R.

Description
The exercise fuze is functionally and dimensionally identical with the HE fuze DM 1002 A1, except for the explosive train.

On receipt of the fuzing signal, the exercise fuze generates an electric impulse of 12 V and 7 A for a duration of 3.5 seconds. This impulse activates a relay or melts a wire which releases a buoy from an exercise charge.

Operational Status

In production and use by the Royal Navy and Royal Australian Navy.

Specifications

Operating range: ≤2000 m
Operating depth: 5-300 m
Total weight: 8 kg
Battery: 12 V/1000 mAh
Relay impulse: 12 V, 4 A, 3.5 s
Reusable with same battery: 20-40 times
Dimensions: 364.5 mm (max length); 200 mm (max diameter)

Contractor

Rheinmetall GmbH, Dusseldorf.

Exercise fuze 158 R

ITALY

CAM

Type

Countermining charge.

Description

Whitehead manufactures this charge for use with automatic mineclearing systems during minehunting missions. The charge comprises:
(a) explosive charge
(b) igniting device
(c) priming device.

These parts are easily and quickly put together; the charge, with its safety devices, can be positioned fully assembled aboard the minehunter.

The charge is contained in a GRP shell capable of withstanding great pressure. The igniter is mounted in the rear of the casing and comprises a receiving hydrophone, electronic circuit, time counter, primary delay, receiving circuit able to receive and process the hydrophone signal and control the firing circuit, firing circuit able to ignite the two detonators of the priming device, and a power supply battery.

Specifications

Max diameter: 360 mm
Length: 830 mm (approx)
Total weight: 110 kg
Main charge weight: 80 kg
Max working depth: 150 m

Contractor

Whitehead, Montichiari (Brescia).

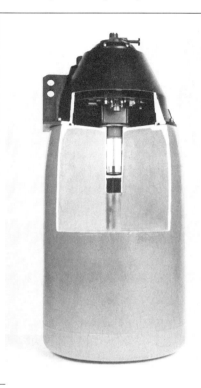

CAM-T

CAM-T

Type

Countermining charge.

Description

This charge is similar to the CAM charge, but with some changes to make it more suitable for mine clearance operations carried out by a diving team.

Operational Status

In production.

Specifications

Max diameter: 360 mm
Length: 830 mm
Total weight: 110 kg
Main charge weight: 80 kg
Max working depth: 150 m

Contractor

Whitehead, Montichiari (Brescia).

CAP

Type

Countermining charge.

Description

The Whitehead CAP charge is particularly suitable for light underwater vehicles for the purpose of detonating mines. The charge is transported by the vehicle by means of a special attachment. When released from the vehicle, hydrostatic pressure activates the system and two detonators automatically advance into a priming position. At the same time a contact is closed, completing the sequence for the command to explode. The fire command is activated by dropping an explosive charge into the water at a safe distance.

Operational Status

In service.

Specifications

Diameter: 250 mm
Length: 615 mm
Weight: 44 kg (in air)
Main charge weight: 28 kg
Maximum operational depth: 150 m

Contractor

Whitehead, Montichiari (Brescia).

CAP underwater explosive charge

MDC

Type

Countermining charge.

Description

The MDC charge is designed for use with the Pinguin ROV. Its charge has the same mode of use as the CAP charge.

Operational Status

In service.

Specifications

Diameter: 360 mm (approx)
Length: 920 mm (approx)
Weight: <120 kg (approx)
Main charge weight: Not less than 80 kg
Maximum operational depth: 150 m

Contractor

Whitehead, Montichiari (Brescia).

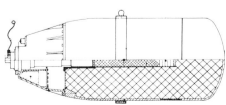

MDC countermining charge

ECP

Type
Explosive cutter.

Description
The ECP is a specially designed explosive cutter device to be installed on the frontal section of remotely controlled vehicles, and is intended for use in cutting the wire or chain of conventional moored mines. It is capable of cutting wires or chains up to a diameter of 30 mm. Two small guidance conveyors allow the device to be fitted on the wire by the pressure of the sensor against the wire or chain to be cut. At the same time the cutter device is disconnected from the vehicle which can then move back freely leaving the cutter locked in position on the wire. A switch is then activated to run an electrical circuit designed to start a time delay from the moment of the vehicle disconnection before the actual cutting operation.

Operational Status
In production and in service.

Specifications
Weight: 1.5 kg (in air)
Dimensions: 360 × 220 × 70 mm
Weight of explosive charge: 150 g (max)
Hydrostatic pressure: >150 m

Contractor
Whitehead, Montichiari (Brescia).

Whitehead explosive cutter device

SWEDEN

MDC 605

Type
Mine disposal charge.

Description
The MDC 605 is designed primarily to detonate and destroy ground influence mines. It can also be placed on the sinker of a buoyant mine to sever its mooring, or be used for various underwater demolition tasks. The charge is normally deployed from an ROV and in particular the SUTEC Sea Eagle or Double Eagle deployed from the Swedish 'Landsort' class MCMV. The MDC 605 can also be carried and placed in position by a diver as it is small enough to be handled and can be made weight neutral in water by fixing a detachable float.

The charge is detonated by acoustic telecommand using a coded (8-bit frequency shift) signal. For safety, firing commands are ignored for a fixed time after laying. The MDC 605 incorporates a hydrostatic device which precludes the charge being armed until it has reached a set depth. The detonator is not fitted until immediately before deployment and the MDC 605 contains only low-sensitivity explosives during storage and transport.

Operational Status
In service with the Swedish Navy.

Specifications
Length: 380 to 615 mm
Diameter: 170 mm
Weight: 4 kg (exclusive of charge)
Charge: 3-10 kg HE (optional)

Contractor
SA Marine, Landskrona.

UNITED KINGDOM

MINE DISPOSAL CHARGE

Type
Countermining charge.

Description
The system comprises a volume of high explosive contained within a charge case which is purpose-designed to be carried by modern remote-controlled mine disposal systems. The charge is compatible with the PAP 104 and Pinguin vehicles.

The fuzing system, manufactured by Rheinmetall, uses an acoustic coded signal transmitted from the MCMV to initiate detonation of the charge. This ensures that the charge can only be detonated by command from the minehunter after any vehicles or divers have moved well away from the danger area.

Specifications
Length: 930 mm
Diameter: 350 mm
Weight: 120 kg
Charge: 80 kg HE

Contractor
BAeSema Marine Division, Bristol.

BAeSema mine disposal charge

CAMERAS

Type
Cameras for ROVs.

Description
The Osprey Electronics OE1356 is a high resolution monochrome CCD camera used principally for high grade inspection and general observation such as tether management.

The OE1362 is a miniature high resolution colour CCD camera for inspection and close-up identification of targets where colour can provide additional information.

The OE1382 is a colour CCD camera with a sensing chip and lens mounted on a gimbal. The camera has the ability to look up and down ±90° and from side to side ±55° without actually moving. In addition the complete internal package can rotate within its own body. All movements are detected and can be displayed on the surface control equipment.

The OE1323 is a highly sensitive SIT camera whose main function is to provide a navigation system for the ROV. The camera operates well in areas of high turbidity using in many cases only ambient light. The resultant pictures are of high quality with little or no self-generating noise.

Specifications

Model	Description	Sensitivity (Lux Face-Plate)	Resolution (TVL/ph)	Angle of view	Dimensions (Θ × L)	Weight air	water
OE1356/7	Monochrome CCD inspection camera	0.1	550	66°	70 × 205	1.2 kg	0.4 kg
OE1323	SIT low light-level	5 × 10⁻⁴	600	110°	101 × 308	3.7 kg	1.1 kg
OE1325	ISIT ultra low light	5 × 10⁻⁶	550	65°	130 × 410	7.0 kg	2.0 kg
OE1362/3	Colour CCD inspection	0.1	460	65°	85 × 200	1.8 kg	0.5 kg
OE1370	Colour CCD wide angle	0.3	280	80°	102 × 222	2.4 kg	0.6 kg
OE1382/3	Colour PTR inspection	0.9	320	60°	104 × 327	3.9 kg	0.8 kg
OE2301	SIT/TVP low light, photographic	5 × 10⁻⁴	600	84°	178 × 467	11 kg	1.3 kg
OE2360/1	Colour TVP inspection photographic	0.9	320	84°	178 × 467	11 kg	1.3 kg

The OE0285 is a highly sensitive ISIT camera providing high quality pictures with very low generated self-noise characteristics. Its two-stage first generation intensifier which is coupled to sophisticated electronics is ideally suited for underwater use.

The OE1326 is a highly light-sensitive ICCD camera which can be used for navigation where low magnetic signature is of paramount importance.

The OE2300 is a series of combined TV/photographic cameras. Using a single lens reflex technique the camera produces both high quality video (colour or intensified monochrome) pictures and high quality 35 mm photographs. A full alphanumeric data chamber overlays the photograph and simultaneously displays characters on the TV screen for 1:1 correlation. Other cameras include the dual colour CCD

and ICCD cameras used on the ECA PAP ROV and Triplex Head colour CCD, SIT and photographic camera on the SPRINT vehicle.

Contractor
Osprey Electronics, Aberdeen.

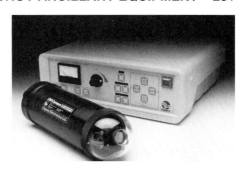

Colour PTR camera and surface control unit

UNITED STATES OF AMERICA

SEAPHIRE

Type
Underwater light.

Description
The Seaphire underwater two-light system is designed to handle the two most significant underwater illumination problems – common volume scattering and hot spots. The light allows the user to select the optimum beam pattern according to the application and operating conditions. The blue-green lens is designed to filter out the majority of red and

yellow light above 600 nanometers wavelength, enabling optimum performance to be achieved from high-gain monochrome TV cameras (that is, SIT and ICCD).

Both long and short housing versions are available allowing direct replacement of existing lamp assemblies without changing mounting configurations. The internal reflector, glass domes and front cowls are also interchangeable, allowing superior beam patterns to be formed. The lights themselves are selectable for flood or spot patterns, and run on 120 or 240 V supply with long life tungsten-halogen lamps of 100, 150 or 250 W.

The lamps are housed in a lightweight aluminium

mounting with a 1000 m depth rating. As an option increased depth ratings are available using stainless steel or titanium housings. Optionally 20° spot and 90° flood beam patterns are available.

Specifications
Length: 21.6 cm (less connector)
Width: 8.9 cm
Depth: 14.6 cm
Weight: Approx 1.8 kg in air
Depth rating: 1000 m

Contractor
Sea Hydro Products, San Diego, California.

SEAVISION

Type
Intensified CCD camera.

Description
Seavision is a low-profile, high performance, intensified, solid-state TV camera with full hemispherical viewing and no external moving parts. This configuration provides a constant hydrodynamic profile and fixed electromechanical connections to the viewing system platform or ROV. When combined with the

Seaphire range of underwater lighting, Seavision enables optimum viewing system performance in resolution, range and reliability.

The system incorporates the Sea Hydro Products Ultravision ICCD imaging system for extreme low light (1×10^{-6} ft candles faceplate illumination) viewing. The camera is mounted on a high speed pan and tilt mounting (45°/s) for optimum real-time viewing.

The camera is mounted in a lightweight aluminium housing rated to 1000 m depth.

Optional features available include: increased depth rating using stainless steel or titanium housing; position feedback for annotation generation; various

input/output command and data formats; 6.5 mm and 12.5 mm lenses.

Specifications
Diameter: 25.8 cm
Length: 28 cm
Weight: 9.5 kg in air
Depth rating: 1000 m

Contractor
Sea Hydro Products, San Diego, California.

NIGHTCOBRA

Description
The Nightcobra camera incorporates Photosea's Nighthawk ICCD solid-state, low light-level camera with the Cobra internal pan and tilt mechanisms. The

ICCD includes an image intensifier coupled to a CCD chip. Although ICCD cameras are unable to provide the high resolution of SIT cameras, they offer several unique features, including instant switch on and low magnetic signature.

The Nightcobra unit is available with a 1000 or 6000 m depth rating. It features low light-level

sensitivity with internal pan and tilt optics, remote lens focus control from 30 cm to infinity and remote lens position readout.

Contractor
Photosea Systems Inc, San Diego, California.

MINESWEEPING SYSTEMS

CHINA, PEOPLE'S REPUBLIC

TYPE 312

Description
China has developed a small, remotely controlled mine countermeasures system believed to be for use in harbours and estuaries. The system uses a 20.94 m long boat with a displacement of 39 tonnes which is controlled by radio signals from a shore station or mother ship up to 3 nm (5.55 km) away.

The boat is powered by a 12V150C 500 hp supercharged diesel engine for transit, giving a maximum speed of 11.6 kts, while for operational use an electric motor is used. It has a range of 108 nm at 9 kts. Acoustic and magnetic sensors are used but it is not clear how the mines it locates are destroyed.

Contractor
China Shipbuilding Trading Company, Beijing.

Chinese remote-control minesweeper Type 312. Key to diagram: (1) steering engine room; (2) special instrument compartment; (3) wheelhouse; (4) main engine compartment; (5) materiel compartment; (6) fuel compartment; (7) auxiliary engine compartment; (8) generator compartment

FINLAND

FIMS

Type
Turnkey magnetic and acoustic minesweeping systems.

Description
FIMS is an integrated package consisting of all components required for minesweeping operations on small to medium sized vessels. The primary components of the system are:
(a) Three electrode magnetic sweep cables, with current capacities of 400 A or 800 A. The sweep cable, with a diameter of either 34 mm or 46 mm, is floating and its low tow resistance enables operation at speeds of up to 10 kts
(b) A depth controllable acoustic source with an output level of up to 150 dB
(c) Power generation units for both the magnetic and acoustic sweep systems. The magnetic sweep current generator can be powered from the ship's supply or by its own diesel generator
(d) A powerful shipboard control computer, backed by an extensive suite of pre- and post-mission analysis tools
(e) Military standard shelters for mobilising the system on COOP ships.

The onboard and mission processing systems are at the heart of the FIMS concept. These systems are located on two 386 PC type computers. One, the TSCPU (Tactical Sweep Coverage and Planning Unit), normally located ashore, runs the planning, database creation, coverage mapping and report generation. The second system, the SSCPU (Shipboard Sweep Control and Positioning Unit), located on the minesweeping ship, runs the sweep control and positioning real-time program.

The onboard SSCPU integrates the magnetic and acoustic sweep controller with the ship's positioning systems. This enables proper control of sweep lines, displays of actual coverage, and position activated generation of sweep footprints. For use with wire sweep systems, FIMS can be interfaced with a sweep wire positioning and depth measurement system.

Both the magnetic and acoustic sweep controllers can generate complex time varying signatures. These signatures are defined interactively, and need not be a precise mathematical or transcendental function. By interfacing to the ship's echo sounder the variation of bottom cover with depth is determined. Also the estimated sweep signal beneath the ship is calculated, again using depth as the control variable. The control algorithms optimise coverage, while maintaining a safe operating environment for the ship.

Graphics displays are provided to show both along track and cross track coverage, and all position and

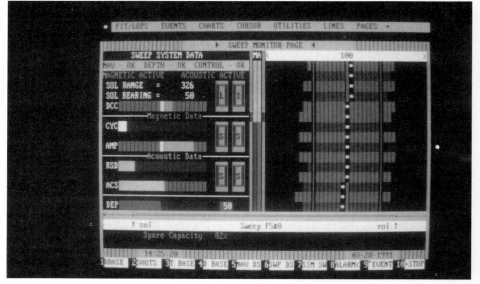

System overview page

coverage data are logged for later analysis by the TSCPU system.

The positions section of the program can calculate position from raw Syledis or Microwave ranges, or accept data from a Differential or Stand-Alone GPS receiver. If two systems are available position comparisons are given. Sweep cycles can be started based on position or time.

Historical quality control information is provided for both positioning and sweep parameters. These can be used to provide both on-line and post-mission evaluations of sweep effectiveness.

While the database is normally created by the TSCPU system all database functions are available on board ship, to increase operational flexibility. The SSCPU software includes a training and simulation mode.

The TSCPU system integrates planning and post-mission analysis functions. The planning system aids the user in the formulation of a set of sweep orders to be dispatched to the minesweeping ships. This is done from an input of the general sweep area, type of mine threat, and resources available. The system would normally be interfaced to some type of digital charting system for this stage of the operation. The output of the system is a data disk for each ship taking part in the operation, the disk contains sweep line co-ordinates, sweep generation parameters, known hazards such as 'friendly' mines and navigation obstructions. Some or all of these parameters can be

modified on the ship depending on changing mission requirements, or changes in the tactical situation.

The sweep coverage subsystem is one of the key features of FIMS. This consists of a database of sweep coverage information organised in small geographical boxes or bins. For each bin, information is stored on various sweep parameters such as type of sweep and number of over-runs. The information is designed to be viewed graphically, which quickly reveals the success of a sweep operation, or areas where further work is required. The database is updated by reading the logging disks generated by the SSCPU system, or in advanced systems by real-time telemetry. Analysis of coverage displays will improve the data produced by the planning part of the system, for example line separation and end of line turn manoeuvres.

Where possible, standard high quality commercial products are used. This results in an excellent price to performance ratio, and improved maintainability. The core technology is well proven and has a clear progress path. The shipboard equipment is small enough to be readily deployed on COOP, and in this mode the integration of the positioning function is particularly useful. Its ability to use several positioning sensors eases the problems of overseas deployment, or non-availability of the primary sensor.

Contractor
Elesco Oy Ab, Espoo.

FRANCE

STERNE 1

Type
Influence minesweeping system.

Description
The Sterne 1 is a towed minesweeping system which simulates both the acoustic and/or magnetic and electromagnetic signatures of various types of ship.

AP 5

Type
Acoustic minesweep.

Description
The AP 5 acoustic sweep (the earlier AP 4 is similar, but with reduced frequency range) comprises: a signal generator which can generate up to 192 acoustic signatures and spectra for various types of ship, signals being pre-programmed within a wide frequency band (in both spectrum and modulation); a very high power, low frequency amplifier with associated impedance matching circuit; an underwater vehicle which is towed by a combined power feed and towing cable and which is fitted with an underwater electrodynamic loud speaker with two symmetrical

The system comprises an AP 5 acoustic sweep and six magnetic bodies. Onboard equipment includes the TSM 2061 tactical display, the TSM 5722 sonar Doppler log, and navigation systems such as the Trident III radio location system and Decca 1226 navigation radar.

The system is towed at speeds of 6 to 10 kts at a minimum distance of 100 m. A mine-avoidance sonar, such as the Petrel, can be integrated into the system.

diaphragms; and an optional piezoelectric transducer for high frequencies. The software programs are designed to meet operational requirements and can be easily developed or modified ashore, with the possibility of exchanging the pre-programmed electronics cards on board. These programs allow the generator to simulate acoustic signatures of various types of ship, to activate mines with known characteristics from a stand-off position, to activate unknown mines by scanning over their probable acoustic spectrum, to inhibit mine detonating devices within an extended area, and to transmit acoustic spectrum in synchronisation with the transmission of the current pulse of a magnetic sweep. The magnetic signature of the ship-borne equipment is very low, even when operating. A pre-programme device enables the transmitted acoustic power to be adjusted according to acoustic propagation conditions over the swept path.

Operational Status
Under development.

Contractor
Thomson Sintra Activités Sous-Marines, Brest Cedex.

The underwater vehicle has successfully passed the impact resistance qualification tests against an underwater explosion of 1 ton of TNT at 80 m abeam.

Minesweeping is performed at speeds of between 3 and 12 kts (8 kts nominal) with a constant immersion of the transmitting vehicle of between 8 and 10 m, this can be extended as an option.

All frequency bands are covered by the sweep and electronically controlled by the computer.

Operational Status
The AP 5 will be carried by the 'Narvik' class mine countermeasures vessels.

Contractor
DCN, Ruelle.

GERMANY

TROIKA

Type
Remote-control minesweep system.

Description
The problem of sweeping mines without endangering men and material has been solved by the new Troika system. The concept is based on using three unmanned minesweepers which are radar-tracked and remotely controlled from a ship-based operations centre, in order to clear specified lanes from acoustic and magnetic influence mines.

The self-propelled minesweepers, designated HFG-F1, are equipped with minesweeping facilities which are accommodated in hollow steel cylinders encased by a ship-like hull. The overall design of the minesweepers is characterised by extremely high shock resistance. Its propulsion system consists of a diesel-driven hydraulic power transmission and a combined rudder propeller. For optimum operation, even in rough seas, an autopilot is provided.

The magnetic sweeping field is generated by two coils, mounted forward and aft on the cylinder. For acoustic minesweeping two medium frequency acoustic hammers and one towed low frequency acoustic displacer are provided.

The Troika system as used by Germany under the navy designation HL 351 uses a converted coastal minesweeper as carrier for the minesweeping operations centre. The system was defined by AEG-Telefunken who also developed the remote-control

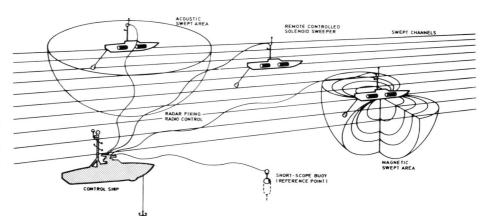

Operating principles of Troika remote minesweeping system

equipment. The latter comprises a horizon-stabilised X/C-band precision navigation radar, a digital target extractor, a master console and three control displays. Each of the latter is associated with one unmanned minesweeper on the one hand, and with a remote action system operator console on the other. The control displays allow the radar image of the swept channel to be greatly magnified so that a high degree of precision is achieved in guiding the unmanned minesweepers along the specified tracks electronically inserted into the radar image.

Control commands are sent to the minesweepers in the form of multiplexed data messages, using a

UHF link, which provides high reliability. Their responses and functions are monitored automatically by an integrated monitoring system. The last main feature is the reference buoy, which provides a geographically stabilised radar image and constitutes a vital component in this Troika configuration.

Operational Status
Delivery since mid-1979.

Contractor
AEG Aktiengesellschaft, Ulm.

GHA

Type
Acoustic minesweeping unit.

Description
The GHA uses the flow of water through a turbine in the body of the vehicle to generate sound for sweeping acoustically sensitive sea mines. The GHA is towed – preferably together with a magnetic sweep – behind a minesweeper and supplies its two electro-dynamic sound generators independently with power

by means of water flowing over its turbines. The beat frequency, sound pressure, operating and pause time, and wobble time of the sound generator are programmable. The beat frequency can be wobbled during towing (that is the beat frequency fluctuates up and down periodically within the adjustment range). Once programmed, the operating mode is independent of the flow speed of the water, and is thus also independent of variable towing speeds. The emission of sound via the electrodynamic system ensures largely wear-free reliability of operation for several thousands of operating hours.

Not requiring a power cable means that the buoy is

easily handled aboard ship, there being no need for generating sets, cable drums, control units and so on and so the system is ideal for use on COOP as well as on purpose-designed minesweepers.

Operational Status
The system is due to enter service with the German Navy following extensive trials.

Contractor
IBAK, Kiel.

DM 19

Type
Explosive minesweeping cutter.

Description
This proven explosive cutter for sweepwire attachment has been in naval service for many years. It uses a linear shaped charge and assures great operational safety even at high sweep speeds and offers a high degree of handling safety. The equipment is lightweight and capable of operation at considerable depths. It is a one-shot system and can be mounted onto sweepwires of different diameters. Neither onboard maintenance nor preparation for use are required.

No residual parts remain on the sweepwire after detonation. The sweepwire is sandwich-shield protected.

Operational Status
In production and used by the former West German Navy and by the Indonesian and Thai navies.

Specifications
Cutting capacity
Stud link chains: 20 mm
Steel wires: 26 mm
Synthetic ropes: 40 mm
Operating depth: 3 to <200 m
Weight in air: 4.4 kg
Weight in water: 2.3 kg
Dimensions: 410 × 470 × 85 mm

Contractor
Rheinmetall GmbH, Dusseldorf.

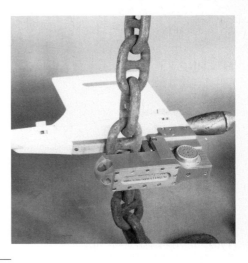

Explosive minesweeping cutter DM 19

166 R

Type
Explosive minesweeping cutter.

Description
This advanced explosive minesweeping cutter complements the DM 19, which it resembles in general appearance, functional principle and operational deployment. The difference lies in the structural separation of the fuze from the shaped charge.

Operational Status
Under development.

Specifications
Cutting capacity
Stud link chains: 20 mm
Steel wires: 26 mm
Synthetic ropes: 40 mm
Operating depth: 5 to <300 m
Weight in air: 4.5 kg
Weight in water: 2.2 kg
Dimensions: 460 × 470 × 85 mm

Contractor
Rheinmetall GmbH, Dusseldorf.

INTERNATIONAL

DOUBLE OROPESA MINESWEEP

Type
Towed minesweeping equipment.

Description
The Double Oropesa Sweep comprises a cable some 200 m in length streamed astern of the minesweeper. At the end of this is a depressor or kite whose weight can be varied to keep the whole system at the required depth. Just above the kite stream two other cables spread out in an inverted Y fashion at the end of which are fitted a spreader or otter board (just as in the fisherman's trawl). The outer positions of these two cables and the otter which form the sweep proper are marked by floats. Attached at suitable intervals to the two spread cables are a number of cutters which operate either mechanically or use an explosive bolt. However, it has been found that not all bolt cutters can handle all types of chain. The German Navy uses a heavy-duty explosive tube cutter capable of severing any type of mine mooring cable.

Sweeping with the Oropesa usually involves a lead ship streaming a Double Oropesa with other minesweepers following *en echelon* astern streaming a single Oropesa and taking up station abeam and inboard of the float streamed by the ship ahead. This ensures that no mines along the intended swept path are missed and also affords each of the minesweepers some measure of immunity, excepting of course the lead ship streaming the Double Oropesa.

The swept path using a Double Oropesa is of the order of 275 m and the maximum depth which can be swept is about 30 m. If the Double Oropesa is streamed as a team sweep between two ships, ocean sweeping depths of 90 m can be achieved with a swept path of 350 m.

Contractors
Various international contractors.

SWEDEN

SAM

Type
Remote-control minesweeping drone.

Description
Developed by Karlskronavarvet, the remote-control SAM uses a diesel-powered rudder propeller to drive the GRP catamaran hull. The craft is fully equipped for power supply, sweeping of acoustic and magnetic mines, navigation and remote-control from a 'Landsort' class MCMV. The MCMV can control up to three of the unmanned sweeps, providing remote-control of the diesel engine, steering, minesweeping equipment, control and monitoring.

The hull of the SAM is fitted with a number of coils which are used for sweeping magnetic mines, and a towed acoustic transmitter is used to counter acoustic mines. A number of buoys are carried to mark the swept passage.

Onboard power is provided by a Volvo Penta TAMD 70D diesel engine with a continuous output of about 159 kW at 2200 rpm, to give the craft a speed of 8 kts.

Power is fed to a Schottel type propeller unit via reversing gear couplings and a shaft. The diesel and reversing gear are mounted on a girder framework which is shock-mounted in an aluminium structure onto the main platform.

Remote-controlled sweeping drones SAM 03 and SAM 05

Although the machinery is normally remotely controlled from the MCMV, it can also be controlled at the engine or from the operating platform. Remote-control functions include diesel engine control, steering control, minesweeping equipment control and monitoring.

Operational Status
In service with the Swedish Navy and the US Navy.

Specifications
Length: 18 m
Beam: 6.2 m
Draught: 1.6 m
Displacement: 19.7 m³
Speed: 8 kts
Range: 330 nm at 7 kts

Contractor
Karlskronavarvet, Karlskrona.

IMAIS

Type
Integrated magnetic and acoustic influence sweep.

Description
The Integrated Magnetic and Acoustic Influence Sweep (IMAIS) incorporates recent advances in technology in the fields of magnetic and acoustic physics as well as signal processing. The sweep is a logical successor to the Combined Influence Sweep which incorporates the magnetic sweep MG 304 and the parallel towed acoustic sweep AS 203.

The integrated sweep comprises the magnetic section, a buoyant cable with three aluminium electrodes, an acoustic generator towed at a constant depth and connected to the buoyant cable at the position of the second electrode, a control unit with keyboard and display, winch system, powerpack for the sweeps – either rectifier or DC-generator, and an optional container for the containerised version.

The buoyant cable's aluminium conductors supply both the electrode and the acoustic generator. The cable also has a cord with a tension load up to 10 kN, allowing sweeping speeds up to approximately 9 kts.

The three-electrode system creates a unique configuration of the magnetic field. It gives a magnetically safe zone (less than 5 nT) of 200 m forward of the first electrode at every depth below the towing ship. It also gives an extensive swept width path against modern mines. With a maximum current amplitude of 700 A (500 A RMS) the swept path is 400 m at a depth of 100 m for mines set at 100 nT.

Compared to earlier sweeps the advantage with this electrode is that the magnetic field can be controlled and automatically kept in a constant and predicted configuration, independent of variations of the water conductivity.

The acoustic sweep is shock resistant (k . 1.0) and has a maximum swept depth of 200 m. Its acoustic generator has a length of 2 m, a diameter of 0.7 m and weighs about 600 kg. It can be programmed for broadband noise and tones from a few Hz and upwards.

The extremely low drag resistance of the towed sweep (less than 10 kN at 6 kts) requires very little propulsion power from the towing ship, thereby generating very little noise disturbance and so increasing the safety of the towing ship.

The total weight of the integrated influence sweep, with streamed as well as ship-mounted equipment, is approximately 2800 kg which makes it suitable for minehunters and small minesweepers as well as converted fishing vessels or other COOP. The system can also be supplied in a containerised version, if requested.

The sweep is very easy to handle and has a high manoeuvring capability in narrow waters because of its small turning radius.

The magnetic and acoustic sweep can be delivered separately.

Contractor
SA Marine, Landskrona.

EXPLOSIVE CUTTER

Type
Explosive minesweeping cutter.

Description
This minesweeping cutter is designed to be integrated with any mechanical sweep system for cutting mooring wires of anchored mines and to eliminate sweep hindering devices. To avoid activating magnetic mines the cutter is constructed of non-magnetic stainless steel and plastic. The cutter is shaped to automatically take up and maintain its ready position, regardless of whether the sweep is carried on the starboard or port side of the ship.

The cutter is a reloadable one shot type. On firing, the explosive device with jaws is blown away, but the fin is retained on the sweep wire and can be fitted with a new explosive device.

The cutter weighs about 7 kg and is fitted with 150 g of TNT+RDX or TNT explosive. It can cut conventional chain with a diameter of 12-14 mm, stainless steel rod of 19 mm diameter, steel rod of 20 mm diameter and mild steel rod of 29 mm diameter. Sweep speed is 4 to 12 kts.

Operational Status
Known to be in service with the Royal Navy aboard 'River' class minesweepers.

Contractor
SA Marine, Landskrona.

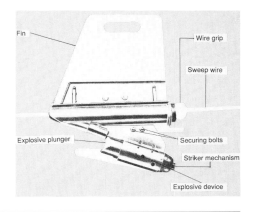

The Explosive Cutter

UNITED KINGDOM

WIRE SWEEP Mk 105

Type
Compact, low magnetic wire sweep system.

Description
The Mk 105 sweep configuration is a development of several Royal Navy wire sweep systems and is available in a range of sizes (including US size 4) to enable a range of naval vessels to undertake minesweeping operations. The smaller variants of the sweeps may be fitted to small patrol craft or minehunting vessels to provide a dual role mine countermeasures capability. The larger variants are generally fitted to dedicated minesweepers or larger Craft of Opportunity where the space and power available is sufficient for the larger equipment.

The design uses a simplified wire configuration enabling relatively small winches to be used to save deck space and in smaller sizes to reduce towing loads. The Mk 105 system may be deployed as either Team, Double Oropesa or Single Oropesa Sweeps and includes all items of minesweeping equipment necessary to fit out a vessel for minesweeping including the wires, winch, floats, cutters, kites and marker buoys. The complete systems are available in either

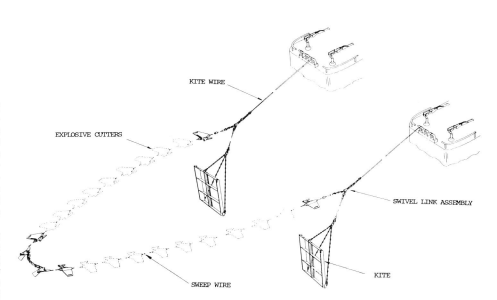

Team Sweep with Wire Sweep Mk 105

low magnetic or ferromagnetic materials. The system is used in conjunction with the BAJ WSME system to achieve good bottom following at depths down to 300 m.

The Mk 105 sweep configuration is a very flexible system that has already been configured to suit several ship designs including minehunters, fast patrol craft, minesweepers and Craft of Opportunity.

Contractor
BAJ Limited (a subsidiary of Meggitt plc), Weston-super-Mare.

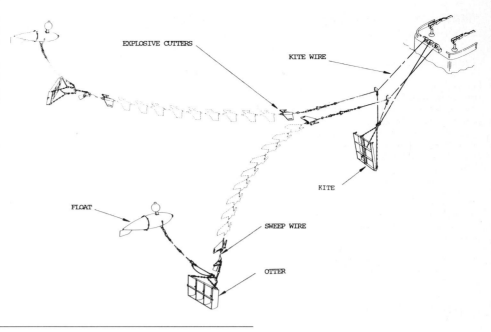

Double Oropesa Sweep with Wire Sweep Mk 105

WIRE SWEEP Mk 8

Type
Low magnetic signature wire sweep system.

Description
The Mk 8 wire sweep system is a development of the wire sweep Mk 3 mod 2 used on the 'Ton' class minesweepers. It has been designed to have a low magnetic signature in order to be compatible with the overall low magnetic characteristics of modern mine countermeasure vessels.

The Mk 8 is designed to sweep most types of buoyant mines that are moored to the seabed with sinkers by severing the mine mooring chain or wires with explosive cutters fitted to the sweep wires. The Mk 8 may be deployed in the following configurations: Single Oropesa, Double Oropesa and Team Sweep.

Operational Status
The Mk 8 system is in service on the Royal Navy 'Hunt' class mine countermeasure vessels.

Contractor
BAJ Limited (a subsidiary of Meggitt plc), Weston-super-Mare.

WIRE SWEEP Mk 9

Type
Controlled, deep wire sweep system.

Development
With the advent of mines that could be laid in much deeper water and mines with much shorter mooring wires, BAJ (who has been principal contractor to the UK MoD since the early 1970s for the development of wiresweeping equipment), together with the Admiralty Underwater Weapons Establishment (AUWE), began in 1974 the development of a sweep system that could be worked much closer to the seabed and at far greater depths. To operate such a system efficiently the minesweeper would have to be able to control both the height of the sweep wire above the seabed and its flatness, so that it does not drag along the seabed and break or prematurely explode the wire cutters connected to it.

It was decided to adapt the existing UK Mk 3 wire sweep system, but use a single wire with the sweep wire attached directly to the end of the kite wire. The kite would then only have to depress one wire instead of two, and allow the sweep to go much deeper with greater control.

Extensive use of computer modelling was followed by lengthy trials, which enabled tables to be drawn up comparing swept depth against length of kite wire and when used in conjunction with Wire Sweep Monitoring Equipment (WSME) it proved possible to heave or veer the kite wire to the exact amount required for the sweep wire to follow the contour of the seabed.

Following extensive trials the system was accepted into RN service in the early 1980s.

Description
Wire Sweep Mk 9 is an effective combat-proven countermeasure against buoyant moored mines in deep waters. The sweep's primary operational mode is Team sweeping but it is also deployed as a double or single Oropesa sweep. It is designed to be

Wire Sweep Mk 9

deployed from either purpose-designed minesweeping vessels or ships taken up from trade – usually stern trawlers. Although usually operated with two ships, several ships have been linked together to form a multiship team sweep providing greater swept paths and allowing an odd number of vessels to be used.

Each vessel carries three lengths of wire on a winch. The outer barrels carry a sweep wire connected to a kite wire and the centre barrel carries a kite wire for Oropesa sweeping.

Using WSME in conjunction with the ship's echo sounder, the sweep can be adjusted to follow the seabed at a constant clearance by heaving or veering kite wire from the winch.

Operational Status
Wire Sweep Mk 9 is combat proven and has been in operation with the Royal Navy since the early 1980s on the 'River' class vessels. This system has also been delivered to Canada for the Canadian Government Naval Reserves Mine Countermeasures Project and is now operational.

Contractor
BAJ Limited (a subsidiary of Meggitt plc), Weston-super-Mare.

WIRE SWEEP MONITORING EQUIPMENT

Type
Wire Sweep Monitoring Equipment (WSME).

Description
WSME was developed primarily for use as part of the UK Wire Sweep Mk 9 to improve its accuracy of sweeping. WSME is fitted on purpose-built MCM vessels and vessels taken up from trade to allow effective minesweeping operations. The purpose of WSME is to provide a flat sweep profile at a steady seabed clearance.

For effective operations against all types of moored mines, it is important to tow the sweep wires just above the surface of the seabed otherwise mines on short moorings will not be swept. To achieve this position the sweep wire must be flat, if the wire sags the sweep will contact the bottom and damage the sweep, if the wire hogs it may miss some mines.

Previous methods of minesweeping used vessel speed to indicate sweep flatness but the errors due to ships' logs, tidal flows and current profiles often lead to sweeps with significant hog or sag.

Sea trials have established that most of the tension measured as the wire passed through the fairlead on the ship to the water could be attributed to the drag of the sweep in the water and the forces acting on the kite. From measurement of this tension the speed of the sweep could therefore be calculated, and hence its flatness. This led to the development of the Wire Sweep Monitoring Equipment (WSME) which displays digital readings of average and peak load on the Sweep Monitor Console. The ship speed is then either manually or automatically adjusted to obtain the correct tow tension.

WSME also electronically measures the length of

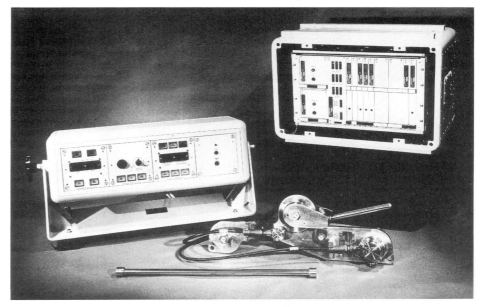

Wire sweep monitoring equipment

kite wire passed through the tension meter to determine the depth of the sweep wire and displays a depth reading which can be directly compared with the digital output of the ship's echo sounder. The winch is then either manually or automatically heaved or veered to maintain the required seabed clearance.

The WSME is also available with sweep location equipment for monitoring actual ground covered by the sweep. Links to the ship's command and control system are also provided.

Operational Status
The WSME has been in service on the 'River' class minesweepers since 1984. In 1989 systems were ordered for the Canadian Government's Naval Reserve Mine Countermeasures Project and the systems are now operational.

Contractor
BAJ Limited (a subsidiary of Meggitt plc), Weston-super-Mare.

MMIMS

Type
Modular multi-influence minesweeping system.

Development
MMIMS has been developed by Marconi Radar and Control Systems in collaboration with the Defence Research Agency at Portland.

Description
This modular system is designed to duplicate four elements of a ship's signature; magnetic, acoustic, underwater electric potential and extremely low frequency electromagnetics, to counter the modern intelligent mine. Design flexibility ensures a system tailored to customer requirement. The sweep is normally configured as either a surface or sub-surface towed array and this, given low overall electrical power consumption and ease of programming from a central controller, allows operation either by MCMV or craft taken up from trade. Given the accuracy with which a ship's signature can be duplicated this sweep is an effective counter to the modern intelligent mine threat.

Prototype MMIMS minesweeping system on trial

Contractor
Marconi Radar and Control Systems, Leicester.

MSSA Mk 1

Type
Acoustic minesweeping system.

Description
MSSA Mk 1 is an advanced acoustic minesweeping system which can generate a wide range of target-like acoustic signatures. Its acoustic output is continually monitored and controlled to ensure that the required amplitude at seabed level is maintained regardless of variations in acoustic propagation conditions. It provides a versatile sweeping system capable of very high power output. It is used to activate all acoustic mines, including those with frequency selective triggering characteristics which are designed to be actuated only by certain types or classes of vessels.

The system comprises a Towed Acoustic Generator (TAG), a Towed Acoustic Monitor (TAM) with a towed hydrophone array, and an onboard control console.

The system has been specially designed to withstand the repeated levels of explosive shock likely to be experienced in operation, and is normally deployed in association with a magnetic sweep to provide a very effective method of sweeping combined influence mines.

Operational Status
MSSA Mk 1 is in service with the Royal Navy on the 'Hunt' class MCMV, and has undergone evaluation trials with the US Navy.

Towed Acoustic Monitor (centre) and Towed Acoustic Generator (left)

Towed Acoustic Monitor of MSSA Mk 1 minesweeping system

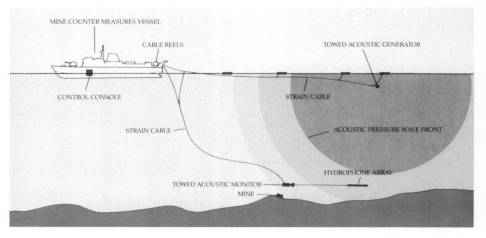

Contractor
BAeSema Marine Division, Bristol.

CIS

Type
Combined influence minesweeping system.

Description
By combining the MSSA Mk 1 TAG with an efficient magnetic sweep BAeSema offers a Combined Influence Sweep (CIS) system which is capable of sweeping the most modern microprocessor-controlled mines. In the CIS both the acoustic and magnetic signatures are controlled by an industry standard PC, giving complete and total control of the generated signature. Both Target Setting Mode and Mine Setting Mode techniques may readily be employed with CIS.

Operational Status
In service with the Royal Navy.

Contractor
BAeSema Marine Division, Bristol.

MINESWEEP POSITIONING AND MONITORING SYSTEM

Type
Monitoring system.

Description
The Minesweep Positioning and Monitoring System is a development from the Simrad Integrated Trawl Instrumentation System, ITI.

It comprises a number of sensors mounted at key locations on the sweep, a transducer mounted on the ship's hull and electronic units mounted on the bridge. The system is designed to interface with standard navigation systems such as gyro-compass, log, sonar and echo sounder.

The Positioning and Monitoring System is capable of being used with either Team Sweep (using two vessels) or Oropesa (single vessel) configurations. In the Team Sweep configuration the system can measure depth and altitude of the depressors, the distance between depressors, depth and altitude of the centre of the sweep, the depth and altitude of other selected points of the sweep and the range and bearing from the guide ship to the depressors and centre of the sweep.

Contractor
Simrad Hydrospace, Bordon.

MINENAV

Type
Acoustic minesweep monitoring system.

Description
The system comprises a number of sensors used to measure the spread of the sweep using transponders, the swept depth using pressure transducers and sweep height using echo sounders, all of which are placed on the sweep wire. Subsea sensors are protected by an impact-resistant plastic outer layer and stainless steel inner cylinder. The receiver is deployed on the wire between the vessel and the sweep kite and is mounted in an attachment frame prior to deployment and suspended from a strain relieving cable, up which received signals are transmitted to the sweep vessel using an FSK acoustic link. It is protected against shock using four rubber universal joints. Computer processing allows graphic display of the information to provide the sweep officer with up-to-the-minute performance data. Data are logged to disk for further analysis.

Operational Status
The system has been used annually for the past three years by the UK MoD DGUW (Director General Underwater Weapons) for trials of minesweeping equipment.

Contractor
Seametrix, Aberdeen.

MINENAV system used in Team Sweep system

LIMPET MINE DISPOSAL EQUIPMENT (LMDE)

Type
Mine disposal equipment.

Description
The LMDE (Limpet Mine Disposal Equipment) has been designed to destroy limpet mines which are difficult to remove from ships because of their powerful magnets and their anti-handling devices.

The system features a stainless steel barrel bomb disruption device similar to those used against terrorist bombs on land. The barrel is mounted in a movable sleeve and trunnion on a stand with flotation magnets to attach it to the ship's hull, and the whole device, which weighs less than 15 kg, has buoyancy blocks to reduce the effective weight in water.

When a limpet mine is discovered the diver assembles the LMDE out of the water, then takes it

Limpet Mine Disposal Equipment

underwater and places it as close as possible to the mine. The barrel is carefully positioned so it faces the heart of the mine, then the diver returns to the surface from where he remotely fires a bridgewire cap cartridge. This propels a powerful slug of water at the mine, smashing open the thin casing and dispersing the main explosive charge.

Operational Status

The system has been purchased by five customers.

Contractor

A B Precision (Poole) Ltd, Poole.

UNITED STATES OF AMERICA

Mk 105 AIRBORNE MINESWEEPING SYSTEM

Type
Magnetic/acoustic influence minesweeping system.

Development
EDO Corporation has developed a highly mobile, helicopter-towed, unmanned MCM system that provides high speed magnetic mine clearance and reduces the risks to personnel inherent in minesweeping operations. The Mk 105 is designed to be towed by the US Navy's MH-53E Sea Dragon helicopter and other heavy lift helicopters of similar performance and size. Minesweeping operations may be either land-based or launched from various amphibious assault ships such as an LPH or LHA.

Description
The Mk 105 Airborne Minesweeping System is a helicopter-towed system carried on a hydrofoil that uses a gas turbine engine generator set (powerpack) to provide electrical power to an array of buoyant sweep cables. The electrical power is programmed and remotely controlled from the helicopter to simulate the magnetic signature of various ships. Magnetic influence mines are detonated by the magnetic field generated by the system. The Mk 105 is composed of a hydrofoil sled, a control programmer that is mounted in the towing helicopter, a tow cable and an array of sweep cables.

The sled consists of a seaborne equipment platform and a generator set. The platform includes two tubular floats of light, metal-alloy construction, connected by an aerofoil wing assembly. Cavitating hydrofoils are attached to the forward and aft sections of each float. The surface piercing tandem foil configuration features two inverted V-shaped foils forward and two aft for balance. High riding pitch control subfoils of similar configuration are located ahead of the two bow foils. For landing and ground manoeuvres, the foils may be rotated up and parallel to the floats by a self-contained hydraulic system. Wheels are attached to the underside of the floats to facilitate ground or deck handling, and a retrieval system is mounted on the platform for helicopter recovery operations. Fuel for the turbine is carried in two centrally located tanks, one in each float. Tow cable attaching points are provided on the forward inward faces of each float. Once launched, the sled is towed by a 1237.16 m tow cable that is capable of carrying both electronic signals and fuel to the gas turbine engine generator set.

Mounted on the wing assembly of the platform is the generator set that includes a gas turbine driven AC generator, a rectifier, a controller containing the waterborne electronics and batteries to power the electronic system. The generator set provides energy for the generation of the magnetic field. The sweep cables are attached to the aft end of the sweep boom,

Mk 105 sled foilborne

which is located on the centreline of the sled, aft of and below the wing assembly. The sweep boom comprises an upper electrode attached to the end of a trailing cable and a lower electrode fitted to the boom fin. Using the water as a conductor, the potential between the electrodes produces a magnetic field that simulates the magnetic signature of a ship.

The control programmer, located in the helicopter,

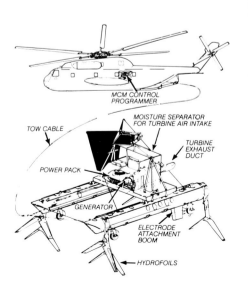

Schematic of Mk 105 system

is the only manned station employed in the system. It contains the airborne electronics and all the controls and instrumentation necessary for operating the Mk 105. The control programmer console contains the fuel transfer control panel, turbine indicators, hydrofoil and boom actuators and the generator controls and indicators. From the control programmer, the operator can start and stop the turbine, raise and lower the foils and control the magnetic influences generated through the sweep cables trailing behind the sled.

The Mk 106 configuration, also towed by the MH-53E, incorporates the Mk 105 magnetic influence sweep and the Mk 104 acoustic influence sweep. The Mk 104 is comprised of a venturi tube and a water-activated turbine that rotates a disc to reproduce a ship's acoustic signature.

Specifications
Length overall: 27 ft 6 in
Beam
    Catamaran structure only: 11 ft 7 in
    Across foils: 21 ft
Height, foils extended
    To top of retrieval rig: 17 ft 3 in
    To top of nacelle: 13 ft 6 in
Height, foils retracted, to base of wheels: 11 ft 6 in
Weights
    Empty: 5522 lb
    Gross: 6432 lb

Contractor
EDO Corporation, Government Systems Division, College Point, New York.

DIVERS' SYSTEMS

This section covers all types of equipment associated with divers and their operations. Diving covers a vast field of activity ranging from sabotage, beach reconnaissance, disarming and removal of unexploded ordnance, to salvage and the inspection of underwater defences and so on. To support these varied underwater operations a wide range of equipment including hand-held sonars, underwater telephones, communications equipment and cameras has been developed, some of which are described here. Future editions of *Jane's Underwater Warfare Systems* will cover more of the equipment and systems used in these underwater operations.

DIVING EQUIPMENT

CANADA

DIVER'S SONAR

Type
Hand-held diver's sonar.

Description
Orcatron has developed an active hand-held diver's sonar equipment which operates down to a depth of approximately 76 m and at ranges of up to 60 m. Transmitter frequency is 200 kHz with a power output of 12 W and a beamwidth of 12°. All frequency and timing information is derived from crystal-controlled digital circuitry for maximum reliability. The built-in display is a three-digit LED type for bright, easy-to-read operation. The display range is selectable for metres or feet by means of an internal selector switch. The equipment is battery-operated by means of a 12 V 0.5 Ah NiCad battery with an estimated life of five hours.

Contractor
Orcatron Manufacturing Ltd, Port Coquitlam, British Columbia.

SCUBAPHONE

Type
Diver underwater communications set.

Description
The Scubaphone wireless underwater communication system is designed for both military and commercial use. It employs crystal-controlled single sideband suppressed carrier ultrasonic techniques. Special filters in the system process the incoming signal to remove background sounds such as the bubble noise caused by the exhalation of the diver.

With the Scubaphone, diver-to-diver communication is independent of the surface unit. Communication from the diver can be performed over ranges up to 1200 m in salt water, while that from the surface ranges up to 3600 m. The diver's microphone and earphones are water- and pressure-proof to 76 m depth in salt water. The set is powered by a 12 V rechargeable battery.

Operational Status
In service.

Specifications
Frequency: 30 kHz
Diver phone
Acoustic power output: 1 W
Audio output: 1 W RMS
Surface phone
Acoustic power output: 3 W
Audio output: 2 W RMS

Contractor
Orcatron Manufacturing Ltd, Port Coquitlam, British Columbia.

Scubaphone diver communications set

SUBPHONE

Type
Submersible communications system.

Description
The Subphone is a dual frequency, single sideband, medium power, through-water communications system. It is a rack-mounted equipment for use in one atmosphere submersibles and can be mounted in a carrying case for surface support use.

The Subphone provides dual frequency operation for both voice and internal pinger. It is intended to provide high quality communications through salt water at nominal ranges up to 10 km.

Subphone is also used as the surface station for the SDU (Special Diver's Unit) which allows clear communications between divers at nominal ranges of 10 km in salt water.

Operational Status
In service.

Specifications
Frequencies: Crystal-controlled user specified: 10, 27 and 30 kHz USB standard.
Transmitter power: 50 W (0.5 W selectable)
Pinger: 15 m pulse/s
Nominal range: 6 km at 27 kHz; 10 km at 10 kHz

Contractor
Orcatron Manufacturing Ltd, Port Coquitlam, British Columbia.

Subphone diver communications set

SPECIAL DIVER'S UNIT

Type
Underwater communications system for divers.

Description
The Special Diver's Unit (SDU) is designed for operations that require long-range voice communications between divers and support vessels. The SDU allows clear voice transmission at a nominal range of 10 km in salt water. It is normally used in conjunction with the Subphone (see separate entry).

Operational Status
In production.

Contractor
Orcatron Manufacturing Ltd, Port Coquitlam, British Columbia.

Special diver's unit

MODEL 1080 COMMUNICATIONS SYSTEM

Type
Diver communication system.

Description
The Model 1080 is a high performance, belt-mounted, voice activated acoustic communications system for use by divers. It operates on a USB frequency of 28.5 kHz, with other frequencies available. Transmitter power is 5 W RMS and the nominal range is 3 km. Depth rating is 76 m.

Contractor
Orcatron Manufacturing Ltd, Port Coquitlam, British Columbia.

VIDEO CAMERA HOUSINGS

Description
A variety of high quality underwater video camera housings enables video cameras to be converted from hand-held models into units capable of operating at depths down to 100 m by plugging the housing to an underwater pluggable umbilical. This feature, coupled with high quality images and low lux capacity (low light required) make the units suitable for use in such diverse applications as reconnaissance, surveillance, the inspection of ships' hulls and so on.

Standard features on the housings include external amphibious microphone and fingertip electronic controls which give the diver access to auto and manual focus, white balance, lens focus control and zoom control as well as record and standby modes.

Operational Status
Units are in service with the US Navy SEALS and submarines and the Royal New Zealand Navy.

Contractor
Amphibico Inc, Dorval, Quebec.

SIVA

Type
Diver rebreather apparatus.

Development
The Siva rebreather apparatus has been developed to provide divers with both shallow water oxygen and deep water mixed gas capability.

Description
The unit features low acoustic signature and with an available non-magnetic form the system is suitable for mine countermeasures duties as well as weapon recovery and other military tasks where minimal signature is important.

The modular Siva system accommodates two interchangeable gas control modules which offer the diver depth adaptability. The '55' gas control module permits the apparatus to be used with oxygen or pre-mixed gases to a depth of 55 m. These are metered to the breathing circuit by a unique mass flow selector and control mechanism. The circuit for the control of pre-mixed gases features four adjustable jets which can be calibrated by the user or supplied factory set to suit the mixtures required for various depths down to 55 m. Once set, the user has the freedom to switch mixtures between dives without recalibrating the unit. In the self-contained mode Siva has an endurance of up to 180 minutes. The '+' gas control module delivers either 100 per cent oxygen or the dynamic mixing of oxygen and diluent gas using a pneumatic control system for dives exceeding 80 m in depth. These depths are made possible by significant gas conservation achieved through the control of the oxygen partial pressure. In the self-contained mode Siva + provides enough gas and CO_2 removal capacity in 0°C water to support a working diver to a depth of 80 m. This includes a 20 min bottom time and allows for in-water decompression. The maximum depth attainable is limited only by the quantity of stored gases.

The use of surface-supplied gases from an optional umbilical permits extended bottom time by employing Siva as a lightweight surface-supplied apparatus or by providing gas for in-water decompression following a free swimming dive. The design and the options extend depth and dive time to expand Siva's operational capabilities.

The ergonomically designed counterlung and breathing loop minimises flow resistance and provides for more comfortable breathing. The counterlung is positioned on the diver to reduce hydrostatic effects while optimising swimming actions. A diver-adjustable valve is employed to establish the best breathing characteristics in all positions. The lung is made of a tough, abrasive-resistant, reinforced polyurethane that will not mildew or rot and is unaffected by ultraviolet light. It comes fitted with integral and jettisonable weights and functions as a flotation device, maintaining a diver face up on the surface.

Specifications
Length: 58 cm
Width: 33 cm
Depth: 17 cm (excluding breathing bag)
Weight: 30 kg (in air)
Gas storage volume: 1177 l
Breathing bag capacity: 8.0 l
Endurance: 3 h (min) at moderate heavy work
Charging pressure: 297 bar

Contractor
Fullerton, Sherwood Engineering Ltd, Mississauga, Ontario.

Siva rebreather apparatus

FRANCE

ERUS-2

Type
Underwater radio communication system.

Description
ERUS-2 equipment is designed to provide underwater radio communication, using HF ultrasonic waves, between divers, divers and surface craft, and between surface craft.

Transmission is omnidirectional. The average range in isothermal conditions is 800 m and down to depths of about 100 m.

Two equipment models have been developed: the ERUS-2-A3 for divers and the ERUS-2-B4 for operation on a surface craft. The 2-A3 equipment includes an enclosed transceiver unit, a mask with a built-in microphone and an earphone. The transceiver unit is provided with one single operation control and a push-to-talk handle drawn back automatically to reception position.

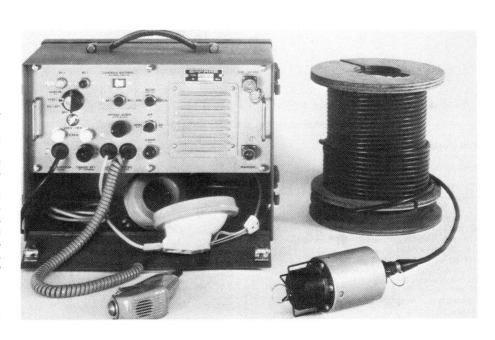

ERUS-2-B4 surface craft equipment

The 2-B4 equipment includes a portable case containing the transceiver unit, the loudspeaker, the 9.6 V storage battery and a battery charger (available also to load the battery of the 2-A3 equipment). The equipment, provided with various adjustment and control features, allows the transmission of modulated telegraphic signals. A hand microphone with a push-to-talk control handle and headphones are housed in a compartment provided in the case.

A 50 m connecting cable is available with a transducer housed inside the cable reel.

Operational Status
Tested by the US Navy.

Contractor
Safare Crouzet, Nice.

RUPG-1A

Type
Diver's homing receiver.

Description
The RUPG-1A receiver is a portable watertight equipment which allows divers either to home towards a submerged transmitter or to hear CW or phone communications from this transmitter.

Received signals are amplified by the equipment, which supplies the audio signal to a bone-conducting earphone. Directivity of the receiver informs the diver of the direction of the transmitter.

The shape of the equipment is similar to that of a portable torch. Metal parts are made out of anodised aluminium alloy to withstand corrosion. A battery charger, designed for the RUPG-1A's storage batteries, can be operated from a 115 to 220 V 50/60 Hz mains or from a 24 V DC supply.

Operational Status
In production for the French Navy.

Specifications
Range: 2500 m in normal propagation conditions with approx 10 W transmitter
Directivity: main lobe angle between ±10° and ±15° for 6 dB attenuation
Power supply: 9 V storage battery 15-20 h continuous operation
Pressure test: 10 bars
Max outer diameter: 191 mm
Overall length: 560 mm
Weight: 2.75 kg (in air)

Contractor
Safare Crouzet, Nice.

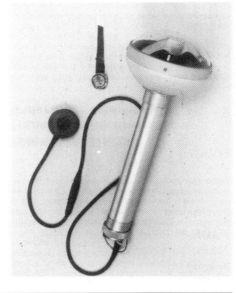
RUPG-1A diver's homing receiver

GERMANY

HAUX-STARCOM

Type
Recompression chambers.

Description
This is a series of compact, modular diver recompression chambers for the treatment of divers, surface decompression and the training of divers. By incorporating the flat bottom principle and front-mounted control panel, chambers can be constructed of varying diameters and the complete utilisation of the entire inner length of the cylinder wall with optimum space being provided. With minimised outer measurements the chambers can be easily transported and with front-mounted control panels they are especially suited for installation in containers, ships or lorries. The front panel houses all the instruments required for control and with an adjacent observation window, supervision of the chamber occupants can be effected from a single point.

The aluminium alloy double-hinged doors are of pancake design and because of their symmetrical design can be operated from both sides.

The use of new Haux-Starvalve and Haux-Ventmaster valves and the integrated electronic Haux-Variomaster for the supervision of pressure changes offer a completely new way of supervising pressure changes enabling precise control over the build-up and maintenance of correct pressures.

With the use of the Haux-Scrubmaster IIIS CO_2 absorber a maximum operating time with a maximum gas consumption can be achieved. The system is low noise in every phase of the operation and Haux-

Haux-Starcom in ISO container on ship's deck

Phonkillers (inside and out) ensure minimal A-sound level. Newly developed Haux-Luxmaster/85D cold-light lamps provide the chamber with anti-dazzle illumination which can be gradually dimmed.

Contractor
Haux-Life-Support GmbH, Karlsbad.

HAUX-SUPER-TRANSSTAR

Type
Transportable one-man diver recompression chamber.

Description
The Haux-Super-Transstar transportable one-man pressure chamber is designed for emergency rescue operations by navy, police, rescue organisations, diving companies and so on. Due to its compact design and low weight, the chamber is easily transportable on lorries, ships and helicopters and can be carried by four people to the site of an emergency.

The pressure chamber is fitted with a standard bayonet flange enabling it to be connected to a treatment chamber.

For short term operations, for example transfer of a patient to another transport system, mating with a treatment chamber and so on, the Haux-Super-Transstar is supplied with its own gas storage and supply systems. The cylindrical design of the pressure chamber in the region of the diver's trunk, and conical design in the leg region provides the occupant with the maximum amount of space. In the chamber the diver lies on a wide, comfortable stretcher with fabric covering and head rest.

The compact control panel is arranged with the observation window adjacent, enabling the operator to maintain permanent visual contact with the occupant. A precision manometer, semi-automatic fresh air ventilation, inlet and outlet valves for pressure control and an intercom system ensure a very high safety level in chamber operation.

Contractor
Haux-Life-Support GmbH, Karlsbad.

Haux-Super-Transstar. Note control panel with observation window to left

HAUX-PROFI-MEDICOM

Type
Two-man diver recompression chamber.

Description
This unit is designed for diver rescue operations by navy, police, or diving companies for example. With its two occupants the chamber can be carried by six to eight people and is easily transportable. The newly developed structural shape allows the chamber to mate with any treatment chamber of any diameter equipped with a DIN/STANAG flange.

For short periods of operation the chamber is fitted with its own gas supply. The inside of the chamber provides the optimum amount of space for the diver who lies on a wide, comfortable stretcher covered with a hard-wearing canvas cloth. The assistant sits upright in the dome of the chamber.

The compact control panel and observation window are sited by each other enabling the operator to view the occupants. The chamber is equipped with a precision manometer, semi-automatic fresh air ventilation and communication system, which includes a battery-powered talkback speaker with volume control. The Medicom is supplied with a newly developed trolley which is easily manoeuvred into position by two people to mate with a treatment chamber.

Contractor
Haux-Life-Support GmbH, Karlsbad.

Haux-Profi-Medicom chamber on trolley ready to mate with large recompression chamber

HAUX-SPACESTAR

Type
Amagnetic recompression chamber for MCMVs.

Description
The Haux-Spacestar recompression chamber is constructed of amagnetic stainless steel for the first aid and treatment of divers on board MCMVs. The unit can also be used as a test and training facility.

The design features a dome mounted on the cylindrical pressure vessel which allows an occupant to stand upright inside the chamber without the need for increased floor space. This feature greatly improves the treatment facilities which medical assistants can provide to injured divers. With neurological problems it is absolutely imperative that the injured diver can stand upright for diagnostic treatment. Increased freedom to move is desirable for the medical assistant and for the practice of a number of life-saving methods. For long-lasting treatment the ability to stand upright improves the physical and psychological

condition of the chamber's occupants. Finally, examination of eyes and ears is much easier if patient and medical assistant are able to face each other.

The Haux-Spacestar consists of a main chamber and an antechamber. Under normal conditions accommodation is provided for two sitting and one lying occupant in the main chamber and one (two) occupants in the antechamber. Pressurisation and fresh air ventilation are achieved using atmospheric air. The diagonal airflow in the chamber provides fresh air for the dome as well as the rest of the chamber. In order to improve decompression treatment, up to three oxygen breathing masks can be connected to a central point of attachment. Should pressurised gas storage begin to run low in long-lasting decompression treatments, safe operation can be assured for a number of days in the closed-circuit operation with the Haux-Scrubmaster CO_2 absorption unit and the addition of a small amount of oxygen from the supply.

The unit is equipped with an intercom system enabling communication between the main chamber, antechamber and control station. Three windows are

fitted in the dome and the cylindrical part of the chamber enabling continual observation of the occupants.

A Masterlock supply lock welded into the main chamber enables the occupants to be supplied with drugs, food and drink without disturbing decompression in the main chamber.

For installation on MCMVs recompression chambers have to meet special requirements, and the Haux flat bottom design offering maximum economy of space and compact design with laterally sited operation centre enables the unit to meet these stringent requirements.

Operational Status
A total of 10 of these chambers will be built to equip the new 'MJ 332' class minehunters of the German Navy.

Contractor
Haux-Life-Support GmbH, Karlsbad.

UNITED KINGDOM

D100 DIVER'S RANGING UNIT

Type
Diver's echo sounder sonar for target location.

Description
The D100 hand-held diver's ranging unit is intended for both naval and commercial diving applications and operating on a frequency of 500 kHz, allows ranging and target location up to 100 m range. It locks on to the strongest target on its bearing. A unique LED dot matrix display gives the equipment a sonar capability by allowing the diver to scan the target zone manually. This displays multi-target data present in a sector, dependent on the rate of manual scan, clearly indicating the range of the target. The rear of the unit contains both a digital display of range in either metres or feet, and a unique rolling display to give a graphic image of the surroundings.

The equipment is manufactured from low magnetic signature materials and is internally powered by two PP3 batteries. The handle base has a fitting for use mounted on a tripod.

Operational Status
In service.

Specifications
Range: 100 m
Resolution: 1 cm
Beam width: 4° conical
Depth rating: 100 m
Weight: 1.3 kg (in air); 0.32 kg (in water)

Contractor
Ulvertech, Ulverston.

D100 scanning sonar head

MODEL 6410B DIVER'S SONAR

Type
Hand-held diver's sonar.

Description
The Model 6410B is a small active sonar unit designed primarily for commercial applications that is also suitable for naval use in the detection of underwater targets. It consists of a transmitter, time-variant gain receiver and electronics which measure the transmission and reception time and calculate the range. A bright LED display gives a readout of the range.

Specifications
Range: 0-10 m; 0-100 m
Operating depth: 305 m
Frequency: 200 kHz
Beam angle: 12°
Weight: 2.4 kg (in air); 0.6 kg (in water)

Contractor
Helle Engineering, Aberdeen.

SEA PIPER

Type
Diver communications system.

Description
The Marconi diver communications system is fully modular and provides underwater communications between divers, between divers and a control platform and between a submersible and surface vessel. The equipment caters for divers working with compressed air or oxy-helium breathing apparatus, and provides line or through-water communications as required. A new design of helium speech converter, which gives increased clarity of speech together with simplified control, is a feature of the system.

In the tethered mode of operation, a four-wire neutrally buoyant cable is used. This doubles as a safety line and gives duplex or round-robin communications which, for divers using oxy-helium mixtures, feeds back an unscrambled version of the diver's own speech.

The through-water mode enables operators to communicate with one or more divers via a simplex link. This facility provides communications in the simplex mode between the diving tender, submersible and divers. In this mode, communications between two or more divers can be achieved without resorting to surface or submersible-carried equipment.

All parts of the system are designed for adverse environments. The control unit can be supplied as a backpack. The diver's module, constructed from plastic, is pressure-proofed to 100 m. The control unit carries its own internal rechargeable battery which has eight hours' duration but, in addition, external power from a ship's 115/240 V AC or 12 V DC supply can be used to power the control unit.

Sea Piper diver communications system

Operational Status
Developed under contract from the UK Ministry of Defence with 150 units purchased by the Royal Navy; also in use with Qatari forces, the Netherlands, Spain and other special forces.

Contractor
Marconi Underwater Systems Ltd, Waterlooville.

SUBCOM 3400 COMMUNICATIONS SYSTEM

Type
Emergency communications system.

Description
Subcom 3400 is an emergency communications system designed primarily for backup to the umbilical wired communications generally used in diving bells. It is a through-water acoustic system which does not rely on the integrity of the bell's umbilical, but operates using the water column between two transducers. With its own battery pack it will operate for a minimum of 20 hours under normal conditions.

Subcom 3400 incorporates a highly reliable single sideband suppressed carrier system.

Specifications
Carrier frequency: 25 kHz ±0.1%
Transmission frequency: 25.3–29.5 kHz

Transmit power: 10 W
Audio power: 1 W

Contractor
Marconi udi (a subsidiary of Marconi Underwater Systems Ltd), Aberdeen.

EL050(2056), EL057, EL067 SERIES

Type
Deep-phone underwater telephones.

Description
These multi-frequency units provide voice communication and pinger indication in the military submarine and diver bands. They are employed as both portable

2056 unit

and installed units in submarines, surface ships and submersibles from helicopters, and as secondary/emergency units in submarines. They are also used for parachute delivery with SUBSUNK emergency assistance teams and are installed in the Royal Navy's crew rescue KR5 submersible.

The units provide diver-to-diver and diver-to-surface communication over ranges of 1000 m. The diver's USB station operates at a carrier frequency of 40.2 or 42 kHz with a bandwidth of 3400 Hz. It measures 78 × 300 mm (excluding helmet or hood microphone and earphones) and its weight in water is 800 g. An on/off switch is water-activated. Powered by nickel/cadmium batteries, the unit provides eight hour operation at a 10 per cent transmit cycle. It is rated to a depth of 450 m.

The surface unit measures 200 × 300 × 280 mm (excluding handset) and weighs 8.65 kg. Its lead acid battery provides 18 hours of operation at a 15 per cent transmit cycle, and it also incorporates a battery charger, a built-in loudspeaker and a tape recorder.

Operational Status
In use with the Royal Navy, the Swedish Navy and several other customers. A further type, the EL067P, is being evaluated by the Swedish Army.

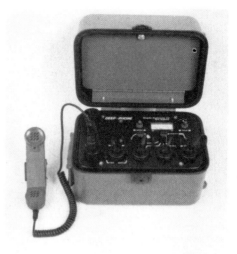

Deep-phone surface unit

Contractor
Slingsby Engineering Ltd, Kirbymoorside.

MODEL 3510 BROADCASTER

Type
Underwater broadcast communication system.

Description
The Model 3510 is a broadcast communication system. No receiver is needed by the diver since the

sound is transmitted directly into the water in the audio region. Ranges of up to nearly 400 m have been obtained.

The system consists of a surface-operated pre-amplifier/power amplifier with a signal processing circuitry which drives a high efficiency, high output speaker. To operate, the speaker is connected to the surface unit and lowered into the water. The operator

can transmit by either voice, tone or siren by using the appropriate switches.

Operational Status
In production.

Contractor
Helle Engineering, Aberdeen.

MODEL 3220B TWO DIVER TELEPHONE

Type
Underwater telephone for diver communications.

Description
The Model 3220B two diver telephone is a two- or four-wire unit. Using two wires to each diver the tender controls communications with one or two divers, or between two divers. Using four wires to each diver up to five divers and the tender can participate in a 'conference call'. A recording device can be

connected when a complete record of all communications is required.

The Model 3220B is a heavy-duty telephone which has a wide range of applications for use with working divers, medical compression treatment chambers, and both naval and commercial operations. Its wide frequency response provides good sound quality for helium atmosphere use to depths of over 60 m.

Operational Status
In production.

Contractor
Helle Engineering, Aberdeen.

Model 3220B two diver telephone

SECURE HELLEPHONE

Type
Secure wireless diver phone.

Description
The Secure Hellephone secure wireless diver phone is based on the Helle standard single sideband (SSB) wireless communication system already in widespread use by navies, marine scientists, police forces and professional scuba divers worldwide.

It is believed to be the first diver through-water communications system with an encryption facility. The system prevents unauthorised listening to diver conversations by scrambling speech, using a proprietary scrambling module whilst maintaining clear communications between divers. The diver can, in addition to encrypted speech, select either 8.075 kHz carrier standard NATO frequency or 25 kHz carrier operation, giving additional security against unauthorised listening.

The encryption system is specifically coded in the factory with a different algorithm for each end user application, thus preventing different customers from understanding each others' communication. The complex circuitry of the Secure Hellephone has been

Secure Hellephone diver's communication system showing encryption device and helmet connection

achieved using state-of-the-art surface mount technology resulting in a compact, portable unit, easily mounted on a diver.

THROUGH-WATER COMMUNICATOR

Type
Untethered diver's communicator.

Description
This compact system developed under contract to the UK MoD provides facilities for diver-to-diver and/or diver-to-surface through-water communications. The equipment comprises a diver's electronic module, headset and attachment belt. The sealed electronic

DS 034-2

Type
Diver speech processor.

Development
Divers operating at great depths use a mixture of oxygen and helium (Heliox) which causes the speech produced to be badly distorted. At depths greater than about 180 m speech becomes virtually unintelligible. This is caused by the increased frequency of the vocal resonances and larynx excitation. At depths of 460 m these resonant frequencies are almost trebled, making speech completely unintelligible.

The DS 034-2 speech processor has been developed to process voice communications from divers breathing Heliox and convert it into coherent speech.

MODEL 6280

Type
Diver-held pinger receiver.

Description
This sensitive, directional acoustic receiver enables a diver or small boat operator to precisely locate underwater pingers. The hand-held, battery-powered ('C' size alkaline batteries) receiver can be used to locate any pinger on a frequency from 8 to 50 kHz. The unit is constructed of corrosion-resistant anodised aluminium and can operate down to 305 m depth. The unit is fitted with a parabolic reflector with a hydrophone mounted at the focal point forming a very narrow listening beam for the precise location of pingers.

A frequency synthesiser allows error-free

MODEL 6281

Type
Surface and diver pinger receiver.

Description
The Model 6281 battery-powered pinger receiver is capable of locating any pinger on a frequency from 8 kHz to 50 kHz. The unit can be used from a small boat and can then be rapidly converted to a diver-held receiver, allowing the diver to swim directly to the located pinger.

The complete system comprises the Model 6515

MODEL 3342A

Type
Diver speech processor.

Description
This two-diver helium unscrambler unit is a two- or four-wire intercom unit with two different unscrambler modes. Using two wires to each diver the surface tender controls communication with one or two divers

Operational Status
In operation with a Far Eastern navy.

module has been designed for use down to 100 m, communication being achieved via acoustic transducers attached to the electronic module. The system uses single sideband, suppressed carrier HF transmitters and receivers. A two-way simplex communication is available, each transmission being preceded by a short tone-burst. The operation of the press-to-talk switch causes the changeover from receive to transmit.

For diver-to-surface communication an adaptive headset is used by the surface operator, whilst the diver uses bone conduction earphones and microphones. When the electronic module is selected for

Description
To convert speech produced under the effect of Heliox, the DS 034-2 writes sections of speech into temporary stores and then reads them out at a lower rate. The sections which are written are those most dominant in each larynx period whilst the remainder are rejected. The frequency compression resulting from the lower replay rate is inversely proportional to the time expansion.

The section length must be less than the shortest larynx period, otherwise the larynx periodicity will be destroyed. A length of 2.5 ms is used which allows operation at larynx frequencies up to 400 Hz. Four temporary stores are used and at any one time while one is being written the remainder are being read. This enables the expansion ratio to be varied over the range 1:1 to 3.5:1 whilst retaining all of the sampled sections.

frequency selection; once the pinger's frequency is selected fine tuning is unnecessary. Since the frequency controls are detented, the diver can change the receiver frequency by 'feel', even in zero visibility.

For operation the sensitivity is set to maximum and the frequency controls adjusted to match the frequency of the pinger. The area is scanned by moving the unit through the water and listening for the pinger's acoustic signal. When a 'ping' is heard the sensitivity is reduced and the operator searches for the strongest signal. By continuing to reduce sensitivity while listening to the 'ping' the operator can easily determine the direction to the pinger by homing in on the strongest signal.

Contractor
Helle Engineering, Aberdeen.

Model 6280 diver-held pinger receiver

antenna, Model 6280 diver-held pinger receiver, a headset for surface operation and an adaptor to convert the system from surface to diver use. The entire system is housed in its own waterproof carrying case.

For small boat operation the antenna sections are fitted together with the directional hydrophone on one end. The cable threads through the connected sections of the antenna and its electrical connector is connected to the adaptor mounted on the receiver.

For operation the sensitivity is set to maximum and the frequency control adjusted to match the frequency of the pinger. The operator scans the area by rotating the antenna through the water and listening for the

or between the divers. Using four wires to each diver, a duplex connection allows the tender and up to five divers to communicate, bypassing the press-to-talk and cross-talk switches.

The two-diver unit uses an audio amplifier with cross-switching to allow communication between the tender and the divers and between divers. The unscrambler uses two complex forms of digital techniques to filter and lower the frequency of a diver's voice resulting in intelligible speech. Mode One is designed for use at normal depths with a normal

Contractor
Helle Engineering, Aberdeen.

long-range operation a minimum range of 1 km can be achieved. A short-range facility is available (less than 100 m) which is dependent on prevailing propagation conditions for use in complex missions.

Any number of divers can be in communication with the controlling surface station. The unit is powered by 10 1.2 V primary cells giving four hours' duration, dependent on usage.

Contractor
Marconi Underwater Systems Ltd, Waterlooville.

Up to four separate voices can be processed simultaneously. The unit comprises a straight audio amplifier from controller to divers, and a delta-sigma helium speech processor with simplex or duplex selection.

The diver's link can be selected for either four-wire duplex or two-wire simplex operation. In the duplex mode the four-wire link routes the processed speech back down to the diver's earpiece enabling diver-to-diver communication. Three settings are available: off/shallow/deep in which selected position the expansion control affects the rate at which data are read into the store.

Contractor
Marconi Underwater Systems Ltd, Waterlooville.

acoustic signal. When a 'ping' is heard the sensitivity is reduced and the operator searches for the strongest signal. By continuing to reduce sensitivity while listening to the 'ping' the operator can easily determine the direction to the pinger by homing in on the strongest signal.

For diver use the hydrophone is connected to the end of the receiver and the headset replaced by the waterproof earphone. The diver operates the unit in the same way as described above.

Contractor
Helle Engineering, Aberdeen.

microphone, Mode Two is intended for use in depths to 610 m using a high quality microphone. A voice-activated switching circuit allows the tender to speak to the diver while leaving his hands free. A remote switch can also be connected allowing a foot pedal to perform the press-to-talk function. A record jack allows a recording device to make complete records of all communications.

Contractor
Helle Engineering, Aberdeen.

MODEL 7370

Type
Breathing noise reducer.

Description
The breathing noise reducer is designed to automatically and independently reduce diver inhalation noise from diver speech for up to three divers. Installed on the surface between diver microphone umbilical pairs

and the communications system, the Model 7370 filters diver inhalation noise in mixed gas and air diving applications.

The unit does not affect speech volume nor inhibit helium speech components essential for good voice reconstruction by unscramblers. The diver supervisor has a single control for each diver which adjusts the inhalation noise cut-off level. The unit also has an in-built bypass for each channel ensuring the integrity of communication from the divers even in the event of failure of the unit.

Contractor
Helle Engineering, Aberdeen.

OE 9020 SEAHAWK

Type
Diver TV system.

Description
The low-cost Seahawk underwater colour television system consists of a wide angle CCD camera

complete with underwater lamp and pistol grip assembly, 50 m of cable and a surface control unit. The control unit incorporates a colour TV monitor, camera and lamp power supplies and a stowage bag for the camera/lamp assembly.

Contractor
Osprey Electronics Ltd, Aberdeen.

Seahawk diver TV system

OE 1211 CYCLOPS

Type
Surface control unit.

Description
A colour monitor, camera and infinitely variable lamp power supply, and data overlay character generator are all contained within the control unit. In addition the unit can accept a wide range of interchangeable

plug-in modules such as divers' communications, pan and tilt control, telemetry control and so on.

Contractor
Osprey Electronics Ltd, Aberdeen.

UNDERWATER LIGHTING

Type
Underwater lamps.

Description
This is a range of lamps specially designed to cope with the divergent operating conditions encountered by operators around the world. The lamps provide a fully variable light output to suit ambient water conditions, and their full light spectrum makes them ideally suited for operation in conjunction with either colour or monochrome cameras.

Specifications

Model	Description	Power	Beam Angle	Dimensions	Weight Air	Weight Water
OE 1130	Underwater floodlight	300 W	100°	152 × 152 mm	0.9 kg	0.35 kg
OE 1131	Heavy-duty underwater floodlight	250 W	100°	152 × 152 mm	1.8 kg	1.15 kg
OE 1132	Compact inspection lamp	300 W	70°	70 × 100 mm	0.63 kg	0.24 kg
Mk 10	Heavy-duty discharge lamp	400 W	100°	222 × 226 mm	2.4 kg	1.4 kg

A miniature model is also available for applications where space is at a premium – mounted on a diver's helmet, or on the manipulator arm of an underwater vehicle.

Contractor
Osprey Electronics Ltd, Aberdeen.

UNITED STATES OF AMERICA

AN/PQS-2A SONAR

Type
Hand-held active/passive sonar.

Description
The AN/PQS-2A is a continuous transmission, frequency modulated, hand-held non-magnetic sonar set intended for use by divers to locate and close on submerged objects such as mines, lost equipment, downed aircraft and sunken vessels in depths to 300 ft (91 m). Because of its non-magnetic construction, the set can be used safely when working near magnetically influenced underwater explosive devices.

The equipment has an active (continuous transmission) mode and a passive (listening) mode. In the active mode, one of three range scales is selected (20, 60 or 120 yd (18, 55 or 110 m)) and an acoustic signal is transmitted over a 30 kHz bandwidth, swept from 114 to 145 kHz. When a submerged object is detected, a ringing tone, proportional in frequency to the distance of the object, is produced in the earphones. The sonar has a 6° beamwidth and can

detect a 12 in (30.5 cm) diameter air-filled sphere at 120 yd (110 m). In the passive mode, the unit detects active sound sources (pingers) in the 24 to 45 kHz range. The sonar can detect a 39 kHz acoustic beacon at a range of at least 2000 yd (1800 m).

Operational Status
In series production for the US Navy. A commercial version is in development.

Specifications
Transmission: Continuous Transmission Frequency Modulated (CTFM)
Readout: Variable audio tone according to range
Active sonar range: 18, 55 or 110 m
Passive range: Up to 1800 m
Weight in water: 0.23 kg (positive buoyancy collar included in accessories)
Size: 31.75 cm long by 11.43 cm diameter

Contractor
General Instrument Corporation, Underseas Systems Division, Westwood, Massachusetts.

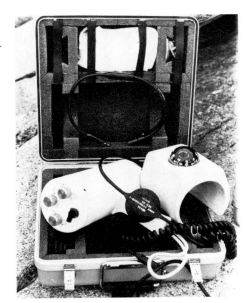

AN/PQS-2A hand-held sonar

MODEL 384A SONAR SYSTEM

Type
Portable diver's sonar system.

Development
The concept of the Model 384A diver's sonar is based on original design data of the US Naval Coastal System Center, Panama City and represents the

efforts of that organisation to support the US Navy in its various diver programmes.

Description
The Model 384A diver's sonar system is a portable, self-contained system designed for use by 'hard hat' or scuba divers. It permits the user to locate suspended or bottom objects, and provides bearing and range information to the target. The Model 384A can also be used to locate pingers or other sound

sources, allowing rapid location of previously marked objects. The small size of the sonar does not interfere with the diver's normal activities, and the configuration and near-neutral buoyancy of the system makes it ideal for search operations by free-swimming divers.

The Model 384A is a Continuous Transmission Frequency Modulation (CTFM) system. In this type of equipment, a continuous signal of varying frequencies is transmitted, and range information is determined by comparing the transmitted frequency to the echo frequency. As a target is located by the equipment, an audio tone is generated in the headphone, a low tone indicating a close target and a high tone a distant target. Moving towards the target will cause the tone to become lower until contact is made. Experience with the 384A allows an operator to determine range and target characteristics from the echo signal. For the purpose of listening to a pinger or other sound source, the passive mode allows selection of the listening frequency to permit selective monitoring of several sources.

Operational Status
In production.

Contractor
EDO Corporation, Electro-Acoustic Division, Salt Lake City, Utah.

Model 384A diver's hand-held sonar

V-COM COMMUNICATIONS SYSTEM

Type
Underwater vehicle communications system.

Description
V-COM is a tri-mode vehicle communications system used by the US Naval Special Warfare Group on its swimmer delivery vehicles. The electronics subsystems consist of an intercom, an underwater telephone and an HF radio, each housed in a pressure-tight aluminium container. The intercom provides communication between the pilot, navigator and the crewman of the vehicle. The underwater telephone is compatible with standard US Navy acoustic communications systems and provides a communication link between the underwater vehicle and its support ship. The HF radio contains a modified man-carried unit (AN/PRC-104) and enables radio communications when the vehicle is surfaced.

Operational Status
In operational service.

Contractor
Allen Osborne Associates, Westlake Village, California.

HELLEPHONE

Type
Underwater communications system.

Description
Hellephone is a single-sideband wireless underwater telephone that provides diver-to-diver, diver-to-surface and surface-to-diver communications. The system consists of the Model 3145 Diver Unit and the Model 3150 Surface Unit or Model 3151 low-cost surface unit. For military applications an 8.0875 kHz version is available, Model 3146, and an optional 'secure communications module' can be incorporated in all versions. Any number of divers and surface personnel can converse in the same area at the same time.

Operational Status
In production.

Specifications
Frequency: 25 kHz USB or 8 kHz
Acoustic output: 2 W
Diver unit
Operating depth: 120 m
Weight: 2.4 kg (in air); 0.5 kg (in water)
Option: Secure communications

Contractor
Helle Engineering, San Diego, California and Aberdeen, UK.

Hellephone underwater communications system

HYDROCOM

Type
Diver speech processor.

Description
Hydrocom is the US Navy's standard helium speech unscrambler system. Using a full duplex connection, Hydrocom allows communication between three divers and a surface tender without the need for a press-to-talk function. The unit operates with either powered or non-powered diver microphones. The power for the microphone can be multiplexed onto the diver's microphone lines and the tender has the option to isolate each of the divers' inputs or outputs in the event of cable failure.

The system provides quality communications in a helium environment using a single audio printed circuit board which uses a differential input stage for each diver to minimise noise and cross-talk picked up in the umbilical cable. Low noise amplifiers are used to amplify divers' voices and a ground plane is incorporated to minimise noise. An optional helium speech unscrambler printed circuit board is available which can be bypassed from the front panel of the control unit for air operation. The unscrambler uses analogue and digital techniques to convert distorted helium speech transmission into intelligible audio. The unscrambling is achieved with audio high frequency boost and time domain sampling and is suitable for operation at depths down to 610 m. At depths of 305 m better than 75 per cent intelligibility has been achieved.

Hydrocom also incorporates inhalation noise attenuation circuitry to reduce breathing noise if an oral/nasal cavity is included in the mask.

Operational Status
In service with the US Navy.

Contractor
SEA-Hydro Products, San Diego, California.

MODEL N30A5B

Type
Diver-held pinger receiver.

Description
The hand-held, battery-powered device is designed for use in conjunction with various types of pinger. The receiving transducer is removable so that it can be used from a small boat. The directional characteristics of the transducer facilitate location of maximum signal area.

With the transducer mounted on the hand-held receiver, and with an underwater earphone, a diver can follow an underwater acoustic signal to the originating pinger. The unit has a visual indicator on the face of the unit enabling the diver to 'see' or hear the signal.

The receiver assembly is designed to withstand pressures down to depths of about 180 m. The battery is housed in a separate sealed compartment in the handle enabling replacement without exposing the interior of the main housing to water.

Contractor
Dukane Corporation, St Charles, Illinois.

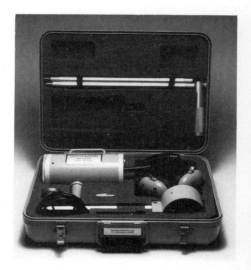

Model N30A5B pinger receiver in carrying case

AN/UQN-5 SANS

Type
Swimmer area navigation system.

Description
The AN/UQN-5 SANS is the swimmer area navigation system used by the United States NSWG Seal Team divers to perform grid searches of large areas of the sea bottom in conditions of reduced visibility. The system employs a diver-held receiver unit synchronised with two acoustic beacons placed at extreme corners of the desired pattern. The diver is given a real-time readout of position and heading and the ability to record the location of significant objects on the seabed. At completion of the dive, the receiver is attached to a printer which produces a hard copy record.

Operational Status
In operational service.

Contractor
Allen Osborne Associates, Westlake Village, California.

SEALITES

Type
Underwater lighting systems.

Description
This range of underwater illumination devices has been developed for a variety of underwater applications.

The standard lightweight SeaLite range offers lamp voltages from 12 V DC to 240 V AC with wattages ranging from 50 to 1000 W. All lamps have positive locking bayonet mounts to reduce the possibility of vibration-related failures. Lamps are replaced simply by unscrewing the front bezel assembly from the light body. The glass-to-metal seal of the pyrex (or quartz) envelope to the bezel is never disturbed during lamp replacement. Modular interchangeable wet reflectors provide beam patterns ranging from a medium spot to a wide angle for almost any conceivable application with both diffuse and specular reflectors being available. The body and bezel are constructed of hard block anodised aluminium. The black finish minimises stray reflections which cause flare in video images. The standard depth rating is 6000 m.

The Max-SeaLite uses the SeaLite body, but with a high efficiency internal reflector head assembly. Three interchangeable reflectors are available offering spot to even, hot spot-free medium and wide angle flood beam patterns. A removable, external shroud protects the borosilicate glass dome and ensures a sharp drop-off at the edge of the light beam. A black neoprene boot provides additional impact resistance. The light has a standard depth rating of 3000 m. An optional deep version rated down to 6000 m is available. A 1000 m depth rated special wet/dry version can burn in air with up to 250 W lamps.

The Multi-SeaLite is available in a variety of configurations using three different reflectors (spot to even, hot spot-free medium, wide angle flood beam patterns), 120 or 240 V AC or DC, multiple wattages and two types of connectors. The light is available in aluminium or stainless steel housings. An external dome-retaining cowl offers protection and acts as a baffle to prevent stray light from entering the water column, which minimises backscatter. The light is rated to 1000 m depth.

Contractor
DeepSea Power & Light, San Diego, California.

DIVER VEHICLES

FRANCE

MARLIN

Type
Divers' vehicle.

Description
Marlin is an electrically powered submersible capable of carrying one or two divers. The craft is a cylindrical body made up of three sections in the single diver version in which the diver lies protected by removable fairings. In the twin diver version an intermediate section has been added in which the two divers lie in tandem. Both positions incorporate vehicle controls. Power is supplied by nickel/cadmium batteries offering continuous speed variation. Underwater steering and navigation are carried out using diving and rudder controls. Navigation equipment includes a horizontal and vertical echo sounder, magnetic compass, depth gauge, standard clock, and control desk displaying the following parameters: horizontal/vertical echo sounder indication by switching, battery voltage and instantaneous intensity.

Operational Status
Marlin is operational.

Specifications

	Single diver version	Twin diver version
Length	3.9 m	6.2 m
Weight	450 kg	700 kg
Speed (max)	6 kts	5 kts
Endurance (at 3 kts)	4 h	6 h

Contractor
Société ECA, Colombes.
  CSI, Colombes (export company).

Marlin underwater vehicle

UNITED KINGDOM

SUBSKIMMER 90

Type
Personnel carrier.

Description
The Subskimmer 90 is a fast, offshore rigid hull inflatable which can convert into a mini-submarine while under way. It enables frogmen to operate at greater distances from their parent surface vessel with more equipment at higher speeds and with much less effort than by other means.

Specifications
Length: 5.35 m
Beam: 1.80 m
Weight: 850 kg
Maximum diving depth: 50 m
Propulsion: 2-stroke outboard motor, 90 hp at 5500 rpm (surface); 2 × 24 V, 1.5 kW DC electric motors (underwater)
Speed: 20-25 kts (surface); 1.5-2.5 kts (underwater)
Range: 70 nm (or more with additional fuel tanks) (surface); up to 9 km (underwater)
All-up payload: 600 kg

Contractor
Defence Boats, Hexham.

SUBSKIMMER 180

Type
Personnel carrier.

Description
The Subskimmer 180 is a fast offshore rigid inflatable personnel carrier developed from the Subskimmer 90, which can quickly transform from a surface craft into a submersible with an underwater endurance of 8 km at 2.5 kts.

The craft is powered by two 90 hp outboard motors mounted on twin transoms and controlled from a common console, or a single 130 hp outboard. Underwater propulsion is achieved using two lead acid batteries to power underwater propulsors which are contained in two glass fibre tubes supporting the deck.

The surface controls are sited at the rear of the buoyancy chamber enabling up to eight personnel and their equipment to be carried. A single steering wheel swivels two forward electric thrusters up and down for vertical underwater manoeuvrability and moves the engines aft to port and starboard for lateral control.

Underwater stability is provided by a small buoyancy control unit mounted amidships. Various parts of the craft and the hull tubes are pressurised from 11 000 litres of compressed air carried in four cylinders mounted on either side of the buoyancy box.

For underwater navigation the hull tubes are deflated and the buoyancy chamber flooded. In the area of operations the vehicle is anchored to the seabed, whilst the personnel proceed with their tasks. On completion, a sonar beacon is remotely activated, the vehicle located, navigated to a safe area where it is brought to the awash condition using compressed air to empty the buoyancy chamber. After careful surveillance to ensure the area is clear, the compressor is started to inflate the side tubes. When at full buoyancy the engine is started and withdrawal proceeds. All vulnerable systems are pressurised when diving and are fitted with non-return valves to allow excess compressed air to escape during ascent without risk.

Specifications
Length: 7 m
Beam: 2.3 m
Weight: 1500 kg
Max diving depth: 50 m
Propulsion: 2 × 2-stroke outboard, 90 hp at 5500 rpm or 1 × 130 hp 2-stroke outboard
Fuel: 70 l per tank
Submerged propulsion: 24 V battery = 1.5 kW at 600 rpm
Speed: 20.25 kts (surface cruise); 30 kts (surface max); 1.5 kts (underwater cruise); 2.5 kts (underwater max)
Range: 70 nm (surface); 4-6 nm (submerged)
All-up payload: 900 kg

Contractor
Defence Boats, Hexham.

YUGOSLAVIA

R 2

Type
Two-man diversionary submersible.

Description
The R 2 is designed for the transport of frogmen and underwater mines, for carrying diversionary equipment and for underwater reconnaissance. The hull is constructed of aluminium alloy resistant to salt water corrosion. The front upper part of the submersible is made of Plexiglas. The spindle-shaped hull can be fully flooded, except for the cylinders housing the storage battery, propulsion unit, navigation equipment, compressed air and ballast tanks.

The built-in ballast system is used for blowing and flooding the ballast tanks to secure static diving in seas of specific gravity 1.01-10.3 t/m³ and at depths of 0-60 m, as well as for compensation of the air consumed by the frogmen's respiratory system.

The vehicle is powered by a DC electric motor which derives its power from a storage battery. The three-bladed propeller is driven via an electromagnetic clutch.

Navigation equipment comprises an aircraft-type

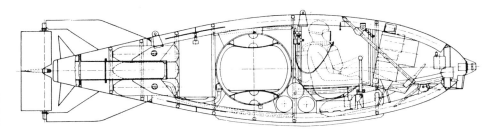

Schematic of the R 2 two-man submersible

gyro-magnetic compass, a magnetic compass, depth gauge with scale 0-100 m, echo sounder, sonar, two searchlights and so on. All navigation equipment is housed in a waterproof container.

Operational Status
In service with the USS, Libya and Sweden.

Specifications
Length: 4.90 (oa)
Diameter: 1.22 m
Breadth: 1.40 m (max over hydroplane)
Height: 1.32 m (max over appendages)

Weight: 1400 kg
Propulsion: 4.5 kW (DC electric motor); 24 V 192 Ah (storage battery)
Speed: 4.4 kts (max); 3.7 kts (cruising)
Range: 18 nm at max speed, 23 nm at cruising speed (with lead-acid batteries); 38 nm at max speed, 46 nm at cruising speed (with silver-zinc batteries 310 Ah)
Max diving depth: 100 m
Armament: two 50 kg underwater mines

Contractor
Brodosplit, Split.

R 1

Type
Diver submersible.

Description
The R 1 submersible is a single-seat underwater frogman vehicle designed for covert tasks such as reconnaissance, harbour protection, minefield surveillance and so on.

The monohull vehicle is built of aluminium alloy and comprises light bow and stern sections that can be flooded. The vehicle can be transported in a submarine torpedo tube and used in both fresh water and seas of specific gravity of 1.000-1.030 t/m³ without reserve updrift.

Navigation instruments are housed in a watertight container and comprise a gyro-magnetic compass, sonar, echo sounder, electric clock and other measuring systems.

Operational Status
In service with the former Yugoslav Navy.

Specifications
Length: 3.72 m (oa)
Breadth: 1.05 m (max)
Height: 0.76 m (max)
Diameter: 0.52 m (max)

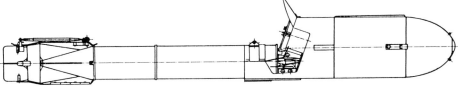

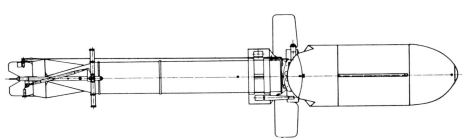

Frogman submersible R 1

Weight: 145 kg (less payload)
Propulsion: DC electric motor, 1 kW; 24 V silver-zinc battery
Speed: 3 kts (max); 2.5 kts (cruising)
Range: 6 nm (at max speed); 8 nm (at cruising speed)
Operating diving depth: 60 m

Contractor
Brodosplit, Split.

Associated Underwater Warfare Systems

Acoustic Management Systems
Environment Measuring Probes
Environment Measuring Buoys
Prediction Systems

Hydrographic Survey Systems
Echo Sounders
Hydrographic Sonars
Ocean Survey Systems

Signature Management
Acoustic Ranges
Magnetic Ranges
Degaussing Systems
Acoustic Control

Training and Simulation Systems
Command Team Trainers
Submarine Trainers
Aircraft Trainers
Sonar Trainers
Targets

Navigation and Localisation Systems

Submarine Radar

Consoles and Displays

Transducers

Miscellaneous Equipment

ACOUSTIC MANAGEMENT SYSTEMS

All forms of warfare are, to a major degree, governed by the surrounding environment, and nowhere is this more true than in the case of underwater warfare. The ocean is the largest of all the environments and covers some 70 per cent of the earth's surface, totalling about 370 000 000 km². At its deepest it reaches down to 11 000 m, with an average depth of 3850 m. The ocean environment is the most difficult of any in which to conduct warfare, and its vagaries are such that even today there is a great deal that still remains unknown about how the sea behaves and how it is affected by the land and air within and above it. Every aspect of underwater warfare – submarine operations, anti-submarine warfare, mine warfare, diver operations and so on – is subject to the behaviour of the sea. It is for this reason that the science of oceanography has become so important to the world's larger navies, and also to many of the smaller ones. The USS for example, maintains a very large fleet of oceanographic vessels designed specifically to study all aspects of the ocean and other environments which affect its behaviour.

The underwater environment differs in many ways from the land and air environments, and is subjected to quite different forces. For example, in sea water, radiation of certain parts of the electromagnetic spectrum is poorly transmitted. This severely restricts the use of visual, radar or other detection systems which use that part of the spectrum. Conversely, sound waves can travel and be detected over very long distances underwater, and are therefore the most important factor relating to underwater warfare, for they form the primary means of detection. Certainly this is true at long range, although at a relatively short range other phenomena such as magnetism are also very important, particularly in mine warfare.

The physical properties of the sea, and how they affect radiation, play a vital role in underwater warfare. For example, geographical considerations, local atmospheric conditions, the season, and the composition of the seabed (is it smooth or rough, does it undulate and by how much, what is its underlying structure, is it rock or a great depth of mud or sand?) all affect very considerably the nature of the environment and how it behaves. Knowledge of how the oceans are conditioned by the atmosphere above, how the masses of ocean water move, and how temperature and salinity are affected, for example, is essential for assessing how the propagation of sound underwater will be affected. Other factors affecting sound propagation underwater relate to boundary layers, both in water movement (in differing parts of the ocean, currents running in opposite directions can be found one above the other) and temperature boundaries (as opposed to gradients), scattering effects, and ambient noise, as noise from shrimps, whales, dolphins and so on, is a constant source of interference.

Knowledge of how sound propagation is affected in the ocean environment can be used to exploit methods of detection, or to avoid detection and to limit the effects of uncontrollable noise generation. Using such knowledge, it is also possible to predict how sound propagation will be affected in a given set of circumstances, hence the probability of detection under those conditions can be predicted.

Various methods and equipment are now becoming available with which to investigate and measure the various physical factors affecting sound propagation underwater.

ENVIRONMENT MEASURING PROBES

CANADA

XSV

Type
Expendable sound velocimeter.

Description
The XSV is designed to provide shipboard ASW operators with a direct measurement of underwater sound velocity. The XSV generates accurate velocity versus depth profiles for the support of ASW operations, target detection and countermeasures and oceanographic research. The XSV's direct measurement capability is particularly useful in waters having a high salinity variability. More traditional methods compute sound velocity data based upon temperature profiles and an assumed salinity value which would be inaccurate under these conditions. Therefore, the XSV is most useful in areas such as the Arctic, the Mediterranean and along coastlines where there may be significant fresh/salt water interfacing.

The expendable sound velocity measuring probe is designed for release from a variety of launchers. The probe is electrically connected through the launcher to the shipboard processor/recorder. Following launch, wire de-reels from the probe as it descends vertically through the water. Simultaneously, wire de-reels from a spool within the probe canister, compensating for any movement of the ship and allowing the probe to freefall from the sea surface unaffected by ship motion or sea state.

The probe measures the speed of sound in water using the well proven 'sing-around' principle. A piezo-electric ceramic transducer within the sensor is pulsed at a high frequency, producing an acoustic signal. An electronic circuit computes the transit time of this acoustic signal through the water. The reflected acoustic pulse initiates a new pulse, causing the signal to 'sing-around' within the sensor at a frequency proportional to the temperature, salinity and pressure of the ocean water present.

The data, telemetered via a single conductor wire to the shipboard processor/recorder, are frequencies directly proportional to sound velocity. Data are recorded and can be displayed in real time as the probe descends. The XSV obtains real-time sound velocity data accurate to ±0.25 m/s at depths of up to 2000 m. Using careful design and manufacturing techniques, each expendable probe has a highly predictable descent rate. From this, probe depth is determined to an accuracy of ±2 per cent. When the probe reaches its rated depth (a function of ship speed and the quantity of wire contained within the shipboard spool) the wire breaks, the profile is completed and the system is ready for another launch.

Three models, each targeted to a specific area of application, are shown in the table. Air-launched and sub-launched configurations are also under consideration (see also *USA* section).

MODEL	XSV-010	XSV-020	XSV-030
Applications	ASW application where salinity varies; naval and civilian oceanographic and acoustic applications	Increased depth for improved ASW operation where salinity varies; naval and civilian oceanographic and acoustic applications	High resolution data for improved mine countermeasures and ASW operations in shallow water; geophysical survey work; commercial oil industry support
Max depth	850 m (2790 ft)	2000 m (6560 ft)	850 m (2790 ft)
Rated ship speed*	15 kts	8 kts	5 kts
Depth resolution	32 cm	32 cm	10 cm

*All probes may be used at speeds above rated maximum. However, there will be a proportional reduction in depth capability.

Operational Status
In service with a number of navies.

Specifications
Description: Expendable sound velocimeter
Function: Direct measurement of sound velocity as the probe descends
Average weight: 1.3 kg

Dimensions: 57 × 394 mm
Activation time: Immediate upon water contact
Data accuracy: ±0.25 m/s
Launchers: Compatible with LM-2A, LM-3A, LM-4A
Processor and display: Compatible with SOC BT/SV processor Model 100, Mk 8, Mk 2A, AN/BQH-7A, Mk 9 digital data acquisition system

Contractor
Sparton of Canada Ltd, London, Ontario.
Sippican Inc, Marion, Massachusetts (USA).
Dowty Maritime Sonar and Communication Systems Ltd, Greenford (UK).

XBT

Type
Expendable bathythermograph probe.

Description
The expendable bathythermograph (XBT) is a surface ship and submarine bathythermograph equipment which uses an expendable probe to obtain accurate and continuous ocean temperature profiles for use by shipboard ASW operators to determine the effect of temperature on sonar propagation and range prediction. The probe may be released over the side from either a hand-held launcher or a deck-mounted launcher. XBTs may also be deployed by means of a

through-hull launcher installed below deck for improved safety and convenience under heavy weather conditions. The probes can be deployed in any sea state at ship speeds up to 30 kts. When the probe contacts the water, the unit is active and begins transmission of the temperature data to the ship.

Upon launch, wire de-reels from inside the probe as well as from a spool inside the launch tube. This allows for a consistent vertical probe descent through the water, without interference from ship movement or sea state.

The probe contains a thermistor located in its nose weight. Changes in temperature are determined by changes in thermistor resistance during the descent. This resistance information is transmitted along the 39 AWG, two-conductor wire to the shipboard processing unit, where it is converted to absolute temperature measurements. The XBT can achieve temperature accuracies of ±0.1°C and a resolution of better than 0.01°C.

The nose of the probe is manufactured to a precise weight and stabilising fins spin the unit to ensure a predictable descent rate. Probe depth is ascertained from this descent rate to an accuracy of ±2 per cent.

Four models are available, each targeted to a specific area of application, as shown in the following table. A cutaway view of the launch container is shown in the figure for comparison of the deep and standard depth probes (see also *USA* section).

Model	XBT-4	XBT-5	XBT-5DB	XBT-6	XBT-7	XBT-7DB	XBT-10	XBT-20
Application	Naval ASW operations	Naval ASW operations	Oceanographic uses	Oceanographic uses	Naval ASW operations	Oceanographic uses	Commercial fisheries	Oceanographic uses
NSN	6655-00 −932-1353	6655-00 −165-0452			6655-00 −162-2479			
Full depth	460 m/1500 ft	1830 m/6000 ft	1830 m/6000 ft	460 m/1500 ft	760 m/2500 ft	760 m/2500 ft	200 m/660 ft	1000 m/3300 ft
Max ship speed*	30 kts	6 kts	12 kts	15 kts	15 kts	20 kts	12 kts	15 kts
Dimensions	7 × 35.5 cm	7 × 48.3 cm	7 × 48.3 cm	7 × 35.5 cm	7 × 35.5 cm	7 × 35.5 cm	7 × 35.5 cm	7 × 35.5 cm

*To achieve rated depth, use of probes at higher ship speeds will result in a proportional depth decrease.

Operational Status
In service with the UK Royal Navy and over 20 other countries.

Specifications
Description: Expendable bathythermal profiler
Function: Aid in sonar propagation and range detection
Average weight: 1.02 kg

Activation time: Immediate upon water contact
Depth accuracy: ±2% of depth
Depth resolution: 65 cm
Thermistor time constant: 100 ms
Temperature resolution: ±0.1°C
Temperature range: −2.2°C to +35.5°C
Operating depth: 460 m
Ship tow speed: 30 kts (max)
Launchers: Compatible with LM-2A, LM-3A, LM-4A

Processor and display: SOC BT/SV processor Model 100, AN/BQH-7A, Mk 2A, Mk 8, Mk 9, SA 810, XBT-ST and other industry standard recorders

Contractor
Sparton of Canada Ltd, London, Ontario.
Sippican Inc, Marion, Massachusetts (USA).
Dowty Maritime Sonar and Communication Systems Ltd, Greenford (UK).

XBT/XSV RECORDING SYSTEMS (BT/SV PROCESSOR, MODEL 100)

Type
Data processing system for expendable probes.

Description
The Expendable Bathythermograph/Expendable Sound Velocimeter Recording System (BT/SV processor, Model 100) is a computer-based bathythermograph/sound velocimeter data acquisition and recording system. Temperature (or sound velocity) data are obtained from expendable probes dropped from standard probe launchers and are displayed for immediate action or recorded for future analysis. The IBM compatible BT/SV processor, Model 100

Processor Card can fit into the expansion slot of a standard PC. Along with user-friendly software, the BT/SV processor, Model 100 Processor Card can provide users with an effective and simple-to-use processor for collecting, storing and displaying XBT/XSV and AXBT data.

Operational Status
In production.

Specifications
System performance is dependent upon type of probe used.
XBT Processing Performance
Temp range: −2.2 to +35.6°C
Accuracy: ±0.15°C (digital); ±0.20°C (paper chart)
Resolution: ±0.01°C (digital); ±0.1°C (paper chart)

Depth: Up to 760 m (with T7 type probe)
Ship speed: Up to 30 kts with T4 type probe; up to 6 kts with T5 probe; up to 15 kts with T7 or T6 probe
Sampling rate: 10 Hz

XSV Processing Performance
Velocity range: 1405-1560 m/s
Depth: Up to 850 m (XSV-01 probe); up to 2000 m (XSV-02 probe)
Ship speed: Up to 15 kts (XSV-01 probe); up to 8 kts (XSV-02 probe)
Velocity accuracy: ±0.25 m/s (on digital interface); ±0.65 m/s (on paper chart)

Contractor
Sparton of Canada Ltd, London, Ontario.

SPARTON ICE THICKNESS TRANSPONDER

Type
Aerially deployed transponder to determine ice thickness.

Development
The Aerially Deployed Ice Thickness Transponder (ADITT) was developed under a partnership agreement between Consolidated Technologies Ltd and Sparton of Canada Ltd. ConTech, a small, independent research and development company, provided expertise on the ice-penetrating thermochemicals and the ice drilling probe. Sparton, a developer and manufacturer of active and passive sonobuoys and underwater acoustic devices, supplied the sonobuoy package including air descent mechanisms and VHF telemetry. Whenever possible, standard sonobuoy components from current production are used, supplemented by special designs developed for ice-penetrator devices. A key focus of the companies' efforts has been the joint development of autonomous thermochemical ice penetrating devices for defence and Arctic research applications.

Description
The ADITT is designed to remotely determine the thickness of ice sheets. Its primary application is as

an aid in the safe landing of aircraft on refrozen leads in the Arctic.

Externally, the device resembles a NATO standard 'A' sized sonobuoy and is deployed from an aircraft via a standard sonobuoy launch tube. Measurement of ice thickness is accomplished by a miniature probe which penetrates the ice, powered by a proprietary ice-melting thermochemical. The ice thickness is determined by detecting the point at which the probe breaks through the bottom ice/water interface.

Operation of the ADITT probe is as follows: the device is pre-armed and launched from the sonobuoy tube of an aircraft, after which the parachute deploys to stabilise and control its descent to the ice surface; upon impact with the surface, the electronic circuitry activates and the package falls to a horizontal orientation with the probe in position to be properly deployed; after a brief delay to allow the orientation to stabilise, the antenna is erected and the ice-penetrating probe is released; the probe begins its travel down through the ice while penetration versus time data are transmitted back to the deploying aircraft via VHF FM radio; the probe's breakthrough of the lower ice/water interface is evident as a sharp rise in the rate of penetration; after a preselected time, the transmitter shuts down to clear the radio channel.

Data are received aboard the deploying aircraft using standard sonobuoy receivers and processed by a purpose-built unit. Ice thickness data can be output on a variety of devices including a strip chart recorder.

Operational Status
Experimental development models being tested.

Specifications
Description: Aerially deployed ice thickness transponder
Function: Remote determination of ice cover thickness
Average weight: 10 kg
Physical size: 12.7 cm (diameter), 91.5 cm (long)
Activation: Pre-arm and impact switch
Transmitter: 31 channels VHF FM
RF power: 1 W
Operating temperature: −40°C (min)
Measurable ice thickness: 1 m (max)
Measurement resolution: ±5 cm
Ice penetration rate: 1-2 cm/s
Auto shutdown: 30 min
Power supply: Alkaline batteries
Shelf life: 2 years (min)
Thermochemical medium: Pyrosolve B

Contractors
Sparton of Canada Ltd, London, Ontario.
Consolidated Technologies Ltd, St John's, Newfoundland.

ICE PENETRATING SENSOR SYSTEM (IPSS)

Type
Sensor system to emplace sonar equipment beneath ice.

Development
Ice penetrating sensor systems are the results of design efforts by Sparton of Canada Ltd working with Consolidated Technologies Ltd (ConTech). Sparton possesses expertise in sonobuoy development and manufacturing, while ConTech developed the ice penetrating thermochemicals and ice drill. In 1986 the two companies agreed to work jointly on aerially deployed ice penetrating systems and instrumentation packages. Also envisaged are upward ice drilling units. In January 1988, Sparton and ConTech were contracted jointly to develop and test the IPSS.

Description
An ice penetrating sensor system (IPSS) is being developed to permit emplacement of oceanographic and tactical anti-submarine and mine countermeasures sensors beneath the thick ice cover of the Arctic Ocean. The system is packaged as a NATO standard 'A' sized sonobuoy and is designed for deployment from aircraft equipped with sonobuoy launch tubes. One half of the package contains the sensor of choice, sonic amplification and VHF transmitter, while the other half is occupied by an ice drill using water-reactive thermochemicals. Through the use of various sensors, the system is configured for acoustic, bathythermal or sound velocity measurements at the time of manufacture. Satellite communications can be achieved by substitution of the VHF transmitter.

Upon air-launch, a parachute stabilises and slows the descent of the system to the ice surface. Upon impact with the ice the parachute and outer housing are ejected, the float is inflated, the antenna is erected, and the ice drill and sensor are deployed. The thermochemical in the ice drill creates a conical crater in the surface. Snow cover will not stall the drill as melt water is produced whether snow or bare ice is

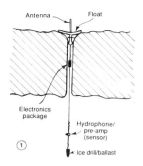
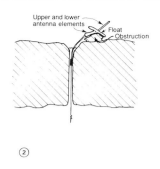
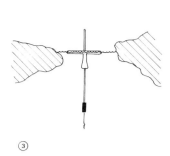

first encountered. The crater provides for the initial downward orientation of the drill and serves as a container for melt water. The drill contains sufficient thermochemical to pull the sensor through about 2.4 m of first year ice in approximately 15 minutes.

The sensor, cable pack and electronics package are pulled into the melt hole by the drill and suspended from the surface by the straddling action of the float/antenna (figure 1). The antenna design provides good performance over the standard sonobuoy VHF band, regardless of whether it is emplaced normally in the melt hole or snagged in a non-vertical orientation on the surface (figure 2). The VHF transmitter, submerged in the melt water, is protected from the climatic extremes on the ice surface. In the sonobuoy configuration, the sonic amplifier, pre-amp and hydrophone provide excellent low noise performance, particularly at low frequencies. This is of paramount importance in the Arctic where ambient acoustic noise levels can be very low.

The thermochemicals used in the IPSS are safe to handle and transport, yet react vigorously with ice. The resultant melt water is environmentally benign anti-freeze. The advantage of this feature, besides lack of environmental impact, is that system components will deploy through the melt hole without freeze-up.

Of concern to airborne ASW crews working in the Arctic is the survival rate of ice penetrating systems deployed on varying thicknesses of ice and snow cover and, at times, in open water. The IPSS is designed to cope with these scenarios. The

float/antenna provides sufficient flotation to support the system in open water and radiate telemetry signals as would a normal sonobuoy (figure 3), yet will radiate equally well when covering the top of a melt hole on thick ice cover (figure 1). Should the system break through a thin ice cover, an additional thermochemical charge on top of the float will ensure that the antenna finds its way to the surface.

Operational Status
Experimental development models are available.

Specifications
Size: NATO 'A' size sonobuoy
Weight: <17.7 kg
Activation time after launch: <30 mins
Transmitter: 31 channels
RF power: 1 W
Power source: Lithium/sulphur-dioxide batteries
Operating life: 4 h
Max launch airspeed: 180 kts
Max launch altitude: 5000 ft (1520 m)
Operating environmental conditions
Launch air temperature: −20°C to +55°C
Seawater temperature: 0 to +35°C
Wind velocity: 30 kts
Ice thickness: 3 m (max)

Contractors
Sparton of Canada Ltd, London, Ontario.
  Consolidated Technologies Ltd, St John's, Newfoundland.

ACCURATE SURFACE TRACKER

Type
Expendable Langrangian drifting tracker.

Description
The accurate surface tracker is specifically designed to track within the top metre of the water column. The lightweight tracker (it weighs approximately 18 kg dry) is configurable with a variety of sensors.

Contractor
Seimac Ltd, Bedford, Nova Scotia.

Barometric Pressure Sensor
Antenna
Air Temperature
Power On/Off
Recovery Beacon
Argos
Electronics
Conductivity Cell
Water Temperature Sensor
Batteries

Accurate Surface Tracker

NEDAL

Type
Navigation data collector.

Description
Nedal provides near real-time data collection of most

shipborne navigation and environmental data at one central site, usually the ship's wheelhouse. Information typically needed for activities throughout the ship are thus available to all users either via a multiple RS 232 connection scheme or via a true network such as the scientific data highway. Data collected include time (GMT), depth, speed, heading, satellite navigation data, air temperature, humidity,

sea temperature, wind speed, wind direction, air pressure and light density.

Contractor
Seimac Ltd, Bedford, Nova Scotia.

RTM

Type
Remote Telemetry Module (RTM).

Description
The RTM is a totally self-contained telemetry module designed such that up to 32 RTM transmitters can be used with the same receiver module. A typical application for the RTM is the monitoring of ocean data, including buoys, where physical links are impractical.

Contractor
Seimac Ltd, Bedford, Nova Scotia.

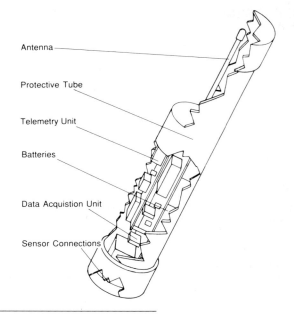

Remote Telemetry Module

AIMS

Type
Arctic Ice Monitoring System (AIMS).

Description
AIMS is capable of measuring air, ice and water column parameters and supporting strings for sensors at depth. The ice penetrator senses ice growth and ice temperature profiles, and can be supplied as a stand-alone sensor element. AIMS units are available with a large range of sensor suites for oceanographic and military applications.

Contractor
Seimac Ltd, Bedford, Nova Scotia.

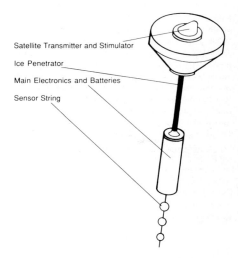

Arctic Ice Monitoring System

ADIB

Type
Air Deployable Ice Beacon (ADIB).

Description
ADIB is one of a family of ice beacons. The ice-deployed version is designed to measure wind and other climate parameters, in addition to Argos satellite position, in Arctic areas. The ice beacon is floatable.

Contractor
Seimac Ltd, Bedford, Nova Scotia.

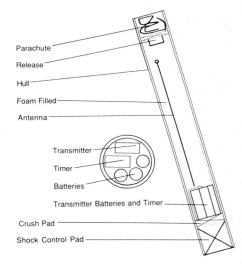

Air Deployable Ice Beacon

GERMANY

BATHYSONDE 2000

Type
Measuring probe for oceanographic applications.

Description
The Bathysonde 2000 HS is an *in situ* measuring instrument for rapid recording of specific electric conductivity (C), temperature (T) and depth/pressure (D) in water. It is designed for use with an internal data recording system. A particular feature is the high measuring rate of 250 CTD values per second, which at a sinking rate of 4 m/s reduces to a minimum the time a ship has to remain in position. The sensors measuring conductivity and temperature are designed to require only 5 ms to respond, with a spatial resolution capacity of 1 cm. The equipment can be operated down to a maximum depth of 6000 m.

The Bathysonde 2000 LS is another high precision *in situ* measuring instrument for recording of the same values. If fitted with fast CTD sensors, as in the 2000 HS, a data rate of 32/64 data sets or more can be achieved.

Operational Status
In production.

Contractor
Salzgitter Elektronik GmbH, Flintbek.

NORWAY

SERIES 7

Type
Series of oceanographic instruments for data collection.

Description
This range of instruments consists of the recording current meter model RCM 7, water level recorder model WLR 7 and temperature profile recorder model TR 7.

The RCM 7 and RCM 8 are self-contained instruments for recording speed, direction, temperature, pressure and conductivity of ocean currents. The RCM 7 can be moored to the seabed and record ocean data. It comprises a recording unit and vane assembly which is equipped with a rod that can be shackled to the mooring line. This arrangement allows the instrument to swing freely and align itself in the direction of the current. The recording unit contains all sensors, monitoring systems, battery and a detachable, reusable solid-state data storing unit. A built-in clock triggers the instrument at preset intervals and a total of six channels are sampled in sequence. The first channel is a fixed reference reading for control purposes. Channels 2, 3 and 4 represent measurements of temperature (three selectable ranges between −2.46 to 36.04°C plus an Arctic range (−2.64 to 5.62°C to special order), conductivity (0-74 mmho/cm standard with two other ranges available on request) and depth (down to 2000 m) respectively; channels 5 and 6 represent measurement of vector averaged current speed and direction since the previous triggering of the

instrument. The data are sequentially fed to the memory unit.

Simultaneously with the taking of the reading, the output pulse keys on and off an acoustic carrier transmitted by a transducer. This allows monitoring of the performance of moored instruments from the surface by a hydrophone and can be used for real-time telemetry of data.

The RCM 8 is identical to the RCM 7 except that it is a high pressure model for measurement to depths of 6000 m.

The RCM unit can be moored in one of two ways: a U-mooring which is best suited for use in relatively shallow water and the I-mooring involving the use of an acoustic release device and which can be used at any depth. It is essential that a sub-surface float is used to avoid wave induced motion on the mooring line.

The WLR 7 water level recorder is a self-contained high precision instrument which is placed on the seabed to record water level by precise measurement of hydrostatic pressure, as well as measuring temperature and conductivity. The WLR averages the pressure over a period of 40 seconds in order to eliminate fluctuation in water level due to waves. When this integration time is completed the data words are recorded. The first data word is a fixed reference reading followed by the temperature of the ambient water. Pressure is recorded as two 10-bit words and finally a 10-bit word for the conductivity (optional sensor). Data are transmitted via an acoustic transducer to a hydrophone receiver hung from a relay buoy at the surface. The buoy then transmits the received data via VHF link to a shore base or parent ship.

The temperature profile recorder TR 7 is a fully

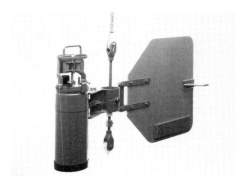

The RCM 7 vector averaging current meter

self-contained instrument for recording temperature profiles in the sea, lakes or fjords. The vertical temperature profiles are used by oceanographers to study internal waves and other physical phenomena. In some applications *in situ* recording of data is required, while for other applications telemetry of data to shore or ship is required, and both functions can be provided by the TR 7. The profile recorder consists of a 12-channel recording unit and a thermistor string. The thermistor string employs 11 temperature sensing thermistors embedded in a polyurethane cable. The thermistors are spaced throughout the cable which can be from 5 up to 400 m maximum in length. Four different temperature ranges are available.

Contractor
Aanderaa Instruments, Bergen.

UNITED KINGDOM

MUMS

Type
Modular Underwater Measurement Systems.

Description
The system comprises a range of interchangeable sensors and components which assemble into underwater measuring systems. A typical MUMS has a number of underwater measurement devices such as temperature, depth, acceleration, proximity and sound sensors, all of which can be plugged into

a polythene cable. This connects to a surface control and display unit via in-cable or bulkhead connectors.

Typical of the range of sensors is the THX/1 hydrophone. This may also be used as a transceiver and has an integral shock absorbing mounting which enables it to operate in harsh environments and to withstand the high shock levels that occur near seismic energy sources. Measuring just 200 mm long by 185 mm diameter, the THX/1 hydrophone has a capacitance of 11.0 nF, receive sensitivity of −200 dB at 0 to 60 kHz, frequency response of 0 to 80 kHz at ±2 dB and a pressure range to 50 bar.

The inexpensive SDI depth indicator system features a rugged transducer 215 mm long by 32 mm diameter and a panel-mounted or hand-held display. Depths to 40 m are displayed in psi, feet or metres.

The MUMS system and its individual components are used for oceanographic research, nearshore and deepwater surveys, seismic exploration and in underwater defence activities.

Contractor
Scientific & Defence Management Associates Ltd.

UNITED STATES OF AMERICA

XCTD

Type
Expendable Conductivity, Temperature and Depth profiling system.

Description
The XCTD is designed to collect salinity profiles up to 1000 m while under way from dedicated research vessels and craft of opportunity. The system comprises an expendable probe that measures conductivity and temperature, the Sippican Mk 9 digital data acquisition system and a launcher.

The system can collect conductivity measurements to 0.03 milli-Siemens (mS) and temperature to 0.03°C. These measurements are sampled at a rate that provides a vertical resolution of 80 cm at a depth accuracy of ±5 m or 2 per cent of depth, whichever is greater.

The XCTD probe contains a conductivity cell, a thermistor, electronics and a BT wire link to the surface. The conductivity cell is a high purity, alumina ceramic tube wired to form a four-electrode conductivity sensor. The cell is potted in a glass-filled, rigid epoxy compound.

The temperature sensor is a very stable glass-encapsulated, fast-response temperature-cycled

thermistor. The cell electronics package converts the measured resistance on the thermistor and conductivity cell into a frequency which is sent up the BT wire to the acquisition circuitry in the Mk 9.

Included in the probe electronics are two precision calibration resistors which are periodically sampled and their output transmitted up the BT wire in conjunction with the temperature and conductivity data to compensate for any induced changes in the electronics during the deployment.

Contractor
Sippican Inc, Marion, Massachusetts.

XCP PROFILER

Type
Horizontal ocean current profiler.

Description
The XCP is an expendable instrument capable of obtaining profiles of horizontal current direction and

speed. The system obtains ocean current data to depths of up to 1500 m with an accuracy of ±1 cm/s RMS. It may be deployed from either a surface or airborne platform since it uses an RF link to transmit data. Three channels are available for RF transmission: 170.5, 172 and 173.5 MHz (standard US Navy sonobuoy channels 12, 14 and 16). A complete XCP system includes expendable probes, a Sippican Mk 10 digital data interface, and a Hewlett Packard

9816S computer. This latter may be substituted by any computer which accepts the IEEE-488 output and which has a minimum of 384 kbytes of memory available for program software.

Contractor
Sippican Inc, Marion, Massachusetts.

RO-308/SSQ-36 DATA RECORDER

Type
Bathythermograph data recorder.

Description
The recorder is an integral part of the US Navy/ Lockheed P-3C Orion aircraft ASW system. The equipment converts seawater temperature information provided by the AN/SSQ-36 bathythermograph buoy-transmitter set and AN/AAR-72 radio receiving set to two output forms:

(a) a permanent record of the vertical temperature profile (temperature versus depth) on a paper strip chart
(b) a parallel mode, eight-bit binary coded data word for delivery to the AN/AYA-8B data processing system.

The buoy is dropped from an aircraft in the target area. Seawater is utilised as the activating agent and after an initial, predictable delay, the buoy releases a temperature sensing probe (TS probe). The TS probe is the variable element in a frequency generation circuit. A radio frequency signal transmitted by the buoy is modulated at a frequency correlated to the temperature of the water. On board the aircraft the ARR-72 radio is tuned to the buoy carrier frequency. Water temperature information is converted to an audio frequency signal and delivered to the recorder.

Contractor
Loral Control Systems, Archbald, Pennsylvania.

XBT/XSV BATHYTHERMOGRAPH SYSTEM

Type
Expendable temperature/sound velocity probe.

Description
The Sippican XBT/XSV system uses inexpensive, expendable probes to obtain vertical profiles of temperature and sound velocity. The temperature data obtained by XBT probes may be used to compute sound velocity, although the XSV probes measure sound velocity directly. Whether computed or measured directly, the sound velocity data obtained assists in the prediction of sound propagation and is crucial to the effective use of sonars. An XBT/XSV system includes XBT or XSV probes, a deck-mounted or through-hull launcher, and a Mk 8 microprocessor-based data processor/recorder.

The temperature data obtained by the XBT may be used to compute sound velocity wherever salinity is constant. In areas where salinity may be highly variable, such as the Mediterranean and the Arctic, temperature data may not be sufficient to calculate velocity and to predict adequately acoustic propagation. In these areas sound velocity must be measured directly with the XSV probe. The XSV uses a unique sing-around transducer to obtain this direct measurement.

Both XBT and XSV probes obtain data in real time. Following launch, data collection begins at the surface and as the probe descends wire unreels from a spool located within the probe, as well as unreeling from another spool in the launch cannister. This enables the ship to use the system without restrictions on speed or manoeuvrability.

The Mk 8 data processing system, US Navy nomenclature AN/BQH-7, is linked to the descending probe by the hard wire link. The processor provides profiles of temperature or sound velocity to three outputs: digital data stored on magnetic tape, analogue chart paper trace, and RS 232 digital interface to external data systems.

XBT/XSV expendable profiling system

Operational Status
In service.

Contractor
Sippican Inc, Marion, Massachusetts.

SSXBT/SSXSV BATHYTHERMOGRAPH

Type
Submarine-launched expendable bathythermograph.

Description
Profiles of temperature and sound velocity versus depth may be obtained by a submerged submarine using the submarine-launched expendable bathythermograph/sound velocity (SSXBT/SSXSV) equipment. This is similar to the XBT/XSV system described previously except that the probes are launched through the signal ejector of the submarine and are carried to the surface by a buoy. The buoy releases the probe just before reaching the surface, and the probe then descends, measuring temperature or sound velocity on the way. The SSXBT expendable probe senses the water temperature profile from the surface to a depth of 760 m and transmits this to the moving submerged submarine. The SSXSV sound velocity probe provides a profile of measured sound velocity down to 850 m. Temperature or sound velocity data are transmitted to the submarine, via the fine wire tether, for recording and display.

The breech door/cable feed-through assembly provides the inner door for the signal ejector. It allows the tether wire to pass through the door and provides a seal around the wire. At the end of probe deployment a guillotine in the assembly cuts the tether wire allowing the SSXBT components to clear the submarine and scuttle.

The AN/BQH-7 and Mk 8 recorders/processors provide temperature and sound velocity profiles. The processor can convert measured temperature to compute sound velocity based on a standard salinity of 35 ppt.

Operational Status
In service.

Specifications
SSXBT
Range: −2°C to +35.6°C
Accuracy: ±0.15°C
Depth range: 0-750 m
Depth accuracy: ±2%
Size: 99.8 × 7.5 cm

SSXSV
Range: 1405-1560 m/s
Accuracy: ±0.25 m/s
Depth range: 0-850 m
Depth accuracy: ±2%
Size: 99.8 × 7.6 cm

Contractor
Sippican Inc, Marion, Massachusetts.

ENVIRONMENT MEASURING BUOYS

CANADA

AN/SSQ-36, AN/SSQ-536 and AN/SSQ-937 BUOYS (AXBT)

Type
Air-launched bathythermograph buoys.

Description
Sparton of Canada has developed a low-cost, lightweight and rugged bathythermograph buoy, responsive to NATO and national requirements and standards. The SSQ-36/536 buoys are air-launched, expendable, temperature sensing devices, packaged as a standard 'A' size buoy. Since the probe and buoy electronics occupy only about one third of the 'A' size package, downsizing to 'G' (A/2) or 'F' (A/3) size is readily achievable. The thermistor probe is capable of providing information on water temperature down to 1000 ft. Deeper depths can be provided on request.

The thermistor, located in the probe, senses the changes in seawater temperatures during its descent from the surface. These changes are transmitted to the launch aircraft by a conventional sonobuoy transmitter where the data are processed and converted to a display of absolute temperature versus depth. These data allow aircrews the capability to evaluate local effect of the seawater temperature on sonar propagation and acoustic range prediction.

The AXBT buoys may be launched at airspeeds between 30 and 370 kts from altitudes of 100-25 000 ft (30-7600 m). Descent is controlled and stabilised by a parachute which deploys when the buoy exits from the vehicle launch tube. Immediately after water entry, the battery is activated and initiates the release of pressurised gas to inflate the flotation bag and erect the internal VHF antenna. The electronics package returns to the water surface, buoyed by the float. The outer housing, float cap and parachute are released and sink. After a short interval to allow stabilisation of the probe temperature, a cutter is activated and the thermistor begins its descent, measuring water temperature. The weight and shape of the probe are closely controlled to ensure a predictable descent rate.

Operational Status
'A' size in production.

Specifications
Description: Passive, bathythermograph

Air-launched Bathythermographs

	Military qualified		Commercial	
	AN/SSQ-36	AN/SSQ-937	Standard AXBT	Deep AXBT
Probe depth	305-800 m (1000-2625 ft)	457 m (1500 ft)	305 m (1000 ft); 457 m (1500 ft); 800 m (2625 ft)	1200 m (3937 ft)
Size	'A'	'F'	'A' or 'F'	'A'
Drop rate	5.0 ft/s (1.52 m/s)	5.0 ft/s (1.52 m/s)	20 ft/s (6 m/s)	20 ft/s (6 m/s)
Depth accuracy	±5%	±5%	±5%	±5 m
Time constant	1 s	1 s	100 m/s	100 m/s
Temperature accuracy	±0.5°C	±0.5°C	±0.18	±0.18
Temperature range	−2°C to +35°C	−2°C to +35°C	−2°C to +35°	−2°C to +35°C

Notes:
Higher accuracies available.
Sizes: 'A' – 124 mm OD × 912 mm length (4.9 in OD × 36 in length)
'F' – 124 mm OD × 305 mm length (4.9 in OD × 12 in length)
Launching: 'F' size buoys are gravity launch only.

Contractor
Sparton of Canada Ltd, London, Ontario.

Function: Bathythermal profile
Application: Standard probe used by US and Canadian Navy for ASW operations
Size/Weight: 'A' size 5.5 kg, 'G' size 4.4 kg, 'F' size 3.8 kg
Activation time (after water entry): Up to 30 s
Transmitter: Channels 12, 14, 16 (or as required)
RF power: 0.25 W
Scuttling: Not longer than 16 h
Max launch altitude: 25 000 ft
Max launch airspeed: 370 kts
Launch envelope: CAN STD

CMOD

Type
Compact Meterological and Oceanographic Drifter.

Description
The CMOD is a low-cost, easily transported and deployed expendable drifting buoy. Constructed of heavy gauge marine grade aluminium, the buoy is compatible with the Argos satellite system. Standard onboard sensors provide for barometric pressure measurement, air temperature measurement, sea surface temperature measurement, battery voltage and data time history. The CMOD is standard 'A' size sonobuoy and can be deployed by hand from virtually any size of aircraft or ship. The buoy is packaged for gravity tube launch from patrol aircraft.

Contractor
Metocean Data Systems Ltd, Dartmouth, Nova Scotia.

ICE DRIFTER

Type
Ice drift measurement buoy.

Description
The Ice Drifter is a robust version of the standard drifter. It is designed and constructed to be a through-ice position measuring platform, capable of supporting a wide variety of optional sensors. Upon ice break-up/melt it functions as a normal drifter. Sensor options include measurement of ice surface temperature, air temperature, sub-surface temperature, barometric pressure, wind speed and direction, and current speed and direction.

Contractor
Metocean Data Systems Ltd, Dartmouth, Nova Scotia.

CALIB

Type
Compact Air-Launched Ice Beacon.

Description
The CALIB is designed to provide a compact, lightweight and low-cost vehicle, capable of providing position information via the Argos satellite system. The CALIB has a sonobuoy 'A' size configuration, which enables air launch from any type of aircraft, both fixed- and rotary-wing. Additional sensors to the standard position package are available. Options include barometric pressure sensor and surface temperature sensor.

Contractor
Metocean Data Systems Ltd, Dartmouth, Nova Scotia.

AMBIENT NOISE DRIFTING BUOY

Type
Acoustic measuring device.

Description
The ambient noise drifting buoy has been developed to provide long-term statistical data on ocean ambient noise. Background noise is monitored in a number of frequency bands, spectral density levels are calculated and the data are telemetred via the Argos satellite system. A sophisticated suspension system ensures that self-generated noise from the buoy is minimised.

Contractor
Seimac Ltd, Bedford, Nova Scotia.

AOB (ARCTIC OCEAN BUOY)

Type
Ambient noise data collection buoy.

Development
The AOB is being designed as a platform for deploying sensors through the ice. Future payloads will include thermistor strings (TZD), conductivity, temperature depth probes (CTD), sound velocity and current profile.

Description
The AOB is an air-launched (from 1000 to 5000 ft at 180-250 kts), 'A' size ice penetrating, passive oceanographic buoy. The thermochemicals used in melt-through are safe to handle and environmentally friendly. The AOB will penetrate 3 m of ice and activate in approximately 30 minutes. The initial operational configuration design calls for the deployment of an omnidirectional hydrophone for collection of broadband (1/3 octave) ambient noise data, estimation of ice thickness, and satellite transmission capability for the data using a service Argos satellite. The buoy has a life capacity of 90 days.

Operational Status
In development.

Specifications
Length: 914 mm
Diameter: 123.82 mm
Weight: 17 kg
Frequency: 401.65 MHz
Sensor range: 5 Hz to 5 kHz
Operating depth: 100 m
Power: Lithium batteries

Contractor
Sparton of Canada Ltd, London, Ontario.

FRANCE

WADIBUOY

Type
Wave Directional Buoy.

Description
The Wadibuoy is a comprehensive meteo/oceanographic real-time instrumentation system. It provides the user with directional spectra of the wave energy, the surface current speed and meteorological parameters such as wind speed, temperature, pressure, solar radiation and so on.

Wadibuoy is a flat-bottom toroidal buoy providing great stability and ability to remain on the surface of the wave. The buoy carries a protected radio transmission system for coastal application at ranges up to 30 km, and a satellite transmission system (Argos/Goes) for oceanic application.

Operational Status
The buoy is in series production following five years of field testing in various oceans around the world. Some 30 buoys are in use in over 10 countries. It is used by the French Navy for determining the environmental conditions during firing trials, calibration of mathematical models which take into account wave frequency, and oceanographic research.

Specifications
Diameter: 2.45 m
Height: 2.5 m
Weight: 800 kg
Buoyancy: 3000 kg

Contractor
Société Nereides, Les Vlis.

Wadibuoy wave directional buoy

UNITED KINGDOM

SSQ-937A BATHYTHERMAL BUOY

Type
Air-launched bathythermal sonobuoy.

Development
The SSQ-937A is a second-generation bathythermal buoy developed by Marconi for the UK Ministry of Defence.

Description
The SSQ-937A uses an innovative approach to provide a cost-effective solution for the air-launched bathythermal requirement. The 'F' size buoy is one-third the size and less than one-third the weight of the standard SSQ-36 'A' size bathythermal buoy, yet it achieves a full 457 m bathythermal profile measurement.

In its packaged 'F' size configuration, the spring steel antenna is folded over the body and secured by a windflap. Immediately after launch, release of the windflap allows the antenna to erect and releases drag fins around the body of the buoy. Air pressure deploys the fins to their flight mode to retard and stabilise the ballistic flight of the buoy into the water.

SSQ-937A bathythermal buoy

A plastic moulded upper unit carries the battery and the transmitter. Being positively buoyant, this forms the flotation system to deploy the antenna.

A seawater entry switch powers the electronic system switching on the RF power on water entry.

After a 20 second delay, when the buoy has stabilised to the surface temperature, timing circuits initiate deployment of the probe unit which descends at a constant velocity to make the bathythermal measurements. On completion of the bathythermal measurements, and transmission of the information, the buoy automatically scuttles.

The SSQ-937A may be launched from a helicopter or fixed-wing ASW aircraft using existing secondary storage and launching systems. With no requirements for a parachute or float inflation system it is extremely safe and reliable in use. The SSQ-937A is fully compliant with the UK MoD specification RAE/RSP 3985.

Operational Status
In production and ordered in large quantity by the UK MoD.

Specifications
Operating depth: 457 m
Descent rate: 1.5 m/s
Modulation frequency: 1360-2700 Hz
Dimensions: 305 × 124 mm (diameter)
Weight: 2 kg

Contractor
Marconi Underwater Systems Ltd, Templecombe.

UNITED STATES OF AMERICA

AN/SSQ-36 BATHYTHERMOGRAPH

Type
Air-launched expendable bathythermograph (AXBT).

Description
This is the standard sonobuoy used by the US Navy to obtain temperature and computed sound velocity from fixed-wing aircraft and helicopters. The AXBT is packaged in a standard sonobuoy configuration and may be launched from all 'A' size sonobuoy launchers. It transmits temperature data over RF frequencies 170.5, 172 and 173.5 MHz (standard US Navy sonobuoy channels 12, 14 and 16).

The AN/SSQ-36 is used to obtain temperature profiles to 300 m (1000 ft), which are used to identify the thermocline and calculate sound propagation. A deep version of the AXBT measures temperatures down to 760 m (2500 ft) to provide information for oceanographic purposes.

In addition to the AXBT, Sippican has developed an aircraft-launched expendable sound velocimeter (ASXV). This measures sound velocity to an accuracy of ±0.82 ft/s at depths to 2800 ft. This sound velocity profile is used to determine the optimum deployment of sonobuoys and dipping sonars in areas where computed sound velocity profiles are unreliable because of variations in seawater characteristics and salinity.

Operational Status
In production.

Contractor
Sippican Inc, Marion, Massachusetts.

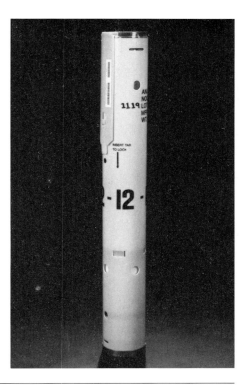

AN/SSQ-36 aircraft-launched expendable bathythermograph (AXBT)

PREDICTION SYSTEMS

CANADA

ACOUSTIC RANGE PREDICTION SYSTEM (ARPS)

Type
Software acoustic prediction system.

Description
The Acoustic Range Prediction System (ARPS) is a software system designed to supply on-scene environmental support to ASW operations. Acoustic models have been developed which, when provided with environmental data such as temperature profiles, wave height and bottom reflectivity, compute acoustic transmission loss and thus predict sensor performance, expected ranges and detection probabilities.

Working from real or recorded data, the operator provides source depth, receiver depth, parameters describing environmental conditions and a selection of acoustic frequencies. ARPS then computes propagation loss as a function of range for each frequency. Propagation loss curves are converted to probability of detection curves to obtain range information. Sound levels of source and ambient noise, detection threshold and standard deviation of signal excess are among the inputs.

These results can then be used to interpret where a sensor can detect a sound and where it is unable to do so because of oceanographic factors. The models have an extension beyond that of ASW applications and could be used to predict performance of any acoustic navigation, control or communication link or MCM operational conditions.

The database which has been integrated into ARPS includes ocean depth on a 5' arc grid, and monthly temperature and salinity profiles on a 1° grid. Shipping density is also available on a 10° grid. The coverage of these is global. Monthly climatology average wind speed is also included on a 10° grid for the North Atlantic and North Pacific.

Operational Status
ARPS was developed for the Department of National Defence, Canada.

Contractor
Oceanroutes Canada Inc (a member of Swire Defence Systems), Halifax, Nova Scotia.

DEPLOYABLE ACOUSTIC CALIBRATION SYSTEM (DACS)

Type
ASW research and calibration device.

Description
DACS consists of an onboard controller and power amplifier module with a deployable calibrated sound source. The system accurately generates low frequency tones at operator specified strengths and frequencies in actual operating conditions. These tones are received by the remote passive sensors being calibrated. From acoustic propagation loss and source strength, the signal level at the receiver can be accurately determined and thus provides a means of calibrating the passive sensor's sensitivity. The standard DACS comprises, at the wet end, a Model 18SA0325 low frequency, wide bandwidth, depth compensated, ringshell projector with a matching network and a projector drive level feedback system.

The shipboard module comprises a winch and a personal computer-based controller and display terminal. The feedback subsystem monitors the projector's output and continuously transmits the formatted data to the control console which then makes the necessary adjustments.

The compact arrangement of DACS makes it suitable for installation on COOP vessels. The system can also be modified for moored applications.

Operational Status
In development with the Canadian Forces.

Specifications
Projector: RSP Model 18SA0325
Display: Display terminal

Input device: Keyboard
Number of tones: 4
Frequency accuracy: ±0.1 Hz
Frequency range (nominal): 60-600 Hz
Frequency separation: 1 Hz
SPL* (nominal): 100-150 dB
SPL accuracy: ±1 dB
Operational depth: 3-300 m (10-1000 ft)
Tow speed: Up to 3 kts
Maximum sea state: 4
Power requirements: 220 V AC, 15 A 1 Phase

*Not all frequency and SPL combinations available at all depths.

Contractor
Sparton of Canada Ltd, London, Ontario.

NETHERLANDS

SPI-04 SOUND RAY PATH ANALYSER

Type
Analyser for determination of sonar conditions.

Description
The SPI-04 is a small special-purpose computer-based equipment providing instantaneous prediction of sound ray paths calculated directly from sound velocity over long, medium and short ranges. All electronics, including power supply, are housed in a single cabinet weighing 70 kg. An 8 in (20.3 cm) diameter CRT display is provided, and sound ray paths are displayed with a spacing of 0.5°, the number depending upon beamwidth selection. The SPI-04 is suitable for both anti-submarine operations and oceanography applications. Sound velocity/depth information is provided by equipment such as the Van der Heem/Sippican expendable sound velocimeter (XSV) or similar equipment.

Main features include: use of microprocessors for high speed and reliability; stable presentation due to high display repetition rate; seven adjustable layers with sound velocity settings; variable bottom depth with reflections on/off switch; profile setting display on LED numerical indicators.

Operational applications include:
(a) on surface ships to provide a clear picture of the

Sound ray path analyser SPI-04

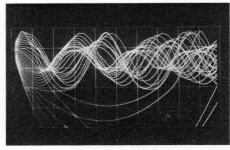
SPI-04 ray path analyser presentation

sonar conditions, enabling the best depth for VDS to be established, and to evaluate the possible submarine tactics
(b) in submarines for the determination of optimum listening and escape depths, and defensive tactics against helicopter sonars and sonobuoys
(c) in ASW aircraft and helicopters to establish the best depth for deploying sonobuoy hydrophones and transducers
(d) minehunting in shallow waters
(e) instant presentation of sonar coverage.

Operational Status
The SPI-04 is in operational service with the Royal Netherlands Navy and various other navies.

Specifications
Angular ray spacing: 0.5°
Function switch: 4.5 m (fixed depth); 0-600 m (variable depth)
Range/depth scales: 5; 9 km/375 m to 72 km/6000 m
Bottom depth setting accuracy: Better than 2 m

Contractor
Signaal Special Products, The Hague.

PRS-02/PATHFINDER

Type
Sound velocity measuring system.

Description
The PRS-02 system is used for two purposes: for measurement of the seawater temperature in relation to depth (by a bathythermograph probe), and for direct measurement of the sound velocity as a function of the depth (by a velocimeter probe).

The system employs expendable probes, made by Sippican Inc, USA. The sensor systems in the sound velocity probes are developed and manufactured by Signaal.

A probe, which can be launched manually or from a fixed launcher, transmits its data via signal wire to the combined processor recorder. The unit then translates the data into temperature/depth or sound velocity/depth profiles in both hard copy or via an RS 232 or 422 interface to other peripherals. In spite of its wide range of uses, the entire system can be operated by just one man. One measurement takes only a few minutes, while neither speed nor course of ship need to be changed.

Pathfinder uses advanced processing for various requirements (for example, recording and playback facilities, extra and/or enlarged copying for better resolution near surface or shallow water).

The PRS-02/Pathfinder system can be digitally interfaced with a sonar or data handling system.

Operational Status
The system is now in production and delivered to several navies and is operational on board submarines and surface ships. In the Royal Netherlands Navy it is aboard the 'M' class frigates and 'Walrus' class submarines; the system is on order for the German Type 123 frigates and the Turkish Navy.

Contractor
Hollandse Signaalapparaten BV, Den Haag.

The PRS-02/Pathfinder processor/recorder

UNITED KINGDOM

MINEHUNTING PERFORMANCE PREDICTION SYSTEM

Type
Computer-based prediction system.

Development
The Marconi minehunting performance and prediction system has been developed to supplement modern minehunting sonars, gathering data on underwater environmental and operating conditions to enable sonar operators to select the optimum operating parameters and to derive a probability factor for detection and classification of perceived mines using those parameters.

Description
The self-contained, computer-based system consists of a variable depth sound velocity measuring probe which is lowered and raised by an outboard winch and jib assembly sited on the upper deck, and which is controlled either remotely from the operations room or from a local control panel on the winch. The computer, plotter and electronic panel and rack are located below deck in the operations room.

The probe measures the velocity of sound, pressure (indicating the depth) and temperature at intervals of time as the probe is raised from the operator-selected depth. An optional lightweight probe is available which can replace the winch and probe assembly.

Using these data together with data keyed in by the operator and data received from minehunting sonar the computer applies pre-programmed algorithms to produce recommendations for the sonar parameters to be used, together with predictions of the classification and detection performance which can be obtained.

The operator interacts with the menu-driven program software in the computer using a keyboard and screen displays to insert parameters, select options, deploy the probe and request recommendations and graphical and tabulated information.

The four-colour X-Y plotter is used to obtain hard copy printout of the computer screen displays.

In computing the recommended sonar parameters and performance predictions, the system takes account of the ship's sonar characteristics (frequency, pulse length and so on); underwater environment

The sound measuring probe and handling system of the minehunting performance prediction system

(reverberation, ambient noise, sound profile and so on); ship characteristics (flow/propulsion noise levels and so on); detection, classification and operator performance models; and specific mission inputs (that is, ship speed, sonar operating mode, target characteristics and deployment). The system computes and displays detection range and depth characteristics (detection contour); probability of detection against range; probability of classification against range for moored and ground mines; real-time monitoring of sonar conditions; performance summary; velocity of sound versus depth; ray trace display; optimum towed body height for the variable depth sonar; and data for mission planning, allowing selection of optimum operating modes.

Operational Status
The system is operational aboard 'Hunt' and 'Sandown' class MCMVs of the Royal Navy.

Contractor
Marconi Underwater Systems Ltd, Templecombe.

SEPADS SYSTEM

Type
Underwater environment prediction system.

Description
The Type 2068 Sonar Environmental Prediction and Display System (SEPADS) has been designed to provide real-time advice to the command on the most effective deployment of units and sensors. This is a particular requirement for towed array sonars which need accurate information so that they can be used to the best effect.

SEPADS interfaces with onboard sensors for the direct input of satellite, radiosonde, navigational, bathythermograph, sonar and meteorological information which is stored on a computer disk. This

information together with other tactical data provides the basis of input to environmental and acoustic models from which tactical decisions are evaluated and displayed. In conjunction with sound-propagation-loss models, it is used to forecast parameters such as the sonar ranges and the optimum depth at which to deploy the towed array. Processing is carried out in a PDP 11/44 computer.

The system also features advanced TMA packages which complement the tactical environmental decision facilities.

Operational Status
A total of 38 in service with the Royal Navy.

Contractor
Dowty-Sema Ltd, Esher.

Typical display from the SEPADS system

HAIS

Type
Acoustic prediction/management system.

Description
HAIS is a state-of-the-art tactical system that was originally developed for the complex shallow water environment of the Baltic Sea, where marked changes in salinity and temperature levels have a significant effect on underwater acoustic propagation. The system has since been modified to work equally well in deep water, making it independent of the operational area.

To exploit the maritime environment effectively demands tactical systems that can accurately model, in a usable timescale, the impact of environmental effects on the performance of sensors and weapons. HAIS incorporates the following key features which enable it to achieve this:

(a) fast display of oceanographic data and hydrographic charts from a comprehensive database
(b) 3D high resolution displays of bottom topography
(c) on-line libraries of climatological data
(d) display and store of *in situ* temperature, salinity and sound speed traces
(e) calculation of acoustic propagation loss using the model most appropriate to the tactical scenario
(f) display of range dependent acoustic ray trace
(g) display of propagation loss and Probability of Detection (POD) in both horizontal and vertical sections
(h) performance assessment of active and passive sonars
(i) assessment of tactical advantage and sonar policy using built-in sonar database
(j) tactical packages giving advice on deployment of sensors and units.

This type of system offers major advantages to the commander at sea, allowing him to deploy his units and sensors to maximum effect. HAIS has the

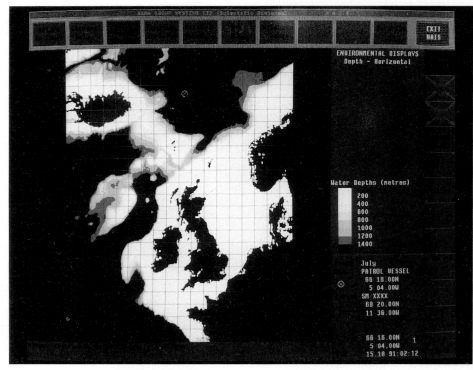

Typical display from the hydroacoustic system HAIS

capability for substantial enhancements in the future as sensor technology evolves and new requirements are formulated.

HAIS is written in the Ada language, using proven workstation hardware, databases and graphics packages. This approach provides a highly reliable system with the benefits of window-based, high resolution graphical displays and safe data storage.

Operational Status
Two systems delivered to the Royal Swedish Navy.

Contractor
Dowty-Sema Ltd, Esher.

LOGICA UNDERWATER TECHNOLOGY

Type
Processing and interpretation services.

Description
Logica is a computer software and consultancy company with considerable experience in the underwater sphere, particularly in acoustic processing and interpretation, and works closely with both government establishments and defence contractors. This covers underwater acoustic modelling signal and data processing, developing of algorithms for autodetection, tracking and classification processes and the conduct of sea trials on operational hardware.

An oceanographic and environmental satellite system, known as ODESSA, is currently operational manipulating oceanographic data. This configuration is in use in NATO countries for research into the use of derived data to assist in the satellite prediction of sea temperature and salinity values and anomalies.

Sonar tracking systems have been developed for the Royal Navy to track targets automatically and derive data from very faint signals. A data logger has also been developed, based on a similar hardware architecture, which interfaces to the tracker so that signals captured at sea can be recorded for subsequent analysis.

Operational Status
In service.

Contractor
Logica Defence and Civil Government Ltd, Cobham.

ODESSA

Type
A software-based Oceanographic Data and Environmental Satellite System Application (ODESSA).

Development
Developed by Logica, ODESSA was derived from a number of UK Ministry of Defence contracts for the Ocean Science Division, Admiralty Research Establishment at Portland.

Description
ODESSA provides the ability to manipulate and display data derived from *in situ* measurements and from remotely sensed satellite images. It can access a large database of *in situ* data in a variety of combinations of spatial, temporal and data type specifications.

The man/machine interface uses an easy-to-use menu-driven system with a mouse to aid certain selections and interactive display enhancement facilities through the use of special function keys. It provides a direct comparison between satellite images and *in situ* data extracted from the database.

The database search area is specified by providing the limits of the required region either through a sequence of menus, or by using a mouse and interactive graphics on the screen. Selected regions can be rectangular, circular (any radius), track or swath in configuration. Any boundaries can be placed on time and the minimum and maximum dates for which data are to be accepted. Different forms of data can be stored on the database and the user can specify the form of datum required, the type of instruments used to obtain it and its source. The system currently operates on oceanographic data, typically comprising temperature/salinity values taken at various depths throughout the region of interest.

Images are obtained either from magnetic tape or directly (over a serial line) from a satellite data receiver. AVHRR and APT images produced by NOAA satellites are those currently processed. The grey levels from different IR bands are used to calculate the sea surface temperature corresponding to each pixel in the image. The image is then mapped from the satellite observed view to a standard map projection. Coastline overlays are used as ground control points, to assist in the accurate location of the image.

Contractor
Logica Defence and Civil Government Ltd, Cobham.

MERMAID

Type
Software and database package for predicting in-water environmental conditions in shallow water areas.

Development
The Mermaid concept has been developed by the Admiralty Research Establishment and the Unit for Coastal and Estuarine Studies at the University of Wales, Bangor.

Description
A forecast model enables Mermaid to make predictions over large or small regions for such parameters as temperature, sound speed, tides and currents. The integral database contains an extensive set of temperature and salinity profiles, tidal components and bathymetry.

In military applications Mermaid can be used in support of shallow water ASW where it provides predictions of thermal structure and sound speed profiles for input to sonar performance models. It also predicts the position and strength of oceanographic features (for example fronts) from database and other considerations. Data on the depth of the surface mixed layer and estimates of diurnal heating effects can be provided together with estimates of environmental factors affecting ASW tactics.

In support of mine countermeasures Mermaid predicts the strength and direction of tidal currents in order to determine safe stand-off distances from hostile objects. It also predicts tidal currents in support of diver and ROV operations. Waves and tides are predicted in order to determine the MCM platform's dynamics and response. It also provides estimates of nearbed tidal and wave induced currents for mine burial prediction and predicts temperatures and sound speed profiles for input to MCM sonar performance prediction models.

Operational Status
The forecast model component of Mermaid is incorporated in the HAIS (Hydro Acoustic Information System) developed by Dowty-Sema Ltd for the Royal Swedish Navy.

Contractor
Admiralty Research Establishment, Portland.

TDA

Type
Prototype Tactical Decision Aid (TDA) system for making optimum use of the ocean environment.

Development
The Admiralty Research Establishment and Sema Scientific jointly developed this unique system which enables the user to make optimum use of the ocean environment for tactical purposes in terms of maximising the detective or evasive capability of a unit.

Description
TDA has been developed as a result of a comprehensive programme of work designed to identify, investigate and demonstrate effective tactical exploitation. It uses an extensive database obtained using actual trials' measurements from a variety of sensors such as satellite remote sensing systems, XBT expendable bathythermographs, environmental sensors (gathering conductivity, temperature and depth data), ODESSA (Oceanographic Data and Environmental Satellite System Application) and a thermistor chain incorporating a high resolution temperature sensor. The system is hosted on a Micro VAX II incorporating advanced computer graphics to create versatile displays. TDA offers both plan and vertical sections of probability of detection on any chosen target. The plan sections allow own and target locations to be superimposed. Ocean depths for any selected depth intervals can be displayed showing bathymetry data and there is a split screen facility allowing the simultaneous viewing of an acoustic ray trace with the corresponding estimates of acoustic propagation loss. This enables the comparison of qualitative and quantitative views of the sound field respectively.

Contractor
Admiralty Research Establishment, Portland.

UNITED STATES OF AMERICA

ATMS TRACK MANAGEMENT SYSTEM

Type
ASW database management system.

Description
The ASW Track Management System (ATMS) receives undersea ocean surveillance information from supporting sensors and sources, and correlates and tracks this information to maintain a database of user-selected surface, sub-surface and air reports.

ATMS provides the user with colour graphics, maps, decision aids and dynamic displays of all platforms and ambiguous contacts within a specified region of the globe using NTDS symbology. Real-time tracking is supported by sophisticated correlation and tracker algorithms. These algorithms produce a database available to the operator. Through this database, the user can access track/contact, library, communications, decision tools, correlator information and automated message generation. The user can edit the information as required.

The baseline ATMS is hosted on a network of DTC II (or Sun 4) workstations over a standard Ethernet local area network. The maximum number of workstations is currently limited to 50. The preferred workstation configuration is a 19 in colour monitor, 32 Mbytes RAM and 250 Mbytes of available disk storage. User interaction is via both keyboard and trackerball or mouse. ATMS is hosted under SunOS version 4.0.3. ATMS is written in C and FORTRAN.

Contractor
Lockheed Missiles & Space Company, Sunnyvale, California.

HYDROGRAPHIC SURVEY SYSTEMS

ECHO SOUNDERS

FRANCE

TSM 5260/5265 ECHO SOUNDERS

Type
Shallow and deep water multi-beam echo sounders.

Description
The TSM 5260 and 5265 are multi-beam echo sounders designed for shallow and deep water operations respectively. These multi-beam equipments are essential tools for effective sea mapping and allow for simultaneous depth measurements on a line perpendicular to the ship's axis. Other features include real-time layout of bathymetric maps in depths over 6000 m and logging of raw data for instant replay or off-line processing.

The TSM 5260 and 5265 incorporate a full range of digital technologies and allow for onboard replay with

Specifications

	TSM 5260	TSM 5265
Operating frequency	100 kHz	12 kHz
Pulse length	1 ± 10 ms	1 ± 10 ms
Source level		
(μPa/m)	210 dB	235 dB
Number of beams	20	60

modification of parameters where required, real-time testing and maintenance of all channels, and optional modification of interfaces and peripherals. Optional equipment includes digital optical disk, side-scan sonar and towed arrays.

The TSM 5260 Lennermor echo sounder uses 20 channels for shallow water (down to below 500 m) sounding on an operating frequency of 100 kHz. The TSM 5265 Nadzomor is a deep water sounder operat-

Maximum ranges

Vertical beams	>500 m	>10 000 m
Lateral beams	at 50°	at 45°
	approx 400 m	approx 9500 m
Angular resolution	5°	2° with
		1.5° intervals
Distance resolution	30 cm	1 m

ing at 12 kHz at depths greater than 10 000 m.

Operational Status
In service.

Contractor
Thomson Sintra Activités Sous-Marines, Valbonne Cedex.

GERMANY

DESO 21/22/25 ECHO SOUNDERS

Type
Survey echo sounders.

Description
The DESO 21 is a single channel, the DESO 22/25 are dual-channel survey sounders with built-in digitisation.

DESO 21/22
The DESO 21/22 are both high precision equipments for shallow and deep water applications with built-in digitiser functions and interface options for heave compensation (DESO 22 only). High precision echograms are annotated with digitiser tick marks, water sound velocity, basic measuring range and phasing, depth threshold line, time, distance and events markers. Operating range is from 1 to 5000 m.

DESO 25
This is a new dual-channel equipment, switchable to

two passive channels as well as manual and fully automatic gain control. Echo strength measurement is supplied with recording and data output functions. Also included are annotation functions together with a built-in clock and calendar. Operating range is from 1 to 15 000 m.

Operational Status
These echo sounders are in widespread use with the navies of Australia, Brazil, France, Greece, India, South Korea, Netherlands, New Zealand, Portugal, Sweden, Thailand, Turkey, USS, UK and USA. The DESO 25 is used by the Royal Navy, French Navy and the Portuguese Navy.

Contractor
Atlas Elektronik GmbH, Bremen.

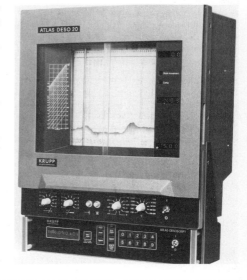

DESO echo sounder

NORWAY

EM100

Type
Multi-beam echo sounder.

Description
This multi-beam swath echo sounder uses computerised post-processing to produce seabed maps in coastal areas and on the continental shelf down to 600 m. It produces records of the total area of the seabed rather than the single lines obtained from conventional echo sounders. It uses three selectable swath widths to give optimum combination of swath coverage and horizontal resolution in water depths to 600 m.

The system uses a simple and easy to understand menu technique based on a joystick control. All other subsystems are controlled from the operator unit, the software carrying out calculations of relative position of depths from each beam, calculation of data for presentation on the VDU, and preparation of data for

storage on tape and/or for input to the real-time contour map processor. The graphic section of the display shows the athwartships depth profile of the swath as well as the along-track profile for any one of the 32 beams for a selected time interval. In addition, the signal strength of each receiving beam is shown below the profiles. Maps can be produced in real time while the echo sounder is in operation, or off-line, using recorded data as input.

The compact design and low weight enable the EM100 to be mounted in small vessels down to launch size. The transducer may be retracted into the hull for protection when not in use.

The main features of the system are the beam forming and interferometric processing within each beam, the unique bottom tracking algorithms, the utilisation of the sound velocity profile for real-time correction for beam deflection and the compact, curved transducer which is lowered below the aerated water during the survey. Post-processing of the bathymetric data is carried out on a DEC MicroVax 3400. The computer, together with peripherals and Simrad's

SP100 post-processing software go to make up the rest of the system. A full ocean depth version, the EM12, has been developed.

Operational Status
Five systems have been delivered to Canadian, Dutch and Norwegian authorities as well as three systems to offshore survey contractors. Two shallow water systems have been ordered for installation in the US Navy's T-AGS 51 and 52 coastal survey vessels currently under construction at Halter Marine. The first of the EM12 versions was delivered to the French ocean research institute IFREMER in 1991.

Contractor
Simrad Subsea A/S, Horten.

EM1000

Type
Multi-beam echo sounder.

Description
Developed from experience with the EM100 multi-beam sounder, the EM1000 uses the same operating frequency, 95 kHz, and is intended for high precision surveys in water depths between 5 and 800 m.

The system produces bathymetric data, and, as an option, a geometrically correct sonar image of the acoustical reflectivity of the seabed. The sounder incorporates full corrections for vessel movements and acoustic raybending in real time.

Like other Simrad echo sounders, the EM1000 uses a common computer and software system, Neptune, to take care of post-processing of data. The system can communicate with other equipments via Ethernet and/or RS 232 connections. A synchronising unit can be added to eliminate problems of interference between different acoustic devices on the ship.

The transducer is based on the EM100 transducer design, using a single transducer for both transmission and reception. The transducer is cylinder-shaped with a radius of 45 cm. It is configured from 128 ceramic staves covering an arc of 160°. Each stave consists of five elements with a fixed weighting in the fore and aft plane, and spaced by 1.25° in the athwartships direction. In the fore and aft direction the fixed beam forming has a 3° opening angle, and the beam centre direction is perpendicular to the transducer surface.

Received beams are formed in real time during the reception phase according to the beam forming mode which has been selected. Each beam is formed as a symmetrically weighted sum of the signal received over a selected group of staves. This eliminates any influence on the beam's direction from the sound velocity at the transducer, and makes it possible to factory calibrate each beam with an accuracy of 0.1°. All receiver beams are electronically roll-stabilised.

Split aperture is used so that measurement within each beam of the instantaneous direction to the point of backscatter is possible. Depths are measured in the centre line of footprint of each beam.

The receive beam forming modes are:
(a) Narrow – intended for deep water surveys typically 600 to 800 m, 48 roll-stabilised beams with a spacing of 1.25° are formed covering a 60° sector and mapping a swath of ×1.15 the depth of water
(b) Wide – intended for medium water depth surveys typically 200 to 600 m, 48 beams with a spacing of 2.5° are formed during each ping. All beams are shifted 1.25° every second ping, so that over a two-ping period the system obtains 96 soundings with a spacing of 1.25°. The measurement sector is 120°, corresponding to a swath width of ×3.4 the depth of water
(c) Ultra Wide – intended for shallow water surveys between 5 and 200 m, 60 beams with a spacing of 2.5° are formed during each ping. All beams are shifted 1.25° every second ping, so that over a two-ping period the system obtains 120 soundings with a spacing of 1.25°. The measurement sector is 150°, corresponding to a swath width of ×7.4 the depth of water.

Split aperture phase detection allows more than one detection per beam, and is to be introduced in future versions of the sounder.

Operational Status
A trial survey was carried out aboard the 65 ft SWATH hydrographic survey ship *Frederick G Creed* of the Canadian Hydrographic Service during the Spring of 1990. At speeds between 14 and 18 kts the sounder produced very successful results.

Contractor
Simrad Subsea A/S, Horten.

EA500

Type
Triple frequency echo sounder.

Description
The EA500 is a modular triple frequency hydrographic echo sounder with totally independent parallel processing within each of the frequency channels. This makes it possible to give separate settings to the different channels to provide optimal solutions for all depths.

Bottom detection is carried out solely by software and separate algorithms for each transducer channel are designed to maintain bottom lock, even for steep variations in depth, and special features have been included to avoid false bottom detection.

A wide range of frequencies is available and transmitted power is adjusted so that most existing transducers can be connected. In connection with the EA500 Simrad has developed a special split beam transducer. Using this transducer it is possible to estimate bottom slope in athwartships direction, detecting or indicating possible shoals to each side of the survey line. The bottom slope is shown in a separate field on the recorder and display echograms.

Depth errors to variations in sound velocity over the water column caused by changes in temperature, salinity and pressures can be automatically compensated for. Bottom depth is computed using a sound velocity profile rather than an average velocity for all depths. This profile consists of up to one velocity value per metre of water depth down to 1000 m. The velocity profile is entered manually from the control panel, automatically from a sound velocity probe, from a remote computer unit or as a combination of these methods.

The sounder also features a sub-bottom expansion feature on the display and recorders which allows the echogram below the detected sea bottom to be shown with a higher resolution than the main echogram. This is of value when studying soft sediment layers and bottom consistency and can be used for all three frequencies independently and simultaneously.

The echo sounder can be connected to a series of external devices including an unlimited number of LCD display units with integrated joystick/keypads, two totally independent colour graphic recorder units, navigation system giving position data to soundings, data logging computer storing all soundings with positioning data for later post-processing, annotation terminal to print comments echogram, vertical reference units to perform compensation, gyro-compass, sound velocity probe, and other external systems, interfacing via an LAN system such as an Ethernet.

A further option is the possibility of integrating the data in the Simrad post-processing system. This software is suitable for implementation in standard desktop computers and enables operators to build up a database of tide levels, digital terrain models, contour maps, 3D plots, depth profiles along specific trajectories and so on.

Operational Status
In service.

Contractor
Simrad Subsea A/S, Horten.

EA300/300P

Type
Compact echo sounders.

Description
This is a new generation of compact depth sounders which provides verified bottom tracking, depth data output digitised for post-operation processing, a menu in six languages, an operating panel with push-button control, an echogram in seven colours, and an input channel for annotation or navigation data for printed reference. A number of options are also available including a UHF link for transferring data to shore or another ship, a laser rangefinder, and various post-processing equipments.

Contractor
Simrad Subsea A/S, Horten.

UNITED KINGDOM

SERIES 700 ECHO SOUNDERS

Type
Echo sounding depth recorders.

Description
AB Precision produces a number of echo sounders to meet modern naval requirements. These include the Type 778, Type 786 and Type 787 systems.

Type 778
This equipment has been designed to meet the requirements of all vessels from coastal minesweepers upwards. It consists of a general service depth recorder operating with a single 10 kHz transducer; a bridge unit with numerical depth display, as an optional fitting, which operates with its own two-element 48 kHz transducer; and a precision depth recorder for oceanographic survey purposes which operates with the transmitter of the depth recorder and its associated transducer.

Type 786
The Type 786 has been developed for use with flat-bottomed landing craft. Main features are that the transducer is extendable below the layer of aerated water which forms beneath a flat-bottomed craft to give good performance, and rapid automatic retraction of the transducer as the craft beaches. The system consists of a main recorder unit with chart facility and digital display which operates with a single 200 kHz transducer, a remote digital display, and an automatic transducer raise and lower mechanism with its control panel.

Type 778 echo sounding equipment

Type 787

The Type 787 is intended to meet the requirements for all vessels in the Royal Fleet Auxiliary Service and the Maritime Auxiliary Service. It consists of a main recorder unit with chart facility and digital readout which operates with a single 50 kHz transducer and remote digital readout(s).

Operational Status

In operational service.

Contractor

AB Precision (Poole) Limited, Poole.

UNITED STATES OF AMERICA

MODEL 9057 (AN/UQN-4) SONAR SOUNDING SET

Type

Depth sounder.

Description

The Model 9057 sonar sounding set (US Navy designation AN/UQN-4) is an echo sounder which incorporates proven design and increased accuracy together with field change upgrades to improve shallow water capability. The Model 9057 has been developed as a replacement for the Model 185 (AN/UQN-1) set which has been the standard equipment on virtually every vessel and submarine in the US Navy, US Coastguard and Coastal Geodetic Service, as well as on domestic and foreign research and offshore vessels, for over 20 years.

The Model 9057 system uses a 12 kHz transducer and visually presents digital and graphic displays of water depths by means of a digital numeric display and permanent strip chart recorder. It may also be used as a passive listening device.

Operational Status

In service with the US Navy.

Contractor

EDO Corporation, Electro-Acoustic Division, Salt Lake City, Utah.

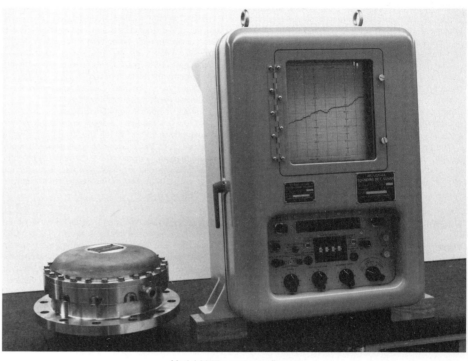

Model 9057 sonar sounding set

DSF-6000

Type

Dual frequency depth sounder.

Description

The DSF-6000 is a portable depth sounding system in which water depth is displayed in a four-digit digital format, visible in full sunlight. Readings are also transferred to an analogue chart recording system which accurately displays the bottom profile. Grid lines are written onto the intially blank chart paper during the bottom recording cycle, avoiding alignment problems encountered with preprinted charts.

The digitiser is programmed to automatically lock-on to the maximum water depth under the control of the system's computer. Once locked onto the bottom, an automatic tracking gate confines the search to the known vicinity of the bottom. Gate marks are printed on the chart, and digital depth data can be communicated to a tape record, data logger or computer through either an RS 232C serial or its parallel BCD I/O port.

The sounder operates at two frequencies: a high frequency of either 100 or 200 kHz for accurate shallow depth recording; or a low frequency of 12, 24 or 40 kHz for greater depths or to measure to firm bottom beneath the mud in shallower water. The DSF-6000 may be operated at either frequency separately, or at both frequencies simultaneously.

Full depth capability, selectable in seven steps via a front panel switch, is from 58 to 5800 ft, fathoms or metres. Full depth is covered on the chart display in seven overlapping phases. A specific phase can be chosen manually for display on the chart, otherwise the automatic mode displays that phase which includes the ocean bottom. Sound speed, tide and drift correction, chart scale, chart speed and a two-digit unit identification selected by the owner is automatically written onto the chart and incorporated into the serial digital data output. Time ticks at one minute intervals are printed along the top and bottom margins of the chart.

Contractor

Ocean Data Equipment Corp, Fall River, Massachusetts.

HYDROGRAPHIC SONARS

UNITED KINGDOM

TYPE 2034 SONAR

Type
Dual side-scan sonar for hydrographic work.

Description
The Waverley sonar Type 2034 is a short-range, high definition, dual side-scan sonar, suitable for: establishing the position of ships, aircraft or other objects lost at sea; for charting obstructions in shipping channels or at sea floor engineering sites; and for evaluating underwater topography. Designed for the utmost reliability and to meet exacting military standards, Type 2034 sonars have been in use with the Royal Navy since 1976. The system is now standard equipment on all ocean-going and coastal vessels of the survey fleet.

The sideways-looking left and right transducers are housed in a towed body, called the 'towfish', the current version of which is the Mk 3. Some of the electronics are contained in the towfish but the majority of this part of the system is housed in the recorder unit which provides for operation and control of the equipment, and display by hard copy, printout chart recordings.

The sonar operating frequency is 110 kHz with a nominal transmitted pulse length of 100 μs. A choice of three range scales, 75, 150 or 300 m, and one of three recording paper speeds, 30, 60 or 150 lines/cm, can be selected to suit the prevailing operating conditions. The transmission repetition rate depends on the selected range scale and is 10 pps for 75 m range, 5 pps for 150 m range or 2.5 pps for 300 m range. For all range settings, scale lines are recorded at 15 m range intervals to aid interpretation of the recordings.

Facilities are provided for putting event markers on the recording.

Operational Status
Over 50 systems are operational with the Royal Navy Hydrographic Service. Older systems are currently being retrofitted with a microprocessor-controlled signal processing system similar to that used in the Waverley 3000 sonar.

Contractor
Dowty Maritime Systems Ltd, Waverley Division, Weymouth.

HYDROSEARCH Mk II

Type
Hydrographic surveying sonar.

Description
Hydrosearch is designed for rapid and precise hydrographic surveying missions and the inspection of underwater targets. The hull-mounted, high definition scanning sonar incorporates motordrives for the stabilised platform which gives it greater precision and response to the ship's movement.

The sonar head, with transmitting and receiving arrays, is mounted on a dynamic, stabilised platform which relates the beam to a fixed spatial reference independent of the vessel's roll, pitch and yaw.

The 60° insonified sector is scanned electronically by a very narrow beam to generate a high definition video image. The scanned sector can be depressed to any angle from the horizontal and can be rotated to any position in azimuth – in either the vertical or horizontal mode. Each echoed pulse represents angle and range data for processing by the computer. The received echoes are digitised and subjected to modern image processing techniques. These greatly enhance the composite video and eliminate flicker.

With the adoption of the standard CCIR TV format, the system can use a wide range of devices such as TV monitors, line scan recorder, video recorder and output printers.

The sonar provides improved performance at speeds up to 10 kts and sea state 4.

Operational Status
The system is in use by the Royal Navy Hydrographic Service.

Specifications
Operating frequency: 180 kHz
Sector: 60°
Bearing resolution: 0.5°
Maximum display range: 600 m

Contractor
Marconi Underwater Systems Ltd, Waterlooville.

UNITED STATES OF AMERICA

MODEL 4065

Type
Side-scan/sub-bottom deep tow survey system.

Description
The Model 4065 integrated deep tow side-scan/bottom penetration system was specifically developed to meet operating requirements in water depths to 3000 m. The deep tow vehicle incorporates a specially designed 10 kW bottom penetration transceiver, side-scan transmitter and pre-amplifier and multiplex system in a unique, positively buoyant tow body.

The tow vehicle tracking option is available to locate the position of the vehicle relative to the surface vessel. The tracking system will locate the towed vehicle with an accuracy of 1° in azimuth and 1 per cent of full scale in range and depth. The short baseline tracking system is multiplexer compatible.

The 4065 system uses a dedicated multiplex system that permits operation over a single 25 000 ft coaxial cable without interference between the three operating channels of sub-bottom and side-scan. The system can be used for side-scan only, bottom penetration only or combined side-scan/bottom penetration. The 4065 deep tow survey system will operate with both the standard three channel recorder and EDO's 706 mapping recorder. The 4065 system includes an emergency location light beacon.

Contractor
EDO Corporation, Electro-Acoustic Division, Salt Lake City, Utah.

The 4065 tow vehicle

MODEL 4175

Type
Side-scan/sub-bottom systems.

Description
The Model 4175 side-scan mapping system consists of a dual frequency side-scan sonar, general-purpose video side-scan display and storage system. It is designed to provide a fully compensated and corrected topographic representation of underwater terrain and targets using the latest in microprocessor, video controller and data storage technologies.

The 4175 is a dual frequency (100/500 kHz) system which allows long-range or high resolution modes of operation in one unit. It can automatically remove the water column, correct for slant range, and correct for ship's velocity to produce a distortion-free display of the sea floor while simultaneously recording the data to optical disk or digital tape. Side-scan data can also be recorded in raw format and corrected in playback.

The 4175 uses a high resolution colour or grey shade monitor for the data display. Up to four channels can be displayed and corrected simultaneously. Targets or sea floor features can be zoomed 1 to 16 times and measured using the included trackerball. The 4175 is ideal for UAV/ROV applications.

Operational Status
In operational service.

Contractor
EDO Corporation, Electro-Acoustic Division, Salt Lake City, Utah.

MODEL 515A HIPACT

Type
Sub-bottom penetration system.

Description
The 515A HIPACT sub-bottom survey system is designed to gather high resolution data concerning the sedimentary layers of the sea floor. The system's High Power Acoustic Coaxial Transducers (HIPACT) produce a high intensity, short duration pulse of a single frequency to maximise both resolution and penetration. The 515A HIPACT offers a wide range of operational characteristics and is the only truly integrated system with all major system components

produced by one company. This total system design concept makes it the most accurate and reliable system in the world, and it has logged more than a million survey miles.

Maximum resolution from a sub-bottom profiling system is dependent on two factors, pulse duration and beamwidth. The pulse length is of prime importance because the system cannot resolve layers or objects less than the sound velocity times the pulse length.

The 515A HIPACT generates an extremely short, high intensity acoustic pulse from a coaxial transducer which is capable of precise repetition of each pulse every time it fires.

EDO's unique transducers produce acoustic beams which are highly directional and operate with extremely high efficiency over the entire ocean depth and frequency range for which they were designed. They are free from excessive side lobes and back responses to prevent interfering reflections from obscuring fine sub-bottom details.

The transducers are more than 10 times as efficient as energy discharge type transducers in gathering short pulse transmissions. This unmatched

performance, combined with the 248E multi-mode transceiver produces unsurpassed high resolution sub-bottom records. These records show maximum detail for the first 100 to 400 ft of the sea floor, with a vertical resolution of 0.5 ms.

Contractor
EDO Corporation, Electro-Acoustic Division, Salt Lake City, Utah.

Model 515A HIPACT

MODEL 706

Type
Side-scan mapping system.

Description
The Model 706 side-scan mapping system produces a plan view to the sea floor using automatically corrected side-scan data. The corrected data present all bottom features in their true scale and shape.

The 706 system automatically removes the water column, corrects for slant range, and correlates paper travel with the ship's speed to produce a distortion free display of the sea floor. The corrected records may be pieced together to form a mosaic of large bottom areas.

The system uses EDO's 602 Tow Fish, a compact unit weighing only 75 lbs including the transducers and sub-electronics. Depth rated at 2000 ft, this portion of the system is available with other psi configurations.

The heart of the system is the 706 recorder, which combines the proven stylus drive mechanism of the EDO Corporation, Electro-Acoustic Division's 606 recorder with a powerful data processor. The 706 system uses multiple microprocessors to provide the required real-time data processing. The 706 system features a keyboard command entry and alphanumeric display that is operator-friendly, with commands entered via the keyboard appearing on the display, which 'prompts' the operator by requesting the next data entry step required. Status information displayed on demand includes time, date, range, tow body altitude, ship speed, channels, shot number and bearing. A system test feature may also be initiated via the built-in keyboard. A self-test is performed every time the recorder is activated.

The 706 recorder is capable of one to four channel

display, permitting simultaneous recording of side-scan and sub-bottom data with selectable channel size (longer side-scan or larger sub-bottom display). The auxiliary channels can also be used to expand any portion of the side-scan data without interrupting the standard side-scan record.

Contractor
EDO Corporation, Electro-Acoustic Division, Salt Lake City, Utah.

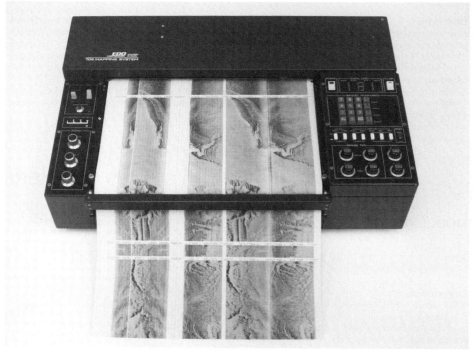

706 mapping system

SIDE-SCAN SONAR

Type
Side-scan sonar for oceanographic research and survey.

Description
The Side-Scan Sonar system provides the operator with the capability to create high resolution images of the sea bottom and sub-bottom for applications such as outer survey for the determination of bottom conditions and the selection of Q routes, the detection of cables and pipelines on and beneath the sea bottom, oceanographic survey for the determination of geological or biological conditions, as well as other applications requiring high resolution images of the seabed.

The design of the Klein sonar equipment permits the operation and real-time display of up to six channels of sonar data, including simultaneous 100 and

500 kHz side-scan sonar for multi-spectral insonification of targets and 3.5 kHz for penetration into the near sub-bottom area of the seabed. These data are displayed in real time correlated on the data medium for ease in operator interpretation. The simultaneous display of the sonar data provides information on both surface and sub-surface targets which could be missed if a single frequency were utilised.

The sonar system utilises a variable depth sonar (VDS) transducer which can be selectively deployed to avoid interference from temperature thermoclines. Various frequencies are available for the VDS transducer including 50 kHz for long-range, medium resolution applications, 100 kHz for high resolution applications, 100/500 kHz for simultaneous insonification of target areas, and 3.5 kHz for penetration of the sea bottom. Transducers are available for operations to depths as deep as 12 000 m.

The sonar images are transmitted up a VDS tow cable to a combination sonar transceiver and graphic recorder for processing and display. Accessories are

available which permit advanced image processing of the sonar data, as well as data reduction. Fully integrated sonar systems which interface with navigation and shipboard equipment are available.

The sonar system is lightweight, weighing less than 100 kg (225 lb) and is easily transported and installed aboard vessels of various sizes down to 5 m in length. The VDS transducer weighs less than 25 kg (55 lb) and is easily deployed by a single crewman without specialised handling equipment.

Full accessories are available for Klein Associates including tow cables, towing winches, as well as various related systems including acoustic positioning, surface navigation, and data management equipment.

Operational Status
In production and in service with hydrographic and oceanographic agencies worldwide, both government and navy.

Contractor
Klein Associates Inc, Salem, New Hampshire.

500 kHz very high resolution side-scan sonar image
of a magnetic sensing device on the ocean bottom
made with Klein high resolution Side-Scan Sonar and
displayed on a Klein graphic recorder

MODEL 260-TH

Type
Dual frequency, image correcting side-scan sonar.

Description
The Model 260-TH microprocessor-based image-correcting side-scan sonar generates plan view images of the seabed depicting the size, shape and location of various sea floor materials and man-made objects. In addition to generating corrected hard copy maps in real time, the system can store minimally processed data on analogue or digital magnetic tape for reprocessing by direct playback through the system.

Sonar images are fully corrected for slant range, ship speed and amplitude. Each pixel is individually corrected which results in a high quality, high resolution image. The acoustics, signal processing, data recording and graphic recording are consistent for a display with a pixel size of 1/800th of the selected range to each side, and an amplitude dynamic range of 64 dB (of acoustic back scattering strength variations).

The system is small, portable and easy to operate and its components include a surface processing and graphics unit, a subsea towfish and a tow cable.

Contractor
EG & G Marine Instruments, Cataumet, Massachusetts.

OCEAN SURVEY SYSTEMS

CANADA

SHOALS

Type
Airborne lidar bathymeter.

Description
Shoals is a laser-based scanning lidar bathymeter that offers hydrographers a cost-effective and rapid means of surveying shallow coastal or inland waters with depth and position accuracy to IHO standards. The equipment is designed to operate in a relatively small fixed-wing aircraft or helicopter, and provides an improvement over acoustic systems both in area coverage and in uniformity of sounding density.

Effective over most coastal areas that are of critical concern to the hydrographer, the system's speed and flexibility in deployment, and convenience in sounding dangerously shallow or highly confined and irregular areas, offers unique advantages in survey planning and deployment strategy.

The system uses two short duration laser pulses, one green and one IR in wavelength, transmitted simultaneously and coaxially from the aircraft down onto the surface of the water. Scattering of the IR pulse from the water surface is detected by a receiver located at the laser source. The green pulse is partially transmitted through the water, and its scattering from the bottom is also detected by the receiver at a later time. The water depth is determined by measuring the elapsed time between receiving the IR and green return signals. This depth value is subsequently corrected for geometric and environmental effects.

To provide wide ground coverage, the consecutive laser pulses are directed by a scanner sequentially across the flight path into preselected orientations that optimise the sounding pattern.

The location of each sounding is derived from data provided by aircraft positioning, attitude reference and scan angle control systems. Depth penetration is strongly dependent on the water quality, varying from several metres in turbid water to greater than 50 m in very clear water. Typically the system has a penetration capability of 20 to 35 m in most coastal areas.

The system is compatible with the Global Positioning System (GPS), which is valuable when surveys have to be conducted in remote areas.

Similar equipment may be used in ASW and MCM applications.

Contractor
Optech Inc, Downsview, Ontario.

FRANCE

HYDRAC-HYTRAI SYSTEM

Type
Hydro-oceanographic data acquisition and processing system.

Description
The Hydrac-Hytrai data acquisition and processing system can be embarked on board hydrographic ships and boats. It is designed to improve the productivity of hydrographic ship missions, and is specifically intended for military use for setting up and updating hydrographic documents used for navigation.

To adapt the different types of hydro-oceanographic vessels two versions of the system are available, designed to equip different types of hydrographic/oceanographic research vessel. The Hydrac launch version is designed for use in small craft employed in surveying the shoreline in shallow water. The Hydrac ship version is for specialised survey ships equipped with real-time computing capacity.

Both systems enable the acquisition of data in bathymetry, bathythermy, currents, gravimetry, magnetometry and tide. Hydrac ship and Hydrac launch are organised around a decentralised network in a loop structure (Hydroboucle) to which all sensors and real-time processors are connected. The primary advantage of this structure over systems organised around a central computer is to provide operational

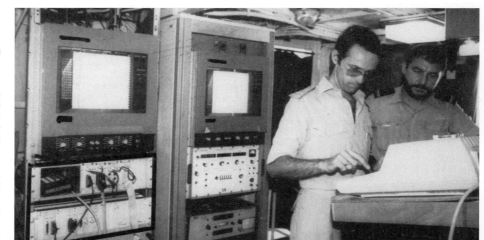

Hydrac-Hytrai system

modularity and flexibility, essential in hydrographic work in the data acquisition process.

The Hytrai system is used on board in survey preparation and data processing, corrections, and correlations to obtain automatically complete and definitive documents such as bathy sheets, profiles, representation in perspective and automatic shading.

Contractor
Société ECA, Colombes.

GERMANY

HYDROSWEEP DS

Type
Multi-beam hydrographic surveying sonar.

Description
Developed for large-scale surveying over depths down to 10 000 m, the multi-beam Hydrosweep system uses advanced electronic beam forming techniques to provide a swath width equivalent to twice water depth. The fully automatic system is self-calibrating, with mean sound velocity applied for depth and slant angle correction in real time. Display of data is in the form of either graduated colour contours or cross-section profiles via a series of high resolution VDUs. Interfaces for heave compensation as well as pre- and post-processing of all data are also available.

Main operating console of Hydrosweep

Operational Status
Various systems are in use by oceanographic institutions. A system is installed in the US Navy's Naval Sea Systems Command *Thomas G Thompson*, a new $20.9 million 263 ft oceanographic research vessel.

Contractor
Atlas Elektronik GmbH, Bremen.

HYDROSWEEP MD

Type
Hydrographic multi-beam surveying sonar with incorporated side-scan imaging function for medium water depth application.

Description
The Hydrosweep MD system covers a maximum swath width of eight times water depth in a depth range covering the continental shelf. The central part of the sound beams reach down to over 1000 m. A novel function is the slant range correc- ted side-scan image which can be produced in real time with an overlay of isobathic depth contour lines.

Besides the hydrographic application, the system is also suitable for object search, for example wrecks, pipelines and the like.

Operational Status
One system has been operational since the beginning of 1991 on the German hydrographic vessel *Wega*.

Contractor
Atlas Elektronik GmbH, Bremen.

PARASOUND

Type
Narrow beam survey system.

Description
Parasound is the world's first hull-mounted combined narrow beam deep sea sounder and sub-bottom profiler with an overall depth sounding capability extending from 3 to 10 000 m. Designed essentially for research and geological survey applications, it comprises a Narrow Beam Sounder (NBS) and a low frequency sediment sounder or Sub-Bottom Profiler (SBP) for high resolution presentation of sediment layers; both generate highly directional sound beams electronically stabilised against ship's motion.

Incorporating an ice-tested, flush-mounted transducer assembly, the system ensures particularly high vertical and horizontal resolution, enabling detection of very fine layers of sediment while maintaining high ground penetration without necessitating traditional costly towed-fish methods of measurement. It has a deep sea sediment penetration capability in excess of 100 m and can be operated at vessel speeds of around 10 kts, even in gale force conditions.

Contractor
Atlas Elektronik GmbH, Bremen.

HYDROMAP 300

Type
Integrated real-time and off-line hydrographic data acquisition and post-processing system specially designed for multi-beam depth sounding data. It provides various types of graphical terrain presentations based on 'digital terrain modelling'.

Operational Status
Various systems operational.

Contractor
Atlas Elektronik GmbH, Bremen.

FANSWEEP

Type
Multi-beam swath surveying system.

Description
A new cost-effective precision multi-beam swath sounding system covering four times water depth for inshore and coastal survey applications. Using electronic beam forming techniques to give 100 per cent bottom coverage via a single, easily fitted transducer assembly, it is designed for operation at depths from 3 to 200 m. It permits optimum levels of versatility and manoeuvrability in narrow and congested waters.

Operational Status
Various systems sold.

Contractor
Atlas Elektronik GmbH, Bremen.

NORWAY

BENIGRAPH

Type
Seabed surveying system.

Description
The high performance Benigraph seabed surveying system uses a high resolution, multi-beam (200 beams) sonar to collect quantitative topographic data of the seabed. The system continuously records survey and navigation data on computer tape and generates a real-time 3D colour display of the seabed. It offers high accuracy fish positioning provided by the Towfish inertial reference system which is updated by the surface and acoustic positioning systems. Bottom reflectivity is retrieved and displayed as a fourth dimension. The system features a full playback facility and interactive processing capability.

Benigraph has an operating frequency of 740 kHz with two other operator-selectable frequencies (1000 and 515 kHz). The range is typically 50 m at 740 kHz and the sonar has a pulse repetition rate of 2 to 7.5 kHz. The towfish can be towed at speeds of 4 to 7 kts at a height of 10 to 40 m above the seabed. Depth resolution is better than 10 cm +0.05 per cent of water depth, and the system has an operating depth of 300 m.

Contractor
Bentech, Tromso.

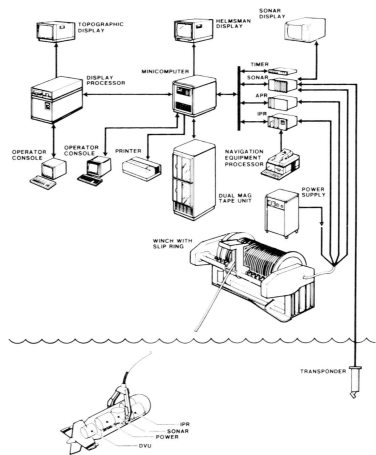

Benigraph seabed surveying system

UNITED KINGDOM

BATHYSCAN

Type
Precision ultra-wide swath depth sounder.

Development
The Bathyscan swath sounder was developed by Bathymetrics Ltd and is distributed by Marconi Underwater Systems Ltd.

The system is available in 100, 200 and 300 kHz variants which offer a maximum swath width of more than 500 m at 100 kHz and an accuracy better than 30 cm at 300 kHz. The towing speed of the survey vessel is normally 6 kts.

Description
The Bathyscan swath sounding system provides high density depth information on both sides of a ship's track. Side-scan signals are transmitted and received on multiple interferometers providing about 1500 soundings per second. These are combined with attitude data of the towed 'fish' to provide real-time display of seabed profiles. With the navigation data, a 'footprint' plot is produced in real time so that survey coverage can be monitored.

Other attitude data, tide corrections and swath merging are applied during post-survey processing.

Sophisticated fast post-processing software merges the depth swaths, taking account of the tidal correction, and provides detailed quality control data about the resulting depth matrix. The latter can be presented on a scale suitable for detailed site investigation or ready for chart compilation.

Side-scan sonar signals are available and can be displayed on a suitable recorder.

Real-time seabed profiles, corrected for roll and heave, are shown on a colour graphics display. About 50 profiles on both sides are shown and scrolled downwards with each transmission. The control VDU displays towed fish attitude, roll, pitch, heave, depth, heading, system condition, mode of operation and system and error messages.

Bathyscan precision ultra-wide swath depth sounder

Operational Status
The hydrographic service of the Royal Netherlands Navy has been operating Bathyscan for more than one year on the survey vessel HrMs *Blommendal* and a second system has been delivered for her sister ship, HrMs *Buyskes*. Marconi Underwater Systems has taken delivery of two systems which are available for lease. A wide range of surveys has been undertaken in the North Sea for the major oil companies and operations are now being extended to overseas surveys.

Contractor
Marconi Underwater Systems Ltd, Waterlooville.

GLORIA

Type
Ocean survey side-scan sonar.

Development
GLORIA is a side-scan sonar for mapping the deep ocean. It was developed by the Institute of Oceanographic Sciences and is operated commercially by Marconi Underwater Systems. GLORIA is able to map a swath up to 60 km wide whilst being towed at 8 kts and can operate in full ocean depths.

Description
GLORIA has been developed to enable coastal states to exercise jurisdiction over their EEZs, resolve boundary disputes and to assess and map their marine resources.

The system enables data to be gathered and compiled for large-scale reconnaissance mapping, covering up to 20 000 km² per day.

The main stages of survey and map production are: data acquisition, onboard mosaic, data processing, image enhancement, overlay preparation, geological interpretation and associated reports.

Operational Status
GLORIA has been selected by the US Geological Survey to map the entire 20 million km² of the US EEZ. GLORIA atlases of the East Coast, the West Coast and the Gulf of Mexico have already been published.

Marconi's GLORIA mapped in less than one month an area of 300 000 km² of the South Pacific seabed in the waters of Fiji, the Solomon Islands, Vanuatu, Western Samoa and Tonga. This was followed by surveys in the deep waters off the coast of southern and western Australia. The object of these surveys was to provide geological information and hydrographic data in these areas and to enhance the understanding of seabed acoustic propagation.

Contractor
Marconi Underwater Systems Ltd, Waterlooville.

Deployment of the GLORIA (Geological Long-Range Inclined Asdic) side-scan sonar mapping the ocean floor

SYSTEM 900

Type
Hydrographic data collection and charting system.

Description
The modular System 900 uses commercially available and well proven equipment, and can be configured to meet a wide range of requirements. It uses distributed processing techniques and in a typical configuration comprises two main elements: the Data Acquisition System (DAS) and the Data Processing System (DPS).

DAS can be rapidly deployed on board a survey vessel or launch to collect and process data gathered from positioning equipment, echo sounders and other sensors. It can be used for survey planning, depth recording, depth editing, track plotting and survey quality control. The keyboard and VDU are used to prepare survey data disks and to control survey area chart preparation in conjunction with the plotter. The survey data disk contains files detailing intended survey routes, basic chain data and operation information, geodetic data (spheroid, projections and grid in use) and any other data required for the planning and conduct of the survey. The system offers a wide range of depth processing routines. A permanent track plot can be maintained on a standard size flat bed or drum plotter. The plot will show the ship's track drawn on prepared trackcharts. The VDU can also be used on-line to display the ship's track, the survey grid, positioning system stations and the ship's speed, heading, and course made good. The DPS is used to edit, adjust and reduce field data collected by the DAS and to prepare high quality sounding overlays or reproducible master charts. Inspection and analysis of sounding data are carried out using a colour graphics display in conjunction with a keyboard and either a digitising table or a trackerball. Soundings are shown in various colours so that certain contours are clearly defined. Soundings can be edited or deleted as required. The surveyor can also add contours on the display using the digitising table and pen. Other features can also be added as required before a hard copy is made.

Contractor
Racal Marine Systems Ltd, New Malden.

UNITED STATES OF AMERICA

SEAMARC™

Type
Towed wide swath bathymetric and backscatter imaging sonar.

Development
The original SEAMARC™ was developed from a high speed exploration system produced for INCO's seafloor mining activity in 1978. The design was improved by International Submarine Technology throughout the 1980s. SEAMARC™ was the first operational seagoing system to combine swath bathymetry and backscatter imagery in a single sensor. Successive models have incorporated current generation digital signal processing, all digital data output, deep tow hardening to 11 000 m, and operating frequencies from 11 kHz to 150 kHz and higher.

Description
The system consists of: a neutrally buoyant towfish; power, tow and control cable (a single, contrahelically armoured coaxial cable with multiplexed power and data) with depressor and handling system; power supplies; system control console; and data displays and processing hardware.

Dual hydrophones mounted on each side of the streamlined towfish produce a highly directional azimuth side-looking sonar pulse. Precise correlation of return-signal phase angle with time allows determination of bathymetric surface within a 120° zone directly beneath the towfish and perpendicular to the track. Quantified absolute signal strength values allow simultaneous backscatter intensity logging. The result is a single-ping, fan-shaped return divided into some 2000 pixels, each showing a bathymetric surface value and a backscatter image value.

Operating speeds vary from 1 to 10 kts and overall swath widths vary with application from less than 100 m to more than 20 km, and the system can be operated in sea state 6, with retrieval in sea state 5.

SEAMARC™ is capable of modification for parametric low frequency operation, as well as multi-beam

Deploying a SEAMARC™ towfish

configurations. It is already proven in such applications as seafloor route surveys and geophysical exploration. Units may be configured for deep towing or high-tow-altitude missions, and for resolutions varying from 5 cm to 10 m.

All towfish provide co-registered bathymetry and backscatter imaging and are controlled from a common shipboard control system and data processing suite.

Operational Status
A total of 10 SEAMARCs™ have been built between 1981 and 1989. Present users include university oceanography departments and geophysical exploration firms. Development and sales continue.

Specifications
Performance: Across track resolution 0.05% of swath width; bathymetric accuracy with respect to towfish is from ½% to 1% of swath width

Contractor
Alliant Techsystems Inc, Marine Systems West, Mukilteo, Washington.

SEA BEAM

Type
Ocean survey system.

Description
Sea Beam is a high resolution bathymetric survey system which combines the Model 853-E narrow beam echo sounder with the Model 875 echo processor for processing sounding data to generate bottom contour charts in real time. The chart is generated by means of soundings from 16 pre-formed beams positioned perpendicular to the ship's axis. Processed or unprocessed sonar echo signals are displayed on a VDU to provide a cross-track profile for each beam. The stabilised vertical beam depth data are displayed in digital form at the echo processor and on multiple depth display repeaters as required. The analogue signal from the vertical beam is displayed on a graphic recorder to provide a survey line directly under the ship's track. A contour plotter provides a continuous 11 in wide plot of the ocean bottom contours in real time. Time, heading and 16 pairs of depth and cross-track sounding co-ordinates are provided in digital form for recording.

Operational Status
There are currently 17 Sea Beam systems in operation worldwide.

Contractor
General Instrument Corporation, Undersea Systems Division, Westwood, Massachusetts.

SEA BEAM 2000

Type
Ocean survey system.

Description
Sea Beam 2000 is an advanced, wide swath bathymetric sonar whose design approach makes it possible to customise the system to meet the user's mission requirements and minimise the individual ship's limitations.

Sea Beam 2000 employs a modular approach to both transducer and electronic design. This creates a sonar which may be modified so as to have a different frequency, beamwidths, number of beams and array sizes.

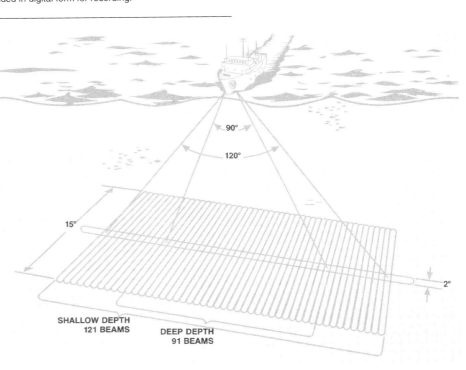

Sea Beam 2000 beam patterns

Sea Beam 2000 utilises software-controlled beam forming. For a given frequency, the sonar beam dimensions are determined by the array lengths, which in turn determine the number of transducers and associated electronics required for the hydrophone and projector arrays. This modular system design provides the flexibility required to optimise frequency and array dimensions for a variety of system requirements and ship sizes. Multiple formed beams of 1 to 3° are accommodated. In all cases, the swath width is 92°.

Special features of the system include: full 92° swath coverage from 10 to 11 000 m; three operational depth modes – shallow, intermediate and deep; a hydrophone array which conforms to hull angles up to 10°; field sheet output showing ship's track against latitude and longitude with depth contour displayed; graphic recorder output of along-track profile; on-track strip chart recording of depth contours; beam intensity displayed with cross-track profile and graphics terminal with 3D waterfall display of most recent cross-track profiles.

Operational Status

Contracts for Sea Beam 2000 have been awarded to General Instrument by the Japan Maritime Safety Agency (JMSA) for the R/V *Meiyo* and also Scripps Institution of Oceanography for the R/V *Melville*.

Specifications

General
Depth (3 modes): 10-11 000 m
Swath width: 92°
Transmit beams: 1 (for 92° swath)
Transmit beamwidth: 100° (athwartship)
Receive beamwidth: 2° (athwartship)
Receive beamwidth: 15° (fore and aft)
Pulse length: 2-20 ms
Frequency: 12 kHz
Ping period: 0.5-22 s
Projectors/power amplifiers: 28
Source level at 30°: 234 dB/μPa/m
Hydrophones/pre-amplifiers: 84

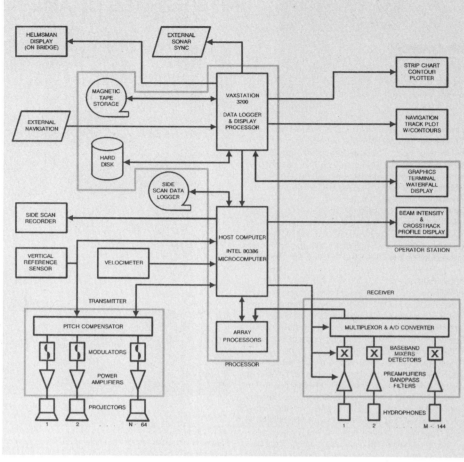

Block diagram of Sea Beam 2000

Depth mode	Shallow	Intermediate	Deep
Depth	10-400 m	200-5000 m	300-11 000 m
Transmit beamwidth (fore and aft)	4°	2°	2°
Receive beams (per ping)	23	46	46
Receive beamwidth (athwartship)	4°	2°	2°
Receive beamwidth (fore and aft)	15°	15°	15°
Pulse length	2 ms	7 ms	20 ms
Ping period	0.5-2 s	1-11 s	4-22 s
Projectors/power amplifiers	14	28	28
Hydrophones/pre-amplifiers	42	84	84

Contractor

General Instrument Corporation, Undersea Systems Division, Westwood, Massachusetts.

HYDROCHART II

Type

Ocean survey system.

Description

The Hydrochart II is a 36 kHz sonar system that employs a new concept in the design of its receive beam patterns. By using narrower beams on the outer edges of the receive beam patterns, significantly enhanced resolution and accuracy is achieved. In addition, the constant track spacing gives a more uniform swath data density. The Hydrochart II provides swath coverage of 2.5 times the depth, up to 1000 m. When normal environmental conditions exist, Hydrochart II delivers full capability at depths in excess of 1500 m.

Designed to satisfy International Hydrographic Standards, the Hydrochart II features constant footprint beam patterns, improved signal processing for echo detection, improved gate depths, distributed microprocessor for real-time signal processing, and user-friendly command and control features. The Hydrochart II system includes a Zeta-8 pen plotter that produces a real-time, full swath contour strip chart of the sea floor. The chart is labelled with fix event number, Julian day, year, time of day, chart scale, contour interval, heading, latitude and longitude.

Hydrochart II utilises the projector arrays, each covering 85° of the athwartship transmitted fan beam for a total of 135°, with 35° overlap. Each array is mounted parallel with the ship's fore and aft centre line. Two hydrophone arrays are mounted athwartship

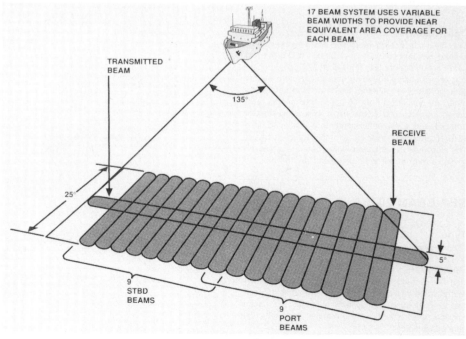

Hydrochart II beam patterns

at an angle of 25° to port or starboard. The transmitter commands port and starboard transmissions on alternate pings and is capable of producing 500 W of electrical output. The transmitter can also attenuate

the drive signal to the projector. The receiver contains pre-amplifier circuits to process hydrophone signals. Port and starboard signals are multiplexed into the pre-amplifiers on alternate ping cycles, which then form nine receive beams. Each beam is processed, and echo envelope detection and log conversion of the detected signals is provided.

Operational Status

Two systems are currently in operation, one aboard the Japan Maritime Safety Agency (JMSA) vessel R/V *Tenyo* and the other aboard the NOAA ship R/V *Whiting*.

Specifications

Frequency: 36 kHz (nominal)
Projectors: 2 arrays; 1 port; 1 stbd
Hydrophones: 2 arrays; 1 port; 1 stbd
Output beams: 17
Power output: 200 dB/μPa/m
Pulsewidth: 1-24 ms
Repetition rate: 100 ms-4 s (nominal)
Depth capability: 10-1000 m
Depth presentation: 0.1 m LSB
Transmit beam: 5°, fore and aft, at the −3 dB points
Receiver beam: 3.5-6.3° (athwartship)
Sidelobes: At least 25 dB from the main lobe for transmitter or receiver
Roll and pitch: Accommodates ±20° of roll and ±10° of pitch
CRT display: X-Y display of cross-track distance versus depth
Data recording: Data are recorded on 9 track magnetic tape at 1600 char/in
Navigation plot: Graphic presentation of survey area showing ship's track and sonar swath coverage
Contour plot: Full contour plot along ship's track
Input power: 115 V, single phase 60 Hz, 30 A

Contractor

General Instrument Corporation, Undersea Systems Division, Westwood, Massachusetts.

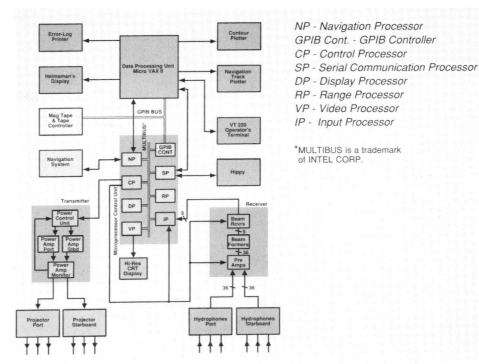

Block schematic of Hydrochart II

NP - *Navigation Processor*
GPIB Cont. - *GPIB Controller*
CP - *Control Processor*
SP - *Serial Communication Processor*
DP - *Display Processor*
RP - *Range Processor*
VP - *Video Processor*
IP - *Input Processor*

*MULTIBUS is a trademark of INTEL CORP.

SIGNATURE MANAGEMENT

Managing and controlling unwanted signatures on warships has become a major consideration for ship designers and operators over the last decade. The advent of new generations of acoustic homing torpedoes and modern acoustic/magnetic mines has led to a major effort being devoted to reducing or modifying various types of signature created by ships and submarines in order to improve their effectiveness.

Some signatures, like a ship's magnetic signature, have been the subject of considerable attention for many years. Even during the Second World War technology was being applied to either reduce to a minimum the magnetic signature, or to install compensation methods which would cancel out the effect of such signatures. Development was spurred on by the increasing losses suffered by merchant shipping from magnetic mines, and the need to build minesweepers which would be virtually immune to the magnetic mine.

With the post-war development of more sensitive magnetic mines, and the increasing use of acoustic mines, together with the development of acoustic homing torpedoes, attention turned towards reducing the effect of a ship's acoustic as well as its magnetic signature, principally through the damping of systems which created noise such as propulsion machinery and so on, by the use of rubber shock absorbers. Using these protective measures had the added effect that if the ship were to suffer an explosion then equipment would be given some measure of protection from effects of the blast and whip.

Stealth Technology

The new science of controlling and modifying these signatures is now referred to as Stealth Technology. Today effort is not only being devoted towards reducing these signatures, but also to adapting them so that a ship is not necessarily made 'invisible' to hostile sensors but appears as something totally different. Using this stealth technology is referred to as signature management.

The relative importance of different types of signatures can be expected to change with time, as new and more sensitive sensors and more sophisticated signal processing techniques become available. The art of signature management lies in the ability to obtain the right balance between the differing signatures. The object of signature management is to achieve the optimum combination of external shape, coatings and internal treatments that will give the desired overall result at an acceptable cost.

Acoustic Signatures

Surface ships and even more importantly submarines emit high levels of underwater radiated noise which can be detected and tracked by passive sonars, possibly hundreds of miles away in the case of towed array sonars. Vessels which strongly reflect incident sound waves are easily detected by active sonar systems. Not only does a high level of radiated noise make a vessel far more susceptible to detection, it also inhibits the vessel's ability to operate its own acoustic sensors.

Underwater acoustic radiation is still the primary means by which submarines betray their presence when they are submerged. Hostile forces also use the underwater noise signature radiated by surface ships for detection and classification. Cavitation and broadband noise are created by propellers and hulls, while various internal pumps, generators and diesel engines produce distinctive sounds at specific frequencies and amplitudes that can both aid and hamper detection. So sophisticated are modern detection methods that, in many instances, not only can the hunter classify a target specifying the class of ship, but even the individual ship within a class can be positively identified from its very own distinctive sound signature (that is, it may have a shaft alignment problem which has been identified by intelligence or a pump with an uneven bearing and the specific sound signature so produced can be recorded for storage in a sound library for future comparison and so on).

Minimum acceptable noise levels are constantly being driven downwards by developments in sonar sensors and signal processing technology. Hence the techniques used to control radiated noise must also become more subtle and refined. One of the most important elements in achieving a 'quiet' design is the ability to predict the noise levels while a ship is still at the design stage. Minimum noise levels are further achieved by the design of high performance machinery isolation systems and methods for reducing cavitation around propellers and so on, and the hull itself.

Noise isolation systems for surface ships and submarines employ a wide range of techniques, including double-elastic mounting systems where, in the case of diesel engines, high intrinsic source levels have to be reconciled with stringent target levels for underwater radiated noise. Increasingly, serious attention is having to be paid to the secondary transmission paths in the design of such advanced isolation systems.

Active control techniques are currently under development to enhance the high degree of isolation which can already be achieved through well balanced, passive isolation treatments. These include systems designed for use with propellers and propulsors and their associated shafting systems, where dynamic interaction with the hull can play an extremely important part in determining the overall level of underwater noise radiation. In rotating machinery, for example, moving parts can be dynamically balanced to reduce the noise from shafts and connections to other machinery. In addition the equipment can be mounted in special acoustically insulated boxes which are, in turn, carried on flexible mounts to isolate them from the hull. In some instances, large items of machinery are additionally flexibly mounted on rafts which are themselves flexibly mounted to isolate the whole machinery installation from the hull. As well as these detailed arrangements, items such as water inlets/outlets and so on, must also be isolated from the hull in flexible mountings to prevent any vibration from the pipe being transmitted to the hull and thence to the surrounding water in the form of noise.

One of the most important areas for noise reduction concerns the propeller. Propeller noise is generally divided into two main types: blade noise which is generated by the thickness of the blade and the loading on the blade as it rotates, leading to low frequency noise; and cavitation noise generated by the implosion of cavitation bubbles which form around the leading edge of the blade, across the low pressure region of the blade and in the vortex behind the hub of the blade. Cavitation is a broadband high frequency noise which can only be avoided by delaying the onset of cavitation.

Coatings applied to hulls in the form of decoupling or damping layers designed to reduce the level of machinery induced noise radiated into the water, or anechoic coatings designed to absorb as much as possible of the incident energy generated by active sonar equipment, are becoming increasingly important in the science of controlling unwanted noise signatures. For some years the hulls of nuclear submarines have been treated with sound-absorbing anechoic tiles. Anechoic tiles are not the same as decoupling tiles, being designed to absorb or scatter incident sound energy from hostile active sonars. They operate on the principle that the material absorbs the incident sound, which is then turned into heat; hence no sound, or only minimal sound, is reflected off the hull. In order not to inhibit own boat sonars, sonar 'windows' are incorporated into the anechoic tiling system.

Other treatments include continuous coatings in which acoustic and anechoic material is applied to large flat or curved surfaces. RhoC (PC) materials are designed to match the impedance and speed of sound in seawater, sonar materials can replace GRP for sonar domes and windows, and encapsulation in which large and small case encapsulation of arrays is achieved with controlled exotherm.

The purpose of using acoustic control measures such as decoupling tiles (whose design is frequency and noise level dependent) and anechoic coatings and so on, is to reduce the detection range and classification capabilities of hostile sensors, at the same time improving the performance of own ship sensors. However, the law of diminishing returns does apply and there comes a time when no amount of treatment will either (a) improve own detection range or (b) significantly reduce enemy detection range.

Air bubble screening has proved to be particularly valuable at higher frequencies and speeds.

Radar Cross-section

Radar cross-section signature control can be approached in a variety of ways ranging from the use of appropriate design techniques (avoiding re-entrant corners, shaping, sloping and reducing the overall size of superstructures and so on), to the addition of temporary or permanent radar absorbent materials (RAMs) and coatings of various types.

The design of structures to avoid specular reflections is an obvious means, but this may not be fully effective on mobile structures and cannot be applied easily to existing vessels. Absorbent coatings represent a versatile solution which can be applied in conjunction with shaping to new designs or retrospectively to existing vessels.

Coatings can be designed to absorb radar signals over a wide or narrow frequency bandwidth, as appropriate to the application. Radar signature can also be controlled by use of structural radar absorbent material (RAM) and materials of this kind are emerging and can be considered at the design stage of new vessels. By appropriate formulation RAM coatings can be endowed with various special properties such as high friction. RAM structures operate on the thickness of the material used according to the angle of incidence of the beam.

Magnetic and Electric Signature

The need to control and manage a ship's magnetic signature has long been the subject of study by naval architects, scientists and operators and so on. Passive countermeasures involve measuring the ferromagnetic mass of the ship itself (including all the individual items of equipment carried by the ship) and reducing the magnetic signature. This is achieved either by magnetically treating the ship in special magnetic treatment facilities (deperming); or by fitting the ship with special active degaussing coils through which electric current is passed, the strength of which can be controlled from within the ship itself, thus ensuring that at any time, in any geographic position the ship's magnetic signature (that is, its own inherent magnetic signature plus that caused by its movement through the earth's magnetic field) is silenced; or by a combination of the deperming and active degaussing processes. Most modern warships are fitted with active degaussing systems, and are also regularly depermed. Merchant ships, on the other hand, rely in the main on deperming only.

Deperming involves putting the ship inside an arrangement of coils or placing an arrangement of coils around the ship, and then passing a powerful electric current through the coils to create a magnetic field in opposition to the magnetic field of the ship, thus cancelling out the ship's own magnetic signature. Alternatively, deperming can be used to create a permanent magnetic field on the ship which is matched to the area in which it will operate. However, the disadvantage of deperming is that it is not permanent, and the vessel will need to be checked periodically in order to maintain the effect desired.

Active degaussing coils, on the other hand, are built into the ship during construction to provide magnetic field correction facilities. The coils are continually fed with electric current provided from special computer-controlled generators to create an opposing magnetic field which is continually matched to the ship's changing magnetic field as it crosses the ocean. Even with this system, however, periodic checks are still necessary and are carried out on special magnetic ranges.

For special-purpose vessels such as MCMVs it is also necessary to attack the basic problem of magnetic content and wherever possible to substitute non-magnetic for magnetic material. In such work it is essential to ensure that the substitution of magnetic materials by those of low permeability does not compromise the performance of the equipment.

Ranges

The sophistication of modern sensors, fire control systems and weapons demands that the various signatures of warships be reduced to the lowest level possible in order to reduce the risk of detection and lower their vulnerability to modern weapons. While considerable attention is paid to reducing a ship's various signatures to a minimum, both at the design stage and during construction, it is an undisputed fact that a ship still retains a definite signature which, like the human fingerprint, enables it to be detected and identified. In order to overcome this limitation, it is essential that the user knows the precise nature of the signatures of his ships, and has available the means to reduce or nullify them, or, as is so in some cases, to actually turn them into an advantage.

But it is not only ships' signatures which are a vital factor in modern naval operations. Weapon performance and behaviour is continuously under close scrutiny, especially as the ranges and speeds of weapons increase. The discriminating user demands to be reassured as to precisely how a weapon behaves under various operating procedures. This is particularly so in the case of underwater-launched weapons. However, the user doesn't only want to know how the weapon behaves, but also how effective it is in various situations against differing types of target, and how well it can overcome countermeasures which might be directed against it.

The question of knowing precisely what a ship's signature is and how an underwater weapon behaves is achieved using the underwater range. These ranges are so specialised, using, as they must, leading edge technology, that only a very few companies are capable of manufacturing the delicate measuring equipment.

The two major signatures about which every user should have precise knowledge concerning his fleet are the acoustic and magnetic signatures. In addition to ranges for measuring these signatures, there are also tracking ranges for weapon system measurement, mine warfare ranges and a very specialised type of calibration range known as FORACS (Fleet Operational Readiness Accuracy and Check System), of which two main ranges for NATO are sited in Norway and Crete.

Noise Ranges

Noise ranges for static and underway measurements of surface vessels and submarines are amongst the largest and most sophisticated systems worldwide. These ranges are primarily designed for the measurement of radiated noise emissions in order to allow reliable and quantitative assessments of signature reduction measures.

There are three main types of noise range:
(a) the static range – in which a vessel is very precisely moored (submarines are held in suspension) between a number of buoys and the acoustic signatures of various equipments transmitted by the ship measured under very carefully controlled conditions

(b) the underway fixed range – used to measure the acoustic signature of ships at various speeds
(c) the portable range – a somewhat less sophisticated version of the fixed range, which can be set up at different sites to measure acoustic signatures.

The portable range comprises a number of hydrophones, suitably deployed from a surface ship or temporarily attached to fixed moorings on the seabed, to measure the noise characteristics of a submarine or surface ship at various ranges, speeds and depths. It is useful for measuring acoustic signatures under different environmental conditions, or when it may not be convenient to send a ship back to a fixed range, or where budgetry restrictions or environmental conditions may preclude the setting up of a fixed range. Portable systems typically cover a circular area of a 2 km radius and can operate suspended from the surface, in depths down to 300 or 400 m. The hydrophones in a portable range must be specially customised to overcome the problem of pressure changes acting on the ceramic head resulting from wave motion which, if severe, can saturate the front end of the pre-amp in the hydrophone.

Tracking Ranges

Modern tracking ranges are capable of tracking up to six vehicles at speeds up to 75 kts with a maximum turn of about 20°/s. Positional accuracy is in the region of 5 m. The hydrophones are laid in a carefully planned arrangement and are calibrated from a surface vessel whose precise position can be fixed with an accuracy of about 2 m.

Magnetic Ranges

Equally as important as a ship's acoustic signature is its magnetic signature. In the case of MCMVs it is vital that this signature be known in detail and checked at frequent and regular intervals. It is also essential that such a signature be measured whenever a vessel is assigned to another theatre of operations hundreds of miles away from its current area of operations. A ship's magnetic signature is not only governed by where it was built and with what materials, but also by the Earth's magnetic field at that point. Even moving a relatively short distance can considerably alter a ship's magnetic signature. It is essential, therefore, to protect ships from the threat of magnetic influence mines. This can be achieved in three ways by:
(a) designing a ship for low magnetic content (of paramount importance in MCMV design)
(b) providing the ship with degaussing coils to compensate for its magnetic field
(c) treating the ship electromagnetically by the application of a powerful external magnetic field to cancel out its magnetic field.

Like acoustic ranges, magnetic ranges are configured to fulfil different functions. There are five main types of magnetic range currently in service around the world:
(a) the open range – used to check the magnetic level of all types of ship and to adjust degaussing coil settings
(b) the fixed range – used to measure the magnetic effects due to eddy currents, stray fields and induced magnetic fields from which the magnetic field components can be separated using a Z-loop
(c) the land-based range – used to carry out magnetic measurements on ship equipments, or for modelling purposes on new designs or modifications to determine the coil system necessary to reduce the magnetic signature
(d) the transportable range – used to process MCMV signatures in the area of operations
(e) the treatment range – used to deperm, or wipe, a ship using an externally created magnetic field.

The tasks of the degaussing range are to measure the ship's magnetic fields and analyse them to provide resolved components on three axes, synthesise the degaussing coil settings required, and record the measured values for archive purposes and reporting.

Early ranges were of the large fixed type and relied on manual tracking and analysis, which required the vessel to make numerous runs over the range in order for the shore-based operator to manually interpret the plotted signatures and to overcome accuracy limitations. These ranges relied on the comparison of signatures in opposing headings to differentiate between induced and permanent magnetisation. They suffered from the limitation that the horizontal field could not be determined. The next generation of range used automated computerised tracking to achieve faster and more accurate tracking of the ship, determining its position within 1 m. Using triaxial sensors the range provided data on the horizontal field. The computer analysis of the results provided, in addition to a much more accurate signature plot than the manual system, contour and 3D plots. A major step forward with these systems was the ability to carry out coil current synthesis, from which the coil current on the ship could be set to counter the ship's magnetism.

Fixed ranges of any description, however, are not easy to put down, requiring a considerable civil engineering effort and extensive surveying. Nor are they cheap, although they are most cost-effective, even if treasury officials consider that money spent on a range has been money spent on a system which cannot prove its value.

Description of specific types of range is almost impossible as ranges are tailor designed, built and laid out to suit each specific requirement. Any description of a system must therefore be in general terms.

ACOUSTIC RANGES

FRANCE

ACOUSTIC RANGES

Acoustic Ranges
Using high dynamic signals input, these ranges are used to locate, identify and measure noise sources on ships. The immersed measurement arrays are made up of hydrophones and/or LF and HF passive antennas.

Contractor
Thomson Sintra Activités Sous-Marines, Brest Cedex.

ITALY

PTM

Type
Tracking range.

Description
The PTM is a mobile 3D underwater tracking range used for testing modern torpedoes in different environmental conditions. It is able to track simultaneously (in real time) the torpedo, the launching unit and the target both in shallow and deep waters and it is capable of off-line analysis of the trajectories.

The PTM consists of a surface vessel (dedicated or fitted for) a multi-hydrophone array, acquisition/recording and analysis equipment, acoustic transmitters and ancillary equipment.

The tracking range acoustic technology is based on a long baseline system with SFSK (space frequency shift keying) acoustic signalling suitable for high speed vehicle tracking. The mobile 3D underwater tracking range can be arranged in different configurations to meet specific requirements and can be laid down in different water depths.

Operational Status
In production.

Contractor
Whitehead, Livorno.

An artist's impression of the PTM

MOBILE RANGE

Type
Transportable acoustic range.

Development
The mobile acoustic range has been developed and built for the Italian Navy to record and analyse radiated noise from ships and submarines.

Description
The range consists of four main subsystems:
(a) the underwater subsystem comprises two hydrophone arrays connected to a surface buoy which is linked to a support ship by radio or by floating cable
(b) a data acquisition subsystem used to collect and record data on board the support ship. The signal recording and control equipment is contained in an easily transportable shelter which can be installed on the support ship or on the surface vessel being measured. It includes an emergency acoustic navigation system. To ensure complete acoustic silencing of the support ship the equipment is completely powered by batteries
(c) a computer-operated navigation subsystem for navigation control of the ship or submarine under test. This is located on the ship being measured and includes compensation for the sound speed profile and deformation of the moorings due to tidal streams. To measure the deformation the moorings are fitted with transducers. If the ship under test is equipped with its own underwater telephone these are not required
(d) a data processing subsystem for signal analysis. This is based on a modern computer system with

Drifting configuration utilising one or two Hydrophone Arrays suspended from the sea surface and a radio link for data transmission to a support ship

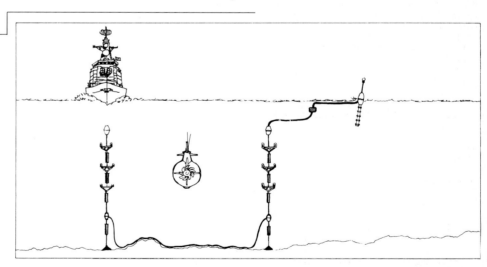

Configuration utilising Hydrophone Stations moored on the seabed and a radio link for data transmission to a support ship

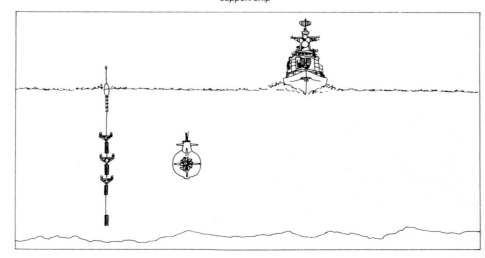

appropriate peripherals and specially developed software. One third octave and line spectral data are compensated for frequency dependent propagation loss and bottom and surface interference, prior to time and space averaging.

The compact, flexible system can be launched and recovered from small support vessels the size of a minesweeper.

Operational Status
The range has been in operation with the Italian Navy since 1987. The flexibility of the design has allowed the system to be used, with minor modifications, for near-field and far-field acoustic measurements of the NATO ASW research ship *Alliance*.

Specifications
Maximum operating depth: 400 m
Frequency range: 10 to 40 000 Hz
Acoustic navigation accuracy: better than 1 m

Contractor
USEA SpA, La Spezia.

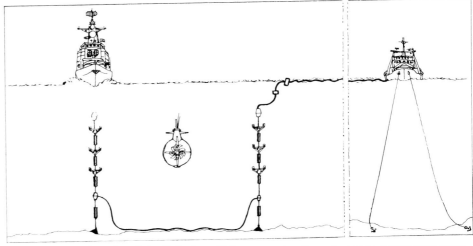

Floating cable configuration utilising Hydrophone Stations moored on the seabed and a floating cable about 1 km long for data transmission to a support ship

UNDERWATER ACOUSTIC RANGE

Type
Range used for positioning and fixing, localisation, navigation and data transmission.

Description
The system comprises three main elements: the Underwater recovery Transponder/Release unit (UTR); the Digital Range Meter (DRM) and the Multi-frequency Transducer (MT).

The DRM is designed to interrogate, control and receive signals from up to 24 UTR units. The DRM is controlled by a 68000 microprocessor and transmission and reception functions are obtained using the MT unit. As input datum the DRM unit requires the measured mean sound velocity (default value 1500 m/s). Enhanced precision can be obtained by the use of a depth/sound velocity table, MT depth and UTR depth. The operating features of the DRM include: standby listening mode, manual transmission command, internal automatic and external triggering,

external acoustic transmission, computations for range calibration, navigation, and optional control by an external computer. The DRM incorporates 128 Kbyte RAM which is expandable to 512 Kbyte. Two serial interfaces are provided together with an optional parallel interface. Data are shown on a liquid crystal display. UTR units are controlled by transmitting an FSK code and interrogation is carried out by transmitting two pulses at different frequencies. Four working bands are available covering 8-16 KHz, 35-44 KHz, 45-54 KHz and 55-64 KHz.

The MT unit operates in conjunction with the DRM and is available either in a hull-mounted or offboard version. Five transducer bands are available covering the same frequencies as the DRM unit and additionally in the 35.64 KHz band. The beam pattern covers 360° in azimuth and 50° in vertical.

The UTR unit is remotely controlled by FSK code and is provided with a serial and parallel port for underwater communications. The unit can also provide depth data via acoustic transmission and replies by a selectable code. Remote-control functions include release, enable/disable transponder

reply, selectable operating mode, transmission frequency selection and depth measurement transmission. Received commands are confirmed by a return code.

In noise range applications two UTR units are positioned and fixed in an appropriate co-ordinate system. At constant time intervals a unit on board the platform sends the interrogation pulses by means of the MT unit. The two UTR units reply with two different frequency pulses. The MT unit receives the two pulses and the DRM unit, after detection, computes the position, velocity and heading of the platform. In the absence of some data, prediction of the position and velocity data are performed by the tracking system. A display presents in graphical and numerical format the infomation relative to the trajectory, and stores the data for post-experimental analysis.

Contractor
USEA SpA, La Spezia.

UNITED KINGDOM

ULTRASHORT BASELINE (USBL)

Type
Torpedo tracking range.

Description
The USBL transportable torpedo tracking range system is designed to track the torpedo, target and attacker in three dimensions. It can be used for equipment development, performance analysis or operator training.

The sensor head or heads (for a multiple range) can be deployed from any ship with the control equipment on board. It provides real-time results for range safety, logs data for post run analysis and provides

hard copy plots for quick look assessment. Tracking frequencies cover the 10 to 25 kHz band and accuracy is 5 m in 1000 m. Coverage is over a 2 to 3 km radius circle. A plot of each object is generated every 1.6 s, with an object speed of 70 kts maximum.

The lightweight USBL range is readily deployed and recovered. Dedicated tracking vessels are not required since the array can be temporarily installed in any vessel.

Ranges are tailored to customers' specific requirements.

Contractor
Marconi Underwater Systems Ltd, Waterlooville.

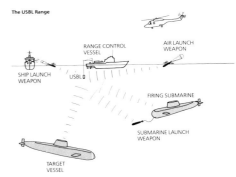

Ultrashort Baseline (USBL)

PATS

Type
Large and transportable acoustic tracking system.

Description
PATS is a high performance system capable of tracking up to four underwater objects simultaneously over a large area. The objects may include the weapon, the target, the launch vessel or any other suitably instrumented vessel in the firing area. The system is fully transportable and may be deployed as a surface range suspended from anchored buoys, or in shallow water laid on the seabed.

In the buoy arrangement, the mooring buoys are anchored in a triangular form approximately 200 m apart. This then becomes the centre of the tracking range and practice firings can take place in an area 4.5 km around the range centre. The three hydrophones are suspended at the selected depth from small flotation buoys, which are in turn attached to each of the mooring buoys when the range is in use. The hydrophones are connected by floating cables to a range support ship normally moored to one of the main buoys. Data are processed on board the range vessel and will normally be transmitted via a radio link to a completely self-contained base station on land, where the tracks are computed and displayed in real time. Alternatively, the base station

may be taken on board the range vessel, allowing an entirely seaborne operation to be carried out.

In shallow water an alternative form of PATS may be used, whereby the hydrophones are placed on the seabed, again in a triangular form. Cables from each hydrophone connect to the range vessel which is moored close to the range centre, and data are again relayed to the base station via a radio link.

Operational Status
In service with the Royal Navy and other navies.

Contractor
THORN EMI Electronics Ltd, Feltham.

SMUTS

Type
Submarine-mounted underwater tracking system.

Description
SMUTS is an in-service Royal Navy system developed for the evaluation of underwater weapon systems. It is a dual mode acoustic tracking system capable of short-range (300 m) high resolution, high accuracy, 3D tracking coupled with a long-range (4.8 km) distance measuring mode. The system may also be used to measure the instant of fuze activation of a suitably instrumented weapon.

SMUTS may be used during the development and proving stages of a new weapon system and for routine in-service weapon effectiveness firings. The 'inboard' system is fully portable and may be rapidly installed on any vessel or installation fitted with an appropriate hydrophone array.

The prime uses are:
(a) to measure and display the approach and three-dimensional trajectory of the terminal phases of an underwater attack on static or moving target submarines, unmanned mobile targets, static targets or fixed installations
(b) to measure and display the discharge trajectory of any underwater weapon (such as a sub-launched missile, stand-off weapon, torpedo, mine or decoy) from a submarine or fixed underwater tubes or dischargers. This may include measurement of the deployment/fall back of torpedo cable dispensers or missile shrouds after water exit separation and so on
(c) to measure and display the deployment and trajectory performance of bottom-launched weapon systems.

SMUTS may be supplied as a 'stand-alone' system, incorporating onboard computing and display together with digital data storage and recording for subsequent off-line processing. Alternatively, it may be supplied as part of a total attack/discharge system, interfacing with the ship's weapon and navigation systems either via discrete interfaces or via a highway.

It has been used extensively by the Royal Navy, both for torpedo attack and discharge measurement, and for sub-launched (Sub-Harpoon) discharge, both in the research and development and fleet exercise role. The system has been used in an open ocean environment and on instrumented weapon ranges where compatibility with the weapon range system is required. It is in operation with, or ordered for, the Royal Navy and other NATO navies.

Contractor
THORN EMI Electronics Ltd, Feltham.

ASR

Type
Acoustic Self-ranging system (ASR).

Description
THORN EMI has developed ASR to meet the need for the acoustic self-ranging of warships to check and evaluate their acoustic signatures during periods between calibration at fixed noise range facilities or locations dictated by operational necessity. ASR is a cost-effective, portable acoustic measurement system. Already in service, the system has proven highly effective in war-zone conditions.

ASR uses data transmitted from an acoustic buoy, or from a bottom-mounted hydrophone, to provide a complete noise range facility. The underwater radiated noise signature is received, recorded and analysed in one-third octaves, LOFARGRAMS or narrow frequency bands to produce results which are directly comparable with those obtained on a fixed calibration noise range, or obtained from other noise measuring systems.

The development of the self-contained acoustic measurement system gives rise to a family of applications which are possible with ASR:
(a) Self-ranging – by installing ASR and deploying an acoustic buoy, a vessel is able to measure its own acoustic signature under varying operational conditions
(b) Transportable ranging – ASR can form the basis of a land-based noise range facility to provide signature measurement and analysis for both surface ships and submarines
(c) Remote ranging – a ship fitted with ASR can provide a floating noise range facility which can be deployed wherever ships or submarines are operational
(d) Covert monitoring – being fully self-contained, easily portable and if required, battery-operated, ASR can rapidly be deployed to monitor sensitive waterways and channels.

Operational Status
In service.

Contractor
THORN EMI Electronics Ltd, Feltham.

ACOUSTIC RANGE EQUIPMENT

Acoustic Transmitters (Pingers)
The weapon to be used must be fitted with an appropriate acoustic pinger. This may take the form of a small hull section for a torpedo or missile, or individual subassemblies for fitting within the weapon system envelope, discharge capsule or shrouds. The pinger may transmit either HF or LF signals, as appropriate, or in certain applications both transmissions will be required. Each transmission will normally comprise two pulses: one to determine range, and one to provide a direct measurement of the weapon operating depth. When operating with a suitable instrumented weapon, the instant of fuze activation may be transmitted by the pinger and registered by the tracking system. The pingers are normally battery-powered giving a completely self-contained package incorporating 'power-up' functions.

Noise Ranging
The measurement and analysis of underwater noise radiated by ships and submarines is a specialised field involving the latest technologies in a most hostile environment. The requirement for surface ships to be as quiet as possible is increasing, firstly due to the advent of advanced acoustic mines and secondly by the need to provide the best possible ratio of signal-to-noise for the surface ships' own sonars. A requirement to ensure that individual ships, particularly the more advanced and complex weapon platforms, are not easily acoustically identifiable at long ranges, is also beginning to be widely appreciated. Ranges for the analysis of underwater radiated noise designed and developed by THORN EMI have included deep water facilities, where submarines may be operated at full speeds, and static facilities, where detailed investigations may be carried out in the absence of noise due to the vessel's movement through the water. Systems for operation in shallow water and which may be combined with magnetic measurement facilities are available. These systems are primarily for use with surface ships, in particular MCMVs.

Tracking And Positioning System (TAPS)
TAPS is a large area 3D fixed tracker for multi-target tracking and forms the basis of many of the available tracking systems. It enables the slant range between a mobile or static acoustic transmitter and a receiving hydrophone to be measured with great precision. The equipment is automatic in operation, requiring minimal operator attention. Modular design enables tracking systems to be configured to meet the user's immediate needs while allowing expansion later if a more complex tracking system is required.

Operational Status
In service with the Royal Navy.

Contractor
THORN EMI Electronics Ltd, Feltham.

NOISE RANGES

Ferranti-Thomson Sonar Systems (UK) Ltd is highly specialised in the design and manufacture of noise ranges. The company has manufactured a complete portable noise range for a European navy which can be deployed in deep water to accurately measure submarine signatures. The data are recorded on IRIG compatible magnetic tape for subsequent analysis ashore using a mainframe computer.

Standard hydrophones specially designed for noise ranging, manufactured by Sonar Systems, produce a very good polar pattern up to 150 kHz. Different gains, acoustic heads and other variations are available on the standard product, which is suitable for either fixed or portable ranges. These hydrophones are used on a number of UK ranges, including the underway noise range at Rona and the static noise range at Loch Goil, as well as portable ranges for two European navies.

The transportable noise range is highly modular and in its standard form provides three noise measuring hydrophones at selectable depths suspended from a surface buoy some 400 m from the deployment boat. An active transducer is suspended from the deployment boat and acts as a pinger for the tracking system. The vessel being ranged carries a self-contained transponder which replies to the active transducer ping. The noise measuring hydrophones receive both the ping and the reply, enabling dynamic position fixing of all elements of the system. The data processing and recording equipment is fitted on board the deployment boat, enabling on- and off-line position line plotting and noise analysis. Software is also included for off-line range normalisation of recorded noise. To assist submariners, the system includes a sonar beacon which may be switched on and off as required.

The fixed underwater acoustic noise range incorporates five variable depth underwater sensors, each connected by its own five-core cable to processing equipment in an offshore, pre-fabricated, air-conditioned building.

The position of vessels being ranged relative to the underwater sensor array is determined by a laser tracking system for surface ships and an acoustic tracking system for submerged submarines, both independent of the normal equipment carried by either platform.

Contractor
Ferranti-Thomson Sonar Systems (UK) Ltd, Stockport.

MAGNETIC RANGES

CANADA

MODELS M-244 and M-234

Type
Recording proton magnetometers.

Description
The Model M-234 is a portable, DC-powered recording proton precision magnetometer. It includes a high speed thermal printer and large area Liquid Crystal Display (LCD) housed in a rugged weatherproof and lightweight case.

The Model M-244 is a rack-mounted recording proton precision magnetometer comprising two standard 19 in rack-mounted units. The electronics, large area liquid crystal display and control keyboard make up one unit, while the second unit contains a high speed thermal printer and power supply for 115/230 V AC operation.

Both instruments can be supplied as shore-based units, airborne or shipborne systems and interface with a variety of the company's sensors and cable lengths enabling them to be used in shallow or deep water areas. Both units use menu-driven software which enables the operator to easily enter or change the operating parameters in a straightforward manner. Parameters controlled via the menu include: cycle rate, time and date, starting value for automatic tuning, line number, manual/external selection, recording format, outputs and LCD viewing angle.

Both Models feature large area LCDs which display a six digit magnetic field reading, supply voltage, signal strength, input values and real-time one- or two-trace analogue representation of every reading. The Models also feature automatic tuning throughout the entire range of the instruments. The operator also has the option to select a starting value of the magnetic field for a fast response of the automatic tuning.

As an option the manufacturers offer a memory PCB with a capacity for storing up to 200 000 readings, which is an important feature for base station applications.

Contractor
Barringer Research Ltd, Rexdale, Ontario.

FINLAND

MGS-900

Type
Transportable seabed measurement system for magnetic, acoustic and pressure signatures.

Description
With high sensitivity and wide dynamic range, the MGS-900 is used in degaussing ranges to measure the magnetic, acoustic and pressure signatures of all classes of ships and submersibles.

The magnetic signature measurement is based on a triaxial fluxgate magnetometer featuring a dynamic range of ±100.000 nT and a noise level down to 0.5 nT RMS. The measuring range of the acoustic signature is −50 dBPa to +80 dBPa with a frequency range of 0.5 Hz to 35 kHz. The pressure signature measurement features a dynamic range of ±15 kPa with a frequency range of 0.005 Hz to 1 Hz and a noise level less than 10 Pa RMS. Each sensor unit contains a microprocessor which controls all measurements. The sensor electronics are housed in a salt water-resistant bronze housing. The sensor units are connected to land-based data processing equipment with a lightweight underwater cable consisting of one twisted pair. Up to 30 sensor units can be connected to a cable with a maximum length of 2000 m.

The graphic operator's interface of the land-based central computer is based on OSF/Motif standard. The magnetic signatures can be displayed in different formats: profile displays, contour displays, vector displays or 3D perspective displays. Acoustic and pressure signatures can be displayed both in frequency and time domain. An optional software package for automatic degaussing coil current synthesis is also available.

Contractor
Elesco Oy Ab, Espoo.

FRANCE

MAGNETIC RANGES

Sea Degaussing Range
The two types of degaussing range manufactured are the open range, in which the vessel follows a track across a line of sensors, and the fixed range, in which the vessel is moored in a stationary position above the sensors. The ranges feature a limited number of sensors and highly sophisticated modelling software. These are used to predict a ship's magnetic signature under varying conditions and to prepare the design of degaussing systems. Ranges are operational in the UK and Norway.

Magnetic Land Range Systems
These systems are used to carry out a full analysis of the magnetic characteristics of all equipments to be installed on a ship. The systems comprise a fixed magnetic range to measure magnetic signatures from which the necessary degaussing corrections can be calculated, deperming equipment studied, and the effects of roll and pitch on magnetic signature together with stray field effects measured.

Contractor
Thomson Sintra Activités Sous-Marines, Brest Cedex.

GERMANY

LAND RANGE

Type
Magnetic measurement system for ships' components.

Description
The magnetic measurement of items of equipment for installation in a ship constitutes part of its proof of performance to appropriate specifications.

The measurement, deperming and degaussing are consecutive operating processes which may need to be repeated until the minimum or acceptable interference field is achieved.

The equipment to be measured is traversed over a specially defined arrangement of sensors on a measuring track arranged in a north-south direction. The non-ferromagnetic test carriage is provided with a platform which can be rotated through 360°. The object-specific permanent fields, the fields induced

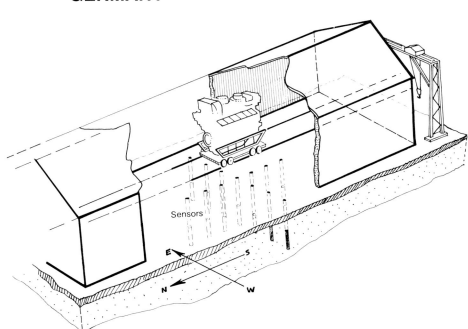

Magnetic measurement system for ships' components

by the earth's field and, where necessary, the electromagnetic stray fields produced by the object under test itself are measured, analysed and recorded. The sensors and the measuring electronic equipment are all components of the Forster Magnetomat magnetic measuring system.

Contractor
Institute Dr Forster, Reutlingen.

STATIONARY RANGE

Type
Stationary magnetic measuring system.

Description
A stationary measuring system with simulation equipment enables all magnetic fields occurring during operation of the ship to be measured and analytically separated in rest conditions. These comprise the ship's own permanent fields, course-dependent induced fields, electromagnetic stray fields, and course and position-dependent eddy current fields resulting from pitching or rolling.

The ship rests above a carpet comprising multi-axial sensors, that is the integral interference field is measured with one, two or three axes in a horizontal plane beneath the ship.

Because of the array of sensors distributed over a large area comparable with the size of the ship and owing to measurement in several axial directions, the interference field analyses and the results of monitoring are more precise, provide more information and are thus more reliable. The advantage of a stationary system as compared with the dynamic overrun measuring system lies in the fact that any particular and temporary change in the earth's fields can be simulated with the aid of large area current coils arranged horizontally and vertically. This also enables the stray field of the entire ship to be monitored in the rest condition for pitching and rolling. It is also possible to simulate known earth field data of intended zones of

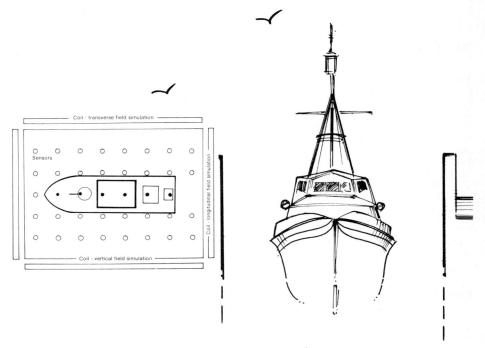

Stationary magnetic measuring system

operation and thus obtain and program the degaussing variables required for self-protection.

Contractor
Institute Dr Forster, Reutlingen.

OPEN RANGE

Type
Dynamic overrun measuring system.

Description
Dynamic overrun measuring systems are permanently installed monitoring stations for strategically important ships endangered by mines.

If the interference field exceeds known critical limit values, degaussing measures must be taken. Further overrun checks qualify the ship for the intended operation.

Following a north-south course and then an east-west course, the ship passes over a perpendicularly arranged chain of magnetic field sensors. This enables measurement to be conducted in two extreme courses with respect to the earth's field. The vertical Z component of the interference field is usually measured with uni-axial sensors. The depth and spacing between the sensors are matched to the type of ship in question.

Measured value acquisition, data processing and representation are carried out using several channels, analogous to the number of sensors or their measuring axes.

This produces, for example, graphical weighted pictorial relief representations in the longitudinal and cross-direction of the ship and these can be assigned geometrically by tracking the ship visually.

Contractor
Institute Dr Forster, Reutlingen.

MOBILE OVERRUN RANGE

Type
Transportable open range.

Description
Open sea ranges are primarily check ranges. The initial object of ranging is to check a vessel's degaussed state before operation and to perform magnetic calibration exercises. Being simple and rapidly installed the transportable degaussing (overrun) range which requires only one range array provides a cost-effective solution to providing facilities to meet logistic and operational requirements in areas where fixed facilities are not available. The transportable range can be set up in a very short time using a limited number of sensors and structures. Such a system can provide measurement, recording, analysis and reporting of actual magnetic states of ships.

The main components of the range are a transportable framework, an ISO container used as the range house, the Forster magnetic measuring system Magnetomat and the tracking system (no separate system is necessary as specialised software and range sensors are used).

Contractor
Institute Dr Forster, Reutlingen.

DEGAUSSING RANGES

Type
Overrun range.

Description
The basic configuration for ranging of modern MCMVs consists of an intercardinal probe array which enables the measurement of the magnetic signature while the vessel is sailing on an intercardinal course (45°) related to magnetic north.

The ranging procedure is conducted on opposite overruns in order to determine the different origins of the ship's magnetisation. During measurement the course of the MCMV is permanently tracked by a shore-based laser course-tracking device and course plots are fed to the evaluation system for further consideration. At the same time course data are transmitted to the MCMV and converted into course information which serves to support the ship's command.

The entire probe array consists of four active triple probes (x-, y-, z-sensors) with 'built-in' electronics. Three probes are installed at the same depth and form the measuring level. The fourth probe is set at a different depth and serves to determine the 'depth law', which enables the vessel's interference field at various depths to be calculated.

Data from the probes are processed by software packages which enable the vessel's induced and permanent magnetisation to be separated. A prerequisite for separation is the ranging of the vessel on opposite intercardinal overruns. The software also enables the required coil settings of the ship's degaussing system to be calculated and the determination of the 'depth law' and conversion of the measured interference field into various depths.

Ranges with reduced numbers of probes on intercardinal alignments offer easy installation and avoid the complex underwater structures associated with stationary ranges. Such systems are suitable for both mobile and permanent systems and containerised versions are available.

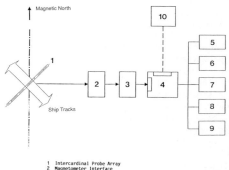

System configuration of overrun range

Type
Stationary range.

Description
With this type of range measurements are carried out while the vessel is moored over an array of probes. The probes are arranged along the longitudinal and transverse axes of the vessel and enable the interference field to be sampled at various points beneath the ship.

The essential feature of the stationary range is its ability to also measure the ship's interference fields caused by eddy currents and to predict the vessel's field at various geographic locations. For this the stationary range is equipped with a coil arrangement which allows the vertical and horizontal magnetic fields to be generated to simulate the roll motions of the ship and its geographic location, including heading. Another important feature of a stationary range is its ability to directly measure and assess stray fields.

Type
Magnetic ranging facilities for components.

Description
This type of range is designed to provide measurement of magnetic conditions and magnetic treatment – if considered necessary – of components such as diesel engines, generators and so on, prior to installation on board the MCMV.

As an additional feature the range provides facilities for the evaluation of magnetic models concerning the ship's magnetic behaviour.

Measurements and treatment activities are computer-controlled and supported by suitable software measures.

Contractor
STN SystemTechnik Nord GmbH, Hamburg and Bremen.

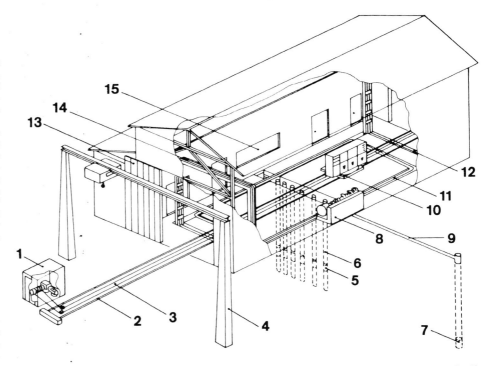

1. **Winch Unit**	7. **ESK-Probe**	12. **Longitudinal Field Coil**
2. **Rail**	8. **Measuring Depth**	13. **Sliding Door**
3. **Rope**	**Adjustment Unit**	14. **Longitudinal Field Coil**
4. **Crane**	9. **Cable Duct**	15. **Measuring Room**
5. **Probe**	10. **Measuring Car**	
6. **Probe Tube**	11. **Vertical Field Coil**	

Main components of measuring range

ITALY

MWR

Type
Mine warfare range.

Description
These systems comprise a group of mobile and fixed equipments that can produce several configurations of range for mine warfare. The ranges are basically used to study mine-target countermeasure interaction and to assess operational data necessary to planners in order to use available resources to achieve the best results.

The MWR system has been designed to operate with mines manufactured by the former Misar Co (now part of Whitehead), and to obtain the maximum information relevant to the desired parameters. This includes: planning the precise location of mines within minefields and their general location; selecting the most effective route for MCMVs as well as targets; recording the responses of the mine; and analytically estimating the required parameters related to mine detonation as well as to units transiting minefields. All this is carried out automatically.

The system comprises the exercise mine linked by cable to a floating buoy, which receives signals from the mine either by the cable or by acoustic detection, and transmits this data by radio to a shore station within a range of 5 km.

Contractor
Whitehead, Livorno.

MINE WARFARE TESTING RANGE

Type
Mine warfare range.

Description
The range is designed to evaluate the effectiveness of mines against naval targets, the effectiveness of MCM operations against various types of mines, and the level of risk for MCM vessels and military and civil traffic.

The range consists of various equipments deployed at sea together with a shore base comprising a master and two remote stations.

At sea a series of exercise mines is acoustically linked with control buoys, and with a mine positioning system installed on board an auxiliary ship and able to compute, before each exercise, the exact position of all the mines in the range.

A communication system links the buoys to the computing centre in the master station. A ship positioning system tracks all the surface ships (targets) with high accuracy using interrogators in the master and remote stations and transponders installed on the target ships.

The computing centre acquires the information from the mines (activation, self-protection) and correlates the data with the time and with the ship positioning information. Data are displayed in real time on a graphic workstation and recorded on magnetic tape for subsequent analysis. Plottings and colour printouts are available to the operators.

The computer system comprises a DEC high resolution graphic workstation equipped with disks and tape units. Peripherals include a printer, colour hard copy, plotter and an alphanumeric terminal. The application software is written in Ada and it has been designed with the HOOD methodology with extensive support of CASE tools (Teamwork).

Operational Status
The range is being installed near the naval base at La Spezia and will be used by the Italian and other NATO navies. It is scheduled to be completed in May 1992.

Contractor
Datamat, Rome.

UNITED KINGDOM

MS 90

Type
Degaussing range.

Description
MS 90 is a new generation of degaussing range that requires just two or four sensors to determine the state of magnetism of a ship, the new sensors generating the same output as the 12 to 18 sensors of the fixed range. It is thus much simpler to deploy, and being transportable (the whole system, comprising two sensors, data transmission system and shore-based computers can be housed and moved in a standard 9 m ISO container) can be moved to the

area of operations, even if this is several hundreds of miles from the ship's normal base, and set up in just two days.

The autonomous, self-surveying, three-axis, flux-gate devices can work out their own orientation and the distance between them, which overcomes the requirement for the precise surveying and alignment required in current ranges and eliminates the need for an accurate ship's tracking system. The twin sensor modules are packaged in a specially designed housing suitable for permanent immersion in depths down to 50 m. The sensor electronics compensate for tilts of more than 30° of arc and geomagnetic field. An automatic calibration facility is included and data can be transmitted using a telemetry buoy up to 5 km offshore. Using mathematical models combined with the new sensors, MS 90 can accurately predict a ship's magnetic field anywhere around it. When a ship sails between the sensors they automatically track it using the ship's own magnetisation, and the data are fed back to the operator onshore either by radio or cable. Using the mathematical models the operator can quickly assess the action required to produce the magnetic stealth ship. The software, also developed by THORN EMI Electronics, is menu-driven and provides range checkout facilities and file protection. The operator is prevented from giving erroneous instructions to the system by error check-ing routines.

Considerable time savings in operation can be achieved with the new system, the three days nor-mally required to complete the degaussing of an MCMV (one degaussed run, one undegaussed run and a final check run) can now be reduced to just half a day. Considerable cost saving can also be realised with this latest generation of range. Conservative estimates put savings in the region of at least a half in terms of capital outlay on a fixed system (costs for which can be in the region of £1-2 million), with considerable additional operational savings.

Not only can MS 90 completely replace an open range without any loss in quality, but it can also be used for harbour entrance ranging, model ranging, roll ranges and check ranges.

For submarines fitted with degaussing coils the MS 90 can be used to measure the near field (enabling the boat to counter mines), or it can be optimised for far field measurement, enabling the submarine to counter magnetic anomaly detection (MAD).

Operational Status

Two systems have been accepted by the UK MoD and were deployed in the Gulf. Earlier successful field trials were held in Australia.

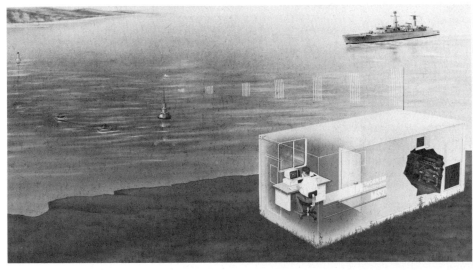
Schematic view of a THORN EMI Electronics MS 90 transportable degaussing range

Inside an MS 90 container showing display presentation of a magnetic signature

A future development with a system such as MS 90 would include the combination of hydrophones to form a combined magnetic and acoustic range.

Contractor

THORN EMI Electronics Ltd, Naval Systems Division, Rugeley.

MAGNETIC SENSORS

THORN EMI Electronics has a complete capability in magnetic detection and measurement with over 20 years' involvement in the design and manufacture of magnetic sensors and systems.

The range of equipment available for military appli-cations includes low power, small dimension fluxgate sensors operating in a fundamental mode to detect magnetic anomalies for fuzing applications. In some examples, the associated electronic circuitry for signal processing is built into the sensor to form a complete magnetometer.

Totally encapsulated, and small in size, they are easy to deploy and accommodate into equipment. Either one, two or three-sensing axis types are avail-able.

THORN EMI magnetic sensors use very little elec-trical power (for instance only 0.5 mW is required for the type LPM2).

There are various sensors in the range for differing applications: for example, a general-purpose sensor which provides an analogue output proportional to magnetic field strength is the type LCM2; the type LPM2 is similar but has exceptionally low power con-sumption. The type LNR1 has very high sensitivity (detecting anomalies of the order of 1 nT) and also low power consumption.

Contractor

THORN EMI Electronics Ltd, Naval Systems Division, Rugeley.

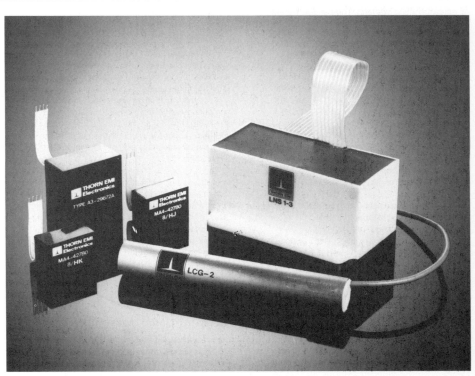
A selection of THORN EMI Electronics magnetic sensors for fuzing applications

MAGNETIC SIGNATURE MEASUREMENT

The fixed range system includes a linear intercardinal array of equispaced three-axis magnetic sensor assemblies individually mounted on either concrete limpets on, or pile-type structures in the seabed. At each end of the line array additional seabed structures incorporate an acoustic transponder for unit position measurement and a depth measuring sensor. Optionally, this unit can also be modified to monitor noise signals of the vessel. The data processing approach used does not demand a high alignment accuracy but does require an accurate knowledge of the actual position of each sensor. Digitised data from the sensors are transmitted by cable to a shore instrumentation facility.

The control instrumentation both supplies the array sensors with power and translates the communication link for use by the range computer. The cable has been designed to ensure very reliable operation after long-term immersion on the seabed. It is of small diameter with single cables being used for each sensor. This minimises the problems in cable laying, recovering and relaying, and enables proven wave zone protection techniques to be used. Avoidance of the use of junction boxes and use of single cables significantly increases range availability, particularly when cable damage occurs.

The computer system is based on the MicroVAX II, as this meets the processing requirements, providing the power necessary to implement the data processing approach and to make available the required signature and coil prediction data to the operator on demand.

Associated with, and digitally linked to the shore instrumentation facility is a laser-based tracking system which also provides data for the helmsman. The data, in the form of distance to go to array and distance off track, are communicated to the vessel by RF link. The laser tracker also requires the temporary addition of two omni-directional prism reflectors to the vessel so that heading information can be obtained. For the transportable system, which would use single axis magnetic field sensors assembled into independent but coupled sub-arrays, and as an option on the permanent range facility, magnetic tracking can be provided.

Contractor
Ferranti-Thomson Sonar Systems UK Ltd, Cheadle Heath.

TRANSMAG 2000

Type
Transportable degaussing range.

Development
Transmag 2000 has been developed as a totally transportable degaussing range system which can be readily moved between locations and deployed with the minimum of effort.

Description
Transmag 2000 can be used with all classes of ship, but is especially adaptable to MCMVs, its two modes enabling it to be used as a ship's magnetic signature check facility or as a ship's calibration facility.

Small and of modular design, Transmag 2000 can be packed into two 3 m standard ISO containers for transport by land, sea or air. Alternatively, the equipment can be mounted in a rugged cross-country vehicle or trailer of the customer's choice. Thus, a task such as helping to keep the Gulf clear of mines can be made safer by carrying a transportable degaussing system to the actual area of operations.

The new system consists of an arrangement of magnetic field sensors located in a line on the seabed. These are connected by a secure cable link to a data acquisition and computer system in a control module, which can be located on shore or on a support vessel. Close by the control module, there is a unit which tracks the ship being ranged.

The computer system acquires magnetic and water depth data along with the ship's track data, and this is processed to advise the ship of the optimum course to steer over the range to resolve the ship's magnetic components and predict coil current settings. It can also predict fields remote from the ship to give safe distance analysis and Magnetic Anomaly Detection effects.

Transmag 2000 does not require the use of heavy mechanical handling equipment, the deployment of sensors being accomplished by manpower alone operating from an inflatable dinghy which is supplied as part of the plan-packed equipment. Three sensors and a junction box are deployed on the seabed, with one lightweight cable returning to the shore. Three sensors form a good compromise between the ease of deployment, adequate ship's navigation width, availability of magnetic information, and a degree of built-in redundancy. Transmag 2000 is a robust system which enables ship's position and heading to be derived throughout its run.

This information is used to provide an onboard navaid, giving assistance to the helmsman to get the ship over the range as quickly and accurately as possible. In the magnetic signature check mode, the ship is sailed over the sensor array on one pair of reciprocal headings. From the data thus acquired, the permanent and induced components of the ship's magnetic signature can be isolated and resolved into longitudinal, athwartship and vertical components.

Should the permanent components be found

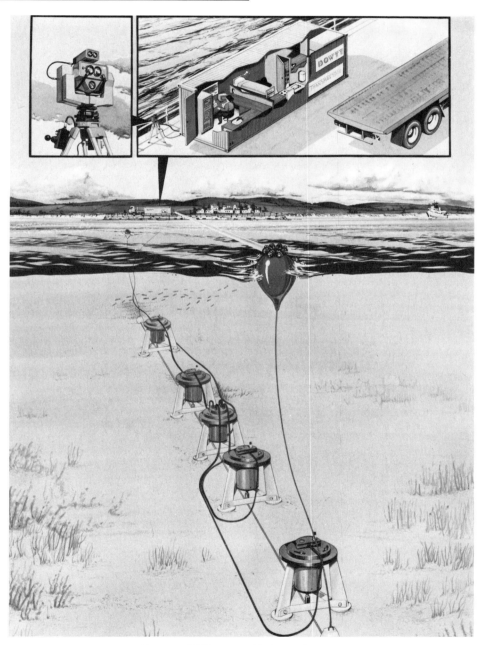

Transmag 2000 transportable degaussing range

greater than that which can be comfortably controlled by the onboard degaussing coils, or they are outside pre-defined limits, then the ship may be instructed to undergo treatment.

In the ship's calibration mode, a coiled ship can carry out a number of passes with its coils either on or off. Providing this is performed in a controlled sequence, the effect on the signature of the ship by energising each coil can be determined. In this way, a ranging can be implemented with onboard degaussing coils currents being predicted so as to reduce the magnetic signature down to the lowest possible value.

Contractor
Dowty Magnetics, Hednesford.

UNITED STATES OF AMERICA

MSMS

Type
Magnetic Signature Measurement System.

Description
This advanced system measures with precision the magnetic signature of ships and gathers, records, processes, stores and displays the data during magnetic ranging and treatment operations.

The system accurately and rapidly ranges both submarines and surface ships to determine the need for magnetic treatment. It also provides critical magnetic measurement readings throughout the deperming process.

With this system, magnetic signals emitted by ships are sensed by underwater magnetometers and recorded simultaneously. Shore-based data acquisition and analysis equipment receives and processes the sensor data to provide, in minimum time, an accurate profile of a ship's magnetic state.

Field measurements are achieved through the use of submerged magnetometers placed in tubes located in the ocean floor. These sensors can also be inserted in a bottom-mounted tube structure, at a customer's option, to give transportability to the system.

Each magnetometer provides a DC voltage that is proportional to the magnitude of the magnetic field component along its magnetic axis.

These sensors are capable of measurement or dynamic range of $\pm$200 000 nT, noise of less than 0.1 nT in a DC to 10 Hz bandwidth, resolution equal to or less than 0.1 nT, and linearity of 0.005 per cent.

Magnetometer outputs can be digital or analogue. These outputs are routed through junction boxes to the interface electronics in the data acquisition console and to a digital computer.

Signatures for keel and athwartship are processed, analysed and displayed. A determination is then made as to whether to record the information.

Raytheon measurement facilities are interactive systems. Software controls the intra-system interface between magnetometers, the interface electronics and standard computer peripherals.

The operator directs the system computer to collect and analyse magnetic field data using appropriate software programs. These programs direct system activity and query the operator for data inputs and desired measurement parameters for ships undergoing treatment.

In response to data and computer commands entered by the operator, the system collects and stores magnetometer data, performs calculations, and displays results relative to the magnetic treatment process.

The system is highly automated and includes data processing and display equipment and specially designed underwater magnetic sensors. It consists of several distinct subsystems.

The Magnetic Range System relies on two arrays of magnetometers, shallow and deep, to range ships. User vessels pass over the sensor arrays at least once in each direction, with signatures continuously recorded via shore-based equipment. This subsystem can also be used to calibrate ship degaussing systems.

The Magnetic Signature Measurement System operates in conjunction with a fixed deperming facility. It uses a keel line of magnetometers to measure signature levels before, during and after deperming operations. User ships are moored over the sensor array and signatures recorded on command and at the appropriate time via shore-based equipment.

These subsystems, although independent, share common data acquisition and analysis equipment. Off-the-shelf hardware is used extensively throughout system designs. A modular approach to design, with key components separate from one another, will enable new features to be incorporated in the future with minimal system upgrade.

Operational Status
MSMS is fully operational and meets US Navy requirements for performance, logistics support and life cycle costs. Early in 1991 Raytheon was awarded a $21.9 million contract to design, build and install an advanced magnetic treatment at the US Navy's Trident Magnetic Silencing Facility in Kings Bay, Georgia. The automated system will rapidly scan the submarines to determine the required magnetic treatment, and then deperm the submarines by removing or altering their permanent magnetism. The system is scheduled to be operational in early 1993. In May 1991 Raytheon won a $29.4 million contract to build four Type IV magnetic silencing ranges at San Diego, California and Mayport, Florida and two options for systems at San Francisco, California and Charleston, South Carolina. The four systems ordered are scheduled to be operational by 1994.

Contractor
Raytheon Company, Submarine Signal Division, Portsmouth, Rhode Island.

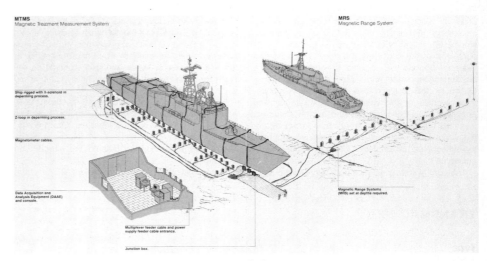

Raytheon MSMS system

DEGAUSSING SYSTEMS

FINLAND

FDS

Type
Onboard degaussing system.

Description
The FDS is an advanced modular degaussing system which can be used aboard a wide range of ships. The solid-state power units range from 270 W up to 30 kW per power module. Each type of power module is based on a microprocessor-controlled switching-mode current amplifier which means high accuracy and efficiency.

The MSO-800 Control Unit is the operator interface for system configuration, system start/stop, the entry of calibration parameters during ranging, alarms with clear descriptions, and manual control. The man/machine interface is menu-driven and based on a 640 × 400 pixel graphic EL display and a 20-key keyboard. The MSO-800 can contain four different calibration parameter sets, for example different load situations, so the operator needs only to select the right pre-calibrated parameter set when the load situation changes. The MSO-800 contains a special approximation program, which calculates earth's local magnetic field from ship's position. The MSO-800 can be connected to a ship's navigation system or the position manually entered.

The automatic control can be based on the MHM-402 masthead magnetometer or ship's gyro-compass combined with built-in inclinometers (pitch and roll) and earth's local magnetic field approximated by the control unit. Both control modes can be in the same system. The manual course control combined with the local field approximation is always available in the control unit.

The FDS features built-in diagnostics at modular level, clear alarm functions and fault-tolerant architecture. All calibration parameters are stored in non-volatile memories both in the control unit and in the power modules. Replacing a power module does not require manual calibration because this is automatically performed by the control unit.

Contractor
Elesco Oy Ab, Espoo.

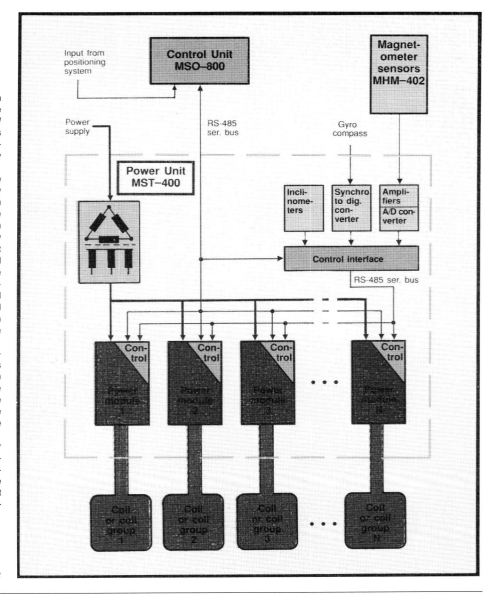

Block schematic of FDS

FRANCE

MAGNETIC TREATMENT SYSTEMS

Magnetic Treatment Systems
Thomson Sintra manufactures magnetic treatment facilities which are used to reduce and/or normalise permanent magnetic signatures using a deperming system. These systems are of particular use for vessels lacking onboard degaussing equipment. Stations can be of the static type, or dynamic, in which the vessel to be depermed makes a run over the range.

Contractor
Thomson Sintra Activités Sous-Marines, Brest Cedex.

GERMANY

DEGAUSSING SYSTEM

Description
In its smallest configuration the degaussing system consists of a triple probe, a miniaturised control cabinet and nine coil systems to control the permanent, induced and eddy current components of the ship's magnetic field in the ship's three axes: vertical, longitudinal and athwartship. The control cabinet includes a three-channel fluxgate type magnetometer and nine channel amplifiers that generate the output power required for non-magnetic vessels.

The structure of the standardised and miniaturised degaussing system is hierarchical, that is, it comprises self-contained non-interlinked components which are provided with their respective input values by the hierarchically superior components. This ensures a high degree of reliability and system availability, or a high degree of probability that the system will function effectively and can be easily maintained even under adverse conditions.

The complete system is composed of independently functioning and testable modules, such as:
(a) degaussing probe with magnetometer
(b) channel amplifiers
(c) operating panel
(d) manual adjustment device
(e) power supply module.
Every module forms a potential island by the separation and screening of power units and by the employment of a new method for floating measurement of DC currents, which has proved itself in this system. Thus there is no mutual influence of the system's individual channels, so that in the event of failures in the coil network the reduction in protection is kept to a minimum.

Provision is made for standby operation of the system with the aid of manual control equipment, which maintains a fixed value for the permanent channels and regulates the induced channels as a function by means of a manual adjustment device.

The system output is sufficient for non-magnetic submarines, small fast patrol boats and mine warfare vessels. Certain components, such as magnetometer and channel amplifier, can also be used for subordinate degaussing systems, such as for separate protection of guns.

However, for larger ships such as frigates and for ferromagnetic submarines, the system output must be increased. STN SystemTechnik Nord supplies the following system components for this purpose:
(a) power supply units (inverters, rectifiers)
(b) additional amplifiers

(c) generator sets
(d) feedback compensation devices
(e) electric compass compensation switchboxes
(f) manual course adjustment devices.
All these components have interfaces compatible with the Degaussing Control Cabinet, so that ergonomical interconnection of the system is ensured.

These components are connected into the control loop according to their function, so that high accuracy

of the system and a correspondingly high degree of ship protection is provided.

Contractor
STN SystemTechnik Nord GmbH, Hamburg and Bremen.

DEG-COMP

Type
Decentralised degaussing system.

Development
This is a future orientated system developed by STN SystemTechnik Nord for use on MCMVs, FACs and submarines. It is designed to meet increased demands for optimum magnetic protection against sea mines.

Description
The system uses common coils for compensating induced magnetism, permanent magnetism and eddy current components. Individual amplifiers are used to feed the decentralised coil. Geomagnetic field data in the ship operating area are synthetically generated. The system carries out digital control and monitoring of data transfer which uses a star configured data transfer structure between the central unit and amplifiers.

The configuration optimises energy requirements by summarising components of compensation and results in reduced cable and installation requirements for the degaussing coils.

The system comprises a central unit and a number of decentralised amplifiers which feed the degaussing

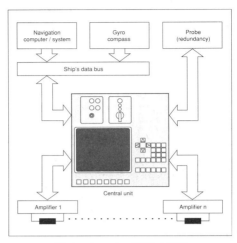

Block schematic of DEG-COMP

coil system with controlled direct currents. The computer-supported central unit is equipped with an internal data memory for archiving the geomagnetic field data occurring in the area of operations.

The central unit is provided with suitable interfaces to the ship's navigation system to receive all relevant data regarding the current operational conditions (for

example ship's position, heading, roll and pitch angle and so on) via the ship's internal databus or via direct interfacing.

The geomagnetic field components affecting the ship are constantly assessed with the aid of the stored geomagnetic field data and taking into consideration the existing operational conditions. These components are converted into set values for the induced current portion of the degaussing amplifiers.

By summarising the set values for the induced, permanent and eddy current portions of magnetic field components, the central unit generates individual set values for each amplifier to effect proportional compensation currents for each coil.

The decentralised amplifiers are directly assigned to the individual coils.

Use of an earthfield triple probe is possible as redundancy to the geomagnetic data memory further increases the system reliability. The man/machine interface relies on a clearly structured user interface incorporating a monitor with associated function keys and hierarchically organised menus for interactive system inputs and outputs within the various operating modes.

Contractor
STN SystemTechnik Nord GmbH, Hamburg and Bremen.

UNITED KINGDOM

CC2

Type
Shipborne degaussing system.

Description
The CC2 onboard degaussing system uses information from an internal geomagnetic database to control ships' magnetic signatures. The fully automatic system correlates position and heading with geomagnetic data and computes the optimum degaussing coil currents for a minimum magnetic signature.

CC2 comprises a system of degaussing coils, coil current drivers, and a bridge control unit which can control the outputs of up to eight separate degaussing power supplies and associated coils.

The control unit is configured for a specific system by the selection of appropriate hardware and software modules. The unit uses microprocessor hardware based on the THORN EMI D3 data distribution equipment. It accepts heading information in synchro data or serial data formats plus inputs of latitude and longitude. Correlating this information with the geomagnetic database, the control unit computes the coil currents for optimum degaussing at the given location. Alternatively the control input can be from a three-axis magnetometer, although this method often suffers complications from the effects of the ship itself. As this information is updated in real time, the degaussing coil drivers continuously change coil current levels to provide minimum magnetic signature.

The coils, installed in the ship's hull during construction, are energised by a DC current from the coil drivers to produce a magnetic field which opposes the existing ship influence caused by permanent or positional effects.

The lightweight coil driver provides an output of

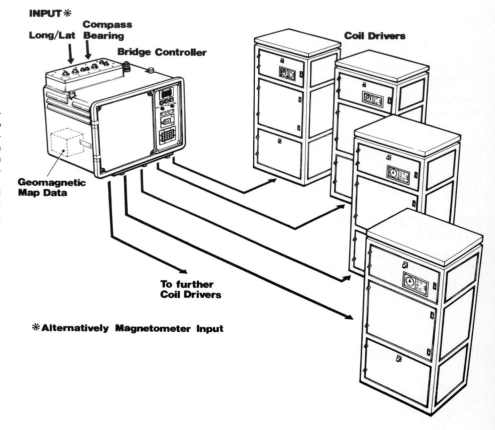

Ship's degaussing control system —CC2

160V 60A (up to 120A with the power supplies connected in parallel). The system is suitable for most types of steel ship or submarine.

Contractor
THORN EMI Electronics Ltd, Naval Systems Division, Rugeley.

DEGAUSSING SYSTEM

Type
Degaussing coil system for surface ships.

Description
Degaussing systems reduce the magnetic signature of a vessel to a specified target level by compensating for the distortion of the earth's magnetic field caused by the presence of the vessel. This is achieved by accurately controlling the currents through a number of coils installed in the vessel. This coiling system can be enhanced by the use of strategically placed Variable Moments Magnets (VMMs).

System Controller
Either analogue or digital control can be provided. The correct degaussing coil current is computed using the local earth's magnetic field and ship's attitude either from a system magnetometer, a magnetic map, ship's gyro system or a combination thereof.

Three part M plus F, Q coils
In this arrangement horizontal coils run around the ship to counter vertical fields. These are generally broken down into three separate coils with independently variable ampere turns to compensate for the unequal distribution of ferrous materials in the ship. Horizontal coils, usually mounted directly below the weather deck, are located forward (F) and aft (Q). These coils counter the vertical component of the longitudinal field.

Three part M, F, Q and A coils
This is a similar system, but with the addition of a pair of athwartships (A) coils to counter the athwartship disturbance. This coil system is the normal fit for steel vessels and allows better signatures to be obtained in the magnetic equatorial zone, where the horizontal component of the earth's field is at its maximum. With the correct control equipment a good magnetic signature should be obtainable in any zone regardless of the ship's heading. Without A coils, some uncompensated athwartship field will be present when the ship departs from a north/south heading.

Multiple M, L and A coils
This system is a full fit. Improved longitudinal compensation is achieved by a series of L coils along the length of the vessel. Normally this fit is used for MCMVs, although steel-hulled surface vessels are also better protected if this fit is used.

Enhanced Coil Fit
Restraints imposed by vessel architecture will, at times, mean that coil positioning is not optimum. This may result in a final signature being outside the design specification. The use of a supplementary VMM fit allows the signature to be brought back within limits.

COOP Degaussing
COOP selected for MCM tasks rarely have a degaussing fit. VMM technology offers a 'carry on board' system which will greatly reduce the vessel's signature.

Contractor
Marconi Radar and Control Systems, Leicester.

SUBMARINE MAGNETIC TREATMENT FACILITY

Description
The design of this equipment provides an innovative application of existing proven techniques to enable the magnetic signatures of all present and future classes of UK submarines to be reduced cost-effectively and without any risk to the submarine or to the environment. Computer-aided design has been used in the calculation of magnetic coil sizes and content to ensure cost-efficient achievement of the requirements.

While vessels are at sea, sailing on various headings, the vessel's permanent magnetic signature alters, which could make it vulnerable to magnetic mines or detection by magnetic sensors. Magnetic measurement on an open sea range will establish the need for a submarine to be treated in this new facility.

Previously, treatment was a lengthy exercise involving wrapping coils around the vessel's hull and taking the vessel out of service for over a week. Now, in much less time, the submarine will be housed accurately inside the treatment berth, which is fitted with electrical coils and equipment to produce homogeneous magnetic fields around the hull. Sensors, instrumentation and control equipment will continuously monitor the submarine's magnetic signature to ensure treatment is carried out fully and accurately. Treatment control can be fully or semi-automatic or manual.

Operational Status
A £50 million contract has been awarded to construct this submarine magnetic treatment facility at the Clyde submarine base, Faslane. It will be the first automatic 'drive-in' unit in the UK.

On-site construction was due to start in Spring 1991 but the UK MoD decided not to proceed with the project for financial reasons.

Contractors
Dowty Magnetics, Hednesford.
  Vosper Thornycroft (UK) Ltd.
  Tarmac Construction Ltd.

ACOUSTIC CONTROL

FRANCE

QSUA SERIES NOISE DETECTORS

Type
Detectors for cavitation and superstructure noise.

Description
The QSUA series is a family of acoustic noise measurement systems which provides detection of propeller cavitation, and the monitoring and localisation of superstructure noise in different parts of the ship.

QSUA-2A
This is an equipment for monitoring and localisation of superstructure noise. Monitoring is performed manually and in wideband only. In service with the French Navy and a number of other navies.

QSUA-2B
Similar to the QSUA-2A but with the additional capability of manual analysis of the received noise.

Operational Status
In operational service with several navies.

Contractor
Safare Crouzet, Nice.

QSUA-4A

Type
Acoustic noise monitoring system.

Description
This equipment provides automatic detection and indication of propeller cavitation, and automatic monitoring and localisation of vibration and superstructure noise along the ship with preset alarm levels. It is an independent system fitted with its own display, but able to be integrated into a sonar system.

The system consists of a processor and display console, one or two remote cavitation indicators and up to 20 hydrophones and accelerometer sensors. The system performs continuously and without manual intervention: the detection and indication of propeller cavitation; monitoring and localisation of superstructure noise and vibration with synthesised display of broadband noise levels throughout the length of the vessel; the triggering of an alarm on detection of propeller cavitation or when a noise level, preset for each sensor, is exceeded; and the optional automatic recording, on any external recorder, of the signals present on the two most noisy channels when one of the preset levels is exceeded. Manual operations available include the precise measurement, analysis and monitoring of noise or vibrations received by two sensors. This is used for determining any correlation between two noises. Other options available include the selection of two channels for external processing (spectral analysis), and the simultaneous external recording of signals present in two or more channels.

Operational Status
In service with the French Navy.

Contractor
Safare Crouzet, Nice.

UNITED KINGDOM

SI 1220 MULTI-CHANNEL SPECTRUM ANALYSER

Type
Underwater surveillance analyser.

Description
The SI 1220 is a portable, versatile commercial analyser, designed and built to AQAP 1 standards, which is tailored to the specific needs of sonar analysis by specially designed software. The instrument incorporates all the functions of an FFT spectrum analyser and includes several features of interest to groups involved in the analysis of underwater noise.

The SI 1220 performs four-channel analysis at frequencies up to 50 kHz, with parallel analysis of cross products between channel pairs. Features include a one million sample buffer for acquisition of transients, time slip between channel for beam forming, simultaneous broadband and Vernier analysis, 16-level greyscale waterfall on a built-in CRT, and generation of chirps or user-defined arbitrary signals.

The equipment is capable of many sonar analysis tasks including noise analysis, LOFARGRAM analysis, transient analysis and noise source correlation. It is suitable for use in trials, operator training, quick-look analysis of on-line tape recordings, calibration of hydrophones, signature analysis and archiving.

Contractor
Schlumberger Instruments Division, Farnborough.

VIMOS MONITORING SYSTEM

Type
Vibration and noise monitoring system.

Description
The Ferranti VIMOS is a computer-controlled vibration and noise monitoring system designed to enable the continuous measurement of radiated noise levels at selected positions throughout the ship. In warships, for which the system was originally designed, this allows the increase in these levels to be detected before they adversely affect the ship's sonar capability or increase the risk of detection. Applications include use by the ship's sonar operators and monitoring of machinery wear and of detection.

A combination of accelerometers and hydrophones, placed at requisite internal and external points, is employed, from which signals are transmitted through amplifiers and switch equipment to one or more central control units. These give data acquisition and signal switch control, fast Fourier transform (FFT) signal processing providing both narrow and broadband analysis and backup facilities.

The control unit incorporates a touch-sensitive

Operator's desk unit of VIMOS

plasma panel which allows the operator to interact with the software database and control the system. Aural listening and communications facilities are provided at the operator position and at various remote points.

Operational Status
Several systems have been ordered by the UK MoD for the Royal Navy Type 23 frigates and for the new

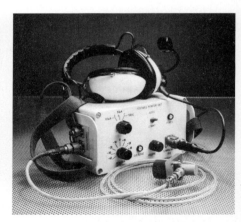

VIMOS portable monitoring equipment

Fleet Auxiliary vessels. Three earlier versions of this system, Hull Vibration Monitoring Equipment (HVME) Mk 1, were delivered to the Royal Navy in 1985.

Contractor
Ferranti-Thomson Sonar Systems UK Ltd, Stockport.

TRAINING AND SIMULATION SYSTEMS

In recent years a great deal of time and effort has gone into the development and production of simulation and training systems for the underwater scene. This has resulted in a much better appreciation of the operational requirements, and has led to a greater efficiency on the part of the operators and designers.

Simulators and trainers in the underwater warfare scenario cover four main areas:
(a) ship handling and manoeuvring
(b) training for machinery control
(c) systems trainers for combat team training
(d) equipment trainers for operator training.
This section deals primarily with system and equipment trainers, in particular those devoted to submarine use.

It should be remembered, however, that many sonar systems incorporate a training facility within the main equipment itself. Details of these will be found in the appropriate section of this book.

COMMAND TEAM TRAINERS

GERMANY

MSG SUBMARINE ATTACK SIMULATOR

Type
Trainer for submarine command teams.

Description
MSG has developed a simulator designed to train submarine command teams in procedures and tactical operations. It is a computer-based modular design consisting of a simulated control room, instructor station and briefing room. The control room uses mainly actual equipments rather than facsimiles, and these are linked to the computer to provide accurate sensor and operational performance parameters. The periscope subsystem is a facsimile with video-generated data. The degree of simulation is created according to individual customers' requirements.

The modular concept ensures that the system can be installed with only a basic subsystem, which may be upgraded as required. The simulator may be mounted in a building, or it may be installed on a motion system with a 6° field of movement.

Operational Status
Fully developed.

Contractor
MSG Marine und Sondertechnik GmbH, Bremen.

AGUS ASW TRAINER

Type
Command team trainer for ASW operations.

Description
AGUS is designed to train and assess command information teams in anti-submarine warfare. It also trains and assesses operators of different grades in the use of sensors and weapon systems, and provides training in tactics. It uses distributed computer architecture techniques employing Atlas MPR/EPR processors which allow modular construction to meet customers' requirements.

The system consists of a number of command information cubicles, an instructor's station, a debriefing auditorium with a large screen display, and an electronics room. It can be equipped with either real operating equipment or with general-purpose consoles which have intelligent graphic displays.

The simulator provides dynamic operating behaviour of own ship's course, speed and mission duration; realistic visual and acoustic presentation for active passive and intercept sonar; presentation and operation on radar and ASW consoles; and weapon data; together with combat situation displays. Also provided is realistic contact association, target motion analysis and data display, and weapon firing tactics, together with simulation of weapons and decoys.

Operational Status
Operational with the German Navy.

Contractor
Atlas Elektronik GmbH, Bremen.

AWU 206/206A ATTACK TRAINER

Type
Attack Submarine Team Trainer (ASTT).

Description
The Attack Submarine Team Trainer (ASTT) is a shore-based computerised simulation system for training all grades of submarine teams in order to achieve maximum combat power at sea. The ASTT simulates tactical situations which can be presented to the trainees who are working in the submarine's Combat Information Centre (CIC) installed at the naval base. The weapon officers and operators are trained in operations and response to specific tactical problems under the supervision of exercise controllers. The ASTT also provides for development and analysis of naval procedures for torpedo firing and guidance methods, and optimum operation of onboard equipment.

Through co-operation between the simulator and the CIC, optimum training can be achieved for:
(a) attack of surface vessels and submarines
(b) defence against attacking surface vessels, submarines, ASW aircraft and helicopters
(c) operations in minefields.

The system consists of an auditorium with large screen displays, instructor consoles and communications equipment, the CIC with real equipment, and a computer room. The latter generates the sonar, tactical data handling, weapon handling, periscope, navigation, ship's controls, radar and ESM inputs for the system.

AWU attack trainer instructor consoles

Operational Status
In service with the German Navy.

Contractor
Atlas Elektronik GmbH, Bremen.

SUBMARINE COMBAT TEAM TRAINER

Type
Trainer for all levels of submarine crews.

Description
The Submarine Combat Team Trainer (SCOTT) is a shore-based simulation system for training all levels of submarine personnel. It provides a submarine's Combat Information Centre (CIC) with all working tasks by using multi-purpose consoles which are independent of any special type of console. The simulation capabilities are, therefore, virtually unlimited.

New hardware concepts, man/machine interfaces and other subjects can be taught and investigated in advance. The available simulators for SCOTT cover all types of sonar, underwater telephony, attack periscope with visual system, fire control systems, radar, ESM, tactical data handling and steering unit.

The officers and operators are trained to operate and combine the available sensors, using the correct tactics and deployment of weapon systems, and optimising and improving procedures. The system is supported by a replay system with detailed briefing/debriefing facilities.

The following training objectives can be achieved with SCOTT:
(a) surveillance and attack of hostile surface vessels, convoys and/or submarines, using detection, classification, threat analysis, tactical navigation, weapon control and weapon deployment
(b) detection and countermeasures against hostile attack by surface vessels, submarines, ASW aircraft and helicopters
(c) minelaying operations and operation in mined areas
(d) tactical manoeuvres.

Operational Status
In service.

Contractor
Atlas Elektronik GmbH, Bremen.

ISRAEL

ASWT-2 ANTI-SUBMARINE WARFARE TRAINER

Type
Trainer for anti-submarine teams.

Description
This system is designed to train anti-submarine warfare teams and provide all aspects of operational experience, including classification of sonar echoes, identifying submarines, attack procedures, and the use of weapons. All types of surface and airborne platforms associated with ASW are represented within the system which is housed in a building.

The ASWT-2 consists of a control room, a computer room, debriefing room, two anti-submarine 'ships' with their own bridge, an operations and sonar room, and a number of multi-purpose cubicles representing various submarine and airborne vehicles. Each bridge has a manoeuvring control console, simulated bridge radar, communications and weapon release equipment. The operations room contains a plotting board, indicators to note target data, and a console for simulated radar and visual detection together with an optional ESM display. The sonar room has only the sonar operator's console and communications.

The exercise covers 100 × 100 nm (185 × 185 km) and realistic maritime environments are created. Three types of torpedo, two types of depth charge, and three other anti-submarine weapons can be simulated.

Operational Status
In service with the Israeli Navy.

Contractor
Elbit Computers Ltd, Haifa.

ITALY

MULTI-APPLICATION SONAR SYSTEM (MAST)

Type
Training system for sonar operators.

Development
MAST is a multi-application sonar which has been developed by Selenia Elsag Sistemi Navali in co-operation with Raytheon, USA, and has been delivered to the Italian Navy training centre.

Description
MAST consists of the following units:
(a) trainee equipment consisting of a set of units identical to those used operationally on board
(b) an instructor control station from which the instructor provides for the insertion of data related to own ship, and the target's course, speed, position and so on
(c) a medium size minicomputer
(d) simulation hardware producing signal and noise to be injected into the front-end transducer array interface. These signals simulate those which would have been generated by the acoustic environment when converted from pressure to voltage by the receiving transducer array.
These include all possible variations due to motion, mutual interference, propagation characteristics, and the environmental conditions simulated.

The MAST system is used for the training of operators and maintenance personnel on the DE 1160 and DE 1164 sonar systems (see entry in the USA section of *Sonar Systems – Surface Ship ASW Systems*). Further improvements will allow training facilities for other passive sonar systems produced by Selenia Elsag Sistemi Navali, and also for the DE 1167 being manufactured under licence from Raytheon.

Operational Status
In service with the Italian Navy.

Specifications
Number of targets: Up to 8
Target types: Own ship, surface ship, submarine, torpedo decoys, biologics, wakes
Own ship simulation: self-noise, spoke noise, VDS simulation
Ocean effects: Propagation loss, sea state, reverberation
Passive simulation: Broadband and narrowband noise
Active simulation: Highlights, target strength

Contractor
Selenia Elsag Sistemi Navali, Genoa.

SUBMARINE TRAINER

Type
Attack team trainer.

Description
The trainer is used to train attack teams of the 'Toti' and 'Sauro' class submarines.

The system comprises real onboard equipment (the command and control system MM/BSN-716(V)2 and the fire control system for A.184 wire-guided torpedoes) and the simulation and control facilities.

The simulation system comprises a computing subsystem for tactical scenario simulation based on a DEC mainframe with magnetic storage units and peripherals (terminals, printer, plotter digitiser). The ship simulation subsystem is based on the Motorola 68020 microprocessor and simulates sensors and equipment interfaces (NTDS, TTL digital I/O, synchro, analogue and so on). A torpedo simulator for A.184 torpedoes is also included. Hardware comprises a number of consoles for controlling sensors and navigation equipment, and includes a video terminal, multi-function keyboard and a control panel for internal and external communication networks. Exercise control is conducted from two instructor consoles based on a DEC Vaxstation. Each includes three video terminals (one high resolution colour), three multi-function keyboards, trackerball and communications control panel. A debriefing system includes a graphics workstation and a large screen projector.

System functions include: scenario preparation with full support provided for storage and retrieval of data including full exercise description, sensors and target characteristics, geographic information and so on; tactical simulation, own ship and target's sensors simulation; data exchange and real onboard equipment; torpedo simulation (logical and electrical); and data recording and playback for debriefing.

Operational Status
The system was installed at the Italian Navy Training Centre at Taranto in June 1991.

Contractor
Datamat, Rome.

NORWAY

MASWT

Type
Anti-submarine warfare trainer for commanding officers and sonar operators.

Description
The MPS Anti-Submarine Warfare Trainer (MASWT) provides realistic and efficient training of commanding officers and sensor and weapon operators on board ASW destroyers, frigates, corvettes, helicopters and so on, and use of active sonars and other sensors. The MASWT also provides training in compilation of information, analysis and decision making in a realistic operational environment.

The MASWT concept gives an advanced simulation of the platforms, weapons, active sonars and other sensors. Consoles and weapon control panels are designed to look like the real consoles. The scenario-definition is done via menu-driven programmes. The concept has an easy-to-use Man/Machine Interface (MMI) and includes full recording and playback/debriefing facilities and a voice communication system.

The sonar equipment consists of the MPS Multi-beam Sonar Simulator (MMSS).

Operational Status
In service with the Norwegian and Swedish navies.

Contractor
Micro Processor Systems A/S, Kongsberg.

A student's console with platforms, weapons and sensors operated via dedicated easy-to-use panels

Norwegian students under instruction in an MASWT

MASTT

Type
Action speed tactical trainer.

Description
The MPS Action Speed Tactical Trainer (MASTT) is a flexible system for building naval tactical trainers and simulators extendable to almost any level of complexity, sophistication and training level. It is designed to fulfil the training requirements of seamen (reporting procedures, sensors and weapon operating procedures, classification training and operator training on real equipment), officers under training (decision making, team work, reporting procedures and weapon employment), and high ranking officers (tactical disposition training, tactical analysis and exercise planning, analysis and debriefing). The system is stated to be very easy to use with students able to operate the cubicles by themselves after one or two one-hour introductory sessions.

MASTT was developed in close co-operation with the Royal Norwegian Navy and installed at the Royal Navy Training Establishment in Bergen in November 1986.

The installation in Bergen consists of 10 cubicle consoles, an extended control section, a standard control section, a system console, two recorder/playback computers, two weapon computers, a communication control, three sonar simulators and two anti-submarine warfare consoles. Each trainer console is software configurable and the user has full control of model and scenario generation. The distributed concept, using an Ethernet network, makes it easy to expand the trainer by adding more cubicles or enhance it by interfacing with the real equipment simulators (such as radar, sonar and weapon control).

The model preparation and storage for the complete game is performed using a commercial computer. The user interacts with the system console via a menu-driven operator dialogue. All game settings and scenarios can be prepared and tested off-line prior to the actual game, so it is possible to build up a library of pre-planned games settings on disk.

Scenarios are based on digitised map contours overlaid with the various optional synthetic information. The actual exercise area is generated from a map library. Two files are made: one for the three-dimensional terrain model and the other for a two-dimensional model for presentation. These files can

be stored for later use.

Each cubicle console can be configured as a surface ship, a submarine, an aircraft, a land-based coastal defence installation, or a coastal radar station. A unit configuration consists of a set of parameters describing the unit type and its dynamic behaviour. In addition, the user specifies the sensors and weapons which the unit will use.

Wind, sea current, precipitation and sea temperature layers are among the data provided.

The exercises are initiated from the same console used for set-up and preparation. The system software is loaded into the different computers in each cubicle via the Local Area Network (LAN) data highway and the complete trainer is started. The user then selects the game set-up to be used, downloads it and the exercise commences.

Following an exercise, a playback programme provides synchronised audio and video playback with freeze and fast wind/rewind functions. During replay the screen in the cubicles and control will contain the same information as during the exercise. Recorded information can be used during a game. The playback programme can also use recordings from real exercises.

Operational Status

The MASTT system has been operational with the Royal Norwegian Training Establishment since 1986.

A MASTT system with 12 cubicles was delivered to the Swedish Navy in 1991.

Contractor

Micro Processor Systems A/S, Kongsberg.

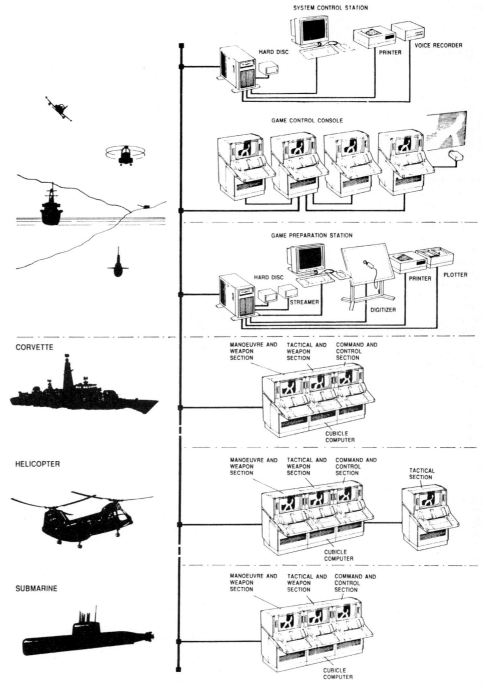

Schematic layout of an MASTT

UNITED KINGDOM

LINK-MILES SUBMARINE COMMAND TEAM TRAINER (SCTT)

Description

This submarine command team trainer is designed to provide effective training in command, target detection and surveillance, weapons control, and the use of combat data systems. The heart of the SCTT is a replica of a submarine control room with microprocessor-driven facsimiles of the command, weapon control and sensors consoles, together with a periscope. Visual simulation for the periscope uses computer-generated imagery. In a total training package for the Royal Norwegian Navy's 'Ula' class, Link-Miles has supplied SCTT, SCS, MCT (Machinery Control Trainer, a comprehensive Technical and Operational Trainer) plus computer-based training to make up 90 per cent of the total requirements.

Operational Status

In September 1987 a £10 million contract was

Instructor's operating station for 'Upholder' class

awarded to provide a system to the Royal Norwegian Navy for use with the 'Ula' class submarine. The system will be installed at the submarine training school at Haakonsvern Naval Base, Bergen, where it will incorporate a Link-Miles SCS.

Contractor
Link-Miles Ltd, Lancing.

UNITED STATES OF AMERICA

LINK ASW TRAINER

Type
Team tactical ASW training device.

Description
The Link Team Tactical Trainer provides ASW team tactical training and evaluation for conventional ASW warships and those equipped with tactical data systems. Simulated sonars, command information centres, bridges and platforms are used to provide trainees with the capability of exercising essential anti-submarine warfare engagement procedures in a simulated multi-threat environment.

These simulated systems will contain general-purpose programmable consoles which can be recognised under software control to represent the anti-submarine capabilities inherent in any of the US Navy's 16 major classes of ASW ships. The system design covers six major categories of simulation: platforms (hostile and friendly), sensor detection, weapons and deployment, tactical environment, probability of interception calculations, and damage assessment. A modular database-driven design provides for future modifications and additions as platforms, weapons and sensors develop in the future.

Operational Status
Two trainers are in manufacture and the first was due for delivery in September 1990. Options on further systems are held by the US Navy.

Contractor
CAE-Link Corporation, Silver Spring, Maryland.

TRIDENT TRAINER

Type
Command and control team trainer and weapons trainer.

Description
The Command and Control Team Trainer (Device 21A42) is an integrated trainer interfacing with a sonar trainer and a radar land mass system/periscope visual scene generator support system to provide training capabilities in all aspects of tactical operation and control. The system includes high fidelity simulation models and equipment on the ocean environment, contacts, weapons, sonar, countermeasures, ship control, periscope visual scene, and ESM environment to provide training in the operational equipment including missile launch, ship control, external countermeasures launch and ESM functions.

The periscope mock-up simulator and image generator are computer-driven providing the periscope with appropriate own ship and contact data. The system provides dynamic surface and air contacts in a 360° field-of-view with sea state, solar illumination, fog, textured sky, foreground/background waves and washover all realistically simulated. The digital acoustic signal generators feature multi-layered ocean modelling, multiple target/platform co-ordinated operations modelling, off-line target generation capabilities and full complement of active and passive special effects.

Contractor
CAE-Link Corporation, Silver Spring, Maryland.

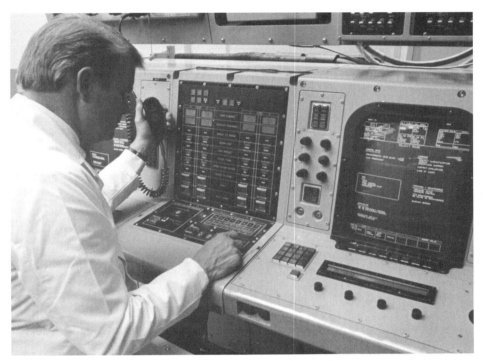

Mk 118 weapon launch console in Trident trainer

Trident command and control trainer

TYPE 20A66 ASW TACTICAL TRAINER

Type
Surface tactical team trainer.

Description
This trainer is designed to provide anti-submarine warfare tactical team training in a multi-platform, multi-threat environment.

The trainer will be equipped to provide ships' officers and crews with essential ASW tactics for a variety of situations for conventional as well as Navy Tactical Data Systems (NTDS) and Combat Direction System (CCDS) equipped ships, aircraft and submarines.

Training is conducted in manned mock-ups which simulate: tactical flag command centre; warfare commander module; aircraft carrier combat information centre; carrier ASW module; surface combatant combat information centre (including those equipped with AEGIS); submarine attack centre and ASW and AEW aircraft crew stations.

Instructors and students can manoeuvre platforms, display and manage sensor data, determine target classification/identification, perform TMA for fire control solutions, engage targets, fire weapons, launch chaff and employ acoustic and EW countermeasures.

Operational Status
Under a $27 million contract, Hughes is scheduled to deliver in 1993 the first trainer for installation in the Fleet ASW Training Center Pacific, San Diego, California. An optional second trainer would be located in the Fleet ASW Training Center Atlantic, Norfolk, Virginia.

Contractor
Hughes Aircraft Company, Manhattan Beach, California.

AN/BSY-2 SCSTT

Type
Submarine combat system team trainer.

Description
This shore-based electronics-based training system will provide individual and team proficiency training and skill reinforcement to sonar and fire control sub-teams, as well as overall combat team training to all personnel manning the US Navy's AN/BSY-2 submarine combat system.

The design is based on distributed modular architecture and unique, hybrid simulation/stimulation and signal processing emulation techniques. The trainer will closely replicate large numbers of complex contacts, the submarine operating environment, the weapons system, periscope, radar, sonar suite and other AN/BSY-2 sensors to provide controlled, tactically realistic training exercises.

New displays, contacts that automatically react to the environment and student performance monitoring are some of the features which will aid the instructor during training exercises.

Operational Status
A competitive contract worth $80.9 million was awarded to Raytheon early in 1991. Under the terms of the contract a full-scale engineering model will be built and installed at Raytheon's Submarine Signal Division at Portsmouth in 1995 to conduct initial training of Navy personnel. Contract options call for Raytheon to produce one additional AN/BSY-2 SCSTT which will be delivered to the US Navy's Submarine School at Groton, Connecticut in 1996 and for the engineering development model to be relocated to the Navy Underwater Systems Center in Newport, Rhode Island in 1997.

Contractor
Raytheon Company, Submarine Signal Division, Portsmouth, Rhode Island.

SUBMARINE TRAINERS

UNITED KINGDOM

LINK-MILES SUBMARINE CONTROL SIMULATOR (SCS)

Description

The Link-Miles Submarine Control Simulator (SCS) incorporates a full-size replica of the forward and port side sections of a control room which is mounted, with the instructor's facilities and computer, on an electrically driven, two-axis motion platform and includes an onboard training management system. The SCS is designed to produce as authentic an environment as possible including sea motion, a simulated intercom system and authentic submarine lighting. A distributed architecture with microprocessors is used in the system.

The instructor's console and management station is separated from the student's compartments, but students can be observed by the instructor through a one-way mirror. Each instructor's position includes consoles for the instructor and the unit operator. The consoles can display detailed schematics which permit the instructor to adapt a wide variety of lesson plans. These facilities permit student-paced training as well as precise monitoring of performance.

Operational Status

A submarine control simulator for the Royal Norwegian Navy's 'Ula' class submarines was delivered in October 1990. The system is being installed alongside the command team trainer at Haakonsvern Naval Base, Bergen. The ship control console is on an electrically driven motion platform providing pitch and roll. Link-Miles has also supplied a control room simulator for the Royal Navy's 'Upholder' class submarines and another control simulator to the Indian Navy. SCSs have been or are being supplied to the Royal Netherlands Navy ('Walrus' class), Royal Navy ('Vanguard' class), Brazilian Navy ('Tupi' class) and Royal Australian Navy ('Collins' class).

Contractor

Link-Miles Ltd, Lancing.

Submarine Control Simulator for Royal Navy's 'Upholder' class showing two-axis electrical motion system

MARCONI PERISCOPE SIMULATOR

Type

Periscope training system.

Description

Training operators in the use of periscopes at sea ties up operational resources and is limited in scope as it can place the submarine in potentially hazardous situations. The periscope trainer enables either initial or continuation training in both recognition and operational procedures related to any type or design of periscope to be simulated in a safe shore environment.

The simulator provides realistic training in search routines and attack procedures, with inshore or open water navigation in a multi-ship scenario, under normal or emergency conditions. To the operator the periscope feels and operates just as it would in a submarine, even rotational inertia being simulated for added realism.

The fidelity of ship-modelling provides sufficient detail of superstructure and major features to enable vessel identification and accurate assessment of range, bearing and angle on the bow. The trainer uses advanced visual systems based on commercially available graphics engines of high performance to provide the best means of day, dusk, night scene representation in a variety of weather and sea state conditions. All aspects of the periscope picture are produced, such as graticule and data displays, providing total control from a single system.

As well as being offered as a stand-alone trainer, the periscope trainer can be integrated with existing command team trainers. In team training, information on own boat, targets and the environment is passed from the command team trainer to the periscope simulator to co-ordinate the periscope visuals with the data received from other sensors.

Operational Status

A periscope simulator has been supplied to the Netherlands Navy for the 'Walrus' class submarine. Periscope simulator visual software has been supplied to the Canadian Navy.

Contractor

Marconi Simulation, Marconi Radar and Control Systems Ltd, Donibristle.

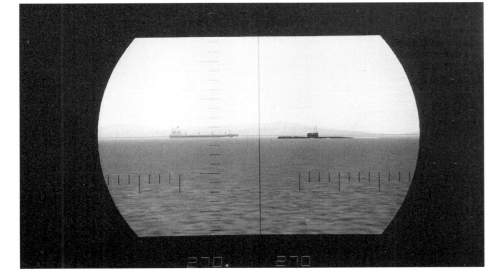

Periscope simulated image

IPT-PERISCOPE

Type

Periscope visual emulator.

Description

The Intelligent Procedure Trainer (IPT) uses the Simulation, Training, Education and Procedural Systems (STEPS) to provide generic training not necessarily dedicated to one form of simulation or training task.

The periscope emulator is a 3D real-time system that operates over 360° providing the user with real-world visuals of background scenery, mast wash over, various naval vessels and airborne platforms in day, night, and other climatic conditions. The emulator also allows for many other facilities within the different scenarios generated by the company.

Contractor

Digital Systems and Design Ltd, Eastleigh.

UNITED STATES OF AMERICA

STVTS PERISCOPE TRAINING SYSTEM

Type
Training device for periscope operation.

Description
Kollmorgen has been designing periscopes and periscope training systems for some years. The most recent example is a Submarine Tactical Visual Training System (STVTS) for the US Naval Training Systems Center at Orlando, Florida. Using a Type 18 periscope, STVTS provides a training device for individual operator, submarine attack centre sub-team or submarine combat system team training in an at-sea environment.

Operational Status
STVTS is in production and service for the US Navy. Kollmorgen has supplied other periscope trainers for the US Navy during the 1980s.

Contractor
Kollmorgen Corporation, Electro-Optical Division, Northampton, Massachusetts.

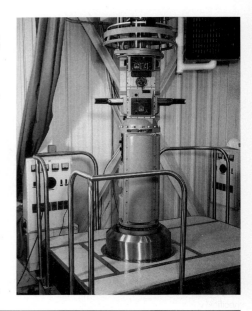

Submarine tactical visual training system

AIRCRAFT TRAINERS

UNITED KINGDOM

AIRBORNE CREW TRAINERS

Type
Anti-submarine training systems.

Description
GEC Avionics' Airborne Crew Trainers (ACTs) provide real-time ASW training for maritime patrol and helicopter crews, offering comprehensive acoustic, tactical and crew co-operational and co-ordination training by simulating sonobuoy, dipping sonar, target and ocean environmental data. Data can be controlled, processed and displayed on the aircraft's acoustic and tactical systems in exactly the same way as real data, providing realistic simulation of surface and sub-surface targets without recourse to the use of a target submarine or the expenditure of sonobuoys. Two basic variants of ACT exist.

ACT 1 is a software-based system developed specifically for the AQS-901 acoustic processing system operational in Royal Air Force Nimrod MR Mk 2 and Royal Australian Air Force P-3C Orion maritime patrol aircraft. It comprises an Exercise Control Unit (ECU) and the computer programme which is downloaded into the AQS-901 from a magnetic tape unit. The crew member acting as the exercise controller uses the ECU to set the target's initial start position, course, speed and depth, and if desired control of subsequent target manoeuvres and ocean conditions. Otherwise, the facilities and functions necessary to simulate fully dynamic ASW scenarios are automatic.

ACT 2 is a hardware-based system which is more flexible than ACT 1. It is capable of being used with any type of airborne acoustic processor, and can simulate the characteristics of any sonobuoy, dipping sonar, target and ocean environment specified by the user.

ACT 2 comprises an Acoustic Simulation Unit (ASU) and Exercise Control Unit (ECU). The ASU, which executes the ASW scenarios and generates synthetic passive and active acoustic data, can be configured to simulate any number of independent sonobuoy and dipping sonar channels. Typical ASU weights range from 12 to 16 kg for 8 and 16 channel systems in a 1 ATR(S) box and 24 kg for 32 channels in a 1+ ATR(S) unit. The ECU is an intelligent control terminal with an integral keyboard and display. In addition to performing all exercise control functions, the battery-supported ECU is used for the temporary storage of exercise scenario data. Two types of ECU are offered, one is a small hand-held unit weighing 3 kg and the other is a laptop unit weighing approximately 6 kg. The latter is more suitable for larger, fixed-wing aircraft.

The most basic ACT 2 system currently available can simulate up to eight independent data channels from any combination of LOFAR, DIFAR, RANGER, DICASS and bathythermal buoys, and generate simultaneously the acoustic signature characteristics of up to three independent targets. Simulation of specific dipping sonars and other types of sonobuoy can be easily accommodated with the existing hardware.

Unlike ACT 1, there are no limitations with ACT 2 on the potential number and variety of exercise scenarios available for training. These are created and maintained by each customer, using an Exercise Support System (ESS), which consists of a software program and an IBM-compatible PC and printer. The ESS can also be used as a classroom trainer and for conducting post-flight debriefs. Scenarios created from target and ocean characteristics defined and stored in the ESS library files by the customer are downloaded into the ECU before the flight. On board the aircraft the scenario data are transferred to the ASU. The ECU is then used to control and record the results of each exercise for subsequent in-flight or post-flight debriefs.

Operational Status
ACT 1 is in sevice with the RAF and RAAF. ACT 2 has completed development and is in full production.

ACT simulation unit and exercise control unit

ACT exercise control unit

Contractor
GEC Avionics Ltd, Maritime Aircraft Systems Division, Rochester.

SONAR TRAINERS

CANADA

SIM-STIM

Type
Onboard sonar simulator-stimulator.

Description
The SIM-STIM injects noise and signal into the front end of a sonar receiver, thereby acting as a substitute for the sonar's transducer outputs. The noise and target signals feed the processing circuits in the sonar receiver to provide video and audio outputs such as operators would experience during a live action.

The SIM-STIM can generate four targets as either ships/submarines/wrecks or torpedoes together with a full range of associated operational noises. If the target simulated is a submarine, an air bubble may also be deployed to simulate an attempt to mask the submarine's echoes. The operator can change or cancel any of the simulated parameters or targets, unless the selected target is a torpedo which is homed in on the ship by the simulation scenario.

The system has four operational modes:
(1) At sea, transmitter on – active and passive targets are simulated and signals superimposed on the ship's noise, flow noise and reverberation received through the transducer from normal operation
(2) At sea, transmitter off – no power is transmitted through the transducer and SIM-STIM signals and reverberation are superimposed on own ship's noise and flow noise from the staves
(3) Harbour operation – SIM-STIM provides all signals, noise, reverberation and target data as well as supplying the receiver with artificial data to represent ship's course and speed
(4) Test set – SIM-STIM performs tests in the sonar receiver on:
(a) pre-formed beam patterns
(b) fine video calibration
(c) Doppler filter
(d) passive and active signal processing gain.

Operational Status
In production. A total of 13 sets is being manufactured for the Canadian Patrol Frigate programme and nine have been delivered.

Contractor
Westinghouse Canada Inc, Burlington, Ontario.

FINLAND

FERS 800 ECHO REPEATER

Type
Sonar training system for surface ships.

Description
The FERS 800 system is primarily designed for navies without a submarine fleet of their own. The Echo Repeater is a turnkey system for training of anti-submarine warfare operators and for performance assessment of sonars. The Echo Repeater is capable of receiving sonar transmissions and echoing the transmissions just like the hull of a submarine, making it suitable for co-operation with surface ships, submarines and fixed- and rotary-wing ASW aircraft.

The Echo Repeater can be adapted to a wide range of sonar equipment (10 kHz to 35 kHz) and environmental conditions within a few minutes. Other features include simulation of target movements even in buoy suspension modes and active depth control in towed mode. During trials in the Gulf of Finland a stable sonar contact was achieved at a distance of 12 km.

The FERS 800 can be used in three operating modes: basic, telemetry and towed mode. As soon as the equipment related to basic and telemetry configuration has been deployed from a surface vessel or from a helicopter, the operation from ship, submarine or aircraft can be started.

Contractor
Elesco Oy Ab, Espoo.

FRANCE

STAR

Type
Artificial towed target.

Description
STAR (an acronym for Submarine TARget) is an artificial towed target designed to replace the use of submarines as torpedo targets. It simulates the acoustic behaviour of a real submarine when under attack from an active torpedo. It has been developed for both the training of naval forces and for the operational evaluation of torpedoes. It can also be used for the technical qualification of new torpedoes, when fitted with specific components, as well as towed decoy under certain conditions.

STAR consists of two main parts, the submerged part and the onboard installation on a surface ship. The submerged part comprises five parts:
(a) the antenna target made up of a linear chain of five of the fifteen acoustic repeater modules and a tail module
(b) the underwater body (fish) containing the electronics, power supplies, navigation sensors and external trajectography sensors
(c) a passive hydrodynamic depressor, weighing about 350 kg in water, which is used to obtain the desired immersion since the antenna, tether cable and fish have a slightly positive buoyancy
(d) a tether cable of approximately 400 m connecting the fish to the depressor
(e) a towing cable of approximately 600 m.

The onboard installation consists of a hydraulic winch, a 15 kW hydraulic power unit and an electronics rack containing the control, checking, visualisation and recording instruments.

STAR is a rugged system which can be launched in sea states up to three or four and will remain operational in sea states up to five. The target can be towed at speeds from 3 to 20 kts and at a depth varying from 20 to 400 m, the depth being set up by the paid out length of the cable. In the event of cable breakage, the tail module ensures hydrodynamic stability of the antenna, as well as its recovery.

Sensors in the fish transmit the heading and depth to the surface ship. The fish can also be equipped with specific sensors to match any external trajectography requirements. When the target antenna is partially or totally illuminated by the torpedo sonar emission, each of the repeater modules retransmits the signal received at the appropriate level, to simulate the submarine volume echo. The simulation operates within a wide frequency band to protect the confidentiality of the torpedo's exact frequency. The torpedo signals received and transmitted by each module are sent via a multiplex cable and are then recorded on board.

Contractors
Société ECA, Colombes (design and manufacture).
Safare Crouzet, Nice (acoustic components).

NETHERLANDS

SSE SONAR SIMULATOR

Type
Training system for sonar operators.

Description
This is one of a number of sonar simulators produced by Signaal Special Products. The SSE sonar simulator, which generates realistic surface and submarine targets in video and audio form, can be used with any passive or active sonar. The system is an onboard trainer and can be used when the ship is berthed or at sea. In the latter situation simulated targets can be inserted into the existing maritime environment.

Several dedicated models are available for use with specific sonar systems including the SSS-03 with the EDO Model 610 sonar, the SSS-05 with the Westinghouse AN/SQS-505, and the SSS-32 with the Signaal PH-32. All the simulators can produce realistic simulation of two targets with accurate representations of the performance envelope in all sonar operational modes. These targets can be ships, submarines, wrecks and false echoes. One simulated target can be a torpedo fired from a target submarine.

Operational Status
In service with the navies of Belgium, Brazil, Canada, France, Iran, Italy and the Netherlands.

Contractor
Signaal Special Products, Hollandse Signaalapparaten BV, The Hague.

NORWAY

MPS MULTIBEAM SONAR SIMULATOR

Type
Training system for commanding officers and sonar operators in the use of active sonar variable depth or dipping.

Description
The MPS Multibeam Sonar Simulator (MMSS) is a computer-based training system for the operators of hull-mounted and variable depth multibeam sonars. The system generates both sonar-image and audio signals in real time so that the student can use the controls of the sonar console, learn to understand the working principles of the equipment and assess the results when the control settings are altered. The system is based on the 32-bit Multibus II real-time computer using the Intel 80386 as the main processor.

Echoes from underwater objects are produced by a digitised representation of reflected sound at different aspect angles. This produces a very accurate generation of the acoustic characteristics of the sound return, which takes into account the size, geometry, location, aspect angle, and Doppler effects. The operational characteristics of the sonar, including the source level, transmission modes, scanning sectors and pulse lengths, together with the actual operator settings such as range scale and tilt, are used as the simulator model.

The system comprises an instructor's position, a student steering console position, a student sonar console (using a specific system), a game preparation console/system control station, laser printer, digitiser for map and terrain-model preparation.

The operator can use either an actual sonar console or an MPS tactical console simulating the man/machine interface, presentation modes and signal processing capabilities of a specific, or a set of real sonar equipments. In order to steer and control the simulated sonar-carrying vessel, a manoeuvre console is connected for setting speed, course and manoeuvre models. The sensor platform and all the targets are controlled by the instructor.

Optional extras include a network interface with an action speed tactical trainer such as MASTT, a plotter for analysis purposes, additional tactical displays and an interface with a video-audio recorder.

Operational Status
In service with the Royal Norwegian Navy (integrated with the MASWT) since 1989, and with the Swedish Navy since 1990.

Contractor
Micro Processor Systems (MPS) A/S, Kongsberg.

UNITED KINGDOM

AS2105 SONAR STIMULATORS

Type
Sonar test and training for ships and submarines.

Description
The AS2105 sonar stimulators have been designed to generate the accurate signals required to test comprehensively modern complex passive sonar systems. They also provide basic training for sonar operators, target motion analysis teams, command teams and maintenance personnel. More comprehensive training facilities are available with the AS2107 sonar trainer (see next entry).

The AS2105 equipments are compact, robust and lightweight, which makes them suitable for use ashore or on board surface ships and submarines, either at sea or in harbour. The AS2105 can emulate a wide range of hull-mounted and towed hydrophone array geometrics, and by generating appropriate electrical signals, stimulate the inboard processing and display equipments.

The AS2105 consists of two parts; a small lap computer which provides the man/machine interface allowing the user to construct scenarios for training and testing, and the electronics unit which generates the acoustic signals.

AS2105 sonar stimulator

Operational Status
A number of systems is in service with the Royal Navy and a NATO navy.

Contractor
Marconi Underwater Systems Limited (formerly GEC Avionics, Sonar Systems Division), Croxley.

AS2107 SONAR TRAINER

Type
Sonar training and system test for surface ships and submarines.

Description
The AS2107 sonar trainer has been developed from the AS2105 sonar stimulator (see previous entry), and will generate set-piece or free-play exercises with either real or simulated own ship parameters.

By synthesising the narrowband, broadband, DEMON and aural emissions of platforms, the AS2107 can describe complex signatures, together with vehicle dynamics, weapon configuration, environmental characteristics and other parameters, which would normally be recalled from the user's data library. Alternatively the AS2107 allows off-line generation to permit the user to develop or modify data. The scenario author/command can perform a fast time review of the scenario using the built-in tactical display, this facility being particularly for the initial construction of scenarios, and for debriefing.

The AS2107 consists of two parts; a laptop computer which provides the man/machine interface allowing the user to construct scenarios for training and testing and the electronics unit which generates the acoustic signals.

The AS2107 is intended primarily for onboard continuation training of sonar operators, target motion analysis teams and command teams, in conjunction with towed and hull-mounted sonar processing systems. It may also be used for extensive testing of sonar systems, particularly within a shore development establishment.

Operational Status
Fully developed and a variant has been installed at the Type 23 Shore Development Facility.

Contractor
Marconi Underwater Systems Limited (formerly GEC Avionics, Sonar Systems Division), Croxley.

AS2107 sonar trainer

MANDARIN SONAR TRAINER

Type
Computer-based sonar training system.

Description
Mandarin is a complete and integrated environment incorporating all the facilities required to author, deliver and manage computer-based training. It is an authoring system designed by Marconi Simulation and sold and supported by local representatives throughout the world. Access to facilities and movement between facilities is readily available by straightforward menu/pointer interfaces, and courseware production is via an interconnected series of on-screen editors for graphics, video, course planning and management.

The Mandarin integrated sonar training system offers four courseware products which represent a comprehensive answer to basic sonar training requirements. The four products provide training in sonar theory, LOFARGRAM analysis, target classification including aural recognition, and equipment-specific operator procedures. Delivery of the courseware is on industry standard IBM-compatible PCs.

Mandarin sonar
This is an interactive course on basic sonar principles consisting of five modular sonar lessons covering all aspects of basic sonar theory. Students interact with the lesson, using mouse and keyboard, through interactive graphics and animated sequences.

Mandarin SOLO-PLUS
This is a unique sonar LOFARGRAM trainer providing analysis and classification. LOFARGRAMs and scenarios can be created in which up to four contacts sail past a passive sonar sensor. The LOFARGRAMs can be displayed in real or accelerated time. The trainer incorporates tools to aid analysis and annotation of LOFARGRAMs.

Mandarin SOLO/PAS
This is an integrated sonar visual and audio training system. The Programmable Audio Synthesiser (PAS) operates from a programmable database and is fully interactive with the dynamic scenarios generated by SOLO.

Mandarin KEYPROC
KEYPROC is a flexible keyboard procedures trainer. Using a touch-screen to provide reconfigurable plasma-panel type keypads, together with high resolution graphics display for the production of sonar pictures, KEYPROC provides instruction on how to use operational equipment in a classroom environment.

Operational Status
In production.

Contractor
Marconi Simulation, Marconi Radar and Control Systems Ltd, Donibristle.

SONAR PRINCIPLES TRAINER

Type
Comprehensive sonar operator training system.

Description
The Ferranti Sonar Principles Trainer has been designed to provide comprehensive facilities for the training of sonar operators in understanding the various elements of complex sonar signals, and the skills associated with detection, evaluation and tracking of targets in a variety of scenarios and environmental conditions.

Consisting of an instructor's console and up to 10 identical trainee operator consoles, the system offers a standardised hardware concept which can be customised by software to meet the specific training needs of the user. Each trainee console consists of a multi-purpose display unit, a sonar control panel, sonar aural facilities and communications facilities. Each trainee has control over his own sonar with respect to mode of operation, frequency, pulsewidth and beam training, and the sonar pictures and aural effects are responsive to individual trainee actions.

The trainer provides the sonar instructor with a flexible training tool which can be used to prepare trainee operators for sonar operations across the range of equipments existing in his particular fleet. Through the means of a user-friendly console, the instructor can prepare, execute, monitor and debrief exercises. The instructor has full control over the ocean environment, movement of own ship, selection of sonar types for training, selection and movement of target vessels and weapon deployment.

A repeater monitor and audio system at the console allows the instructor to select any trainee operator's sonar picture and associated sonar aural for on-line monitoring of individual sonar performance. Each trainee operator's use of the relevant functional controls is also monitored as part of the performance analysis.

Operational Status
In service.

Contractor
Ferranti-Thomson Sonar Systems UK Ltd, Stockport.

SAINT TRAINING SYSTEM

Type
Initial training system for sonar operators.

Description
The Sonar Analysis Initial Trainer (SAINT) is intended to train operators in the principles of underwater noise generation, signal display presentation and narrowband analysis from basic principles to advanced levels. The system consists of a dual position instructor's console linked to as many as 12 students' positions, each for two students. As a result of this organisation each instructor can carry out independent exercises with six positions.

Each instructor position has two video monitors, one for monitoring the trainer control menu and the other to repeat any picture seen by the students.

The Ferranti SAINT (Sonar Analysis Initial Trainer) in operational mode

There is also a joystick-controlled pointer to identify important features on any video seen by students, a digitiser pad, and a stylus for repositioning of frequency lines in a LOFARGRAM, or hand-free graphics 'sketching' on to video pictures. Video record facilities and communications with the students are also provided.

Each student position has a paper chart recorder with numeric display to identify the frequency window covered by the LOFARGRAM being recorded. The positions also have a video monitor and cursor generation electronics. Cursors can be superimposed on a picture relayed by the instructor to allow the student to identify both frequencies and harmonic relationships. Numeric display can also be incorporated.

Operational Status
Fully developed.

Contractor
Ferranti-Thomson Sonar Systems UK Ltd, Stockport.

FERRANTI ONBOARD SONAR TRAINER

Type
Onboard trainer for the Type 2050 sonar.

Description
This simulator has been designed to provide training facilities for the Type 2050 sonar system being fitted to surface ships of the Royal Navy. The trainer is intended to provide high-fidelity acoustic contacts to stimulate the operational sonar processor. Controlled by an intelligent graphics terminal, it will incorporate a comprehensive ocean environment model to ensure realistic detection scenarios under different thermal and acoustic conditions.

Several contacts can be injected simultaneously into the sonar processor. Exercises can be constructed to include either a real-world, real-time environment, or alternatively contacts can be generated together with a large variety of synthetic operational scenarios for continuation training either at sea or in harbour. A data recorder is included in the system.

Operational Status
In May 1989, Ferranti received a £2 million contract to supply an onboard trainer to the Royal Navy.

Contractor
Ferranti-Thomson Sonar Systems UK Ltd, Stockport.

AS 1092 SONAR RADAR TRAINER

Type
Shipborne sonar and radar training system.

Development
The AS 1092 is a development of the AS 1077 marketed by THORN EMI.

Description
The AS 1092 is the latest in a series of multi-sensor onboard trainers produced by THORN EMI which is designed to train command team sensor operators within their operational environment using their normal equipment. The systems are modular and compact to fit in the limited space normally available. They can be used to provide procedures training, or as part of command team training with simulated weapons firing.

The AS 1092 consists of a main control console, a sonar simulation cabinet and initial detection and classification trainer (IDCT), a second control unit, weapon system and navigation radar interface units, an IDCT remote-control unit, a second control position and a hand-held helmsman's control unit.

The THORN EMI AS 1092 shipborne radar and sonar trainer

Synthetic targets can be inserted onto screens receiving real data or an exercise may be carried out using simulated targets and simulated background, with the latter covering an area of 2048 × 2048 nm (3788 × 3788 km) to heights of 80 000 ft (24 384 m) and depths of 20 000 ft (6096 m). A ship movement simulator is included with high integrity changeover devices for switching between real and simulated data. Up to 70 vehicles (surface ships, submarines, aircraft, torpedoes) can be simulated simultaneously from a library with a capacity for 254 classes.

DIGITAL SYSTEMS AND DESIGN SONAR PROCEDURES TRAINER

Type
Basic and advanced trainer for sonar operators.

Description
These are intelligent procedure trainers which provide students with a full range of training facilities from basic to complete tactical and weapons operations. Each student is guided automatically by verbal and visual commands in accordance with a pre-defined training programme. The system features high resolution animated graphic presentations, full touch-screen control operations, complete emulation of main system displays and operating control panels, together with emulation of targets and the environment.

Two dedicated products have been produced; the IPT-32 and the IPT-2022. The first of these is an anti-submarine warfare sonar simulator for training operators of the PHS-32 hull-mounted sonar, while the IPT-2022 is a minehunting sonar and mine disposal system trainer for operators of the TSM 2022 sonars. Both systems have been supplied to Dornier for use with their training systems.

Operational Status
In production and in service with the Nigerian Navy.

Contractor
Digital Systems and Design Ltd, Eastleigh.

Operational Status
In full service with the Royal Navy.

Contractor
THORN EMI Electronics Ltd, Feltham.

The IPT-32 anti-submarine sonar trainer

IPT-ASW

Type
ASW sonar emulator.

Description
The Intelligent Procedure Trainer (IPT) uses the Simulation, Training, Education and Procedural Systems (STEPS) to provide generic training not necessarily dedicated to one form of simulation or training task.

This IPT-ASW emulates many discrete functions and facilities within an ASW system. The real-time emulation is realistic to the degree of unpredictability (expert system behaviour).

It provides a wide range of training scenarios from basic operator training to exercises involving complex tactical situations. The system features acoustic noise emulation, full touch-screen control, synthesised speech for commands and student management facilities.

Contractor
Digital Systems and Design Ltd, Eastleigh.

IPT-MHS

Type
Minehunting sonar emulator.

Description
The IPT-MHS has been designed to provide a comprehensive sonar training facility for mine warfare sonar operators at various levels of expertise in the principles, theory and operation of specific-to-type minehunting sonars and mine disposal methods. Training in the following disciplines is provided: sonar environment, high definition sonar, mine detection, mine classification, mine identification, operating drills and procedures, and operating of ROVs.

It emulates many of the discrete systems and functions within the mine warfare scenario.

The MHS operates on a self-teach principle starting with basic sonar operating and classification procedures and increases in degrees of difficulty up to advanced tactical levels including mine disposal. Two levels of training are provided: basic and advanced. Each module comprises instructional lessons and simulated minehunting exercises. On completion of each lesson/exercise, the training/supervising officer is provided with a hard copy printout indicating student errors and the progress. All simulated mine warfare exercises performed by students are recorded for replay and debriefing purposes.

The trainer displays instructions and prompts, reacts and advises on correct and wrong actions performed by the trainee. In addition, all commands normally associated with minehunting drills and actions are communicated to the operator through synthesised speech. During each exercise, own ship's course and speed is adjusted in response to actions performed by the trainee operator. Each exercise level increases in difficulty requiring the operator to detect, classify and subsequently identify mines (IPT-ROV), mine-like and non-mine targets from different aspects in varying environmental conditions. Targets identified as mines are neutralised by the use of an ROV.

In addition to the prime task of training, the IPT-MHS provides a true simulation system for the evaluation of new warfare tactics, platforms, devices, and sensors under controlled conditions using known and empirical data.

Contractor
Digital Systems and Design Ltd, Eastleigh.

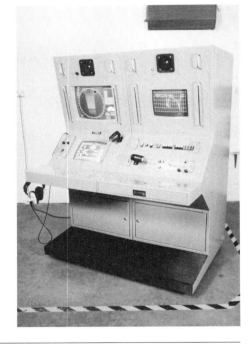

IPT double console with minehunting sonar TSM 2022 operating

UNITED STATES OF AMERICA

AN/SQQ-T1 SONAR TRAINING SET

Type
Training equipment for ASW sonar operators.

Description
Designed to provide passive and active acoustic training for teams and sub-teams, the AN/SQQ-T1 stimulates sonar systems to generate high-fidelity acoustic signals resembling those of actual targets under various sea and environmental conditions. It provides sonar operators with the ability to improve their skills in target identification, analysis and classification during long periods at sea or at ground-based stations.

The system can be operated on an individual basis using recorded scenarios, or on an interactive team basis with an instructor. Up to 64 pre-programmed scenarios may be stored within the system and displayed on the instructor control unit. Two modes are available to the instructor; one with all simulated data and the other with live sensor data overlaid with simulated data. In addition, the instructor can select from 16 pre-programmed ocean areas, and sea state and shipping density levels. All databases are stored on a removable hard disk unit to facilitate updates.

The AN/SQQ-T1 simulates up to four high-fidelity acoustics simultaneously, selected from the US Government-based Common Acoustic Database, or from other acoustic databases. This extensive library is expandable and can accommodate up to 40 acoustic target types. Each acoustic contact has both passive and active signatures, which can be modified by the instructor for added realism.

The set supports onboard training by providing high-fidelity simulated data separately or simultaneously to the AN/SQR-17A and AN/SQR-18A. The inputs to the AN/SQR-17A are two selectable sonobuoy channels, eight beams for the AN/SQS-26CX PEC and a DIMUS channel. In addition, the AN/SQQ-T1 provides hydrophone group data for injection into the beam former of the AN/SQR-18A.

The instructor interfaces with the AN/SQQ-T1 via a graphics terminal, keyboard and colour display.

The set can simulate own ship's data under pre-programmed scenario or instructor control, or use live own ship's data. Information available to the instructor for training scenarios includes own ship course and speed and AN/SQR-18A array depth.

Operational Status
DRS received a $6 million contract from the US Navy in 1986 to design, develop and manufacture three trainers. As of June 1991 three units have been delivered.

Contractor
Diagnostic/Retrieval Systems Inc, Oakland, New Jersey.

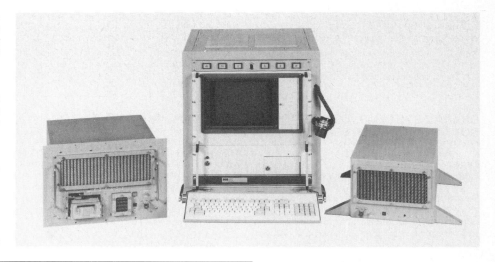

AN/SQQ-T1 sonar training set

MASTER STIMULATOR/TRAINER

Type
Acoustic analysis trainer and tester.

Description
MASTER (Modular Acoustic Stimulator/Emulator) is based on a Ferranti design and is a modular hardware/software system which can be tailored to meet surface ship, submarine or airborne ASW sensor system requirements, including acoustic analysis, training and sonar equipment testing. The hardware consists of target generators, inverse beam forms and sonobuoy simulation modules with a software package that can be used to alter the target scenario, and ocean model parameters to provide a realistic environment with the generation of high-fidelity signatures, including composite narrowband/broadband. The system can be used to provide individual or full team training at all proficiency levels.

A PDP 11/44 host mini-computer operates the software with a simple instructor monitor. MASTER can be embedded in a larger system, or it can be provided as a stand-alone stimulator. A typical installation consists of the instructor facilities, the controller computer, an acoustic signal generator, an interface with the trainer computer, and assorted peripherals.

Operational Status
Two systems are in operation with the US Navy.

Contractor
Diagnostic/Retrieval Systems Inc, Oakland, New Jersey.

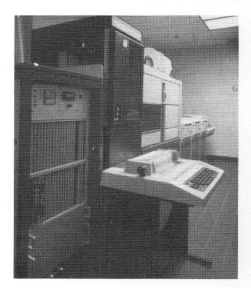

The Diagnostic/Retrieval Systems MASTER acoustic stimulator

RAYTHEON SONAR TRAINERS

Type
Training systems for both surface ships and submarines.

Development
The prototype DS1200 system was first used by the US Navy's submarine forces to successfully demonstrate an onboard sonar training capability for a passive subsystem. The production version, designated the DS1210, now provides ASW target training capability on 31 fleet ballistic missile submarines.

Two production DS1210 trainers, modified and designated DS1213, were later applied by US Navy submarine forces to totally integrated sonar suites. Although installed for test purposes and on a temporary basis, overall performance proved so effective that both trainers remain in service to this date aboard US Navy ships.

The Submarine Active Detection System – Transmit Group (SADS-TG) was developed and produced by Raytheon for the US Navy's advanced submarine combat system. SADS-TG contains a fully integrated onboard trainer which supports training for all shipboard sonar arrays and provides a moving contact for fire control team training.

In addition to the SADS-TG OBT capability, the company has developed a shore-based Multiple Array Test Set (MATS). It uses the same technology applied to the onboard trainers and will interface with the acoustic sensor group for these new submarine combat systems. MATS provides acoustic sensor stimulation and simulation of the high frequency array, top and bottom sounders, sound velocity and noise monitoring systems, and acoustic communications during system level performance and operability testing.

The Generalised Simulation/Stimulation (GSS) system was developed to facilitate the US Navy's programme of pre-planned product improvements to the fast attack submarine Combat Control System (CCS). GSS necessitated development of facilities and tools to support the evolution, certification, life cycle support, and training for major CCS components.

Description
Raytheon Company is a leader in advanced 'stimulation' technologies, with many trainer systems incorporating distinct simulation techniques to provide controlled, tactically 'real' training that is dynamic and highly effective.

The trainers provide operational and maintenance support to shipboard sonar and combat systems. They function at sea, in port, on shore or in the air, and address the training needs of surface ship, submarine and airborne platforms.

These trainers are engineered for versatility. They have successfully interfaced with the acoustic sensors of widely varying systems, both aboard multiple ship platform classes and in shore-based training facilities. They rely on advanced techniques for sonar stimulation and target simulation to generate for both individuals and teams the most realistic ASW training exercises available anywhere.

The entire ASW trainer series is based on established concepts of proven value. Expandable system architecture; the use of common hardware and software modules; programmable, functional partitioning of system processing – these techniques enhance realism and ensure design flexibility and manageable system upgrades in the future.

Raytheon trainers have high-fidelity, electronically synthesised active and passive contacts that are manoeuvrable, three-dimensional and ultra-realistic. The multiple targets found in systems are sensor coordinated. Onboard trainers feature 'front end' target injection and sonar system stimulation, a technique pioneered by Raytheon and conducted aboard ship for unrivalled, at-sea realism and training effectiveness. Essentially off-the-shelf proven systems, Raytheon trainers typically present a lower risk in technology, performance and schedule than many other competitive training systems.

Operational Status
In service. Over 40 trainers from the company's DS1200 Trainer Series, representing four system generations, have collectively logged in excess of 80 000 operating hours.

Raytheon has delivered nine GSS laboratories for use with CCS components and two trainers which allow a shore-based CCS to function as it would aboard an operating submarine.

Contractor
Raytheon Company, Submarine Signal Division, Portsmouth, Rhode Island.

AN/SQQ-89(V) TRAINER

Type
Onboard sonar trainer.

Description
The trainer is designed to provide individual and team proficiency training and skill reinforcement to embarked combat system personnel. The trainer uses simulated targets to stimulate the hull, towed array, sonobuoys and electronic sensors of the AN/SQQ-89 combat system. The trainer allows personnel to carry out realistic training exercises both in port and while at sea.

The trainer uses electronic synthesis techniques to

generate realistic high-fidelity active and passive, three-dimensional moving acoustic contacts. The system features front end injection and sonar system stimulation.

The trainer interfaces with the AN/SQS-53B/C active and passive sonar, the AN/SQR-19(V) towed array, the AN/SQQ-28(V) and LAMPS Mk III systems and the Mk 116 Mod 5, 6 or 7 ASW combat system.

Realistic passive and active targets stimulate these sensors while under way and without interfering with the ability of the sensors to detect, classify and track naturally appearing targets.

The trainer provides stimulation to the LAMPS Mk III sensors with the helicopter in the air or on deck.

The system can also simulate the helicopter's navigation information for heading, altitude and speed. The trainer is provided in four configurations to support the different AN/SQQ-89(V) configurations aboard the various ship platforms. Together with the ASW training capability, the device supports training activities for AN/SLQ-32(V) operators, AN/ALQ-142 operators, EW supervisors and other ESM personnel.

Operational Status
Raytheon won the original development and production contract for the AN/SQQ-89(V)-T trainer in 1985 and delivered 52 systems under direct contract to the US Navy. In 1988 General Electric led a team which won a contract for a complete AN/SQQ-89 combat

system suite, including trainers. Under this team Raytheon's Submarine Signal Division became the team member responsible for producing and supporting 35 trainer systems. Recently the General Electric team won the FY91 production buy for AN/SQQ-89 combat system suites and awarded Raytheon a contract for the production of 17 trainers, the first to be delivered in October 1992 with final deliveries being made by February 1993.

Contractor
Raytheon Company, Submarine Signal Division, Portsmouth, Rhode Island.

TSMT FES

Type
Shore-based trainer.

Description
TSMT FES (Trident Sonar Maintenance Trainer Front End Simulator) will offer personnel basic, intermediate and advanced instruction in sonar system troubleshooting, repair and calibration. Realism will be

achieved through the injection of TSMT-generated signals into the front end of AN/BQQ-6 and AN/WLR-17 sonar systems.

The system is totally integrated and will initiate and control the exercises, providing the required signal generation, propagation processing and sonar interfacing.

The trainer has four electronically synthesised manoeuvrable, active and passive three-dimensional, realistic targets.

The system can be modified to provide personnel

with operational training in addition to maintenance training. It can be further developed into a compact, onboard trainer for all submarine classes or can be used to establish a realistic fully capable standard evaluation facility.

Contractor
Raytheon Company, Submarine Signal Division, Portsmouth, Rhode Island.

ACOUSTIC OPERATOR TRAINER

Type
Acoustic operator trainer for the AN/SQQ-89 system.

Description
This trainer, also known as Device 14E35, has been developed as a trainer for the AN/SQQ-89(V) combat suite. The latter has four components; the AN/SQQ-28 LAMPS Mk III sonobuoy signal processor, the AN/SQR-19 towed array system (TACTAS), AN/SQS-53B hull-mounted sonar and the AN/UYQ-25 assessment system.

The trainer simulates the performance of the operational systems and can also provide a variety of exercises from a simple target location to complex

tactical operations. Each system comprises an instructor station, four trainee stations, a computer system and an acoustic signal generator. Each system is configured to a particular class of ship.

Later versions of the trainer will incorporate an improved AN/SQS-53B sonar, together with an interface with the Mark 116 underwater fire control system.

Operational Status
A number of these trainers is in service with the US Fleet Anti-Submarine Training Center in San Diego, California. Six systems are being supplied to San Diego under a contract costing over $100 million.

Contractor
CAE-Link Corporation, Silver Spring, Maryland.

The CAE-Link Device 14E35 acoustic operator trainer

DEVICE 14G1 SONAR TRAINER

Type
Sonar operator and diagnostic basic trainer.

Description
The sonar and diagnostic trainer, Device 14G1, is designed to provide student sonar operators with familiarisation training on the AN/SQQ-89(V) sonar.

It consists of four student stations, each of which simulates the appearance and operations of the AN/UYK-21 console associated with the AN/SQQ-89(V) sonar. The students receive six hours of fully automated, self-paced, computer-controlled instruction. The instructor selects either individual exercises or a common exercise for all the students. Each includes recorded narrations and visual presentations of the selected exercise.

Operational Status
In service with the US Navy.

Contractor
EMS Development Corporation, Farmingdale, New York.

HONEYWELL SONAR TRAINERS

Type
Range of sonar operator trainers.

Description
Honeywell designs and manufactures a number of training systems intended for use by sonar operators and teams. Each system is dedicated to a particular sonar equipment; 14E36 for the AN/SQR-18A, 14E19 for the AN/SQS-26CX, 14E23 for the AN/SQS-35, 14E24 for the AN/SQQ-23, and 14E25 for the AN/SQS-53.

Depending on the operational system the student stations are either actual sonar consoles or facsimiles. In either case, video and acoustic signals are fed to the student stations. The underwater acoustic environment is modelled in all operator trainers. Simulated sonar propagation is generated based on ocean variables, such as regional seasonal velocity profiles, layer depth and sea state as well as bottom type, depth and slope. These simulations use complex algorithms to account for the effects of propagation loss, reverberation, surface duct, time delays, convergence zones, multi-path returns and shallow water operation.

The instructor consoles are used to set up problem exercises, to monitor student performance and to modify the tactical situation in real time. The instructor is also able to insert malfunctions and replay the exercise.

Operational Status
In service with the US Navy.

Contractor
Honeywell Inc, Training and Control Systems Division, West Covina, California.

PASSIVE ACOUSTIC ANALYSIS TRAINERS

Type
Analysis and classification trainers for sonar operators.

Description
The US Navy Passive Acoustic Analysis Trainers

(Devices 14E40 and 21H14) are designed for use by sonar operators for both surface vessels and submarines. They are used for training in analysis and classification of various surface ship, submarine and commercial ship contacts, both aurally and visually. The displays include LOFARGRAM, DEMON and Vernier frequency data, together with a student worksheet. The displays are time synchronised with raw and processed audio signal presentation. The worksheet data entered by each student provide a record

of performance for training feedback and course grading.

The trainer generates realistic sonar audio and visual displays by pre-processing user-supplied acoustic analogue tape recordings from a variety of user-deployed tactical sonar systems. The pre-processing facility produces VHS tape cassettes for audio signals and computer disk files of synchronised generic visual displays for replay on the trainer. The trainer includes hardware for up to 12 student stations, a single instructor station, and a pre-processing facility.

Operational Status
In service with the US Navy.

Contractor
Ship Analytics, North Stonington, Connecticut.

Two Passive Acoustic Analysis (PAA) trainer student displays

ROCKWELL ONBOARD TRAINER

Type
Onboard sonar training system.

Description
Rockwell has designed an onboard trainer, based on a trainer developed for the US Navy, which simulates sonar hydrophones and array geometry prior to beam forming. It is therefore not affected by changes to hardware or software in the processing or display part of the system. The trainer can operate either under way with the contacts generated by the trainer overlaid on the environmental background noise from the array, or in harbour with the trainer generating independent background noise to simulate at-sea conditions.

The trainer design incorporates commercial VME-based processors programmed with software to provide maximum flexibility for adaptation to various sonar and array conditions, growth features which allow for multiple contacts to satisfy the most demanding training scenarios, and programmable contact signature generation to allow the user to create a library of specific contacts.

The Rockwell trainer can be tailored to meet the most basic or the most complex training needs of the customer.

Operational Status
In service.

Contractor
Rockwell International Corporation, Autonetics Marine Systems Division, Anaheim, California.

SONAR/ARTIFICIAL INTELLIGENCE LABORATORY

Type
Simulator system for emulation and research.

Description
The Sonar/Artificial Intelligence Laboratory (SAIL) is a shore-based Rockwell facility designed to provide broad support for simulation and research. SAIL encompasses the original Rockwell sonar system simulator and an artificial intelligence laboratory. It is equipped with both unique and commercially available hardware and software, written in FORTRAN and LISP, to allow algorithms and architectures to be implemented easily.

The sonar system simulator facility is a signal processing equipment designed for flexibility, and provides support for many types of emulation and research activities.

The combination of onboard acoustic data acquisition systems and the SAIL allows complete sensor signal acquisition at sea and exhaustive near-real-time analysis in the laboratory ashore. In the SAIL environment, existing sonar systems can be emulated to develop and verify improvements, or new and improved systems can be developed.

At the core of the SAIL system are three pieces of Rockwell hardware:
(a) a high speed ring bus capable of eight million 64-bit words per second per node. The bus provides communication among up to 64 units at speeds supporting faster-than-real-time sonar system emulation
(b) a programmable beam former which provides 213 million weight-delay-and-sum operations per second for up to 64 independent beams from as many as 52 hydrophones
(c) a reconfigurable signal processor which provides five million two-pole, two-zero filter stages per second simultaneously with 2.5 million butterflies per second and 512 000 adaptive filter taps per second.

The beam former and the processor provide the massive 'number-crunching' capability required for near-real-time sonar system emulation.

Operational Status
In operational service at the Rockwell facility in California.

Contractor
Rockwell International Corporation, Autonetics Marine Systems Division, Anaheim, California.

TARGETS

UNITED KINGDOM

SOUNDTRAK TYPE 2058

Type
ASW target simulator.

Description
The SoundTrak ASW target simulator is designed for use with surface ships, submarines and aircraft. It provides realistic acoustic target signatures for use in training sonar operators, and for performance

assessment of ASW weapon systems and platforms and other operational applications which require a versatile, reliable underwater acoustic source. It can be configured to provide an ASW target with a wide range of passive sonars and may be adapted for active operations.

SoundTrak comprises an acoustic generator, installed in a body that is towed behind a surface ship, and an inboard signal generator. An operator can enter the desired noise characteristics manually from tape or by downloading from a computer. The system can store up to eight pre-programmed scenarios. It can simultaneously transmit a mixture of narrowband tones and broadband noise signals, fully programmable in frequency, modulating amplitude and time, covering the full sonar frequency bands. A submarine-mounted version of the system is also available.

The SoundTrak target simulator reduces the need for the deployment of ships and submarines in exercise roles and is ideally suited to the training of ASW operators and command teams of surface ships and ASW aircraft.

Among the operational applications for SoundTrak are:
(a) acoustic countermeasures
(b) own signature confusion
(c) acoustic mine sweep
(d) force countermeasure/decoy.

SoundTrak provides a reliable acoustic source for the assessment of performance and calibration of the following: hull sonars, towed array sonars, flank sonars, intercept sonars, dipping sonars, passive

Type 2058 towed sonar source

ranging sonars, sonobuoys, noise ranges and ocean surveillance systems. SoundTrak is easy to deploy on, and recover from, a variety of vessels including fleet auxiliaries and warships, utilising standard winches.

Operational Status
In service with the Royal Navy and the French Navy and is under assessment by several other navies.

Contractor
THORN EMI Electronics Ltd, Feltham.

RASAT sonar target

RASAT SONAR TARGET

Type
Underwater sonar target.

Description
RASAT (Radar And Sonar Alignment Target) is a programmable, universal sonar target which is compatible with a wide range of modern, high technology sonars, including hull-mounted and dipping types. It can be used for the alignment of surface ship or helicopter systems, can act as both a transponder for

active sonar and as a triggered noise source for passive sonar, and it can be deployed from a ship or helicopter. Operating depth is selectable from 15 to 50 m and bandwidth is 3 to 12.5 kHz for the basic model; other operating depths and bandwidths are available.

Contractor
Dowty Maritime Systems Ltd, Waverley Division, Weymouth.

SLUTT TYPE G 733

Type
Programmable ship-launched underwater transponder target.

Description
SLUTT is designed to simulate underwater targets enabling both active and passive functional testing of a wide range of surface ship, submarine and helicopter sonar systems checking sonar range and bearing accuracies. The transponder can also be used for sonar/radar alignment checks and operator training and as a sea acceptance trials target.

It features user-programmable sonar parameters,

multiple output signals in FM or CW for realistic sonar simulation, provides a noise source for checking passive sonars and may be used as a transponder or a free-running pinger.

SLUTT is a self-contained unit suspended at depths of between 10 and 120 m beneath a free-floating buoy fitted with a radar reflector. This method of deployment allows interrogation of the transponder by both surface duct and variable depth sonars.

All incoming acoustic signals are monitored and SLUTT responds to signals which have the correct frequency and continuity. A suitable time simulated output echo pulse is then transmitted. SLUTT is immune to the reverberations set up by the transponder pulse and to incorrect inputs. SLUTT simulates a typical beam aspect submarine target or an incoming

torpedo, by providing suitable FM, CW and pulsed noise outputs to enable both active and passive sonars to be checked for range and bearing accuracy against relevant radar readings.

Optional extras include remote VHF/UHF data link to programme the transponder while deployed at sea.

The system is provided with a battery charger, hand-held programming unit and lead, handbook and stowage case.

Operational Status
In service use with the Royal Navy and many other navies.

Contractor
Graseby Marine Ltd, Watford.

TOPAT

Type
Towed passive torpedo target.

Description
The ultimate capability of a torpedo can only be

assessed by conducting in-water trials and exercises against realistic targets. The availability of ships and submarines which can be suitably armour-clad for this purpose is severely limited and such trials can be extremely dangerous. Free-swimming target vehicles have limited endurance, little real-time command and control and may not be cost-effective.

The TOPAT system is designed to overcome these

disadvantages. It provides a simple, adaptable set of equipment which will meet torpedo trials requirements.

The system emulates the acoustic returns expected from real targets. This is achieved by placing two acoustic reflectors in the same relative positions, and of the same strength, as the expected acoustic highlights from a real target. Two towed bodies carry the acoustic reflectors to obtain the required towing configuration and provide a 'moving target' capability.

The bodies can be deployed and recovered from a range of towing vessels, including stern trawlers, without any need for specialised handling equipment.

Operational Status
In service with the Royal Navy.

Contractor
BAeSema Marine Division, Bristol.

TOPAT passive torpedo target

IPT-ROV

Type
Remote-operating vehicle emulator.

Description
This trainer, which can either be shore-based or installed on board the MCMV, provides operators with a real-time, 3D visual emulation of complex underwater scenarios. Operators can be trained to become accustomed to the different types of sub-surface mines and the multitude of environmental conditions in which the ROV may have to operate.

Contractor
Digital Systems and Design Ltd, Eastleigh.

FAST

Type
Fuze Activating Static Target.

Description
An alternative to using a target submarine for underwater weapon evaluation is to use the FAST frame. The frame is covered with wire mesh panels to simulate the magnetic field of a submarine and is also fitted with an acoustic transmitter to provide a signal for the torpedo's homing system. The system's four hydrophones are mounted on the frame, which is lowered to a maximum of 100 m below the surface. All the processing equipment for tracking the torpedo is fitted inside the ship supporting the FAST frame. The frame is designed to fold for easy storage and deployment.

Operational Status
In service with the Royal Navy.

Contractor
THORN EMI Electronics Ltd, Feltham.

THOR TORPEDO SIMULATION

Type
Torpedo weapon system assessment system.

Description
THOR is a generic, multi-application Monte Carlo simulation model for the statistical evaluation of torpedo weapon system performance and the assessment of anti-torpedo and anti-submarine countermeasure effectiveness.

THOR may simulate an engagement involving attacking platforms which fire torpedoes, and target platforms which may become alerted to the threat weapon and attempt to evade by manoeuvres and/or by the employment of countermeasures. The attacking platforms may be submarines, surface vessels or aircraft in any number or combination. Target detection, closure, TMA refinement, weapon launch and guidance are simulated. Torpedo firing on the target vessels may be in single shot or salvo, involving wire-guided or unguided acoustic homing torpedoes, other homing torpedoes, or simple straight/pattern runners. There is a capability to simulate the deployment of missiles with a lightweight torpedo payload, the direct deployment of lightweight torpedoes from aircraft and tube launch from submarines and surface ships.

THOR incorporates the following representations:
(a) platform sensors – active and passive hull-mounted sonar, towed array, sonobuoy and dipping sonar systems for target or torpedo detection
(b) weapon sensors – active and passive sonar and other sensors
(c) combat information systems – data fusion, target or threat torpedo tracking, weapon and countermeasure targeting and guidance and tactical decision making
(d) acoustic countermeasures – towed, thrown and directly launched (static or mobile) jammers and decoys operating at weapon or platform frequencies
(e) other anti-torpedo 'soft-kill' countermeasures
(f) 'hard-kill' countermeasures
(g) physical characteristics – broadband and narrowband radiated noise, target strength, manoeuvrability and device deployment error and limitations
(h) environment and interference effects, including masking.

THOR has been developed as a combined continuous and discrete event based simulation model and has been designed in accordance with rigorous programming standards and development methodology to form a fully self-consistent and robust product. The facilities are currently hosted on the VAX range of mini- and super mini-computers and operate under VMS. The software is programmed largely in ANSI standard FORTRAN 77, but with some VAX specific management and graphics functions to enhance the user interface.

THOR has an established and growing user base currently comprising government research establishments, defence contractors and navies in the UK and in other NATO countries. Its applications are wide ranging from current system tactical development to the assessment of desirable options for future systems.

Operational Status
THOR was originally developed under contract to the UK Admiralty Research Establishment, and early versions of it have been supplied to UK industry and to US defence establishments.

Contractor
SD-Scicon, Camberley.

Mk-38

Type
Miniature Mobile Target.

Description
The Mk-38 Miniature Mobile Target (MMT) is a self-propelled submersible acoustic transponder. It is normally launched over the side of a surface ship and glides to its preset operating depth, where the propulsion motor turns on. The Miniature Mobile Target maintains its operating depth within ±60 ft and travels at a speed of 5 kts on a random course. The acoustic receiver and signal processor is programmed to recognise different acoustic interrogation frequencies and will respond with either an acoustic CW pulse or an FM pulse at the interrogation frequencies desired by the user. During periods when the target is not interrogated, it transmits two CW tones simultaneously. The Mobile Target can be used as a training device with any underwater acoustic detection system which has a transmission frequency within the receiving frequency band of the target and includes the Mk 6 torpedo.

Contractor
Hazeltine Corporation, Greenlawn, New York.

Mk 39 EMATT

Type
Expendable Mobile ASW Training Target.

Description
The Mk 39 Expendable Mobile ASW Training Target (EMATT) is a sophisticated, cost-effective training target easily deployed from surface ships or ASW aircraft. EMATT offers ASW forces the opportunity to practice the complete ASW problem from detection to attack in the open ocean in any area of operations.

The capability for launch from ASW aircraft gives EMATT an added dimension previously lacking in ASW training. The Mk 39 also has the capability to simultaneously interact with all sonars, towed arrays, dipping sonars and sonobuoys, both active and passive. Operational commanders have the opportunity to conduct co-ordinated exercises with several different air and surface ASW platforms.

EMATT will contribute to fleet readiness by:

(a) increasing ASW tactical proficiency in anticipation of actual submarine operations
(b) providing a means to complete readiness exercises when submarines or trainers are not available, especially on deployment
(c) supplementing retrievable targets on underwater ranges when unavailability or weather may cancel operations
(d) giving operational commanders the option to conduct ASW training anytime, anywhere, under most operating conditions.

EMATT's innovative design is flexible and versatile.

The Mk 39 is the same size as a standard 36 in sonobuoy and may be loaded in any sonobuoy chute for deployment. For surface ships, it may be slipped over the side by hand.

Prior to launch, one of three pre-programmed run geometries may be selected. Each run may be set for operation with or without the MAD capability.

After deployment, the vehicle runs for up to three hours. Acoustically, it generates four discrete frequencies. It has a unique echo repeat system which receives, stores, then re-transmits active sonar signals enhanced to simulate the echo from an actual submarine. It also has a transponder for use with the Mk 46 torpedo to provide acquisition and homing response.

For use with MAD-equipped aircraft, EMATT deploys a 100 ft wire through which DC power is pulsed to create a magnetic field around the vehicle. The signature is unique and provides excellent tactical training for aircrews.

Operational Status
Sippican Inc was awarded a contract in late 1991 for 3200 EMATT systems. The targets are all due to be delivered by December 1993.

Contractor
Sippican Inc, Marion, Massachusetts.

NAVIGATION AND LOCALISATION SYSTEMS

FRANCE

NUBS-8A/NUUS-8A

Type
Navigation echo sounder.

Description
This navigation echo sounder is designed to measure the distance of the seabed, on a vertical line beneath the ship, for normal navigation purposes.

The sounder provides a graphic recording of the seabed, and is equipped with a minimum depth alarm device.

The difference between the two versions lies in the type of transducer used.

Maximum range is 1400 m.

Operational Status
In service with the French Navy.

Contractor
Safare Crouzet, Nice.

ITALY

RS-100

Type
Underwater transponder.

Description
The RS-100 underwater transponder has been designed for installation on submarines as an emergency localisation device. Should a damaged vessel be lying on the seabed, the RS-100 will automatically respond to suitable acoustic interrogation pulses received.

Reception and transmission of acoustic pulses is accomplished by means of a cylindrical piezo-electric transducer which is on or installed on the outside of the hull.

The RS-100 is sensitive to acoustic pulses at three frequencies in the 6 to 10 kHz band. When a pulse at one of these frequencies and of the expected duration is received, a corresponding pulse at the same frequency is sent back by the transponder.

RS-100 underwater transponder

Operational Status
Currently installed on board 'Sauro' class submarines.

Contractor
USEA, La Spezia.

APLF SIGNAL

Type
Submarine marker device.

Description
This submarine marker device is designed to be launched at a maximum depth of 600 m to produce a multicolour light and smoke display on the surface. Due to its flare and smoke composition it can be used by both day and night. The body is anodised aluminium and contains two pyrotechnic charges, the first being a flare and the second a smoke composition, which burn consecutively. Burn time is 60 seconds for both flare and smoke. Overall weight of the device is 3 kg.

Operational Status
Over 25 000 devices have been produced to date.

Contractor
Stacchini Sud SpA, Milan.

UNITED KINGDOM

MASTER YEOMAN

Type
Electronic chart table – navigation integrator.

Description
Master Yeoman integrates electronics and conventional navigation on the nautical chart, combining the labour saving of electronics with the safety of manual plotting and pilotage.

Having fixed and referenced his chart to the chart table the navigator has all the vessel's electronic aids output data at his fingertips on the chart.

The user interface is the Puck which acts as the system control and display unit. Around the window of the Puck are four cardinal point lights which indicate the direction in which the Puck needs to be moved on the chart to locate the navigation position or other locations depending in which mode the Puck is activated. To assist speed of plotting two colours are used. Red when larger distances are involved and green when the Puck is close to the target radius. When the lights are extinguished the target position is located and plotted.

Master Yeoman operates on latitude and longitude and in its basic form takes output data from a selectable single navigation aid, GPS, Loran, Decca and so on, which is displayed in the Puck window. Whenever the Puck is moved across the chart the position is displayed and with the press of a key,

The intelligent Puck of Master Yeoman

range and bearing to and from navigation position, a mark position or waypoint is displayed.

Master Yeoman provides the navigator with all information required for position plotting, route planning, waypoint acquisition, range and bearing calculations and multiple chart navigation.

Master Yeoman can be expanded to take in speed, log and gyro outputs. range and bearing from an ARPA radar and is able to download all information to video plotter or printer.

An alarm card can be included in the system together with a pilot card to display port and starboard guidances. Also a GPS facility with its own antenna unit and a monitor card can be included as an integral part of the system.

Operational Status
Entering service with the Royal Navy, Commonwealth navies, NATO, Middle East and Far East navies, as well as many commercial users.

Contractor
Qubit UK Ltd, Passfield.

TYPE 639 SUBMARINE INDICATOR BUOY

Type
Submarine indicator buoy.

Description
The Type 639 Submarine Indicator Buoy (SIB) has been designed and developed for the Royal Navy to replace the current in-service Type 609. It operates at 8.364 and 243 MHz. The role of the SIB is to alert search and rescue to the plight of a submarine unable to reach the surface.

Two units are installed on each submarine. Each can be released independently from within the vessel; one is tethered from the forward end and one from the aft. Once on the surface, the SIB antenna is erected automatically, the HF and VHF distress transmissions are initiated, and a visual flasher is activated.

The buoy transmits at HF (distress message and

Type 639 indicator buoy

DF mark), and at VHF (SARBE homing signal), while also displaying the high intensity flashing light for a period of 72 hours.

The SIB is supplied either as a one-for-one

replacement item or as a complete system for first fit during submarine construction or refit.

Operational Status
In use with the Royal Navy, the Brazilian Navy, the Indian Navy and the Royal Netherlands Navy. It will shortly enter service with the Canadian Navy.

Specifications
HF transmitter
Frequency: 8364 kHz ±800 Hz
Effective radiated power: 15 mW (min)
Transmission mode: A1A (CW) Morse
Antenna: Helical whip
UHF transmitter
Frequency: 243 MHz ±7.5 kHz
Effective radiated power: 250 mW (min)
Transmission mode: Amplitude shift keying
Antenna: 4 monopole

Contractor
Marconi Underwater Systems Ltd, Addlestone.

HONEYPOT BEACON

Type
VHF location beacon.

Description
Honeypot is a VHF location beacon operating at standard sonobuoy VHF frequencies which provides the ASW aircraft with a direction finding location capability to its home base of ship.

The beacon is indistinguishable from normal LOFAR sonobuoy transmission to provide a tactical navigation aid in support of anti-submarine warfare operations. It is a low-cost system, designed for high reliability and minimum maintenance, and gives the ship's ASW commander flexibility in the choice of navigation aid to recover the ASW helicopter.

Honeypot operates over the VHF frequency band 136 to 173.5 MHz, to transmit 1 W as an omnidirectional beacon. A channel switch on the transmitter unit allows transmission on any of the 99 sonobuoy

channels. The unit is powered from the standard 24/28 V DC supplies.

Honeypot is a portable system, the transmitter unit is designed for installation in a shipborne, naval enclosed environment. The unit is a sealed splash-proof housing which may be fitted in any suitable compartment. The system can be supplied complete with a coaxial feed cable and antenna, which may be clamped to any suitable point on the ship's structure.

Operational Status
In production and in service with the Royal Navy.

Contractor
Marconi Underwater Systems Ltd, Templecombe.

Honeypot VHF location beacon

MODEL 2433A

Type
Submarine warning beacon.

Description
The Model 2433A submarine warning beacon is a heavy-duty underwater acoustic beacon designed to meet MoD specifications to protect oil/gas rigs operating in NATO authorised submarine transit lanes.

The pinger is activated by water immersion or manually using a shorting link, and emits sufficient power to alert dived submarines of the rig's position allowing safe submersible navigation past the rig without risk of collision.

The beacon operates for up to 60 days.

Specifications
Operating depth: 1200 m
Frequency: 9.5 kHz

Coverage: omnidirectional
Pulse length: 115 ms
Pulse repetition rate: 0.5 pps
Dimensions: 89.5 cm (long) × 13.03 cm (diameter)
Weight: 16 kg (in air)

Contractor
Helle Engineering, Aberdeen.

CORRELATION ELECTRO-MAGNETIC LOG CML 001

Type
Two-axis correlation electromagnetic ships' speed log.

Description
This is a new concept in speed measurement in which a correlation sonar log provides highly accurate two-axis speed over the ground and is combined in a single 150 mm diameter transducer with a two-axis electromagnetic log with probes set at 45° to the ship's axis to give speed through the water.

Correlation speed values are derived independently of the speed of sound and are only dependent on manufacturing tolerances (which are checked

during initial ship calibration trials) and time measurement which is easily calculated. The correlation system is much less sensitive to pitch and roll effects and can tolerate large vertical velocity components with no problems. The system is therefore suitable for applications in relatively rough waters and the resulting high integrity measurements of ship speed and tide represent a major advance in ship navigation and position aids. The log accepts synchro or digital heading input and computes the tide with reference to north. This tidal value is displayed in polar co-ordinates.

As a by-product of the ground referenced speed measurement, the depth of water is also available and this can be fed to a hard copy printer to form a standard echo sounding system.

The log incorporates its own position integrators and these provide positional information in latitude

and longitude format, or in transverse mercator projection.

When the depth of water exceeds the bottom tracking limit of the system, internally stored position referenced currents may be used to generate ground speed from the water speed measurements.

Serial data outputs are available to feed the speed information to inertial navigation, fire control and other systems.

Operational Status
In production for autonomous underwater vehicle applications for clients in Australia and the USA. A similar system has been developed for the Royal Navy 'Sandown' class minehunters.

Contractor
Sonar Research and Development Ltd, Beverley.

SONAR LOCATOR BEACON

Type
Locator beacon for downed aircraft.

Description
The Sonar Locator Beacon will withstand the impact of any aircraft crashing into the sea, the mechanical design being such that it will remain undamaged when subjected to impact shocks of up to 600*g* and crush forces of more than 2200 kg. On submersion in water the beacon is switched on automatically when a depth of 5 to 10 m is reached. It then transmits acoustic signals which can be received by ship or airborne sonar equipment. The beacon is designed to

withstand pressure corresponding to a water depth of 600 fathoms (1100 m).

Operational Status
In service.

Specifications
Operating frequency: 9.5 kHz
Operating life: 10 days minimum
Size: 75 × 75 × 150 mm
Weight: 1.2 kg

Contractor
Dowty Maritime Systems Ltd, Communications Division, Greenford.

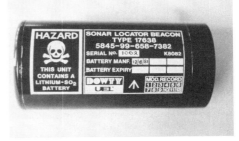

Sonar locator beacon

PINGERS

Type
Locating devices.

Description
Helle Engineering has developed several types of pinger for locating various underwater devices such as ROVs, divers and so on.

ROV power loss pingers comprise a family of 2 W and ¼ W pingers operating at frequencies of 27 kHz and 37 kHz with battery lives ranging from four days to three months. The ROV power loss pinger starts pinging when power to the vehicle is interrupted.

The Model 6280 pinger receiver, introduced in the last two years and in service with several navies around the world, is a versatile receiver capable of being tuned to any acoustic source between 8 kHz and 50 kHz.

The Model 6281 is suitable for surface or diver use.

Contractor
Helle Engineering, Aberdeen.

UNITED STATES OF AMERICA

3951 UNDER-ICE SONAR SYSTEM

Type
Dual beam, high resolution side-scan for under-ice applications.

Description
This is a high frequency, dual beam side-scan sonar system for use in imaging the underside of the polar ice pack. The transducers utilise an inverted horizontal beam pattern which images the underside of the ice pack providing the sonar operator with a high resolution image.

The transducers, being less than 50 cm (20 in) in length and 13 cm² (2.2 sq in) in cross-sectional area, are sufficiently small to permit installation on board the submarine in any available location. Pre-processing electronics are located exterior to the pressure hull and required penetrations into the pressure hull are minimal. The processing electronics can be located at locations up to 500 m (1500 ft) from the sonar transceiver for ease in installation.

The sonar transceiver is 19 in rack-mountable and requires only 10 in of vertical rack space. The processed sonar data are displayed on a video display unit (VDU) which can be located in a location remote from the sonar transceiver.

Three frequencies are available for operation with the sonar system to optimise the installation for the specific mission requirements: 50 kHz provides long-range with medium resolution; 100 kHz provides high resolution and simultaneous dual frequency; 100/500 kHz provides simultaneous insonification of the ice pack for maximum resolution of the ice/water interface.

Operational Status
The under-ice sonar system is currently in production and in operation aboard various units of the US Navy.

Contractor
Klein Associates Inc, Salem, New Hampshire.

390 video display unit for under-ice sonar

MODEL 3040 DOPPLER VELOCITY LOG

Type
Doppler velocity log.

Description
The Model 3040 Doppler Velocity Log is EDO Electro-Acoustic Division's fifth generation Doppler system.

Digital signal processing results in the least variance estimate and eliminates the need for periodic alignment and calibration. The system is microprocessor-controlled, and automatically selects its operating parameters for optimum bottom tracking performance.

The Model 3040's main features are:
(a) Full MIL parts selection, environmental testing (EMI/RFI, shock, vibration) and electrical interface
(b) Simultaneous, two axes, water track and bottom track, providing measurement of platform set and drift
(c) BIT functions to circuit board level
(d) Phased array beam forming provides automatic compensation for sound velocity variations. The transducer face forms a flat surface which installs flush with the hull.

Model 3040 Doppler velocity log

The system is crystal-controlled and completely coherent. All system frequencies are derived from a common source. Minimum velocity variance is achieved through the use of an optimal (maximum likelihood) digital spectral estimator (called the Rummler estimator). Bias errors are suppressed through the use of a phased array transducer, an unbiased Doppler frequency discriminator and platform trim compensation. High reliability is achieved through a substantial reduction in card count as compared to analogue designs.

Operational Status
Systems under contract for Australian, Turkish and Swedish submarines. Also used with various US Navy applications.

Contractor
EDO Corporation, Electro-Acoustic Division, Salt Lake City, Utah.

MODEL 4208 CORRELATION VELOCITY LOG (CVL)

Type
Full ocean depth velocity log.

Description
The Model 4208 provides a method of measuring absolute velocity relative to the ocean floor. Its ability to make these measurements in deep water, with practical acoustic power levels, high accuracy and with total autonomy, make it an ideal sensor for commercial, military, surface ship, submarine and UUV applications. The CVL is capable of establishing its velocity in motion or stationary at system power-up. The transmitted waveform, power level and duty cycle

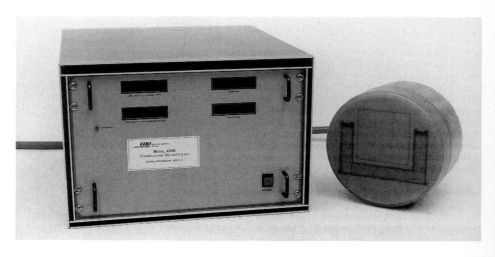

Model 4208 correlation velocity log

are dynamically adjusted with depth and platform velocity. Everything needed for operation is continued in EPROM embedded software.

Unlike a Doppler velocity measuring system which operates in the frequency domain, the correlation velocity log operates in the time domain. Relative velocity between the vehicular-mounted sonar and the bottom is determined by correlating sonar returns in time and position. The CVL's measurement is not dependent on the speed of sound.

The acoustic arrays are housed in a single unit, requiring a gate valve installation of only 12 in. The transducer generates one broad acoustic beam directed downward, which returns to the hydrophone after striking the bottom. The CVL is most sensitive to positional changes that are parallel to the transducer array plane. In this respect, the CVL provides a better match to the propagation geometry and platform constraints than other sonar velocity measuring techniques.

Operational Status
In production.

Contractor
EDO Corporation, Electro-Acoustic Division, Salt Lake City, Utah.

MODEL MRQ-4015 DUAL AXIS SPEED LOG

Type
Doppler speed log.

Description
MRQ-4015 provides both fore/aft and port/starboard velocities in addition to distance travelled and depth. Speed range on both axes is 0-39.9 knots, with bottom tracking to 200 m depth. The system can be forced to track the water column rather than the bottom at any time from a front panel control. Construction is of a non-ferrous material providing low magnetic signature required on vessels involved in mine warfare operations.

A standard system consists of: main display and control unit, Main Electronics Unit (MEU), transducer junction box, and transducer.

The system design allows cable lengths of up to 365 m between transducer, J-Box and MEU, and between MEU and main display and control unit.

Operational Status
In service with US and South Korean minesweepers.

Contractor
EDO Corporation, Electro-Acoustic Division, Salt Lake City, Utah.

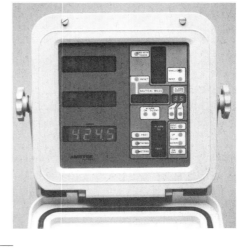

Model MRQ-4015 speed log display and control unit

HONEYWELL POSITION SENSING UNIT

Type
Proximity position sensing system for submarines.

Description
A deep sea proximity sensing system for accurate position sensing outside submarine hulls has been developed by Honeywell. Typical applications include checking the deployment of towed array sonar dispensers, ECM towers, auxiliary propulsion units, and the position of 'wet side' doors and valves.

The system comprises two parts, a passive sensor developed for extreme reliability under deep sea conditions, and a hermetically sealed electronic module for installation as part of the control electronics within the pressure hull.

Sensors are designed for installation without adjustment. They contain two identical coils, an active (sensing) coil and a balance coil, forming two arms of a bridge circuit. Two resistors within the separate electronic module provide the other two arms of the bridge circuit, which is driven by an audio frequency oscillator.

Contractor
Honeywell Inc, Underseas Systems Division, Hopkins, Minnesota.

C-100 SENSOR

Type
Miniature digital compass sensor.

Description
The C-100 miniature heading sensor is a very small, extremely accurate heading sensor ideal for use in systems where space is at a premium and where direction information is essential. Typical applications include oceanographic equipment, underwater warfare systems, vehicle navigation and mapping systems, night vision devices, and weapon aiming systems. The C-100 provides heading information with ±0.5° accuracy.

The digital fluxgate compass sensor, which uses no moving parts, measures the earth's magnetic field electronically, thus eliminating card movement which makes conventional compasses inaccurate and unstable. It comprises a remote sensor coil and electronics board featuring a microprocessor-controlled calibration system. The sensor is compatible with multiple floating ring coil sensors which allow gimballing from ±8° to ±45° of tilt. The modular design enables sensor coils and electronics to be separately mounted in different locations. The sensor communicates with other electronics with a variety of analogue and digital interfaces.

Contractor
KVH Industries Inc, Middletown, Rhode Island.

C-100 miniature digital compass sensor

SUBMARINE RADAR

CANADA

CMR-85 RADAR SYSTEM

Type
Lightweight air surveillance system.

Description
Although designed for operation in a submarine environment, the CMR-85 can be used in any role requiring a lightweight/portable system, either as the primary radar in a small craft or as a secondary radar in larger vessels. It is another variant of the LN66 series.

The CMR-85 is a 6 kW radar comprising only two units: a scanner unit incorporating the transmitter/receiver and the antenna (either a 4 ft slotted waveguide or a 3 ft antenna enclosed in a radome), and an 8 in display unit. With the 4 ft unit, the antenna can be removed from the transmitter/receiver unit by means of a quick-release mechanism, for easy transportation through hatches.

Operational Status
Various versions of the LN66 family are in widespread use with the US Navy and other customers.

Specifications
Frequency: 9410 ±30 MHz
Peak power: 6 kW
Rotation speed: 30 rpm
Pulse width: 0.15 and 0.12 μs
PRF: 1200 and 3000 Hz
Topside weight: 13.6 kg (including antenna)

Contractor
Canadian Marconi Company, Naval and Ground Systems Division, Kanata, Ontario.

FRANCE

TRS 3100 (CALYPSO III) SUBMARINE RADAR

Type
Submarine air and surface surveillance radar.

Description
Calypso III is a higher performance radar than Calypso II, but with the same general characteristics and same functions (navigation, surveillance, target designation) with a greater detection range and the additional capability of accurate range measurement. It is also of compact design and well fitted to the severe environment characteristics of a submarine.

Calypso III comprises four main parts: an antenna (the Calypso II antenna); a non-rotating periscopic mast (hoisting is achieved using the ship's hydraulic system pressure); a transmitter/receiver cabinet (secured at the foot of the mast) and an operational console.

The transmitter is a conventional magnetron transmitter with a frequency adjustable magnetron and the same I/J-band transmitter employed by Calypso II. The receiver is fitted with RF elements of long service life, a modern mixer and a logarithmic anti-clutter chain. The operational console comprises the control panel for radar and antenna, the display (16 in (406 mm) CRT) with an accurate digital rangefinding device and an operational panel. The optical-

periscope direction ESM and sonar data can be displayed on the PPI. The accurate range measurement of a detected target can be made during short time transmission.

The detection range capability of Calypso III for a typical ASW aircraft at 2500 m altitude is 18 nm but depends on the radar horizon for a surface vessel.

Operational Status
In service with three navies. No longer in production.

Contractor
Thomson-CSF, Division Systèmes, Défense et Contrôle, Meudon-la-Forêt.

TRS 3110 (CALYPSO IV) SUBMARINE RADAR

Type
Submarine air and surface surveillance radar.

Description
The TRS 3110 Calypso IV radar is an I/J-band equipment designed to provide surveillance and navigation facilities for submarines. It is comprised of three units: an antenna, a transmitter/receiver cabinet and an operating console. The antenna and operating console are similar to those of Calypso III. To fulfil its functions, the Calypso IV is built around a simple navigation radar equipment, and modular design enables

it to carry out additional functions merely by the addition of appropriate functional circuit cards, without changing the basic equipment.

In using a 1 ms pulse it has a good detection range, limited only by the radar horizon (at least 10 nm in free space), and using the narrow (50 ns) pulse it has very high resolution over shorter detection ranges (down to about 15 m). The receiver design has been developed to give good capabilities and performance in the presence of clutter.

The transmitter is a 25 kW (peak) klystron-driven unit providing several transmission options (continuous, sectoral, burst, or receive only – radar silence) giving maximum operational flexibility to meet specific requirements. The overall radar transmission parameters have been chosen to enhance the

submarine's security by arranging for them to be similar to those of merchant ship radars, frequency and pulse length in particular.

The operating console and its data processor permit the presentation, in either raw radar or synthetic form, of information such as target echoes, range and bearing labels, transmission sectors and so on, as well as providing for operational control of the radar equipment itself.

Operational Status
In service but no longer in production.

Contractor
Thomson-CSF, Division Systèmes, Défense et Contrôle, Meudon-la-Forêt.

ISRAEL

EA-20/SRD RADAR ANTENNA

Type
Submarine navigation radar antenna.

Description
The EA-20/SRD radar antenna has been developed for integration into a submarine navigation and surveillance radar system. The dimension restraints led to the design of a hog horn feeding a cylindrical parabolic dish. The complete antenna will withstand a pressure of 40 atmospheres and is constructed of anti-corrosive stainless steel.

Operational Status
Fully developed.

Specifications
Frequency range: 9.44 GHz ±0.02 GHz
Polarisation: Horizontal
Peak power: 200 kW
Average power: 200 W
Gain: >25 dBi
Azimuth beamwidth: <3° (3 dB)§
Elevation beamwidth: >15°
VSWR: 1.2:1
Dimensions: 175 mm (high) × 800 mm (wide)
Weight: 32 kg

Contractor
Elta Electronics Ltd (a subsidiary of Israel Aircraft Industries Ltd), Ashdod.

EA-20/SRD submarine radar antenna

ITALY

MM/BPS-704 SUBMARINE RADAR

Type
Submarine search and navigation radar.

Description
The MM/BPS-704 is a naval search and navigation radar for use aboard submarines. The general characteristics are similar to the MM/SPN-703 surface ship version, using the same 20 kW transmitter and

pulse length/PRF combinations. A specially designed high pressure resistant antenna (up to 60 kg/cm²) mounted on a telescopic mast, is part of the equipment. The principal differences are indicated in the specifications.

Operational Status
Italian Navy 'Sauro' class submarines are fitted with the MM/BPS-704.

Specifications
Antenna span: 1 m
Beamwidth: 2.2° (horizontal); 11° (vertical)
Gain: At least 27 dB
Noise figure: Better than 11 dB

Contractor
SMA – Segnalamento Marittimo ed Aereo, Florence.

MM/BPS-704 antenna

NETHERLANDS

ZW07 SUBMARINE RADAR

Type
Surface search and navigation radar.

Description
The ZW07 is one of the family of the ZW series radars produced by Signaal. This particular model is designed for use by submarines for navigation, distance measuring, surface search and limited air warning. A single shot mode has been adapted for a range of surface targets. Other operational features are surface coverage up to the radar horizon, limited air warning and high resolution for navigation. Anti-clutter measures and ECCM provisions include sector-scan facilities, logarithmic receiver with pulse length discriminator, suppression of non-correlated pulses, and a tunable transmitter.

Operational Status
In operational service.

Specifications
Frequency: I-band
Polarisation: Horizontal
Transmitter: Tunable magnetron
Power output: 60 kW (peak); 100 W (average)
Maximum range: 22 km (against small surface craft)

Contractor
Hollandse Signaalapparaten BV, Hengelo.

SWEDEN

SUBFAR 100 SUBMARINE RADAR ANTENNA

Type
Submarine air and surface search radar antenna.

Description
The Subfar 100 equipment is an air and surface search radar antenna for submarines.

The 9GA 300 antenna is designed for mounting with its hydraulic turntable on top of the submarine radar mast. This way of mounting eliminates bearing error sources from torsion effects in the radar mast. The hydraulic turntable gives a very low noise level as well as flexible control of antenna rotation speed. The antenna direction can be manually set or slaved to external sources such as the periscope.

Sector transmission as well as short time transmission down to emitting a single pulse can be selected.

Operational Status
In service.

Specifications
9GA 300 antenna
Frequency range: 8.5-9.6 GHz
Rotation speed: 0.5-24 rpm
Beamwidths
Horizontal: 2.4°
Vertical: 16°
Gain: 26 dB
Sidelobes: 6-18 dB
Antenna aperture: 1000 × 140 mm
Height above mast tube: 600 mm
Rotating circle diameter: 1040 mm
Antenna system weight: 95 kg
Dimensions
Height: 1100 mm
Width: 665 mm
Depth: 600 mm plus desk 250 mm
Weight: 135 kg

Subfar 100 submarine antenna

Contractor
NobelTech Systems AB, Järfalla.

UNION OF SOVEREIGN STATES

SNOOP TYPE SUBMARINE SURVEILLANCE RADARS

Type
Submarine surface search radars.

Description
A number of surveillance radars has been fitted to USS submarines since the Second World War, some of which have been given a NATO designation starting with Snoop. The early submarines were equipped with a small antenna designated Snoop Plate. This was apparently fitted to all boats up to the 'Golf'/'Foxtrot'/'Romeo' classes and was reputed to have a range of 25 nm against aircraft and 12 nm against surface vessels. Later, 'Foxtrot' class vessels and the 'Hotel' generation were fitted with a larger and more powerful I-band radar known as Snoop Tray which entered service in the early 1960s.

Since the early 1960s, nearly all new USS submarines have been fitted with Snoop Tray, with the exception of the 'Juliett' and 'Echo' classes of cruise missile submarines which are equipped with a much larger antenna known as Snoop Slab. This antenna is twice the size of the Snoop Tray unit and must be capable of greater definition at longer ranges. Both the 'Juliett' and 'Echo' classes are armed with the surface-launched SS-N-3 'Shaddock' missile, and it is assumed that the Snoop Slab radar will also provide backup cover for the Front Door/Front Piece missile guidance equipment.

Two other Snoop class radars are the Snoop Pair fitted to 'Typhoon' and 'Akula' class submarines, and Snoop Head fitted to the 'Oscar' and 'Alfa' classes.

Operational Status
In operational service. Virtually all USS submarines are fitted with one or other type of Snoop class surveillance radar.

UNITED KINGDOM

TYPE 1006 RADAR

Type
Navigation and surface search radar.

Description
The Type 1006 was the standard I-band navigational radar of the Royal Navy until the arrival of the Type 1007.

Antenna Outfit AZJ
The surface role antenna outfit consists of a 2.4 m slotted waveguide linear array rotated at 24 rpm by a turning mechanism. The array has a horizontal beamwidth of 1° and low sidelobe levels to give good bearing discrimination. The vertical beamwidth of 18° gives good performance in conditions of roll and pitch.

Antenna Outfit AZK
This antenna outfit consists of a 3.1 m slotted waveguide linear array rotated by the same type of turning mechanism as outfit AZJ. It is used when the improved bearing discrimination given by the 0.75° beamwidth is required.

Transmitter/Receiver (Surface)
The equipment is non-thermionic with the exception of the magnetron, and operates at a frequency of approximately 9445 MHz.

Transmitter/Receiver (Submarine)
In order to be compatible with the existing submarine antenna system a variant of the transmitter/receiver operating at 9650 MHz is available.

Display Unit JUD
This unit is non-thermionic with the exception of the cathode ray tube. The display unit uses a rotating scanning coil system to provide range scales from 0.5 to 96 data miles.

Operational Status
The Type 1006 radar has been in series production

for the Royal Navy and other navies since 1971. It is now being superseded by the Type 1007.

Contractor
Kelvin Hughes Ltd, Hainault.

TYPE 1007 RADAR

Type
Navigation and surface search radar.

Description
The Type 1007 is an I-band navigation radar for naval use, consisting of an antenna unit, transmitter/receiver and a display.

Three versions of the antenna outfit are available for surface vessels: a 2.4 m single array, a 3.1 m single array, and a 2.4 m dual array for use with helicopter transponders and outfit RRB. All are horizontally polarised end feed slotted line arrays, incorporating vertical polarisation filters to give low sidelobe and back radiation levels. The 2.4 m array has a horizontal beamwidth of 0.75°. Surface ship antenna outfits can operate in winds up to 100 kts and withstand funnel gas temperatures up to 120°C. De-icing systems are available. A fully pressure-tested antenna is also available for fitting to a variety of submarine masts.

The transmitter/receiver is solid-state, with the exception of the magnetron, and operates at a frequency of 9410 MHz with a transmitter power output of 25 kW. A wide dynamic range logarithmic receiver is provided. A built-in monitoring system is included to check that the equipment is operating at peak performance. A low leakage dummy load allows for system testing during periods of radar silence. Centralised emission control circuitry enables command to inhibit transmission immediately. Sector transmission is also

incorporated with direct control from the main display. Blanking pulses are incorporated to safeguard sensitive ESM equipment.

The 16 in main display has nine range scales from ⅜ to 96 nm and displays relative motion, true motion pictures in either head-up, north-up or course-up orientation. Crystal-controlled range rings and range marker, fixed or floating electronic bearing line, with digital readout of range, bearing and ship's head, form the basic controls and operational features. Additional features include target track history, electronic plotting, target labelling, guard zones, electronic mapping, sector zones, helicopter track and sector transmission. Auxiliary video inputs are also included to allow inputs from simulators and helicopter transponders.

Additional displays are also available, including an NHR high resolution and colour tactical displays. The NHR display uses a combination of advanced microprocessor technology and dedicated hardware to give high resolution (1360 × 1024 pixels), with a wide selection of operational facilities. The colour tactical display is designed principally for small warships and gives the operator a clear sharp colour surface tactical picture with ability to label radar tracks with ships' names/tactical call signs or track numbers. Track hostility is easily defined by the use of soft keys to change the track colouring to indicate hostile/friendly/neutral/unknown.

Two types of weatherproof auxiliary display are available as options: a polar scan giving a high definition picture with standard operational facilities, and a raster scan high resolution display which provides

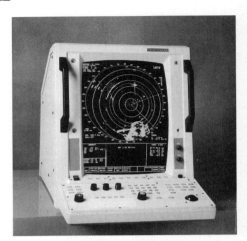

The Type 1007 NHR display

flicker-free daylight viewing. Both units are portable and are designed for use on an open bridge or submarine fin.

Operational Status
Series production started in 1986 for the Royal Navy and other navies using the displays Type JUF and JQA.

Contractor
Kelvin Hughes Ltd, Hainault.

UNITED STATES OF AMERICA

AN/BPS-15 SUBMARINE RADAR

Type
Submarine surface search and navigation radar.

Description
The AN/BPS-15 is a submarine radar designed for surface search, navigation and limited air warning facilities. It operates in I-band using a horn array antenna.

Operational Status
In widespread use on US Navy submarines.

Specifications
Frequency: I-band
Peak power: 35 kW
Pulse width: 0.1; 0.5 μs
PRF: 1500; 750 pps
Scan rate: Up to 9.5 rpm
Antenna dimensions: 101 cm (aperture)
Antenna weight: 76 kg

Range resolution: 10 m (in short pulse mode); 30 m (in long pulse mode)

Contractor
Sperry Marine, Charlottesville, Virginia.

AN/BPS-16 SUBMARINE RADAR

Type
Submarine navigation and surface search radar.

Description
The AN/BPS-16 (previously known as the AN/BPS-XX) is a submarine radar designed to provide nuclear-powered fast attack and ballistic missile submarines with a navigation and search

radar capability when operating on the surface. The AN/BPS-16 features a new 50 kW frequency-agile transmitter in I-band and the latest in signal processing techniques to enhance operational performance in heavy weather. Unlike the AN/BPS-15, the new system is supplied with a new unique radar mast assembly to more effectively and reliably raise and retract a state-of-the-art antenna.

Operational Status
The AN/BPS-16 has completed First Article Testing

and Navy Operational Evaluation. The equipment to be furnished under Production Options to the present contract will be installed in 'SSN-688', 'SSBN-726' and 'SSN-21' ('Seawolf') class submarines. At the end of 1991 Sperry Marine was awarded a contract for 10 AN/BPS-16 radars for 'Ohio' and 'Seawolf' class submarines and training sites. Work is expected to be completed by December 1994.

Contractor
Sperry Marine, Charlottesville, Virginia.

CONSOLES AND DISPLAYS

CANADA

AN/UYQ-501(V) SHINPADS DISPLAY

Type
Multi-sensor display system.

Description
The MSD-7001, developed by Computing Devices for the Canadian Department of National Defense as part of a distributed processing system, is a true multi-function display that accepts inputs from all shipboard sensor systems and computers. In its naval configuration it is designated AN/UYQ-501(V), and provides both sensor information and complex graphical overlays on high resolution full-colour television monitors. It interfaces with any general-purpose NTDS-capable computer functioning as a display processor, and hence communicates over any standard databus.

Standardisation of hardware, software and interfacing has been achieved to the point where this display satisfies all the requirements for operator interface with any sensor, weapon or machinery control function. It is a powerful tactical and command situation display providing the command and control team with instant access to all data available on board.

The AN/UYQ-501(V)'s unique video features optimise the match between sensor and display for all sensors on board ship. Careful human engineering, a high resolution display and judicious use of colour combine to give the operator all the information required to carry out his tasks effectively on a single display CRT.

Video processing
(a) 4-megabit RAM video memory
(b) 1024 × 1024 pixels each having four bits for intensity
(c) video area controlled in size and position anywhere within complete raster area.

Radar characteristics
(a) accepts normal radar PPI range and bearing, and converts these to X-Y co-ordinate display
(b) presents section of PPI radar in B-scan format
(c) presents PPI and fire control (A + R-scan) radar simultaneously
(d) offset centre of PPI outside display field-of-view
(e) any scan rate 0 to 60 rpm
(f) video pulse width, 50 ns minimum.

Line scan characteristics
High speed serial interface using STANAG 4153 format accepts pre-formatted data from sensors such as sonar and infra-red at 10 Mb/s rate directly into memory.

Television characteristics
(a) accepts TV composite video via RS 343A or RS 170 interface at standard 525, 625, 875 and 1075-line rates.
(b) TV video stored digitally in video memory permitting annotation, magnification, freeze-frame and image processing.

Graphic processing
(a) 8-megabit RAM graphics memory
(b) covers full extent of 1225-line raster scan format. 1152 active lines at 1536 pixels/line with 4 bits/pixel for colour and blink
(c) eight colours selectable from palette of 4096
(d) graphics repertoire includes ASCII characters, NTDS characters, vectors, arcs, circles, in both positive and inverted video format

(e) graphics are overlaid without loss of information
(f) graphics scrolling is independent of video presentation
(g) graphics addressing is user-friendly and simple; built-in generator does the housekeeping.

Application processing
(a) interfaces with any TDS-capable computer of the user's choice by means of a 16-unit NATO Standard NTDS parallel interface (STANAG 4146)
(b) specifically designed for universal application in the combat system environment with distributed processing interconnected by a standard high speed serial databus.

Operational Status
In service with the Canadian Navy (Canadian Patrol Frigate and Trump). Special variants have been developed for the AN/SQR-501 (CANTASS) and AN/SQS-510 (see separate entries).

Contractor
Computing Devices Company (a division of Control Data Canada Ltd), Ottawa, Ontario.

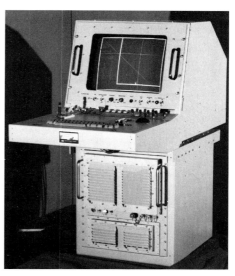

AN/UYQ-501(V) SHINPADS display

MAGIC 2 COMMAND AND CONTROL DISPLAY SYSTEM

Type
Multi-mode graphics and sensor imaging system for command and control.

Description
MAGIC 2 (Multi-mode Advanced Graphics and Imaging system for Command and control) offers a

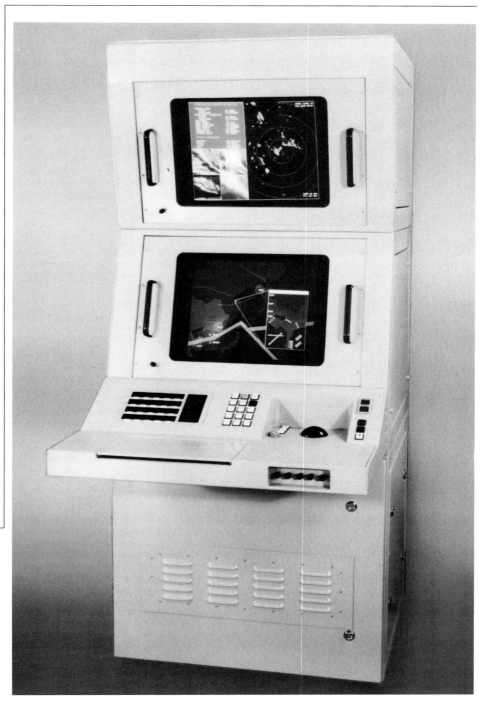

Ruggedised MAGIC 2 graphics and sensor imaging system

solution to the requirements for a multi-sensor display system. It provides a major step forward in the ability to interface with the wide variety of sensors currently employed, plus those envisaged for the future. It does this through the use of a totally modular architecture. This approach allows simultaneous display of up to four sensor outputs with one multi-window display, providing visual comparison for information 'fusing'. The design possesses enough inherent growth capabilities to cater for all foreseeable requirements, both near- and long-term.

The display consists of a high resolution, colour raster TV with 1075 line resolution, refreshed at 60 Hz. This provides a bright, crisp display format which reduces operator fatigue.

The MAGIC 2 system is a single bay, 24 in (63.5 cm) wide, containing the embedded electronics and displays. Mounted on the front is an ergonomically designed operator control panel. This control unit is

removable to enable convenient relocation of the system during installation or refit. The operator control unit provides the appropriate man/ machine interface, and includes keyboard, trackerball and electro-luminescent panel with soft keys.

The main electronics bay has one or two (optional) stacked, angled 19 in colour CRT screens at the top. Each screen provides high resolution raster TV images. Below the displays are the graphics and sensor data display electronics in a standard VME chassis, complete with main power supply and power conditioning module.

The MAGIC 2 system features:
(a) fast traditional tactical display capability
(b) modern workstation environment including windowing of both sensor and graphics data
(c) display manager functions provided by a 68020 based card
(d) single or dual monitor configurations

(e) multi-sensor capability (for example two radars and a TV picture simultaneously)
(f) international standard based VME architecture
(g) provision of space and power for additional computing capability within the console
(h) advanced operator control panel with multi-function keyboard and programmable EL panel with tactile overlay
(i) optional water-cooling to minimise noise and heat
(j) compact modular and rugged construction.

Operational Status
Advanced development model on trial.

Contractor
Computing Devices Company (a division of Control Data Canada Ltd), Ottawa, Ontario.

ISIS

Type
Integrated sensor information system.

Description
ISIS is a computer-based, integrated command, control, communications and display system for harbour surveillance, vessel traffic management, process monitoring and control, video monitoring and other C³D applications.

The system is designed to interface with, and display and control, a large variety and number of

sensors such as radar, sonar, TV and IR cameras. Based on predetermined criteria, ISIS automatically responds to intrusion, vessel movement or other activities detected by the interfaced sensor systems. It assists the operator with initial and rapid assessment by selecting and positioning appropriate imaging devices to further investigate the situation.

The system also automatically computes relative ranges and bearings between designated locations to provide other resources with the data required to take appropriate action.

ISIS comprises two main elements: the operator's workstation and an electronics equipment rack.

The workstation includes a large, high resolution

colour graphics display for presentation of a map of the local site under surveillance, and two smaller displays for imaging sensors such as TV and IR video.

The basic ISIS uses 68030 microprocessors capable of 30 mips but the distributed architecture allows additional computing power to be added via the internal LAN. This also enables a number of ISISs to be linked together to accommodate very large system requirements.

Contractor
Computing Devices Company (a division of Control Data Canada Ltd), Ottawa, Ontario.

UNITED STATES OF AMERICA

MRCS

Type
Militarised reconfigurable console system.

Description
The MRCS is a tactical console capable of being reconfigured for a wide range of applications. The hardware has been designed and tested in accordance with stringent US Navy requirements for shock, vibration, EMI, humidity, sand, dust, saltfog, fungus and acoustic noise.

Up to four MC68020/68881 processors are built around the VME bus architecture, providing a large, commercial support base which is expandable in both processing power and memory. Processing speeds up to 6 mips are possible and the system has a 4-24 Mbyte RAM memory. Up to 12 interfaces are available in the single display, depending on the configuration selected.

It provides a flexible I/O configuration which can be expanded or modified to fit the application. A powerful embedded graphics generator can provide multiple, independent graphics output and input video overlays and the system supports two independent monitor

drivers with four graphics planes per driver (expandable to eight).

Operational Status
The consoles are being supplied under contract to the US Navy for the MHC-51 programme where they are used in the navigation/command and control system, the AN/SYQ-13.

Contractor
Unisys Defense Systems Inc, Reston, Virginia.

MMC

Type
Modular multi-function console.

Description
This is a real-time, tactical, general-purpose computer and high resolution colour graphics system. The console is capable of operation as a single integrated unit or being split into major sub-units allowing the monitor and keyboard to be operated remotely from the base assembly.

The open architecture of the console is based on the VME bus enabling the system to be easily reconfigured by the integration of extra printed circuit cards in Eurocard form.

The console is modularly constructed allowing it

to meet almost any customer requirement. It is configured in three basic units (the processor, base assembly, and operator entry panel and monitors). From these units a variety of configurations can be formed. These include installation and operation as a personal workstation, as a unit interfaced as a node in a distributed system, or interfaced to a central mainframe.

Single or dual monitors can be fitted in any of these configurations as well as systems where the monitor and operator entry panels are remotely located from the processor.

The MMI is further enhanced through an optional touch-screen mounted on the monitor CRT. Multiple processors are interconnected via the internal VME-VSB bus or LAN. The central processor uses the 680X0 processor family combined with an optional floating point co-processor. The processor memory is

downloaded from an internal EPROM assembly during power-up or at operator command.

Contractor
Unisys Defense Systems Inc, Reston, Virginia.

OJ-653/UYQ-21(V) INTEGRATED DISPLAY CONSOLE

Type
Raster video processor for multi-sensor functions.

Description
The Advanced Video Processor (AVP) has been designed to provide a standardised approach to the requirements of multi-platform environments. It is a militarised processor and display system for multi-sensor functions, capable of processing sensor data from sonar, radar, ECM, infra-red, strategic C³I and

tactical C³I inputs, and meets requirements for shipborne, submarine, land-based and airborne applications. It incorporates flexible system architecture with high resolution colour raster technology for the most demanding display requirements.

A dual display station configuration of AVP is being integrated into the AN/SQQ-89 sonar suite. Designated the OJ-653/UYQ-21(V) Integrated Display Console for shipboard application, the AVP is expected to become the standard colour display system on board most US Navy surface ships.

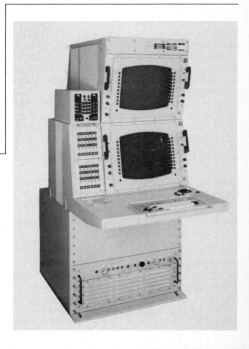

Integrated display console

AVP provides high speed, high data rate processing for improving target detection, resulting in increased operator effectiveness. The system combines functionally independent hardware and software modules, a standard VME bus and multiple input/ output interfaces for maximum flexibility. The display formats are fully programmable and feature multi-image windowing, on-line fault detection and localisation, and GKS graphics.

The OJ-653/UYQ-21(V) Integrated Display Console is built in accordance with MIL-E-16400.

Operational Status
Already selected for the AN/UYQ-21 display station in the AN/SQQ-89 sonar suite and SURTASS. Under contract to the US Naval Sea Systems Command (NAVSEA).

In late 1991 it was announced that Hughes Aircraft Company was being awarded a contract for AN/UYQ-21(V) display systems for various classes of ship, combining purchases for the US Navy, Japan and Spain.

Contractor
Diagnostic/Retrieval Systems, Inc, Oakland, New Jersey.

AN/ASA-66 DISPLAY

Type
Tactical data display for the P-3C Orion.

Description
The AN/ASA-66 is a multi-purpose cockpit display for the P-3C Orion maritime patrol aircraft. A 9 in (23 cm) diameter CRT provides pilot and co-pilot with real-time presentation of tactical situations. Graphic and alphanumeric data are presented on the high brightness and contrast screen for viewing in either high

ambient or controlled lighting conditions. The display is designed to be driven from a remote display generator to provide analogue deflection and video drive from computer data or direct from aircraft sensors.

Operational Status
In operational service.

Contractor
Loral Electronic Systems, Yonkers, New York.

AN/ASA-66 tactical data display system

AN/ASA-66

AN/ASA-82 DISPLAY SYSTEM

Type
Tactical data display system for the S-3A.

Description
The AN/ASA-82 tactical data display system (TDS) is the primary data display for the US Navy S-3A Viking maritime aircraft. It serves as a real-time link between the four-man crew and the various electronic sensors which the aircraft carries for its task of maritime reconnaissance and anti-submarine warfare. High speed, high density data in the form of alphanumeric symbols, vectors, conic projections and other appropriate display formats from both acoustic and non-acoustic sensors are presented to the crew. Information is stored, updated and refreshed by the onboard general-purpose digital computer and selectively displayed. Tactical and tabular data, controlled by the computer, are routed to the display via the display generator unit (DGU), which also provides the computer with display fault status information on a priority basis, controls the routeing of display information, and generates the system built-in test functions.

The equipment consists of five CRT displays in addition to the DGU. The TACCO and SENSO (tactical co-ordinator and sensor operator) are each provided with identical multi-purpose display units in their respective consoles. In addition, the SENSO has an

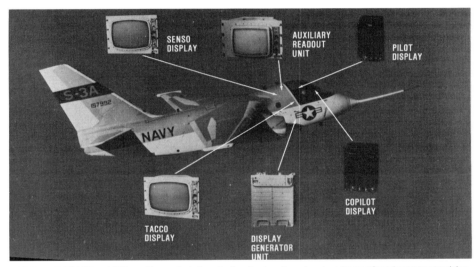

US Navy S-3A Viking ASW maritime patrol aircraft, showing location of various items of equipment comprising AN/ASA-82 tactical data display system

auxiliary readout unit (ARU), the co-pilot has a multi-purpose display, and the pilot has a display which presents a summary tactical plot. The DGU provides all of the displays with digital computer data except the ARU which receives acoustic data direct from the acoustic data processor.

Operational Status
In operational service.

Contractor
Loral Electronic Systems, Yonkers, New York.

C-12151/AYQ PROGRAMMABLE ENTRY PANEL

Type
Programmable airborne plasma panel.

Description
The Programmable Entry Panel represents highly reliable, low maintenance mission avionics carried by the US Navy P-3C Update IV ASW patrol aircraft. These software-configurable, menu-driven command devices provide the means for controlling and monitoring the P-3C's avionics suite. The ability to reconfigure panels via software allows any station to perform any function. In addition to the host computer interface, operator interfaces include an alphanumeric keyboard, numeric keypad, trackerball assembly and a touch-sensitive screen on the plasma panel. Built-in test and fault location are included in overall panel design.

Operational Status
In development for the US Navy P-3C Update IV.

Contractor
Dowty Avionics, Arcadia, California.

Programmable entry panel for US Navy P-3C Update IV

ADVANCED VIDEO PROCESSOR

Type
High resolution video processing and display system.

Description
The Advanced Video Processor (AVP) is a militarised raster video processor for multi-sensor applications capable of processing sensor data from sonar, radar, ECM, IR, strategic C^3I and tactical C^3I inputs. It incorporates flexible system architecture with high resolution colour raster technology for the most demanding display requirements.

The AVP provides high speed high data-rate processing for improving target detection. It combines functionally independent hardware and software modules, a standard VME bus, and multiple input/output interfaces with high speed processing that provides maximum flexibility and user support for a diversity of display applications.

In addition to the above the system features: high speed multiple graphics processing; open architecture; fully programmable display formats; standard VME bus, VRTS operating system and GKS graphics; and multi-image windowing.

A dual-display configuration of the AVP is being integrated into the AN/SQQ-89(V) sonar suite to replace the existing equipment. This configuration features:

(a) a command set which is electronically switchable to either an emulation of the AN/UYQ-21 or a GKS-based command set

(b) a high speed NTDS and/or fibre optic communication link with external host application computers

(c) a flexible operator interface which permits direct interaction with the display without the need for host-computer intervention.

Operational Status
Planned for service with the US Navy. In production. DRS received a $29.5 million production contract in May 1991 to produce 148 AVP units. Previous AVP contract awards were received for critical design definition, full-scale engineering development and logistic support services, and totalled approximately $60 million as at June 1991.

Contractor
Diagnostic/Retrieval Systems Inc, Oakland, New Jersey.

TRANSDUCERS

CANADA

DEPTH COMPENSATED RINGSHELL PROJECTOR (RSP)

Type
ASW research and training device.

Development
The ringshell underwater sound projector was originally developed by the Defence Research Establishment Atlantic in Dartmouth, Nova Scotia, Canada. Transfer of the technology to Sparton of Canada Ltd was initiated in 1981. Sparton is the sole licensee for production.

Models of various sizes, weights and performance characteristics have been produced incorporating industrial design improvements to increase the overall efficiency. Design for models resistant to underwater explosive shock are in development.

Description
The depth compensated RSP is a Class V flextensional transducer that can efficiently produce a high acoustic output over a broad operating band. With its passive pressure compensation system, a ringshell projector has an operating depth exceeding 400 m and a safe immersion depth exceeding 500 m.

RSPs are being used in a variety of applications where high power and efficiency are required at relatively low frequencies. As a research tool the projectors have been employed as single units or as part of large arrays. Further, passive ASW training is available without the use of a target submarine by generating signals of desired frequencies and levels. The projector can also be adapted for use as a separate active adjunct element to current passive arrays, or as part of a low frequency, high power, deep submergence sonar. Applications also involve using RSP-based systems to calibrate and evaluate the performance of towed arrays.

Because the Class V RSP has only two

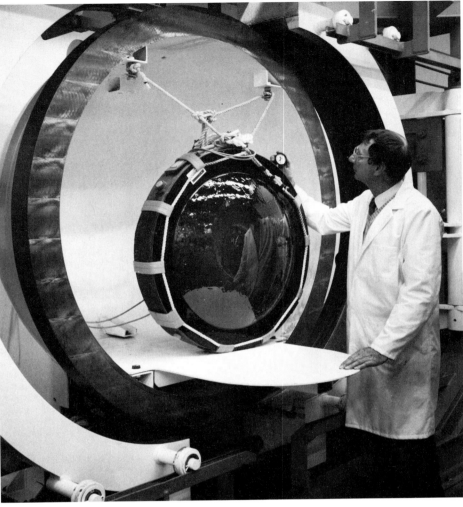

Depth compensated ringshell projector

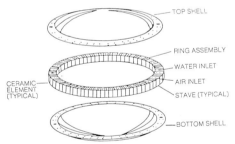

fundamental modes of vibration (ring and shell modes), its element-to-element interactions are minimised when compared to other technologies. Reduction of these interactions is extremely significant when operating projectors in large arrays, and allows for a much tighter projector spacing. In addition the design results in low mechanical stresses in the shell and ring.

The RSP consists of a glass fibre-wrapped, piezoelectric ceramic ring fastened between two convex-domed spherical metals (see diagram). Stimulation of the ceramic ring by an electrical signal causes vibration of the ring in the radial plane. The radial motion of the ceramic ring drives the shells in a flexure mode with a significant enhancement of the volume velocity at the shell surfaces.

Model Number	Resonance Frequency*	Source Level†	Bandwidth at Resonance‡	Ring size diameters A	B	C	D	E Min	Comments
09A2700	2700 Hz	206 dB	635 Hz	9.2 in (23.4 cm)	12.2 in (31.0 cm)	3.8 in (9.6 cm)	2.8 in (7.1 cm)	24.0 in (61.0 cm)	
12A1950	1950 Hz	207 dB	440 Hz	11.7 in (30.0 cm)	14.7 in (37.3 cm)	5.5 in (13.97 cm)	3.0 in (7.6 cm)	24.0 in (61.0 cm)	§
18A0325	325 Hz	205 dB	32 Hz	18.6 in (47.2 cm)	21.6 in (54.9 cm)	3.8 in (9.6 cm)	2.9 in (7.4 cm)	24.0 in (61.0 cm)	219 dB possible at 3.6 KHz
18A0400	400 Hz	205 dB	45 Hz	18.6 in (47.2 cm)	21.6 in (54.9 cm)	4.2 in (10.7 cm)	2.9 in (7.4 cm)	24.0 in (61.0 cm)	219 dB possible at 3.6 KHz
18A1000	1000 Hz	207 dB	235 Hz	18.6 in (47.2 cm)	21.6 in (54.9 cm)	6.5 in (16.5 cm)	3.1 in (7.9 cm)	24.0 in (61.0 cm)	217 dB possible at 3.6 KHz
34A0400	400 Hz	211 dB	85 Hz	34.6 in (87.9 cm)	37.6 in (95.5 cm)	9.3 in (23.6 cm)	4.5 in (11.4 cm)	24.0 in (61.0 cm)	220 dB possible at 2.5 KHz
34A0610	610 Hz	213 dB	160 Hz	34.6 in (87.9 cm)	37.6 in (95.5 cm)	12.2 in (30.1 cm)	4.7 in (11.9 cm)	24.0 in (61.0 cm)	220 dB possible at 2.5 KHz
34A1000	1000 Hz	217 dB	147 Hz	34.6 in (87.9 cm)	37.6 in (95.5 cm)	19.4 in (49.3 cm)	5.0 in (12.7 cm)	24.0 in (61.0 cm)	
73A0155	155 Hz	211 dB	30 Hz	73.0 in (185 cm)	76.0 in (193 cm)	19.6 in (50.0 cm)	5.0 in (12.7 cm)	24.0 in (61.0 cm)	

Nominal technical data and design parameters provided for guidance only. Contact Sparton of Canada Ltd for specific requirements.

Notes:
* at nominal 30 m depth
† at 3000 V RMS drive
‡ at 3 dB points
§ Designed for shallow water applications.

The projector design is flexible and can be tailored to specific requirements for frequency, bandwidth and power. The resonant frequency can be adjusted for specific ring diameters, over a three octave band, by altering the shell curvature and thickness. Acoustic power for a given ring size is controlled by the ceramic ring cross-sectional area. Bandwidth is adjusted by modifying the acoustic radiating surface.

Passive depth compensation is effected through the use of a bladder housed within the inner of the RSP. The bladder is open to the sea and is allowed to expand or contract within the pressurised chamber of the RSP. Because of the depth compensating system and the pressurised air cavity, the depth capability of the RSP exceeds three times the range of one without these features.

The design of ringshell projectors by computer is the result of extensive finite element modelling, of which the accuracy and correlation with test results has been confirmed over several years.

Good thermal contact between sea water and the ceramic ring, both outside and inside the projector, allows the use of projectors for extended durations. The standard operating drive voltage of 3000 V provides a duty cycle of greater than 10 per cent.

The table in this entry lists some of the characteristics of units manufactured and delivered to various customers. These units tracked computer data with extremely good fidelity (model number codes list the diameter in inches as the first two digits and resonant frequency as the last four digits. Resonant frequencies are listed at 30 m depth and source levels are based on a 3000 V RMS drive).

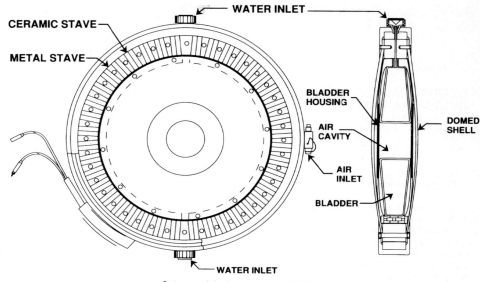

Cutaway of depth compensated RSP

Operational Status

In production. RSPs have been deployed from surface ships and submarines as hull-mounted or VDS transmitters, in single or multi-element arrays.

Contractor

Sparton of Canada Ltd, London, Ontario.

BARREL STAVE PROJECTOR (BSP)

Type

Acoustic projector.

Description

The BSP is a class III flextensional transducer capable of providing high power and broad bandwidth in a very small unit. The unit comprises a stack of PZT ceramic rings bolted to end plates. Staves mounted in an inverted barrel configuration are also attached to the end plates. The expansion and contraction of the stack is transmitted to the staves and results in an enhanced volume velocity. With diameters of 10 cm to 30 cm, BSPs are able to efficiently radiate at frequencies below 1 kHz. The transducer is ideal for air-deployable applications and for use in multi-element towed or volumetric arrays.

Operational Status

In development with Canadian Forces.

Contractor

Sparton of Canada Ltd, London, Ontario.

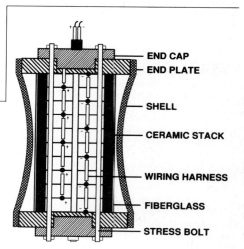

Cutaway of BSP

FREE FLOODING RING (FFR)

Type

Acoustic projector.

Description

The FFR consists of a ring of piezoelectric ceramics contained within a neoprene boot, which produce a toroidal beam pattern around the horizontal plane. The unit has exceptional properties for low frequency active sonar applications, very wide bandwidth ($Q~1$), very high efficiency (greater than 90 per cent), and unlimited depth capability. The FFR operates in the 500 Hz to 8 kHz frequency band and can produce source levels in excess of 220 dB per element at 1 kHz.

Operational Status

In production.

Contractor

Sparton of Canada Ltd, London, Ontario.

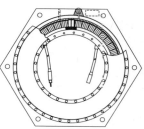

Cutaway of FFR

RESONANT PIPE PROJECTOR (RPP)

Type

Acoustic projector.

Description

The RPP consists of a ring of piezoelectric ceramics attached to the centre of a large pipe. The ring excites the pipe, which acts as an acoustic wave guide. Because of its free flooding design, the RPP has unlimited depth capability and can operate at or below 250 Hz. The unit was developed specifically for the oceanographic community and is being used as a source for SOFAR (Sound Ocean Fixing and Ranging) experiments.

Operational Status

In production.

Contractor

Sparton of Canada Ltd, London, Ontario.

Specifications

	Frequency range*	Source level†	Dry weight	Efficiency	Maximum operating depth	Cavitation depth	Bandwidth at Resonance‡	Outside diameter	Thickness
Free flooded ring									
13FA2000	2000-4000 Hz	216 dB		90%	Unlimited		2000 Hz	31 cm	15 cm
27FA200	1000-2000 Hz	222 dB		90%	Unlimited		1000 Hz	68 cm	
Resonant pipe projector									
28PA0260	260 Hz	205 dB	242 kg	75%	Unlimited	111 m	4 Hz	71 cm	178 cm
Barrel stave projector									
03BA1100	1100 Hz	191 dB	2 kg	70%	200 m		260 Hz	8 cm	18 cm
06BA0550	550 Hz	200 dB	10 kg	70%	250 m		110 Hz	8 cm	26 cm

Nominal technical data and design parameters provided for guidance only. Contact Sparton of Canada Ltd for specific requirements.

Notes: * at nominal 30 m depth
† at 3500 volts RMS drive
‡ at 3 dB points

MOVING COIL TRANSDUCER (MCT)

Type
ASW research and calibration device.

Description
The MCT has been developed to meet the requirement for high power and low frequency tones for the accurate calibration of towed arrays, sonobuoys and sonobuoy receivers, dipping sonar and fixed hydrophone arrays.

The main components of the system are: a towfish body which houses two acoustic transducers, sensor package and depth compensation system; tow cable; winch; and shipboard electronics and power amplifier. Two diametrically opposed moving coil transducers are housed within the towed body and driven out of phase and cooled using a fan incorporated into the MCT tow body. The MCT generates very low frequency, high power acoustic tones at various depths over a long period of time. These tones are received by the remote passive sensors under calibration. The depth compensation system can operate for over 50 hours in high sea states. Where lower powers are required the MCT can be used as a single element.

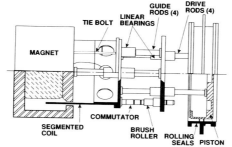

Cutaway of moving coil transducer

The multi-sensor monitoring package ensures all critical data on the towed body are available. In high power level applications a cooling system is provided to dissipate the heat produced in the moving coils.

Operational Status
In development.

Specifications
Standard
Length: 4.3 m
Width: 1.4 m
Height: 0.9 m

Weight: 2050 m
Frequency range: 3-500 Hz
Number of tones: 4
Harmonic content: less than -35 dB
Source level dB/μPa/m: 180-190
Tow speed (knots): up to 10 in Sea State 4; up to 7 in Sea State 5
Max sea state: Typically Sea State 5
Endurance (typical): 50 hours at Sea State 3
Turnaround time (mins): 45 for air replenishment
Depth: greater than 150 m (500 ft)
Depth compensation system: 8 tanks
Number of transducers: 2
Cooling system: Yes

Options
Length: 2.5 m
Width: 1.3 m
Height: 1.6 m
Weight: 1300 m~
Number of tones: 8
Source level dB/μPa/m: 170-180
Depth compression system: 4 tanks
Cooling system: No

Contractor
Sparton of Canada Ltd, London, Ontario.

SUBEX

Type
Acoustic noise augmenter.

Description
The SUBEX system has been designed to provide training for passive sonar operators in submarine detection, localisation and tracking techniques in actual submarine depths and speed, by imitating and augmenting the sound generated by submarines. The signal generated is used during naval training exercises to enhance or disguise the boat's radiated noise. The level and type of augmentation can be adjusted as the scenario requires. The standard SUBEX system comprises a Model 18SA0350 RSP housed within the sail of the submarine and a small, remote, lightweight display console which interfaces with an RSP controller. The RSP is a depth compensated, low frequency, wide bandwidth acoustic source.

The flexible mounting requirements of the standard SUBEX system enable it to be easily modified to fit most existing submarines with minimum intrusion in the main command centre area. An event recorder allows future recall of settings for comparison with other sensor information. Analogue inputs to the control unit allow the system to generate signals other than discrete tones. Lock-outs on the main control unit ensure only authorised use of the system.

Specifications
Number of tones: 4
Frequency range: 60-600 Hz (nominal)
Frequency separation: 1 Hz
Operational depth: 2-300 m

Contractor
Sparton of Canada Ltd, London, Ontario.

UNITED KINGDOM

FERRANTI ACOUSTIC TRANSDUCERS

Description
Ferranti-Thomson Sonar Systems is engaged in the production of a variety of transducers, arrays and a variety of acoustic devices.

The former Universal Sonar Ltd at Bridlington, Yorkshire specialises in the development and manufacture of a variety of underwater acoustic transducers, including towed arrays. The range includes passive arrays, self-noise monitoring hydrophones, parametric arrays, high resolution scanning arrays, intercept arrays and broadband countermeasures transducers. Many of the latest elements are modular devices designed to be configured into a variety of array shapes to meet individual needs.

The company is a leader in the application of materials technology to transducer design, with particular emphasis placed upon the development of specialist polymer and ceramic techniques to improve the performance, reliability and cost of sonar devices.

The former DBE Technology at Aldershot, Hampshire specialises in the design and manufacture of sonar systems and acoustic devices. These include large hull-mounted ship and submarine active/passive transducer arrays, as well as flank and towed arrays for the Royal Navy. The company manufactures a wide range of underwater transducers covering the frequency range from 1 Hz to 6 MHz, including barrel staves, piston stacks, tubular and spherical ceramic transceivers and omnidirectional wideband hydrophones.

Ferranti-Thomson Sonar Systems towed arrays being assembled

The company also develops and produces active and passive acoustic targets for a variety of customers, for calibration, fleet training and system testing functions. For minehunting, the group has produced high resolution sonars for target identification and classification, together with miniature marker beacons. New generation mine sensors have been developed.

Contractors
Ferranti-Thomson Sonar Systems UK Ltd, Stockport.

BAe FLEXTENSIONAL TRANSDUCERS

Type
Underwater acoustic transducer devices.

Description
British Aerospace has developed a flextensional transducer for underwater applications consisting of a piezoelectric ceramic stack operating in piston mode and enclosed in an elliptical shell. These transducers can be built for operation at low frequencies from 300 Hz to 3000 Hz, either from a range of standard designs, or to suit individual requirements. They can be assembled into staves or arrays to achieve specific directivity requirements. Research is continuing, both on new types of transducers, particularly those using magnetostrictive rare earth metals, and on increasing the depth performance of the present range of transducers.

Operational Status
BAeSema has supplied low frequency flextensional transducers to the US Naval Underwater Systems Center which have now been successfully evaluated. Similar transducers have also been supplied to France, Japan and Norway.

Contractor
BAeSema Marine Division, Bristol.

UNITED STATES OF AMERICA

LOW FREQUENCY TRANSMIT SUBSYSTEM (LTS)

Type
Flextensional transducer acoustic source array.

Description
The LTS is a large, hydrodynamically stabilised array of low frequency, active, high-powered projectors with associated power amplifiers, handling equipment, software, data processing and a unique transmit control beam former.

The LTS is claimed to be the largest flextensional system ever built. LTS forms part of the US Navy's SURTASS system, the total system including a towed hydrophone array, processing and display equipment, communications and navigation segments and a shore facility (qv).

Operational Status
Two LTS sets have been built under a contract issued in October 1990. One set is for technical evaluation and the other for operational evaluation. Hardware endurance testing commenced in May 1991 and was completed in the Summer of 1991. The first production module was delivered early in July 1991 and is undergoing test prior to the start of full-scale production.

Contractor
Lockheed Sanders ASW Division, Manchester, New Hampshire.

MODEL 30

Type
High-powered projector.

Description
The Model 30 is a high-powered, low frequency, broadband, flextensional transducer with an uncompensated operating pressure of 300 psi. The transducer weighs 96 lb in air with the coil, and has a minimum efficiency rating of 80 per cent.

Operational Status
The Model 30 has been used by the US Navy in a variety of exercises and operational tests.

Contractor
Lockheed Sanders ASW Division, Manchester, New Hampshire.

MODEL 40

Type
Flextensional transducer.

Description
The Model 40 flextensional transducer is a high power, low frequency, broadband flextensional transducer. It has an uncompensated operating pressure of 300 psi, weighs 185 lb in air with its coil, and has a minimum efficiency rating of 80 per cent.

Operational Status
The Model 40 has been used by the US Navy in a variety of exercises and operational tests.

Contractor
Lockheed Sanders ASW Division, Manchester, New Hampshire.

MODEL TC-12/34

Type
High power transducer.

Description
The Model TC-12/34 was developed as a ceramic version of the AN/UQN-1 sonar transducer used by the US Navy in depth measurement systems and is designed for use with bathymetric and near-bottom survey systems. The active element of the transducer is made up of an array of mechanically resonant, mass-loaded, lead-zirconate-titanate ceramic elements arranged to produce an effective radiating piston 9 in in diameter. Ceramic assemblies are contained within a castor oil filled cylindrical nickel aluminium bronze housing. A rubber boot provides acoustic coupling to the water.

Contractor
Ocean Data Equipment Corp, Fall River, Massachusetts.

MODEL TR-109

Type
High power transducer.

Description
The Model TR-109 transducer is designed for use in systems used in bathymetric and near sub-bottom surveys. It is a high power unit operating at a frequency of 3.5 kHz and 600 W maximum power input for up to 30 per cent duty cycle and 200 W maximum input for continuous operation.

In shallow water the maximum output is cavitation limited. Minimum depths of approximately 30 and 100 ft are recommended for output powers of 200 and 600 W respectively, to avoid cavitation.

The active radiating portion of the transducer is a circular piston, driven by lead-zirconate-titanate elements.

The modular design of the TR-109 facilitates its assembly into multiple arrays in any configuration to achieve the beam patterns and source levels desired. Arrays from 2 to 16 units are normally recommended, depending on the application required.

Contractor
Ocean Data Equipment Corp, Fall River, Massachusetts.

MISCELLANEOUS EQUIPMENT

GERMANY

SHORE-BASED INTELLIGENCE CENTRE

Type
Analysis system for sonar information.

Description
The Shore-Based Intelligence Centre (SOBIC) has been developed and manufactured by Atlas to enable naval forces to get the maximum use from the passive sonar sensors installed in surface ships, submarines, ASW aircraft, helicopters and shore-based facilities.

Intercepted noise and signals of target vessels not only reveal the bearing but also carry numerous items of information regarding the source of sound. This information can be extracted and understood only when it has been processed, evaluated and classified. To perform this SOBIC acts in a central role as follows:

(a) raw acoustic recordings from all available sensors are handed in to the SOBIC, accompanied by a detailed description of the classified source of noise

(b) the analysis personnel of SOBIC evaluate the recordings with all the different kinds of signals processing available, such as LOFAR (with different algorithms of pre- and post-FFT processing), DEMON, spectrum analysis, audio analysis, transient analysis and sonar signal analysis. The optimum output is the identification after correlation of the classified signals with information from other sources

(c) SOBIC establishes and updates an acoustic database, including all vessels possible in the operational area

(d) SOBIC distributes user adapted databases to all vessels, shore stations and training centres equipped with passive acoustic sensors

(e) feedback reports and recordings are used to provide continuous updating of the SOBIC-controlled database.

Operational Status
In operational service.

Contractor
Atlas Elektronik GmbH, Bremen.

NETHERLANDS

BATTERIES

Type
Batteries for mines, depth charges and sonobuoys.

Description
Reserve lithium batteries for mines and depth charges are hermetically sealed and provide a guaranteed storage period of 10 years and an expected storage period of 30 years. During storage the batteries can be left unattended. Several methods can be used for activation, which occurs in less than one second. They are fully operable in extreme temperatures ranging from −40°C to +63°C.

Operational Status
In production and in use for mines, fuzes and underwater applications.

Specifications
Nominal current: From mA to several A

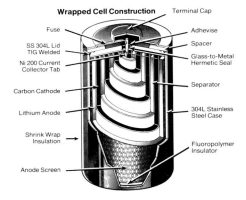

Wrapped Cell Construction — Terminal Cap
Fuse — Adhevise
SS 304L Lid TIG Welded — Spacer
Ni 200 Current Collector Tab — Glass-to-Metal Hermetic Seal
Carbon Cathode — Separator
Lithium Anode — 304L Stainless Steel Case
Shrink Wrap Insulation — Fluoropolymer Insulator
Anode Screen

Nominal voltage: 2.5 V to 40 V
Activation time: <1 s
Shelf life: >10 years
Operating temperature: −40°C to +63°C

Active lithium batteries use various different lithium chemistries to suit a wide range of requirements and applications. They have very high voltage per cell and can operate in extreme temperatures. Special designs are available for very high sustained current drains, including pulsed loads. The batteries are available in standard sizes and as battery packs of series and parallel combinations.

Operational Status
In production.

Specifications
Capacity: ranging from 0.8 to 30 Ah per cell*
Rated current: ranging from 1 mA to 4 A per cell*
Storage life: up to 10 years
Operating temperature: −40°C to +63°C per cell*
*depending on chemical system used.

Contractor
Signaal-USFA, Eindhoven.

SWEDEN

SAFE NET BARRIER

Type
Intruder security system.

Description
Developed in collaboration with the Swedish Navy, Safe Net Barrier is a physical alarm system designed to protect vital installations against both surface and underwater intruders. The system can be adapted to suit the anticipated type of intruders whether divers, underwater vehicles or small surface vessels.

The system consists of alarm and prevention nets, vertically adjustable gates which allow ships to pass, strong anchorings and a complete alarm system. Each net section is equipped with alarm functions activating a visual and/or audible warning.

The net consists of a series of panels comprising a continuous matrix of steel wire rope coated with thermoplastic. The knots are of solid polyurethane. Mesh sizes, wire diameters, bottom anchorage weights and flotation tubes can be varied to meet individual requirements depending on the anticipated threat.

If the net is penetrated using a mechanical cutter or explosives an alarm is automatically activated. If the protective covering around the steel wire rope is damaged so that contact occurs between the steel wire rope and the surrounding water, a first alarm is activated. If the steel wire rope is cut (or tampered with) a second alarm will be activated.

Regardless of the method, reaction time will be about the same. The time period between the alarm activating and the point when the intruder reaches the target will be adequate for other action to be taken against the intruder.

The gate is lowered by evacuating the air in the flotation tube. To lift the gate, compressed air is fed into the flotation tube. Constant pressure is automatically maintained in the tube. Operation can be tele-controlled.

If required, future enhancements may be implemented in the total defence system such as sonar, underwater TV cameras and so on.

If required another net can be added above water and connected to the underwater net and flotation tube.

Operational Status
The system is operational in Sweden and the first system has been supplied to the USA for the protection of a nuclear power plant.

Contractor
Safe Bridge Scandinavia AB, Vaestervik.

UNITED KINGDOM

EASAMS UNDERWATER ANALYSIS

Type
Range of underwater studies and analysis.

Description
EASAMS has been engaged in underwater engineering for a number of years and undertakes a wide range of studies in this field. This includes effectiveness of missions, vehicles, sensors, weapon detection, weapon evasion and attack profiles. Analysis of sensors and weapon assessment covers missiles and torpedoes; the development of sonar arrays, signal processing systems, sonar system performance and prediction, and reconstruction of operations and trials.

In the field of acoustic signal processing EASAMS offers a complete digital signal processing service which can involve project planning, collection of data, calibration analysis, and interpretation. The various processing systems are based on a wide range of processor technologies including DEC VAX, Hewlett Packard computers, SUN workstations and FPS array processors, plus real-time desktop processing based on a transputer network with associated DSP components.

Operational Status
In service.

Contractor
EASAMS Ltd, Camberley.

MIDAS (MARINE INCLINATION DIFFERENTIAL ALIGNMENT SYSTEM)

Type
Precision alignment system.

Description
MIDAS carries out alignment operations on ships afloat or other structures. It allows clinometers and theodolites to be used on a vessel in sheltered water and obviates the need for dry docking, and hence any possibility of any induced distortion of the vessel.

Reference surfaces can be accurately coplaned during tilt tests. Bearing references can also be transferred without line-of-sight, simply by heeling the vessel through a small angle in each direction. Recorders, printers, data loggers and repeater indicators can be added to enhance the basic system. A dedicated computer with software for a wide range of applications can also be supplied.

MIDAS comprises a central display unit carrying analogue and digital displays and function controls, to which are connected several tilt transducers which may be used independently or associated with conventional surveying and measuring instruments. The system may be specified with conventional force-balance transducers or for high resolution with Electrolevel tilt transducers. MIDAS uses a tilt sensor as a reference and compares other sensors to it. The ship's motion is therefore common to both and can virtually be discounted.

Operational Status
In service with Australian Defence Industries, and in France, Canada, Denmark and the UK.

Contractor
Tilt Measurement Ltd, Baldock.

SD-SCICON UNDERWATER TECHNOLOGY

Type
Processing and interpretation services.

Description
SD-Scicon is a computer systems and software services company with considerable work and services in the underwater scene. This includes studies into the effectiveness of ASW, analytic modelling support, development of weapon systems, wake homing studies, torpedo simulation programmes, and sonar system software design.

A new sonar detection and tracking system that uses transputer technology and artificial intelligence (AI) to assist in the analysis of data from towed arrays has been developed. This new technology is built into a product called SATAID (Sonar Auto-detection and Tracking, AI and Data fusion). Further developments planned for this programme include the actual identification of vessel types through the use of AI.

A Minewarfare Operational Analysis Tool (MOAT), programmed in FORTRAN 77 and hosted on the DEC VAX range of mini- and super mini-computers, has been developed to improve the effectiveness of MCM systems. The system provides facilities for the display and annotation of MCM vessels' operational plots. The operator may interactively display selected vessel positions, distances and bearings for time or event-based replay. Minefield data may be created, edited and plotted, and MOAT can perform MCM vessels' performance analysis using nominated mine targets. Performance data are displayed in either graphical form or as a tabular summary. A feature of MOAT is a package that permits the operator to recall a complete MCM operation on either a time or event basis.

Operational Status
In service.

Contractor
SD-Scicon, Fleet.

UNITED STATES OF AMERICA

COMBAT CONTROL CONCEPT EVALUATION SYSTEM (CCCES)

Type
Acoustic testbed for research and development.

Description
The CCCES is a technological testbed based on an innovative architecture that integrates acoustic sensor models, hardware and a comprehensive ASW platform database. The system is flexible and supports the rapid prototyping of a variety of ASW systems for engineering and algorithm development.

CCCES is designed for the development, evaluation, validation and demonstration of new ASW command and control concepts for both existing and future systems. It also supports the validation and demonstration of new concepts using recorded data or actual at-sea testing.

The hardware configuration is based on the VAX/VMS computer system. It can use a single VAX computer or multiple computers networked together without impacting the applications. Rapid prototyping of combat systems is achieved via the software structure. This structure provides processing 'shells' and a well-defined database architecture.

The man/machine interface consists of dual colour CRTs, a plasma panel display with numeric entry keyboard, fixed function buttons, a touch-screen overlay and trackerball encoders. A wide range of RAMTEK display devices can be supported and the CCCES system can be expanded to other display makes and models.

Operational Status
In service.

A technology testbed system for assisting in the development and rapid prototyping of advanced ASW combat control systems

Contractor
Raytheon Company, Submarine Signal Division, Portsmouth, Rhode Island.

Addenda

Mine Warfare

COMMAND AND CONTROL AND WEAPON CONTROL SYSTEMS
COMBAT INFORMATION SYSTEMS

ITALY

MINE DATA CENTRE

Type
MCM mission planning centre.

Description
The Mine Data Centre (MDC) has been developed to support the operational activities of MCM vessels.

It enables operators to plan MCM missions through the analysis of data stored in the system files and the extraction of information pertinent to the missions. MCM missions can also be evaluated through the analysis of information acquired during the MCM mission. It also enables MCM missions to be fully exploited by the updating of information stored in the system files.

The input to the centre consists primarily of data recorded by MCMVs during missions using their combat information system (in the Italian Navy the MM/SSN-714(V)2).

Input data are recorded both manually and automatically on magnetic support. Various types of outputs can be obtained including reports on missions, statistics, content of the database and planning information for future missions. Various output formats can be used including map display on graphic workstations or large screen projector, plots and listings.

Planning data are delivered to the MCMV on magnetic support, the data being directly input to the combat information system.

The system is based on a DEC mainframe (VAX 8250) equipped with disk and tape units and running the ORACLE Rdbms. User interaction is via graphic workstations (VAX stations) connected through Ethernet LAN, PCs and terminals. Peripherals include colour plotter, printers, digitiser and large screen projector.

All the application software is written in Ada, and it has been designed with the HOOD methodology; database design used the Entity-Relationship approach.

Operational Status
The system has been in use at the Italian Navy mines and mine countermeasures centre since July 1991 and was used to support Italian MCMVs deployed in the Persian Gulf.

Contractor
Datamat, Rome.

NORWAY

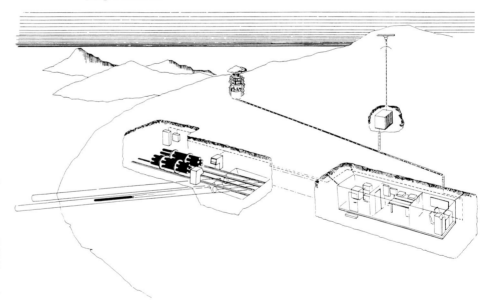

IDA

Type
Command, weapon control and information system for controllable minefields and torpedo batteries.

Development
In 1990 it was decided to undertake a comprehensive modernisation of existing minefields and torpedo batteries which form part of Norway's coastal defences. These comprise a series of fortresses which form an essential part of the country's anti-invasion defence system. The fortresses consist of a combination of controllable minefields, fortified torpedo batteries and artillery installations.

Description
The modernised minefields and torpedo batteries will have identical and completely new weapons control systems and sensors. The existing torpedoes and

Torpedo battery

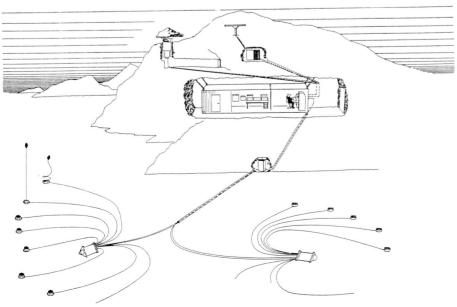

mines will be upgraded and new mines are to be developed.

The central part of the system is the operator console which is located in the fire control centre. The console accommodates two operators – one fire control operator and one sensor operator. The console is equipped with one colour raster scan monitor for display of the radar picture with computer-generated overlay, and one monochrome monitor for display of a TV or IR picture.

Operational functions such as target tracking, track correlation, threat evaluation and weapon allocation are fully automatic but provide for operator vetoes at certain preselected stages.

The sensor assembly consists of a radar and an electro-optical package which contains a laser rangefinder, IR camera and a daylight TV camera. All the sensors are remotely controlled from the fire control centre. Radar video is processed in an advanced radar data extractor prior to feeding the fire control computer.

The stationary parts of the sensors are housed in containers below ground. The electro-optical package

Controllable minefield

is fitted with a mushroom-shaped top cover which is elevated prior to use.

Signals and data are transmitted between the fire control centre and the sensors via fibre optic cables. This transmission also incorporates video from the radar and video from the IR and TV cameras.

The minefields consist of both ground and pop-up mines. Supervision and control are executed via a new concealed distribution system. The mines are operated individually or in groups. Extensive safety measures are employed in order to prevent unintended firings. These measures also imply new safety and arming units in each mine.

Microprocessor technology is applied to the torpedoes in order to improve torpedo guidance and incourse positioning. The torpedoes enter the water below sea level. A new loading and launching mechanism and wire guiding arrangement will reduce the reloading time and enable the launching of multiple salvos.

Operational Status

IDA is under contract for the Royal Norwegian Navy and is halfway through the development phase. A pilot system will be in operation by 1993.

Contractor

Norsk Forsvarsteknologi A/S, Kongsberg.

SONAR SYSTEMS

ROV SONARS

DENMARK

SEA BAT 6012

Type
ROV scanning sonar.

Description
The SEA BAT 6012 is a high definition, real-time electronically scanning sonar specifically designed for use in ROVs. The hydrophone array is mounted in a forward-looking configuration and is used to locate objects such as mines and swimmers at distances up to 100 m.

The system comprises a low-weight sonar head and a surface unit featuring a high resolution colour display and trackerball 'point and click' operation.

On-screen commands include three different colour ranges, grid on/off control, distance measurement, zoom and freeze frame. The display features 256 colours and high resolution output to an optional RGB monitor. Display recording is possible in either PAL or NTSC video format. An optional 3.5 in disk drive allows importing of images to word processors for documentation. The sonar scans a forward sector of 90° using 60 individual beams that are simultaneously updated seven times per second at the 100 m range and 30 times per second at the 10 m range. Seven range settings from 2.5 m to 200 m are available offering an actual increase in sonar display resolution. The sub-surface/surface links comprise uplink a black and white video channel and downlink a single RS232 channel. By averaging up to 16 frames/s random noise and clutter is reduced. Undistorted real-time images are displayed with the unit moving at up to 7 kts in water.

Specifications
Operating frequency: 450 kHz
Bandwidth: 20 kHz
Range settings: 2.5, 5, 20, 25, 50, 100, 200 m
Range resolution: 5 cm
Horizontal beamwidth: 1.5° (receive, each beam); 100° (transmit)
Vertical beamwidth: 15° (receive); 15° (transmit)

Contractor
Reson Systems A/S, Slangerup.

Analysis Tables

Italic type indicates that the equipment is neither in production nor in widespread operational use. Details of these systems can be found in previous editions of Jane's Weapon Systems.

Sonar Equipment

Note: This section covers surface ship, submarine and airborne sonar systems, including processors, sonobuoys and sonobuoy receivers

Designation	Description	Contractor(s)
AUSTRALIA		
Barra (SSQ-801)	Project Barra. RAAF/RAN advanced sonobuoy and airborne detection system	Sonobuoys Australia
Kariwara	Towed array sonar for surface ship/submarine use. In development	
Mulloka	Hull-mounted sonar for RAN	THORN EMI
SSQ-801	Barra system passive directional sonobuoy	Sonobuoys Australia
CANADA		
AN/AQS-503	*Helicopter acoustic processing system*	*CDC*
AN/UYS-503	Sonobuoy processor	CDC
AQA-801	Side processor for Barra sonobuoys	CDC
CMAS-36	Mine avoidance sonar	C-Tech
CSAS-80	Harbour surveillance sonar	C-Tech
CSS-80AS	Harbour surveillance sonar	C-Tech
CTS-36	LHMS omni-sonar	C-Tech
Hermes	Low-cost micro sonobuoy	Hermes
HS-1000	Lightweight search/attack sonar. Hull-mounted and towed versions	Westinghouse Canada
IPSS	Ice penetrating sensor system	Sparton/Contech
MAGIC 2	Multimode graphic and imaging system	CDC
Model 971/977	Colour imaging sonar for minehunting	Mesotech
Model 972	Side-scan sonar system	Mesotech
SCAN 500	Series of shallow water sonars for patrol boats	Westinghouse Canada
SQQ-504	Towing condition monitor	Westinghouse Canada
SQR-501	CANTASS (Canadian towed array sonar system) for surface ships	CDC
SQS-505	Medium search/attack sonar	Westinghouse Canada
SQS-507 (Helen)	*Lightweight variable depth towed sonar*	*Westinghouse Canada*
SQS-509	Lower frequency version of SQS-505	CAE
SQS-510	Digital sonar for surface ships	CDC
SSQ-36	Air-launched bathythermograph sonobuoy	Sparton
SSQ-517	*Passive sonobuoy*	*Sparton*
SSQ-518	*Passive sonobuoy. Long-life version of SSQ-517*	*Sparton*
SSQ-522	Active sonobuoy. Canadian version of SSQ-47	Sparton
SSQ-523	*CANCASS, Canadian command-active sonobuoy system*	*Sparton*
SSQ-527B	Improved version of SSQ-517B	Sparton
SSQ-529	*Directional sonobuoy*	*Sparton*
SSQ-530	Directional passive sonobuoy	Sparton
SSQ-531	*Directional sonobuoy*	*Sparton*
SSQ-536	Air-launched bathythermograph sonobuoy	Sparton
Type 5051	Surface ship attack sonar	Westinghouse Canada
UYS-501	Digital signal processor	CDC
UYQ-501(V)	Multi-sensor display system	CDC
FINLAND		
FERS 800	Sonar training system for surface ships	Elesco
FHS-700	*Fixed passive surveillance system*	*Elesco*
FRANCE		
Albatros	Torpedo alert system	Thomson Sintra
Amethyste	Version of the Eledone sonar	Thomson Sintra
Anaconda	Hull-mounted and tactical towed array series (TACTAS)	Thomson Sintra
Argonaut	Sonar for RN Type 2400 submarine (Eledone family)	Thomson Sintra
ATAS	Active towed array sonar	Thomson Sintra
DHAX-1	*Magnetic anomaly detection system*	*Crouzet*
DHAX-3	*Magnetic anomaly detection system*	*Crouzet*
Diodon	Surface vessel sonar. Active	Thomson Sintra
DSBV 61A	Surface ship towed linear array sonar	Thomson Sintra
DST A 3	*Active sonobuoy used with DSAA-4 system*	*Thomson Sintra*
DSTV-4M/DSTV-7Y	Passive sonobuoys	Thomson Sintra
DUAV-4	Helicopter active/passive sonar	Thomson Sintra
DUAV-18	*Helicopter sonar. Superseded by HS.70*	*Thomson Sintra*
DUBA-3A	*Surface vessel attack sonar*	*Thomson Sintra*
DUBA 25	Surface vessel sonar (Tarpon)	Thomson Sintra
DUBM-20A	*Active minehunting sonar*	*Thomson Sintra*
DUBM 21B	Mine countermeasures sonar, (IBIS). TSM 2019/2021/2022 variants also (TSM 2021)	Thomson Sintra
DUBM-40	*Active minehunting sonar, towed*	*Thomson Sintra*
DUBM-41B	High resolution side-looking sonar	Thomson Sintra
DUBM-42 (IBIS 42)	Side-looking sonar for minehunting	Thomson Sintra
DUBV 23D	Active surface vessel search/attack sonar	Thomson Sintra
DUBV-24C	*Low frequency panoramic search/attack sonar*	*Thomson Sintra*
DUBV 43B/C	Variable depth towed sonar. Active or passive. Used with DUBV 23	Thomson Sintra
DUBY-24C	*Active submarine sonar. Panoramic, sector or tracking modes*	*Thomson Sintra*
DUUA-1	*Submarine sonar. A, B and C versions*	*Thomson Sintra*
DUUA-2A	Simultaneous search and attack sonar for modernised 'Daphne' class submarines	Thomson Sintra
DUUA-2B	Development of DUUA-2A	Thomson Sintra
DUUX-2	*Passive submarine detection system*	*Thomson Sintra*

Designation	Description	Contractor(s)
FRANCE cont		
DUUX-5	Successor to DUUX-2. See Fenelon	Thomson Sintra
Eledone	Passive sonar for submarines	Thomson Sintra
Fenelon	Passive rangefinder for submarines	Thomson Sintra
FLASH	Helicopter dunking sonar	Thomson Sintra
HS 12	Helicopter version of SS 12 sonar	Thomson Sintra
HS 70	*Helicopter sonar*	*Thomson Sintra*
HS 71	Helicopter sonar	Thomson Sintra
HS 312S	Integrated ASW equipment for helicopters, including HS 12 and Lamparo processor	Thomson Sintra
IBIS 43	Towfish-mounted side-looking sonar system	Thomson Sintra
Lamparo	Airborne sonobuoy processing and display systems	Thomson Sintra
Lamproie (TSM 2930)	Linear towed array for surface ships and submarines	Thomson Sintra
MAD Mk III	Inboard magnetic anomaly detector	Sextant Avionique
Octopus	Version of the Eledone sonar	Thomson Sintra
Pascal	*Surveillance and tracking sonar for small and medium sized ships*	*Thomson Sintra*
Piranha (TSM 2140)	*Small ship attack sonar*	*Thomson Sintra*
Premo	*Panoramic search/attack sonar*	*Thomson Sintra*
Sadano	Airborne acoustic processing system	Thomson Sintra
Salmon	Hull-mounted sonar	Thomson Sintra
Scylla	Version of the Eledone sonar	Thomson Sintra
SLASM	High performance integrated sonar system	Thomson Sintra
SPDT-1A	Passive sonar for torpedo detection	Sextant Avionique
SQS-17A	*Panoramic sonar. Superseded by Premo*	*Thomson Sintra*
SS 12	*Panoramic sonar for small/medium ships*	*Thomson Sintra*
SS 24	*Panoramic sonar; more powerful version of SS 12*	*Thomson Sintra*
SS 24LF	*Passive panoramic sonar*	*Thomson Sintra*
SS 48	*Panoramic surface ship sonar*	*Thomson Sintra*
TSM 2019	Version of DUBM-21 minehunting family	Thomson Sintra
TSM 2021B	DUBM 21B minehunting sonar for Tripartite Minehunter (France, Netherlands, Belgium)	Thomson Sintra
TSM 2022	Derivative of TSM 2021 (above)	Thomson Sintra
TSM 2023	Hull-mounted, VDS or ROV sonar	Thomson Sintra
TSM 2054	Side-scan minehunting sonar	Thomson Sintra
TSM 2600	*Lightweight version of TSM 2630*	*Thomson Sintra*
TSM 2630	Improved version of Diodon sonar	Thomson Sintra
TSM 2633	Spherion version of TSM 2630	Thomson Sintra
TSM 2640	Passive sonar for patrol craft	Thomson Sintra
TSM 2820	Active panoramic sonar (Tarpon)	Thomson Sintra
TSM 8050	Active omnidirectional sonobuoy	Thomson Sintra
TSM 8200	Airborne sonar processing equipment	Thomson-CSF
Velox M7	Sonar intercept receiver	Safare Crouzet
GERMANY		
AIS	Submersible minehunting active sonar	Atlas Elektronik
AS 90 Series	Hull-mounted and variable depth sonar	Atlas Elektronik
ASES	Active harbour surveillance sonar	Atlas Elektronik
ASO4-2 Mod	Active sonar for small vessels. Hull-mounted	Atlas Elektronik
ASO 83-86	See DSQS-21 series	Atlas Elektronik
Bathysonde 2000	Oceanographic probe	Salzgitter
CSU 3	Passive/active/intercept sonar for submarines	Atlas Elektronik
CSU 83	Submarine active/passive sonar	Atlas Elektronik
CSU 90	Submarine attack, intercept and flank sonar	Atlas Elektronik
DSQS-11A	Hull-mounted mine avoidance sonar for minesweepers	Atlas Elektronik
DSQS-11H	Hull-mounted minehunting sonar	Atlas Elektronik
DSQS-21B/C/D	Series of active panoramic anti-submarine sonar for surface ships. Hull-mounted and towed versions	Atlas Elektronik
FAS-3	Submarine flank array sonar	Atlas Elektronik
OSID	Submarine integrated passive panoramic sonar	Atlas Elektronik
PRS 3	Passive ranging sonar for submarines	Atlas Elektronik
PSU 1-2	Passive submarine sonar	Atlas Elektronik
PSU-83	Submarine passive surveillance and detection sonar	Atlas Elektronik
SIP 3	Sonar processor for passive classification	Atlas Elektronik
TAS 83	Towed array sonar	Atlas Elektronik
TAS 90	Passive towed array sonar	Atlas Elektronik
TASS 3-2	Passive towed array sonar	Atlas Elektronik
INTERNATIONAL		
Bottle Cap	*Static surveillance systems*	*Saab/Simrad/Mesotech*
GETAS	Passive surveillance sonar for towed and bottom-laid array applications (GECO A/S provides 'wet end')	Marconi/GECO A/S
HELRAS	*Long-range helicopter dunking sonar*	*Bendix/BAe/FIAR*
Osprey	*Dipping sonar for helicopter use*	*Saunders/GEC Avionics/Thomson Sintra*
ITALY		
ASWAS	Underwater surveillance area system	WELSE
BI68	*Sonobuoy. Double number of RF channels*	*ELSAG*
BIR	Miniature sonobuoy for helicopter use	Servomeccanismi
BIT-3	Passive sonobuoy	Servomeccanismi
BIT-8	Passive sonobuoy	Servomeccanismi
FALCO	*Submarine detection and location system*	*ELSAG*
IP64/MD64	Submarine sonars	Selenia
IPD70/S	Integrated sonar for 'Sauro' submarines	Selenia
MD 100/S	Passive submarine sonar	Selenia
MLS/1A	*Sonar*	*ELSAG*
MSR-810	Passive sonobuoy	Whitehead
P2072	Upgraded version of GE AN/SSQ-14	FIAR
SARA	Spectral analysis and classification system	Selenia
SQQ-14IT	Mine detection and classification sonar	FIAR

Designation	Description	Contractor(s)
JAPAN		
QQS-4	**Solid-state, low frequency surface ship sonar**	
NETHERLANDS		
HSS-15	**Compact panoramic sonar for small ships**	**Signaal**
LW-30	**Passive sonar/intercept system**	**Signaal**
PHS-32	**Search/attack sonar**	**Signaal**
PHS-34	*Long-range panoramic sonar*	*Signaal*
PHS-36	Active panoramic sonar	Signaal
SIASS-14-2	Submarine attack and surveillance sonar	Signaal
SPI-04	Sound path ray analyser	Signaal
NORWAY		
SA 950	Mine detection and avoidance sonar	Simrad Subsea
SH 100	Mine detection and classification sonar	Simrad Subsea
SK3D	*Higher frequency version of SQ-D*	*Simrad Subsea*
SQ-D	*Hull-mounted sonar*	*Simrad Subsea*
SQ3D/SF	*Medium-range 'searchlight' sonar*	*Simrad Subsea*
SS105	Coastguard 360° scanning sonar	Simrad Subsea
SS240	Active omni multibeam sonar for small ships	Simrad Subsea
SS245	Active multibeam sonar	Simrad Subsea
SS304	Small ship scanning sonar	Simrad Subsea
SS575	Hull-mounted medium-range active sonar	Simrad Subsea
ST240 Toadfish	Dipping and variable depth sonar	Simrad Subsea
ST	*Bulkhead-mounted sonar*	*Simrad Subsea*
SU	*Bulkhead-mounted version of SU-R*	*Simrad Subsea*
SU-R	*Similar to SU-RS without transducer stabilisation*	*Simrad Subsea*
SU-RS	*Hull-mounted long-range sonar*	*Simrad Subsea*
UNION OF SOVEREIGN STATES		
Bull Horn	MF active/passive surface ship sonar	State
Bull Nose	MF bow sonar (surface ships)	State
Elk Tail	MF variable depth sonar	State
Fez	Underwater telephone system	State
Foal Tail	MF dipping sonar	State
Herkules	*HF scanning sonar for surface ships and submarines*	*State*
Horse Jaw	LF surface ship bow sonar	State
Horse Tail	LF variable depth sonar	State
Lamb Tail	MF dipping sonar	State
Mare Tail	MF variable depth sonar	State
Moose Jaw	LF hull sonar	State
Pegas	*HF searchlight sonar system*	*State*
Perch Gill	HF submarine searchlight attack sonar	State
Pike Jaw	Searchlight attack sonar - probably obsolescent	State
Rat Tail	HF variable depth and helicopter dipping sonar	State
Sail Plates	Believed to be intercept receiver transducers	State
Seal Skin	Submarine sonar	State
Shark Fin	Submarine medium frequency sonar	State
Shark Teeth	Submarine low frequency bow sonar	State
Steer Hide	Variable depth sonar	State
Tamir	*HF searchlight sonar - obsolescent*	*State*
Trout Cheek	Submarine passive array bow sonar	State
UNITED KINGDOM		
3000 Series	ROV and small boat minehunting sonars	Ulvertech
3025	Search sonar	Ulvertech
AA34030	Sonobuoy command transmitter	GEC Sensors
AD 130	Sonar homing and DF receiver	GEC Sensors
AQS 901	Acoustic data processing and display system	GEC Avionics
AQS 902/920	Lightweight acoustic processing and display system series	GEC Avionics
AQS 903/930	Acoustic processing and display system series	GEC Avionics
ARMS	Advanced remote minehunting system for ROVs	Marconi
ARR 901	Airborne sonar receiver	Marconi
AS360	Mechanically scanned sonar for use on ROVs	Marconi udi
AS370	Seabed surveillance sonar to detect swimmers and small submersibles	Marconi udi
AS380	Active and passive dunking sonar - can be displayed from ships or helicopters	Marconi udi
AS2105	Sonar stimulator	Marconi
AS2107	Training and test system	Marconi
ATAS	Active towed array sonar for small surface vessels	BAeSema
COMTASS	Compact towed array systems (COMTASS 1, 2 & 3)	Marconi
Cormorant	Passive/active helicopter sonar	Marconi
Dolphin 100	Processing and display upgrading equipment for submarines	Marconi
D100	Diver's ranging unit	Ulvertech
FIS 3	Integrated submarine sonar suite	Ferranti-Thomson Sonar
FMS Series	Family of active and passive sonar systems for surface vessels	Ferranti-Thomson Sonar
G738	Towed system for decoying active and passive homing torpedoes	Graseby
G750	Multipurpose all-round active/passive sonar for corvettes and above. Based on RN Type 184. Auto-tracking	Graseby
G768	Derived from G750 series for use in smaller ships	Graseby
G777	*Compact sonar for patrol craft down to 100 tons*	*Graseby*
G780	Passive sonar for very small submarines	Graseby
GS 7110	Multi-channel FSK receiver	Marconi
Guillemot	*Dipping sonar*	*Marconi*
Hi-Scan	Surveillance, navigation and target location sonars	Smiths
Hydra	Surface ship and submarine sonar systems	Marconi
Minescan	Mine detection and route surveillance sonar	Dowty

Designation	Description	Contractor(s)
UK cont		
Model 1256	Scanning sonar series	Marine Electronics
Model 3025	Search sonar for ROVs and harbour surveillance	Ulvertech
MOSAIC	*ASW avionics integration system*	*GEC Avionics*
MS70	Solid-state version of Type 193 minehunting sonar	Marconi
PMS 26/27	Lightweight search/attack sonar	Marconi
PMS 32	*Active panoramic sonar*	*Marconi*
PMS 35	*Small frigate digital sonar*	*Marconi*
PMS 40	Series of modular sonar systems	Marconi
PMS 56	Lightweight search/attack sonar - successor to PMS 26	Marconi
PMS 58	Minehunting sonar	Marconi
PMS 75	Side-scan sonar equipment	Marconi
PSC 8400	*Passive sonar surveillance system for coastal defence*	*Marconi*
R 612	Sonobuoy receiving set	Dowty
RASAT	Programmable universal sonar target	Dowty
SADE	*Sensitive Acoustic Detection Equipment. Intruder detection system*	*Marconi*
Sea Hunter	Medium-range sonar for small boats	Graseby
Sea Searcher	Active and passive ASW sonar	Graseby
Sonar 3000 Plus	Side-scan sonar developed from the Type 2034	Dowty
SP 2104	Airborne acoustic processor	GEC Avionics
SP2110	Passive towed array sonar	Marconi
SSQ-904A	Miniature passive (Jezebel) sonobuoy	Dowty/Marconi
SSQ-905A	Miniature passive sonobuoy	Marconi/Dowty
SSQ-906	Miniature passive sonobuoy	Dowty/Marconi
SSQ-907	Miniature passive sonobuoy	Marconi/Dowty
SSQ-937	'F' size version of SSQ-36	Dowty
SSQ-937A	Air-launched miniature bathythermal buoy	Marconi
SSQ-947B	Omnidirectional active sonobuoy	Dowty
SSQ-954B	'G' size miniature directional passive sonobuoy	Dowty
SSQ-963	Command Active Multi-Beam Sonobuoy	Dowty
SSQ-981	Second generation Barra sonobuoy derived from SSQ-801	Marconi
SSQ-991	Air-launched sonar transponder	Marconi
STS 3000	*Submarine integrated sonar system*	*Marconi*
STS 3100	*Surface ship/submarine towed array sonar*	*Marconi*
SWS-35	Miniature towed array sonar	Marconi
T17164	*Mk IC active sonobuoy*	*Marconi*
TATTIX	Tactical processing and display system	GEC Avionics
TG-1	High frequency sonar system	Ferranti-Thomson Sonar
Type 843	Sonobuoy signal receiver for helicopters	Marconi
Types 864, 865, 866	*Airborne sonobuoy telemetry systems*	*Marconi*
Type AN/ARR-XX	Sonobuoy data receiver	Dowty
Type T613	Sonobuoy data receiver	Dowty

A certain amount of confusion has existed in the past over the nomenclatures used by the Royal Navy for its acoustic equipment. The consolidated list printed below may help to clear up queries. Inevitably there are a few gaps in the list, due partly to security restrictions and partly to numbers allocated to systems which were cancelled or abandoned. The list covers surface, submarine and airborne sonars, as well as other acoustic equipment. The more recent systems are described in detail in the respective surface, submarine and airborne sonar sections. References and descriptions have been obtained from already published sources.

Type	Description	Prime contractor
162M	Surface ship high frequency, submarine bottom profile system using transducers in the bow area, port/starboard and keel	Kelvin Hughes
170	Short-range active attack sonar - normally associated with the Mortar Mk 10	Graseby
176	Passive protection sonar for surface ships	Graseby
177	Surface ship medium-range searchlight sonar	Graseby
182	Surface ship towed noise maker	Graseby
183	Emergency submarine underwater telephone fitted in each escape compartment and battery operated	Graseby
184	Medium-range active surface ship RDT/Omni sonar which preceded the Type 2016. Associated with the Ikara weapon system	Graseby
185	Medium-range underwater telephone - preceded the Type 2008. Used on submarines and some surface ships	Graseby
186	Submarine sonar system	THORN EMI
187	Passive/active sonar fitted to 'Oberon' class submarines - preceded the Type 2051. Can be integrated with the Type 719 and underwater telephones. Fitted to 'Oberon' class submarines not equipped with the Type 2051	THORN EMI
189	Cavitation intercept - fitted to all submarines	
189(P)	Portable Type 189 used for noise monitoring. Fitted to submarines and some surface ships	
193M	Short-range high definition active minehunting sonar for surface ships	Marconi
195	Medium-range active sonar - sector scanning helicopter dipping sonar - fitted to Sea King aircraft	Marconi
197	Active sonar intercept on 'Oberon' class submarines (Safare Crouzet Velox system) - preceded Type 2019. Direction and frequency sensors mounted on the fin	Safare Crouzet
199	Medium-range active variable depth sonar in use in Canadian, New Zealand, Australian and South African surface ships	THORN EMI
719	Short-range passive sonar on some 'Oberon' class submarines - being removed from all SSKs during refit	THORN EMI
776	Shallow echo sounder	Kelvin Hughes
778	Combined shallow/deep echo sounder	Kelvin Hughes
780	Shallow echo sounder	Kelvin Hughes
2001(AA)	Long-range low frequency active/passive sonar with a conformal array around the bow in the 'eyebrow' position. Fitted to 'Valiant' class submarines and SSBNs	Graseby
2001(BC)	Long-range low frequency active/passive sonar with a conformal array around the bow in the 'chin strap' position. Fitted to most 'Swiftsure' class submarines	Graseby
2004	Velocity meter fitted to all submarines - sound head mounted in the keel and fin	Graseby
2007	Low frequency broadband passive sonar with two flank arrays. Each array contains 24 hydrophones. Fitted to all submarines except 'Valiant' class	Graseby

Type	Description	Prime contractor
UK cont		
2008	Medium-range active/passive underwater telephone system with three fixed transducers, and two small trainable transducer arrays in the fin of the submarines. Fitted to surface ships and submarines	Marconi
2009	Acoustic recognition (IFF) integrated with Types 2008 and 185. Transmits challenge and reply in six-tone combinations. Fitted to some submarines and surface ships	Marconi
2010	Acoustic rapid automatic teletype (RATT) integrated with Type 2008 underwater telephone. Fitted to some surface ships and submarines	
2014	Acoustic ray tracer indicator producing ray path plot which can be photographed with a polaroid camera. Fitted to some submarines and surface ships	
2015	Surface ship expendable bathythermograph	Marconi
2016	Medium-range active sonar - three selectable transmission frequencies. Superseded Type 184 and preceded Type 2050. Fitted to aircraft carriers, and Type 22 and Batch 3A 'Leander' class frigates	Marconi
2017	LOFAR/DEMON frequency analyser using a drum correlator. Was fitted to submarines but now phased out	
2018	Submarine fitted DEMON frequency analyser (0 to 80 Hz) - stylus driven with electro-sensitive paper. Previously known as 'Diana'. Can also be used with a Freelator	
2019	Active sonar intercept, commonly known as PARIS (Passive/Active Range and Intercept Sonar). Joint Anglo/French/Dutch project. Submarines but not 'Oberon' class	Thomson Sintra/Marconi
2020	Low frequency active/passive computer-assisted sonar. Superseded 2001(BC). Fitted to 'Trafalgar' class and two 'Swiftsure' submarines. Being updated (2020 EX)	Marconi/Ferranti-Thomson Sonar
2023	SSBN reeled towed array. RN nomenclature for the United States AN/BQR-15	Western Electric
2024	Clip-on towed array for submarines. A Type 2024 suite combines Type 2035 and the towed array	Marconi
2026	Double octave towed array system which encompasses both towed array and inboard processing. 'Trafalgar' and 'Upholder' class submarines	Marconi
2027	Submarine passive ranging system which uses 2001/2020 array	Marconi
2028	Surface ship sonar	Graseby
2030	Processor using waterfall and A-scan display formats. US in origin (AN/BQR-22)	Spectral Dynamics
2031(I)	Reeled towed array (five-octave) system	Marconi
2031(Z)	As 2031(I) but uses 'Curtis technology'	Dowty
2031(Y)	2031(Z) testbed	Dowty
2032	Consists of nominally 15 de-tuned hydrophones within the main bow array (2001/2020) to provide a low frequency response - bow array narrowband capability. Fitted to most SSBNs	Ferranti-Thomson Sonar
2033	Hydrographic survey sonar equipment	Marconi
2034	Short range side-scan bottom profiling system. Fitted to hydrographic survey ships	Dowty
2035	A 16-channel waterfall display processor which forms the processing part of the Type 2024 suite. US in origin (AN/BQR-23)	AT&T
2038	Cancelled submarine towed array (superseded by Type 2046)	
2040	Combined active/passive/intercept suite, also known as 'Argonaut'. For 'Upholder' class submarines	Thomson Sintra
2041	Passive ranging system for 'Upholder' class submarines	
2042	Self-protection system for submarines (probably decoy system)	
2043, 2044, 2045	Bow, towed array and sonar intercept parts of Type 2054 sonar. These designations are no longer in use	Marconi
2046	Submarine 3-octave narrowband sonar system combining both towed array and inboard processing (clip-on towed array)	Ferranti-Thomson Sonar
2047	12-channel narrowband processor - replaces Type 2035	Marconi
2047(AC)	Submarine 16-channel narrowband processor, associated with a Type 2024 towed array	Marconi
2048	Add-on electronics for the Type 193M	Marconi
2050	Surface ship medium-range active sonar system with improved passive broadband and narrowband capability. Replaces the Type 2016. Type 42 destroyers, and Types 22 and 23 frigates	Ferranti-Thomson Sonar
2051	Passive/active/intercept narrowband sonar suite for surface ships	Marconi
2052	Submarine passive broadband sonar associated to Type 2024 array fitted units which do not have a Type 2007	Marconi
2053	Hydrographic side-scan sonar for survey ships?	Dowty
2054	Submarine combined sonar suite for 'Vanguard' class	Marconi
2057	Low frequency passive towed array system for Type 23 frigates	Ferranti-Thomson Sonar
2058	Surface ship towed LF noise projector sonar for calibration purposes	THORN EMI
2059	Acoustic tracking system for PAP 104 ROV	EDO
2060	Surface ship expendable bathythermograph system replacing Type 2015	
2061	SSBN submarine interim narrowband sonar processing system which is being replaced by the Type 2046	Dowty
2062	SSBN submarine interim towed array system being replaced by the Type 2046	
2063	Submarine noise vibration measuring equipment	Ferranti-Thomson Sonar(?)
2064	Minehunter tracking system?	
2065	Experimental reeled wet end thin-line towed array	
2066	Surface ship anti-torpedo countermeasure	Dowty
2067	Underwater tracking system (noise source mounted on submarines)	THORN EMI
2068	Surface ship environmental sonar ray path range prediction system	Dowty-SEMA
2069	Modernised Type 195M helicopter dunking sonar	Marconi/Ferranti-Thomson Sonar
2071	Noise decoy system fitted to submarines	Gearing & Watson
2072	Submarine flank array processor	Marconi
2074	Long-range active/passive sonar to replace Type 2001	Marconi
2075	Submarine active/passive/intercept/underwater telephone broadband and narrowband sonar to replace Type 2040. Cancelled	Ferranti-Thomson Sonar
2076	Mid-life update for 'Trafalgar' class submarines	Ferranti-Thomson Sonar/Marconi
2077	HF sonar system for nuclear submarines (probably underside obstacle avoidance sonar)	Marconi
2078	Comprehensive SSN 20 future sonar suite	
2080	LF active sonar system	Joint UK/French study
2082	Passive intercept	THORN EMI
2093	Surface ship high frequency minehunting sonar	Marconi

Designation	Description	Contractor(s)
UNITED STATES OF AMERICA		
ALFS	New helicopter dunking sonar	Hughes/Thomson Sintra
AMSS	Advanced minehunting sonar system development	Raytheon/Thomson-CSF

Designation	Description	Contractor(s)
USA cont		
AQS-13	Helicopter sonar	Bendix
AQS-14	Helicopter towed side-looking sonar for mine clearance	Westinghouse
AQS-18	Helicopter panoramic sonar	Bendix
ARR-72	Sonobuoy receiving system	Flightline Electronics
ARR-75	Sonobuoy receiving system	Flightline Electronics
ARR-76	Sonobuoy receiving system	Dowty
ARR-84	Airborne and surface vessel sonobuoy receivers	Flightline Electronics
ARR-146	Airborne sonobuoy receiver	Flightline Electronics
ARR-502	Airborne sonobuoy receiver	Flightline Electronics
ASQ-208(V)	MAD system	Texas Instruments
BQG-1/4	Submarine passive fire control sonars	Sperry/Raytheon
BQQ-1	Search and fire control sonars	Raytheon
BQQ-2	Sonar for SUBROC system	Raytheon
BQQ-5	Nuclear attack submarine sonar	IBM/Raytheon
BQQ-6	Active/passive sonar for Trident submarines	IBM
BQR-2	Submarine passive sonar	EDO/Raytheon
BQR-3	Submarine passive sonar	Raytheon
BQR-7	Passive sonar. Part of BQQ-2 system. Fitted FBM submarines	EDO/Raytheon
BQR-15	Towed submarine sonar. Fitted FBM submarines	Western Electric
BQR-19	Submarine sonar	Raytheon
BQR-21	Submarine passive detection and tracking set (DIMUS). Fitted FBM submarines	Honeywell
BQS-4	Adds active sonar capability to BQR-2B	EDO
BQS-6	Active submarine sonar. Part of BQQ-2 system	Raytheon
BQS-8	Under-ice navigation sonar	Hazeltine/EDO
BQS-13	Submarine search sonar. Passive/active	Raytheon
BQS-14A	Submarine navigation and mine detection sonar for under-ice passage	Hazeltine
BQS-15	Submarine sonar for mine detection in heavily mined areas. Active and passive	Amtek
BRT-1	*Sonar radio transmitting buoy*	*Sparton*
BSY-1(V)	Sonar/fire control integrated combat system	IBM
BSY-2	Sonar/fire control integrated combat system	General Electric
DE1164	SQS-56 plus variable depth sonar	Raytheon
DE1167	Small ship sonar based on SQS-56	Raytheon
DE1191	Upgrade suite for SSQ-23 and SQS-23 sonars	Raytheon
Dwarf	Reduced length, standard diameter miniature passive sonobuoy	Sparton
Dwarf DIFAR	DIFAR version of Dwarf	Sparton
Dwarf DIFAR(VLAD)	Vertical line array DIFAR version of Dwarf	Sparton
Dwarf omni	*Dwarf omnidirectional passive sonobuoy*	*Sparton*
FDS	Static surveillance system (Fixed Distribution System)	
HIPAS	Helicopter active/passive dunking sonar	Martin Marietta
Hydroscan	Towed side-scan seabed mapping sonar	Klein
MicroPUFFS	Submarine passive ranging sonar	Sperry
Mk 24 Mod 0	Underwater ordnance location sonar	Klein
Model 258	ROV scanning sonar	Ketema
Model 260	Dual frequency side-scan sonar	EG & G Marine Instruments
Model 384A	Diver's sonar system	EDO
Model 984	Precision altitude sonar system	EDO
Model 1110	Passive flank array system	EDO
Model 1550	Obstacle avoidance sonar	Ketema
Model 4200	Obstacle detection sonar	EDO
Model 4235	Short-range Doppler sonar for ROVs	EDO
Model 4235A	ROV Doppler sonar	EDO
Model 6410	Hand-held diver sonar	Helle
Moored sensor buoy	Surveillance sonobuoy	Sparton
Pilot Fish	Static seabed sonar system	
PQS-2A	Hand-held active/passive diver sonar	General Instrument Corp
RDSS	Seabed deployable sonar system	
Seaprobe 120	Static surveillance system	Ketema
SOSUS	Large-scale fixed sonar surveillance system	AT&T
SQA-10	Variable depth sonar	Litton
SQA-13	Variable depth sonar	EDO/Litton
SQA-14	'Searchlight' sonar	Raytheon
SQA-16	'Searchlight' sonar	Raytheon
SQA-19	Variable depth sonar	Litton
SQG-1	Anti-submarine attack sonar	Raytheon
SQQ-14	Minehunting and classification sonar	General Electric
SQQ-23	Sonar for A/S patrol ships	
SQQ-30	Minehunting sonar	General Electric
SQQ-32	Minehunting sonar	Raytheon
SQR-14	Surface sonar	
SQR-15	Surface ship towed array sonar	Martin Marietta
SQR-17A	Four-channel processor forming part of LAMPS Mk I	Diagnostic/Retrieval
SQR-18A(V)	TACTAS - Tactical Towed Array Sonar	EDO
SQR-19	Tactical towed array sonar	Martin Marietta
SQS-4	Short-range active sonar	Sangamo/GE
SQS-23	Long-range active sonar	Sangamo
SQS-26	Bow-mounted, 'bottom bounce' mode sonar to replace SQS-23	EDO/General Electric
SQS-29/32	Surface vessel active sonars. Numbers relate to differing frequencies. 'B' models are associated with variable depth sonar	Sangamo
SQS-35	Variable depth sonar	EDO
SQS-36	Medium-range hull sonar	EDO
SQS-38	Medium-range hull-mounted version of SQS-35	EDO
SQS-53	Development of SQS-26	General Electric
SQS-56	Lightweight sonar for USN PF (Patrol Frigate) ships and other navies (DE1160B)	Raytheon/Sparton
SSQ-36	*Bathythermograph sounding buoy*	*Sparton*
SSQ-41B	Sonobuoy, passive	Magnavox/Sparton

Designation	Description	Contractor(s)
USA cont		
SSQ-47B	Sonobuoy, active	Sparton
SSQ-50	*CASS - Command Active Sonobuoy System*	*Sparton*
SSQ-53A	Sonobuoy, passive directional (DIFAR)	Sparton/Magnavox/Sanders
SSQ-53B	Successor to SSQ-53A	Sparton
SSQ-53C	Dwarf version of SSQ-53B	Sparton
SSQ-53D	Improved version of SSQ-53B	Sparton
SSQ-57A	Sound reference sonobuoy	Sparton
SSQ-58A	Moored sensor buoy	Sparton
SSQ-62B	Sonobuoy, directional version of SSQ-50 (DICASS)	Sparton/Raytheon/Magnavox
SSQ-71	ATAC, air transportable acoustic communication buoy	Sparton/Sanders
SSQ-75	Long-range, low frequency active sonobuoy	Allied/ERAPSCO
SSQ-77A	VLAD - vertical line array DIFAR passive sonobuoy	Sparton/Magnavox
SSQ-77B	Passive directional search and surveillance sonobuoy	Sparton
SSQ-79	Steered vertical line array sonobuoy. Development	Hazeltine
SSQ-86	Downlink communications sonobuoy	Sparton
SSQ-102	Tactical surveillance sonobuoy	Magnavox/Sippican/Hazeltine
SSQ-103	Low-cost passive sonobuoy	Sippican
SST	Advanced sonar standard transmitter	Raytheon
STRAP	Experimental programme to use sonobuoys for beam forming	US Navy/Lockheed/Magnavox
SURTASS	Mobile towed array sonar	Hughes/TRW
TB-16	Towed array	Martin Marietta
TB-23	Thin-line towed array	Martin Marietta
TVLAD	Tuned vertical array omni sonobuoy	Sparton
UQS-2	Minehunting sonar	General Electric
Widetrac	*Sonar communications buoy*	*Sparton*
610	Long-range hull sonar	EDO
700 Series	Medium-range. Hull and variable depth versions	EDO
780 Series	Variable depth sonar. 13 kHz	EDO
786 Series	Hull-mounted sonar. 13 kHz	EDO
795 Series	Hull-mounted sonar. 5 kHz	EDO
796 Series	Hull-mounted sonar. 7 kHz	EDO
900 Series	*Submarine mine avoidance*	*EDO*
910 Series	*Surface ship mine avoidance*	*EDO*
1102/1105 Series	*Submarine active/passive medium range*	*EDO*
7860 Series	Combined 13 kHz hull sonar with 13 kHz VDS	EDO
7950 Series	Combined 5 kHz hull sonar with 13 kHz VDS	EDO
7960 Series	Combined 7 kHz hull sonar with 13 kHz VDS	EDO

Torpedoes

Designation	Dimensions (d × l)	Description	Contractor(s)
FRANCE			
E14	550 mm × 4.29 m	Passive acoustic homing. Contact and magnetic fuze	DCN/CIT-ALCATEL
E15	550 mm × 6 m	As E14 but 300 kg charge instead of 200 kg	DCN/CIT-ALCATEL
F17	533 mm × 5.9 m	Wire-guided or automatic homing. Surface or submarine targets. F17P multimode version	DCN
L3	550 mm × 4.32 m	Active acoustic homing. Contact and magnetic	DCN/CIT-ALCATEL
L3 Mod 1	533 mm × 4.3 m	fuze. 200 kg charge 21 in version available for manufacture but not in production	DCN/CIT-ALCATEL
L4	533 mm × 3.13 m	Circular search. Active acoustic homing. Contact and acoustic proximity fuze. Air-launched. Used in Malafon	DCN
L5 Mod 1	533 mm × 4.4 m	Active/passive acoustic homing. Direct attack programmed search. Surface launch weight 1000 kg	DCN
L5 Mod 3	533 mm × 4.4 m	As Mod 1 but submarine launch weight 1300 kg	DCN
L5 Mod 4	533 mm × 4.4 m	Anti-submarine model	DCN
L5 Mod 4P	533 mm × 4.4 m	Multi-purpose model	DCN
Z16	550 mm × 7.2 m	Preset course and depth. Runs zig-zag pattern if no target encountered after predetermined distance run. Magnetic and contact fuze. 300 kg charge. Obsolescent	
Murène	324 mm × 2.96 m	Lightweight torpedo for air- and surface-launch	DCN/ECAN
GERMANY			
DM2A3	533 mm	Anti-submarine and anti-surface wire-guided heavyweight torpedo	STN
DM2A4	533 mm	Planned development of DM2A3	
Seal	533 mm × 6.15 m	Wire-guided heavy surface target torpedo. Surface launch. Magnetic and impact fuzes. 260 kg warhead	STN/Atlas Elektronik
Seeschlange	533 mm × 6.08 m	Wire-guided, surface launch, submarine targets. 100 kg warhead	STN/Atlas Elektronik
SST-4	533 mm × 6.5 m	Wire-guided; active/passive acoustic homing. Torpedo sonar linked to shipborne FCS. 260 kg warhead	STN/Atlas Elektronik
SUT	533 mm × 6.1 m	Wire-guided; active/passive acoustic homing. Similar to SST-4 but dual-purpose	STN/Atlas Elektronik
ITALY			
A.184	533 mm × 6 m	Wire-guided, automatic acoustic homing	Whitehead
A.244	324 mm × 2.7 m	Acoustic homing. Shallow water and anti-reverberation capabilities	Whitehead
A.244/S	324 mm × 2.7 m	A.244 with more sophisticated CIACIO homing head	Whitehead
A290		Lightweight anti-submarine homing torpedo	Whitehead
JAPAN			
GRX-2		Future development	
GRX-3		Torpedo in development	
SWEDEN			
Type 41	*400 mm × 2.44 m*	*Passive homing in azimuth and depth. Shallow water capability. Impact and proximity fuze. Electric propulsion. All-up weight 250 kg*	*Förenade Fabriksverken (FFV)*

Designation	Dimensions (d × l)	Description	Contractor(s)
Sweden cont			
Type 42	400 mm × 2.44 m	Successor to Type 41 with helicopter-launch capability and optional wire guidance (length then 2.6 m)	Swedish Ordnance
TP 43XO	400 mm × 2.645 m	Lightweight ASW torpedo	Swedish Ordnance
Torpedo 43X2	400 mm × 2.8 m	Multi-purpose lightweight torpedo	Swedish Ordnance
TP 421/422	–	Swedish Navy versions of Type 42	Swedish Ordnance
TP 427	–	Export version of Type 42	Swedish Ordnance
TP 61	533 mm × 6.98 m	Long-range heavyweight passive homing	Swedish Ordnance
TP 613	–	Swedish Navy version of Type 61	Swedish Ordnance
TP 617	–	Export version of Type 61	Swedish Ordnance
Torpedo 2000	533 mm × 5.75 m	Long-range active/passive heavyweight	Swedish Ordnance
UNION OF SOVEREIGN STATES			
21 in Torpedo	21 in (533 mm)	Standard fit for submarines. Alternatives available for surface vessel launching	State
Airborne Torpedoes	–	–	State
Heavyweight Torpedoes	–	–	State
Light Torpedoes	406 mm × 5 m approx	Training deck launchers	State
Type 65	660 mm × 9.144 m	Reported large diameter torpedo	State
UNITED KINGDOM			
Mk 8	*533 mm × 6.7 m*	*1930s design. Compressed air propulsion, free-running. In RN service until 1973.*	*UK MoD*
Mk 24 Mod 0	533 mm	Wire-guided active/passive acoustic homing	UK MoD
Tigerfish (Mk 24)	533 mm × 6.46 m	Long-range wire-guided active/passive acoustic homing. Anti-surface, anti-submarine. Computer allows autonomous homing and re-attack. Entered RN service 1980; Mod 2 operational 1986	Marconi
Spearfish	533 mm × 6 m approx	Advanced heavyweight sub-launched torpedo. Entering service	Marconi
Stingray	324 mm × 2.6 m	Lightweight air- and surface-launched torpedo. In service with the Royal Navy	Marconi
UNITED STATES OF AMERICA			
Mk 14	*533 mm × 5.25 m*	*1930s design. Free-running with preset depth and course angles. Compressed air propulsion. Weight 1780 kg. Withdrawn from USN service in 1973 but still in service elsewhere*	
Mk 27 Mod 4	*483 mm × 3.23 m*	*Passive acoustic homing. Electric propulsion. Used in USN service as training round prior to Mk 37 introduction but some warshot torpedoes sold elsewhere*	
Mk 32	*483 mm × 2.08 m*	*Acoustic anti-submarine. Operational but obsolescent*	
Mk 37 Mod 0	483 mm × 3.52 m	Active/passive acoustic homing. Electric dual speed propulsion. Warshot weight 648 kg. Mod 3 is improved version	Westinghouse
Mk 37 Mod 1	483 mm × 4.09 m	Wire-guided. Electric dual speed propulsion. Warshot weight 766 kg. Mod 2 is improved version	Westinghouse
Mk 37 Mod 2	483 mm × 4.09 m	Updated version (minor modifications) of Mod 1	Westinghouse
Mk 37 Mod 3	483 mm × 3.52 m	Updated version (minor modifications) of Mod 0	Westinghouse
NT 37C	*483 mm × 3.52 m*	*Improved versions of Mk 37 Mods 2 and 3. Better speed,*	*Northrop/Honeywell*
Mods 2 & 3	*483 mm × 4.09 m*	*range, and acoustic performance including shallow water capacity. Sold to Canada and Netherlands*	
NT 37D	*483 mm × 4.5 m (Mod 2) 483 mm × 3.8 m (Mod 3)*	*Further improvement of NT 37 with Mk 46 OTTO engine and enhanced guidance*	*Honeywell*
NT 37E	483 mm × 4.5 m	Latest modernised model of Mk 37	Alliant Techsystems
NT 37F	450 mm × 3.84 m	Upgraded thermochemical propulsion and new long-range sonar	Alliant Techsystems
Mk 44	324 mm × 2.56 m	2 versions differing slightly in length and both weighing about 233 kg. Active acoustic homing. Electric propulsion. Replaced by Mk 46 in US and UK service but still in service elsewhere	Several in USA, also licence-built overseas
Mk 46 Mod 0	324 mm × 2.67 m	Deep-diving, high speed, active/passive acoustic homing. Weight about 258 kg. First US torpedo with solid-fuel propulsion	Aerojet-General and others
Mk 46 Mod 1	324 mm × 2.59 m	As Mod 0 but slightly larger and with liquid mono-propellant (OTTO) motor	Aerojet-General and others
CAPTOR	324 mm × 3.7 m	Mk 46 Mod 4 torpedo inserted in a mine casing and released when target detected	Goodyear
Mk 48 Mod 0	533 mm	AS-only version. Superseded during development by Mod 2	Westinghouse
Mk 48 Mod 1	533 mm × 5.8 m	Deep-diving (914 m) high speed (93 km/h) active/passive acoustic homing, wire-guided, long-range (46 km) weapon. Weight about 1600 kg. Said to be most complex torpedo ever designed	Gould (formerly Clevite)
Mk 48 Mod 2	21 in (533 mm)	Competing against Mod 1 which was selected after comparison at pilot production stage	Westinghouse
Mk 50	324 mm × 2.896 m	Formerly EX-50 Advanced Lightweight Torpedo (ALWT). Successor to Mk 46 (NEARTIP). In initial production phase	Honeywell
Freedom Torpedo	*19 in × 4.83 mm*	*Private-venture development. Wire-guided, pattern-running, electrically propelled. Contact fuze. Warshot weight 1237 kg. Charge 295 kg. Range 11 km. Adapts to 21 in tubes*	*Westinghouse*

Mines, Depth Charges and Underwater ASW Weapons

Designation	Description	Contractor(s)
AUSTRALIA		
Branik	Version of Ikara for Brazilian frigates	BAeSema
Ikara	Torpedo carrying missile system	BAeSema
Super Ikara	Torpedo carrying missile system - development has now ceased	BAeSema
BRAZIL		
MCF-100	Moored contact mine	Consub

Designation	Description	Contractor(s)
CHILE		
AS-228	Air- and/or surface-launched depth charge	Cardoen
MS-C	Magnetic influence ground mine	FAMAE
MS-L	Magnetic ground mine	FAMAE
CHINA		
CY-1	Anti-submarine surface-launched ballistic missile	China Precision Machinery
DENMARK		
MTP-19	Cable controlled mine	Nea-Lindberg
FRANCE		
Malafon	Torpedo carrying winged vehicle	Latecoere
SM 39	Submarine-launched Exocet	Aerospatiale
SM Polyphem	Submarine self-defence missile system against ASW aircraft	Euromissile
TSM 3510 (MCC 23)	Ground influence mine	Thomson Sintra
TSM 3530 (MCT 15)	Ground influence mine	Thomson Sintra
GERMANY		
DMT 211/221	Anti-frogman depth charge and underwater signal charge	Rheinmetall
FG 1	Surface- or submarine-launched ground influence mine	Faun-Hag
SAT/AIM	Anti-invasion mine	STN
SGM-80	Ground influence mine	Atlas Elektronik
SM G2	Ground influence mine	Atlas Elektronik
INTERNATIONAL		
Milas	Torpedo carrying missile system	OTO Melara/Matra
IRAQ		
Al Kaakaa/16	Floodable submersible mine	
SAGEEL/400	Ground mine	
ITALY		
ERP/2.5	Small anti-frogmen/limpet mine	Tecnovar
Manta	Shallow water ground influence mine	Whitehead
MAL/17	Moored anti-invasion mine	Tecnovar
MAS/22	Ground anti-invasion mine	Tecnovar
MP-80	General-purpose ground mine	Whitehead
MR-80	Ground influence mine	Whitehead
MRP	General-purpose ground influence mine	Whitehead
Seppia	Moored influence mine	Whitehead
NORWAY		
Terne III	Rocket-propelled depth charge	
SWEDEN		
Bunny	Ground influence mine	NobelTech
Elma	Lightweight anti-submarine projectile	Saab
GMI 100	Anti-invasion ground influence mine	NobelTech
MMI 80	Moored influence mine	NobelTech
SAM 204	Air-launched depth charge	SA Marine
Type 375	ASW rocket system	NobelTech
TAIWAN		
M89A2	Magnetic limpet mine	Military Technology Consulting
UNION OF SOVEREIGN STATES		
AMD 500	Bottom influence mine	State
AMD 1000	Bottom influence mine	State
AMG-1	Contact mine, chemical horn	State
Cluster Bay	Rising influence mine, acoustic	State
Cluster Gulf	Rising influence mine, acoustic	State
FRAS-1	Anti-submarine rocket	State
KMD 500	Bottom influence mine	State
KMD 1000	Bottom influence mine	State
M-08	Moored contact mine, chemical horn	State
M-12	Moored contact mine, chemical horn	State
M-16	Moored contact mine, chemical horn	State
M-26	Moored contact mine, inertial	State
M-AG	Moored antenna mine	State
MBU 1200	Anti-torpedo rocket launcher	State
MBU 1800	Anti-submarine rocket launcher - probably obsolescent	State
MBU 2500	Anti-submarine rocket launcher	State
MBU 4500A	Anti-submarine rocket launcher	State
MBU 6000	Anti-submarine rocket launcher	State
Mirab	Magnetic bottom influence mine	State
M-KB	Moored contact mine, chemical horn	State
M-KB-3	Moored contact mine, chemical horn	State
MYaM	Moored contact mine, chemical horn	State
MZ-26	Obstructor mine	State
PLT	Moored contact mine, impact-inertial	State
PLT-3	Moored contact mine, chemical horn	State
SS-N-14 'Silex'	Torpedo carrying missile system	State
SS-N-15	Anti-submarine depth charge carrying missile	State
SS-N-16	Version of SS-N-15 carrying a torpedo instead of a depth charge	State
UEP?	Moored influence mine, electrical	State
UNITED KINGDOM		
Dragonfish	*Anti-invasion ground influence mine*	*Marconi*
Limbo	Anti-submarine mortar	Marconi

Designation	Description	Contractor(s)
UK cont		
Mark 11	Air-launched depth charge	BAeSema
Sea Urchin	Programmable influence mine	BAeSema
Stonefish	Ground influence mine	Marconi
VEMS	Exercise ground mine	BAeSema
UNITED STATES OF AMERICA		
ASM	Advanced sea mine (development)	
ASROC	Anti-submarine missile system	Honeywell/Loral
Captor	Encapsulated Mk 46 torpedo	Loral
Mk 36	Aircraft-laid bottom mine - magnetic/seismic	Loral
Mk 40	Aircraft-laid bottom mine - magnetic/seismic	Loral
Mk 41	Aircraft-laid bottom mine - magnetic/seismic	Loral
Mk 52	Aircraft-laid bottom mine - magnetic, acoustic and/or pressure	Loral
Mk 53	Aircraft-laid bottom mine - magnetic, acoustic and/or pressure	Loral
Mk 55	Aircraft-laid bottom mine	Loral
Mk 56	Aircraft-laid moored mine - magnetic dual channel	Loral
Mk 57	Submarine/ship laid moored mine - magnetic dual channel	Loral
Quickstrike	Shallow water bottom mine series - (Marks 62, 63, 64 and 65)	Loral
Sea Lance	Submarine-launched anti-submarine missile	Boeing
SLMM	Submarine-launched mobile mine	Boeing
SUBROC	Submarine-launched anti-submarine missile	Loral
Type 115A	Aircraft-laid surface mine - magnetic/seismic	Loral
YUGOSLAVIA		
M66	Distraction mine	SDPR
M70	Ground influence mine	SDPR
M71	Limpet mine	SDPR

Mine Countermeasures

Designation	Description	Contractor(s)
DENMARK		
FOCUS 400	Towed inspection vehicle	MacArtney A/S
FINLAND		
FIMS	Integrated minesweeping control system	Elesco
FMS	*Magnetic minesweeping system*	*Elesco*
FRANCE		
AP 5	Acoustic minesweep	DCN
EVEC 20	Data handling system for minehunters	Thomson Sintra
IBIS	Mine countermeasures data handling systems	Thomson Sintra
PAP Mk 5	ROV mine disposal system	Société ECA
Sterne 1	Towed minesweeping system	Thomson Sintra
GERMANY		
DM 19	Explosive minesweeping cutter	Rheinmetall
DM 801 B1	Mine disposal charge	Rheinmetall
DM 1002 A1	Remote-controlled fuze	Rheinmetall
E 67/E Mod 1	Firing transmitter	Rheinmetall
GHA	Acoustic minesweeping unit	IBAK
MWS 80	Minehunting data handling system	Atlas Elektronik
Pinguin	ROV mine disposal system	MBB
Troika	Remote-control minesweeping system	AEG
96 R	Exercise fuze	Rheinmetall
165 R	Remote-controlled explosive cutter	Rheinmetall
166 R	Explosive minesweeping cutter	Rheinmetall
INTERNATIONAL		
ARMS	*Remote minehunting system*	*Marconi*
Double Oropesa	Towed minesweeping system	Various
Trailblazer	ROV mine disposal vehicle	ISE
ITALY		
CAM	Countermining charge	Whitehead
CAM-T	Countermining charge	Whitehead
CAP	Countermining charge	Whitehead
ECP	Explosive cutter	Whitehead
IMICS	Integrated minehunting combat system	Datamat
MDC	Countermining charge	Whitehead
MIN	ROV mine disposal system	Selenia/Riva Calzoni
MM/SSN-714	Minesweeper data processing system	SMA/Datamat
Pluto	Remote-operated mine disposal vehicle	Gaymarine
Pluto Plus	Remote-control mine disposal vehicle	Gaymarine
SWEDEN		
Double Eagle	ROV mine disposal vehicle	SUTEC
IMAIS	Integrated minesweeping system	SA Marine
MDC 605	Mine disposal charge	SA Marine
SAM	Remote-control minesweeping drone	Karlskronavarvet
Sea Eagle	ROV mine disposal vehicle	SUTEC
9MJ 400	Navigation and combat information minehunting system	NobelTech
UNITED KINGDOM		
Dragonfly	ROV minehunting vehicle	GEC Avionics/OSEL

Designation	Description	Contractor(s)
UK cont		
FIMS	Integrated mine countermeasure system	Ferranti-Thomson Sonar
LMDE	Limpet mine disposal equipment	AB Precision
MAINS	Minehunting combat information and navigation system	Racal
MC500	Mine countermeasures command and control system	Ferranti International
MINENAV	Acoustic sweep monitoring system	Seametrix
Minnow	*ROV minehunting vehicle*	*Marconi*
MMIMS	Magnetic minesweeping system	Marconi
NAUTIS-M	Integrated minehunting combat information system	Marconi/Sema
Osborn	Acoustic minesweeping system	BAeSema
QMSS	Mine surveillance system	Qubit
SATAM	Integrated minehunting system	Ferranti
System 880	Minesweeping/minehunting control system	Racal
Tow Taxi	Towed underwater research vehicle	BAeSema
TUMS	*Towed unmanned submersible*	*BAeSema*
WS Mk 9	Wire sweep system	BAJ Vickers
UNITED STATES OF AMERICA		
ML 105	Airborne magnetic/acoustic influence sweep	EDO
MNS	Remotely operated mine disposal system	Honeywell
Phantom	ROV series for mine countermeasures	Deep Ocean Engineering
PINS	Integrated navigation system for MCMVs	Magnavox
RCV-425	*ROV minehunting vehicle*	*Hydro Products*
Recon IV	ROV minehunting vehicle	Perry Tritech
TP Scout	ROV mine surveillance vehicle	Ketema
Viper Mk 1	ROV mine countermeasures vehicle	Allen Osborne

Acoustic and Electronic Countermeasures

Designation	Description	Contractor(s)
CHINA		
921-A	Submarine ESM direction-finder	CNEIEC
FRANCE		
DR 2000	ESM receiver (TMV 434)	Thomson-CSF
DR 3000U	Submarine ESM surveillance system	Thomson-CSF
DR 4000U	Submarine ESM surveillance system	Thomson-CSF
INTERNATIONAL		
SSTD	Joint US/UK surface ship torpedo defence system	
ISRAEL		
ATC-1	Towed torpedo decoy	Rafael
ES Series	Submarine ESM antennas	Elbit
NS-9034	Submarine ESM/ELINT system	Elisra
Scutter	Expendable torpedo decoy	Rafael
ITALY		
C303	Expendable acoustic countermeasures system	Whitehead
RQN-5	ESM/ELINT system	Selenia
Thetis	Submarine ESM surveillance system	Elettronica
UNITED KINGDOM		
ATAAC	Anti-torpedo acoustic countermeasures	Ferranti
Bandfish	Expendable torpedo countermeasures system	Dowty
G 738	Towed torpedo decoy equipment	Graseby
Manta	Submarine ESM system	THORN EMI
Porpoise	Submarine ESM system	Racal
RDL Series	*Submarine ESM system*	*Racal*
Sealion	Submarine ESM search and warning system	Racal
Sea Siren	Towed torpedo decoy system	Graseby
UA Series	UAB/UAC/UAJ/UAP ESM systems for the Royal Navy	Racal
UNITED STATES OF AMERICA		
ADC Mk 1	5-inch/6-inch expendable acoustic countermeasures devices	Conax
ADC Mk 2	3-inch expendable acoustic countermeasures	Hazeltine
ADC Mk 3	6-inch expendable acoustic countermeasures	Bendix
ADC Mk 4	Expendable acoustic countermeasures	Hazeltine
ADC Mk 5	Expendable acoustic countermeasures device, probably mobile (development)	Bendix
BLD-1	Submarine ESM DF system	Litton
CCCU	Submarine countermeasures control unit	
Guardian Star	Family of ESM systems	Sperry
NAE Mk 3	5-inch sonar beacon countermeasures device	Pique Engineering
Phoenix	Submarine ESM radar warning system	ARGO
S-2150	Submarine ESM radar warning system	EM Systems
S-3000	Submarine ESM radar warning system	EM Systems
Sea Sentry	Submarine ESM surveillance system	Kollmorgen
SLQ-25	Torpedo countermeasures device (Nixie)	Frequency Engineering
WLQ-4	Submarine SIGINT system	GTE
WLR-8	Submarine ELINT receiver	GTE

Underwater Communications

Note: The following list includes those systems devoted to submarine-to-surface, surface-to-submarine, and submarine-to-submarine communications systems, plus those used by divers. It should be remembered, however, that many surface ship and submarine sonar systems also include an underwater telephone facility. Details of these will be found in the appropriate entry in the main sections of this yearbook.

Designation	Description	Contractor(s)
CANADA		
Model 1080	Diver's communications system	Orcatron
Scubaphone	Diver's underwater communications	Orcatron
SDU	Diver's underwater communications	Orcatron
Subphone	Submersible communications system	Orcatron
FRANCE		
ERUS-2	Diver's communications system	Safare Crouzet
MCA 30	Buoyant cable antenna	CSI
RUPG-1A	Diver's homing receiver	Safare Crouzet
TSM 5152A/B	Underwater telephony system	Thomson Sintra
TUUM-2C/D	Underwater wireless telephone	Safare Crouzet
TUUM-4A	Underwater multi-channel telephone	Safare Crouzet
ISRAEL		
SACU-200	Communications unit	Elbit
ITALY		
APLF	Submarine marker device	Stacchini
RS 100	Underwater transponder	USEA
TS-200	Underwater telephone system	USEA
TS-300	Emergency underwater telephone	USEA
TS-400	Modular underwater telephone	USEA
TS-500	Underwater telephone system	USEA
UNION OF SOVEREIGN STATES		
Fez	Underwater telephone	State
UNITED KINGDOM		
DS 034-2	Speech processor	Marconi
ECB-680	Expendable communications	Marconi
EL050/057/067	Deep-phone underwater telephone	Slingsby
G732 Mk II	Underwater telephone	Graseby
GS7110	Submarine VLF/LF receiver	Marconi
Honeypot	VHF location beacon	Marconi
Model 6281	Pinger receiver	Helle
Model 3342A	Speech processor	Helle
Model 3220B	Two-diver telephone	Helle
Model 3510	Underwater broadcast system	Helle
R 1800	VLF receiver	Marconi
Sea Piper	Diver communication system	Marconi
Subcom 3400	Throughwater emergency communications system	Marconi udi
Type 183	Submarine emergency underwater telephone	Graseby
Type 185	Medium-range underwater telephone	Graseby
Type 639	Submarine indicator buoy	Marconi
Type 2008	Medium-range underwater telephone	Marconi
Type 2010	Acoustic rapid automatic teletype	Marconi
Type 2073	Underwater telephone	AB Precision
Type 3200	Submarine communications system	Marconi udi
VLF MSK	VLF MSK system	Marconi
UNITED STATES OF AMERICA		
BRC-6	Expendable communications buoy	Sippican
BRT-1	Communications buoy	Sippican
BRT-6	UHF satellite communication system	Hazeltine
Hellephone	Underwater communications system	Helle
Hydrocom	Speech processor	Hydro OHT
Model 6280	Pinger receiver	Helle
Model 5400	Underwater telephone	EDO
Model N30AA5B	Pinger receiver	Dukane
SSQ-71	Expendable acoustic communications set	Sparton
SSQ-86 (XN-1)	One-way surface-to-submarine communications system	Sparton
SUBTACS	Submarine tactical communications system	GTE
SUS Mk 84	Air-to-submarine communications device	Sippican
V-COM	Underwater vehicle communications system	Allen Osborne
WQC-2A	Sonar underwater communications set	GI Corporation

Miscellaneous Underwater Detection Equipment

Designation	Description	Contractor(s)
CANADA		
ASA-64	Anomaly detection signal processor	CAE
ASA-65	Compensator group adapter for ASA-65(V)	CAE
ASA-65(V)	Nine-term compensator for MAD system	CAE
ASQ-502	Submarine detecting set (MAD)	CAE
ASQ-504(V)	Submarine detecting set (MAD)	CAE
Models M-244, H-234	Recording magnetometers	Barringer
OA 5154/ASQ	Automatic MAD compensation system	
FRANCE		
DHAX-1	Airborne MAD equipment	

Designation	Description	Contractor(s)
FRANCE cont		
DHAX-3	ASW helicopter MAD equipment	
TSM 5260/5265	Multibeam echo sounder	Thomson Sintra
TSM 9310	*Sound ray tracer for analysis of sonar equipment performance*	*Thomson-CSF*
GERMANY		
DESO 21/25	Deep channel echo sounder	Atlas Elektronik
Polartrack	3D tracking and positioning system	Atlas Elektronik
ITALY		
CIACIO	Torpedo self-homing system	Selenia-Elsag
ELT/810	Sonar prediction system	Elettronica
P MICCA	Remote-control of deep sea mines	Selenia-Elsag
SFM-A	Underwater target identification system	Selenia-Elsag
NETHERLANDS		
SP1-04	Sound ray path analyser	Signaal
XSV-01	Expendable sound velocimeter for sound ray path measurement	Signaal
NORWAY		
EM 100/300/500	Echo sounders	Simrad Subsea
HPR-300	Hydro-acoustic positioning system	Simrad Subsea
MICOS	MCM command system	Simrad Subsea
UNITED KINGDOM		
D100	Diver's ranging unit	Ulvertech
Deep Mobile target	Instrumented ship-launched target	THORN EMI
G733	Underwater transponder target	Graseby
G740	Transportable long-range calibration system	Graseby
Hydrosearch Mk II	Surveying sonar	Marconi
LRN-3-3	Magnetic sensor	THORN EMI
Series 700	Echo sounders	AB Precision
SI 1220	Underwater surveillance analyser	Schlumberger
TRAC	Data acquisition and processing system	Qubit
UNITED STATES OF AMERICA		
Active processing system	Airborne system for processing sonobuoy data	Sparton
AKT-22(V)4	Telemetry data transmitting set	Flightline Electronics
AQH-8	Mission recording and playback system	Diagnostic/Retrieval
ARR-78(V)	Advanced sonobuoy communication link (ASCL)	Hazeltine
ASA-66	P-3C Orion ASW tactical data cockpit display	Loral
ASA-82	S-3A Viking ASW tactical data display system	Loral
ASPRO	Parallel processing computer system	Loral
ASQ-81(V)	Airborne MAD system	Texas Instruments
AYA-8B	P-3C Orion ASW data processing system	General Electric
DSP Sona-Graph	Analysis station for acoustic signals	Kay
Masscomp 5600	Onboard multiprocessing system	Concurrent Computers
Model 4068	Ultra short-baseline tracking system	EDO
Model 4268	Ultra short-baseline tracking system	EDO
OE 259	Ultra short-baseline tracking system	Osprey
OL-5003(–)/ARR	Sonobuoy signal processor	Sparton
Q-MIPS	Sonar image processing system	Triton
R-1651/ARA	VHF radio receiver for sonobuoy location	Flightline Electronics
RO 308/SSQ-36	P-3C Orion bathythermograph data recorder	Western Components
Seamark	Bathymetric sonar	Alliant Techsystems
Spectrum analyser	Airborne digital processor for real-time spectral analysis	Sparton
TD 1135A	Demultiplexer processor/display	Sparton
UQN-4	Digital depth sounder	EDO
UQN-5	Swimmer area navigation system	Allen Osborne

NATO Designations of USS Underwater Warfare Systems and Equipment

Designation	Description	Designation	Description	Designation	Description
Boat Sail	Submarine radar	**Herkules**	Submarine and surface ship HF scanning sonar	**Serb**	Submarine-launched ballistic missile (SS-N-6)
Brick Group	Submarine ECM system incorporating Brick Pulp and Brick Spit	**Horse Jaw**	Low frequency surface ship sonar	**Shark Fin**	Submarine medium frequency sonar
		Horse Tail	Variable depth sonar	**Shark Teeth**	Submarine LF bow sonar
Brick Pulp	Submarine EW antenna	**Lamb Tail**	Variable depth sonar	**Silex**	SS-N-14 ASW missile
Brick Spit	Submarine EW antenna	**Mare Tail**	Variable depth sonar	**Skiff**	Submarine-launched ballistic missile (SS-N-23)
Buck Toe	Searchlight sonar system	**Moose Jaw**	Low frequency hull sonar		
Bull Horn	Active/passive surface ship sonar	**Park Lamp**	Submarine mast direction finding loop antenna	**Snipe**	SS-N-17 SLBM
Bull Nose	Medium frequency surface ship bow sonar			**Snoop Plate**	Submarine surveillance radar
		Perch Gill	Submarine HF attack sonar	**Snoop Slab**	Submarine surface search radar
Cluster Bay	'Rising' sea mine	**Pike Jaw**	Searchlight attack sonar	**Snoop Tray**	Submarine surveillance radar
Clusterguard	Anti-sonar hull coating for submarines	**Port Spring**	Submarine mast conical spiral antenna	**Stag Ear**	HF searchlight sonar
				Stag Hoof	HF searchlight sonar
Cluster Gulf	'Rising' sea mine for deep water	**Punch Bowl**	Submarine mast radome	**Starbright**	SS-N-7 submarine-launched anti-ship missile
Cod Eye A/B	Submarine sensors	**Rat Tail**	Variable depth helicopter dipping sonar		
Dustbin	Submarine mast sensor			**Steer Hide**	Variable depth sonar
Elk Tail	Variable depth sonar	**Sampson**	SS-N-21 intermediate-range SLBM	**Stingray**	SS-N-18 SLBM
Feniks	Submarine sonar	**Sark**	SS-N-5 submarine ballistic missile	**Sturgeon**	Submarine-launched ballistic missile (SS-N-20)
Fez	Underwater telephone	**Sawfly**	Submarine-launched ballistic missile (SS-N-8)		
Foal Tail	Variable depth sonar			**Tamir**	Submarine sonar
Golf Ball	Submarine mast sensor	**Seal Skin**	Submarine sonar	**Trout Cheek**	Submarine passive array bow sonar

Contractors

Australia

AWA Defence Industries Pty Ltd
Endeavour House
Module 3
Fourth Avenue Technology Park
South Australia 5095
Australia
Tel: 8 349 8009
Fax: 8 260 8938

British Aerospace Australia
Head Office
PO Box 180
Salisbury
South Australia 5108
Australia
Tel: 8 343 8211
Tx: 88 342

Fairey Australasia Pty Ltd
2-6 Ardtornish Street
Holden Hill
South Australia 5088
Australia
Tel: 8 266 0666
Tx: AA89029

Nautronix Limited
13 Corkhill Street
North Freemantle
Western Australia 6159
Australia
Tel: 619 430 5900
Fax: 619 430 5901

Plessey Australia Pty Ltd
Electronic Systems Division
Faraday Park Railway Road
Meadowbank
New South Wales 2114
Australia
Tel: 2 807 0400
Tx: AA21471

Sonobuoys Australia Ltd
Faraday Park Railway Road
Meadowbank
New South Wales 2114
Australia
Tel: 2 807 0400
Tx: 121 339

Canada

Amphibico Inc
9563 Côte de Liesse
Dorval
Quebec H9P 1A3
Canada
Tel: 418 636 9910

Barringer Research Ltd
304 Carlingview Drive
Metropolitan Toronto
Rexdale
Ontario
Canada
Tel: 416 675 3870
Tx: 06 989183

CAE Electronics Ltd
PO Box 1800
St Laurent
Montreal
Quebec H4L 4X4
Canada
Tel: 514 341 6780
Tx: 05824856

Computing Devices Co
PO Box 8508
Ottawa
Ontario K1G 3M9
Canada
Tel: 613 596 70 00; 596 70 59
Tx: 0534139
Twx: 610 563 1632

Consolidated Technologies Ltd
St John's
Newfoundland
Canada

C-Tech Ltd
PO Box 1960
525 Boundary Road
Cornwall
Ontario K6H 6N7
Canada
Tel: 613 933 7970
Tx: 811538

Fullerton Sherwood Engineering Ltd
6450 van Deemter Court
Mississauga
Ontario L5T 1S1
Canada
Tel: 416 670 0656
Fax: 416 670 8318

Hermes Electronic Ltd
40 Atlantic Street
PO Box 1005
Dartmouth
Nova Scotia B2Y 4A1
Canada
Tel: 902 466 7491
Tx: 1921 744

Indal Technologies Inc
3570 Hawkestone Road
Mississauga
Ontario L5C 2V8
Canada
Tel: 416 275 5300
Fax: 416 273 7004

International Submarine Engineering Ltd
2601 Murray Street
Port Moody
British Columbia V3H 1X1
Canada
Tel: 604 937 3421
Tx: 4353 554

Metocean Data Systems Ltd
PO Box 2427 DEPS
40 Fielding Avenue
Dartmouth
Nova Scotia
Canada
Tel: 902 468 2505
Fax: 902 468 4442

Oceanroutes Canada Inc
Halifax
Nova Scotia
Canada
Tel: 902 468 3008

Optech Inc
701 Petrolia Road
Downsview
Ontario M3J 2N6
Canada
Tel: 416 661 5904
Fax: 416 661 4168

Orcatron Manufacturing Ltd
86 North Bend Street
Port Coquitlam
British Columbia V3K 6H1
Canada
Tel: 604 941 7909
Tx: 04352848

Scannar Industries Inc
PO Box 5009
777 Walkers Line
Burlington
Ontario L7R 4B3
Canada
Tel: 416 528 8811
Tx: 618 401

Seimac Ltd
1378 Bedford Highway
Bedford
Nova Scotia B4A 1E2
Canada
Tel: 902 825 9680
Tx: 091 22888

Simrad Mesotech Systems Ltd
2830 Huntington Place
Port Coquitlam
British Columbia V3C 4T3
Canada
Tel: 604 464 8144
Tx: 4353 637

Sparton of Canada Ltd
PO Box 5125
99 Ash Street
London
Ontario N6A 4N2
Canada
Tel: 519 455 6320
Tx: 0645876

Westinghouse Canada Inc
Government Projects Dept
Information & Defence Technology Division
PO Box 5009
Burlington
Ontario L7R 4B3
Canada
Tel: 416 333 6006
Tx: 061 8401

Chile

Fabricas & Maestranzas de Ejerato (FAMAE)
Santiago
Chile

Industrias Cardoen Ltda
AvDa Providencia 2237
660 Piso
Santiago
Chile
Tel: 2321081/2321082/2515884
Tx: 340997 INCAR CK

China, People's Republic

China National Electronics Import & Export
Corporation
49 Fuxing Road
Beijing
People's Republic of China
Tel: 810910
Tx: 22475

China Precision Machinery
Import & Export Corp
17 Wenchang Hutong Xidan
PO Box 845
Beijing
People's Republic of China
Tel: 895012
Tx: 22484 CPMC CN

Dalian Shipbuilding Industry Corporation
16 Zhuqingje
Dalian
People's Republic of China
Tel: 26277
Tx: 86171

Denmark

MacArtney ApS
Guldagervej 48
DK-6710 Esbjerg V
Denmark
Tel: 05 11 66 77
Tx: 54 272

Nea-Lindberg A/S
Industriparken 39-43
Post Box 226
DK-2750 Ballerup
Denmark
Tel: 42 97 2200
Tx: 35338

Terma Elektronik AS
Hovmarken 4
DK-8520 Lystrup
Denmark
Tel: (6) 222000
Tx: 68109

Finland

Elesco Oy AB
Luomannotko 4
PO Box 128
SF-02201 Espoo
Finland
Tel: 420 8600
Fax: 420 8610

France

Aerospatiale
Division Engins Tactiques
2 rue Beranger BP 84
F-92322 Chatillon Cedex
France
Tel: 47 46 21 21
Tx: 250 881 AISPA F

R Alkan & Cie
rue de 8 Mai 1945
F-94460 Valenton
France
Tel: 43 89 39 90
Tx: 203876

Creusot-Loire Mecanique
Immeuble Ile de France
Paris La Defense
France
Tel: 49 00 60 50
Tx: 615638
Fax: 49 00 57 30

Crouzet SA
25 rue Jules Vedrines
F-26027 Valence Cedex
France
Tel: 75 79 85 95
Tx: 345807
Twx: 75 55 22 50

Direction des Construction Navales (DCN)
2 rue Royale BP1
F-75200 Paris Naval
France
Tel: 42 60 33 30

Engins Matra
Matra SA
17 rue Paul Dautier
F-78140 Velizy
France
Tel: 39 46 97 86
Tx: 698130

Etablissment des Constructions
et Armes Navales
(ECAN de St Tropez)
F-83990 St Tropez
France

Euromissile
12 rue de la Redoute
F-92260 Fontenay-aux-Roses
France
Tel: 46 61 73 11
Tx: 204691
Fax: 46 61 64 67

Safare Crouzet SA
98 Avenue Saint Lambert
PO Box 171
F-06105 Nice Cedex 2
France
Tel: 93 84 72 79
Tx: 460 813

SAGEM
27 rue Leblanc
F-75512 Paris Cedex 15
France
Tel: 40 70 63 63
Fax: 40 70 66 40

Sextant Avionique
5/7 Rue Jeanne Braconnier
F-92366 Meudon-la-Forêt Cedex
France
Tel: 46 29 88 00
Tx: SEXTANT 631155
Fax: 40 94 02 51

Société ECA
17 Avenue de Chateau
Boite Postale 16
F-92194 Meudon Cedex
France
Tel: 46 26 71 11
Tx: ECABLVU 200 336F

Societe Industrielle d'Aviation Latecoere
(SILAT)
79 avenue Marceau
F-75116 Paris
France
Tel: 47 20 01 05
Tx: 631712

Société Nereides
4 avenue des Indes
ZA de Courtaboeuf
F-91969 Les Ulis Cedex 13
France
Tel: 69 07 20 48
Fax: 69 07 19 14

Sopelem
19 Boulevard Ney
BP 264
F-75866 Paris Cedex 18
France
Tel: 42 02 89 80
Tx: 620 111F

Thomson-CSF
Aerospace Group
51 Esplanade du General de Gaulle
Cedex 67
F-92045 Paris La Defense
France
Tel: 49 07 80 00
Tx: THOM 616780
Fax: 49 07 83 00

Thomson-CSF
Division Systèmes
Defense et Contrôle
18 avenue du Marechal Juin
F-92363 Meudon-la-Forêt Cedex
France
Tel: 40 94 34 21
Tx: 270375

Thomson Sintra Activites Sous-Marines
1 avenue Aristride-Briand
F-94117 Arcueil Cedex
France
Tel: 49 85 35 35
Tx: 204780

Thomson Sintra Activites Sous-Marines
Rte de Sainte Anne du Porzie
F-29601 Brest Cedex
France
Tel: 98 45 38 20
Tx: 204780

Thomson Sintra Activites Sous-Marines
525 route des Dolines
BP 138
06561 Valbonne Cedex
France
Tel: 92 96 30 00

Germany

Atlas Elektronik GmbH
Post Box 44 85 45
D-2800 Bremen 44
Germany
Tel: 0421 4570
Fax: 0421 457 2900

Faun-Werke
PO Box 8
D-8560 Lauf ad Pegnitz
Germany
Tel: 9123 1-85-0
Tx: 626093
Fax: 9123 7-53-20

Haux-Life Support GmbH
Descostrasse 19
D-7516 Karlsbad-Iltersbach
Germany
Tel: 072 48 1050
Tx: 782950

IBAK Helmut Hunger GmbH
Wehdenweg 122
PO Box 6260
D-2300 Kiel 14
Germany
Tel: 431 72 70-0
Tx: 292824
Fax: 431 72 62 20

Institut Dr Forster
Postfach 1564
D-7410 Reutlingen
Germany
Tel: 07121 140-0
Tx: 729781
Fax: 07121 140 488

Rheinmetall GmbH
Ulmenstrasse 125
PO Box 6609
D-4000 Dusseldorf 30
Germany
Tel: (211) 44 71; 447 20 66
Tx: 8584 963

Salzgitter Elektronik GmbH
PO Box 160
D-2302 Flintbek
Germany
Tel: 04347/908-0
Tx: 292976

SystemTechnik Nord GmbH
Postfach 107845
D-2800 Bremen 1
Germany
Tel: (421) 538-1
Tx: 240210
Fax: (421) 538-3320

Telefunken SystemTechnik
Radio and Radar Systems
Theodor-Stern-Kai 1
D-600 Frankfurt 70
Germany
Tel: 69 6003759
Tx: 411 064

Israel

Elbit Computers Ltd
Advanced Technology Center
PO Box 5390
Haifa 31053
Israel
Tel: 04 524222
Tx: 46774

Elisra Electronic Systems Ltd
(a subsidiary of Tadiran Ltd)
48 Mivtza Kadesh Street
51203 Bene Beraq
Israel
Tel: (3) 754 5111
Tx: 33553

Rafael
Israel Armament Development Authority
PO Box 2082
Haifa 31021
Israel
Tel: 708174
Tx: 45173 VERED

Italy

Datamat Ingegneria Dei Sistemi SpA
Via Simone Martini 126
I-00143 Rome
Italy
Tel: (396) 50451
Tx: 613436
Fax: (396) 504 3057

Elettronica San Giorgio - ELSAG SpA
Via Puccini 2
PO Box 125
I-16154 Genova-Sestri
Italy
Tel: (6) 84 14 41
Tx: 621273

Elettronica SpA
Via Tiburtina Km 13 700
I-00131 Rome
Italy
Tel: (6) 43641
Tx: 611024

Elmer SpA
Vialle dell'Industria 4
I-00040 Pomezia
Italy
Tel: (6) 912971
Tx: 610112

FIAR
Via Montefeltro 8
I-20156 Milano
Italy
Tel: 02 357901
Tx: 331140 FIARMO I

Gaymarine Srl
Via Giovanni XXIII 39
I-20090 Trezzano Sul Naviglio MI
Italy
Tel: 2 4455347
Tx: 313 539

Riva Calzoni SpA
via Emilia Ponente 72
I-40133 Bologna
Italy
Tel: 051 527511
Fax: 051 437233

Selenia Elsag Sistemi Navali
Defence Systems Division
Via Tiburtina Km 12 400
I-00131 Rome
Italy
Tel: (6) 43601
Tx: 613690

SEPA
Societa di Elettronica
per L'Automazione SpA
Corso Giulio Cesare 294
I-10154 Turin
Italy
Tel: 011 205 3371
Tx: 221527 SEPA 1

Servomeccanismi
Via Mediana Km 29.3
I-00040 Pomezia
Italy

Sistemi Subacquei WELSE SpA Consortile
Via L Manara 2
I-16154 Genova-Sestri
Italy
Tel: 010 6511321
Fax: 010 6512147

SMA
Segnalatmento Marittimo ed Aereo SpA
via del Ferrone 5
I-50124 Florence
Italy
Tel: 055 27501
Fax: 055 714934

Tecnovar Italiana SpA
Via Argiro 95
I-70121 Bari
Italy
Tel: (80) 21 17 44
Tx: 810 345

USEA SpA
Via G Matteotti 63
I-19030 Pugliola di Lerici
La Spezia
Italy
Tel: (187) 96 71 25
Tx: 281216
Fax: (187) 96 54 45

Whitehead
(a division of Gilardine SpA)
Via di Levante 48-50
I-57128 Salviano
Livorno
Italy
Tel: 586 84 01 11
Tx: 500 192

Whitehead
Commercial Office
Str Statale 236 Goitese
Loc. Fascia D'oro
I-25018 Montichiari (Brescia)
Italy
Tel: 030 962561
Tx: 302280 MISAR I
Fax: 030 962561

Netherlands

Hollandse Signaalapparaten BV
Zuidelijke Havenweg 40
PO Box 42
NL-7550 GD Hengelo
Netherlands
Tel: 074 488111
Tx: 44310 SIGN NL
Fax: 074 425936

Norway

Anderaa Instruments
Fanaveien 13 B
N-5050 Bergen
Norway
Tel: (05) 132500
Tx: 40049
Fax: (05) 13 79 50

Micro Processor Systems A/S
Banaveien 32
N-3600 Kongsberg
Norway
Tel: (473) 73 57 66

Norsk Forsvarsteknologi A/S
PO Box 1003
N-3601 Kongsberg
Norway
Tel: (473) 73 82 00
Tx: 71491 VAAPN N

Simrad Subsea A/S
PO Box 111
N-3191 Horten
Norway
Tel: 47 033 44250
Tx: 70391 Simh n
Twx: 47 033 44424

Sweden

AB Bofors
S-691 80 Bofors
Sweden
Tel: 46 586 81000
Tx: 73210 BOFORS S
Fax: 46 586 58145

Ericsson Radar Electronics AB
PO Box 1001
S-431 26 Molndal
Sweden
Tel: 031 671000
Tx: 20905 ERICRAS

Ericsson Radar Electronics AB
Computer Systems Division
Torshamnsgatan 21-23
S-16380 Stockholm
Sweden
Tel: (8) 7570000
Tx: 15872

Karlskronavarvet AB
S-37182 Karlskrona
Sweden
Tel: 455 19440
Tx: 8395018

NobelTech Systems AB
S-175 88 Järfalla
Sweden
Tel: 0758 10000
Tx: 12688

Saab Missiles AB
S-58188 Linkoping
Sweden
Tel: (13) 286000
Tx: 50001

Safe Bridge Scandinavia AB
Box 3003
S-59303 Vaestervik
Sweden
Tel: 490 362 40
Fax: 490 197 46

SA Marine Aktiebolag
PO Box 627
S-261 02 Landskrona
Sweden
Tel: 418 2 40 10
Tx: 72 063

Sutec AB
PO Box 7073
Linkoping
Sweden
Tel: 13158060
Tx: 50 150

Swedish Ordnance
S-631 87 Eskilstuna 1
Sweden
Tel: 16 155000
Tx: 46075

Taiwan

Military Technology Consulting
PO Box 11-006
Pei-Tou
Taipei
Taiwan
Tel: 8216805
Tx: 17326

United Kingdom

AB Precision (Poole) Ltd
Stanley Green Road
Poole
Dorset BH15 3AL
UK
Tel: 0202 673185
Tx: 417 102

Admiralty Research Establishment
Ministry of Defence (Procurement Executive)
Southwell
Portland
Dorset DT5 2JS
UK
Tel: 0305 820381

BAeSema
PO Box 5
Filton
Bristol BS12 7QW
UK
Tel: 0272 693831 x3709
Tx: 449452

BAJ Ltd
Banwell
Weston-super-Mare
Avon BS24 8PD
UK
Tel: 0934 822251
Tx: 44 259

Bennico Ltd
Unit B1
Kirkhill Place
Kirkhill Industrial Estate
Dyce
Aberdeen AB1 0ES
UK
Tel: 0224 772266
Fax: 0224 771294

Bridport Aviation Products Ltd
The Court
West Street
Bridport
Dorset DT6 3OU
UK
Tel: 0308 56666
Tx: 41132
Fax: 0308 56605

Defence Equipment & Systems Ltd
Salhouse Road
Norwich
Norfolk NR7 9AY
UK
Tel: 0603 484065
Tx: 975067
Fax: 0603 415649

Digital Systems and Design Ltd
18 Shakespeare Business Centre
Hathaway Close
Eastleigh
Hampshire SO5 4SR
UK
Tel: 0703 620499
Tx: 477575

Dowty Maritime Systems Ltd
Sonar and Communications Division
419 Bridport Road
Greenford Industrial Estate
Greenford
Middlesex UB6 8UA
UK
Tel: 081 578 0081
Tx: 934512

Dowty Maritime Systems Ltd
Gresham Division
Twickenham Road
Feltham
Middlesex TW13 6HA
UK
Tel: 081 894 5511
Tx: 27419

Dowty Maritime Systems Ltd
Waverley Division
Hampshire Road
Granby Estate
Weymouth
Dorset DT4 9XD
UK
Tel: 0305 784738
Tx: 41477 WAVLEC G
Fax: (0305) 777904

Dowty-SEMA Ltd
Biwater House
Portsmouth Road
Esher
Surrey KT10 9SJ
UK
Tel: 0372 466660
Fax: 0372 466566

EASAMS Limited
Lyon Way
Frimley Road
Camberley
Surrey GU16 5EX
UK
Tel: 0276 63377
Tx: 858115
Fax: 0276 683468

Fairey Hydraulics Ltd
Claverham
Bristol BS19 4NF
UK
Tel: 0934 835224
Fax: 0934 835337

Ferranti International
Bracknell Division
Western Road
Berkshire RG12 1RA
UK
Tel: 0344 483232
Fax: 0344 861003

Ferranti-Thomson Sonar Systems UK Ltd
Bird Hall Lane
Cheadle Heath
Stockport SK3 0XQ
UK
Tel: 061 491 4001
Tx: 665106

Gearing & Watson (Electronics) Ltd
South Road
Hailsham
East Sussex BN27 3JJ
UK
Tel: 0323 846464
Tx: 877563

GEC Avionics Ltd
Airport Works
Rochester
Kent ME1 2XX
UK
Tel: 0634 844400
Fax: 0634 827332

GEC Sensors Ltd
Electro-Optical Surveillance
Christopher Martin Rd
Basildon
Essex SS14 3EL
UK
Tel: 0268 522822
Tx: 99225

Graseby Marine Ltd
Park Avenue
Bushey
Watford
Hertfordshire WD2 2BW
UK
Tel: 0923 28566
Tx: 923 010

Helle Engineering Ltd
Howe Moss Ave
Kirkhill Industrial Estate
Dyce
Aberdeen AB2 0GP
UK
Tel: 0224 724663
Tx: 739216

Kelvin Hughes Ltd
New North Road
Hainault
Ilford
Essex IG6 2UR
UK
Tel: 081 500 1020
Tx: 896 401

Logica Aerospace & Defence Ltd
Cobham Park
Downside Road
Cobham
Surrey KT11 3LX
UK
Tel: 071 637 9111
Tx: 27200

Marconi Radar and Control Systems Ltd
Weapon Systems Division
New Parks
Leicester LE3 1UF
Tel: 0533 871481
Tx: 34551

Marconi Simulation
Marconi Instruments Ltd
Napier Building
Donibristle Industrial Park
Nr Dunfermline
Fife KY11 5JZ
UK
Tel: 0383 822131
Tx: 727779
Fax: 0383 824227

Marconi udi
Denmore Road
Bridge of Don
Aberdeen
Scotland AB2 8JW
UK
Tel: 0224 703551
Tx: 73361
Fax: 0224 821339

Marconi Underwater Systems Ltd
Elettra Avenue
Waterlooville
Hampshire PO7 7XS
UK
Tel: 0705 264466
Tx: 869233

Marconi Underwater Systems Ltd
Wilkinthroop House
Templecombe
Somerset BA8 0DH
UK
Tel: 0963 70551

Marine Electronics Ltd
Houmet House
Rue des Houmets
Catel
Guernsey, Channel Islands
Tel: 0481 53181
Tx: 4191 501

NEI Clarke-Chapman Ltd
Victoria Works
Gateshead
Tyne & Wear
NE8 34S
UK
Tel: 091 477 2271
Tx: 53239

Osprey Electronics
Campus 1
Aberdeen Science & Technology Park
Balgownie Road
Bridge of Don
Aberdeen AB2 8GT
UK
Tel: 0224 826464
Fax: 0224 826363

Pilkington Optronics
Caxton Street
Anniesland
Glasgow
G13 1H2
UK
Tel: 041 954 9601
Tx: 778114

Qubit UK Ltd
Lynchborough Road
Passfield
Hampshire GU30 7SB
UK
Tel: 0252 33 14 18
Tx: 858 593

Racal Marine Systems Ltd
Burlington House
118 Burlington Road
New Malden
Surrey KT3 4NR
UK
Tel: 081 942 2464
Tx: 22 891

Racal Radar Defence Systems
Davis Road
Chessington
Surrey KT9 1TB
UK
Tel: 081 397 5281
Tx: 27720

Redifon Limited
Newton Road
Crawley
West Sussex RH10 2TU
UK
Tel: 0293 518855
Tx: 877131
Fax: 0293 561096

Schlumberger Instruments Division
Victoria Road
Farnborough
Hampshire GU14 7PW
UK
Tel: 0252 544433
Tx: 8858245
Fax: 0252 543854

Science Defence Management
Associates Ltd (SDMA)
Unit 8
Murrills Industrial Estate
East Street
Portchester
Hampshire PO16 9RD
Tel: 0705 384124
Fax: 0705 324125

SD-Scicon
127/147 Fleet Road
Fleet
Hampshire GU13 8PD
UK
Tel: 0252 622171
Tx: 859921
Fax: 0252 615990

SEMA
Naval Systems
20-26 Lamb's Conduit Street
London WC1N 3LF
UK
Tel: 071 404 0911
Tx: 28863
Twx: 071 405 9469

Siemens Plessey Defence Systems
Grange Road
Christchurch
Dorset BH23 4JE
UK
Tel: 0202 486344
Fax: 0202 404221

Slingsby Engineering Ltd
Kirkbymoorside
York YO6 6EZ
UK
Tel: 0751 31751
Tx: 57911

Smiths Industries
Aerospace and Defence Systems
Bishops Cleeve
Cheltenham
Gloucestershire GL52 4SF
UK
Tel: 0242 673333
Tx: 43172

Sonar Research & Development Ltd
Unit 1B
Grovehill Industrial Estate
Beverley
North Humberside HU17 0JW
UK
Tel: 0482 869559
Tx: 592126
Fax: 0482 872184

STC Defence Systems
Cable Systems Division
Wednesbury Street
Newport
Gwent NP9 0WS
UK
Tel: 0633 244244
Tx: 498368

STC Defence Systems
Marine Systems Division
Chester Hall Lane
Basildon
Essex SS14 3BW
UK
Tel: 081 945 5000
Tx: 99 101

Strachan & Henshaw
Ashton Vale Rd
PO Box 103
Bristol BS99 7TJ
UK
Tel: 0272 664677
Tx: 44170

THORN EMI Electronics Ltd
Manor Royal
Crawley
West Sussex RH10 2PZ
UK
Tel: 0293 28787
Tx: 87 267

THORN EMI Electronics Ltd
Defence Group
1 Forrest Road
Feltham
Middlesex TW13 7HE
UK
Tel: 081 751 6464
Fax: 081 751 4774

THORN EMI Electronics Ltd
Naval Systems Division
PO Box 4
Rugeley
Staffordshire WS15 1DR
UK
Tel: 08894 5151
Tx: 361351

Ulvertech Ltd
33-49 Farwing Lane
Bromley
Kent BR1 3RE
UK
Tel: 081 290 0200
Tx: 8951912

Winchester Associates Ltd
Unit 16
Denmore Industrial Estate
Denmore Road
Bridge of Don
Aberdeen
UK
Tel: 0224 822833
Tx: 94011419
Fax: 0224 702469

United States of America

Note: Many area codes in the United States
need to have a 1 dialled before the area code

Allen Osborne Associates
756 Lakefield Road, Bldg J
Westlake Village
California 91361-2624
USA
Tel: 805 495-8420
Twx: 910 494 1710
Fax: 805 373 6067

Alliant Techsystems Inc
6500 Harbour Heights Parkway
Everett
Washington 98275
USA
Tel: 206 356 3000

Allied Signal Inc
Bendix Oceanics Division
15825 Roxford Street
Sylmar
California 91342
USA
Tel: 818 367 0111
Tx: 662900

ARGO Systems Inc
PO Box 3452
Sunnyvale
California 94088-3452
USA
Tel: 408 737 2000
Tx: 6711100
Fax: 415 737 9236

Bathy Systems Inc
Gardner Road
West Kingston
Rhode Island 02892
USA
Tel: 401 294 2190
Tx: 467443

BBN Systems and Technology Corporation
10 Moulton Street
Cambridge
Massachusetts 02238
USA
Tel: 617 873 3000
Tx: 921470
Fax: 617 873 3776

Boeing Aerospace
PO Box 3999, M/S 85-19
Seattle
Washington 98124
USA
Tel: 206 773 2816

CAE-Link Corporation
1180 Tech Road
Silver Spring
Maryland 20904
USA
Tel: 301 622 4400
Twx: 710 825 9786

Concurrent Computer Corporation
One Technology Way
Westford
Massachusetts 01886
USA
Tel: 508 692 6200

Deep Ocean Engineering Co
1431 Doolittle Drive
San Leandro
California 94577
USA
Tel: 415 562 9300
Tx: 705816

Diagnostic/Retrieval Systems Inc
8 Wright Way
Oakland
New Jersey 07436
USA
Tel: 201 337 3800
Tx: 710 988 4191

Dowty Avionics
300E Live Oak Avenue
Arcadia
California 91006-5617
USA
Tel: 818 445 5955
Fax: 818 447 0880

Dukane Corporation
Seacom Division
2900 Dukane Drive
St Charles
Illinois 60174
Tel: 708 584 2300

EDO Corporation
2001 Jefferson Davis Highway
Suite 1000
Arlington
Virginia 22202-3688
USA
Tel: 703 415 1560
Fax: 703 415 1564

EG & G Inc
1396 Piccard Drive
Rockville
Maryland 20850
USA
Tel: 301 840 3000

Emerson Electric Co
Electronics & Space Division
8100 W Florissant Ave
St Louis
Missouri 63136
USA
Tel: 314 553 3232
Tx: 44 869

EMS Development Corp
100 Sea Lane
Farmingdale
New York 11735
USA
Tel: 510 293 7900
Fax: 510 244 6480

EM Systems Inc
Textron
45757 W Northport Loop
Fremont
California 94538
USA
Tel: 415 657 9960
Tx: 497 0260

Flightline Electronics Inc
7500 Main St
PO Box 750
Fishers
New York 14453-0750
USA
Tel: 716 924 4000
Tx: 510 254 2896

GE Aerospace
Electronic Systems Division
French Road
Utica
New York 13503
USA
Tel: 315 793 7708

General Electric Company
Government Electronics Systems Division
3135 Easton Turnpike
Fairfield
Connecticut 06431
USA
Tel: 203 373 2121

General Electric Company
Ordnance Systems Division
100 Plastics Avenue
Pittsfield
Massachusetts 01201
USA
Tel: 413 494 3634
Fax: 413 494 3791

General Electric Company
Undersea Systems Division
Electronics Park
Building 67
Syracuse
New York 13221
USA
Tel: 315 456 0123

General Instrument Corp
767 Fifth Avenue
New York
New York 10153
USA
Tel: 212 207 6333

General Instrument Corp
Underseas Systems Division
33 Southwest Park
Westwood
Massachusetts 02090
USA
Tel: 617 326 7815

GTE Government Systems Corp
77 'A' Street
Needham Heights
Massachusetts 02194-2892
USA
Tel: 617 449 2000
Tx: 922497

GTE Government Systems Corp
100 Ferguson Drive
Mountain View
California 94039
USA
Tel: 415 966 2000
Tx: 910 3796 948

Hazeltine Corporation
Cuba Hill Road
Greenlawn
New York 11740
USA
Tel: 516 351 4000
Tx: 510 221 2113

Helle Engineering Inc
7198 Convoy Court
San Diego
California 92111-1093
USA
Tel: 619 278 3521
Tx: 697961

Honeywell Inc
Honeywell Plaza
Minneapolis
Minnesota 55408
USA
Tel: 612 870 6442
Twx: 910 576 2692

Honeywell Inc
Training & Control Systems
1200 East San Bernardino Road
West Covina
California 91790
USA
Tel: 213 331 0011
Tx: 670452
Fax: 818 915 9255

Honeywell Inc
Underseas Systems Division
600 Second St NE
Hopkins
Minnesota 55343
USA
Tel: 612 931 5160
Tx: 910 756 3437

Hughes Aircraft Co
Ground Systems Group
PO Box 3310
Fullerton
California 92634
USA
Tel: 714 871 3232
Tx: 685504

IBM
Federal Systems Division
6600 Rockledge Drive
Bethesda
Maryland 20817
USA
Tel: 301 493 1321

Kay Elemetrics Corporation
12 Maple Avenue
Pine Brook
New Jersey 07058-9798
USA
Tel: 201 227 2000
Twx: 710 734 4347
Fax: 201 227 7760

Klein Associates Inc
Undersea Search & Survey
Klein Drive
Salem
New Hampshire 03079
USA
Tel: 603 893 6131
Tx: 947439

Kollmorgen Corp
Electro-Optical Division
347 King Street
Northampton
Massachusetts 01060
USA
Tel: 413 586 2330
Tx: 955409

KVH Industries Inc
110 Enterprise Center
Middletown
Rhode Island 02840
USA
Tel: 401 847 3327
Tx: 382051

Librascope Corporation
833 Sonora Avenue
Glendale
California 91201-2433
Tel: 818 244 6541
Tx: 715620
Twx: 910 497 2266

Litton Systems Inc
Amemcom Division
5115 Calvert Road
College Park
Maryland 20740
USA
Tel: 301 864 5600
Tx: 710 826 9650

Lockheed Sanders Inc
Daniel Webster Highway South
PO Box 868
Nashua
New Hampshire 03061-0868
USA
Tel: 603 885 4321
Tx: 943430

Loral Control Systems
Archbald
Philadelphia 18403
USA
Tel: 717 876 1500
Tx: 290432

Loral Systems Group
1210 Massillon Road
Akron
Ohio 44315-0001
USA
Tel: 216 796 2121
Tx: 986 439

Loral Systems Group
Defence Systems Division
PO Box 85
Litchfield Park
Arizona 85340-0085
USA
Tel: 602 925 3232

Magnavox Advanced Products and
Systems Co
2829 Maricopa Street
Torrance
California 90503
Tel: 213 618 1200
Tx: 674373; 696101
Twx: 910 349 6657
Fax: 213 618 7001

Magnavox Electronics Systems Co
1313 Production Road
Fort Wayne
Indiana 46808
USA
Tel: 219 429 6000
Tx: 232 478

Martin Marietta Corporation
6801 Rockledge Drive
Bethesda
Maryland 20817
USA
Tel: (301) 897 6000
Tx: 898 437; 248 934
Twx: 710 8249 050
Fax: (301) 897 6252

Martin Marietta Corporation
Aero and Naval Systems
103 Chesapeake Plaza
Baltimore
Maryland 21220
Tel: 301 338 5000
Twx: 710 239 9049

Martin Marietta Corporation
Ocean Systems Division
6711 Baymeadow Drive
Glen Burnie
Maryland 21061
USA
Tel: 301 760 31 00
Tx: 908 310

McDonnell Douglas Astronautics Co
PO Box 516
St Louis
Missouri 63166
USA
Tel: 314 232 0232
Tx: 44857
Twx: 7620635

Morton Thiokol Inc
PO Box 241
Elkton
Maryland 21921
USA
Tel: 301 398 3000

Numax Electronics
720 Old Willets Path
Hauppage
New York 11787
USA

Ocean Data Equipment Group
473 Pleasant Street
PO Box 2557
Fall River
Massachusetts 02721
USA
Tel: 508 679 5284
Fax: 508 672 9780

Osprey Sub-Sea Inc
PO Box 1749
1225 Stone Drive
San Marcos
California 92069
USA
Tel: 619 471 2223
Tx: 695409
Fax: 619 471 1121

Perry Tritech Inc
821 Jupiter Park Drive
Jupiter
Florida 33458
USA
Tel: 407 743 7000

Raytheon Company
141 Spring Street
Lexington
Massachusetts 02173
USA

Raytheon Company
Submarine Signal Division
1847 West Main Road
PO Box 360
Portsmouth
Rhode Island 02871
USA
Tel: 401 847 8000
Tx: 927787

Rockwell Corporation
Autonetics Maritime Systems Division
PO Box 4921
3370 Miraloma Avenue
Anaheim
California
Tel: 714 762 3327
Fax: 714 762 0146

Sachse Engineering Associates Inc
5550 Oberlin Drive
San Diego
California 92121-1717
USA
Tel: 619 453 8200
Fax: 619 453 7100

Ship Analytics
North Stonnington Professional Center
Route 184
North Stonnington
Connecticut 06359
USA
Tel: 203 535 3092
Tx: 643732
Fax: 203 535 0560

Sippican Inc
Seven Barnabas Road
Marion
Maryland 02738
USA
Tel: 617 748 1160
Tx: 200189 S051 UR

Sparton Corporation
Electronics Division
2400 East Ganson Street
Jackson
Mississippi 49202
USA
Tel: 517 787 8600

Spears Associates Inc
249 Vanderbilt Avenue
Norwood
Massachusetts 02062
USA
Tel: 617 769 6900
Fax: 617 769 9725

Sperry Marine
1070 Seminole Trail
Charlottesville
Virginia 22906
USA
Tel: 804 974 2000
Tx: 822411
Twx: 510 587 5463

Texas Instruments Inc
Defense Systems Electronics Group
PO Box 660246 M/S 3134
Dallas
Texas 75266
USA
Tel: 214 480 6867
Fax: 214 480 3281

Triton Technology Inc
12 Westridge Drive
Watsonville
California 95076
USA
Tel: 408 722 7373
Fax: 408 722 1405

TRW Inc
Systems Division
7600 Colshire Drive
McLean
Virginia 22102-7603
USA
Tel: 703 734 6000
Twx: 710 831 0030

Unisys Corporation
Surveillance and Fire Control Systems Divison
Marcus Avenue & Lakeville Road
Great Neck
New York 11020
USA
Tel: 516 574 0111
Tx: 960 167
Twx: 510 223 0401

United Technologies Corporation
Norden Systems
Norden Place
Box 5300
Norwalk
Connecticut 06856
USA
Tel: 203 852 5000
Tx: 240005

Western Electric
Guildford Centre
PO Box 20046
Greenboro
North Carolina 27420
USA
Tel: 919 697 6770

Westinghouse Electric Corp
Defense & Electronic Centre
PO Box 1693, M/S 4610
Baltimore
Maryland 21203
USA

Manufacturers Index

Alphabetical Index

Printed in Great Britain by
Butler & Tanner Ltd, Frome and London